CLYMER®

YAMAHA

OUTBOARD SHOP MANUAL

115-250 HP TWO-STROKE • 1999-2002 (Includes Jet Drives)

The World's Finest Publisher of Mechanical How-to Manuals

PRIMEDIA
Business Magazines & Media

P.O. Box 12901, Overland Park, KS 66282-2901

FIRST EDITION
First Printing June, 2003

Printed in U.S.A.

CLYMER and colophon are registered trademarks of PRIMEDIA Business Magazines & Media Inc.

This book was printed at Von Hoffmann an ISO certified company.

ISBN: 0-89287-846-0

Library of Congress: 2003106773

AUTHOR: Mark Rolling.

TECHNICAL PHOTOGRAPHY: Mark Rolling.

TECHNICAL ILLUSTRATIONS: Mike Rose.

WIRING DIAGRAMS: Bob Meyer.

EDITOR: Jason Beaver.

PRODUCTION: Susan Hartington.

COVER: Mark Clifford Photography at www.markclifford.com. Boat courtesy of Wholesale Marine, Canyon Country, CA.

CLYMER PUBLICATIONS
PRIMEDIA Business Magazines & Media

The following product lines are published by PRIMEDIA Business Directories & Books.

The Electronics Source Book

More information available at *primediabooks.com*

Contents

CHAPTER TEN
POWER TILT/TRIM 407

CHAPTER ELEVEN
MIDSECTION 424

CHAPTER TWELVE
OIL INJECTION SYSTEM 432

CHAPTER THIRTEEN
REMOTE CONTROL 448

INDEX 461

WIRING DIAGRAMS 465

Quick Reference Data

ENGINE INFORMATION

MODEL:______________________ YEAR:______________________

VIN NUMBER:______________________

ENGINE SERIAL NUMBER:______________________

CARBURETOR SERIAL NUMBER OR I.D. MARK:______________________

GENERAL TIGHTENING TORQUE

Screw or nut size	N•m	in.-lb.	ft.-lb.
Metric			
M5 bolt	5	44	–
M6 bolt	8	70	–
M8 bolt	18	156	13
M10 bolt	36	–	27
M12 bolt	43	–	32
M14 bolt	81.3	–	60
8 mm nut	5	44	–
10 mm nut	8	70	–
12 mm nut	18	156	13
14 mm nut	36	–	27
17 mm nut	43	–	32

MAINTENANCE SCHEDULE

After each use	Check for loose nuts, bolts, spark plugs Check the propeller Check the oil reservoir level* Flush the cooling system Lubricate the jet drive bearings* Check for and correct leaking fluids Wash the exterior of gearcase and drive shaft housing Touch up paint damage on external surfaces
Before each use	Check for and correct fuel leakage Check the steering and controls for proper operation Check the oil reservoir level* Check for a proper cooling system operation (water stream) Check for proper operation of the neutral only start system
Initial 10 hours or one month	Lubricate the swivel tube, tilt tube and steering system Check throttle operation Check shift linkages for proper operation Check tightness of all accessible nuts and bolts

(continued)

MAINTENANCE SCHEDULE (continued)

Initial 10 hours or one month (continued)	Check power tilt and trim operation* Check choke lever operation* Inspect fuel filter for contamination Inspect fuel hoses and connections Adjust the idle speed* (Chapter Four) Inspect mid-section components (Chapter Eleven) Inspect the spark plug(s) Adjust the oil pump linkage (Chapter Four)* Inspect oil reservoir for water or contamination* Check electrical wiring and connections Check power head for water and exhaust leakage Check gearcase lubricant level and condition Check condition and charge level of battery Check carburetor synchronization and adjustments Check cylinder compression (Chapter Two)
Initial 50 hours or 90 days	Lubricate the swivel tube, tilt tube and steering system Adjust the carburetor(s) Inspect fuel filter for contamination Check spark plug condition and gap Check and adjust the ignition timing Check the oil injection system* Check electrical wiring and connections Check power head for water and exhaust leakage Check gearcase lubricant level and condition Inspect the water pump impeller (Chapter Eight) Check the propeller Check propeller nut for tightness Clean and inspect sacrificial anodes Check all accessible nuts and bolts for tightness Check cylinder compression (Chapter Two)
Each 100 hours of usage or 180 days	Lubricate the swivel tube, tilt tube and steering system Check carburetor synchronization and adjustments Inspect fuel filter for contamination Check fuel hoses and clamps for leakage Check the spark plug condition and gap Check the power tilt/trim fluid level (Chapter Ten) Inspect the mid-section components. Check the oil injection system* Check electrical wiring and connections Check power head for water and exhaust leakage Check gearcase lubricant level and condition Check the condition and charge level of the battery* Clean and inspect sacrificial anodes Check all accessible nuts and bolts for tightness Check the propeller nut for tightness Check cylinder compression
Each 200 hours of usage or one year	Inspect fuel tank, hoses and clamps Clean or replace the fuel filter Replace the water pump impeller Check the fuel pump oil level* (HPDI models) Adjust the throttle position sensor (EFI and HPDI models) Inspect the fuel pump drive belt* (HPDI models)
Each 1000 hours or five years	Change the fuel pump oil* (HPDI models) Replace the fuel pump drive belt* (HPDI models)
*This maintenance item does not apply to all models.	

FLUID CAPACITIES

Model	Capacity (approximate)
Gearcase	
115 and 130 hp	
Standard RH rotation	760 ml (25.7 oz.)
Optional LH rotation	715 ml (24.2 oz.)
150-200 hp (2.6 liter models)	
Standard RH rotation	980 ml (33.1 oz.)
Optional LH rotation	870 ml (29.4 oz.)
Twin counter rotating propellers	900 ml (30.4 oz.)
200-250 hp (3.1 liter models)	
Standard RH rotation	1.15 L (38.9 oz.)
Optional LH rotation	1.0 L (33.8 oz.)
Oil Reservoir	
115 and 130 hp	
On board reservoir	10.5 L (11.1 qt.)
Engine mounted reservoir	0.9 L (0.95 qt.)
150-200 hp (2.6 liter models)	
On board reservoir	10.5 L (11.1 qt.)
Engine mounted reservoir	0.9 L (0.95 qt.)
200-250 hp (3.1 liter models)	
On board reservoir	10.5 L (11.1 qt.)
Engine mounted reservoir	1.2 L (1.27 qt.)

OIL AND FUEL MIXING RATIOS

Quantity of fuel	Oil for 50:1 ratio	Oil for 25:1 ratio
3.8 L (1 gal.)	76 cc (2.6 oz.)	152 cc (5.2 oz)
7.6 L (2 gal.)	152 cc (5.2 oz)	304 cc (10.4 oz.)
11.4 L (3 gal.)	228 cc (7.8 oz.)	456 cc (15.6 oz.)
15.4 L (4 gal.)	304 cc (10.4 oz.)	608 cc (20.8 oz.)
18.9 L (5 gal.)	380 cc (12.8 oz.)	760 cc (25.6 oz.)
22.8 L (6 gal.)	456 cc (15.6 oz.)	912 cc (31.2 oz.)
26.6 L (7 gal.)	530 cc (18.2 oz.)	1060 cc (36.4 oz.)
30.8 L (8 gal.)	608 cc (20.8 oz.)	1216 cc (41.6 oz.)
34.2 L (9 gal.)	684 cc (23.4 oz.)	1368 cc (46.8 oz.)
37.8 L (10 gal.)	760 cc (25.6 oz.)	1520 cc (51.2 oz.)
41.6 L (11 gal.)	832 cc (28.2 oz.)	1664 cc (56.4 oz.)
45.6 L (12 gal.)	912 cc (31.2 oz.)	1824 cc (62.4 oz.)

SPARK PLUG SPECIFICATIONS

Model	NGK plug	Champion plug	Gap
115 hp	BR8HS-10	QL78C	1.0 mm (0.039 in.)
130 hp	BR9HS-10	QL77CJ4	1.0 mm (0.039 in.)
150 hp			
C150	B8HS-10	L82C	1.0 mm (0.039 in.)
D150	BR8HS-10	QL82C	1.0 mm (0.039 in.)
P150, S150, L150, DX150, LX150, PX150 SX150, VX150	BR7HS-10	RL82C	1.0 mm (0.039 in.)
HPDI			
With black coil connector	BKR6E-S1-10	*	1.0-1.1 mm (0.039-0.043 in.)
With gray coil connector	BKR6E-KU-10	*	0.6 mm (0.024 in.)

(continued)

SPARK PLUG SPECIFICATIONS (continued)

Model	NGK plug	Champion plug	Gap
175 hp			
Carburetor and EFI models	B8HS-10	QL78C	1.0 mm (0.039 in.)
HPDI models			
With black coil connector	BKR7E-S1-10	*	1.0-1.1 mm (0.039-0.043 in.)
With gray coil connector	BKR7E-KU-10	*	0.6 mm (0.024 in.)
200 hp			
2.6 Liter models with carburetors	B8HS-10	QL78C	1.0 mm (0.039 in.)
2.6 Liter models with EFI	BR8HS-10	QL78C	1.0 mm (0.039 in.)
3.1 Liter models (1998)	BR9HS-10	*	1.0 mm (0.039 in.)
3.1 Liter models (1999-on)	BR8HS-10	QL78C	1.0 mm (0.039 in.)
HPDI model			
With black coil connector	BKR7E-S1-10	*	1.0-1.1 mm (0.039-0.043 in.)
With gray coil connector	BKR7E-KU-10	*	0.6 mm (0.024 in.)
225 and 250 hp (except HPDI)	BR9HS-10	*	1.0 mm (0.039 in.)

*The manufacturer does not provide a Champion plug part No. for this model.

BATTERY CHARGE PERCENTAGE

Specific gravity reading	Percentage of Charge Remaining
1.120-1.140	0
1.135-1.155	10
1.150-1.170	20
1.160-1.180	30
1.175-1.195	40
1.190-1.210	50
1.205-1.225	60
1.215-1.235	70
1.230-1.250	80
1.245-1.265	90
1.260-1.280	100

BATTERY CAPACITY

Accessory draw	Provides continuous power for:	Approximate recharge time
80 amp-hour battery		
5 amps	13.5 hours	16 hours
15 amps	3.5 hours	13 hours
25 amps	1.6 hours	12 hours
105 amp-hour battery		
5 amps	15.8 hours	16 hours
15 amps	4.2 hours	13 hours
25 amps	2.4 hours	12 hours

Chapter One

General Information

This detailed and comprehensive manual covers Yamaha two-stroke outboard engines (115-250 hp).

This manual can be used by anyone from a first time do-it-yourselfer to a professional mechanic. The text provides step-by-step information on maintenance, tune-up, repair and overhaul. Hundreds of illustrations guide the reader through every job.

A shop manual is a reference that should be used to find information quickly. Clymer manuals are designed with this in mind. All chapters are thumb tabbed and important items are indexed at the end of the manual. All procedures, tables, photos and instructions in this manual are designed for the reader who may be working on the machine or using the manual for the first time.

Keep the manual in a handy place such as a toolbox or boat. It will help to better understand how the boat runs, lower repair and maintenance costs and generally increase enjoyment of the boat.

Frequently used specifications and capacities from individual chapters are summarized in the *Quick Reference Data* at the front of the book. Specifications concerning specific systems are at the end of each chapter.

Tables 1-4 are at the end of this chapter.

Table 1 lists the engines identification codes.

Table 2 lists technical abbreviations.

Table 3 lists metric tap and drill sizes.

Table 4 lists conversion formulas.

Table 5 lists general torque specifications.

MANUAL ORGANIZATION

All dimensions and capacities are expressed in U.S. standard and metric units of measurement.

This chapter provides general information on shop safety, tool use, service fundamentals and shop supplies. The tables at the end of the chapter include general engine information.

Chapter Two provides methods and suggestions for quick and accurate diagnosis and repair of problems. Troubleshooting procedures discuss typical symptoms and logical methods to pinpoint the trouble.

Chapter Three explains all periodic lubrication and routine maintenance necessary to keep the outboard operating well. Chapter Three also includes recommended tune-up procedures, eliminating the need to constantly consult other chapters on the various assemblies.

Subsequent chapters describe specific systems, providing disassembly, repair, assembly and adjustment procedures in simple step-by-step form.

Some of the procedures in this manual specify special tools. When possible, the tool is illustrated in use. Well-equipped mechanics may be able to substitute similar tools or fabricate a suitable replacement. However, in some cases, the specialized equipment or expertise may make it impractical for the home mechanic to attempt the procedure. When necessary, such operations are identified in the text with the recommendation to have a dealership or specialist perform the task. It may be less expensive to have them perform these jobs, especially when considering the cost of the equipment. This is true with machine work for power head rebuilds, as machinists spend years perfecting their trade and even professional mechanics will often rely upon their services.

WARNINGS, CAUTIONS AND NOTES

The terms WARNING, CAUTION and NOTE have specific meanings in this manual.

A WARNING emphasizes areas where injury or even death could result from negligence. Mechanical damage may also occur. WARNINGS *are to be taken seriously.*

A CAUTION emphasizes areas where equipment damage could result. Disregarding a CAUTION could cause permanent mechanical damage, though injury is unlikely.

A NOTE provides additional information to clarify or make a procedure easier. Disregarding a NOTE could cause inconvenience, but would not cause equipment damage or injury.

SAFETY

Professional mechanics can work for years and never sustain a serious injury or mishap. Follow these guidelines and practice common sense to safely service the engine.

1. Do not operate the engine in an enclosed area. The exhaust gasses contain carbon monoxide, an odorless, colorless, and tasteless poisonous gas. Carbon monoxide levels build quickly in small, enclosed areas and can cause unconsciousness and death in a short time. Make sure the work area is properly ventilated or operate the engine outside.
2. *Never* use gasoline or any extremely flammable liquid to clean parts. Refer to *Handling Gasoline Safely* and *Cleaning Parts* in this chapter.
3. Never smoke or use a torch in the vicinity of flammable liquids, such as gasoline or cleaning solvent.
4. After removing the engine cover, allow the engine to air out before performing any service work. Review *Safety Precautions* at the beginning of Chapter Five.
5. Use the correct tool type and size to avoid damaging fasteners.
6. Keep tools clean and in good condition. Replace or repair worn or damaged equipment.
7. When loosening a tight or stuck fastener, always consider what would happen if the wrench should slip. In most cases, it is safer to pull on a wrench or a ratchet than

it would be to push on it. Be careful; protect yourself accordingly.
8. When replacing a fastener, make sure to use one with the same measurements and strength as the old one. Refer to *Fasteners* in this chapter for additional information.
9. Keep your work area clean and uncluttered. Keep all hand and power tools in good condition. Wipe greasy and oily tools after using them. Unkept tools are difficult to hold and can cause injury. Replace or repair worn or damaged tools. Do not leave tools, shop rags or anything that does not belong in the hull.
10. Wear safety goggles during all operations involving drilling, grinding, or the use of a cold chisel, or *anytime* the safety of your eyes is in question (when debris may spray or scatter). *Always* wear safety goggles when using solvent and compressed air.
11. Do not carry sharp tools in clothing pockets.
12. Always have an approved fire extinguisher available. Make sure it is rated for gasoline (Class B) and electrical (Class C) fires.
13. Do not use compressed air to clean clothes, the boat/engine or the work area. Debris may be blown into eyes or skin. *Never* direct compressed air at yourself or someone else. Do not allow children to use or play with any compressed air equipment.
14. When using compressed air to dry rotating parts, hold the part so it cannot rotate. Do not allow the force of the air to spin the part. The air jet is capable of rotating parts at extreme speed. The part may become damaged or disintegrate, causing serious injury.

Handling Gasoline Safely

Gasoline is a volatile flammable liquid and is one of the most dangerous items in the shop.

Because gasoline is used so often, many people forget that it is hazardous. Only use gasoline as fuel for gasoline internal combustion engines. Do not use it as a cleaner or

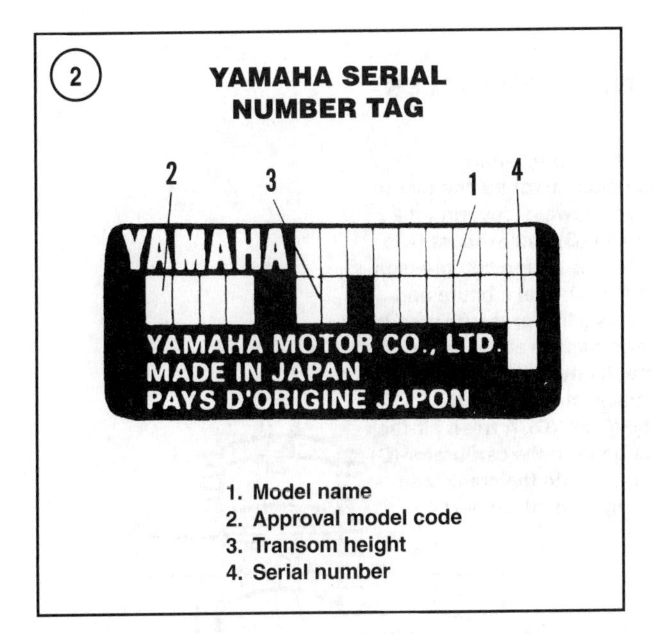

degreaser. Keep in mind, when working, gasoline is always present in the fuel tank, fuel lines and carburetor or fuel rail. To avoid a disastrous accident when working around the fuel system, carefully observe the following precautions:

1. *Never* use gasoline to clean parts. See *Cleaning Parts* in this chapter.
2. When working on the fuel system, work outside or in a well-ventilated area.
3. Do not add fuel to the fuel tank or service the fuel system while the boat is near an open flame, sparks or where someone is smoking. Gasoline vapor is heavier than air. It collects in low areas and is much more easily ignited than liquid gasoline.
4. Allow the engine to cool completely before working on any fuel system component.
5. When draining a carburetor or fuel fitting, catch the fuel in a plastic container and then pour it into an approved gasoline storage device.
6. Do not store gasoline in glass containers. If the glass breaks, a serious explosion or fire may occur.
7. Immediately wipe up spilled gasoline with rags. Store the rags in a metal container with a lid until they can be properly disposed of, or place them outside in a safe place so the fuel can evaporate.
8. Do not pour water onto a gasoline fire. Water spreads the fire and makes it more difficult to extinguish. Use a class B, BC or ABC fire extinguisher to extinguish the fire.
9. Always turn OFF the engine before refueling. Do not spill fuel onto the engine components. Do not overfill the fuel tank. Leave an air space at the top of the tank to allow room for the fuel to expand because of temperature fluctuations.

Cleaning Parts

Cleaning parts is one of the more tedious and difficult service jobs performed in the home garage. There are many types of chemical cleaners and solvents available for shop use. Most are poisonous and extremely flammable. To prevent chemical exposure, vapor buildup, fire and serious injury, observe each product warning label and note the following:

1. Read the entire product label before using any chemical. Always know what type of chemical is being used and whether it is poisonous and/or flammable.
2. Do not use more than one type of cleaning solvent at a time. If mixing chemicals is called for, measure the proper amounts according to the manufacturer.
3. Work in a well-ventilated area.
4. Wear chemical-resistant gloves.
5. Wear safety glasses.
6. Wear a vapor respirator if the instructions call for it.
7. Wash hands and arms thoroughly after cleaning parts.
8. Keep chemical products away from children and pets.
9. Thoroughly clean all oil, grease and cleaner residue from any part that must be heated.
10. Use a nylon brush when cleaning parts. Metal brushes may cause a spark.
11. When using a parts washer, only use the solvent recommended by the equipment manufacturer. Make sure the parts washer is equipped with a metal lid that will lower in case of fire.

MODEL IDENTIFICATION

Before troubleshooting the engine, verify the model name, model number, horsepower and serial number of the engine. It is absolutely essential that the model be correctly identified before performing any service on the engine. In many cases, the tables list specifications by horsepower, type of fuel system, type of lubrication system, gear case rotational direction, drive shaft length, engine displacement and/or model name. The identification tag is located on the port side clamp bracket (**Figure 1**). Refer to **Figure 2** to review the various forms of information on the tag.

The first numerals in the model name (1, **Figure 2**) indicate the horsepower rating. The letters following the numbers indicate the drive shaft length and if the engine is a jet drive model. All models covered in this manual have let-

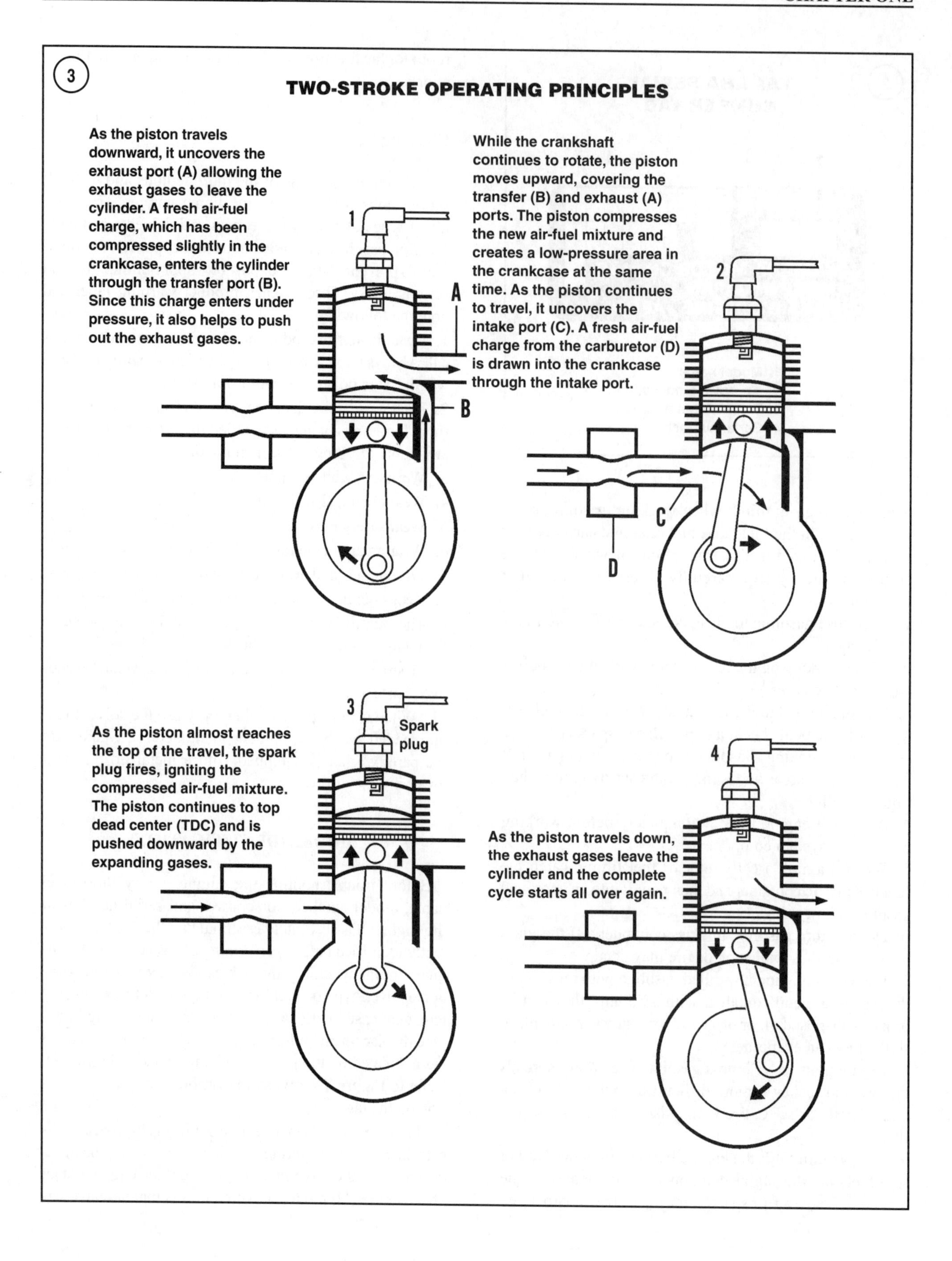
3
TWO-STROKE OPERATING PRINCIPLES
As the piston travels downward, it uncovers the exhaust port (A) allowing the exhaust gases to leave the cylinder. A fresh air-fuel charge, which has been compressed slightly in the crankcase, enters the cylinder through the transfer port (B). Since this charge enters under pressure, it also helps to push out the exhaust gases.
1
A
B
While the crankshaft continues to rotate, the piston moves upward, covering the transfer (B) and exhaust (A) ports. The piston compresses the new air-fuel mixture and creates a low-pressure area in the crankcase at the same time. As the piston continues to travel, it uncovers the intake port (C). A fresh air-fuel charge from the carburetor (D) is drawn into the crankcase through the intake port.
2
C
D
As the piston almost reaches the top of the travel, the spark plug fires, igniting the compressed air-fuel mixture. The piston continues to top dead center (TDC) and is pushed downward by the expanding gases.
3
Spark plug
As the piston travels down, the exhaust gases leave the cylinder and the complete cycle starts all over again.
4

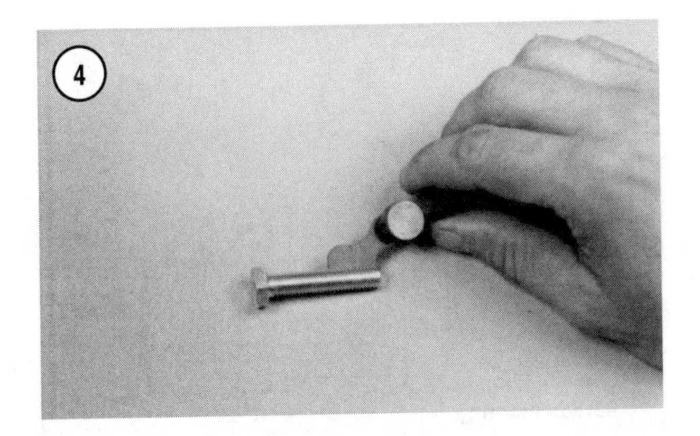

4

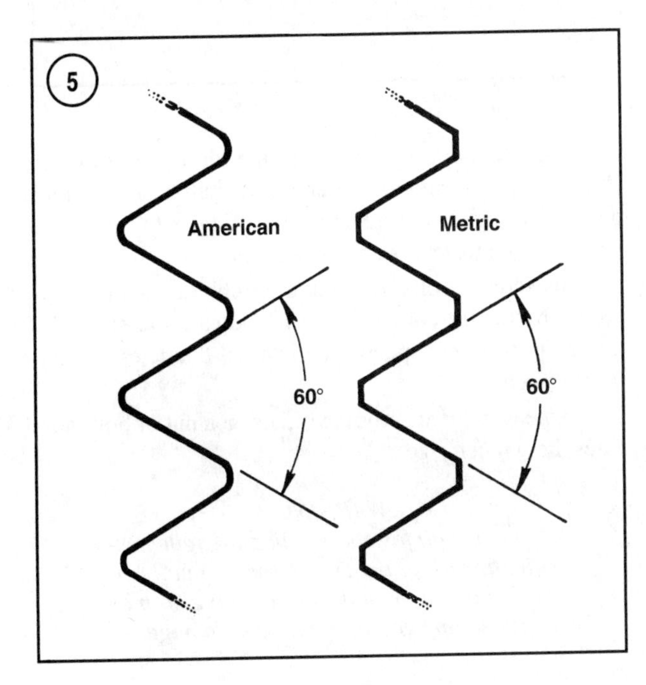

5

ters preceding the horsepower rating number. This letter also identifies the engine characteristics.

The final letter in the model name is the year model code. Refer to **Table 1** and the information on the tag to identify the horsepower, year model and engine characteristics.

The serial number (4, **Figure 2**) identifies the engine from others of the same model. Record this number and keep it in a safe place. In the event of theft, this number may be needed to identify and reclaim the engine.

The approval model code (2, **Figure 2**) and transom height number (3) are factory model identification codes.

All the numeral and letter codes on the tag are required when purchasing replacement parts for Yamaha outboards.

ENGINE OPERATION

All marine engines, whether two- or four-stroke, gasoline or diesel, operate on the Otto cycle intake, compression, power and exhaust phases. All Yamaha engines covered in this manual are of two-stroke design. **Figure 3** shows typical gasoline two-stroke engine operation.

FASTENERS

Proper fastener selection and installation is important to ensure that the engine operates as designed and can be serviced efficiently. The choice of original equipment fasteners is not arrived at by chance. Make sure replacement fasteners meet all the same requirements as the originals.

Threaded Fasteners

Threaded fasteners secure most of the components on the boat and engine. Most are tightened by turning them clockwise (right-hand threads). If the normal rotation of the fastener being tightened loosens, the fastener may have left-hand threads. If a left-hand threaded fastener is expected, it is noted in the text.

Nuts, bolts and screws are manufactured in a wide range of thread patterns. To join a nut and bolt, the diameter of the bolt and the diameter of the hole in the nut must be the same and the threads must be properly matched.

The best way to tell if the threads on two fasteners match is to turn the nut on the bolt (or the bolt into the threaded hole in a piece of equipment) with fingers only. Make sure both pieces are clean; remove Loctite or other sealer residue from threads if present. If force is required, check the thread condition on each fastener. If the thread condition is good but the fasteners jam, the threads are not compatible. A thread pitch gauge (**Figure 4**) can also be used to determine pitch.

NOTE

To ensure the fastener threads are not mismatched or cross-threaded, start all fasteners by hand. If a fastener is hard to start or turn, determine the cause before tightening it with a wrench.

Two dimensions are required to match the thread size of the fastener: the number of threads in a given distance and the outside diameter of the threads.

Two systems are currently used to specify threaded fastener dimensions: the U.S. Standard system and the metric system. Although fasteners may appear similar, close inspection shows that the thread designs are not the same (**Figure 5**). Pay particular attention when working with unidentified fasteners; mismatching thread types can damage threads.

NOTE
Most Yamaha engines are manufactured with predominantly International Organization for Standardization (ISO) metric fasteners, though some models may be equipped with components using U.S. Standard fasteners depending on the model and application.

U.S. Standard fasteners are sorted by grades (hardness/strength). Bolt heads are marked to represent different grades; no marks means the bolt is grade zero, two marks equal grade two, three marks equal grade five, four marks equal grade six, five marks equal grade seven and six marks equal grade eight. It is important when replacing fasteners to make sure the replacements are of equal or greater strength than the original.

U.S. Standard fasteners generally come in two pitches: coarse and fine. The coarse bolts/screws have fewer threads per inch than the fine. They are normally referred to by size such as 1/2-16 or 3/8-24. In these names the first number, 1/2 or 3/8 in the example, represent the measurement of the bolt diameter from the top of the threads to the top of the other side. The second number represents the number of threads per inch (16 or 24 in the case of the examples).

International Organization for Standardization (ISO) metric threads come in three standard thread sizes: coarse, fine and constant pitch. The ISO coarse pitch is used for most common fastener applications. The fine pitch thread is used on certain precision tools and instruments. The constant pitch thread is used mainly on machine parts and not for fasteners. The constant pitch thread, however, is used on all metric thread spark plugs.

The length (L, **Figure 6**), diameter (D) and distance between thread crests (pitch) (T) classify metric screws and bolts. The numbers 8–1.25 × 130 identify a typical bolt. This indicates the bolt has diameter of 8 mm. The distance between thread crests is 1.25 mm and the length is 130 mm.

NOTE
*When purchasing a bolt from a dealership or parts store, it is important to know how to specify bolt length. The correct way to measure bolt length is to measure the length, starting from underneath the bolt head to the end of the bolt (**Figure** 7). Always measure bolt length in this manner to avoid purchasing or installing bolts that are too long.*

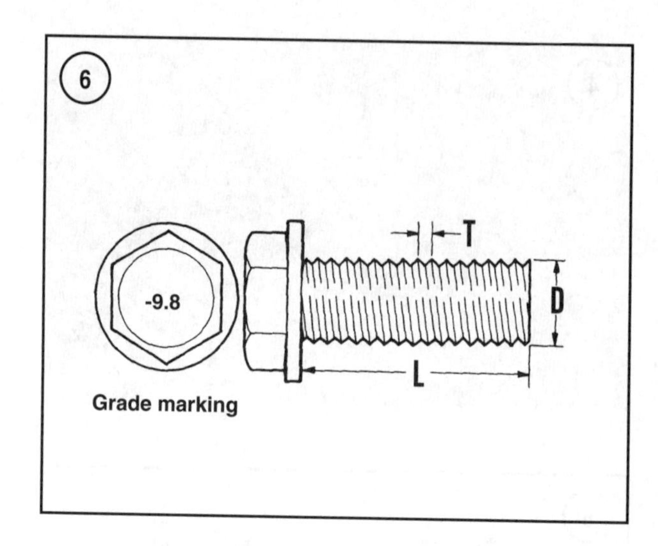

The grade marking located on the top of the fastener (**Figure 6**) indicates the strength of metric screws and bolts. The higher the number, the stronger the fastener. Unnumbered fasteners are the weakest.

Many screws, bolts and studs are combined with nuts to secure particular components. To indicate the size of a nut, manufacturers specify the internal diameter and the thread pitch.

The measurement across two flats on a nut or bolt indicates the wrench size.

WARNING
Do not install fasteners with a strength classification lower than what was originally installed by the manufacturer. Doing so may cause equipment failure and/or damage.

Torque Specifications

The materials used during the manufacturing of the engine may be subjected to uneven stresses if the fasteners of the various subassemblies are not installed and tightened correctly. Fasteners that are improperly installed or work loose can cause extensive damage. It is essential to use an accurate torque wrench, described in this chapter, with the torque specifications in this manual. Torque specifications are listed at the end of each chapter. If a torque is not listed, use the general torque specifications in **Table 5** of this chapter.

Specifications for torque are provided in Newton-meters (N•m), foot-pounds (ft.-lb.) and inch-pounds (in.-lb.). Torque specifications for specific components (including all critical torque figures) are at the end of the appropriate chapters. Torque wrenches are covered in the *Basic Tools* section.

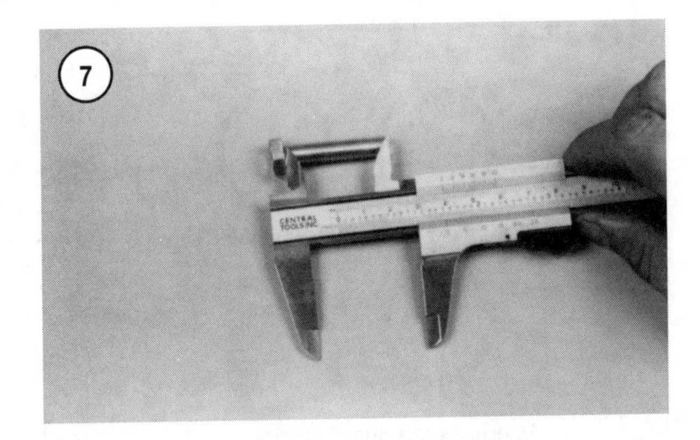

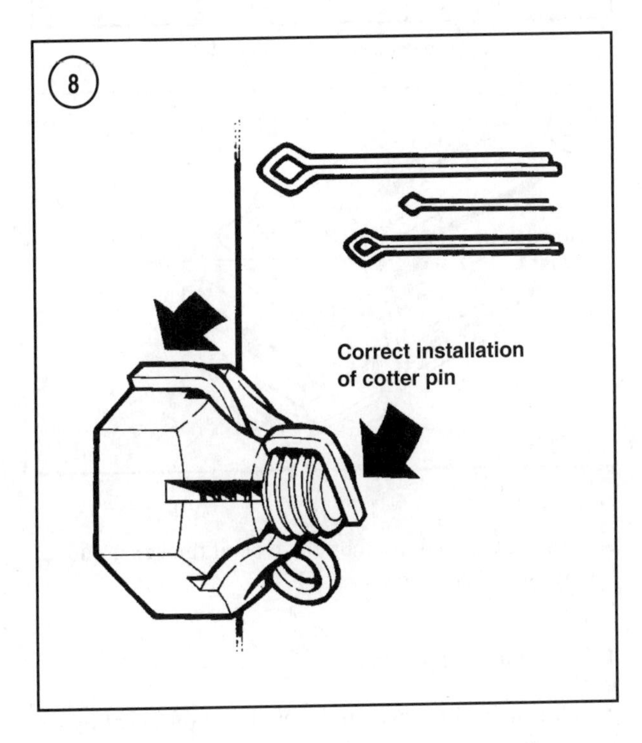

Self-Locking Fasteners

Several types of bolts, screws and nuts incorporate a system that creates interference between the two fasteners. Interference is achieved in various ways. The most common type is the nylon insert nut and a dry adhesive coating on the threads of a bolt.

Self-locking fasteners offer greater holding strength than standard fasteners, which improves their resistance to vibration. Most self-locking fasteners cannot be reused. The materials used to form the lock become distorted after the initial installation and removal. It is a good practice to discard and replace self-locking fasteners after their removal. Do not replace self-locking fasteners with standard fasteners.

Washers

There are two basic types of washers: flat washers and lockwashers. Flat washers are simple discs with a hole to fit a screw or bolt. Lockwashers prevent a fastener from working loose. Washers can be used as spacers and seals, or to help distribute fastener load and to prevent the fastener from damaging the component.

As with fasteners, when replacing washers make sure the replacement washers are the same design and quality.

NOTE

Give as much care to the selection and purchase of washers as given to bolts, nuts and other fasteners. Avoid washers that are made of thin and weak materials. These will deform and crush the first time they are used in a high torque application, allowing the nut or bolt to loosen.

Cotter Pins

A cotter pin is a split metal pin inserted into a hole or slot to prevent a fastener from loosening. In certain applications, the fastener must be secured in this way. For these applications, a cotter pin and castellated (slotted) nut are used.

To use a cotter pin, first make sure the diameter is correct for the hole in the fastener. After correctly tightening the fastener and aligning the holes, insert the cotter pin through the hole and bend the ends over the fastener (**Figure 8**). Cut the arms to a suitable length to prevent them from snagging on clothing, or worse, skin; remember that exposed ends of the pin cut flesh easily. When the cotter pin is bent and the arms cut to length, it must be tight. If it can be wiggled, it is improperly installed.

Unless instructed to do so, never loosen a torqued fastener to align the holes. If the holes do not align, tighten the fastener just enough to achieve alignment.

Cotter pins are available in various diameters and lengths. Measure length from the bottom of the head to the tip of the shortest pin.

Do not reuse cotter pins as the ends may break, causing the pin to fall out and allowing the fastener to loosen.

Snap Rings

Snap rings (**Figure 9**) are circular-shaped metal retaining clips. They help secure parts and gears in place such as shafts, pins or rods. External type snap rings retain items on shafts. Internal type snap rings secure parts within housing bores. In some applications, in addition to securing the component(s), snap rings of varying thickness also determine endplay. These are usually called selective snap rings.

There are two basic types of snap rings: machined and stamped snap rings. Machined snap rings (**Figure 10**) can be installed in either direction, since both faces have sharp edges. Stamped snap rings (**Figure 11**) have a sharp edge and a round edge. When installing a stamped snap ring in a thrust application, install the sharp edge facing away from the part producing the thrust.

Observe the following when installing snap rings:

1. Remove and install snap rings with snap ring pliers. See *Snap Ring Pliers* in this chapter.
2. In some applications, it may be necessary to replace snap rings after removing them.
3. Compress or expand snap rings only enough to install them. If overly expanded, they lose their retaining ability.
4. After installing a snap ring, make sure it seats completely.
5. Wear eye protection when removing and installing snap rings.

E-rings and circlips (**Figure 9**) are used when it is not practical to use a snap ring. Remove E-rings with a flat blade screwdriver by prying between the shaft and E-ring. To install an E-ring, center it over the shaft groove and push or tap it into place.

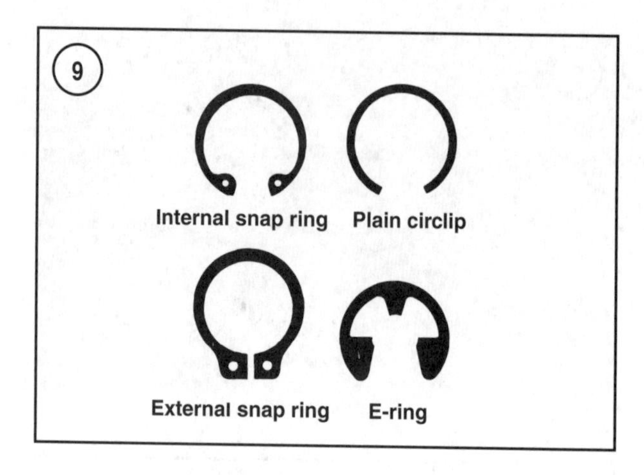

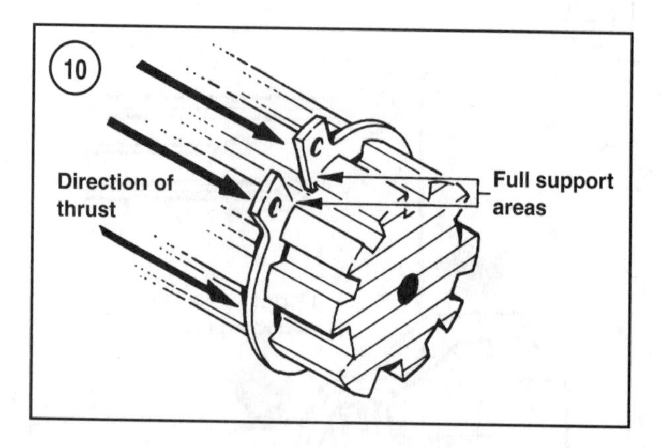

SHOP SUPPLIES

Lubricants and Fluids

Periodic lubrication helps ensure long life for any type of equipment. The *type* of lubricant used is just as important as the lubrication service itself, although in an emergency the wrong type of lubricant is usually better than no lubricant at all. The following information describes the types of lubricants most often used on marine equipment. Be sure to follow the manufacturer's recommendations for lubricant types.

NOTE
*For more information on Yamaha recommended lubricants, please refer to **Quick Reference Data** at the beginning of this manual or the information and tables in Chapter Three.*

Generally, all liquid lubricants are called *oil*. They may be mineral-based (including petroleum bases), natural-based (vegetable and animal bases), synthetic-based or emulsions (mixtures). *Grease* is oil to which a thickening base was added so that the end product is semi-solid. Grease is often classified by the type of thickener added; lithium soap is commonly used.

Two-Stroke Engine Oil

Lubrication for a two-stroke engine is provided by oil mixed into the incoming air-fuel mixture. Some of the oil mist settles in the crankcase, lubricating the crankshaft, bearings and lower end of the connecting rod. The rest of the oil enters the combustion chamber to lubricate the piston, rings and the cylinder wall. This oil is burned with the air-fuel mixture during the combustion process.

Engine oil must have several special qualities to work well in a two-stroke engine. It must mix easily and stay in suspension with gasoline. When burned, it cannot leave behind excessive deposits. It must also withstand the high operating temperature associated with two-stroke engines.

The National Marine Manufacturer's Association (NMMA) has set standards for oil used in two-stroke, water-cooled engines. This is the NMMA TC-W (two-cycle, water-cooled) grade (**Figure 12**). It indicates the oil's performance in the following areas:

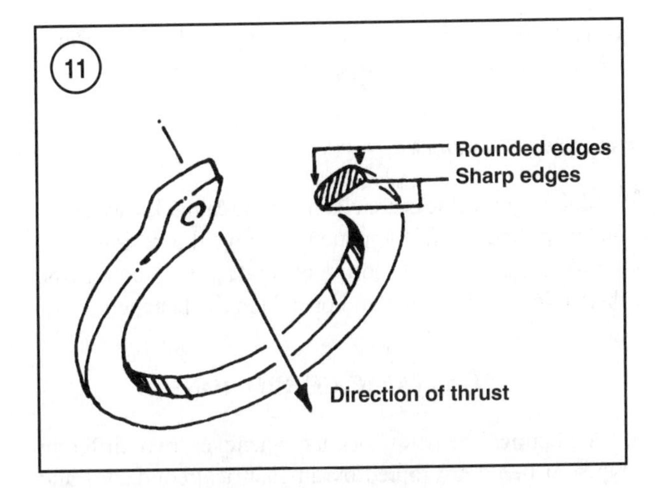

1. Lubrication (prevention of wear and scuffing).
2. Spark plug fouling.
3. Piston ring sticking.
4. Preignition.
5. Piston varnish.
6. General engine condition (including deposits).
7. Exhaust port blockage.
8. Rust prevention.
9. Mixing ability with gasoline.

In addition to oil grade, manufacturers specify the ratio of gasoline to oil required during break-in and normal engine operation.

Gearcase Oil

Gearcase lubricants are assigned SAE viscosity numbers under the same system as four-stroke engine oil. Gearcase lubricant falls into the SAE 72-250 range. Some gearcase lubricants, such as SAE 85-90, are multigrade.

Three types of marine gearcase lubricant are generally available: SAE 90 hypoid gearcase lubricant is designed for older manual-shift units; Type C gearcase lubricant contains additives designed for the electric shift mechanisms; High viscosity gearcase lubricant is a heavier oil designed to withstand the shock loading of high performance engines or engines subjected to severe duty use. Always use a gearcase lubricant of the type specified by the gearcase manufacturer.

Greases

Grease is lubricating oil with thickening agents added to it. The National Lubricating Grease Institute (NLGI) grades grease. Grades range from No. 000 to No. 6, with No. 6 being the thickest. Typical multipurpose grease is NLGI No. 2. For specific applications, manufacturers may recommend water-resistant type grease or one with an additive such as molybdenum disulfide (MoS^2).

Cleaners, Degreasers and Solvents

Many chemicals are available to remove oil, grease and other residue.

Before using cleaning solvents, consider how they will be used and disposed of, particularly if they are not water-soluble. Local ordinances may require special procedures for the disposal of many types of cleaning chemicals. Refer to *Safety and Cleaning Parts* in this chapter for more information on their use.

Use electrical contact cleaner to clean wiring connections and components without leaving any residue. Carburetor cleaner is a powerful solvent used to remove fuel deposits and varnish from fuel system components. Use this cleaner carefully, as it may damage finishes.

Generally, degreasers are strong cleaners used to remove heavy accumulations of grease from engine and frame components.

Most solvents are used in a parts washing cabinet for individual component cleaning. For safety, use only nonflammable or high flash point solvents.

Gasket Sealant

Sealants are used in combination with a gasket or seal and are occasionally used alone. Follow the manufacturer's recommendation when using sealants. Use extreme care when choosing a sealant different from the type originally recommended. Choose sealants based on their resistance to heat, various fluids and their sealing capabilities.

One of the most common sealants is RTV, or room temperature vulcanizing sealant. This sealant cures at room temperature over a specific time period. This allows the repositioning of components without damaging gaskets.

Moisture in the air causes the RTV sealant to cure. Always install the tube cap as soon as possible after applying RTV sealant. RTV sealant has a limited shelf life and does not cure properly if the shelf life has expired. Keep partial tubes sealed and discard them if they have surpassed the expiration date.

Applying RTV sealant

Clean all old gasket residue from the mating surfaces. Remove all gasket material from blind threaded holes; it can cause inaccurate bolt torque. Spray the mating surfaces with aerosol parts cleaner and then wipe with a lint-free cloth. The area must be clean for the sealant to adhere.

Apply RTV sealant in a continuous bead 2-3 mm (0.08-0.12 in.) thick. Circle all the fastener holes unless otherwise specified. Do not allow any sealant to enter these holes. Assemble and tighten the fasteners to the specified torque within the time frame recommended by the RTV sealant manufacturer (usually within 10-15 minutes).

Gasket Remover

Aerosol gasket remover can help remove stubborn gaskets and prevent damage to the mating surface that may be caused by using a scraping tool. Most of these types of products are very caustic. Follow the gasket remover manufacturer's instructions.

Threadlocking Compound

A threadlocking compound is a fluid applied to the threads of fasteners. After tightening the fastener, the fluid dries and becomes a solid filler between the threads. This makes it difficult for the fastener to work loose from vibration, or heat expansion and contraction. Some threadlocking compounds also provide a seal against fluid leaks.

Before applying threadlocking compound remove any old compound from both thread areas and clean them with aerosol parts cleaner. Use the compound sparingly. Excess fluid can run into adjoining parts.

Threadlocking compounds come in different strengths. Follow the particular manufacturer's recommendations regarding compound selection. Two manufacturers of threadlocking compound are ThreeBond and Loctite, which offer a wide range of compounds for various strength, temperature and repair applications.

Applying threadlock

Make sure surfaces are clean. If a threadlock was previously applied to the component, remove this residue.

Shake the container thoroughly and apply to both parts, then assemble the parts and/or tighten the fasteners.

GALVANIC CORROSION

A chemical reaction occurs whenever two different types of metal are joined by an electrical conductor and immersed in an electrolytic solution such as water. Electrons transfer from one metal to the other through the electrolyte and return through the conductor.

The hardware on a boat is made of many different types of metal. The boat hull acts as a conductor between the metals. Even if the hull is wooden or fiberglass, the slightest film of water on the hull provides conductivity by acting as electrolyte. This combination creates a good environment for electron flow. Unfortunately, this electron flow results in galvanic corrosion of the metal involved, causing one of the metals to be corroded or eroded away. The amount of electron flow, and therefore the amount of corrosion, depends on several factors:

1. The types of metal involved.
2. The efficiency of the conductor.
3. The strength of the electrolyte.

Metals

The chemical composition of the metal used in marine equipment has a significant effect on the amount and speed of galvanic corrosion. Certain metals are more resistant to corrosion than others. These electrically negative metals are commonly called *noble*; they act as the cathode in any reaction. Metals that are more subject to corrosion are electrically positive; they act as the anode in a reaction. The more noble metals include titanium, 18-8 stainless steel and nickel. Less noble metals include zinc, aluminum and magnesium. Galvanic corrosion becomes more excessive as the difference in electrical potential between the two metals increases.

In some cases, galvanic corrosion can occur within a single piece of metal. For example, brass is a mixture of zinc and copper, and, when immersed in an electrolyte, the zinc portion of the mixture will corrode away as a galvanic reaction occurs between the zinc and copper particles.

Conductors

The hull of the boat often acts as the conductor between different types of metal. Marine equipment, such as the engine/gearcase of the outboard can act as the conductor. Large masses of metal, firmly connected together, are more efficient conductors than water. Rubber mountings and vinyl-based paint can act as insulators between pieces of metal.

Electrolyte

The water in which a boat operates acts as the electrolyte for the corrosion process. The more efficient a conductor is, the more excessive and rapid the corrosion will be.

Cold, clean freshwater is the poorest electrolyte. Pollutants increase conductivity; therefore, brackish or saltwater is an efficient electrolyte. This is one of the reasons that most manufacturers recommend a freshwater flush after operating in polluted, brackish or saltwater.

Protection from Galvanic Corrosion

Because of the environment in which marine equipment must operate, it is practically impossible to totally prevent galvanic corrosion. However, there are several ways in which the process can be slowed. After taking these precautions, the next step is to *fool* the process into occurring only in certain places. This is the role of sacrificial anodes and impressed current systems.

Slowing Corrosion

Some simple precautions can help reduce the amount of corrosion taking place outside the hull. These precautions are not substitutes for the corrosion protection methods discussed in *Sacrificial Anodes* and *Impressed Current Systems* in this chapter, but they can help these methods reduce corrosion.

Use fasteners made of metal more noble than the parts they secure. If corrosion occurs, the parts they secure may suffer but the fasteners are protected. The larger secured parts are able to withstand the loss of material. Also major problems could arise if the fasteners corrode to the point of failure.

Keep all painted surfaces in good condition. If paint is scraped off and bare metal exposed, corrosion rapidly increases. Use a vinyl- or plastic-based paint, which acts as an electrical insulator.

Be careful when applying metal-based antifouling paint to the boat. Do not apply antifouling paint to metal parts of the boat or the outboard engine/gearcase. If applied to metal surfaces, this type of paint reacts with the metal and results in corrosion between the metal and the layer of paint. Maintain a minimum 1 in. (25 mm) border between the painted surface and any metal parts. Organic-based paints are available for use on metal surfaces.

Where a corrosion protection device is used, remember that it must be immersed in the electrolyte along with the boat to provide any protection. If the outboard is raised out of the water when the boat is docked, any anodes on the engine will be removed from the corrosion process rendering them ineffective. (Of course, the engine requires less protection when raised out of the water/electrolyte.) Never paint or apply any coating to anodes or other protection devices. Paint or other coatings insulate them from the corrosion process.

Any change in boat equipment, such as the installation of a new stainless steel propeller, changes the electrical potential and may cause increased corrosion. Always consider this fact when adding equipment or changing exposed materials. Install additional anodes or other protection equipment as required to ensure the corrosion protection system is up to the task. The expense to repair corrosion damage usually far exceeds that of additional corrosion protection.

Sacrificial Anodes

Sacrificial anodes are specially designed to do nothing but corrode. Properly fastening such pieces to the boat causes them to act as the anode in any galvanic reaction that occurs; any other metal in the reaction acts as the cathode and is not damaged.

Anodes are usually made of zinc, a less noble material. Some anodes are manufactured of an aluminum and indium alloy. This alloy is less noble than the aluminum alloy in drive system components, providing the desired sacrificial properties. The aluminum and indium alloy is more resistant to oxide coating than zinc anodes. Oxide coating occurs as the anode material reacts with oxygen in the water. An oxide coating insulates the anode, dramatically reducing corrosion protection.

Anodes must be used properly to be effective. Simply fastening anodes to the boat in random locations does not do the job.

First determine how much anode surface is required to adequately protect the equipment surface area. A good starting point is provided by the Military Specification MIL-A-818001, which states that one square inch of new anode protects either:

1. 800 sq. in. of freshly painted steel.
2. 250 sq. in. of bare steel or bare aluminum alloy.
3. 100 sq. in. of copper or copper alloy.

This rule is valid for a boat at rest. If underway, additional anode areas are required to protect the same surface area.

The anode must be in good electrical contact with the metal that it protects. If possible, attach an anode to all metal surfaces requiring protection.

Quality anodes have inserts around the fastener holes that are made of a more noble material. Otherwise, the anode could erode away around the fastener hole, allowing the anode to loosen or possibly fall off, thereby losing needed protection.

Impressed Current System

An impressed current system can be added to any boat. The system generally consists of the anode, controller and reference electrode. The anode in this system is coated with a very noble metal, such as platinum, so that it is almost corrosion-free and can last almost indefinitely. The reference electrode, under the boat waterline, allows the control module to monitor the potential for corrosion. If the module senses that corrosion is occurring, it applies positive battery voltage to the anode. Current then flows from the anode to all other metal components, regardless of how noble or non-noble these components may be. Essentially, the electrical current from the battery counteracts the galvanic reaction to dramatically reduce corrosion damage.

Only a small amount of current is needed to counteract corrosion. Using input from the sensor, the control module provides only the amount of current needed to suppress galvanic corrosion. Most systems consume a maximum of 0.2 Ah at full demand. Under normal conditions, these systems can provide protection for 8-12 weeks without recharging the battery. Remember that this system must have constant connection to the battery. Often the battery supply to the system is connected to a battery switching device causing the operator to inadvertently shut off the system while docked.

An impressed current system is more expensive to install than sacrificial anodes, but considering the low maintenance requirements and the superior protection it provides, the long-term cost may be lower.

PROPELLERS

The propeller is the final link between the boat drive system and the water. A perfectly maintained engine and hull are useless if the propeller is the wrong type, damaged or deteriorated. Although propeller selection for a specific application is beyond the scope of this manual, the following provides the basic information needed to make an informed decision. A professional at a reputable marine dealership is the best source for a propeller recommendation.

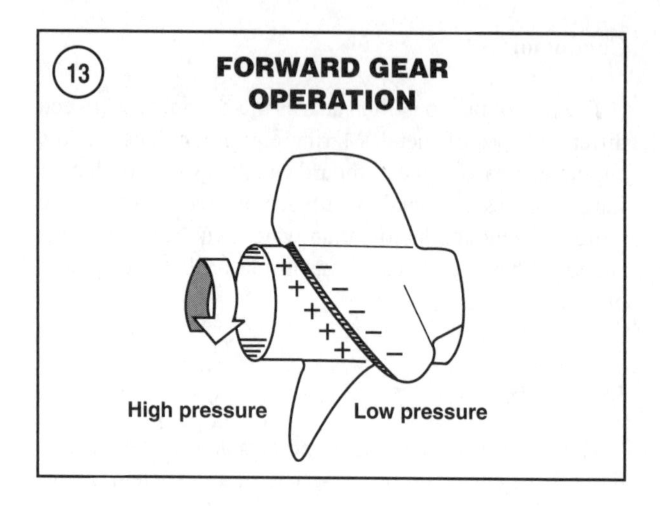

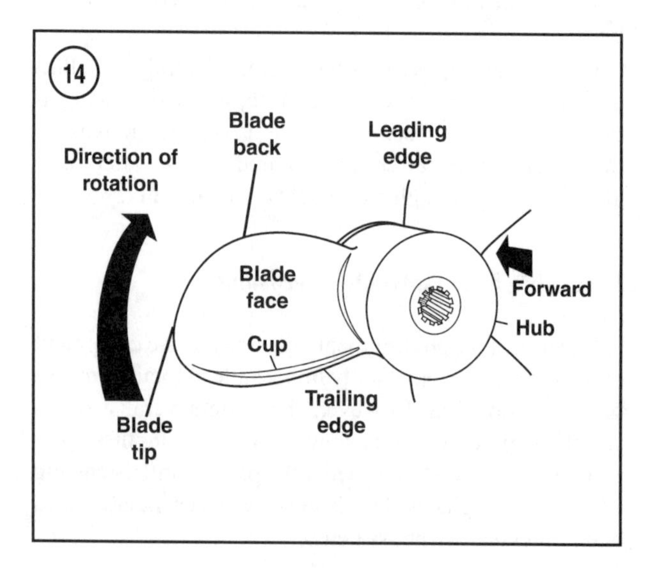

How a Propeller Works

As the curved blades of a propeller rotate through the water, a high-pressure area forms on one side of the blade and a low-pressure area forms on the other side of the blade (**Figure 13**). The propeller moves toward the low-pressure area, carrying the boat with it.

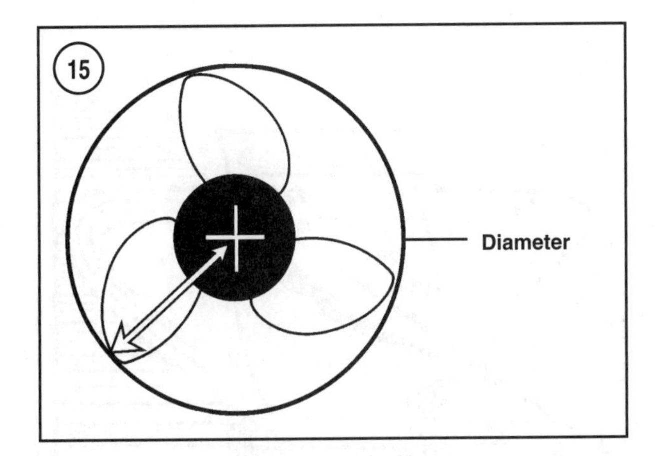

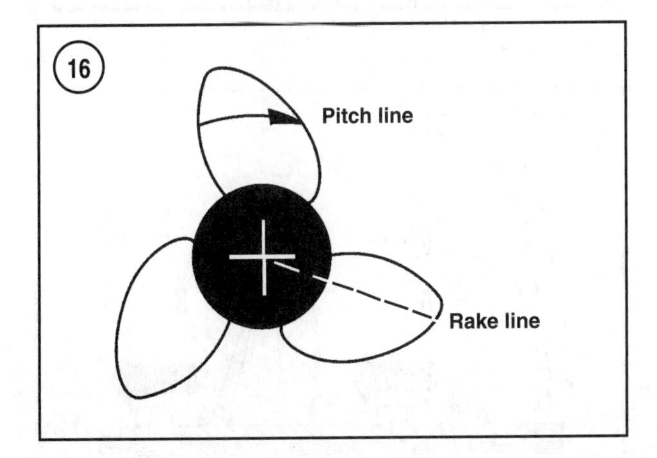

Propeller Parts

Although a propeller is usually a one-piece unit, it is made of several different parts (**Figure 14**). Variations in the design of these parts make different propellers suitable for different applications.

The blade tip is the point of the blade furthest from the center of the propeller hub or propeller shaft bore. The blade tip separates the leading edge from the trailing edge.

The leading edge is the edge of the blade nearest the boat. During forward operation, this is the area of the blade that first cuts through the water.

The trailing edge is the surface of the blade furthest from the boat. During reverse operation, this is the area of the blade that first cuts through the water.

The blade face is the surface of the blade that faces away from the boat. During forward operation, high-pressure forms on this side of the blade.

The blade back is the surface of the blade that faces toward the boat. During forward gear operation, low-pressure forms on this side of the blade.

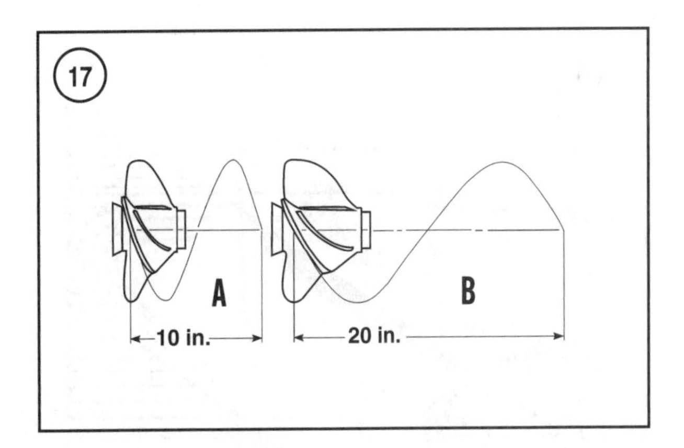

The cup is a small curve or lip on the trailing edge of the blade. Cupped propeller blades generally perform better than non-cupped propeller blades.

The hub is the center portion of the propeller. It connects the blades to the propeller shaft. On most drive systems, engine exhaust is routed through the hub; in this case, the hub is made up of an outer and inner portion, connected by ribs.

A diffuser ring is used on through-hub exhaust models to prevent exhaust gasses from entering the blade area.

Propeller Design

Changes in length, angle, thickness and material of propeller parts make different propellers suitable for different applications.

Diameter

Propeller diameter is the distance from the center of the hub to the blade tip, multiplied by two. Essentially it is the diameter of the circle formed by the blade tips during propeller rotation (**Figure 15**).

Pitch and rake

Propeller pitch and rake describe the placement of the blades in relation to the hub (**Figure 16**).

Pitch describes the theoretical distance the propeller would travel in one revolution. In A, **Figure 17**, the propeller would travel 10 in. in one revolution. In B, **Figure 17**, the propeller would travel 20 in. in one revolution. This distance is only theoretical; during typical operation, the propeller achieves only 75-85% of its pitch. Slip rate describes the difference in actual travel relative to the pitch. Lighter, faster boats typically achieve a lower slip rate than heavier, slower boats.

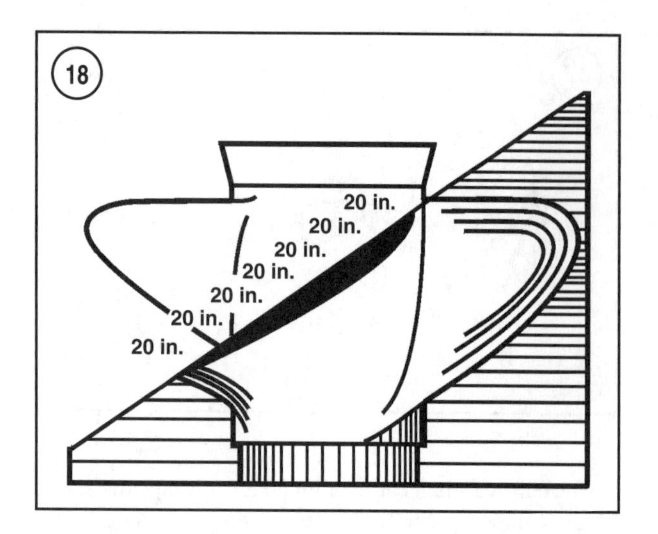

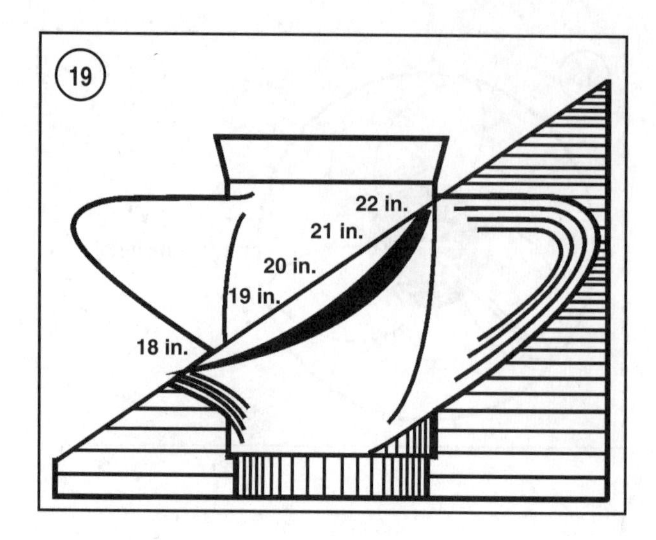

Propeller blades can be constructed with constant pitch (**Figure 18**) or progressive pitch (**Figure 19**). On a progressive propeller, the pitch starts low at the leading edge and increases toward the trailing edge. The propeller pitch specification is the average of the pitch across the entire blade. Propellers with progressive pitch usually provide better overall performance than constant pitch propellers.

Blade rake is specified in degrees and is measured along a line from the center of the hub to the blade tip. A blade that is perpendicular to the hub (A, **Figure 20**) has 0° rake. A blade that is angled from perpendicular (B, **Figure 20**) has a rake expressed by its difference from perpendicular. Most propellers have rakes ranging from 0-20°. Lighter, faster boats generally perform better using a propeller with a greater amount of rake. Heavier, slower boats generally perform better using a propeller with less rake.

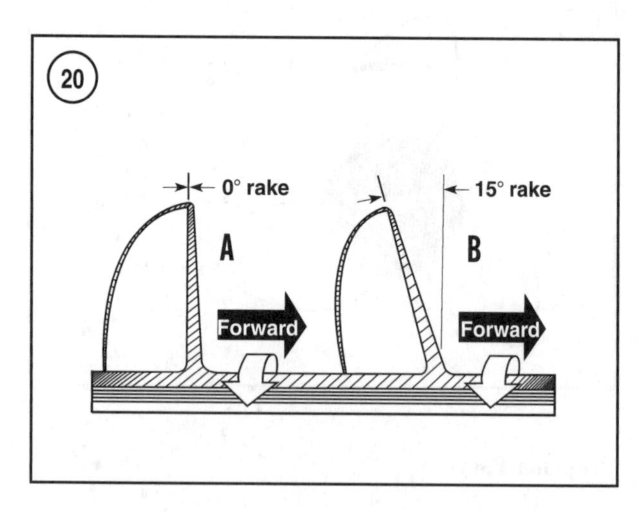

Propellers that run only partially submerged, as in racing applications, may have a wedge shaped cross-section (**Figure 21**). The leading edge is very thin and the blade thickness increases toward the trailing edge, where it is thickest. If a propeller such as this type is run totally submerged, it is very inefficient.

Blade thickness

Blade thickness is not uniform at all points along the blade. For efficiency, blades are as thin as possible at all points while retaining enough strength to move the boat. Blades are thicker where they meet the hub and thinner at the blade tips. This construction is necessary to support the heavier loads at the hub section of the blade. Overall blade thickness is dependent on the strength of the material used.

When cut along a line from the leading edge to the trailing edge in the central portion of the blade, the propeller blade resembles an airplane wing. The blade face, where high-pressure exists during forward rotation, is almost flat. The blade back, where low-pressure exists during forward rotation, is curved, with the thinnest portions at the edges and the thickest portion at the center.

Number of blades

The number of blades on a propeller is a compromise between efficiency and vibration. A one-bladed propeller would be the most efficient, but it would create an unacceptable amount of vibration. As blades are added, efficiency decreases, but so does vibration. Most propellers have three or four blades, representing the most practical trade-off between efficiency and vibration.

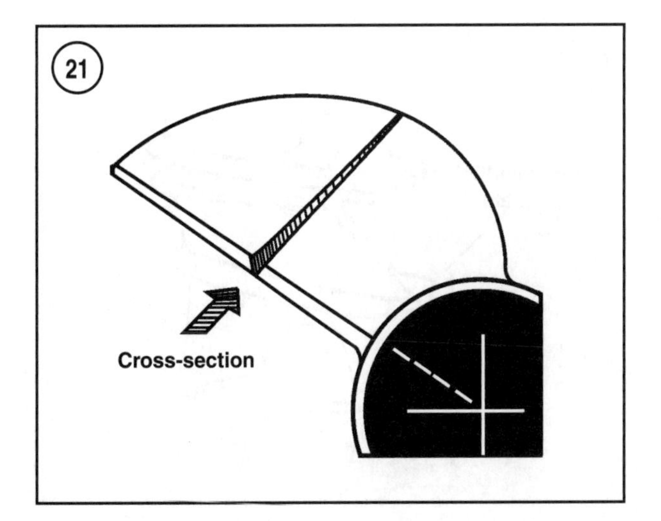

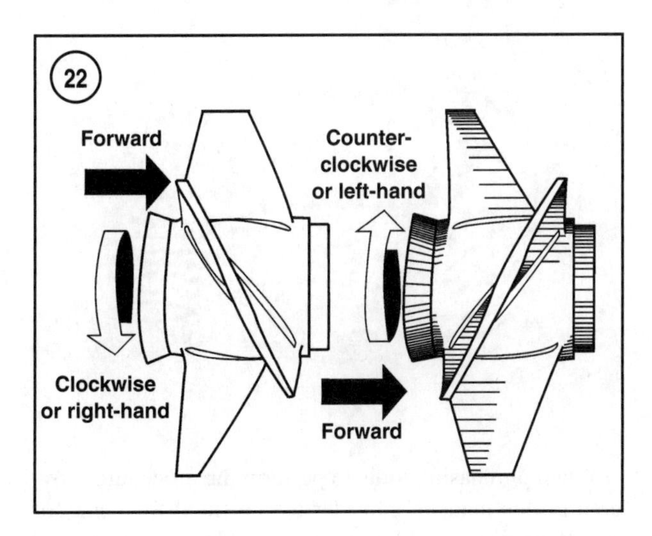

Material

Propeller materials are chosen for strength, corrosion resistance and economy. Stainless steel, aluminum, plastic and bronze are the most commonly used materials. Bronze is quite strong but rather expensive. Stainless steel is more common than bronze because of the combination of strength and lower cost. Aluminum alloy and plastic materials are the least expensive but usually lack the strength of stainless steel. Plastic propellers are more suited for lower horsepower applications.

Direction of rotation

Propellers are made for both right-hand and left-hand rotations although right-hand is the most commonly used. As viewed from the rear of the boat while in forward gear, a right-hand propeller turns clockwise and a left-hand propeller turns counterclockwise. Off the boat, the direction of rotation is determined by observing the angle of the blades (**Figure 22**). Right-handed propeller blades slant from the upper left to the lower right; left-handed propeller blades are opposite.

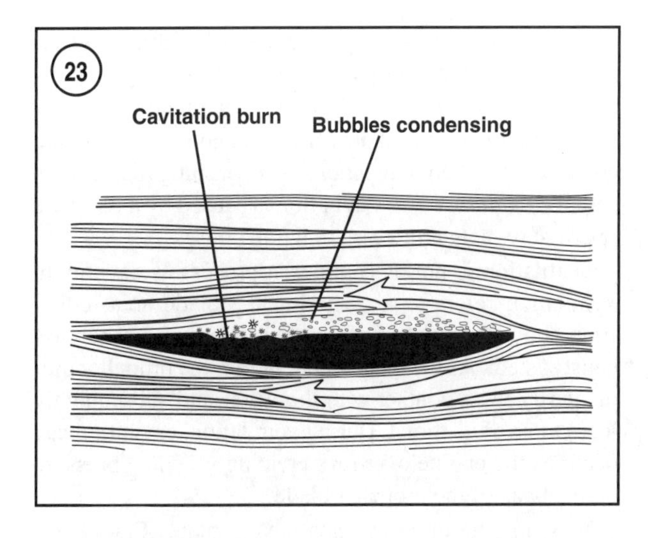

Cavitation and Ventilation

Cavitation and ventilation are *not* interchangeable terms; they refer to two distinct problems encountered during propeller operation.

To help understand cavitation, consider the relationship between pressure and the boiling point of water. At sea level, water boils at 212° F (100° C). As pressure increases, such as within an engine cooling system, the boiling point of the water increases—it boils at a temperature higher than 212° F (100° C). The opposite is also true. As pressure decreases, water boils at a temperature lower than 212° F (100° C). If the pressure drops low enough, water will boil at normal room temperature.

During normal propeller operation, low pressure forms on the blade back. Normally the pressure does not drop low enough for boiling to occur. However, poor propeller design, damaged blades or using the wrong propeller can cause unusually low pressure on the blade surface (**Figure 23**). If the pressure drops low enough, boiling occurs and bubbles form on the blade surfaces. As the boiling water moves to a higher pressure area of the blade, the boiling ceases and the bubbles collapse. The collapsing bubbles release energy that erodes the surface of the propeller blade.

Corroded surfaces, physical damage or even marine growth combined with high-speed operation can cause low pressure and cavitation on outboard gearcase surfaces. In such cases, low pressure forms as water flows over a protrusion or rough surface. The boiling water forms bubbles that collapse as they move to a higher pressure area toward the rear of the surface imperfection.

This entire process of pressure drop, boiling and bubble collapse is called *cavitation*. The ensuing damage is called *cavitation burn*. Cavitation is caused by a decrease in pressure, not an increase in temperature.

Ventilation is not as complex a process as cavitation. Ventilation refers to air entering the blade area, either from above the water surface or from a through-hub exhaust system. As the blades meet the air, the propeller momentarily loses contact with the water and subsequently loses most of its thrust. During ventilation, cavitation can occur as the engine over-revs creating very low pressure on the back of the propeller blade.

Most marine drive systems have a plate (**Figure 24**) above the propeller designed to prevent surface air from entering the blade area. This plate is an *anti-ventilation plate*, although it is often incorrectly called an *anti-cavitation plate*.

Most propellers have a flared section at the rear of the propeller called a diffuser ring. This feature forms a barrier, and extends the exhaust passage far enough aft to prevent the exhaust gases from ventilating the propeller.

A close fit of the propeller to the gearcase is necessary to keep exhaust gasses from exiting and ventilating the propeller. Using the wrong propeller attaching hardware can position the propeller too far aft, preventing a close fit. The wrong hardware can also allow the propeller to rub heavily against the gearcase, causing rapid wear to both components. Wear or damage to these surfaces allows the propeller to ventilate.

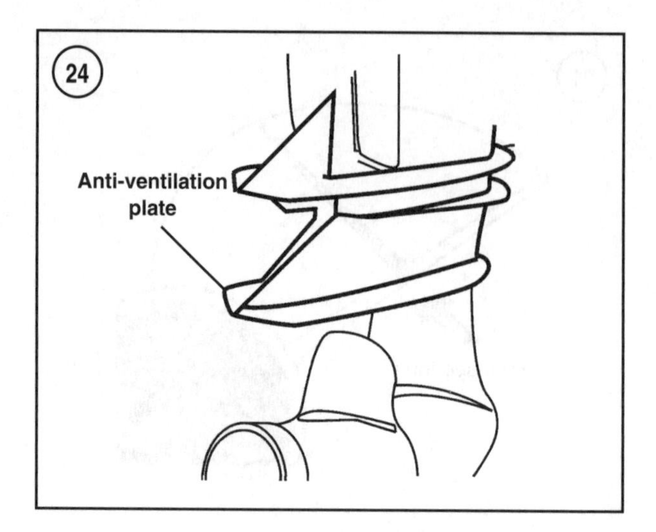

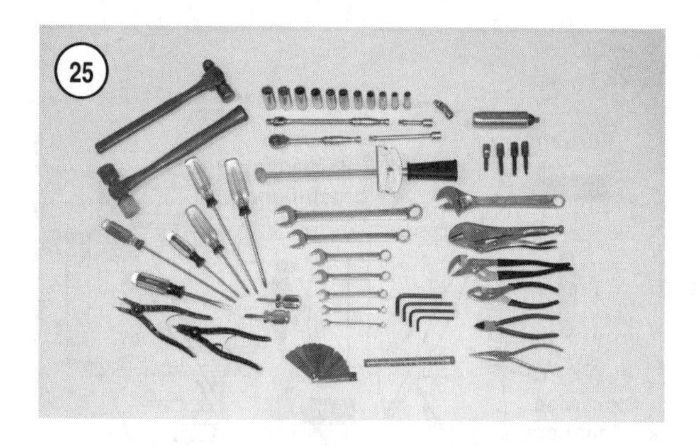

BASIC TOOLS

Most of the procedures in this manual can be carried out with simple hand tools and test equipment familiar to the home mechanic. Always use the correct tools for the job at hand. Keep tools organized and clean. Store them in a tool chest with related tools organized together.

After using a tool, wipe off dirt and grease with a clean cloth. Wiping tools off is especially important when servicing the craft in areas where they can come in contact with sand. Sand is very abrasive and causes premature wear to engine parts.

Quality tools are essential. The best are constructed of high-strength alloy steel. These tools are light, easy to use and resistant to wear. Working surfaces are devoid of sharp edges and the tool is carefully polished. They have an easy-to-clean finish and are comfortable to use. Quality tools are a good investment.

When purchasing tools to perform the procedures covered in this manual, consider the potential frequency of use. If starting a tool kit, consider purchasing a basic tool set (**Figure 25**) from a large tool supplier. These sets are available in many tool combinations and offer substantial savings when compared to individually purchased tools. As work experience grows and tasks become more complicated, specialized tools can be added.

Screwdrivers

Screwdrivers of various lengths and types are mandatory for the simplest tool kit. The two basic types are the slotted tip (flat blade) and the Phillips tip. These are available in sets that often include an assortment of tip sizes and shaft lengths.

As with all tools, use a screwdriver designed for the job. Make sure the size of the tip conforms to the size and shape of the fastener. Use them only for driving screws. Never use a screwdriver for prying or chiseling metal. Re-

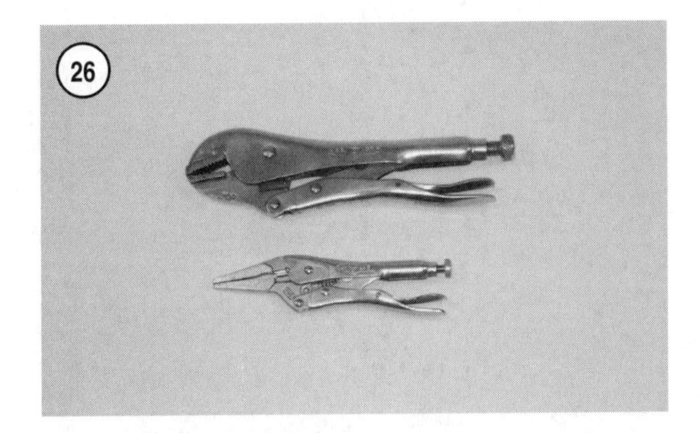

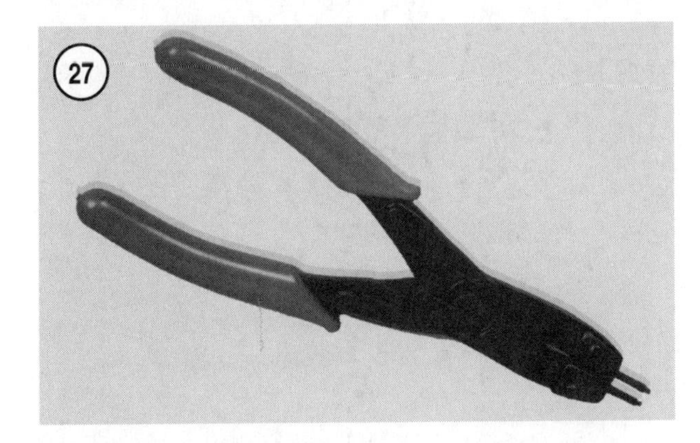

pair or replace worn or damaged screwdrivers. A worn tip may damage the fastener, making it difficult to remove.

Phillips screwdrivers are sized according to point size. They are numbered one, two, three and four. The degree of taper determines the point size; the No. 1 Phillips screwdriver is the most pointed. The points are more blunt as the number increases.

Pliers

Pliers come in a wide range of types and sizes. Though pliers are useful for holding, cutting, bending and crimping, they should never be used to turn bolts or nuts.

Each design has a specialized function. Slip-joint pliers are general-purpose pliers used for gripping and bending. Diagonal cutting pliers are needed to cut wire and can be used to remove cotter pins. Needlenose pliers are used to hold or bend small objects. Locking pliers (**Figure 26**), sometimes called vise-grips, are used to hold objects very tightly. They have many uses ranging from holding two parts together to gripping the end of a broken stud. Use caution when using locking pliers, as the sharp jaws damage the objects they hold.

Snap Ring Pliers

Snap ring pliers (**Figure 27**) are specialized pliers with tips that fit into the ends of snap rings to remove and install them.

Snap ring pliers are available with fixed action (either internal or external) or convertible (one tool works on both internal and external snap rings). They may have fixed tips or interchangeable ones of various sizes and angles. For general use, select convertible type pliers with interchangeable tips.

WARNING

Snap rings can slip and fly off when removing and installing them. Also, the snap ring plier tips may break. Always wear eye protection when using snap ring pliers.

Hammers

Various types of hammers are available to fit a number of applications. A ball-peen hammer is used to strike another tool, such as a punch or chisel. Soft-faced hammers are required when a metal object must be struck without damaging it. *Never* use a metal-faced hammer on engine components, as damage does occur in most cases.

Always wear eye protection when using hammers. Make sure the hammer face is in good condition and the handle is not cracked. Select the correct hammer for the job and make sure to strike the object squarely. Do not use the handle or the side of the hammer to strike an object.

When striking a hammer against a punch, cold chisel or similar tool, the face of the hammer should be at least 1/2 in. larger than the head of the tool. When it is necessary to strike hard against a steel part without damaging it, use a brass hammer. Brass will give when used on a harder object.

Wrenches

Box-end, open-end and combination wrenches (**Figure 28**) come in a variety of types and sizes.

The number stamped on the wrench refers to the distance between the work areas. This size must match the size of the fastener head.

The box-end wrench is an excellent tool because it grips the fastener on all sides. This factor reduces the chance of the tool slipping. The box-end wrench is designed with either a six or 12-point opening. For stubborn or damaged fasteners, the six-point provides superior holding ability by contacting the fastener across a wider area at all six edges. For general use, the 12-point works well. It allows the wrench to be removed and reinstalled without moving the handle over such a wide arc.

An open-end wrench is fast and works best in areas with limited overhead access. It contacts the fastener at only two points, and can slip under heavy force, or if the tool or fastener is worn. A box-end wrench is preferred in most instances, especially when breaking loose and applying the final tightness to a fastener.

The combination wrench has a box-end on one end, and an open-end on the other. This combination makes it a very convenient tool.

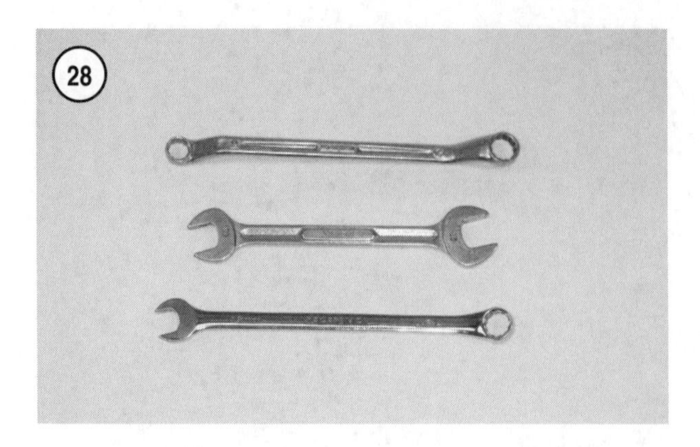

Adjustable Wrenches

An adjustable wrench or Crescent wrench can fit nearly any nut or bolt head that has clear access around the entire perimeter. Adjustable wrenches are best used as a backup wrench to keep a large nut or bolt from turning while the other end is being loosened or tightened with a box-end or socket wrench.

Adjustable wrenches contact the fastener at only two points, which makes them more subject to slipping off the fastener. The fact that one jaw is adjustable and may loosen only aggravates this shortcoming. Make certain the solid jaw is the one transmitting the force.

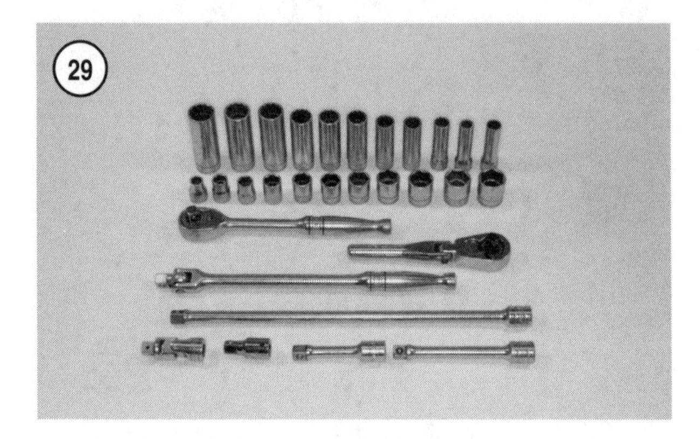

Socket Wrenches, Ratchets and Handles

Sockets that attach to a ratchet handle (**Figure 29**) are available with six-point or 12-point openings and different drive sizes. The drive size (1/4, 3/8, 1/2 and 3/4 in.) indicates the size of the square hole that accepts the ratchet handle. The number stamped on the socket is the size of the work area and must match the fastener head.

As with wrenches, a six-point socket provides superior-holding ability, while a 12-point socket needs to be moved only half as far to reposition it on the fastener.

Sockets are designated for either hand or impact use. Impact sockets are made of thicker material for more durability.

WARNING
Do not use hand sockets with air or impact tools, as they may shatter and cause injury. Always wear eye protection when using impact or air tools.

Various handles are available for sockets. The speed handle is used for fast operation. Flexible ratchet heads in varying lengths allow the socket to be turned with varying force, and at odd angles. Extension bars allow the socket setup to reach difficult areas. The ratchet is the most versatile. It allows the user to install or remove the nut without removing the socket.

Sockets combined with any number of drivers make them undoubtedly the fastest, safest and most convenient tool for fastener removal and installation.

Impact Driver

An impact driver provides extra force for removing fasteners, by converting the impact of a hammer into a turning motion. This makes it possible to remove stubborn fasteners without damaging them. Impact drivers and interchangeable bits (**Figure 30**) are available from most tool suppliers. When using a socket with an impact driver, make sure the socket is designed for impact use. Refer to *Socket Wrenches, Ratchets and Handles* in this chapter.

WARNING
Do not use hand sockets with air or impact tools as they may shatter and cause injury. Always wear eye protection when using impact or air tools.

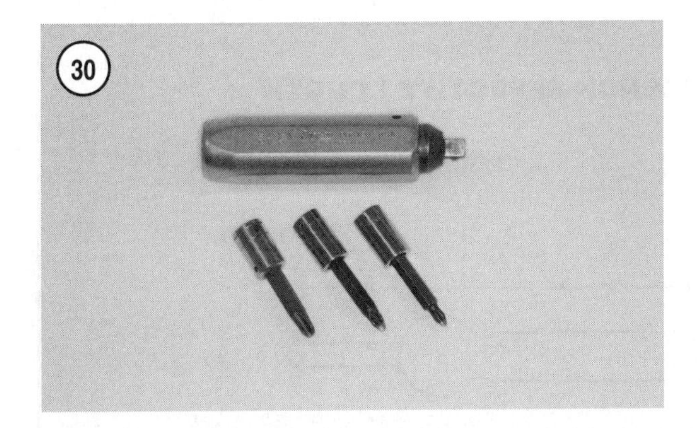
30

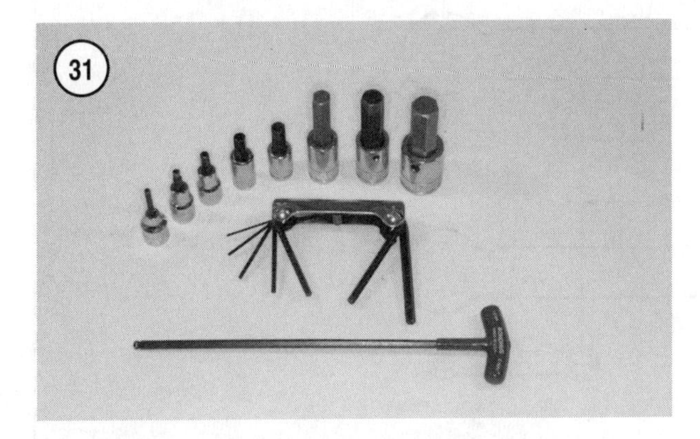
31

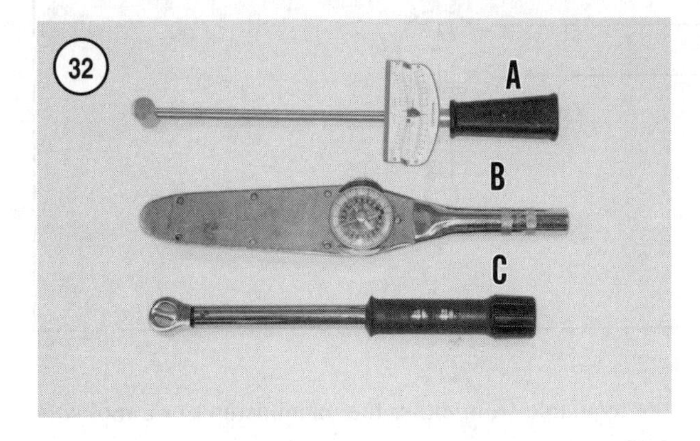

32

Impact drivers are great for the home mechanic as they offer many of the advantages of air tools without the need for a costly air compressor to run them.

Allen Wrenches

Allen or setscrew wrenches (**Figure 31**) are used on fasteners with hexagonal recesses in the fastener head. These wrenches come in L-shaped bar, socket and T-handle types. Allen bolts are sometimes called socket bolts.

Torque Wrenches

A torque wrench is used with a socket, torque adapter or similar extension to measure torque while tightening a fastener. Torque wrenches come in several drive sizes (1/4, 3/8, 1/2 and 3/4) and have various methods of reading the torque value. The drive size is the size of the square drive that accepts the socket, adapter or extension. Common types of torque wrenches include the deflecting beam (A, **Figure 32**) the dial indicator (B) and the audible click (C).

When choosing a torque wrench, consider the torque range, drive size and accuracy. The torque specifications in this manual provide an indication of the range required.

A torque wrench is a precision tool that must be properly cared for to remain accurate. Store torque wrenches in cases or separate padded drawers within a toolbox. Follow the manufacturer's care and calibration instructions.

Torque Adapters

Torque adapters or extensions extend or reduce the reach of a torque wrench. The torque adapter shown on the top of **Figure 33** is used to tighten a fastener that cannot be reached because of the size of the torque wrench head, drive, and socket. If a torque adapter changes the effective lever length the torque reading on the wrench will not equal the actual torque applied to the fastener. It is necessary to recalibrate the torque setting on the wrench to compensate for the change of lever length. When using a torque adapter at a right angle to the drive head, calibration is not required, since the effective length has not changed.

To recalculate a torque reading when using a torque adapter, use the following formula, and refer to **Figure 33**.

$$TW = \frac{TA \times L}{L + A}$$

TW is the torque setting or dial reading on the wrench.

TA is the torque specification (the actual amount of torque that should be applied to the fastener).

A is the amount that the adapter increases (or in some cases reduces) the effective lever length as measured along the centerline of the torque wrench (**Figure 33**).

L is the lever length of the wrench as measured from the center of the drive to the center of the grip.

The effective length of the torque wrench measured along the centerline of the torque wrench is the sum of **L** and **A** (**Figure 33**).

Example:

TA = 20 ft.-lb.

A = 3 in.

L = 14 in.

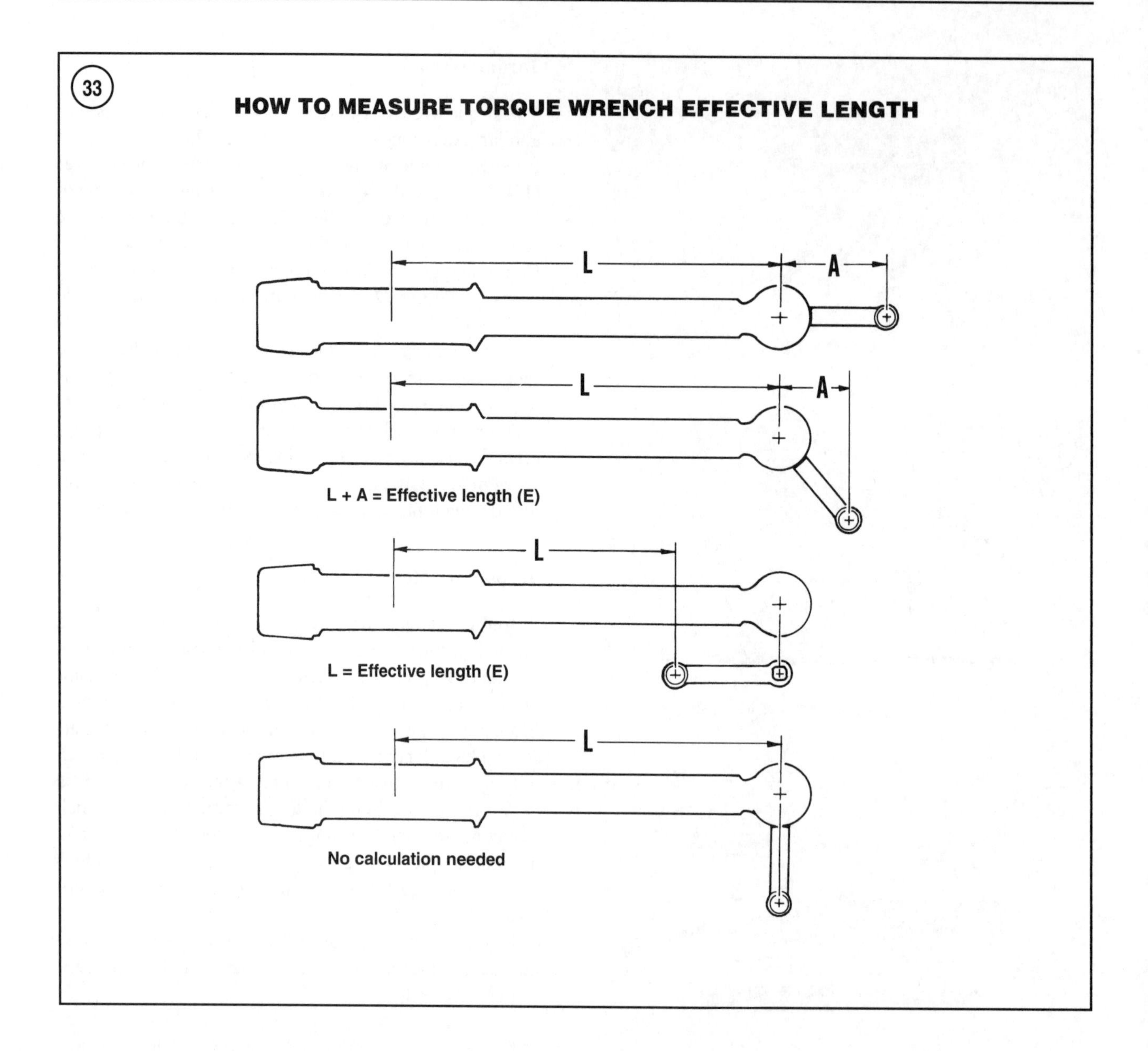

$$TW = \frac{20 \times 14}{14 + 3} = \frac{280}{17} = 16.5 \text{ ft. lb.}$$

In this example, the torque wrench would be set to the recalculated torque value (TW = 16.5 ft.-lb.). When using a beam-type wrench, tighten the fastener until the pointer aligns with 16.5 ft.-lb. In this example, although the torque wrench is pre-set to 16.5 ft.-lb., the actual torque is 20 ft.-lb.

SPECIAL TOOLS

Some of the procedures in this manual require special tools. These are described in the appropriate chapter and are available from either the manufacturer or a tool supplier.

In many cases, an acceptable substitute may be found in an existing tool kit. Another alternative is to make the tool. Many schools with a machine shop curriculum welcome outside work that can be used as practical shop applications for students.

PRECISION MEASURING TOOLS

The ability to accurately measure components is essential to successfully rebuilding an engine. Equipment is manufactured to close tolerances, and obtaining consis-

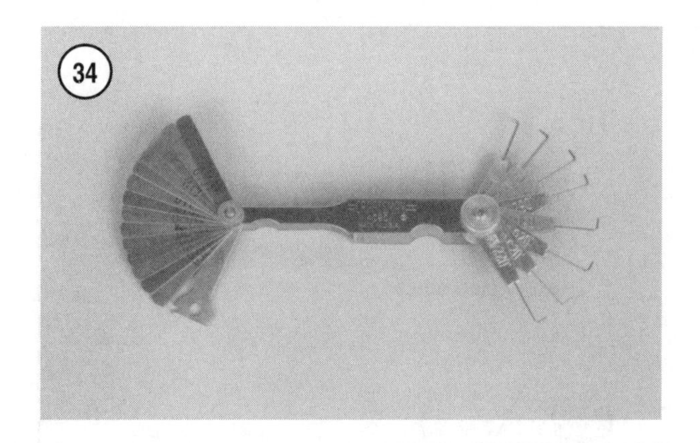

tently accurate measurements is essential to determining which components require replacement or further service.

Each type of measuring instrument is designed to measure a dimension with a certain degree of accuracy and within a certain range. When selecting the measuring tool, make sure it is applicable to the task.

As with all tools, measuring tools provide the best results if cared for properly. Improper use can damage the tool and result in inaccurate results. If any measurement is questionable, verify the measurement using another tool. A standard gauge is usually provided with measuring tools to check accuracy and calibrate the tool if necessary.

Precision measurements can vary according to the experience of the person performing the procedure. Accurate results are only possible if the mechanic possesses a feel for using the tool. Heavy-handed use of measuring tools produces less accurate results than if the fingertips grasp the tool gently so the point at which the tool contacts the object is easily felt. This feel for the equipment produces more accurate measurements and reduces the risk of damaging the tool or component. Refer to the following sections for specific measuring tools.

Feeler Gauge

The feeler or thickness gauge (**Figure 34**) is used for measuring the distance between two surfaces.

A feeler gauge set consists of an assortment of steel strips of graduated thickness. Blades can be of various lengths and angles for different procedures.

A common use for a feeler gauge is to measure valve clearance. Wire (round) type gauges are used to measure spark plug gap.

To obtain a proper measurement using a feeler gauge, make sure the proper-sized blade passes through the gap with some slight drag. The blade should not need to be forced through, and should not have any play up-and-down between the surfaces being measured.

Calipers

Calipers are excellent tools for obtaining inside, outside and depth measurements. Although not as precise as a micrometer, they allow reasonable precision, typically to within 0.05 mm (0.001 in.). Most calipers have a range up to 150 mm (6 in.).

Calipers are available in dial, vernier or digital versions. Dial calipers have a dial gauge readout that provides convenient reading. Vernier calipers have marked scales that are compared to determine the measurement. Most convenient of all, the digital caliper uses an LCD to show the measurement.

To help ensure accurate readings, properly maintain the measuring surfaces of the caliper. There must not be any dirt or burrs between the tool and the object being measured. Never force the caliper closed around an object; close the caliper around the highest point so it can be removed with a slight drag. Some calipers require calibration. Always refer to the manufacturer's instructions when using a new or unfamiliar caliper.

To read a vernier caliper refer to **Figure 35**. The fixed scale is marked in 1 mm increments. Ten individual lines on the fixed scale equal 1 cm. The movable scale is marked in 0.05 mm (hundredth) increments. To obtain a reading, establish the first number by the location of the 0 line on the movable scale in relation to the first line to the left on the fixed scale. In this example, the number is 10 mm. To determine the next number, note which of the lines on the movable scale align with a mark on the fixed scale. A number of lines will seem close, but only one aligns exactly. In this case, 0.50 mm is the reading to add to the first number. The result of adding 10 mm and 0.50 mm is a measurement of 10.50 mm.

Micrometers

A micrometer is an instrument designed for linear measurement using the decimal divisions of the inch or meter (**Figure 36**). While there are many types and styles of micrometers, most of the procedures in this manual call for an outside micrometer. The outside micrometer is used to measure the outside diameter of cylindrical forms and the thickness of materials.

Micrometer size indicates the minimum and maximum size of a part that it can measure. The usual sizes are 0-1 in. (0-25 mm), 1-2 in. (25-50 mm), 2-3 in. (50-75 mm) and 3-4 in. (75-100 mm).

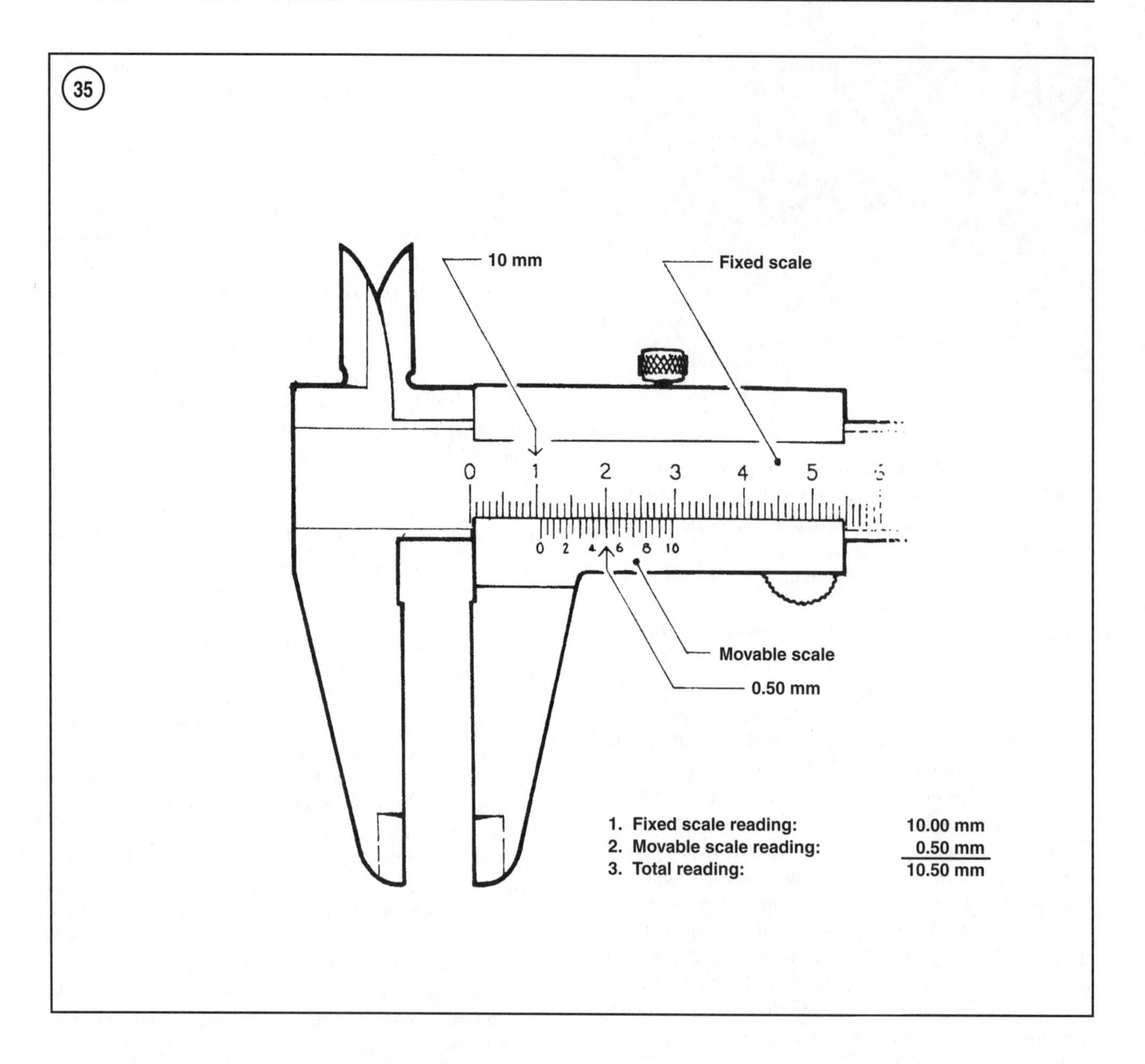

Micrometers covering a wider range of measurement are available and use a large frame with interchangeable anvils of various lengths. This type of micrometer offers a cost savings; however, the overall size make it less convenient.

Reading a micrometer

When reading a micrometer, numbers are taken from different scales and added together. The following sections describe how to take measurements with various types of outside micrometers.

For accurate results, properly maintain the measuring surfaces of the micrometer. There cannot be any dirt or burrs between the tool and the measured object. Never force the micrometer closed around an object. Close the micrometer around the highest point so it can be removed with a slight drag. **Figure 37** shows the markings and parts of a standard inch micrometer. Be familiar with these terms before using a micrometer in the following sections.

Standard inch micrometer

The standard inch micrometer is accurate to one-thousandth of an inch or 0.001. The sleeve is marked in 0.025 in. increments. Every fourth sleeve mark is numbered 1, 2, 3, 4, 5, 6, 7, 8, 9. These numbers indicate 0.100 in., 0.200 in., 0.300 in., and so on.

DECIMAL PLACE VALUES*

0.1	Indicates 1/10 (one tenth of an inch or millimeter)
0.010	Indicates 1/100 (one one-hundreth of an inch or millimeter)
0.001	Indicates 1/1,000 (one one-thousandth of an inch or millimeter)

*This chart represents the values of figures placed to the right of the decimal point. Use it when reading decimals from one-tenth to one one-thousandth of an inch or millimeter. It is not a conversion chart (for example: 0.001 in. is not equal to 0.001 mm).

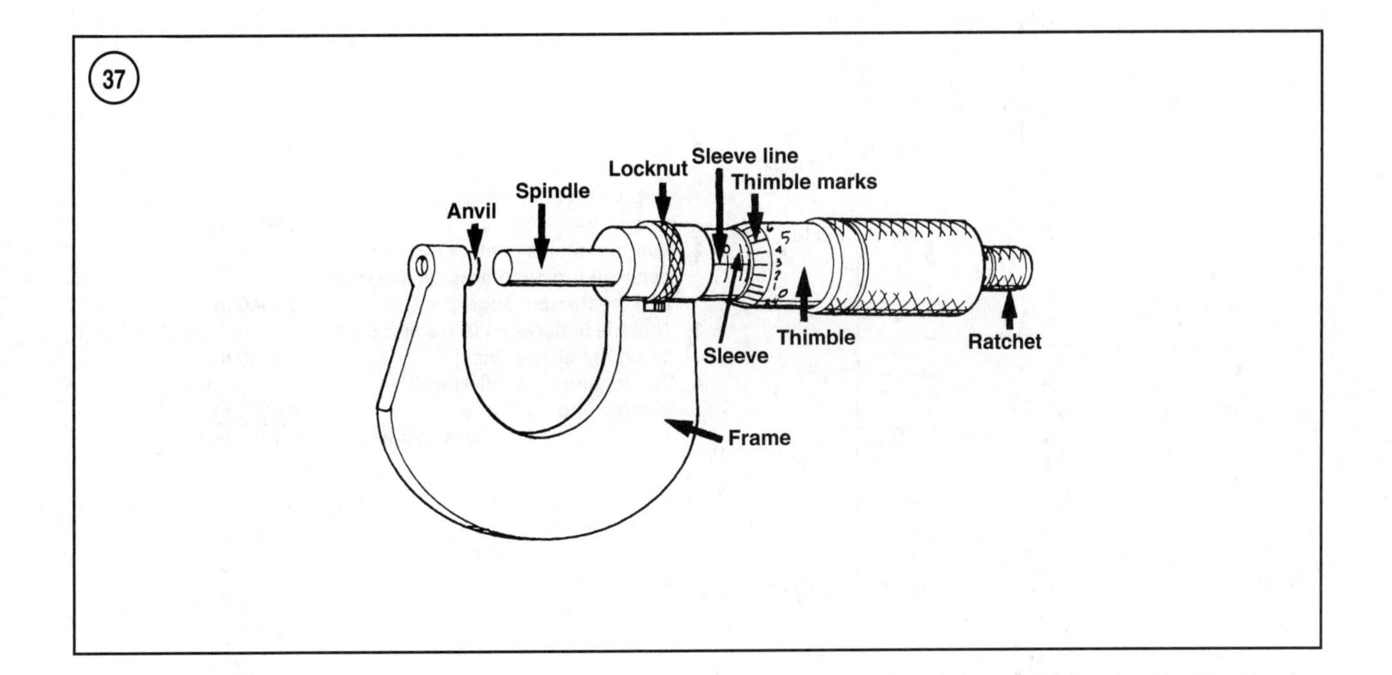

The tapered end of the thimble has twenty-five lines marked around it. Each mark equals 0.001 in. One complete turn of the thimble will align the zero mark with the first mark on the sleeve or 0.025 in.

When reading a standard inch micrometer, perform the following steps while referring to **Figure 38**.

1. Read the sleeve and find the largest number visible. Each sleeve number equals 0.100 in.
2. Count the number of lines between the numbered sleeve mark and the edge of the thimble. Each sleeve mark equals 0.025 in.
3. Read the thimble mark that aligns with the sleeve line. Each thimble mark equals 0.001 in.

NOTE

If a thimble mark does not align exactly with the sleeve line, estimate the amount between the lines. For accurate readings in ten-thousandths of an inch (0.000 in.), use a vernier inch micrometer.

4. Add the readings from Steps 1-3.

Vernier inch micrometer

A vernier inch micrometer is accurate to one ten-thousandth of an inch or 0.0001 in. It has the same marking as

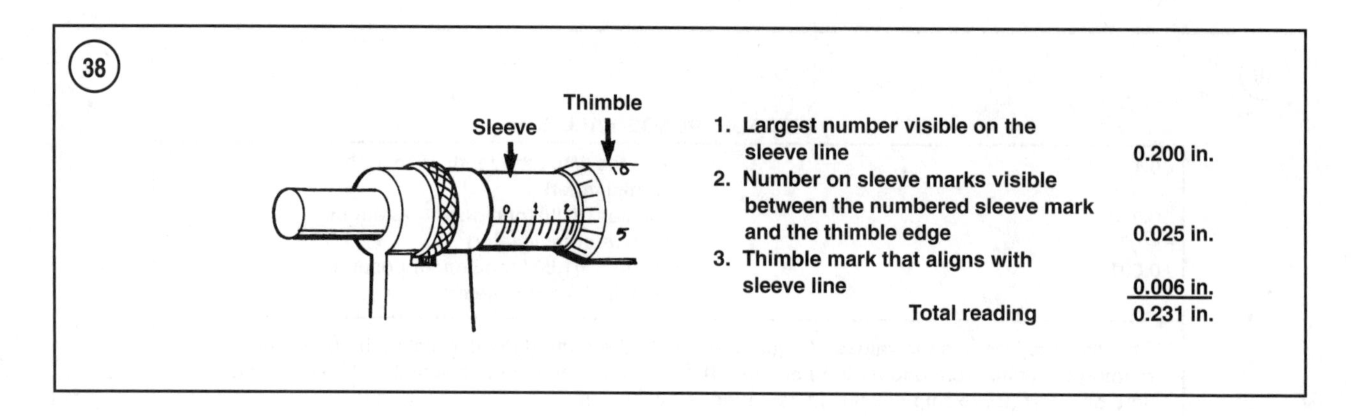

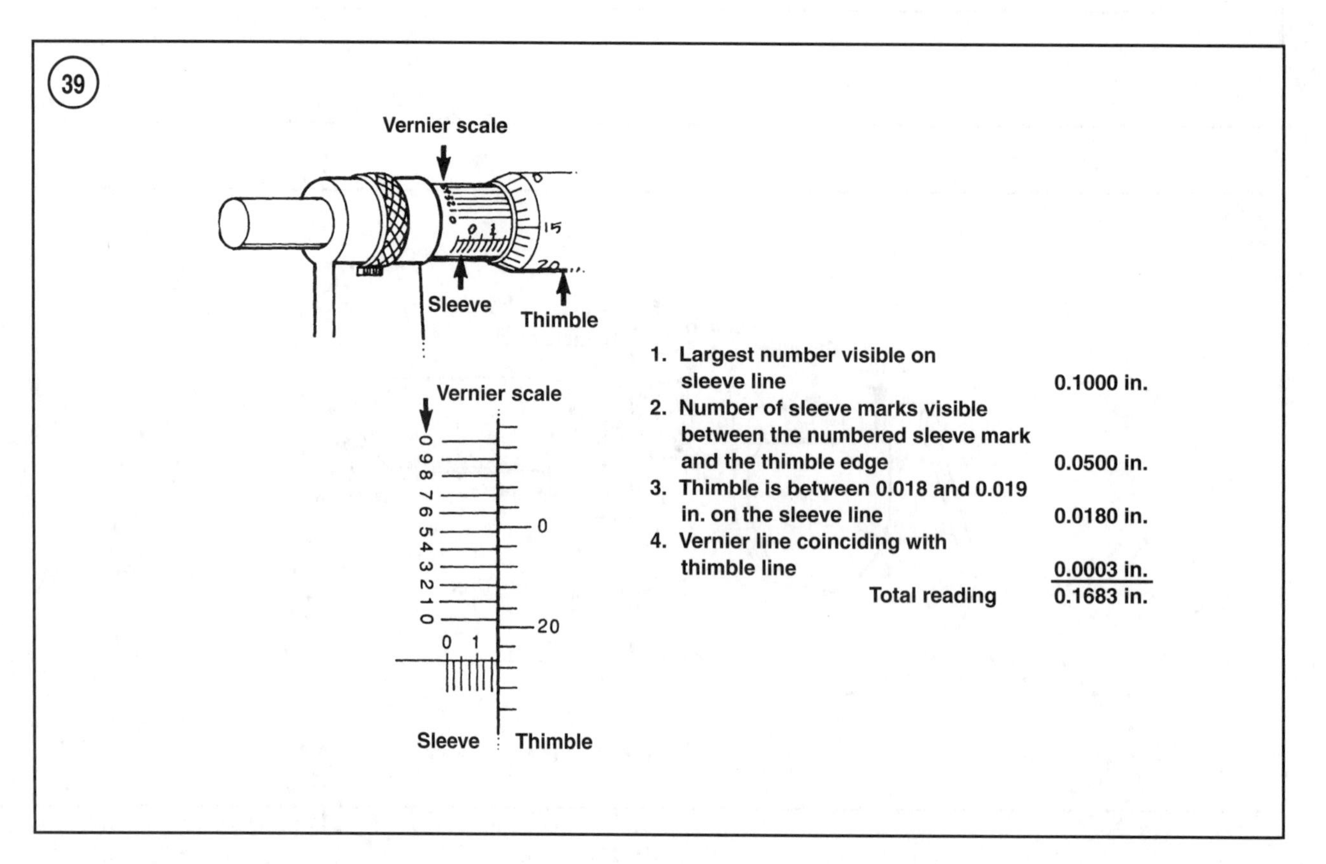

a standard inch micrometer with an additional vernier scale on the sleeve (**Figure 39**).

The vernier scale consists of 11 lines marked 1-9 with a 0 on each end. These lines run parallel to the thimble lines and represent 0.0001 in. increments.

When reading a vernier inch micrometer, perform the following steps while referring to **Figure 39**.

1. Read the micrometer in the same way as a standard micrometer. This is the initial reading.
2. If a thimble mark aligns exactly with the sleeve line, reading the vernier scale is not necessary. If they do not align, read the vernier scale in Step 3.
3. Determine which vernier scale mark aligns with one thimble mark. The vernier scale number is the amount in ten-thousandths of an inch to add to the initial reading from Step 1.

Metric micrometer

The standard metric micrometer (**Figure 40**) is accurate to one one-hundredth of a millimeter (0.01 mm). The sleeve line is graduated in millimeter and half millimeter increments. The marks on the upper half of the sleeve line

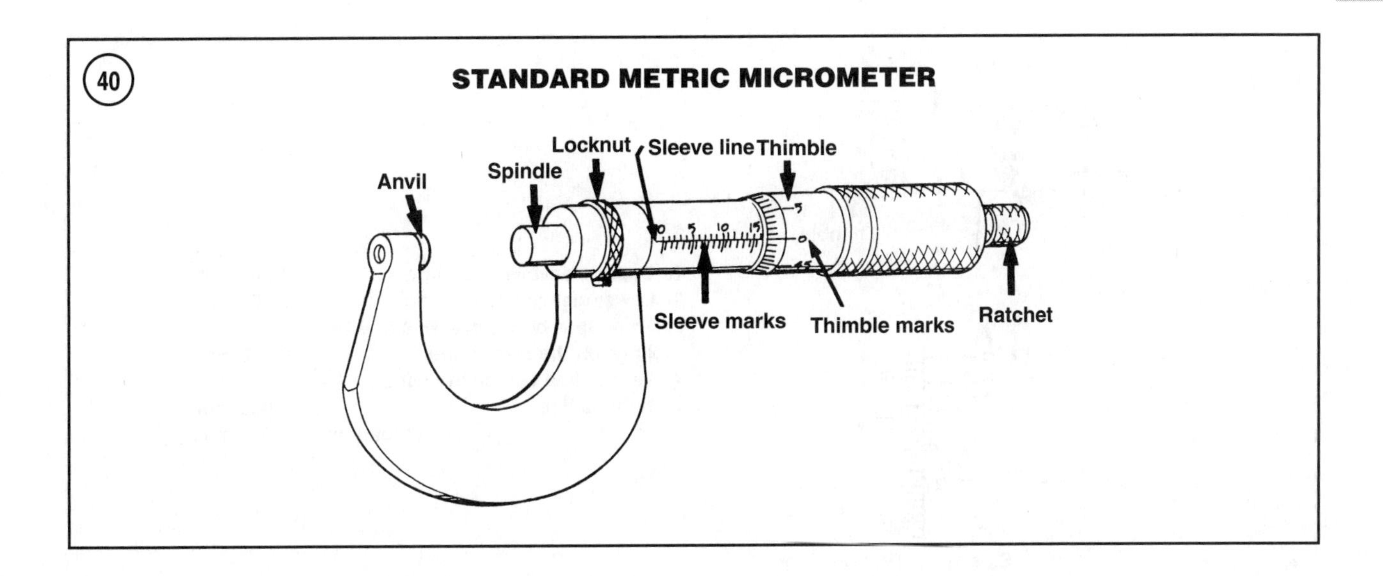

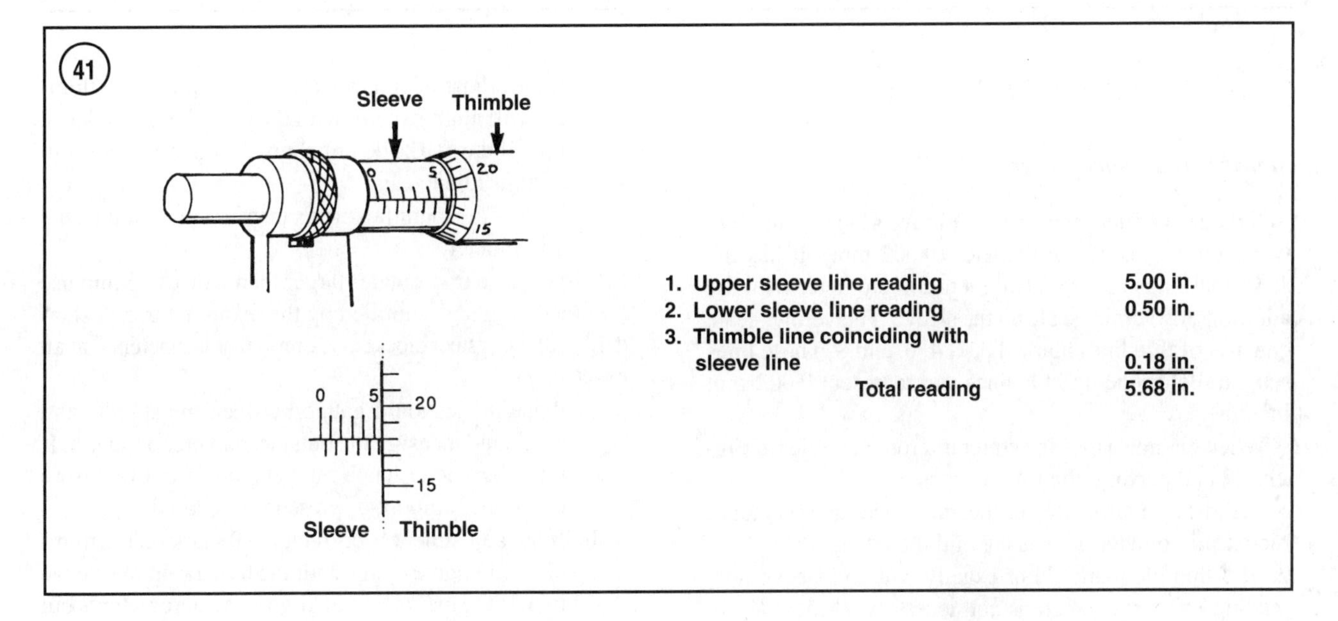

equal 1.00 mm. Every fifth mark above the sleeve line is identified with a number. The number sequence depends on the size of the micrometer. A 0-25 mm micrometer, for example, has sleeve marks numbered 0 through 25 in 5 mm increments. This numbering sequence continues with larger micrometers. On all metric micrometers, each mark on the lower half of the sleeve equals 0.50 mm.

The tapered end of the thimble has fifty lines marked around it. Each mark equals 0.01 mm.

One complete turn of the thimble aligns the 0 mark with the first line on the lower half of the sleeve line or 0.50 mm.

When reading a metric micrometer, add the number of millimeters and half-millimeters on the sleeve line to the number of one one-hundredth millimeters on the thimble. Perform the following steps while referring to **Figure 41**.

1. Read the upper half of the sleeve line and count the number of lines visible. Each upper line equals 1 mm.
2. See if the half-millimeter line is visible on the lower sleeve line. If so, add 0.50 to the reading in Step 1.
3. Read the thimble mark that aligns with the sleeve line. Each thimble mark equals 0.01 mm.

NOTE

If a thimble mark does not align exactly with the sleeve line, estimate the amount between the lines. For accurate readings in two-thousandths of a millimeter (0.002 mm), use a metric vernier micrometer.

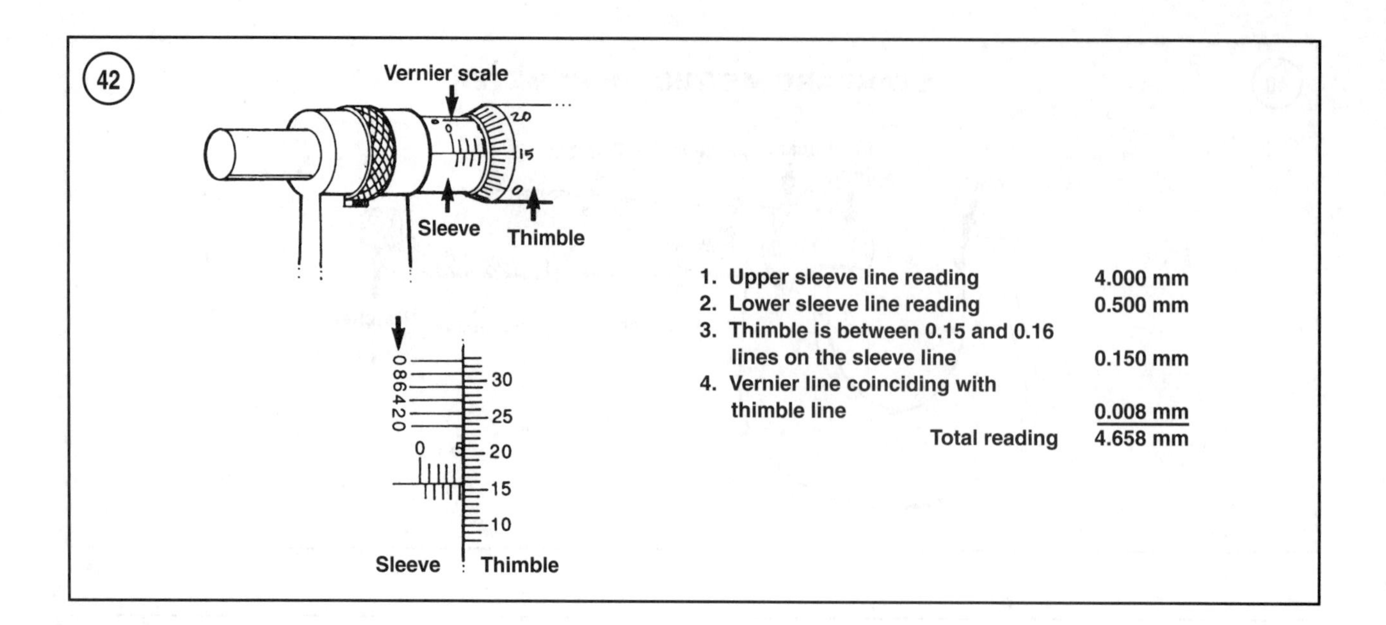

4. Add the readings from Steps 1-3.

Metric vernier micrometer

A metric vernier micrometer (**Figure 42**) is accurate to two-thousandths of a millimeter (0.002 mm). It has the same markings as a standard metric micrometer with the addition of a vernier scale on the sleeve. The vernier scale consists of five lines marked 0, 2, 4, 6, and 8. These lines run parallel to the thimble lines and represent 0.002-mm increments.

When reading a metric vernier micrometer, refer to **Figure 42** and perform the following steps.

1. Read the micrometer in the same way as a standard metric micrometer. This is the initial reading.
2. If a thimble mark aligns exactly with the sleeve line, reading the vernier scale is not necessary. If they do not align, read the vernier scale in Step 3.
3. Determine which vernier scale mark aligns exactly with one thimble mark. The vernier scale number is the amount in two-thousandths of a millimeter to add to the initial reading from Step 1.

Micrometer Calibration

Before using a micrometer, check the calibration as follows.

1. Clean the anvil and spindle faces.

2A. To check a 0-1 in. or 0-25 mm micrometer:

   a. Turn the thimble until the spindle contacts the anvil. If the micrometer has a ratchet stop, use it to ensure the proper amount of pressure is applied.
   b. If the adjustment is correct, the 0 mark on the thimble will align exactly with the 0 mark on the sleeve line. If the marks do not align, the micrometer is out of adjustment.
   c. Follow the manufacturer's instructions to adjust the micrometer.

2B. To check a micrometer larger than 1 in. or 25 mm use the standard gauge supplied by the manufacturer. A standard gauge is a steel block, disc or rod that is machined to an exact size.

   a. Place the standard gauge between the spindle and anvil, and measure the outside diameter or length. If the micrometer has a ratchet stop, use it to ensure the proper amount of pressure is applied.
   b. If the adjustment is correct, the 0 mark on the thimble will align exactly with the 0 mark on the sleeve line. If the marks do not align, the micrometer is out of adjustment.
   c. Follow the manufacturer's instructions to adjust the micrometer.

Micrometer care

Micrometers are precision instruments. Use and maintain them with great care.

Note the following:

1. Store micrometers in protective cases or separate padded drawers in a toolbox.
2. When in storage, make sure the spindle and anvil faces do not contact each other or another object. If they do, temperature changes and corrosion may damage the contact faces.

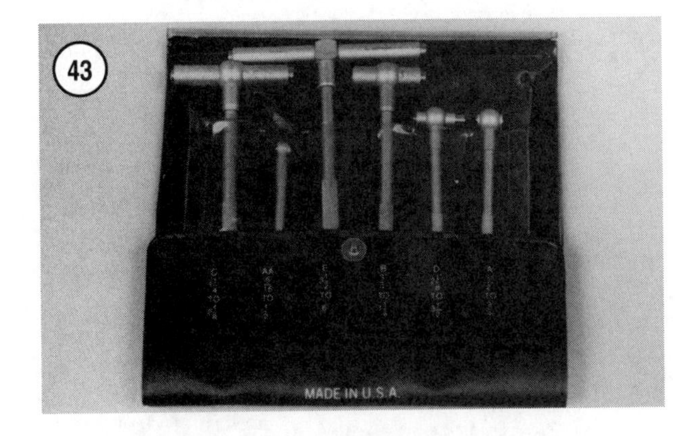

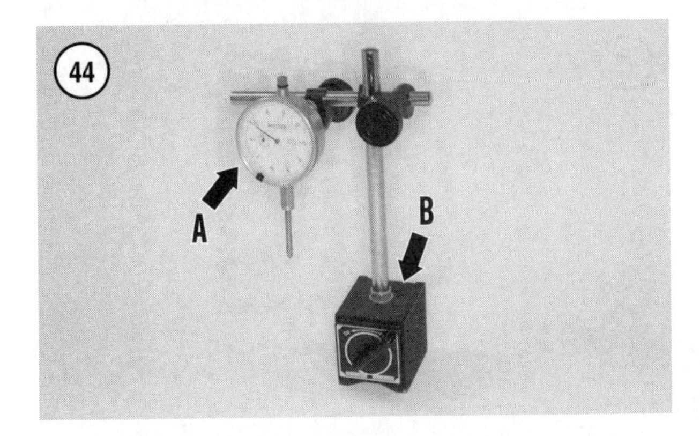

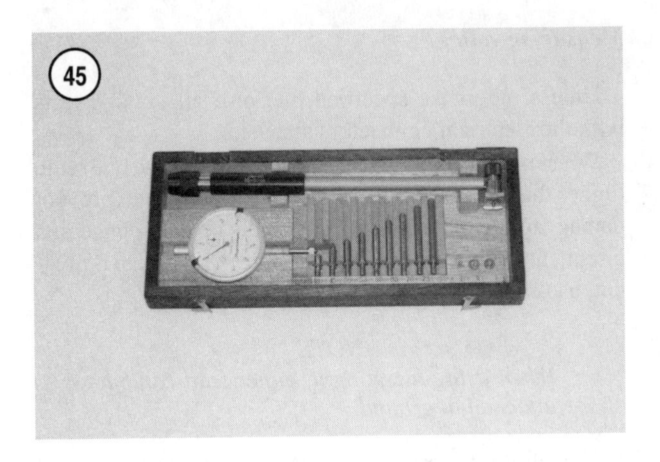

3. Do not clean a micrometer with compressed air. Dirt forced into the tool causes wear.

4. Lubricate micrometers with WD-40 to prevent corrosion.

Telescoping and Small Bore Gauges

Use telescoping gauges (**Figure 43**) and small hole gauges to measure bores. Neither gauge has a scale for direct readings. Use an outside micrometer to determine the reading.

To use a telescoping gauge, select the correct size gauge for the bore. Compress the movable post and carefully insert the gauge into the bore. Carefully move the gauge in the bore to make sure it is centered. Tighten the knurled end of the gauge to hold the movable post in position. Remove the gauge and measure the length of the posts. Telescoping gauges are typically used to measure cylinder bores.

To use a small-bore gauge, select the correct size gauge for the bore. Carefully insert the gauge into the bore. Tighten the knurled end of the gauge to carefully expand the gauge fingers to the limit within the bore. Do not overtighten the gauge, as there is no built-in release. Excessive tightening can damage the bore surface and the tool. Remove the gauge and measure the outside dimension. Small hole gauges are typically used to measure valve guides.

Dial Indicator

A dial indicator (**Figure 44**) is a gauge with a dial face and needle used to measure variations in dimensions and movements, such as crankshaft and gear shaft runout limits.

Dial indicators are available in various ranges and graduations and with three basic types of mounting bases: magnetic, clamp, or screw-in stud. When purchasing a dial indicator, select the magnetic stand type (B, **Figure 44**) with a continuous dial (A).

Cylinder Bore Gauge

The cylinder bore gauge is a very specialized precision tool that is only needed for major engine repairs or rebuilds. The gauge set shown in **Figure 45** is comprised of a dial indicator, handle and a number of different length adapters (anvils) used to fit the gauge to various bore sizes. The bore gauge can be used to measure bore size, taper and out-of-round. When using a bore gauge, follow the manufacturer's instructions.

Compression Gauge

A compression gauge (**Figure 46**) measures the combustion chamber (cylinder) pressure usually in psi or kg/cm^2. An engine is capable of mechanically generating on the compression stroke. The gauge adapter is either inserted or screwed into the spark plug hole to obtain the reading. Disable the engine so it does not start and hold the throttle in the wide-open position when performing a compression test. An engine that does not have adequate compression cannot be properly tuned.

Multimeter

A multimeter (**Figure 47**) is an essential tool for electrical system diagnosis. The voltage function indicates the voltage applied or available to various electrical components. The ohmmeter function tests circuits for continuity, or lack of continuity, and measures the resistance of a circuit.

Some less expensive models contain a needle gauge and are known as analog meters. Most high-quality (but not necessarily expensive) meters available today contain digital readout screens. Digital multimeters are often known as DVOMs. When using an analog ohmmeter, the needle must be zeroed or calibrated according to the meter manufacturer's instructions. Some analog and almost all digital meters are self-zeroing and no manual adjustment is necessary.

Some manufacturers' specifications for electrical components are based on results using a specific test meter. Results may vary if using a meter not recommended by the manufacturer.

ELECTRICAL SYSTEM FUNDAMENTALS

A thorough study of the many types of electrical systems used in engines today is beyond the scope of this manual. However, a basic understanding of electrical basics is necessary to perform simple diagnostic tests.

Voltage

Voltage is the electrical potential or pressure in an electrical circuit and is expressed in volts. The more pressure (voltage) in a circuit, the more work that can be performed.

Direct current (DC) voltage means the electricity flows in one direction. All circuits powered by a battery are DC circuits.

Alternating current (AC) means the electricity flows in one direction momentarily then switches to the opposite direction. Alternator output is an example of AC voltage. This voltage must be changed or rectified to direct current to operate in a 12-volt battery powered system.

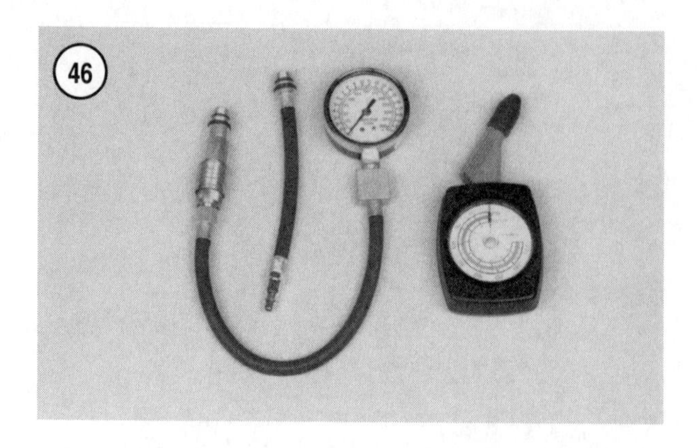

46

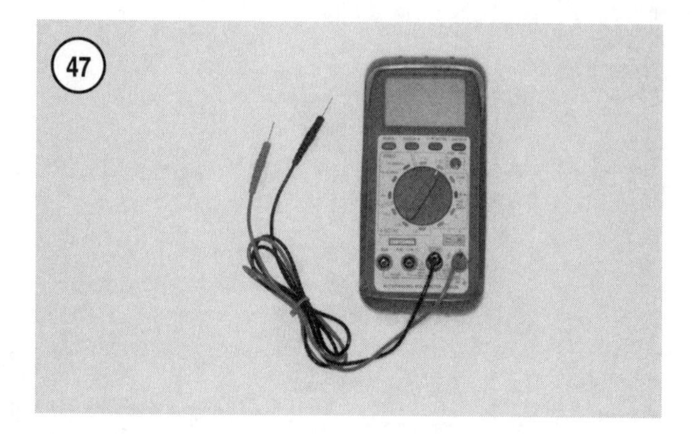

47

Measuring voltage

Unless otherwise specified, perform all voltage tests with the electrical connectors attached.

When measuring voltage, select a meter range one scale higher than the expected voltage of the circuit to prevent damage to the meter. To determine the actual voltage in a circuit, use a voltmeter. To simply check if voltage is present, use a test light.

NOTE
When using a test light, either lead can be attached to ground.

1. Attach the negative meter test lead to a good ground (bare metal). Make sure the ground is not insulated with a rubber gasket or grommet.
2. Attach the positive meter test lead to the point being checked for voltage (**Figure 48**).
3. If necessary for the circuit being checked, turn on the ignition switch. This will be necessary if the point being checked only has power applied when the ignition switch is turned ON, but the example in **Figure 48** shows a measurement at the positive battery terminal, which should always have voltage if the battery is charged. The test light should light or the meter should display a reading. The

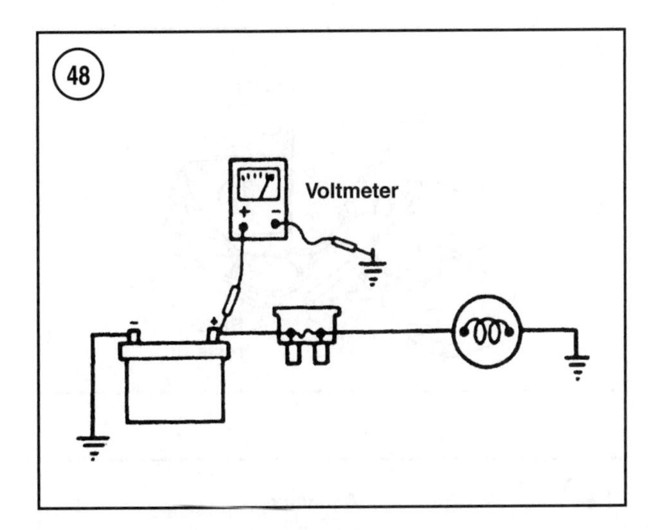

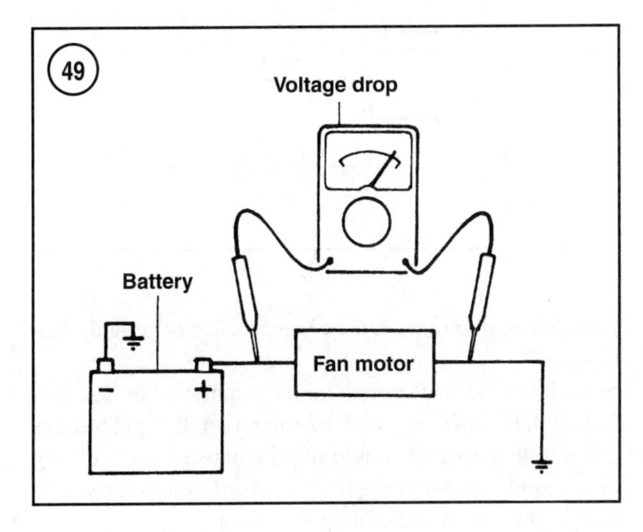

reading should be within one volt of battery voltage. If the voltage is less, there is a problem in the circuit.

Voltage drop test

Resistance causes voltage to drop. This resistance can be measured in an active circuit by using a voltmeter to perform a voltage drop test. A voltage drop test compares the difference between the voltage available at the start of a circuit to the voltage at the end of the circuit. But it does so while the circuit is operational. If the circuit has no resistance, there will be no voltage drop. The greater the resistance, the greater the voltage drop will be. A voltage drop of one volt or more usually indicates excessive resistance in the circuit.

1. Connect the positive meter test lead to the electrical source (where electricity is coming from).
2. Connect the negative meter test lead to the electrical load (where electricity is going). See **Figure 49**.
3. If necessary, activate the component(s) in the circuit.
4. A voltage reading of 1 volt or more indicates excessive resistance in the circuit. A reading equal to battery voltage indicates an open circuit.

Resistance

Resistance is the opposition to the flow of electricity within a circuit or component and is measured in ohms. Resistance causes a reduction in available current and voltage.

Resistance is measured in an inactive circuit with an ohmmeter. The ohmmeter sends a small amount of current into the circuit and measures how difficult it is to push the current through the circuit.

An ohmmeter, although useful, is not always a good indicator of the actual ability of the circuit under operating conditions. This fact is due to the low voltage (6-9 volts) that the meter uses to test the circuit. The voltage in an ignition coil secondary winding can be several thousand volts. Such high voltage can cause the coil to malfunction, even though it tests acceptable during a resistance test.

Resistance generally increases with temperature. Perform all testing with the component or circuit at room temperature. Resistance tests performed at high temperatures may indicate high resistance readings and result in the unnecessary replacement of a component.

Resistance and continuity test

CAUTION

Only use an ohmmeter on a circuit that has no voltage present. The meter will be damaged if it is connected to a live circuit. Remember, if using an analog meter, it must be calibrated each time it is used or the scale is changed.

A continuity test can determine if the circuit is complete. This type of test is performed with an ohmmeter or a self-powered test lamp.

1. Disconnect the negative battery cable.
2. Attach one test lead (ohmmeter or test light) to one end of the component or circuit.
3. Attach the other test lead to the opposite end of the component or circuit (**Figure 50**).
4. A self-powered test light will come on if the circuit has continuity or is complete. An ohmmeter will indicate either low or no resistance if the circuit has continuity. An

open circuit is indicated if the meter displays infinite resistance.

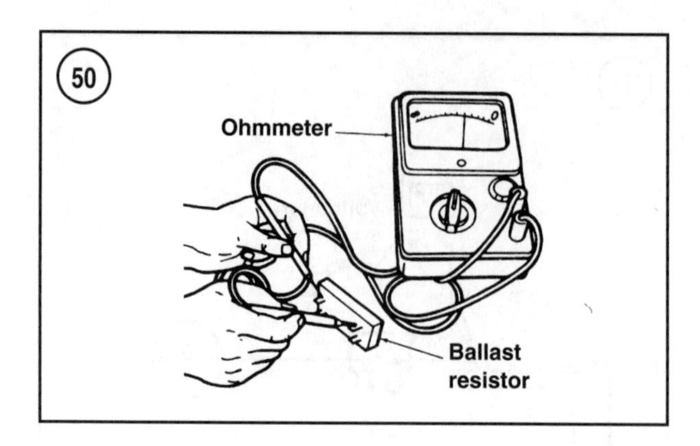

Amperage

Amperage is the unit of measure for the amount of current within a circuit. Current is the actual flow of electricity. The higher the current, the more work that can be performed up to a given point. If the current flow exceeds the circuit or component capacity, the system will be damaged.

Amperage measurement

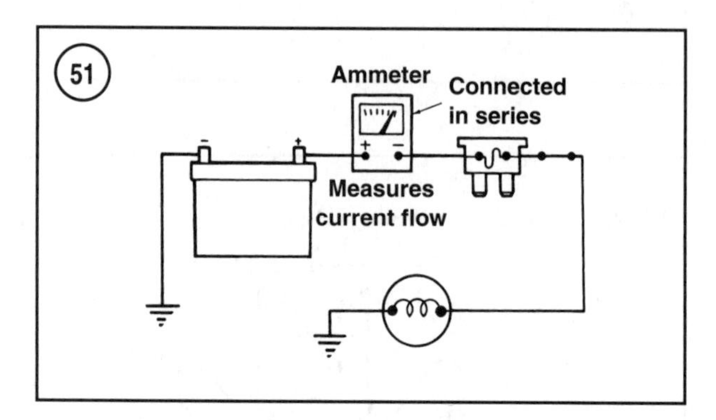

An ammeter measures the current flow or amps of a circuit (**Figure 51**). Amperage measurement requires that the circuit be disconnected and the ammeter be connected in series to the circuit. Always use an ammeter that can read higher than the anticipated current flow to prevent damage to the meter. Connect the red test lead to the electrical source and the black test lead to the electrical load.

BASIC MECHANICAL SKILLS

Most of the service procedures covered are straightforward and can be performed by anyone reasonably handy with tools. It is suggested, however, to consider your own capabilities carefully before attempting any operation involving major disassembly.

1. *Front*, as used in this manual, refers to the front of the engine or the side of the engine facing the boat; the front of any component is the end closest to the front of the engine or boat. The *left* and *right* sides refer to the position of the parts as viewed by the boat operator sitting and facing forward. These rules are simple, but confusion can cause a major inconvenience during service.
2. When disassembling engine or drive components, mark the parts for location and mark all parts that mate together. Placing them in plastic sandwich bags can identify small parts, such as bolts. Seal the bags and label them with masking tape and a marking pen. Because many types of ink fade when applied to tape, use a permanent ink pen. If reassembly will take place immediately, place nuts and bolts in a cupcake tin or egg carton in the order of disassembly.
3. Protect finished surfaces from physical damage or corrosion. Keep gasoline off painted surfaces.
4. Use penetrating oil to free frozen or tight bolts, then strike the bolt head a few times with a hammer and punch. (Use a screwdriver on screws.) Avoid the use of heat where possible, as it can warp, melt or affect the temper of parts. Heat also ruins finishes, especially paint and plastics.
5. Unless otherwise noted, no parts removed or installed (other than bushings and bearings) in the procedures given in this manual should require unusual force during disassembly or assembly. If a part is difficult to remove or install, find out why before proceeding.
6. Cover all openings after removing parts or components to prevent things like dirt or small tools from falling in.
7. Read each procedure *completely* while looking at the actual parts before starting a job. Make sure you *thoroughly* understand what is to be done and then carefully follow the procedure, step by step.
8. For the Do-it-Yourselfer, recommendations are occasionally made to refer service or maintenance to a dealership or a specialist in a particular field. In these cases, the work will be done more quickly and economically than performing the job yourself.
9. In procedural steps, the term *replace* means to discard a defective part and replace it with a new or exchange unit. *Overhaul* means to remove, disassemble, inspect, measure, repair or replace defective parts, reassemble and install major systems or parts.
10. Some operations require the use of a hydraulic press. If a suitable press is not available, it would be wiser to

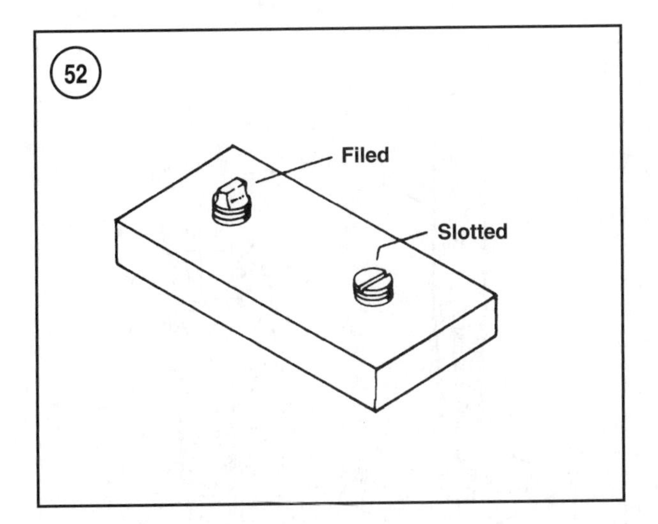

have these operations performed by a shop equipped for such work, rather than to try to do the job yourself with makeshift equipment that may damage the engine.

11. Repairs go much faster and easier if the machine is clean before beginning work.

12. If special tools are required, make arrangements to get them before starting. It is frustrating and time-consuming to start a job and then be unable to complete it.

13. Make diagrams or take a picture wherever similar-appearing parts are found. For instance, crankcase bolts often are not the same length. You may think you can remember where everything came from—but mistakes are costly. You may also get sidetracked and not return to work for days or even weeks—in which time, carefully laid out parts may have become disturbed.

14. When assembling parts, be sure all shims and washers are put exactly where they came out.

15. Whenever a rotating part butts against a stationary part, look for a shim or washer. Use new gaskets if there is any doubt about the condition of the old ones. A thin coating of silicone sealant on non-pressure type gaskets may help them seal more effectively.

16. If it becomes necessary to purchase gasket material to make a gasket for the engine, measure the thickness of the old gasket (at an uncompressed point) and purchase gasket material with the same approximate thickness.

17. Heavy grease can be used to hold small parts in place if they tend to fall out during assembly. However, keep grease and oil away from electrical components, unless otherwise directed.

18. Never use wire to clean out jets and air passages. They are easily damaged. First remove the diaphragm before using compressed air when cleaning the carburetor.

19. Take your time and do the job right. Do not forget that a newly rebuilt engine must be broken in just like a new one.

Removing Frozen Fasteners

If a fastener cannot be removed, several methods may be used to loosen it. First, apply penetrating oil such as Liquid Wrench, WD-40 or PB Blaster. Apply it liberally and let it penetrate for 10-15 minutes. Rap the fastener several times with a small hammer. Do not hit it hard enough to cause damage. Reapply the penetrating oil if necessary.

For frozen screws, apply penetrating oil as described, then insert a screwdriver in the slot and rap the top of the screwdriver with a hammer. This loosens the corrosion so the screw can be removed in the normal way. If the screw head is too damaged to use this method, grip the head with locking pliers and twist the screw out.

Avoid applying heat unless specifically instructed, as it may melt, warp or remove the temper from parts.

Removing Broken Fasteners

If the head breaks off a screw or bolt, several methods are available for removing the remaining portion. If a large portion of the remainder projects out, try gripping it with locking pliers. If the projecting portion is too small or a sufficient grip cannot be obtained on the protruding piece, file it to fit a wrench or cut a slot in it to fit a screwdriver (**Figure 52**).

If the head breaks off flush, use a screw extractor. To do this, centerpunch the exact center of the remaining portion of the screw or bolt. Drill a small hole in the screw and tap the extractor into the hole. Back the screw out with a wrench on the extractor (**Figure 53**).

NOTE

Broken screw extraction sometimes fails to remove the fastener from the bore. If this occurs, or if the screw is drilled off-center and the threads are damaged, a threaded-insert will be necessary to repair the bore. Check for one at a local dealership or supply store and follow the manufacturer's instructions for installation.

Repairing Damaged Threads

Occasionally, threads are stripped through carelessness or impact damage. Often the threads can be repaired by running a tap (for internal threads on nuts) or die (for external threads on bolts) through the threads (**Figure 54**). Use only a specially designed spark plug tap to clean or repair spark plug threads.

If an internal thread is damaged, it may be necessary to install a Helicoil or some other type of thread insert. Fol-

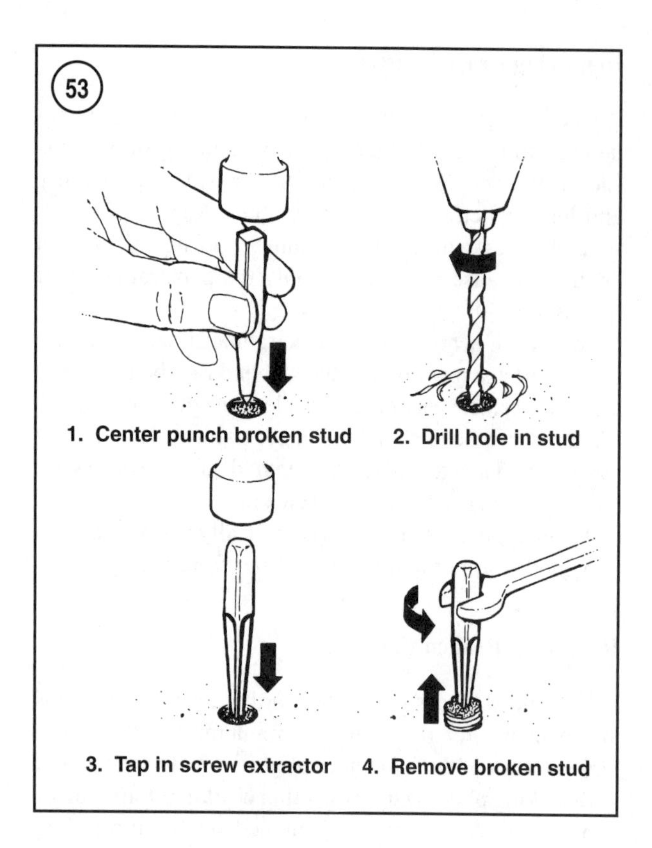

low the manufacturer's instructions when installing the insert.

If it is necessary to drill and tap a hole, refer to **Table 3** for metric tap drill sizes.

Stud Removal/Installation

A stud removal tool that makes the removal and installation of studs easier is available from most tool suppliers. If one is not available, thread two nuts onto the stud and tighten them against each other to lock them in place, then remove the stud by turning the lower nut (**Figure 55**).

NOTE
If the threads on the damaged stud do not allow installation of the two nuts, it is necessary to remove the stud with a pair of locking pliers or a stud remover.

Removing Hoses

When removing stubborn hoses, do not exert excessive force on the hose or fitting. Remove the hose clamp and carefully insert a small screwdriver or pick tool between the fitting and hose. Apply a spray lubricant under the hose and carefully twist the hose off the fitting. Clean the

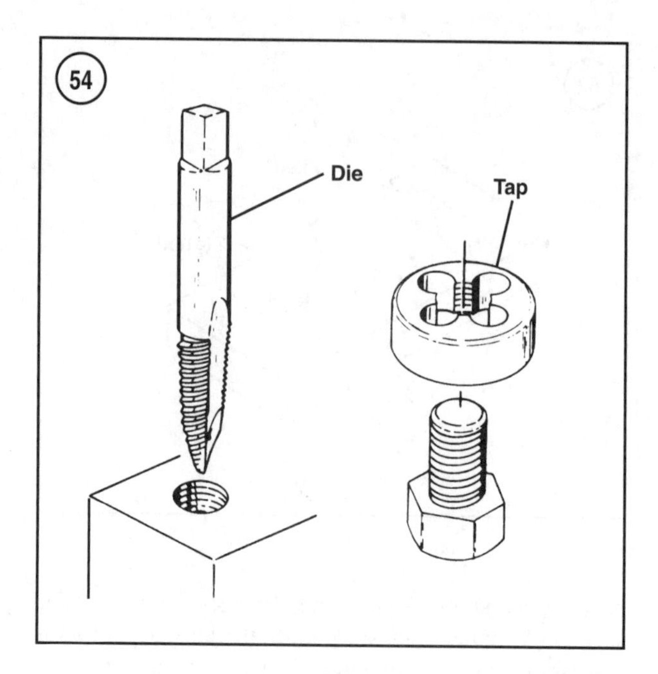

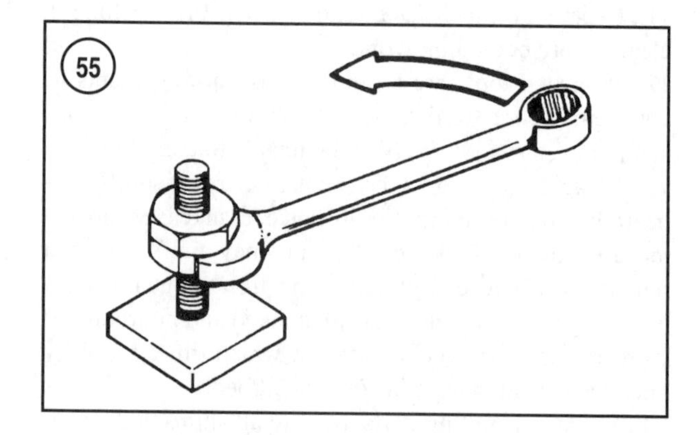

fitting of any corrosion or rubber hose material with a wire brush. Clean the inside of the hose thoroughly. Do not use any lubricant when installing the hose (new or old). The lubricant may allow the hose to come off the fitting, even with the clamp secure.

Bearings

Bearings are used in the engine and gearcase assembly to reduce power loss, heat and noise resulting from friction. Because bearings are precision parts, it is necessary to maintain them with proper lubrication and maintenance. If a bearing is damaged, replace it immediately. Bearing replacement procedures are included in the individual chapters where applicable; however, use the following sections as a guideline.

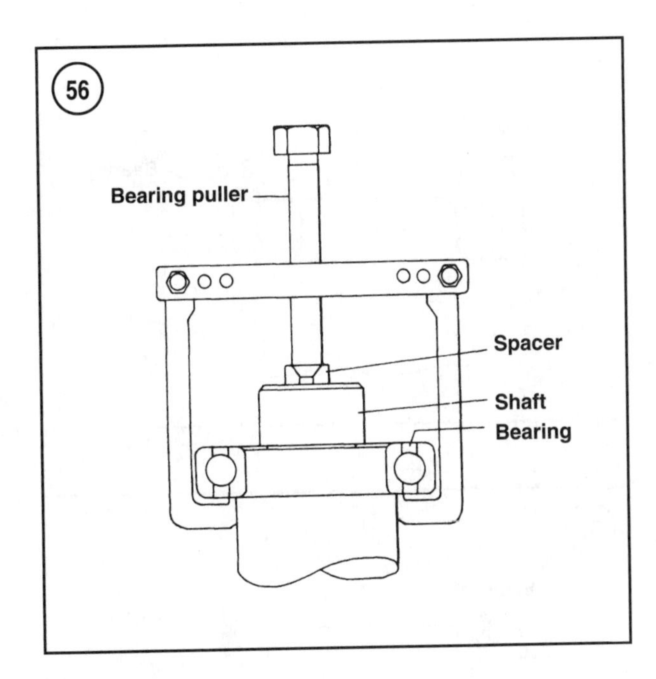

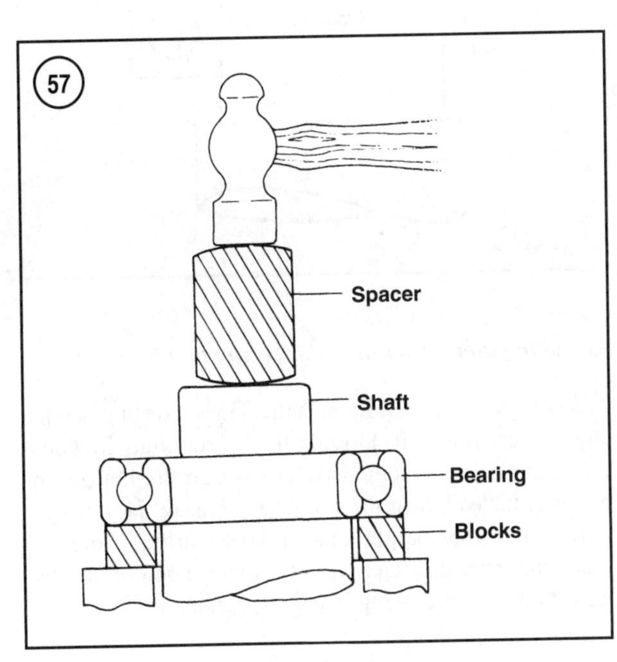

NOTE
Unless otherwise specified, install bearings with the manufacturer's mark or number facing outward.

Removal

While bearings are normally removed only when damaged, there may be times when it is necessary to remove a bearing that is in good condition. However, improper bearing removal will damage the bearing, shaft, and/or case half. Note the following when removing bearings.

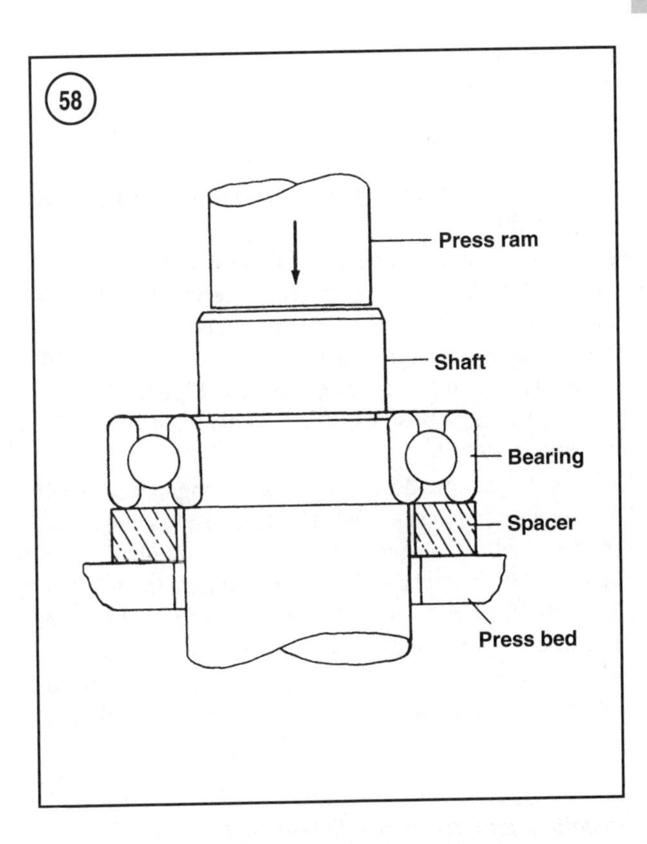

1. When using a puller (**Figure 56**) to remove a bearing from a shaft, take care not to damage the shaft. Always place a piece of metal between the end of the shaft and the puller screw. In addition, place the puller arms next to the inner bearing race.
2. When using a hammer to remove a bearing from a shaft, do not strike the hammer directly against the shaft. Instead, use a brass or aluminum spacer between the hammer and shaft (**Figure 57**) and make sure to support both bearing races with wooden blocks as shown.
3. The ideal method of bearing removal is with a hydraulic press. In order to prevent damage to the bearing and shaft or case, note the following when using a press:
   a. Always support the inner and outer bearing races with a suitable size wooden or aluminum spacer ring (**Figure 58**). If only the outer race is supported, pressure applied against the balls and/or the inner race will damage them.
   b. Always make sure the press ram (**Figure 58**) aligns with the center of the shaft. If the ram is not centered, it may damage the bearing and/or shaft.
   c. The moment the shaft is free of the bearing, it will drop to the floor. Secure or hold the shaft to prevent it from falling.

Installation

1. When installing a bearing in a housing, apply pressure to the *outer* bearing race (**Figure 59**). When installing a bearing on a shaft, apply pressure to the *inner* bearing race (**Figure 60**).
2. When installing a bearing as described in Step 1, use some type of driver. Never strike the bearing directly with a hammer or the bearing will be damaged. When installing a bearing, use a driver, a piece of pipe or a socket with a diameter that matches the bearing race. **Figure 61** shows the correct way to use a socket and hammer to install a bearing on a shaft.
3. Step 1 describes how to install a bearing in a case half or over a shaft. However, when installing a bearing over a shaft *and* into a housing at the same time, a tight fit is required for both outer and inner bearing races. In this situation, install a spacer underneath the driver tool so that pressure is applied evenly across both races. See **Figure 62**. If the outer race is not supported as shown, the balls or rollers will push against the outer bearing race and damage it.

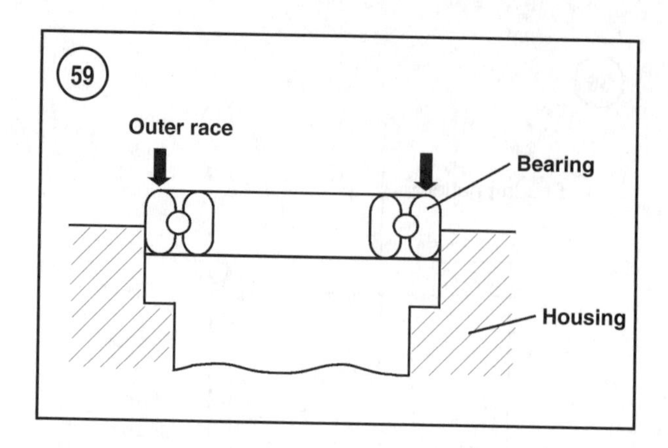

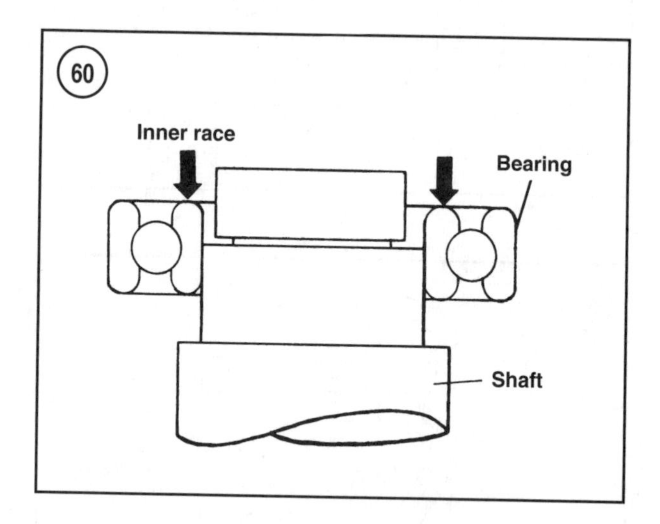

Installing an interference fit bearing over a shaft

When a tight fit is required, the bearing inside diameter will be smaller than the shaft. In this case, driving the bearing on the shaft using normal methods may cause bearing damage. Instead, heat the bearing before installation. Note the following:

1. Secure the shaft so that it is ready for bearing installation. While the parts are still cold, determine the proper size and gather all necessary spacers and riders for installation.
2. Clean the bearing surface on the shaft of all residues. Remove burrs with a file or sandpaper.
3. Fill a suitable pot or beaker with clean mineral oil. Place a thermometer rated higher than 120° C (248° F) in the oil. Support the thermometer so it does not rest on the bottom or side of the pot.
4. Remove the bearing from its wrapper and secure it with a piece of heavy wire bent to hold it in the pot. Hang the bearing so it does not touch the bottom or sides of the pot.
5. Turn the heat on and monitor the thermometer. When the oil temperature rises to approximately 120° C (248° F), remove the bearing from the pot and quickly install it. If necessary, place a socket on the inner bearing race and tap the bearing into place. As the bearing chills, it will tighten on the shaft so work quickly when installing it. Make sure the bearing is installed completely.

Replacing an interference fit bearing in a housing

Bearings are generally installed in a housing with a slight interference fit. Driving the bearing into the housing using normal methods may damage the housing or bearing. Instead, heat the housing (to make the inner diameter of the bore larger) and chill the bearing (in order to make the outer diameter slightly smaller) before installation. This makes bearing installation much easier.

CAUTION

Before heating the crankcases in this procedure to remove the bearings, wash the cases thoroughly with detergent and water. In order to prevent a possible fire hazard, rinse and rewash the cases as required to remove all traces of oil and other chemical deposits.

1. While the parts are still cold, determine the proper size and gather all necessary spacers and drivers for installation.
2. Place the new bearing in a freezer to chill it and slightly reduce the outside diameter.

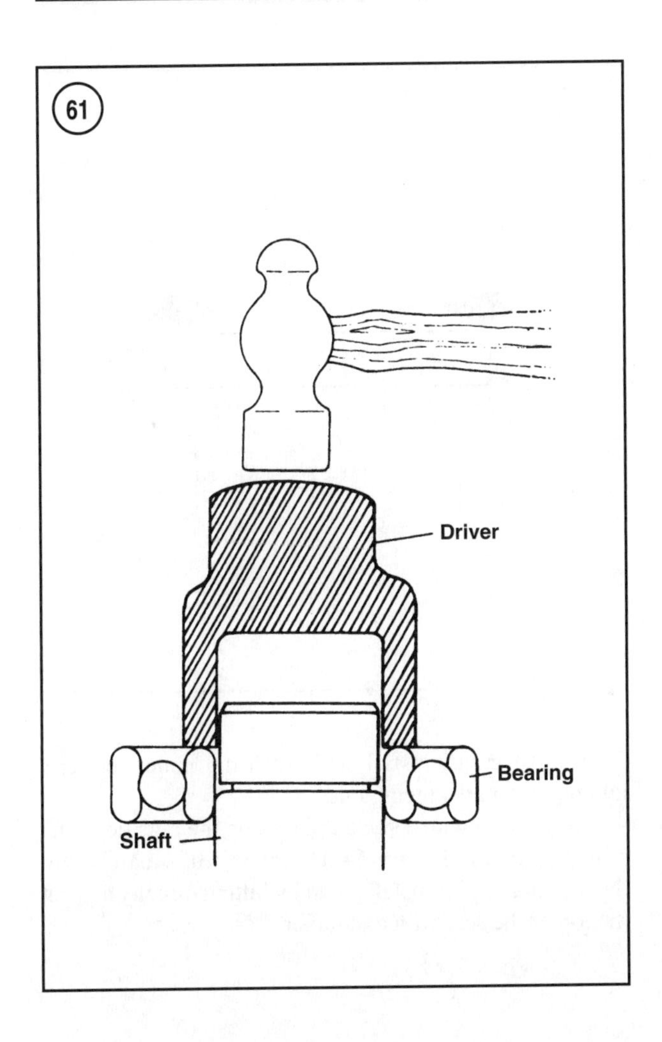

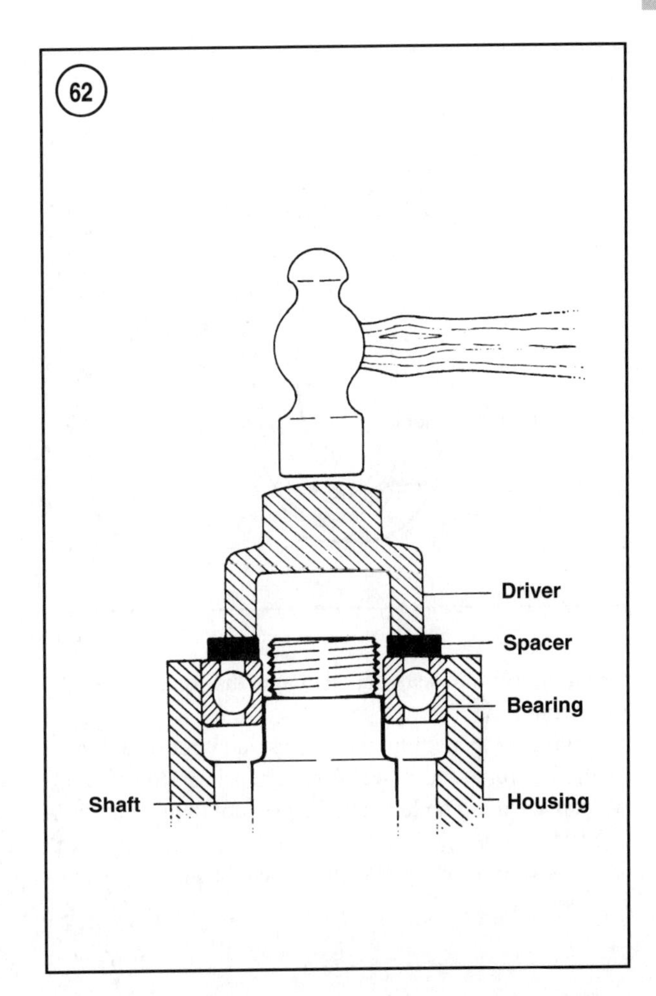

3. While the bearing is chilling, heat the housing to a temperature of about 100° C (212° F) in an oven or on a hot plate. An easy way to check if the housing is hot enough is to drop tiny drops of water on the case; if they sizzle and evaporate immediately, the temperature is correct. Heat only one housing at a time.

CAUTION

Do not heat the housing with a propane or acetylene torch. Never bring a flame into contact with the bearing or housing. The direct heat will destroy the case hardening of the bearing and will likely warp the housing.

4. Remove the housing from the oven or hot plate using a thick kitchen potholder, heavy protective gloves or heavy shop cloths.

NOTE

A suitable size socket and extension works well for removing and installing bearings.

5. Hold the housing with the bearing side down and tap the bearing out. Repeat for all bearings in the housing.

NOTE

Always install bearings with the manufacturer's mark or number facing outward.

6. While the housing is still hot, install the chilled new bearing(s) into the housing. Install the bearings by hand, if possible. If necessary, lightly tap the bearing(s) into the housing with a socket placed on the outer bearing race. *Do not* install new bearings by driving on the inner bearing race. Drive each bearing into the bore until it seats completely.

Seal Replacement

Seals (**Figure 63**) are used to contain oil, water, grease or combustion gasses in a housing or shaft. Improper removal of a seal can damage the components. Improper in-

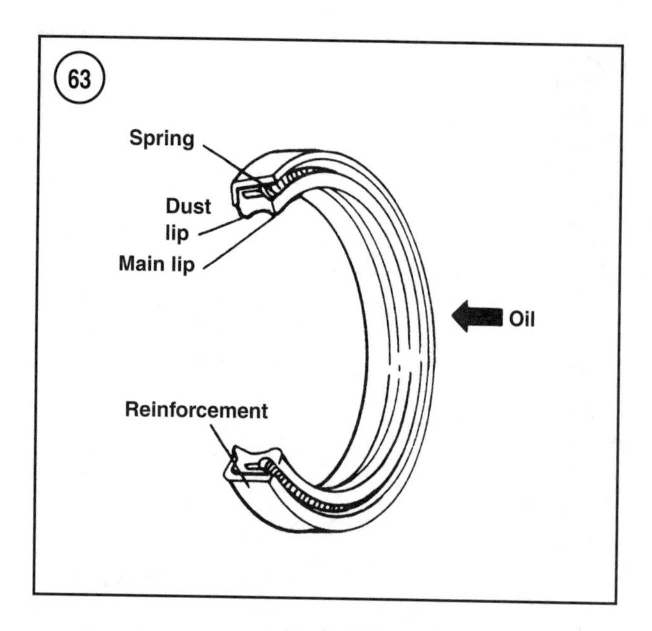

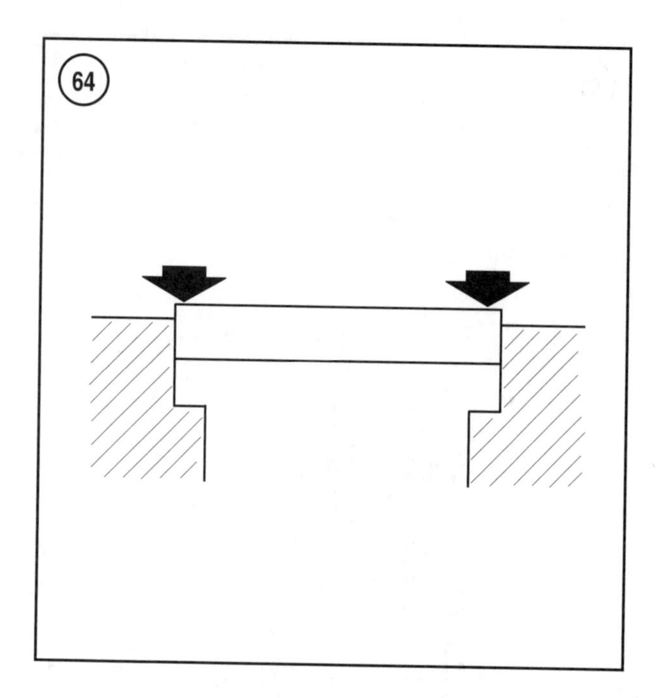

stallation of the seal can damage the seal. Note the following:

1. Prying is generally the easiest and most effective method of removing a seal from a housing. However, always place a rag underneath the pry tool to prevent damage to the housing.
2. Pack waterproof grease in the seal lips before installing the seal.
3. In most cases, install seals with the manufacturer's numbers or marks facing out.
4. Install seals with a socket placed on the outside of the seal as shown in **Figure 64**. Drive the seal squarely into the housing. Never install a seal by hitting directly against the top of the seal with a hammer.

Table 1 ENGINE IDENTIFICATION CODES

Model characteristic codes	
Letters preceding the horsepower rating	
C	C series models, premix, carburetor equipped
D	Twin counter rotating propellers
X	Electronic fuel injection (EFI)
P	Pro series
L	Counter rotation (left-hand) gearcase
S	Saltwater series
V	V-Max series
Z	Direct fuel injection (HPDI)
Letters following the horsepower rating	
J	Jet drive model
S	Short drive shaft model (15 in.)
L	Long shaft drive model (20 in.)
X	Extra long drive shaft model (25 in.)
U	Double extra long drive shaft model (30 in.)
T	Power tilt and trim
	(continued)

Table 1 ENGINE IDENTIFICATION (continued)

Model characteristic codes	
Year model codes	
X	1999 year model
Y	2000 year model
Z	2001 year model
A	2002 year model
B	2003 year model

Table 2 TECHNICAL ABBREVIATIONS

ABDC	After bottom dead center
ATDC	After top dead center
BBDC	Before bottom dead center
BDC	Bottom dead center
BTDC	Before top dead center
C	Celsius (centigrade)
cc	Cubic centimeters
CDI	Capacitor discharge ignition
CKP	Crankshaft position
CMP	Camshaft positioin
CT	Cylinder temperature
CTP	Closed throttle position
cu. in.	Cubic inches
DOHC	Dual-overhead camshafts
ECU	Electronic control unit
EFI	Electronic fuel injection
EM	Exhaust manifold
F	Fahrenheit
ft.-lb.	Foot-pounds
g	Gram
gal.	Gallons
hp	Horsepower
HPDI	High Pressure Direct Injection
IAC	Idle air control
IAT	Intake air temperature
in.	Inches
kg	Kilogram
kg/cm^2	Kilograms per square centimeter
kgm	Kilogram meters
km	Kilometer
L	Liter
m	Meter
MAG	Magneto
MAP	Manifold absolute pressure
mm	Millimeter
MPEFI	Multi-Port Sequential Electronic Fuel Injection
N.A.	Not available
N·m	Newton-meters
oz.	Ounce
OHC	Overhead camshaft
psi	Pounds per square inch
pto	Power take off
pts.	Pints
qt.	Quarts
RFI	Radio Frequency Interference
rpm	Revolutions per minute
WOT	Wide open throttle

Table 3 METRIC TAP AND DRILL SIZES

Metric tap (mm)	Drill size	Decimal equivalent	Nearest fraction
3 × 0.50	No. 39	0.0995	3/32
3 × 0.60	3/32	0.0937	3/32
4 × 0.70	No. 30	0.1285	1/8
4 × 0.75	1/8	0.125	1/8
5 × 0.80	No. 19	0.166	11/64
5 × 0.90	No. 20	0.161	5/32
6 × 1.00	No. 9	0.196	13/64
7 × 1.00	16/64	0.234	15/64
8 × 1.00	J	0.277	9/32
8 × 1.25	17/64	0.265	17/64
9 × 1.00	5/16	0.3125	5/16
9 × 1.25	5/16	0.3125	5/16
10 × 1.25	11/32	0.3437	11/32
10 × 1.50	R	0.339	11/32
11 × 1.50	3/8	0.375	3/8
12 × 1.50	13/32	0.406	13/32
12 × 1.75	13/32	0.406	13/32

Table 4 CONVERSION TABLES

Multiply	By	To get equivalent of
Length		
Inches	25.4	Millimeter
Inches	2.54	Centimeter
Miles	1.609	Kilometer
Feet	0.3048	Meter
Millimeter	0.03937	Inches
Centimeter	0.3937	Inches
Kilometer	0.6214	Mile
Meter	3.281	Mile
Fluid volume		
U.S. quarts	0.9463	Liters
U.S. gallons	3.785	Liters
U.S. ounces	29.573529	Milliliters
Imperial gallons	4.54609	Liters
Imperial quarts	1.1365	Liters
Liters	0.2641721	U.S. gallons
Liters	1.0566882	U.S. quarts
Liters	33.814023	U.S. ounces
Liters	0.22	Imperial gallons
Liters	0.8799	Imperial quarts
Fluid volume (continued)		
Milliliters	0.033814	U.S. ounces
Milliliters	1.0	Cubic centimeters
Milliliters	0.001	Liters
Torque		
Foot-pounds	1.3558	Newton-meters
Foot-pounds	0.138255	Meter-kilograms
Inch-pounds	0.11299	Newton-meters
Newton-meters	0.7375622	Foot-pounds
Newton-meters	8.8507	Inch-pounds
Meters-kilograms	7.2330139	Foot-pounds
Volume		
Cubic inches	16.387064	Cubic centimeters
Cubic centimeters	0.0610237	Cubic inches

(continued)

Table 4 CONVERSION TABLES (continued)

Multiply	By	To get equivalent of
Temperature		
Fahrenheit	(F – 32) × 0.556	Centigrade
Centigrade	(C × 1.8) + 32	Fahrenheit
Weight		
Ounces	28.3495	Grams
Pounds	0.4535924	Kilograms
Grams	0.035274	Ounces
Kilograms	2.2046224	Pounds
Pressure		
Pounds per square inch	0.070307	Kilograms per square centimeter
Kilograms per square centimeter	14.223343	Pounds per square inch
Kilopascals	0.1450	Pounds per square inch
Pounds per square inch	6.895	Kilopascals
Speed		
Miles per hour	1.609344	Kilometers per hour
Kilometers per hour	0.6213712	Miles per hour

Table 5 GENERAL TORQUE SPECIFICATIONS

Screw or nut size	in.-lb.	ft.-lb.	N•m
U.S. Standard			
6-32	9	–	1.0
8-32	20	–	2.3
10-24	30	–	3.4
10-32	35	–	4.0
12-24	45	–	5.1
1/4-20	70	–	7.9
1/4-28	84	–	9.5
5/16-18	160	13	18
5/16-24	168	14	19
3/8-16	–	23	31
3/8-24	–	25	34
7/16-14	–	36	49
7/16-20	–	40	54
1/2-13	–	50	68
1/2-20	–	60	81
Metric			
M5	36	–	4
M6	70	–	8
M8	156	13	18
M10	–	26	35
M12	–	35	48
M14	–	60	81

Chapter Two

Troubleshooting

The most successful way to troubleshoot an outboard is to take an orderly and logical approach. Taking a haphazard approach may eventually find the problem, but it can be costly in terms of wasted time and unnecessary parts replacement. Follow the step-by-step instructions provided in this chapter to perform troubleshooting in an efficient and timely manner.

The first step in any troubleshooting procedure is to define the symptoms and then to localize the problem. Subsequent steps involve testing and analyzing those areas causing the symptoms.

Never assume anything. Do not overlook the obvious. If the engine does not start, is the safety lanyard attached? Is the fuel system primed? Is the engine cranking slowly because the battery is discharged? Check for disconnected wiring or hoses. Is the proper starting procedures being followed?

If nothing obvious turns up in a quick check, look a little further. Learning to recognize and describe the symptoms will make diagnosis easier. Gather as many symptoms as possible to aid in diagnosis. Note whether the engine lost power gradually or all at once, and what color of smoke came from the exhaust. Remember that the more complicated a machine is, the easier it is to troubleshoot because symptoms point to specific problems.

After defining the symptoms, test and analyze areas that could cause problems. Guessing at the cause of a problem may provide a solution, but it can easily lead to frustration, wasted time and a series of expensive, unnecessary parts replacement.

It is not necessary to have expensive equipment or complicated test gear to determine whether repairs can be attempted at home. A few simple checks could save a large repair bill and lost time while the boat sits in a dealership service department. On the other hand, be realistic. Do not attempt repairs beyond your abilities. Service departments tend to charge heavily for putting together a disassembled engine; some do not even take on such a job, so use common sense and do not get in over your head.

This chapter provides sections covering model identification, preliminary inspection, operating requirements, starting difficulty, followed by systems or component testing.

Tables 1-3 list common problems related to the starting, fuel and ignition systems. The probable cause(s) and corrective action are also listed. **Tables 1-3** may refer the reader to the *Component Testing* section(s) of this chapter for further testing. **Tables 4-15** list specifications for various components and systems on the engine.

Refer to the wiring diagrams at the end of the manual during electrical system testing. Due to changes made by the manufacturer, in some cases the wire colors called out in the text may not match a particular model. If there is a differrence, make sure the correct component(s) is identified.

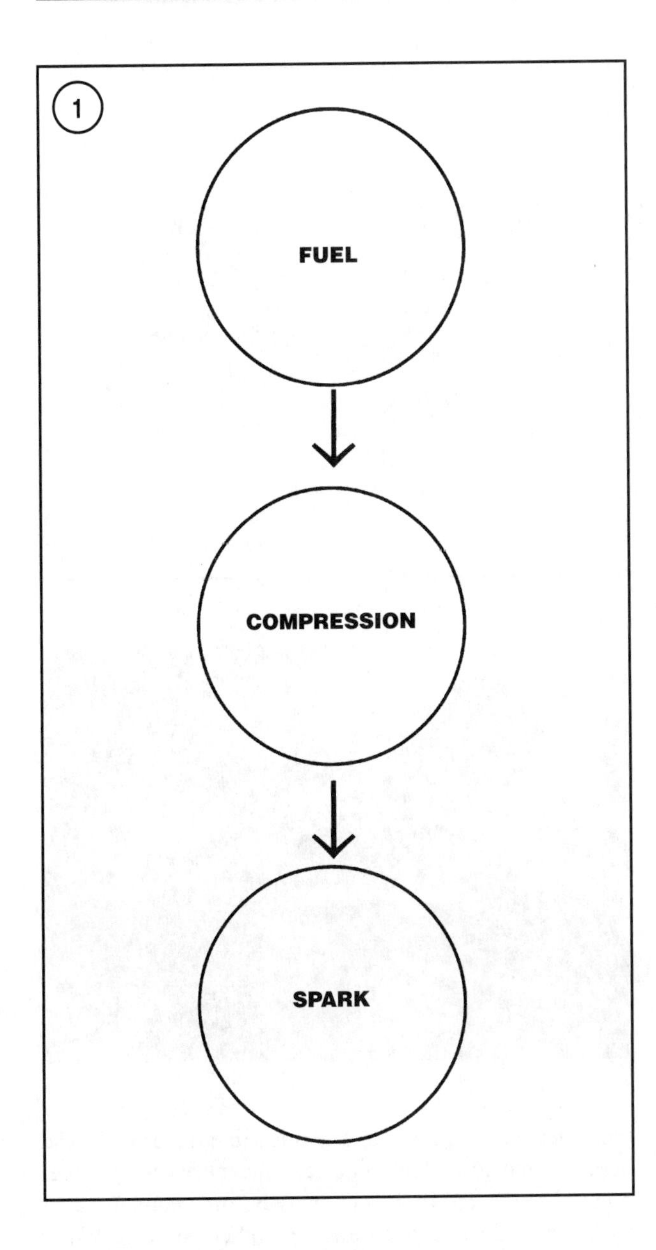

PRELIMINARY INSPECTION

Most engine malfunctions can be corrected by checking and correcting a few simple items. Check the following and if problems still exist refer to **Tables 1-3** for starting, fuel and ignition system troubleshooting. Additional troubleshooting tips are provided in this chapter for the specific system or component.

1. Place the lanyard safety switch into the run position.
2. Inspect the engine for loose, disconnected, dirty or corroded wiring.
3. Provide the engine with fresh fuel and the proper mix of Yamaha outboard engine oil (premix models).
4. Check the battery for proper and clean connections.
5. Fully charge the battery if needed.
6. Test for spark at each cylinder with a spark gap tester (part No. YM34487/908906754 or equivalent).
7. Check spark plug condition and gap settings.
8. Verify that the boat hull is clean and in good condition.

OPERATING REQUIREMENTS

An internal combustion engine requires three basic things to run properly: a fresh supply of fuel and air mixed in the proper proportion, adequate compression in the combustion chamber and a source of ignition at the proper time (**Figure 1**). If any of these are missing, the engine will not run. If any of these are weak or lacking, the engine will not run properly.

STARTING DIFFICULTY

Starting Procedures

If starting the engine becomes difficult, the problem may be with the engine or may be related to the starting procedure. The owner's manual is the best source for finding the proper starting procedure. Refer to **Table 1** to troubleshoot problems with the electric starting system.

Determining a Fuel or Ignition Fault

It can be quite difficult to determine if a starting problem is related to fuel, ignition or other causes. If the engine does not start, it is easiest to first verify that the ignition system is operating. Check the fuel system if the ignition system is operating properly. Use a spark gap tester to indicate output from the ignition system at cranking speeds. Spark gap testers are available from a variety of sources including Yamaha (part No. YM34487/90890-06754).

Spark test

WARNING

High voltage is present in the ignition system. Never let wires or connections touch any part of the body. Never perform ignition system testing in wet conditions. Electric shock can be fatal or cause serious injury. Never perform electrical testing when fuel or fuel vapor is present as arcing can cause a fire or explosion.

NOTE

The flywheel must rotate at a minimum of 300 rpm to generate adequate spark. Weak

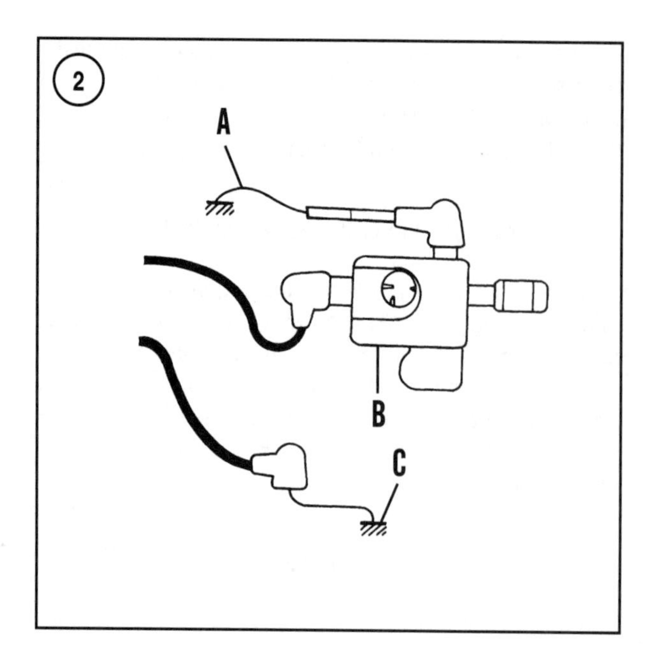

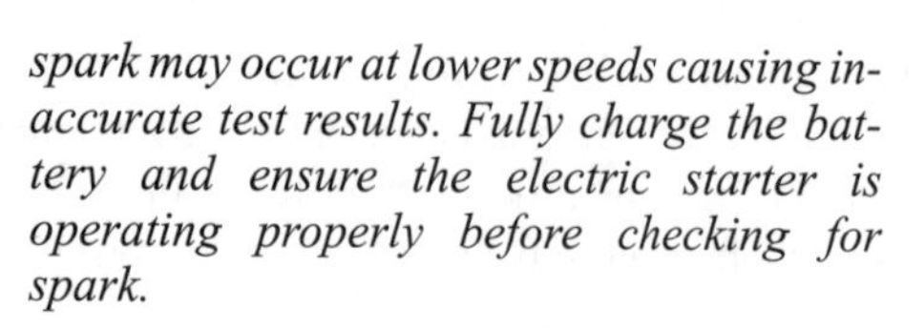
spark may occur at lower speeds causing inaccurate test results. Fully charge the battery and ensure the electric starter is operating properly before checking for spark.

1. Remove the spark plug(s) to prevent accidental starting.
2. Connect the alligator clip lead of the spark tester onto a suitable engine ground (A, **Figure 2**).
3. Attach the selected spark plug lead onto the spark gap tester (B, **Figure 2**). Adjust the spark gap on the tester to 9 mm (0.35 in.).
4. Connect the remaining spark plug leads to the engine ground to prevent arcing (C, **Figure 2**).
5. Crank the engine while observing the spark gap tester (**Figure 3**). A strong blue spark that jumps a 9 mm (0.35 in.) gap indicates adequate spark.
6. Repeat Steps 3-5 for the remaining cylinders.
7. Install the spark plug(s) and connect the lead(s) when testing is complete. Refer to **Table 3** for ignition system testing if spark is lacking or weak on any of the cylinders. Refer to **Table 2** for fuel system troubleshooting if the ignition system is working but the engine does not start.

Fuel test

Fuel-related problems are common with outboard engines. Fuel has a relatively short shelf life. Fuel becomes sour if stored for longer than 14 days. A gummy deposit may form in the carburetor fuel bowl or vapor separator tank and fuel passages as the fuel evaporates. Over time

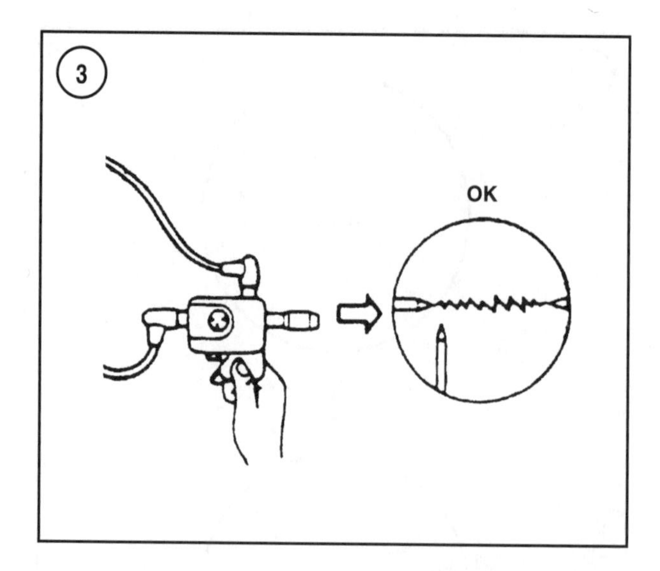

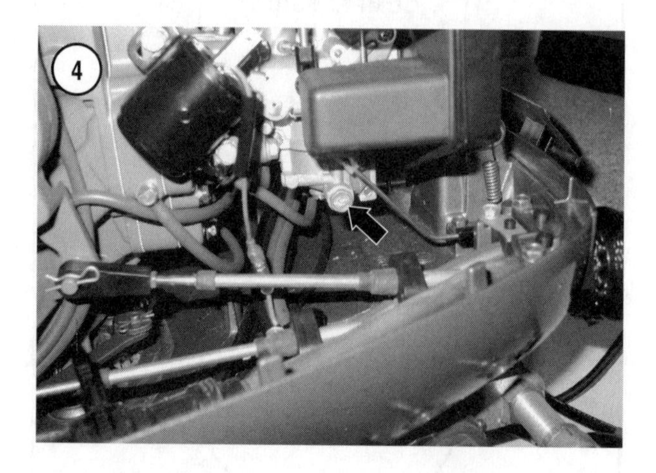

these deposits harden into a varnish-like material. These deposits may also clog fuel filters, fuel hoses and the fuel pump. Fuel stored in the tank may become contaminated with water from condensation or other sources. Water contamination causes the engine to run poorly or not at all.

Check the fuel condition if the engine has been stored for some time and does not start. Carefully drain a sample of the fuel into a suitable container. Chapter Five provides drawings that indicate the specific location of the carburetor bowl drain screw (**Figure 4**) or vapor separator tank drain screw (**Figure 5**). Unusual odor, debris, cloudy appearance or presence of water is a sure sign of a fuel related problem. If any of these conditions are noted, dispose of the old or contaminated fuel in an environmentally responsible manner. Contact a local marine dealership or automotive repair facility for information on the proper disposal of fuel. Clean and inspect the entire fuel system if contaminants are found in the carburetor float bowl or vapor separator tank. Repeating engine problems

are inevitable if the entire system is not thoroughly cleaned. Replace all fuel filters.

If fuel cannot be drained from the carburetor float bowl or vapor separator tank, the carburetors or vapor separator tank must be disassembled along with the fuel pump and hoses. Typically, the fuel inlet needle is stuck closed or blocked with debris. Refer to Chapter Five for all disassembly and assembly procedures of fuel system components.

A faulty choke valve or fuel injection system component can cause starting difficulty. Test these components or systems if there is no problem with the ignition system or fuel system and hard starting persists.

STARTING SYSTEM

An electric starting system is used on all Yamaha outboards covered in this manual.

The common components of the electric starting system include the battery, ignition, key switch, starter relay, starter solenoid (HPDI models), neutral safety switch, starter motor and wiring.

The electric starter motor (**Figure 6**) is similar in design to what is commonly used on automotive applications.

The electric motor portion of the starter is capable of producing a tremendous amount of torque, but only for a short period of time. Never operate the starter motor for over 20 seconds at a time and allow a two minute cooling period between each attempt to start the engine.

Weak or undercharged batteries are the leading cause of starting system problems. Battery requirements are listed in the *Quick Reference Data* section at the beginning of the manual. Battery maintenance and testing procedures are described in Chapter Six.

Two different types of starters are used on the Yamaha outboards covered in this manual. They use a different method for engaging and disengaging the flywheel ring gear.

Carburetor equipped and EFI models—An inertia engagement type starter (**Figure 6**) is used. The rotating armature shaft causes the starter drive gear (A, **Figure 7**) to move upward on the helical shaft splines and engage the flywheel ring gear while compressing the return spring (B). When the starter is disengaged, the flywheel moves the drive gear downward on the helical splines to disengage the ring gear with assistance from the return spring (B, **Figure 7**).

HPDI models—When the starter is engaged, an electromagnetic coil and plunger in the starter solenoid (**Figure 8**) physically moves a connecting lever that moves the starter drive gear upward to engage a flywheel ring gear. When the starter is disengaged, a spring in the solenoid

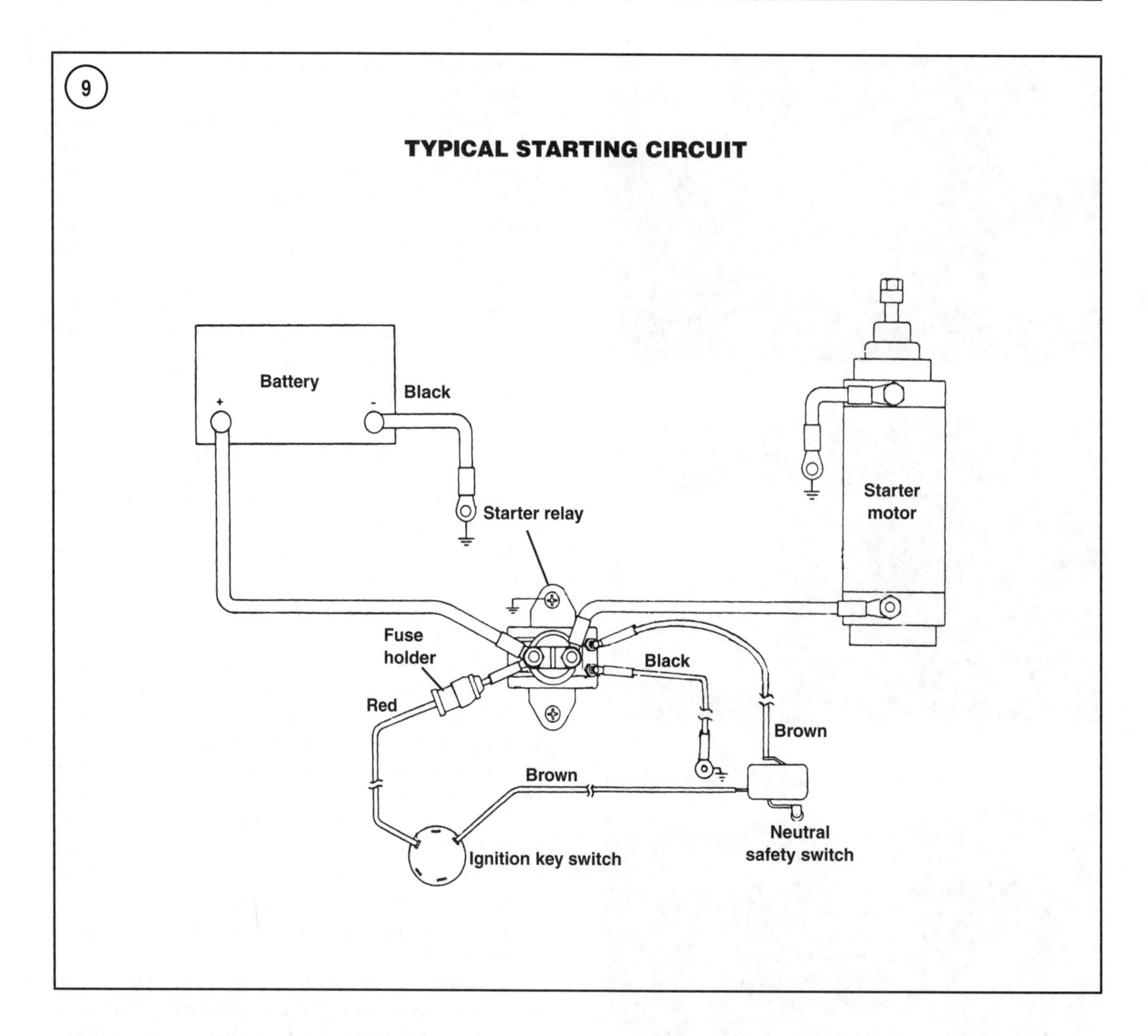

physically moves the drive gear downward to disengage the ring gear. **Figure 9** is a typical starting circuit.

Most marine dealerships and parts suppliers will not accept the return of any electrical part. If you cannot determine the exact cause of any electrical system malfunction, have a Yamaha dealership retest that specific system to verify the test results. If you purchase a new electrical component(s), install it, and then find that the system still does not work properly, you will probably not be able to return the unit for a refund.

Consider any test results carefully before replacing a component that tests only *slightly* out of specification, especially resistance. A number of variables can affect test results dramatically. These include: the test meter internal circuitry, ambient temperature and conditions under which the machine has been operated. All instructions and specifications have been checked for accuracy; however, successful test results depend to a great extent upon individual accuracy.

Starter Cranking Voltage Test

This test measures the voltage delivered to the starter motor during cranking. The battery must be tested and fully charged prior to performing this test. Refer to Chapter Six for battery testing, maintenance and charging procedures. A seized power head, gearcase or jet pump must be ruled out before repairing or replacing the starter motor. If at all possible, use an analog meter for this test. The

10

11

readings may fluctuate on a digital meter and prevent accurate voltage measurement.

1. Disconnect and ground the spark plug leads to prevent accidental starting. Do not remove the spark plugs.
2. Calibrate a voltmeter to the 20 or 40 VDC scale.

3A. *Carburetor equipped and EFI models*—Connect the positive test lead onto the large terminal on the starter motor (**Figure 10**). Carefully scrape the neoprene coating from the terminal nut to ensure a good connection. Connect the negative test lead to an engine ground.

3B. *HPDI models*—Connect the positive test lead to the large solenoid terminal closest to the starter motor frame (**Figure 11**). Push the rubber boot to the side to access the terminal. Connect the negative test lead onto an engine ground.

4. Crank the engine while observing the voltmeter.
   a. *No voltage is present*—Test the starter relay as described in this chapter. On HPDI models, test the starter relay and starter solenoid as described in this chapter.
   b. *Voltage is 9.5 VDC or greater*—The starter relay is supplying adequate current for the starter. Replace or repair the starter if it does not rotate or it rotates slowly.
   c. *Voltage reading is 9.4 volts or lower*—Check the battery terminals and cables, then replace or fully charge the battery. Repeat the test.
5. Reconnect the spark plug leads.

Starter Relay

The starter relay allows a large amount of current to pass from the battery to the starter motor or starter solenoid (HPDI models). When the ignition key switch activates, current flows through the neutral safety switch using the brown wire and into the relay. This closes the relay contacts, which completes the circuit and the staring system operates. The first test checks the voltage supply and the ground wire for the relay. The relay function test verifies proper operation of the relay.

Starter relay voltage test

Refer to the information in Chapter Six to locate the starter relay and wires.

1. Disconnect and ground the spark plug leads to prevent accidental starting.
2. Disconnect the large diameter cable from the starter terminal (**Figure 10**). Thoroughly cover the cable terminal to prevent it from contacting any components during testing. Route the cable to position the terminal away from all components.
3. Calibrate a volt/ohm meter onto the 20 or 40 VDC scale.
4. Connect the negative meter test lead to a good engine ground.
5. Touch the positive meter test lead to the battery terminal of the relay. The meter should indicate battery voltage. If not, check the wiring and terminals of the large diameter cable connecting the battery to the starter relay. If the wiring and terminals check satisfactorily, check for a faulty battery ground terminal or cable.
6. Repeat Step 5 with the negative meter test lead touching the ground terminal of the starter relay. Repair the faulty relay ground wire or terminal if battery voltage is only present with the negative meter test lead connected to engine ground.
7. Disconnect the brown wire from the relay. Connect the positive meter test lead to the brown wire harness terminal. Connect the negative meter lead to a good engine ground. Turn the ignition key switch to all three positions. The meter should indicate battery voltage each time the switch reaches the start position and 0 volt with the key in the off or run positions. If it does not, test the fuses, neutral safety switch, key switch and related wiring.

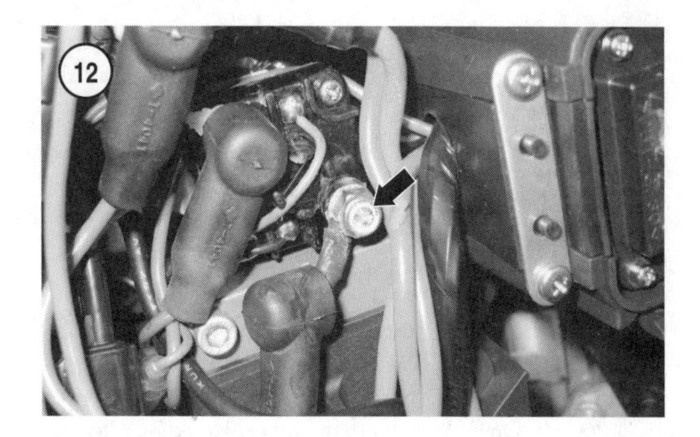

8. Reconnect the brown engine harness wire onto the brown relay wire. Route the wiring to avoid interference with other components.

9. Connect the positive meter test lead to the starter cable terminal at the starter. Turn the ignition key switch to the start position. Repeat the test with the positive meter test lead connected to the starter cables terminal at the relay (**Figure 12**). The meter should indicate battery voltage at both test points.

   a. *Voltage only at the starter relay terminal*—The cable or cable terminals are faulty. Replace or repair the starter cable and retest.

   b. *No voltage at either test point*—Perform a relay function test as described in this chapter.

10. Reconnect the large diameter starter cable to the starter terminal (**Figure 10**). Securely tighten the terminal and apply a coating of liquid neoprene onto the terminal to prevent arcing and corrosion.

11. Connect the spark plug leads.

Starter relay functional test

This test requires a fully charged battery, an analog multimeter and suitable jumper wires.

1. Remove the relay from the engine as described in Chapter Six.

2. Calibrate the multimeter to the R × 1 scale. Connect the negative meter test lead to one of the large terminals on the relay (**Figure 13**). Connect the positive test lead to the other large terminal.

3. Using a jumper wire, connect the black relay wire to the negative battery terminal (**Figure 13**).

4. Connect a jumper wire to the positive battery terminal. Observe the multimeter while repeatedly touching the jumper wire to the brown wire terminal of the relay. The meter should indicate *continuity* each time the connection in made and *no continuity* each time the connection is broken. The relay should make a clicking noise each time the connection in made.

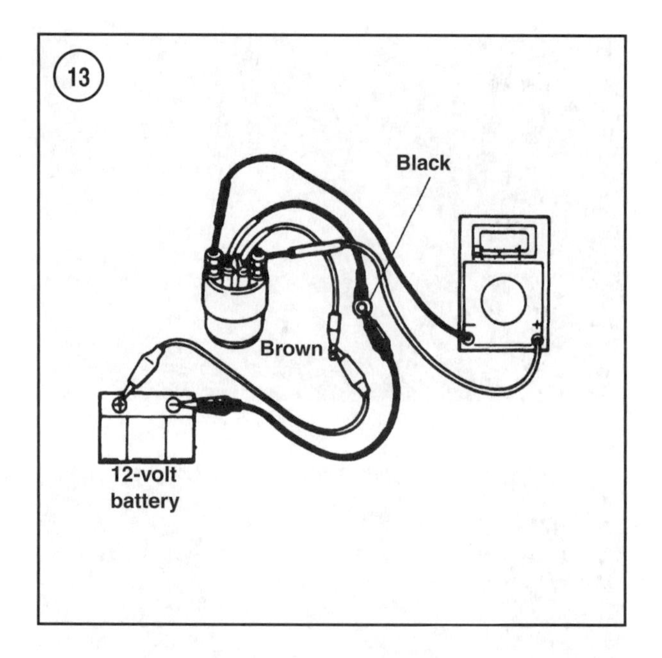

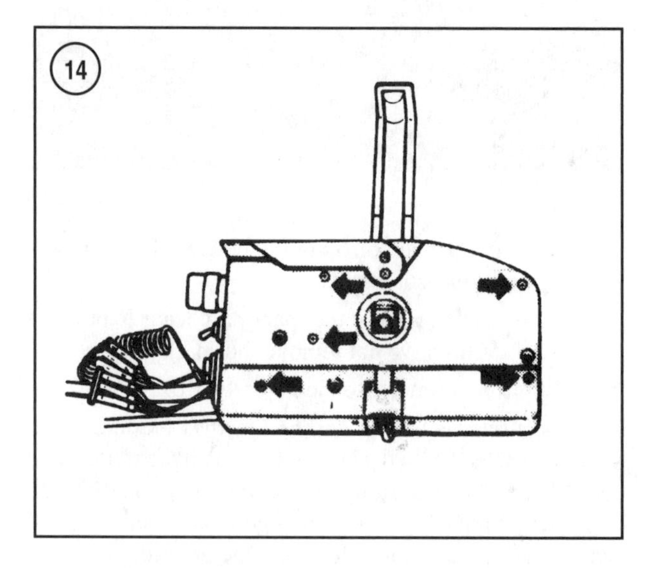

5. Replace the starter relay if it fails to perform as described.

6. Install the starter relay as described in Chapter Six.

Neutral Safety Switch Test

The neutral safety switch prevents the electric starter from operating when the engine is in forward or reverse gear. The switch is mounted within the remote control. A common multimeter is required for this test.

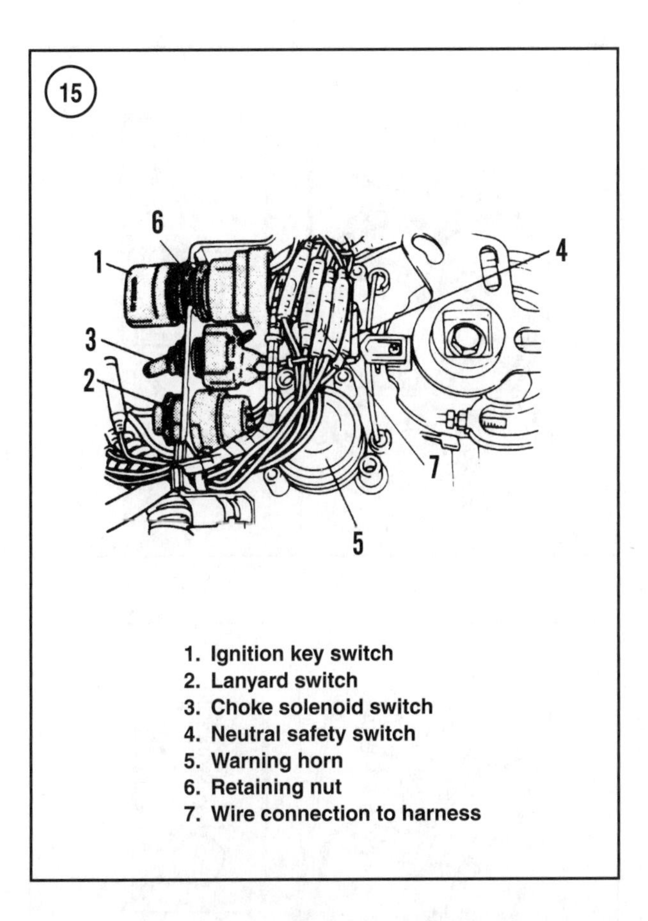

1. Ignition key switch
2. Lanyard switch
3. Choke solenoid switch
4. Neutral safety switch
5. Warning horn
6. Retaining nut
7. Wire connection to harness

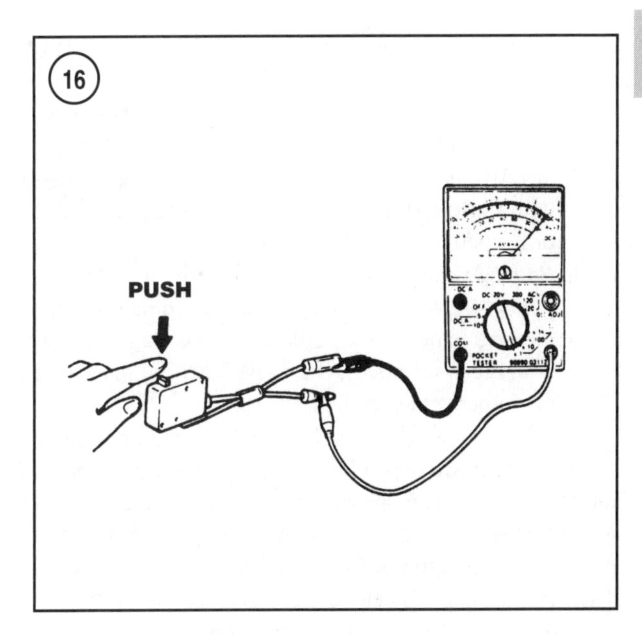

1. Remove the remote control and remove the rear covers (**Figure 14**) as described in Chapter Thirteen. Do not remove the control handle.
2. Locate the neutral safety switch (4, **Figure 15**), then trace the switch leads to the connections onto the remote control harness (7). Disconnect both switch leads.
3. Calibrate the multimeter to the R × 1 scale. Connect the meter test leads to each of the neutral safety switch wires.
4. Move the control lever to the NEUTRAL gear position. The meter should indicate continuity. Observe the meter while alternately moving the control handle into the FORWARD and REVERSE gear positions. The meter should indicate no continuity each time the engine is shifted into forward or reverse gear. Repeat this step several times to check for intermittent switch failure. Test the key switch and related wiring if the meter readings test as specified. Proceed to Step 5 if the neutral safety switch tests incorrectly.
5. Disassemble the control and remove the neutral safety switch as described in Chapter Six. Test the switch as follows:
   a. Connect the ohmmeter to each of the switch wires (**Figure 16**).
   b. Observe the ohmmeter while repeatedly depressing the switch lever (**Figure 16**). The meter should indicate continuity with the switch depressed and no continuity with the switch released. Replace the neutral safety switch if it fails either test.
   c. Install the switch into the remote control as described in Chapter Six.
6. Reconnect the neutral safety switch wires to the control harness. Route the wiring to prevent contact with moving components.
7. Reassemble and install the remote control as described in Chapter Thirteen.

Starter Solenoid Test

The starter solenoid allows current to pass from the battery to the starter motor.

This test checks the voltage supply to the solenoid.

1. Disconnect and ground the spark plug leads to prevent accidental starting.
2. Calibrate a volt/ohm meter onto the 20 or 40 VDC scale.
3. Connect the positive test lead to the positive battery terminal of the solenoid. Connect the negative test lead to a good engine ground. Note the voltage the negative test lead to the negative battery terminal. Note the voltage. Next, move the positive test lead connected to the positive battery terminal.
   a. *Battery voltage for all three tests*—The cables and terminals test correctly. Go to Step 4.
   b. *No or low voltage for all three tests*—Test the battery as described in Chapter Six. Clean and inspect

the battery cables and terminals if the battery test is satisfactory.

c. *Battery voltage only with the negative test lead to battery*—Clean and inspect the terminals on the negative battery cable. If low or no voltage persists with clean terminals, the cable has high resistance or is open and must be replaced.

d. *Battery voltage only with the positive test lead to battery*—Clean and inspect the terminals on the positive battery cable. If low or no voltage persists with clean terminals, the cable has high resistance or is open and must be replaced.

4. Connect the negative test lead onto an engine ground. Connect the positive test lead to the brown wire terminal on the solenoid. The brown wire connects to the *S* terminal on the solenoid. Note the meter reading while repeatedly turning the ignition key switch to all three positions. The meter should indicate battery voltage in the start position and 0 volt in the on or run positions. Test the starter relay, neutral safety switch, ignition key switch and related wiring if the tests fail.

5. Connect the negative test lead to an engine ground. Connect the positive test lead to the starter terminal on the solenoid (**Figure 13**). Note the meter reading while turning the ignition key to all three positions. The meter should indicate battery voltage and click noise each time the switch reaches the start position and 0 volt each time the switch reaches the on or run positions. Test the starter solenoid as described in Chapter Six if the solenoid fails these tests.

6. Connect the spark plug leads.

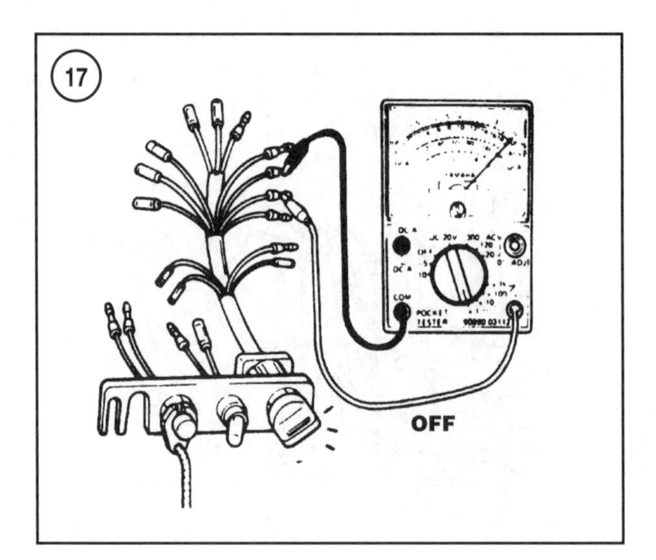

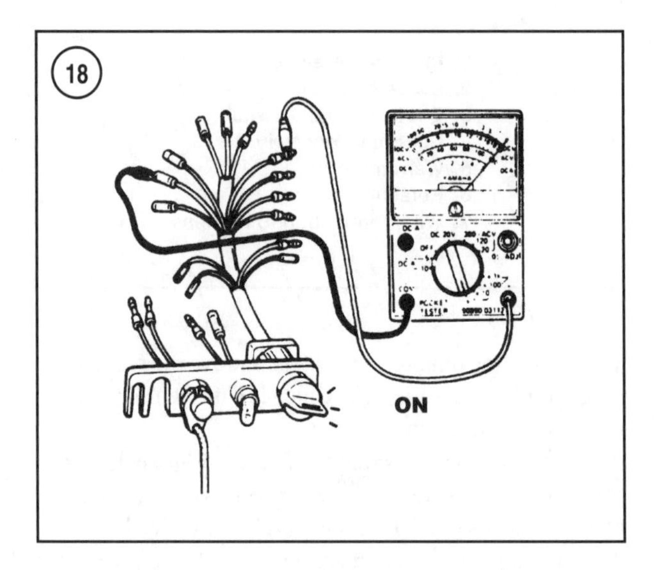

Ignition Key Switch Test

The key switch is mounted in the remote control or on the control station dashboard.

1. Disconnect the battery cables. Disconnect and ground the spark plug leads to prevent accidental starting.

2. Remove the ignition key switch as described in Chapter Six.

3. Calibrate a multimeter to the R × 1 scale.

4. Connect the positive test lead to the white switch wire and the negative test lead to the black switch wire. Observe the meter while repeatedly turning the ignition key switch to the OFF (**Figure 17**), ON (**Figure 18**) and START (**Figure 19**) positions. Correct test results:

a. *Switch in the off position*—Continuity.

b. *Switch in the on position*—No continuity.

c. *Switch in the start position*—No continuity.

5. Connect the positive test lead to the red switch wire and the negative test lead to the yellow switch wire. Observe the meter while turning the key switch to the OFF, ON and START positions. Correct test results:

a. *Switch in the off position*—No continuity.

b. *Switch in the on position*—Continuity.

c. *Switch in the start position*—Continuity.

6. Connect the positive test lead to the red switch wire and the negative test lead to the brown switch wire. Observe the meter while turning the key switch to the OFF, ON and START positions. Correct test results:

a. *Switch in the off position*—No continuity.

b. *Switch in the on position*—No continuity.

c. *Switch in the start position*—Continuity.

7. Replace the switch if any incorrect test results are noted.

8. Install the key switch as described in Chapter Six.

9. Connect the battery cables and spark plug leads.

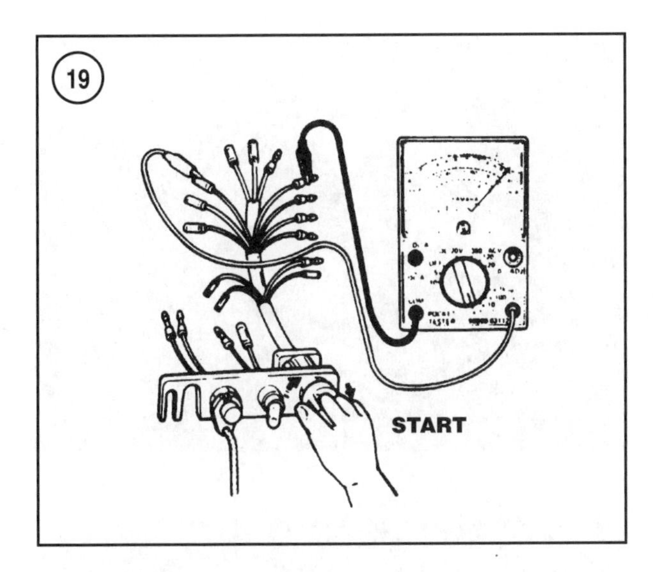

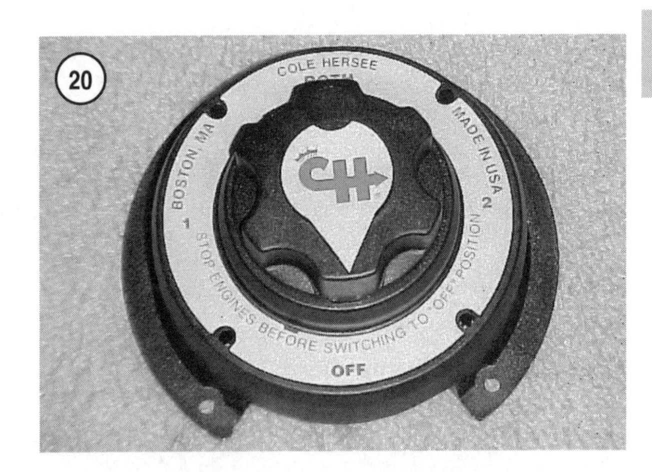

Choke Switch Test

The choke switch is used on carburetor equipped models. The switch is either integrated into the ignition key switch or separately mounted in the remote control (3, **Figure 15**).

1A. *Remote control mounted ignition key switch*—Remove the remote control and remove the rear cover (**Figure 14**) as described in Chapter Thirteen. Do not remove the control handle.

1B. *Dash mounted ignition key switch*—Locate the ignition key switch wire connections to the instrument wiring harness.

2A. *Choke switch separately mounted in the remote control*—Locate the choke switch (3, **Figure 15**), then trace the switch wires to the remote control (7). Disconnect both switch wires.

2B. *Choke switch integrated into the key switch*—Disconnect the blue and black ignition key switch wires from the remote control (7, **Figure 15**).

3. Calibrate the multimeter to the R × 1 scale. Connect the positive test lead to the blue switch wire. Connect the negative test lead to the black switch wire.

4A. *Choke switch mounted in the remote control)*—Turn the ignition key switch to the RUN position, then toggle the choke switch toward the ignition key switch.

4B. *Choke switch integrated into the ignition key switch*—Turn the switch to the ON or START position while pushing straight in on the ignition key switch.

5. Observe the meter while repeatedly activating then releasing the choke switch. The meter should indicate continuity each time the choke switch is activated and no continuity each time the choke switch is released.

6. Replace the choke switch or ignition key switch if there are any incorrect test results.

7. *Remote control mounted ignition key switch*—Assemble and install the remote control as described in Chapter Thirteen.

CHARGING SYSTEM

The function of the charging system is to maintain the battery charge level. Accessories (depth sounders, stereos and the like) can draw considerable current from the battery. The increased use of electrical accessories has placed a greater demand on the typical outboard charging system.

Determine the total amperage consumed by the onboard accessories and compare the total with the charging system output specification in **Table 4**. Keep in mind that the charging system delivers considerably less output when operating the engine at lower speeds.

Charge the battery at frequent intervals if the amperage load is near or exceeding the charging system output. Consider adding additional batteries or installing a battery with greater capacity. If two batteries are used, install a battery switch (**Figure 20**) and connect the accessories to one of the batteries. Battery maintenance and charging procedures are described in Chapter Six.

The components of the charging system include the flywheel (**Figure 21**), battery/ignition charging coil (**Figure 22**), rectifier/regulator (**Figure 23**), the wiring and the battery. Test all components of the charging system if the battery fails to maintain a charge.

The battery charging coil and ignition charging coil are integrated into a single assembly mounted under the flywheel. Replace the complete assembly if either coil fails.

As the flywheel magnets pass near the coil it produces alternating current (AC). The current is directed to the rectifier/regulator where it is converted into direct current. The rectifier/regulator limits the charging system output when necessary to prevent overcharging.

NOTE
In addition to charging the battery, the charge coil produces the electrical pulses that operate the dash mounted tachometer. If the tachometer is operating, the charge coil is producing alternating current.

Charging System Output Test

The battery must be fully charged for this test.

WARNING
Stay clear of the propeller shaft while running an engine on a flush/test adapter. The propeller must be removed before running the engine. Disconnect the battery and all spark plug leads before removing or installing a propeller.

CAUTION
Never run an outboard without first providing cooling water. Use either a flush/test adapter if the engine cannot be operated under normal conditions or in a suitable test tank. Install a test propeller to run the engine in a test tank.

1. Calibrate an ohmmeter to measure 10-15 VDC.
2. Connect the positive test lead to the battery positive terminal. Connect the negative test lead to a suitable engine ground. Record the battery voltage.
3. Start the engine and run at fast idle until the engine reaches normal operating temperature. Turn off all accessories.
4. Advance the throttle to approximately 2500 rpm. Note and record the voltage reading. If so equipped, note if the tachometer is functioning. Compare the engine running voltage with the engine off voltage.
   a. *Engine running voltage equal to or below the engine off voltage*—The charging system is not operating properly. Test all charging system components as described in this chapter.
   b. *Engine running voltage exceeds 14.0 volt*—The charging system is overcharging. Check for a faulty wiring or a low electrolyte level in the battery. Test the rectifier/regulator as described in this chapter.
   c. *Engine running voltage is 0.3 volt or higher than the engine off voltage*—The charging system is charging the battery. If the battery fails to maintain a charge, check for excessive accessory load or a faulty battery.

21

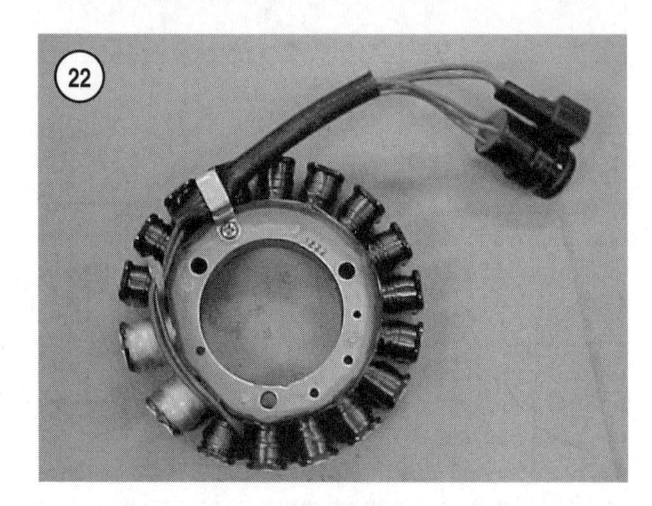
22

   d. *Tachometer is working, engine running voltage is equal to or below engine off voltage*—Test the rectifier/regulator as described in this chapter.
   e. *Tachometer is not working, engine running voltage reading is equal to or below the engine off voltage*—Test the battery charge coil and rectifier/regulator as described in this chapter.
   f. *Tachometer is not working, engine running voltage is 0.3 volt or higher than engine off voltage reading*—Substitute a known good tachometer and retest. If the tachometer still does not operate, test the rectifier/regulator.

Battery Charge Coil Tests

This section tests the battery charge coil resistance and battery charge coil output. Battery charge coil resistance specifications are provided only for 80 Jet and 115-130 hp models (except C115). Battery charge coil output specifications are provided for all models. On 80 Jet and 115-130 hp models (except C115) perform the resistance and out-

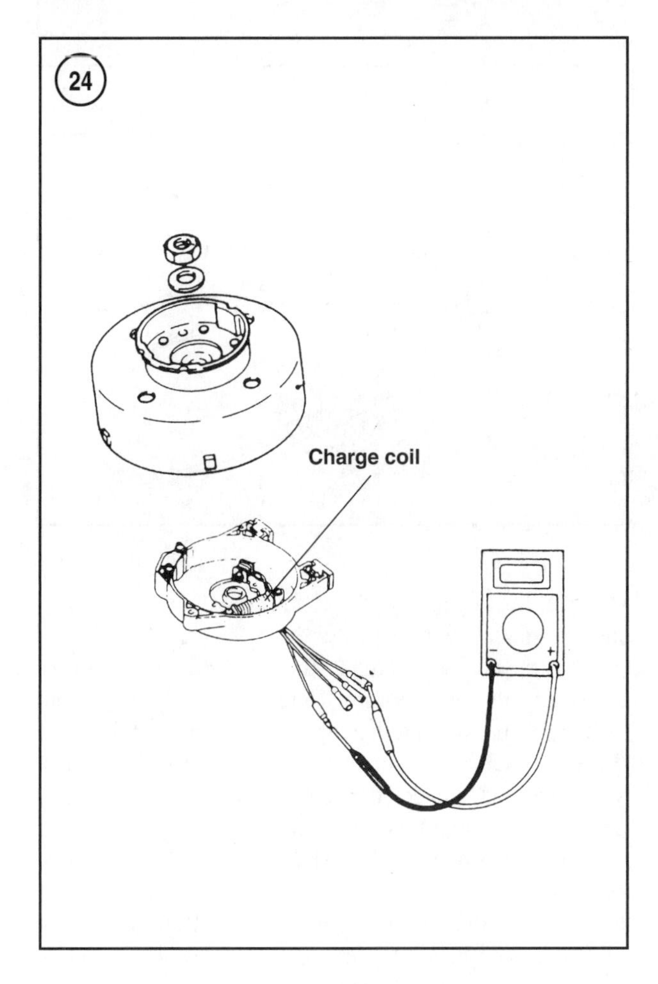

put test. On all other models, perform only the output test to determine the condition of the coil.

NOTE
Replace the rectifier/regulator if the charging system is not operating and the battery charge coil tests correctly.

Battery charge coil resistance test

Use a digital volt/ohm meter for this test. The battery charge coil can be tested without removing it from the engine. The ambient temperature will affect the measured resistance. Perform this test with the engine at approximately 20° C (68° F).

1. Disconnect the battery cables and ground the spark plug leads to prevent accidental starting.
2. Calibrate a digital volt/ohm meter onto the R × 1 scale.
3. Locate the green and green/white battery charge coil wires. The wires lead to the coil mounting location under the flywheel. Disconnect the wires from the rectifier/regulator harness connectors.
4. Connect the positive test lead to the green battery charge coil wire (**Figure 24**). Connect the negative test lead to the green/white coil wire. Do not inadvertently connect the test leads to the rectifier/regulator wires. Record the resistance reading.
5. The resistance reading must be between 0.4-0.6 ohm. If not, replace the battery charge coil as described in Chapter Six.
6. Reconnect the green and green/white wires. Route the wiring to prevent interference with moving components.
7. Connect the battery cables and spark plug leads.

Battery charge coil output test

CAUTION
Never run an outboard without first providing cooling water. Use either a flush/test adapter if the engine cannot be operated under normal conditions or in a suitable test tank. Install a test propeller to run the engine in a test tank.

WARNING
Stay clear of the propeller shaft while running an engine on a flush/test adapter. The propeller must be removed before running the engine. Disconnect the battery and all spark plug leads before removing or installing a propeller.

Use a common meter with peak output voltage capability (Yamaha part No. J-39299/90890-06752 and YU-39991/90890-03169) along with an accurate shop tachometer for this test. Use test harnesses on all models except C115 and C150.

Use the three pin Yamaha test harness part No.YB-06770/90890-06770 on all models except C115, C150 and HPDI models. On 150-250 hp HPDI models use the single pin test harness part No. YB-06788/90890-06788 and the two pin test harness test harness part No. YB-06787/90890-06787. A test harness for C115 and C150 models is not required.

1. Connect a shop tachometer to the engine. Follow the tachometer manufacturer's instructions.
2. Locate the three green and green and green/white battery charge coil wires. The wires lead to the coil mounting location under the flywheel. Trace the wires to their connection to the rectifier/regulator wires.
   a. *80 Jet, 105 Jet, 115-130 hp (except C115)*—A single connector connects the three green battery charging coil wires to the rectifier/regulator wires.
   b. *C115 and C150 (premix models)*—Two bullet connector connect the green and green/white wires to the rectifier/regulator leads.
   c. *150-250 hp (except HPDI models)*—A single connectors connects the three green battery charging coil wires to the rectifier/regulator wires.
   d. *150-250 HPDI models*—Two connectors are used. One connector contains a single green wire. The other connectors contain two green wires. The connectors are located behind the electrical component cover on the starboard side of the power head.
3. Refer to **Table 5** for the test specifications, then calibrate the multimeter to the correct peak AC voltage scale. Scale changes may be required when testing at different engine speeds.

4A. *C115 and C150*—Connect the positive test lead to the green battery charge coil lead (**Figure 25**). Connect the negative test lead to the green/white wire. Do not disconnect the wires. Slip the test leads between the bullet connector sleeves to contact the terminals as shown in **Figure 26**.

4B. *All models except for C115, C150 and HPDI models*—Disconnect the three green wire connectors from the rectifier/regulator and battery charge coil harnesses. Connect the test harness (Yamaha part No. YB-06770/90890-0670) to the charge coil and rectifier connectors. Connect the positive test lead to one of the green wire terminals on the test lead and the negative test lead to one of the remaining green wire terminals.

4C. *150-250 hp (HPDI models)*—Disconnect the two green wire and single green wire connectors from the battery charge coil and rectifier/regulator harnesses. Connect the single wire test harness (Yamaha part No. YB-06788/90890-06788) to the single green wire connector of the charge coil and rectifier/regulator. Connect the two-pin test wire harness (Yamaha part No. YB-06787/90890-06787) to the two green wire connectors of the coil and rectifier/regulator. Connect the positive test lead to the of the single-pin test wire harness.

25

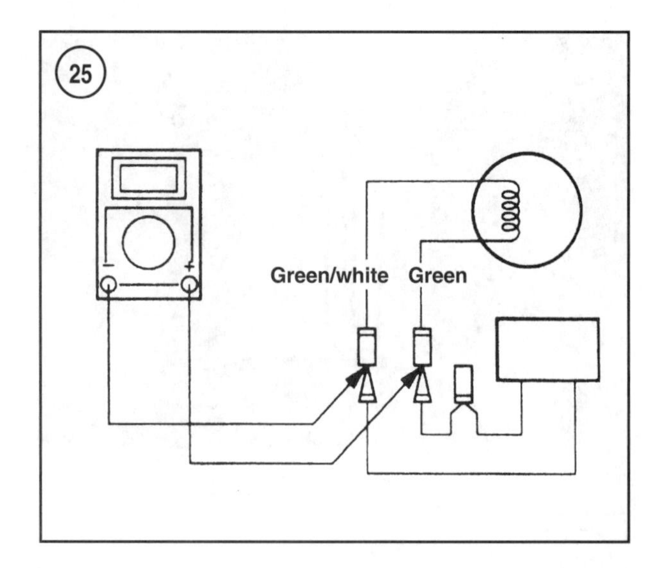

26

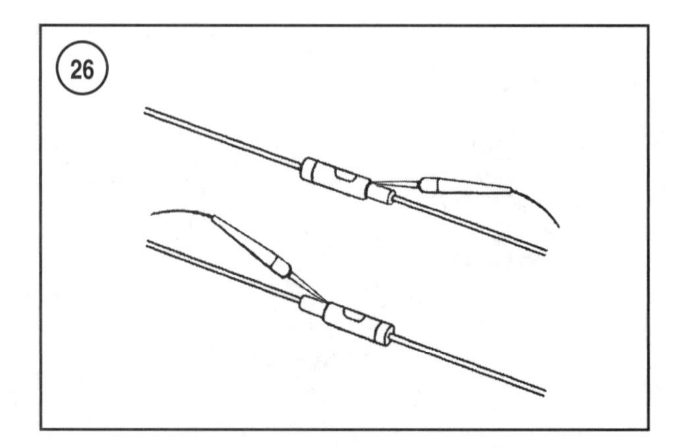

Connect the negative test lead to one of the test leads of the two-pin test wire harness.

5. Start the engine and run at fast idle until it reaches normal operating temperature. Note and record the output voltage while running the engine at the speed(s) listed in **Table 5**. Change the meter scale as required for the test specification. To check the voltage at cranking speed, disconnect and ground the spark plug leads. Note the meter readings while operating the electric starter during normal starting.
6. *All models except C115 and C150*—Repeat this test with the negative test lead connected to the remaining green test wires. This tests the remaining coil winding.
7. All charge coil output tests must meet or exceed the minimum specification in **Table 5** at each designated engine speed. If not, replace the battery charge coil as described in Chapter Six.
8. Stop the engine and disconnect the shop tachometer.
9. Disconnect the test harness and test leads. Reconnect all battery charging coil wires and connectors to the recti-

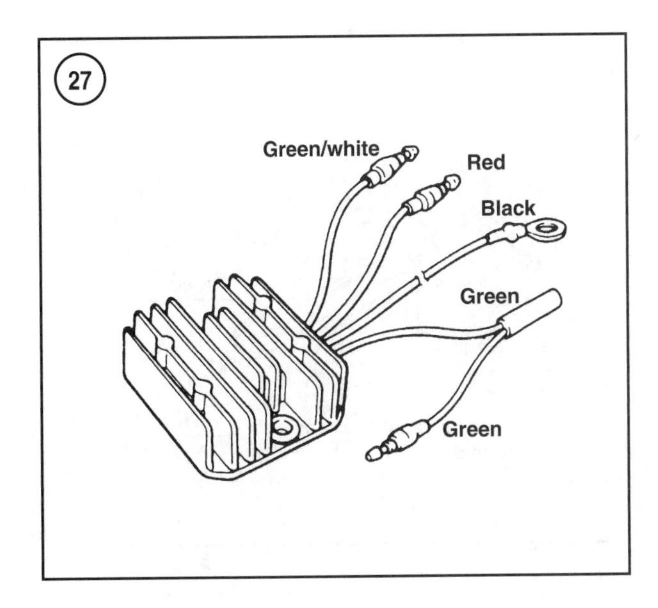

fier/regulator harness. Route all wiring to prevent interference with moving components.

Rectifier/Regulator Tests

Use a common multimeter for rectifier/regulator testing. An analog meter is recommended. Perform the resistance test on all carburetor equipped models. Perform the output test on EFI and HPDI models.

NOTE
Internal circuitry may vary by the model and brand of the test meter. This may result in opposite test results when selecting the ohm function for testing diodes. If incorrect test results occur, perform the testing with the polarity opposite to what is designated in the procedures.

Rectifier/regulator resistance test (models C115 and C150)

Refer to **Figure 27** for this procedure.

1. Remove the rectifier/regulator from the power head as described in Chapter Six.
2. Calibrate and ohmmeter to the R × 1 scale.
3. Connect the positive test lead to the red rectifier/regulator wire. Connect the negative test lead to the green/white lead to the rectifier/regulator. Note the meter reading, and then repeat the test with the negative test lead connected to the green rectifier/regulator wire. If the meter indicates no continuity at both test points, the rectifier/regulator has failed open and should be replaced.
4. Connect the positive test lead to the red rectifier/regulator wire. Connect the negative test lead to the black rectifier/regulator wire. If the meter indicates no continuity, the rectifier/regulator has failed open and should be replaced.
5. Connect the positive test lead to the black rectifier/regulator wire. Connect the negative test lead to the green rectifier/regulator wire. If the meter indicates continuity, the rectifier/regulator has shorted and should be replaced.
6. Connect the positive test lead to the black rectifier/regulator wire. Connect the negative test lead to the red rectifier/regulator wire. If the meter indicates continuity, the rectifier/regulator has shorted and should be replaced.
7. Connect the positive test lead to the black rectifier/regulator wire. Connect the negative test lead to the green/white wire to the rectifier/regulator. If the meter indicates continuity, the rectifier/regulator has failed open and should be replaced.
8. Connect the positive test lead to the green rectifier/regulator wire. Connect the negative test lead to the black rectifier/regulator wire. Note the meter reading, then repeat the test with the negative test lead connected to the green/white wire to the rectifier/regulator. If the meter indicates no continuity for both test points, the rectifier/regulator has failed open and should be replaced.
9. Connect the positive test lead to the green rectifier/regulator wire. Connect the negative test lead to the red rectifier/regulator wire. Note the meter reading, then repeat the test with the positive test lead connected to the green/white wire to the rectifier/regulator. If the meter indicates continuity for both test points, the rectifier/regulator has shorted internally and should be replaced.
10. Connect the positive test lead to the green rectifier/regulator wire. Connect the negative test lead to the green/white wire to the rectifier/regulator. If the meter indicates continuity, the rectifier/regulator has failed open and should be replaced.
11. Connect the positive test lead to the green/white wire of the rectifier/regulator. Connect the negative test lead to the green rectifier/regulator wire. If the meter indicates continuity the rectifier/regulator has shorted and should be replaced.
12. Install the rectifier/regulator as described in Chapter Six.

Rectifier/regulator resistance test (carburetor equipped models except C115 and C150)

Refer to **Figure 28** for this procedure.

1. Remove the rectifier/regulator from the power head as described in Chapter Six.
2. Calibrate an ohmmeter to the R × 1 scale.

3. Connect the positive meter test lead to the red rectifier/regulator wire in the single terminal connector. Connect the negative meter test lead to the black rectifier/regulator wire. If the meter indicates continuity, the rectifier/regulator has shorted and should be replaced.

4. Connect the positive test lead to the red rectifier/regulator wire in the single terminal connector. Observe the meter while touching the negative test lead to each of the green wires in the three-terminal connector. If the meter indicates continuity at each connection point, the rectifier/regulator has shorted and should be replaced.

5. Connect the negative test to the black rectifier/regulator wire. Observe the meter while touching the positive test lead to each of the green wires in the three-terminal connector. If the meter indicates continuity at each connection point, the rectifier/regulator has shorted and should be replaced.

6. Connect the positive test lead to the black rectifier/regulator wire. Observe the meter while touching the negative test lead to each of the green wires in the three-terminal connector. If the meter indicates no continuity at each connection point, the rectifier/regulator has failed open and should be replaced.

7. Connect the positive test lead to the black rectifier/regulator wire. Connect the negative test lead to the single terminal connector. If the meter indicates no continuity, the rectifier/regulator has failed open and should be replaced.

8. Connect the negative test lead to the single terminal connector. Observe the meter while touching the positive test lead to each of the green wires in the three-terminal connector. If the meter indicates no continuity at each connection point, the rectifier/regulator has failed open and should be replaced.

9. Install the rectifier/regulator as described in Chapter Six.

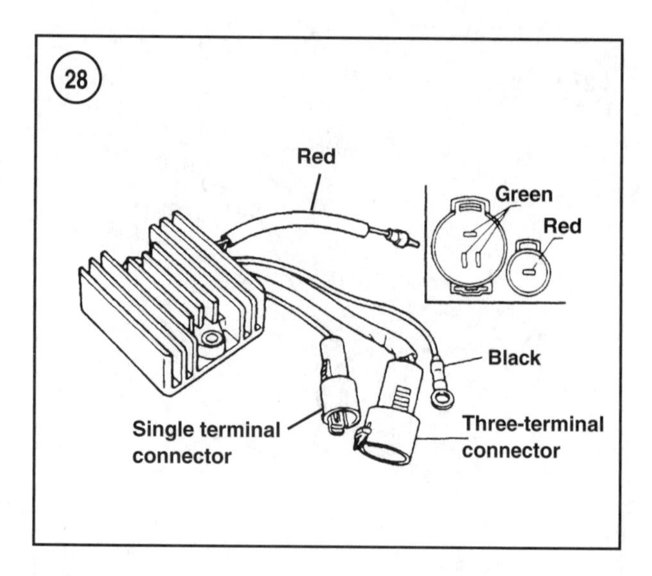

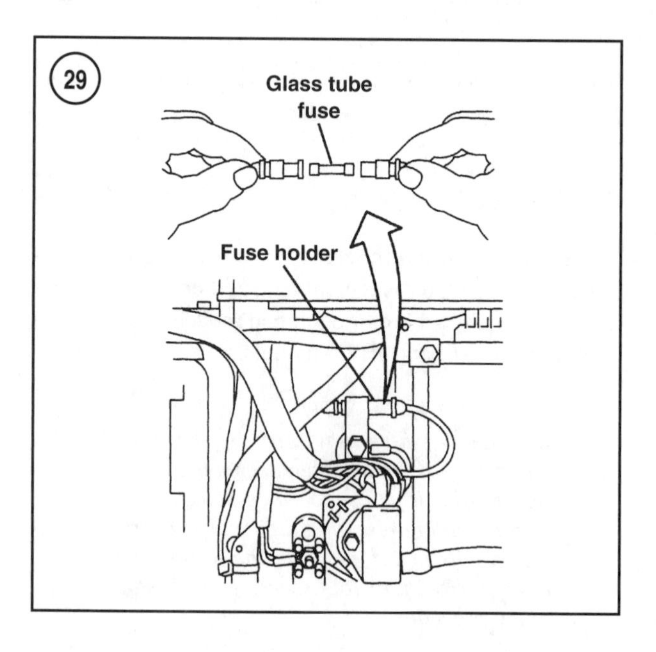

Rectifier/regulator output test

CAUTION
Never run an outboard without first providing cooling water. Use either a flush/test adapter if the engine cannot be operated under normal conditions or in a suitable test tank. Install a test propeller to run the engine in a test tank.

WARNING
Stay clear of the propeller shaft while running an engine on a flush/test adapter. The propeller must be removed before running the engine. Disconnect the battery and all spark plug leads before removing or installing a propeller.

NOTE
A meter with peak output voltage measuring capability should be used when measuring peak output voltage from the battery charging coil. Low test results are likely if the coil is tested using the AC voltage scale.

1. Connect a shop tachometer to the engine. Follow the tachometer manufacturer's instructions.

2. Trace the three green wires from the battery charging coil to the rectifier/regulator harness. Connect the positive test lead to the red rectifier/regulator wire terminal. Do

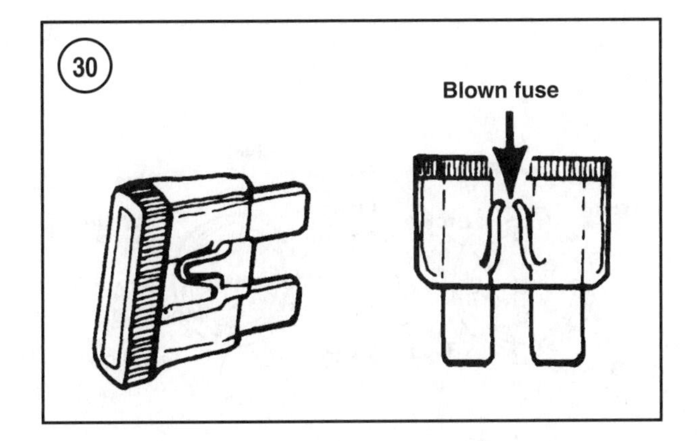

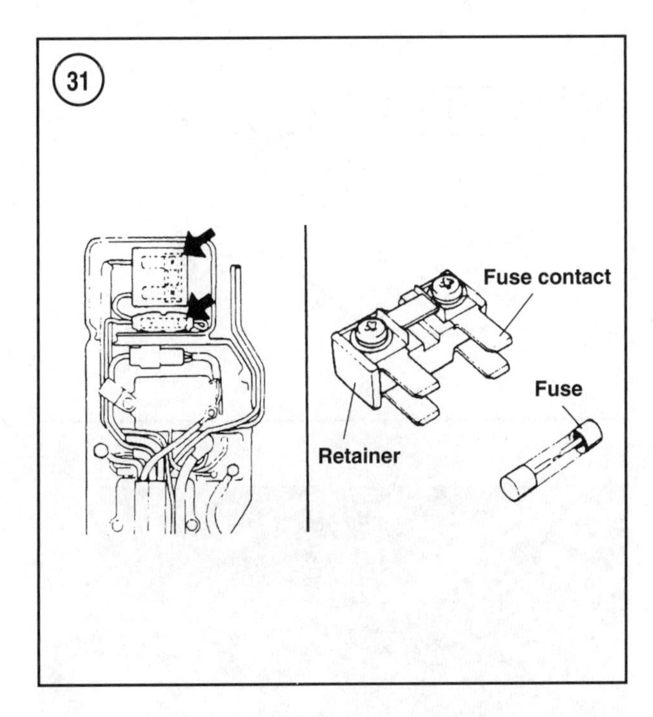

not disconnect the red wire. If the terminal is inaccessible, touch the test lead to the exposed 20 amp fuse terminal. Connect the negative test lead to the engine ground.
3. Calibrate the multimeter to 20 or 40 peak AC voltage scale.
4. Start the engine and run the engine at a fast idle until it reaches normal operating temperature. Record the output voltage while running the engine at the speeds listed in **Table 6**. To check the voltage at cranking speed, disconnect and ground the spark plug leads. Note the meter readings while operating the electric starter during normal starting.
5. All rectifier/regulator tests should meet or exceed the minimum specification in **Table 6** at each designated engine speed. Replace the rectifier/regulator as described in Chapter Six if it does fail to meet the specifications.
6. Stop the engine and disconnect the shop tachometer.

FUSE AND WIRE HARNESS

Fuse Test

Fuses are used to protect the electrical system in the event of a short circuit or overload. Never replace a fuse without performing a thorough inspection of the electrical system. Keep in mind that fuses are designed to open the circuit if an overload occurs. For safety, never bypass a fuse or install a fuse with greater capacity than specified. The most common symptom of a blown fuse is the electric starter not operating and the dash mounted gauges not working. In some instances, the first symptom is the tilt/trim system not working. Use a common multimeter to test fuses. Never rely solely on a visual inspection to test fuses. Different types of fuses and mounting arrangement are used on the various models covered in this manual. **Figures 29-31** show the most commonly used types and mounting arrangements.
1. Refer to the wiring diagrams at the end of the manual to identify the wire colors and capacity of the fuse(s).
2. Disconnect the battery cables.
3. Disconnect and ground the spark plug leads to prevent accidental starting.
4. Trace the indicated wire color to the fuse location. The fuse is located under the electrical component cover or in another easily accessible location on the power head.
5. Remove the fuse from the retainer. Visually check for a blown fuse (**Figure 30**).
6. Calibrate the meter to the R × 1 scale. Touch the meter test leads to each fuse terminal. If the meter indicates no continuity, the fuse has failed open and should be replaced. If the fuse tests correctly, proceed to Step 9.
7. Thoroughly inspect the electrical system and correct the overload or short before replacing the fuse.
8. Clean the fuse terminals and contacts in the fuse holder, then install the fuse into the holder.
9. Route all wiring to prevent contact with moving components.
10. Connect the battery cables and spark plug leads.

Wire Harness Test

Due to the harsh operating environment, problems with the wiring harness are common. A problem may occur continuously or only intermittently. When an electrical problem is evident and all components test correctly, the wire harness is suspect. Check both the engine and instrument harnesses on remote control models. Gently twist, bend and pull on wire harness connectors when checking the wires for continuity. Often this is the way an intermittent fault is located.

Use a multimeter or self powered test light to test the wire harness.

1. Disconnect the battery cables and ground the spark plug leads to prevent accidental starting.
2. Disconnect the engine harness (C, **Figure 32**) from the instrument wire harness. Disconnect the wire harness wires from the engine components or instruments.
3. Calibrate the meter to the R × 1 scale.
4. Connect the meter or test light lead to one of the harness wires (A, **Figure 32**). Touch the other test lead or test light probe to the corresponding harness pin connector (B, **Figure 32**) If the meter indicates no continuity or the test light does not illuminate, the circuit is open. Check for and repair the wiring or faulty terminal.
5. Refer to the wiring diagrams at the end of the manual to determine connection points for each of the wires. Verify that continuity exists between the chosen wire and other wires only if the diagrams indicate a connection.
6. Replace the wiring harness if the open or shorted wire or connection cannot be properly repaired.
7. Install the wire harness. Connect the wires to the corresponding electrical components or instruments. Route all wiring to prevent interference with moving components. Secure the wiring with suitable clamps where needed.
8. Connect the battery cables and spark plug leads.

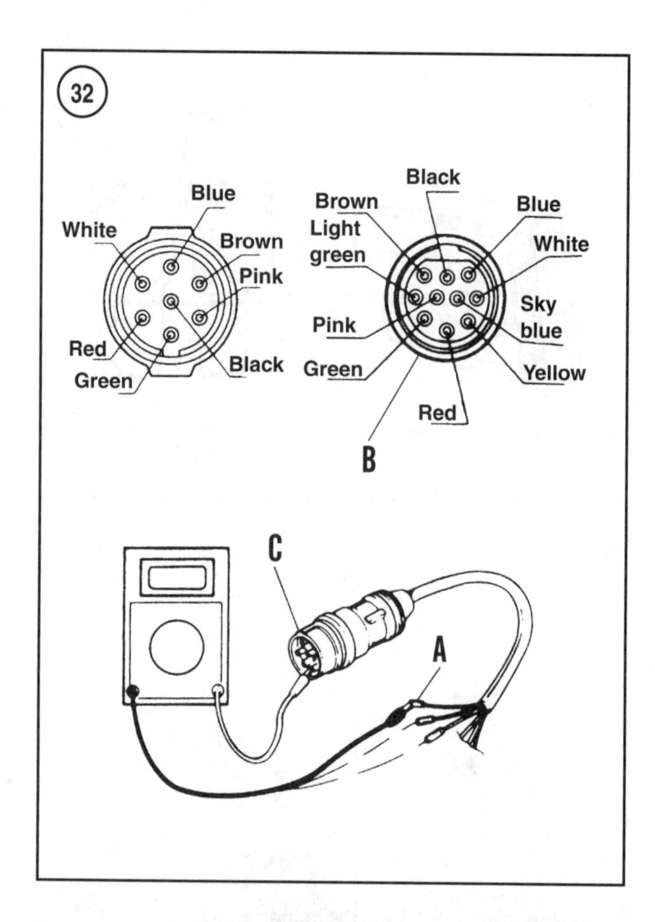

WARNING SYSTEM

A warning system is used on all models to warn the operator of a problem with the engine. Continued operation with the warning system activated can lead to serious and expensive engine damage.

Warning System Operation

Warning system operation and component usage varies by model and the type of lubrication system. Overheat warning is present on all models. Low oil level warning is used on all oil injected models. Major components of the warning system include:

1. Overheat switch.
2. Oil level sensor.
3. Emergency switch.
4. CDI unit or electronic control unit.
5. Warning horn.

Read the following descriptions to determine if the warning system is operating correctly. Test all applicable warning system components if there is a false warning or lack of warning.

Overheating

CAUTION

Never operate the engine while overheated. The power head will suffer serious damage or seize if operated with insufficient cooling.

All models covered in this manual use two overheat switches (**Figure 33**). The switches mount onto each cyl-

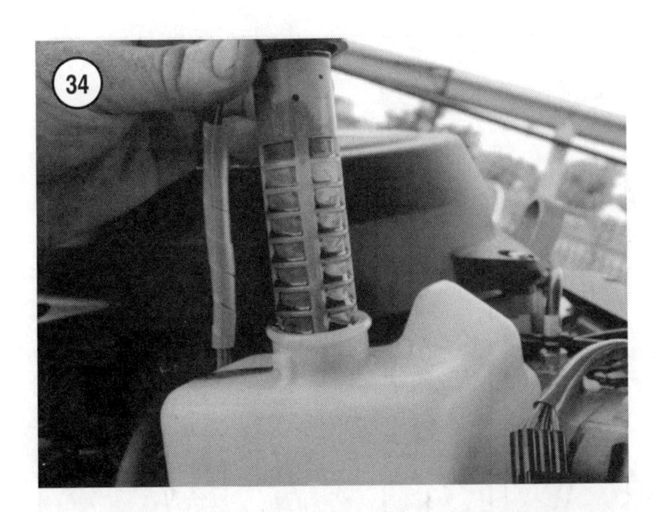

inder head or cylinder bank water jacket. The black wire of the overheat switch connects to an engine ground. The pink lead connects to the CDI unit or engine control unit.

If the engine should reach the predetermined temperature, the overheat switch grounds the pink lead connecting to the CDI unit or engine control unit. The CDI unit or engine control unit then completes the circuit to ground and the warning horn sounds. Grounding the pink wire also signals the CDI unit or engine control unit to initiate power reduction. To initiate power reduction, the CDI unit or engine control unit initiates a controlled ignition misfire, which retards ignition timing or reduces fuel delivery until the engine speed is reduced to approximately 2000 rpm. Full power is only restored after correcting the overheating problem.

Low oil level

CAUTION

Never operate the engine with the warning horn sounding or the low oil level light illu-

minated. The power head will fail if the engine operates with insufficient oil.

A warning system continuously monitors the oil level in the engine mounted oil reservoir using the oil level sensor (**Figure 34**). As oil is consumed, the sensor mounted float drops until it activates the oil reservoir refill switch in the sensor assembly. The switch signals the CDI unit, engine control unit or oil injection module (80 Jet, 115 hp and 130 hp models) to switch on the electric oil pump in the onboard mounted oil reservoir (**Figure 35**). Oil is then pumped from the onboard reservoir until the engine mounted reservoir is refilled, at which point the oil pump is switched off. If the system fails the oil level will drop below the minimum level, causing the sensor to switch on the low oil level warning. The CDI unit or engine control unit sounds the warning horn and/or illuminates a dash mounted light to notify the operator. If not corrected, the oil level will drop below a critical level, causing the CDI unit or engine control unit to initiate power reduction. To initiate power reduction, the CDI unit or engine control unit initiates a controlled ignition misfire, retards ignition timing or reduces fuel delivery until the engine speed is reduced to approximately 2000 rpm. Full power is only restored after reducing the engine to idle speed and refilling the engine mounted oil reservoir.

If the oil level sensor, oil injection module, CDI unit, engine control unit fails to switch on the pump, the emergency switch (**Figure 36**) is used to switch on the electric oil pump to refill the reservoir. This feature allows manual control of the oil reservoir level.

Overheat Warning Circuit Test

CAUTION

Never run an outboard without first providing cooling water. Use a test tank if the en-

gine cannot be operated under normal conditions. Install a test propeller to run the engine in a test tank.

Run the engine under actual conditions or in a test tank for this procedure.

1. Start the engine and operate at a fast idle until the engine reaches normal operating temperature.
2. Advance the throttle to approximately 2000 rpm.
3. Locate one of the overheat switches on the cylinder head or cylinder block. Disconnect the pink wire at the switch. Using a suitable jumper wire, connect the pink wire to an engine ground.
4. If the warning horn does not sound with the pink wire grounded, stop the engine and test the warning horn as described in this chapter. If the horn tests correctly, the fault is with the power supply to the dash mounted horn, wiring, wire connections, CDI unit or engine control unit. Replace the CDI unit or engine control unit only if all other components and circuits test correctly.
5. Reconnect the pink wire. Route the wiring to prevent interference with moving components. Secure the wiring with plastic locking type clamps as necessary.

Overheat Switch Test

Use a common multimeter, liquid thermometer and a container of water that can be heated to test the overheat switch.

1. Remove the overheat switches as described in Chapter Six. Allow the switches to cool to room temperature before testing.
2. Calibrate the multimeter to the R × 1 scale. Test one overheat switch at a time.
3. Connect the positive test lead to one of the overheat switch wires. Connect the negative test lead to the other overheat switch wire.
4. If the meter indicates continuity with the switch at room temperature, the switch is shorted and must be replaced. Continue to Step 5 if the overheat switch tests correctly.
5. Fill the container with cool tap water. Suspend the overheat switch in the water so the tip is below the surface. Place the thermometer in the container with the overheat switch (**Figure 37**).
6. Begin heating and gently stir the water while observing the meter and thermometer. Immediately note the temperature when the meter switches to a continuity reading. Discontinue the test if the water begins to boil before the meter reading changes.
7. Allow the water in the container to slowly cool while observing the meter and thermometer. Immediately note the temperature when the meter changes to a no continuity reading.
8. The overheat switch must switch from *no continuity* to *continuity* reading at a temperature of 84°-90° C (183°-194° F) and switch back to *no continuity* at a temperature of 60°-74° C (140°-165° F). Replace the overheat switch if the switching does not occur within the specified ranges.
9. Repeat this procedure for the remaining overheat switch.
10. Install the overheat switches as described in Chapter Six.

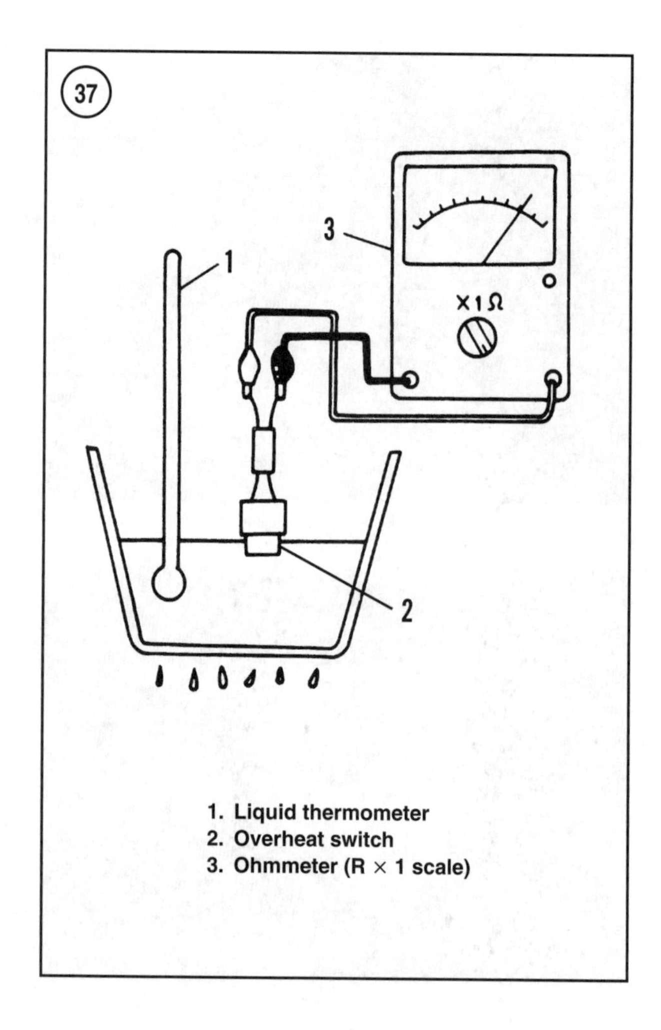

1. Liquid thermometer
2. Overheat switch
3. Ohmmeter (R × 1 scale)

Oil Level Sensor Test

Use a multimeter and measuring ruler to test the sensor.

1. Remove the oil level sensor as described in Chapter Twelve. Carefully dry the sensor.
2. Calibrate the multimeter to the R × 1 scale.

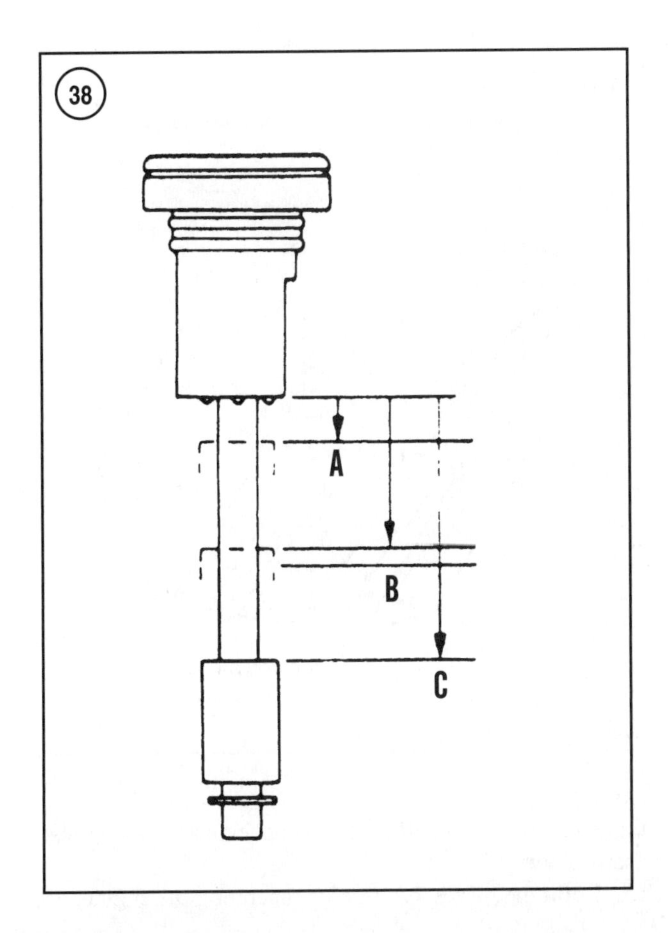

3. Connect the positive test lead to the white or blue/white sensor wire terminal. Connect the negative test lead to the black sensor wire terminal. Move the float up the sensor shaft until it contacts the sensor body. If the meter indicates no continuity with the float in this position, one of the sensor switches has stuck or failed open. Replace the sensor.
4. With the test leads connected as described in Step 3, observe the meter while slowly moving the float downward on the sensor shaft. The meter reading should switch to no continuity when the float is 3.0-6.0 mm (0.12-0.24 in.) from the sensor body (A, **Figure 38**). Replace the sensor if the meter reading does not switch at the specified distance.
5. Connect the positive test lead to the brown or blue/green sensor wire terminal. Connect the negative test lead onto the black sensor wire terminal. Move the float up the sensor shaft until it contacts the sensor body. If the meter indicates continuity with the float in this position, one of the sensor switches is stuck or shorted and the sensor should be replaced.
6. With the test leads connected as described in Step 5, observe the meter while slowly moving the float downward on the sensor shaft. The meter reading should switch to continuity when the float is 33.0-36.0 mm (1.30-1.42 in.) from the sensor body (B, **Figure 38**) and switch back to no continuity when the distance exceeds 36.0 mm (1.42 in.). Replace the sensor if the meter reading does not switch at the specified distance.

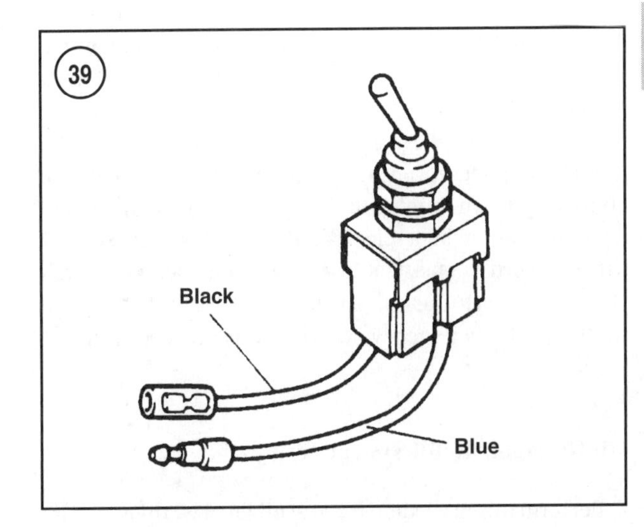

7. Connect the positive test lead to the blue/red or red sensor wire terminal. Connect the negative test lead to the black sensor wire terminal. Move the float up the sensor shaft until it contacts the sensor body. If the meter indicates continuity with the float in this position, one of the sensor switches is stuck or shorted and the sensor must be replaced.
8. With the test leads connected as described in Step 7, observe the meter while slowly moving the float downward on the sensor shaft. The meter reading should switch to continuity when the float is 53.0-56.0 mm (2.09-2.20 in.) from the sensor body (C, **Figure 38**). Replace the sensor if the meter reading does not switch at the specified distance.
9. Install the oil level sensor as described in Chapter Twelve.

Emergency Switch Test

Perform this test only on oil injected 105 Jet and 150-250 hp models. On all other models, the emergency switch is integrated into the oil injection module and cannot be tested alone. A multimeter is required for this procedure.

1. Remove the emergency switch as described in Chapter Twelve.
2. Calibrate a multimeter to the R × 1 scale.
3. Connect the positive meter test lead to the blue switch lead terminal (**Figure 39**). Connect the negative test lead to the black switch lead terminal.

4. If the meter indicates continuity with the switch in the released position, the switch is stuck or shorted and must be replaced.
5. Observe the meter while toggling the switch from the ON to the released position. The meter should indicate continuity when toggled to the ON position and indicate no continuity when released. Repeat this step several times to verify consistent test results. Replace the switch if it fails to perform as described.
6. Install the emergency switch as described in Chapter Twelve.

Oil Reservoir Refill System Tests

Perform this procedure if a low oil level warning occurs and sufficient oil is present in the on-board reservoir.

CAUTION
Never run an outboard without first providing cooling water. Use either a flush/test adapter if the engine cannot be operated under normal conditions or in a suitable test tank. Install a test propeller to run the engine in a test tank.

WARNING
Stay clear of the propeller shaft while running an engine on a flush/test adapter. The propeller must be removed before running the engine. Disconnect the battery and all spark plug leads before removing or installing a propeller.

1. Refer to *Oil Reservoir Removal/Installation* in Chapter Twelve and drain the oil reservoir until approximately one-half full.
2. Prepare the engine for operation on a flush test adapter, in a test tank or with the boat in the water.
3. Start the engine and run at idle speed. Observe the oil level in the reservoir while toggling the emergency switch (**Figure 36**) to the RUN position. The oil level should slowly rise while operating the switch.
   a. *Oil level rises*—The on-board electric oil pump is operating. Test the oil level sensor as described in this chapter. Check for faulty wiring or connections if the sensor tests correctly.
   b. *Oil level does not rise*—Test the emergency switch and the on-board electric oil pump as described in this chapter.
4. Thoroughly inspect all related wiring and connections if the oil refill system fails to automatically refill the engine mounted reservoir. A faulty oil injection module, CDI unit or engine control unit is indicated if the system fails to operate and all other components and wiring test satisfactorily.
5. Refill the oil reservoir before operating the engine.

40

ON-BOARD OIL RESERVOIR

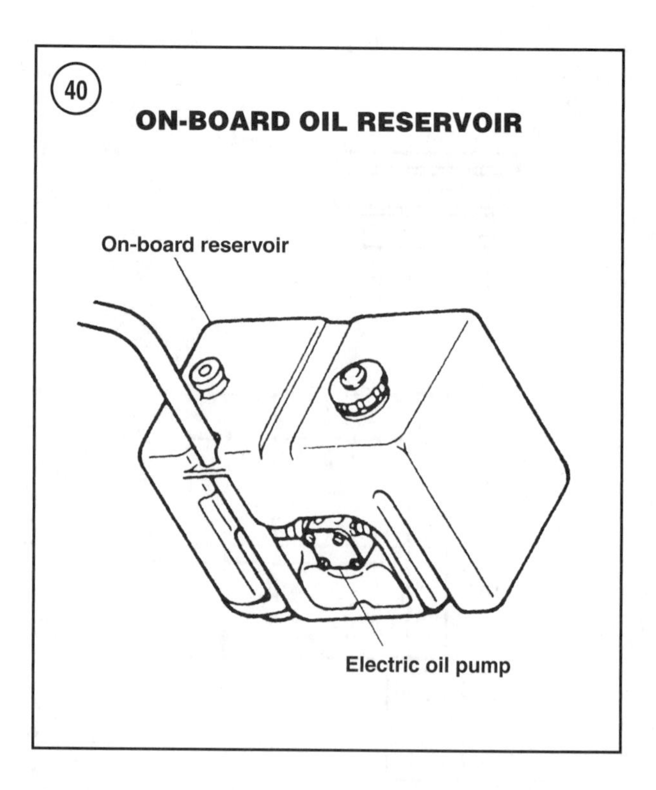

Testing the on-board electric oil pump

This procedure requires a 12 volt battery and jumper wires.
1. Trace the oil hose from the on-board tank (**Figure 40**) to the connection to the engine mounted reservoir. Carefully disconnect the oil hose from the engine mounted reservoir. Direct the disconnected hose into a reservoir suitable for holding oil.
2. Disconnect the wire harness from the on-board oil pump harness. Using a jumper wire to connect the negative terminal of the battery to the brown wire terminal on the oil pump harness.
3. Observe the disconnected hose. Use a jumper wire to connect the positive battery terminal to the blue wire terminal on the oil pump harness.
   a. *Oil flows from the hose*—The pump is operational. Check for faulty wiring if all other components test correctly and the system fails to operate.
   b. *The oil pump does not run*—The oil pump is faulty. Replace the on-board reservoir pump as described in Chapter Twelve.
   c. *Oil pump runs but oil does not flow from the hose*—Remove the on-board reservoir pump as de-

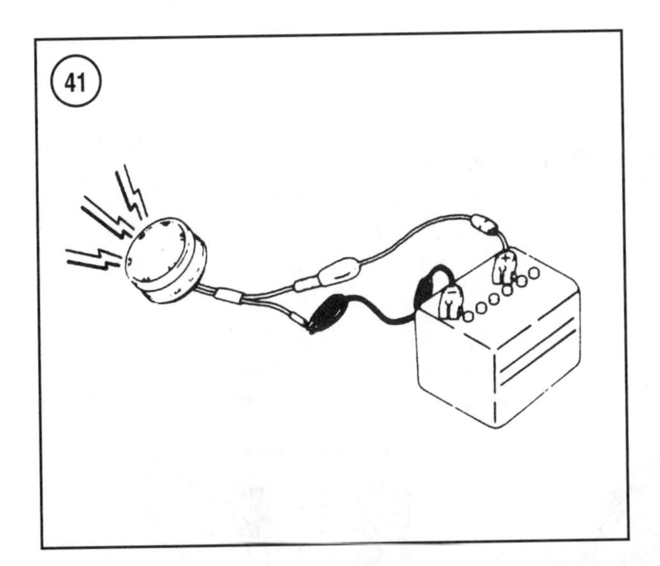
41

scribed in Chapter Twelve and inspect the inlet filter. Replace the on-board reservoir pump if the filter is not plugged and the pump does not pump oil.

4. Remove the jumper wires. Connect the wire harness to the on-board oil pump harness. Connect the oil hose to the reservoir.

Warning Horn

WARNING
When performing tests using a battery, never make the final connection of a circuit at the battery terminal. Arcing may occur and ignite the explosive gasses that are emitted near the battery.

This test requires a fully charged battery and suitable jumper leads.

1. Remove the warning horn as described in Chapter Six.
2. Use jumper wires, to connect the warning horn wire to the battery terminals as indicated in **Figure 41**. Replace the warning horn if it fails to emit a load tone.
3. Install the warning horn as described in Chapter Six.

IGNITION SYSTEM

The ignition system used on Yamaha outboards is composed of solid-state components. Commonly used components include the flywheel, ignition charge coil, pulser coil, CDI unit or engine control unit, ignition coil, crankshaft position sensor, spark plugs and the wiring. Except for the spark plugs very little maintenance or adjustment is required. All components, except the CDI unit or engine control unit, can be accurately tested using resistance or peak voltage output tests. The best means to test the CDI unit or engine control unit is to use a process of elimination. If all other components of the ignition system test correctly, the source of an ignition system malfunction is likely the CDI unit or engine control unit.

Troubleshooting Notes and Precautions

Several troubleshooting precautions should be observed to avoid damaging the ignition system or injuring oneself.

1. Do not reverse the battery connections. Reverse battery polarity damages electronic components.
2. Do not spark the battery terminals with the battery cable connections to determine polarity.
3. Do not disconnect the battery cables while the engine is running.
4. Do not crank or run the outboard if any electrical components are not grounded to the power head.
5. Do not touch or disconnect any ignition components while the outboard is running.
6. Do not rotate the flywheel when performing ohmmeter tests. The meter will be damaged.
7. If a sudden unexplained timing change is noted:
   a. Check the flywheel magnets for damage or a possible shift in magnet position. If the magnets are cracked, damaged or have shifted position, replace the flywheel. See Chapter Seven.
   b. Check the flywheel key for a sheared condition. See Chapter Seven.
8. The ignition system requires that the electric starter crank the engine at normal speed in order for the ignition system to produce adequate spark. If the starter motor cranks the engine slowly or not at all, go to the *Starting System* in this chapter and correct the starting system problems before continuing.
9. The spark plugs must be installed during the troubleshooting process. The ignition system must produce adequate spark at normal cranking speed. Removing the spark plug(s) artificially raises the cranking speed and may mask a problem in the ignition system at lower cranking speeds.
10. Check the battery cable connection for secure attachment to both battery terminals and the engine. Clean any corrosion from all connections.
11. Check all ignition component ground wires for secure attachment to the power head. Clean and tighten all ground leads, connections and fasteners as necessary.

42

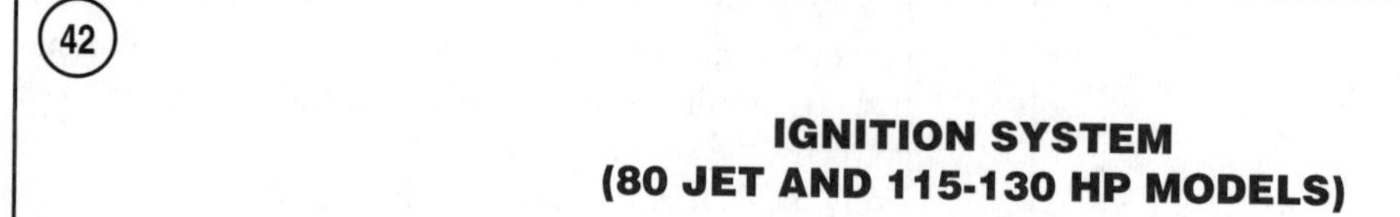

IGNITION SYSTEM
(80 JET AND 115-130 HP MODELS)

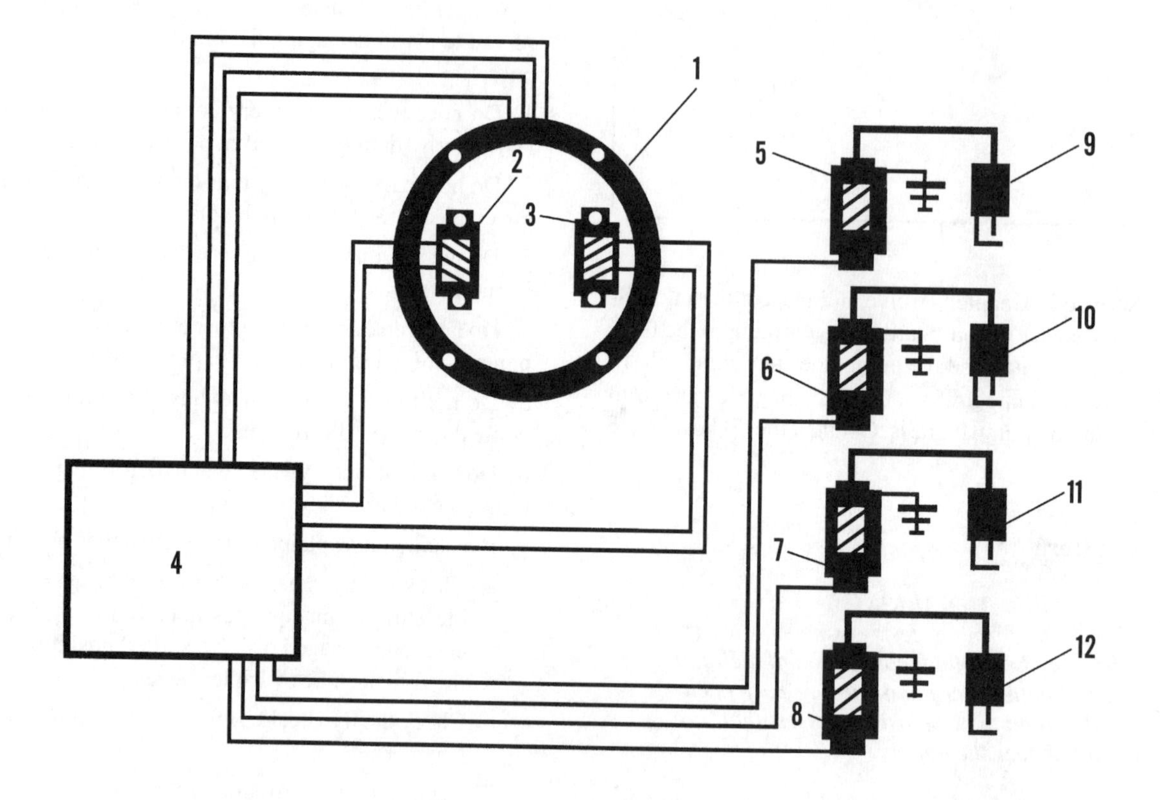

1. Ignition charge coil
2. Pulser coil No. 1
3. Pulser coil No. 2
4. CDI unit
5. Ignition coil No. 1
6. Ignition coil No. 2
7. Ignition coil No. 3
8. Ignition coil No. 4
9. Spark plug No. 1
10. Spark plug No. 2
11. Spark plug No. 3
12. Spark plug No. 4

Resistance (Ohmmeter) Tests

The resistance values specified in the following test procedures are based on tests performed at a normal room temperature or 20° C (68° F). Actual resistance readings obtained during testing will generally be slightly higher if checked on hot components and lower on cold components. In addition, resistance readings may vary depending on the manufacturer of the ohmmeter. Therefore, use discretion when failing any component that is only slightly out of specification. Many ohmmeters have difficulty reading less than 1 ohm accurately. If this is the case, specifications of less than 1 ohm generally appear as a very low (continuity) reading. If at all possible, use an accurate digital ohmmeter to measure values of less than 1 ohm.

System Description and Troubleshooting Sequence

Ignition system operation, components used and testing procedures vary by model. Refer to the appropriate system description to gain an understanding of the system prior to performing any test. The troubleshooting sequence follows the description.

Peak Output Voltage Tests

Peak output voltage tests are designed to check the voltage output of the ignition charge coil, pulser coil, CDI unit or engine control unit and the crankshaft position sensor. The test procedures check voltage output only at normal cranking speed. If an ignition misfire or failure occurs only when the engine is running and cranking speed tests do not show any defects, perform the output tests at the engine speed at which the ignition symptom or failure occurs.

If checking peak voltage output of a component, observe the meter for fluctuations, which indicate erratic voltage output. The voltage output of the ignition charge coil and CDI unit or engine control unit may change with engine speed, but should not be erratic.

CAUTION
Do not run the engine without an adequate water supply and do not exceed 3000 rpm without an adequate load.

The term *Peak Volts* is used interchangeably with *DVA (Direct Volts Adapter)*. The Yamaha recommended meter and adapter (Yamaha part No. J-39299/90890-06752 and YU-39991/90890-03169) or other accurate meter with peak voltage capability should be used whenever the specification is in *DVA or Peak Volts*.

WARNING
High voltage is present during ignition system operation. Do not touch ignition components, leads or test leads while cranking or running the engine.

CAUTION
Unless otherwise noted, perform all peak volt testing with the leads connected, but with the terminals exposed to accommodate test lead connection. All electrical components must be securely grounded to the power head any time the engine is cranked or started or the components will be damaged.

80 Jet and 115-130 hp models

Refer to **Figure 42**.

This ignition system uses a flywheel, ignition charge coil, two pulser coils, CDI unit and four ignition coils.

Alternating current is generated as the flywheel magnets pass by the ignition charge coil. The charge coil incorporates two coil windings. One coil is wound to provide maximum output at lower engine speed. Conversely, the other coil provides maximum output at higher engine speeds. Full output from both coils is required for proper ignition system operation. Alternating current from the ignition charge coil is directed to the CDI unit to be converted to direct current and stored in a capacitor for later release.

This ignition system incorporates two pulser coils. Each pulser coil controls the firing of two cylinders. The triggering magnets in the flywheel pass by the pulser coils as the corresponding piston approaches the firing position. This creates an electrical pulse that is directed to a corresponding switch in the CDI unit. The switch then releases the capacitor charge to the corresponding ignition coil. The ignition coil increases the capacitor voltage to the level needed to jump the spark plug gap. A linkage connects the throttle linkages to the pulser coil assembly. Moving the throttle rotates the pulser coil assembly relative to the triggering magnets in the flywheel. This mechanically advances or retards the ignition timing. The remote control stop circuit prevents ignition system operation by shorting the ignition charge current to ground. Perform the following to troubleshoot the ignition system.

1. Spark test.
2. Stop circuit test.
3. Ignition charge coil output test.

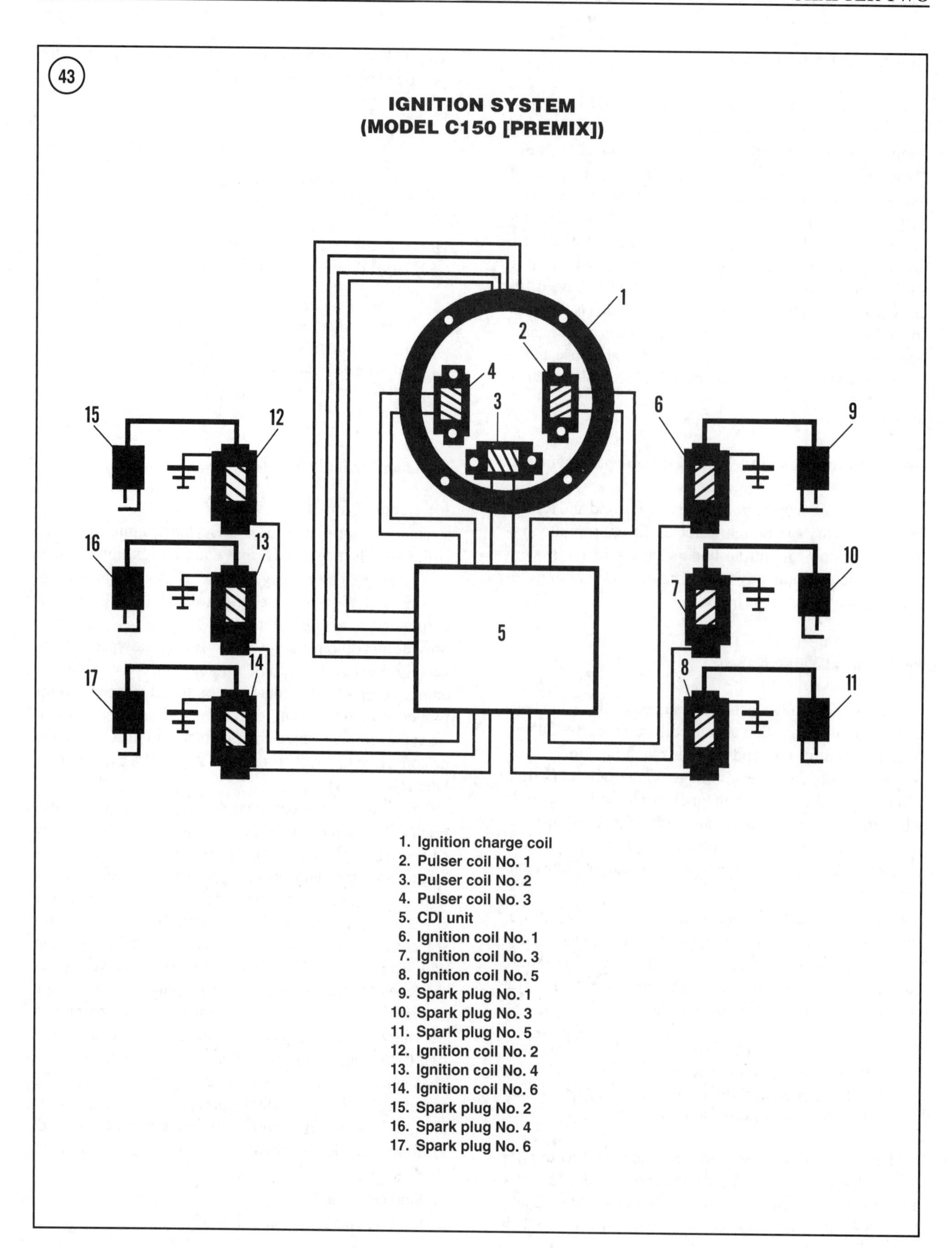

1. Ignition charge coil
2. Pulser coil No. 1
3. Pulser coil No. 2
4. Pulser coil No. 3
5. CDI unit
6. Ignition coil No. 1
7. Ignition coil No. 3
8. Ignition coil No. 5
9. Spark plug No. 1
10. Spark plug No. 3
11. Spark plug No. 5
12. Ignition coil No. 2
13. Ignition coil No. 4
14. Ignition coil No. 6
15. Spark plug No. 2
16. Spark plug No. 4
17. Spark plug No. 6

4. Pulser coil output test.
5. Spark plug cap resistance.
6. CDI unit output test.

Model C150 (premix)

Refer to **Figure 43**.

This ignition system uses a flywheel, ignition charge coil, three pulser coils, CDI unit and six ignition coils.

Alternating current is generated as the flywheel magnets pass by the ignition charge coil. The charge coil incorporates two coil windings. One coil is wound to provide maximum output at lower engine speed. Conversely, the other coil provides maximum output at higher engine speeds. Full output from both coils is required for proper ignition system operation. Alternating current from the ignition charge coil is directed to the CDI unit to be converted to direct current and stored in a capacitor for later release.

This ignition system incorporates three pulser coils. Each pulser coil controls the firing of two cylinders. The triggering magnets in the flywheel pass by the pulser coils as the corresponding piston approaches the firing position. This creates an electrical pulse that is directed to a corresponding switch in the CDI unit. The switch then releases the capacitor charge to the corresponding ignition coil. The ignition coil increases the capacitor voltage to the level needed to jump the spark plug gap. A linkage connects the throttle linkages to the pulser coil assembly. Moving the throttle rotates the pulser coil assembly relative to the triggering magnets in the flywheel. This mechanically advances or retards the ignition timing. The remote control stop circuit prevents ignition system operation by shorting the ignition charge current to ground. Perform the following to troubleshoot the ignition system.

1. Spark test.
2. Stop circuit test.
3. Ignition charge coil output test.
4. Pulser coil output test.
5. Spark plug cap resistance.
6. CDI unit output test.

105 Jet and 150-200 hp models (carburetor equipped [oil injected])

Refer to **Figure 44**.

This ignition system uses a flywheel, ignition charge coil, crankshaft position sensor, two pulser coils, the engine control unit, engine temperature sensor and six ignition coils.

Alternating current is generated as the flywheel magnets pass by the ignition charge coil. This current is directed to the engine control unit to be converted to direct current and stored in a capacitor for later release. The triggering magnets in the flywheel pass by the pulser coils creating an electrical pulse that is directed to the engine control unit. An electrical pulse is also generated as the flywheel gear teeth pass by the crankshaft position sensor. These pulses are also directed to the engine control unit where they are used to determine the piston position for all six cylinders.

As each piston reaches the firing position, the circuits in the engine control unit activate a switch that releases the stored charge to the corresponding ignition coil. The ignition coil increases the charge voltage to the level needed to jump the spark plug gap. Moving the throttle rotates the pulser coil assembly relative to the triggering magnets in the flywheel. This mechanically advances or retards the ignition timing. The remote control stop circuit prevents ignition system operation by shorting the ignition charge current to ground. The engine temperature sensor provides input to the engine control unit where it is used to electronically advance the timing when the engine is cool. This increases idle speed for smoother engine operation during warm-up. The remote control stop circuit prevents ignition system operation by shorting the ignition charge current to ground. Perform the following to troubleshoot the ignition system.

1. Spark test.
2. Stop circuit test.
3. Diagnostic code check.
4. Ignition charge coil output test.
5. Pulser coil output test.
6. Crankshaft position sensor output test.
7. Spark plug cap resistance.
8. Engine control unit output test.
9. Engine temperature sensor test.

150-250 hp EFI and HPDI models

Refer to **Figure 45**.

This ignition system uses a flywheel, ignition charge coil, crankshaft position sensor, six pulser coils (in one assembly), the engine control unit, engine temperature sensor and three ignition coils.

Alternating current is generated as the flywheel magnets pass by the ignition charge coil. This current is directed to the engine control unit to be converted to direct current and stored in a capacitor for later release. The triggering magnets in the flywheel pass by the pulser coils creating an electrical pulse that is directed to the engine control unit. An electrical pulse is also generated as the

(44)

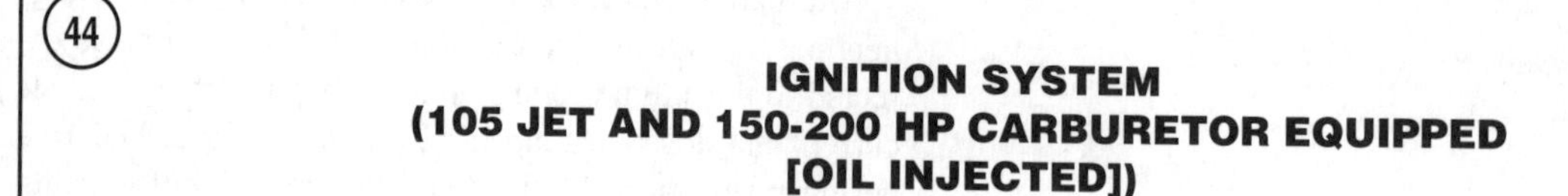

IGNITION SYSTEM
(105 JET AND 150-200 HP CARBURETOR EQUIPPED [OIL INJECTED])

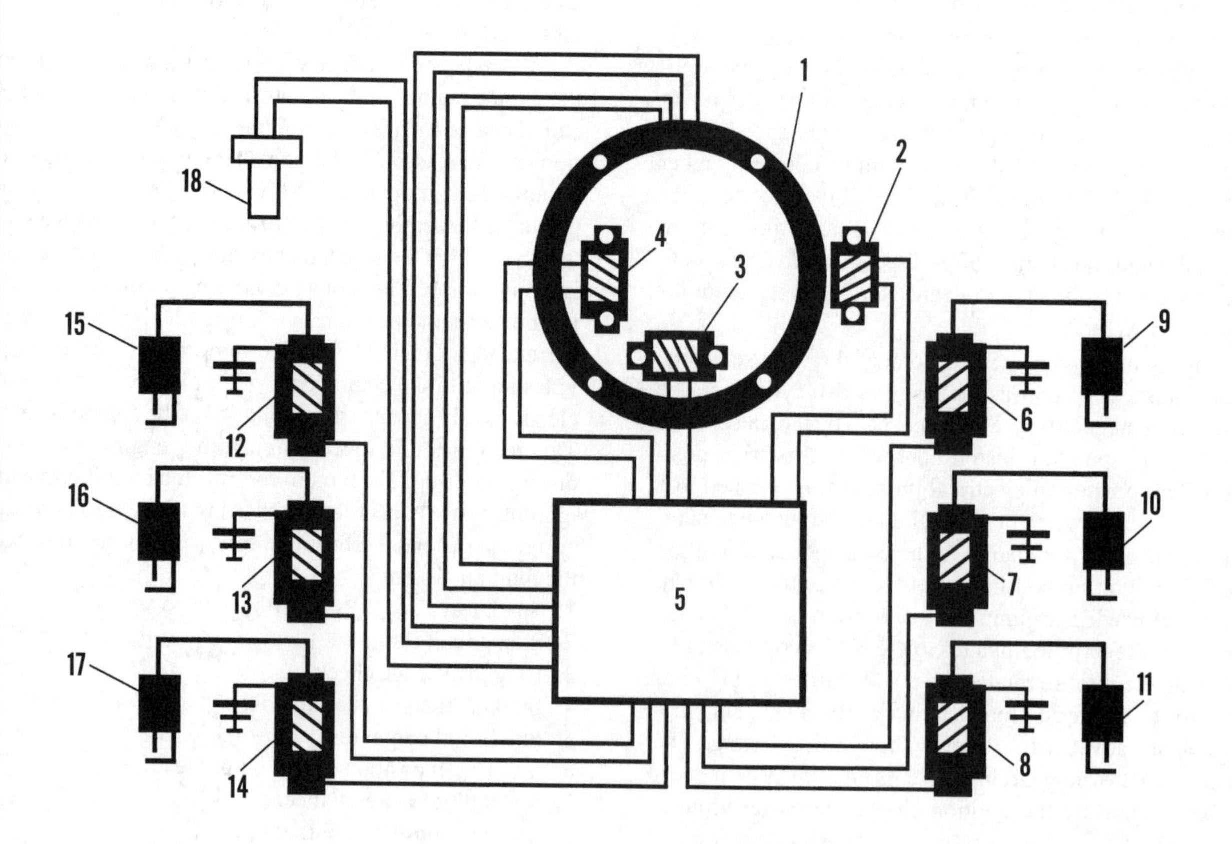

1. Ignition charge coil
2. Crankshaft position sensor
3. Pulser coil No. 1
4. Pulser coil No. 2
5. Engine control unit
6. Ignition coil No. 1
7. Ignition coil No. 3
8. Ignition coil No. 5
9. Spark plug No. 1
10. Spark plug No. 3
11. Spark plug No. 5
12. Ignition coil No. 2
13. Ignition coil No. 4
14. Ignition coil No. 6
15. Spark plug No. 2
16. Spark plug No. 4
17. Spark plug No. 6
18. Engine temperature sensor

(45)

IGNITION SYSTEM (150-250 HP [EFI AND HPDI MODELS])

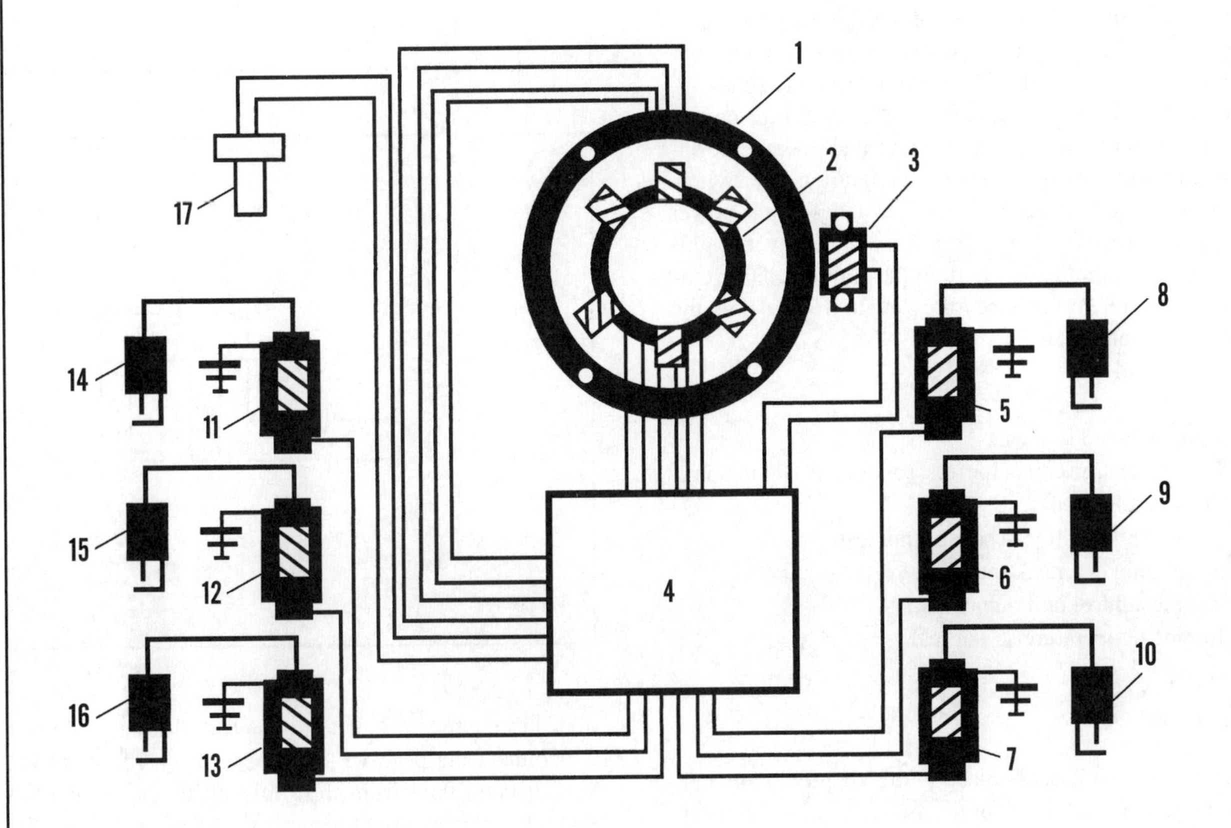

1. Ignition charge coil
2. Pulser coils
3. Crankshaft position sensor
4. Engine control unit
5. Ignition coil No. 1
6. Ignition coil No. 3
7. Ignition coil No. 5
8. Spark plug No. 1
9. Spark plug No. 3
10. Spark plug No. 5
11. Ignition coil No. 2
12. Ignition coil No. 4
13. Ignition coil No. 6
14. Spark plug No. 2
15. Spark plug No. 4
16. Spark plug No. 6
17. Engine temperature sensor

flywheel gear teeth pass by the crankshaft position sensor. These pulses are also directed to the engine control unit where they are used to determine the piston position for all six cylinders.

As each piston reaches the firing position, the circuits in the engine control unit activate a switch that releases the stored charge to the corresponding ignition coil. The ignition coil increases the charge voltage to the level needed to jump the spark plug gap. The engine temperature sensor provides input to the engine control unit where it is used to electronically advance the timing when the engine is cool. This increases idle speed for smoother engine operation during warm-up. The ignition timing is electronically adjusted for optimum ignition timing for the engine operating conditions. Electrical current to operate the engine control unit is provided by the ignition key switch. Turning the ignition key switch off switches off the power supply. Also, the remote control stop circuit prevents ignition system operation by shorting the ignition charge current to ground. Perform the following to troubleshoot the ignition system.

1. Spark test.
2. Stop circuit test.
3. Diagnostic code check.
4. Ignition charge coil output test (except HPDI models).
5. Pulser coil output test.
6. Crankshaft position sensor output test.
7. Spark plug cap resistance.
8. Engine control unit output test.
9. Engine temperature sensor test.

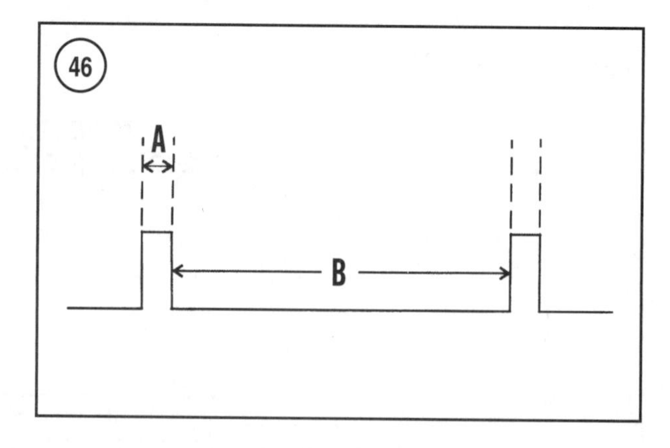

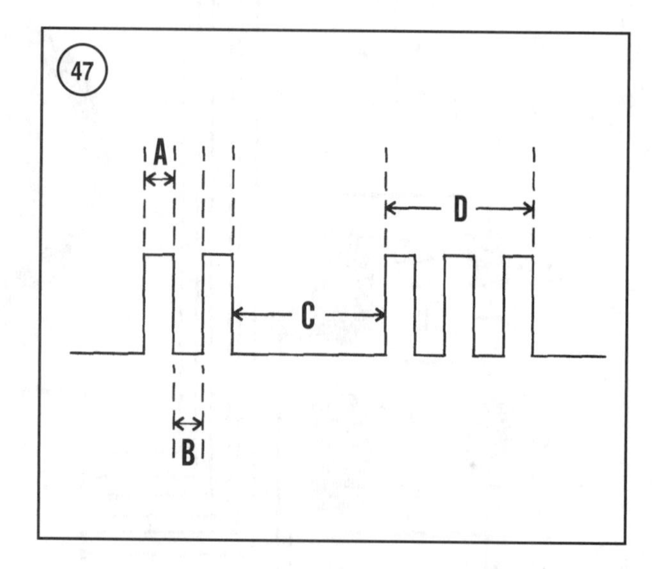

Spark Test

Check for spark as described in this chapter. If any cylinders have a weak spark or no spark, perform the specified test procedures to determine the cause(s).

Stop Circuit Test

When activated, the stop circuit connects the white wire to engine ground. A failure in the stop circuit can cause the engine to lose spark or the engine not being able to stop. Use a multimeter for this test.

1. Disconnect the battery cables and ground the spark plug leads to prevent accidental starting.
2. Refer to the wiring diagrams at the end of the manual to identify the connection points for the white ignition stop wire. The wire connects to the ignition key switch and the CDI unit or engine control unit.
3. Disconnect the harness connector or plug that contains the white wire from the CDI unit or engine control unit.
4. Calibrate the multimeter to the R × 1 scale.
5. Connect the positive test lead to the white wire that was disconnected from the CDI unit or engine control unit. Do not inadvertently connect the test lead to the CDI unit or engine control unit wiring. Connect the negative meter test lead to a good engine ground.
6. The meter should indicate continuity in the following conditions:
   a. The key switch is in the OFF position.
   b. The lanyard cord is removed from the lanyard safety switch.
7. The meter should indicate no continuity for the following conditions:
   a. The key switch is in the ON and START position.
   b. The lanyard cord is properly attached to the lanyard safety switch.
8. If the circuit fails to test as described, disconnect the wiring for one of the switches and repeat the test. Replace the disconnected switch if the circuit now tests correctly. See Chapter Six. If isolating the switches cannot identify

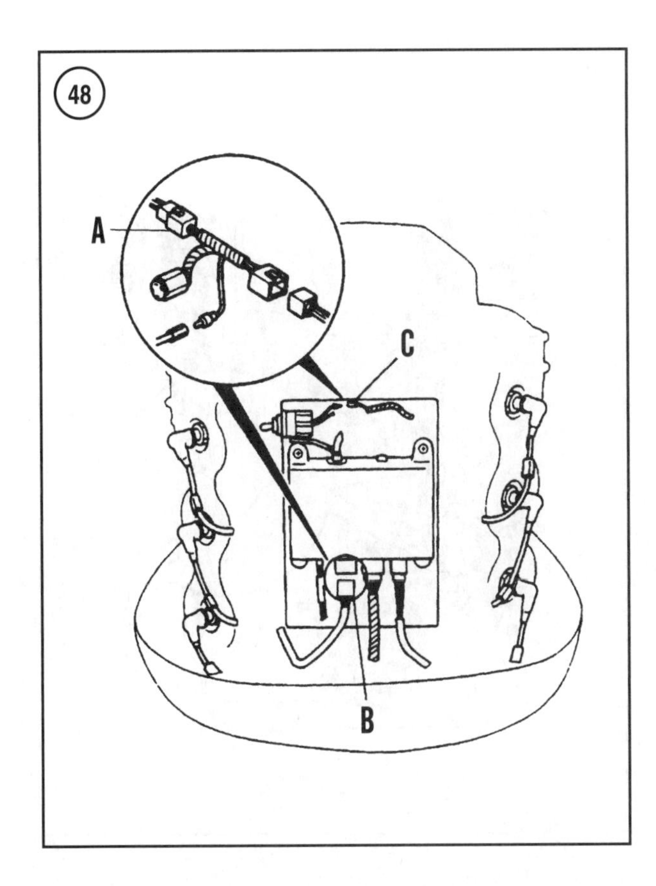

the fault, repair or replace the shorted or open white lead in the engine or instrument wiring harness.

9. Reconnect the harness connector to the CDI unit or engine control unit. Route the wiring to prevent contact with moving components.

10. Reconnect the battery cables and spark plug lead(s).

Diagnostic Code Check

Perform this test on all carburetor equipped and EFI models using an engine control unit. During operation, the engine control unit continuously monitors the various ignition and fuel injections system sensors for faults. Faults are identified by reading the flashing light codes displayed on the diagnostic flash harness (Yamaha part No. YB-06765/90890-06765). Read the codes while running or cranking the engine.

Check all connecting wiring and connections prior to replacing a suspect engine control unit. Actual failure of the engine control unit is extremely rare and replacement seldom corrects the malfunction. Refer to **Table 7** for diagnostic code description, reason and corrective action.

Reading the codes

The engine control unit alternately switches the test harness light *on* and *off* to display the diagnostic codes. A light *on* for 0.3 seconds (A, **Figure 46**), followed by a light *off* for 5 seconds and so on indicates a Code 1. A light *on* for 0.3 seconds (A, **Figure 47**) followed by a light *off* for 0.3 seconds (B, **Figure 47**), then turning back *on* before a 1.3 second pause (C) indicates a Code 2. A series of flashing lights (D, **Figure 47**) after the pause indicates the second digit of the code. Additional codes may be displayed after a pause (light off) starting with the lowest number, then the second number and so on.

CAUTION

Never run an outboard without first providing cooling water. Use either a flush/test adapter if the engine cannot be operated under normal conditions or in a suitable test tank. Install a test propeller to run the engine in a test tank.

WARNING

Stay clear of the propeller shaft while running an engine on a flush/test adapter. The propeller must be removed before running the engine. Disconnect the battery and all spark plug leads before removing or installing a propeller.

NOTE

A repeating diagnostic code 1 indicates a normal operating condition and is always displayed.

1A. *Carburetor equipped models*—Connect the test harness as follows:

a. Remove the cover from the engine control unit. Then locate and unplug the harness connector (B, **Figure 48**) from the bottom of the engine control unit.
b. Connect the test harness (A, **Figure 48**) to the engine and engine control unit harnesses.
c. Unplug the blue wire bullet connector leading to the emergency switch. Connect the test lead harness bullet connector to the disconnected female bullet connector (C, **Figure 48**).

1B. *EFI and HPDI models*—Connect the test harness as follows:

a. Locate the test harness connection point on the port side of the power head. The harness connectors are located just aft of the vapor separator tank. The connectors are easily identified by the short wire with a plugged female bullet connector (A, **Figure 49**) extending from the connector.

b. Disconnect the harness connectors. Connect the test harness (B, **Figure 49**) to the harness connectors (C).

c. Remove the plug, then connect the single wire of the test harness to the female bullet connector (A, **Figure 49**) of the engine harness.

2. Start the engine and run it at idle speed. Observe the test harness light to read the codes. Record any displayed codes. The warm up code should display until the engine reaches normal operating temperature.

3. Refer to **Table 7** to determine if a fault is present. Test, adjust or replace any components displayed by the diagnostic codes.

NOTE
Higher than normal idle speed occurs when a fault occurs with the crankshaft position sensor, engine temperature sensor, knock sensor, oxygen density sensor or throttle position sensor. No code is displayed in the event of oxygen density sensor failure.

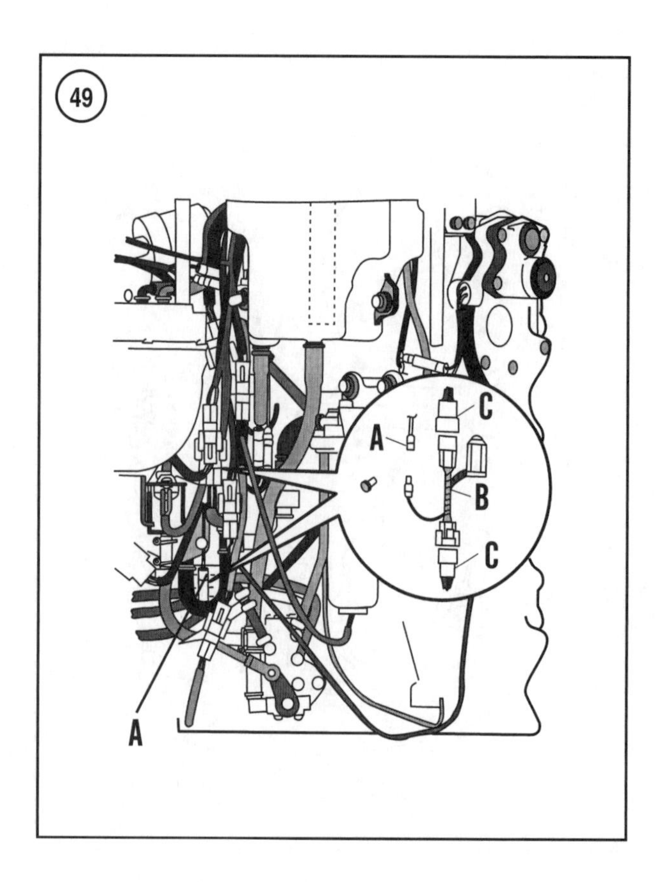

Ignition Charge Coil Output Test

Perform this test on all carburetor equipped and EFI models. This test is not required on HPDI models. The ignition charge coil powers the ignition system. A fault with this component can cause an intermittent spark, weak spark or no spark. A faulty charge coil can cause the engine to run properly at one speed range and misfire at another. The ignition charge coil is located under the flywheel. Perform the test carefully to avoid misdiagnosis and unnecessary flywheel removal. Flywheel removal is not required for testing, as the coil wires are accessible. Ignition charge coil output specifications are listed in **Table 8**. Perform this test under actual operating conditions or with the engine running is a suitable test tank. Use a test propeller if running the engine in a test tank.

Use a multimeter with peak output voltage capability (or Yamaha part No. J-39299/90890-06752 and YU-39991/90890-03169) along with an accurate shop tachometer for this test.

Perform the output test with all wires connected. Use the recommended test harness as follows:

80 Jet and 115-130 hp models—Use Yamaha part No. YB-38831/90890-06771.

105 Jet and 150-250 hp models (oil injected models)—Use Yamaha part No. YB-38831/90890-06771.

150-200 hp models (premix models)—Use Yamaha part No. YB-38832/90890-06772.

WARNING
The ignition system produces very high voltage current. Never touch the test lead of any engine wiring while the flywheel is rotating. Never stand in water while testing or working around wet wiring or connections. Electrical shock can cause serious injury or death.

CAUTION
Never run an outboard without first providing cooling water. Either use a test tank or run the engine under actual conditions. Install a test propeller to run the engine in a test tank.

1. Connect a shop tachometer to the engine. Follow the tachometer manufacturer's instructions.

2. Disconnect the ignition charge coil wire harness from the engine wire harness and CDI unit or engine control unit. Connect the test harness to the charge coil harness and engine harness, CDI unit or engine control unit as shown in **Figure 50**. Connect the multimeter test leads to the test harness wire colors specified in **Table 8**.

3. Refer to **Table 8** for the test specifications, then calibrate the multimeter to the correct peak AC voltage scale.

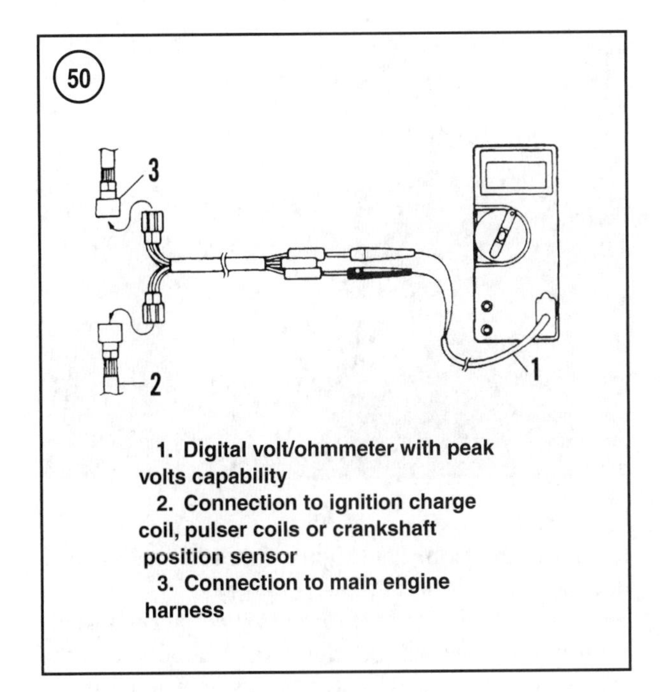

1. Digital volt/ohmmeter with peak volts capability
2. Connection to ignition charge coil, pulser coils or crankshaft position sensor
3. Connection to main engine harness

4. Start the engine and run it at fast idle until it reaches normal operating temperature. Record the output voltage while running the engine at the speeds listed in **Table 8**. Change the meter scale as required for the test specification. To check the voltage at cranking speed, disconnect and ground the spark plug leads. Note the meter readings while operating the electric starter as during normal starting. Repeat the test using the second set of wire colors listed in **Table 8**.
5. The coil output must meet or exceed the specification in **Table 8** at each designated engine speed. If it does not, replace the ignition charge coil as described in Chapter Six.
6. Disconnect the shop tachometer.
7. Remove the test harness and reconnect the charge coil harness.

Pulser Coil Output Test

The pulser coil creates a pulsating current that is used to initiate spark at the plug. A faulty pulser coil can cause an intermittent spark or no spark on one or all coils. Pulser coils are located under the flywheel. Follow the test carefully to avoid misdiagnosis and unnecessary flywheel removal. Pulser coil output specifications are listed in **Table 9**. Perform this test under actual operating conditions or with the engine running in a suitable test tank. Use a test propeller if running the engine in a test tank.

Use a multimeter with peak output voltage capability (or Yamaha part No. J-39299/90890-06752 and YU-39991/90890-03169) along with an accurate shop tachometer for this test.

Perform the output test with all wires connected. Use the recommended test harness as follows:

80 Jet and 115-130 hp models—Use Yamaha part No. YB-38831/90890-06771.

105 Jet and 150-200 hp models (carburetor equipped)—Use Yamaha part No. YB-38831/90890-06771 for oil injected models. Use Yamaha part No. YB-38832/90890-06772 for premix models.

150-250 hp models (HPDI models)—Use Yamaha part No. YB-06779/90890-06779.

150-250 hp models (EFI models)—Use Yamaha part No. YB-38832/90890-06772.

WARNING

The ignition system produces very high voltage current. Never touch the test lead of any engine wiring while the flywheel is rotating. Never stand in water while testing or working around wet wiring or connections. Electrical shock can cause serious injury or death.

CAUTION

Never run an outboard without first providing cooling water. Either use a test tank or run the engine under actual conditions. Install a test propeller to run the engine in a test tank.

1. Connect a shop tachometer to the engine. Follow the tachometer manufacturer's instructions.
2. Disconnect the pulser coil wire harness from the engine wire harness, CDI unit or engine control unit. Connect the test harness onto the pulser coil harness and engine wire harness, CDI unit or engine control unit as shown in **Figure 50**. Connect the multimeter test leads to the test harness wire colors specified in **Table 9**. Test lead polarity does not affect the test results.
3. Refer to **Table 9** for the test specifications, then calibrate the multimeter to the correct peak AC voltage scale. Scale changes may be required when testing at different engine speeds.
4. Start the engine and run it at fast idle until it reaches normal operating temperature. Record the output voltage while running the engine at the speeds listed in **Table 9**. Change the meter scale as required for the test specification. To check the voltage at cranking speed, disconnect and ground the spark plug leads. Note the meter readings while operating the electric starter as during normal starting.

5. Repeat the test and record the voltage for each wire color combination listed in **Table 9**.
6. The pulser coil output must meet or exceed the minimum specification in **Table 9** at each designated engine speed. If not, replace the pulser coil as described in Chapter Six.
7. Disconnect the shop tachometer.
8. Remove the test harness and reconnect the pulser coil harness.

Crankshaft Position Sensor Output Test

The crankshaft position sensor creates a pulsating current as raised protrusions or the flywheel teeth pass next to the sensor coil. The pulsating current is directed to the CDI unit to provide an rpm reference and crankshaft positioning for each cylinder relative to TDC. A faulty sensor can cause no spark or intermittent spark at the coil.

Follow the test carefully to avoid misdiagnosis. Crankshaft position sensor output specifications are listed in **Table 10**. Perform this test under actual operating conditions or with the engine running in a suitable test tank. Use a test propeller if running the engine in a test tank.

Use a common multimeter with peak output voltage capability (or Yamaha part No. J-39299/90890-06752 and YU-39991/90890-03169) along with an accurate shop tachometer for this test.

WARNING
The ignition system produces very high voltage current. Never touch the test lead of any engine wiring while the flywheel is rotating. Never stand in water while testing or work around wet wiring or connections. Electrical shock can cause serious injury or death.

CAUTION
Never run an outboard without first providing cooling water. Either use a test tank or run the engine under actual conditions. Install a test propeller to run the engine in a test tank.

1. Connect a shop tachometer to the engine. Follow the tachometer manufacturer's instructions.
2. Locate the crankshaft position sensor next to the flywheel (**Figure 51**). Trace the sensor wires to the connection to the engine control unit or engine wire harness.
3A. *150-200 hp models (carburetor equipped models)*—Connect the test leads to the two green sensor harness connectors. The engine control unit and crankshaft position sensor wire harnesses must remain connected for this test. Insert the test leads between the bullet connectors

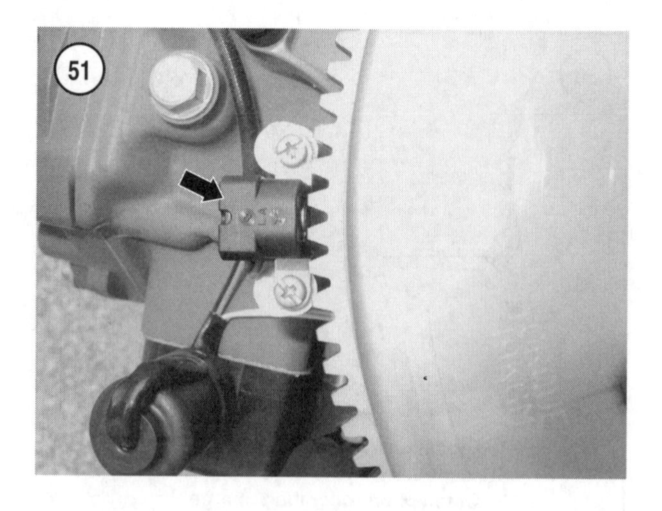

to contact the terminals (**Figure 52**). Test lead polarity does not affect the test results.
3B. *150-250 hp models (EFI and HPDI models)*—Disconnect the crankshaft position sensor harness from the engine wire harness. Connect the two-pin test harness (Yamaha part No. YB-06767/90890-06767) to the crankshaft position sensor harness and engine wire harness. Connect the test leads to the two test harness wires as shown in **Figure 50**.
4. Refer to **Table 10** for the test specifications, then calibrate the multimeter to the correct peak AC voltage scale. Scale changes may be required when testing at different engine speeds.
5. Start the engine and run it at fast idle until it reaches normal operating temperature. Record the output voltage while running the engine at the speeds listed in **Table 10**. Change the meter scale as required for the test specification. To check the voltage at cranking speed, disconnect and ground the spark plug leads. Note the meter readings while operating the electric starter.
6. If the crankshaft position sensor output does not meet or exceed the minimum specification in **Table 10** at each designated engine speed. Replace the sensor as described in Chapter Six.
7. Disconnect the shop tachometer.

CDI Unit and Engine Control Unit

Failure of the CDI unit or engine control unit is relatively rare and replacement seldom corrects an ignition system malfunction. Replace the CDI unit or engine control unit only if it fails the output test or if a constant or intermittent misfire is verified with a spark tester and all other ignition system components and wiring test satisfactorily. On occasion, the engine will experience a misfire only at higher engine speeds and when all ignition compo-

52

nents test correctly. Many times, the overspeed prevention circuits in the CDI unit or engine control unit causes the misfire. The circuits initiate a misfire at a designated speed and higher. Normal operation resumes when the throttle is reduced enough to bring the engine into the normal engine speed range. Full throttle engine speed specifications are listed in **Table 11**.

Overspeed prevention circuit test

If the overspeed prevention circuit maybe causing a misfire, perform the following procedure:

1. Connect a shop tachometer to the engine following the manufacturer's instructions.
2. Observe the tachometer while an assistant operates the engine at the speed in which the misfire occurs.
3. Compare the engine speed with the specification in **Table 11**. Refer to the following for recommendations:
   a. *The engine is exceeding the maximum rated rpm*—Install a larger or higher pitch propeller to reduce engine speed (See Chapter One).
   b. *The engine is well within the maximum rated rpm—The engine is misfiring*—Test all ignition and fuel system components. Replace the CDI unit or engine control unit only after all fuel and ignition system components test satisfactorily.
   c. *The engine is near the top of the maximum operating speed range—Misfires intermittently*—The engine is intermittently exceeding the maximum rated speed. Install the next larger or higher pitch propeller to prevent intermittent overspeed prevention.
   d. *The engine is near the bottom or below the maximum operating range—Engine is misfiring*—Test all ignition and fuel system components. Replace the CDI unit or engine control unit only after all fuel and ignition system components test satisfactorily.

CDI unit or engine control output test

The CDI unit or engine control unit converts alternating current to direct current and stores the current from the ignition charge coil, then releases the charge to the correct ignition coil using input from the pulser coil and/or crankshaft position sensor. A faulty CDI unit or engine control unit can cause no spark on one or more of the cylinders. This test measures the peak voltage delivered to the ignition coil(s). Low output to one or more of the coils indicates a faulty ignition charge coil, crankshaft position, pulser coil or stop circuit. Test all these components prior to performing this procedure.

Follow the test carefully to avoid misdiagnosis and part replacement. CDI unit and engine control unit output specifications are listed in **Table 12**. Perform this test under actual operating conditions or with the engine running in a suitable test tank. Use a test propeller if running the engine in a test tank.

Use a multimeter with peak output voltage capability (or Yamaha part No. J-39299/90890-06752 and YU-39991/90890-03169) along with an accurate shop tachometer for this test. A two-pin test harness (Yamaha part No. YB-06767/90890-06767) is also required on HPDI models.

Perform the output test with all ignition coil leads or harness plugs connected.

WARNING

The ignition system produces very high voltage current. Never touch the test lead of any engine wiring while the flywheel is rotating. Never stand in water while testing or working around wet wiring or connections. Electrical shock can result in serious injury or death.

CAUTION

Never run an outboard without first providing cooling water. Either use a test tank or run the engine under actual conditions. Install a test propeller to run the engine in a test tank.

NOTE

On some models, the engine control unit does not provide spark to the coil on all cylinders until the engine speed exceeds 1500 rpm. This enhances low speed operation and does not indicate a fault with the engine control unit. On such models, the test specifications list a 0 volt on the effected cylinders at cranking and 1500 rpm speed.

1. Connect a shop tachometer to the engine. Follow the tachometer manufacturer's instructions.

2A. *80 Jet, 105 Jet and 115-250 hp models (except HPDI models)*—Connect the negative test lead to the black ground terminal for the ignition coil. Connect the positive test lead to one of the coil wire colors listed in **Table 12**. Do not disconnect any wires for this test. Insert the test leads between the bullet connectors to contact the terminal as shown in **Figure 52**.

2B. *150-250 hp models (HPDI models)*—Disconnect the engine harness connector from the ignition coil harness. Connect the two-pin test harness (Yamaha part No. YB-06767/90890-06767) onto the engine harness and ignition coil harnesses. Touch the test leads to the pins in the test harness to connect to the coil wire colors listed in **Table 12**.

3. Refer to **Table 12** for the test specifications, then calibrate a multimeter to the correct peak AC voltage scale. Scale changes may be required when testing at different engine speeds.

4. Start the engine and run it at fast idle until it reaches normal operating temperature. Record the output voltage while running the engine at the speeds listed in **Table 12**. To check the voltage at cranking speed, disconnect and ground the spark plug leads. Note the meter readings while operating the electric starter. Repeat the test for each cylinder using each pair of coil leads. The test harness and the coil wire colors are listed in **Table 12**. Record all the meter readings.

5. The CDI unit or engine control output voltage must meet or exceed the minimum specifications in **Table 12** at each designated engine speed. Refer to the following for repair recommendations.

   a. *Correct voltage measurement—No spark at the coil*—Check for faulty coil connections. Replace the coil and retest if the connections are satisfactory.

   b. *No voltage measurement*—Test all applicable ignition system components and circuits. Replace the CDI unit or engine control unit only if no spark is present and all other components and circuits test correctly.

   c. *Voltage measurement is low*—Test the ignition charge coil output and stop circuit as described in this chapter. Replace the CDI unit or engine control unit only if all other ignition system components and circuits test satisfactorily.

6. Disconnect the shop tachometer.

7. *150-250 hp HPDI models*—Disconnect the test harness and reconnect the engine harness connectors onto the ignition coil harness.

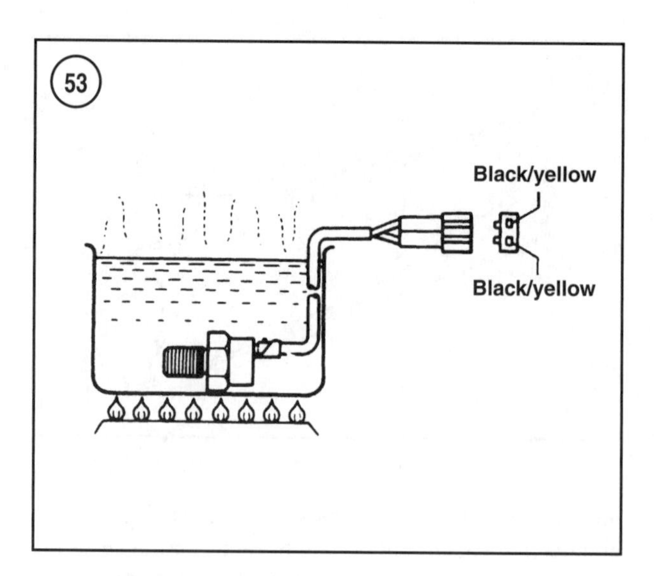

Engine Temperature Sensor Test

The engine temperature sensor is used by the engine control unit or CDI to determine the temperature of the cooling water. This allows the engine control unit to adjust the ignition timing and/or fuel system for optimum performance under varying conditions. Use a digital multimeter, liquid thermometer and a container of water that can be heated for this procedure.

1. Remove the engine temperature sensor as described in Chapter Six.

2. Place the sensor into the container. Do not immerse the sensor harness connection into the water.

3. Select the ohms function on the multimeter. Select the auto scale range. Select the range for the specified temperature (**Table 13**) on meters without this feature.

4. Connect the positive test lead to one of the black/yellow wire terminal of the sensor. Connect the negative test lead to the other black/yellow wire terminal. Suspend the sensor in a container of cool water (**Figure 53**).

5. Place the thermometer in the container next to the sensor. Note the temperature after allowing a few minutes for the thermometer to stabilize.

6. Refer to **Table 13** to determine the specified testing temperatures. Heat or cool the water to the listed temperatures. Note the meter reading and stop the heating when the water begins to boil. Record the resistance at the listed temperatures.

7. Compare the resistance reading and temperatures with the specifications listed in **Table 13**. Replace the temperature sensor if it does not perform as specified. Repeat this test for models using two engine temperature sensors.

8. Install the sensor as described in Chapter Six.

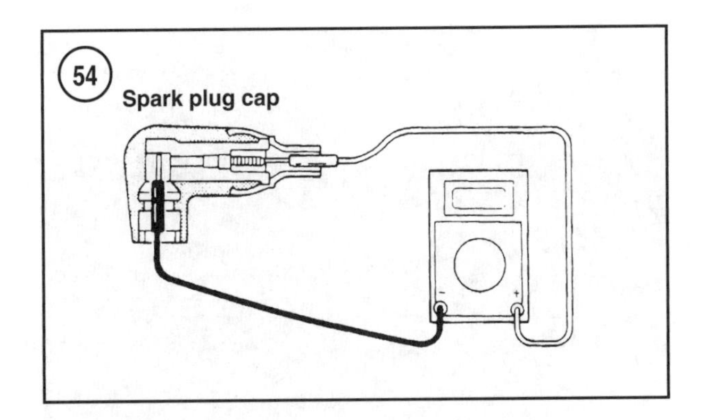

Spark Plug Cap Test

This test measures the resistance in the spark plug cap. Perform this test only on resister spark plug caps. A faulty cap can cause a constant or intermittent ignition misfire. Handle the spark plug cap with care. The resistor in the cap can be damaged if the cap is dropped. Use a multimeter for this test.

1. Disconnect the spark plug cap from the spark plug. Turn the cap counterclockwise to remove the cap from the spark plug lead.
2. Select the scale on the multimeter for measuring 4000-6000 ohms of resistance.
3. Touch the multimeter test leads to the contacts in the cap as shown in **Figure 54**. Note the meter reading.
4. Replace the spark plug cap if it is not within 4000-6000 ohms resistance.
5. Turn the cap clockwise to thread it fully onto the spark plug lead.
6. Repeat this procedure for the remaining caps.

FUEL SYSTEM (CARBURETOR EQUIPPED MODELS)

Fuel Tank, Fuel Supply Hose and Diaphragm Type Fuel Pump

The fuel tank, fuel supply hose or diaphragm type fuel pump can cause engine surges at high speeds but operate properly at lower speeds. Boats equipped with built-in fuel tanks are equipped with an anti-siphon device to prevent fuel from siphoning from the fuel tank if a leak occurs in the fuel hose. These devices are necessary from a safety standpoint but can cause problems if they malfunction. Temporarily run the engine using a portable fuel tank with a fresh fuel or a fuel/oil mixture on premix models. Inspect the fuel tank pickup, anti-siphon device and fuel tank vent if the engine performs properly on the temporary attached fuel tank. Always replace corroded or plugged anti-siphon devices.

To check for a problem with the diaphragm type fuel pump, try gently squeezing the primer bulb when the symptom occurs. Perform a complete inspection of the fuel pump and fuel hoses if the symptom improves while squeezing the primer bulb. Fuel system repair is described in Chapter Five. Always check for and correct fuel leaks after working with fuel system components.

CAUTION
Never run an outboard without providing the engine with cooling water. Use either a test tank or flush test adapter if the engine cannot be operated under actual conditions. Remove the propeller before running the engine on a flush test adapter. Install a test propeller if operating the engine in a test tank. Refer to ***Safety Precautions*** *in Chapter Four for information on test propellers.*

Carburetor Malfunction

A rough running engine that smokes excessively usually indicates a rich fuel/air mixture. The typical causes are a flooding carburetor, faulty recirculation system or stuck choke valve. The most common cause is a flooding carburetor or improper float level adjustment. Weak spark or faulty spark plugs can also cause rough running and excessive smoking.

Other faults with the carburetor(s) can cause a lean running condition that leads to poor performance or bogging on acceleration.

Flooding carburetor

1. Disconnect the battery cables and ground the spark plug lead(s) to prevent accidental starting.
2. Remove the silencer cover (**Figure 55**, typical) as described in Chapter Five. Do not remove the carburetor(s) from the engine.
3. Look into the front of the carburetor (**Figure 56**, typical) while gently squeezing the primer bulb.
4. If fuel flows in the opening of the carburetor, remove the carburetor and repair as described in Chapter Five.
5. Install the silencer cover as described in Chapter Five.
6. Connect the battery cables and spark plug leads.

Plugged carburetor passages

Blocked jets, passages, orifices or vent openings can cause either a rich or lean condition. Operating the engine

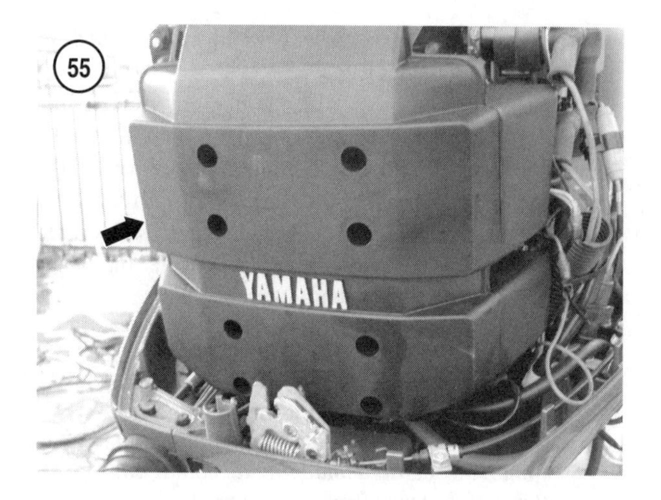

55

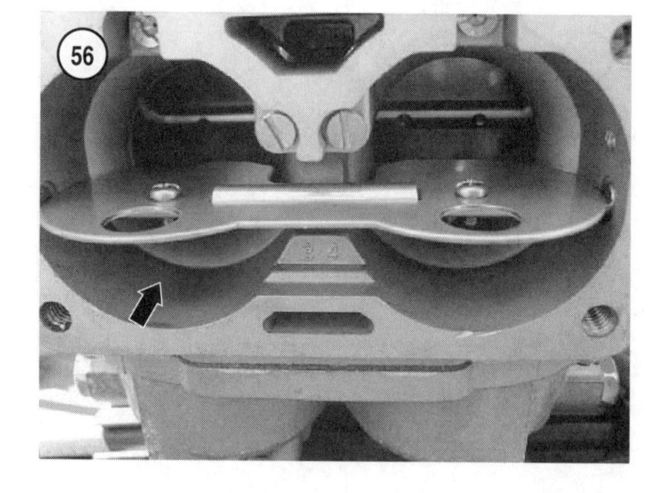

56

under a lean condition can lead to serious power head damage. Symptoms of inadequate or excess fuel include bogging down during rapid acceleration, rough idle, poor performance at high speed or surging at any engine speed.

If the engine is bogging on acceleration, push in on the key switch or operate the choke switch when the bogging occurs. The engine is operating under a lean condition if the symptoms improve with the enriched fuel. The engine is operating under a rich condition if the symptoms become much worse. In either case, clean and inspect the carburetor(s) as described in Chapter Five.

Altitude adjustments

In some instances, changes to carburetor jets or carburetor adjustments are required to correct engine malfunctions while operating at high elevation. Operation in extreme climates, hot or cold, may also require jet or adjustment changes. If operating in these conditions, contact a Yamaha dealership in the area where the engine will be operated for recommendations.

Damaged Reed Valve(s)

WARNING

Use extreme caution when working with the fuel system. Avoid damage to property and potential injury or death. Never smoke around fuel or fuel vapor. Make sure no flame or source of ignition is present.

A chipped or broken reed valve (**Figure 57**) can cause poor idle quality and rough running, primarily at lower engine speeds. With damaged reed valves, the engine may run satisfactorily at higher engine speeds.

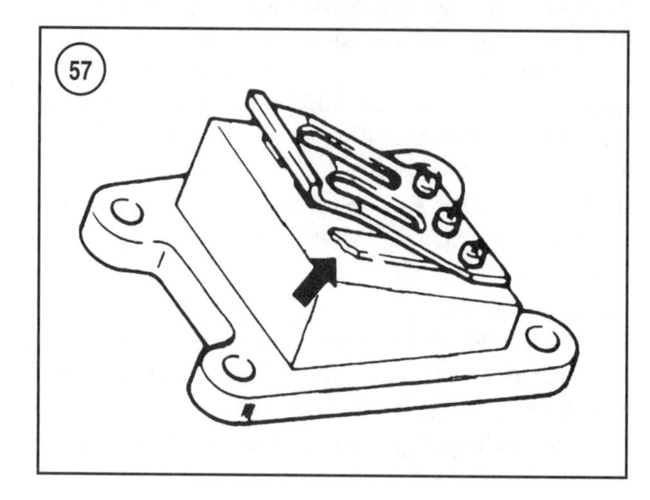

57

Test for damaged reed valves while running the engine at low speed in the water, in a test tank or on a flush/test adapter. Although many variations exist in the number of carburetors and mounting arrangements for the silencer cover and carburetors, the testing procedures are similar.

1. Remove the silencer cover (**Figure 55**, typical) as described in Chapter Five.
2. Observe the carburetor opening (**Figure 56**, typical) while running the engine at idle speed. Fuel spitting out of the carburetor opening indicates problems with the reed valve. If so noted, remove and inspect the reed valve and related components as described in Chapter Five.
3. Install the silencer cover as described in Chapter Five.

Fuel Enrichment System

All carburetor equipped models have a solenoid actuated choke valve. The solenoid (**Figure 58**) moves a plunger that closes the choke valve(s). To activate the choke, simply push in the ignition key switch or operate

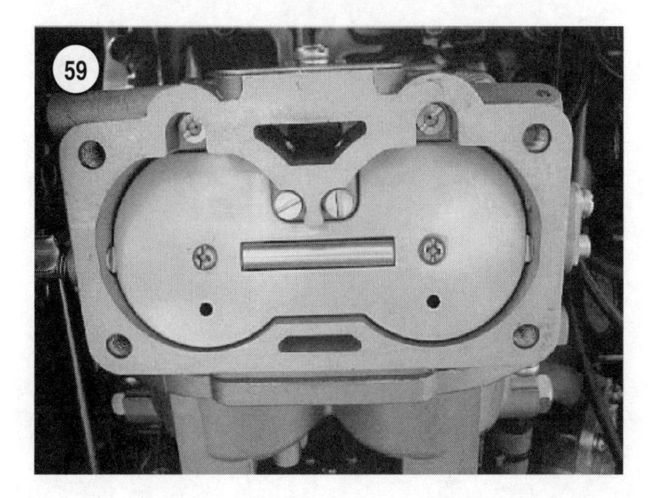

the control mounted choke switch while starting the engine.

Test the choke solenoid if hard starting occurs.

Choke solenoid test

The battery must be fully charged for this procedure.

1. Remove the silencer cover (**Figure 55**, typical) as described in Chapter Five.
2. Turn the key switch to the ON position. Do not activate the electric starter.
3. Observe the choke valves while repeatedly pushing in the key switch or flipping the choke switch. The choke valve must close (**Figure 59**) with a smooth, brisk motion each time the switch is activated and open (**Figure 60**) each time the switch is released.
   a. *The valve opens as specified*—The cause of hard starting is related to other cause(s). Refer to **Table 2** and **Table 3**.

   b. *The choke valve opens and closes slowly*—Check for binding linkages and clean corrosion or other contaminants from the choke plunger. Proceed to Step 4 if slow operation persists.
   c. *The choke valve does not move when the switch is activated*—Proceed to Step 4.
4. Calibrate a multimeter to the 20 VDC scale.
5. Disconnect the blue wire bullet connector from the choke solenoid.
6. Connect the positive test lead to the blue choke solenoid wire connector. Do not inadvertently connect the test lead onto the solenoid lead. Connect the negative test lead to the black wire of the solenoid.
7. Observe the meter while repeatedly pushing in on the key switch or tripping the choke switch. The meter should indicate battery voltage each time the switch activates and 0 volt each time the switch is released. If it does not repeat the test with the negative test lead connected to an engine ground.
   a. *Battery voltage only with the negative test lead connected to an engine ground*—A fault is present in the solenoid black ground wire or terminal. Repair the wire or terminal and retest.
   b. *Low voltage with either negative test lead connection point*—Test the key switch and related wiring as described in this chapter.
   c. *Battery voltage indicates voltage does not move*—Check for binding choke valve linkages. Replace the choke solenoid as described in Chapter Five if the battery voltage is correct and the choke does not operate.
8. Test the solenoid winding resistance as follows:
   a. Remove the choke solenoid as described in Chapter Five.
   b. Calibrate a digital mulitmeter to the R × 1 scale.

c. Connect the positive test lead to the blue solenoid wire (**Figure 61**). Connect the negative test lead to the black solenoid wire.
d. The meter should indicate 3.7-4.0 ohms. If it does not, replace the solenoid.
e. Install the choke solenoid as described in Chapter Five.

9. Reconnect all wiring, then install the silencer cover as described in Chapter Five.

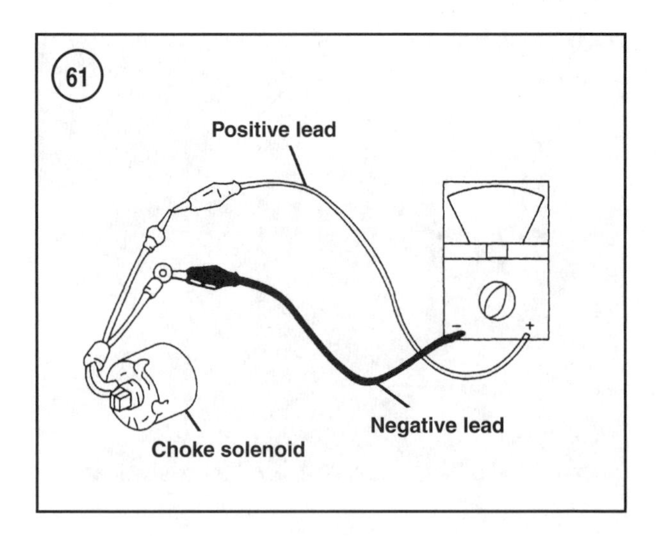

Fuel Supply Hose and Primer Bulb

NOTE
Run the engine at full throttle for several minutes to verify a faulty fuel supply hose or primer bulb.

A faulty fuel supply hose or primer bulb (**Figure 62**) can cause fuel starvation and lean operating conditions at higher engine speeds or cause the engine to simply run out of fuel at idle speed.

Faulty check valves can prevent the primer bulb from pumping fuel or restrict fuel flow to the engine. Leakage at the check valve and hose connections can cause fuel and air leakage, which allows air to be drawn into the fuel while the engine is running.

The most effective method for troubleshooting a suspect fuel supply hose is to operate the engine using a suitable hose from an engine that is operating correctly. Test the primer bulb as described in Chapter Five if the symptoms disappear with the replacement hose. Inspect the fuel supply hose connectors (**Figure 63**) for loose clamps or other defects and replace the hoses if the primer bulb tests correctly.

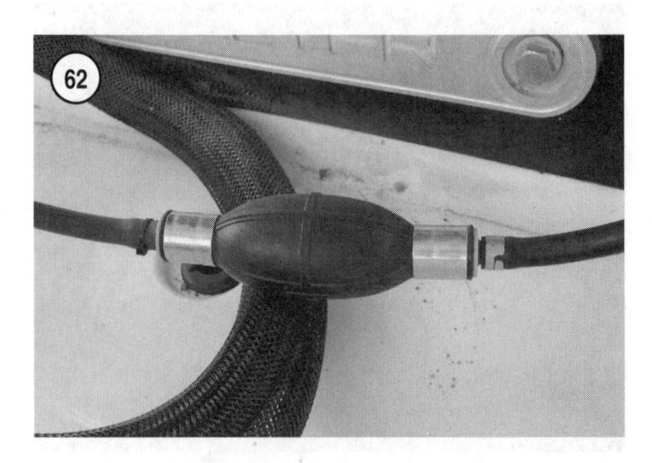

FUEL SYSTEM (EFI AND HPDI MODELS)

Compared to traditional carburetor equipped outboards, advanced electronic fuel injection (EFI) provides quicker starting, automatic altitude compensation, improved fuel economy and smoother overall operation. High pressure direct injection (HPDI) provides these same benefits with even better fuel economy and very low exhaust emissions. The primary difference between EFI and HPDI outboards is the mounting location of the fuel injectors.

On EFI outboards, the fuel injectors are mounted on the front of the crankcase. The fuel is injected into the throttle valve openings. Similar to carburetor equipped outboards, the fuel flows through the crankcase and into the combustion chambers.

On HPDI outboards, the fuel injectors are mounted into the cylinder head. The fuel is injected directly into the combustion chamber. Air only flows through the throttle valve openings where it mixes with the lubricating oil before entering the combustion chamber. The direct injected approach allows fuel to enter the combustion chamber after the exhaust port closes to prevent unburned fuel flow from exiting the exhaust.

Troubleshooting either system is relatively simple if one system at a time is checked. Remember to first perform a preliminary inspection as described in this chapter. Much time and expense can be wasted testing the fuel injection system only to find a fouled plug, low cylinder compression or a reed valve is damaged.

Troubleshooting either system is divided in to two sections the fuel supply system and the electronic control system. Always check the fuel supply system first. Test the electronic control system if all components of the fuel supply system test correctly.

WARNING
The ignition system produces very high voltage current. Never touch the test lead to any

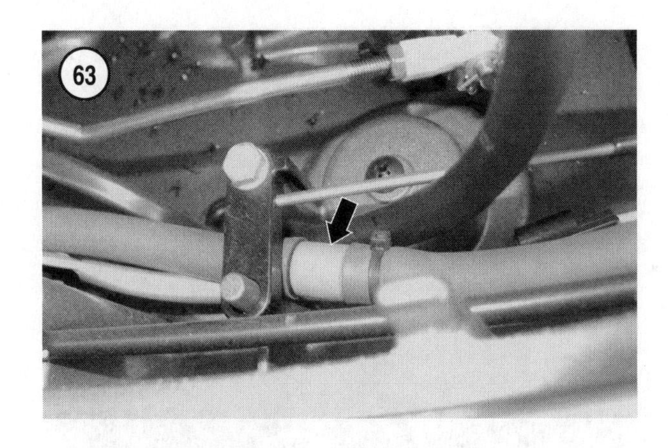
63

engine wiring while the flywheel is rotating. Never stand in water while testing or working around wet wiring or connections. Electrical shock can result in serious injury or death.

WARNING

Use extreme caution when working with the fuel system. Fuel can spray out under high pressure. Always use required safety gear. Never smoke or perform any test around an open flame or other source of ignition. Fuel vapors can ignite or explode causing damaged property, and/or serious injury or death.

CAUTION

Never run an outboard without first providing cooling water. Either use a test tank or run the engine under actual conditions. Install a test propeller to run the engine in a test tank.

Fuel System Test (EFI and HPDI)

This section covers troubleshooting medium and high pressure fuel systems. The high pressure components include the vapor separator tank, fuel rail, fuel injectors, high pressure mechanical pump (HPDI) and fuel pressure regulator.

The low pressure side of the system includes the diaphragm type fuel pump, fuel lines and fuel filters. These components are similar to the components used on carburetor equipped models. Refer to *Fuel Tank, Fuel Supply Hose and Diaphragm Type Fuel Pump* in this chapter for test procedures.

EFI tests and inspectons

1. Check for electric fuel pump operation.
2. Check for a flooding vapor separator tank.
3. Check for fuel in the vapor separator tank.
4. Test the electric fuel pump pressure.
5. Test the electric fuel pump pressure regulator.
6. Check for fuel rail leakage.
7. Check for fuel injector operation.

HPDI tests and inspections

1. Test the electric fuel pump pressure.
2. Check for a flooding vapor separator tank.
3. Check for electric fuel pump operation.
4. Check for fuel in the vapor separator tank.
5. Test the electric fuel pump pressure regulator.
6. Test the fuel pressure sensor.
7. Check for fuel rail leakage.
8. Check for fuel injector operation.

Vapor Separator Tank

The vapor separator tank provides a means to supply liquid fuel to the enclosed electric fuel pump.

On EFI models, the electric fuel pump supplies the high fuel pressure required by the fuel injectors.

On HPDI models, the electronic fuel pump supplies fuel under medium pressure to the mechanical high pressure fuel pump. The mechanical pump increases the fuel pressure to the high level required by the direct fuel injectors. Verify proper operation of the ignition system as described in this section. Then verify proper operation of the low pressure fuel system as described in this chapter (see *Fuel Tank, Fuel Supply Hose and Diaphragm Type Fuel Pump*). If these systems test correctly, check for fuel in the vapor separator tank and for vapor separator tank flooding as described in this chapter.

Electric fuel pump test

Perform this procedure if the engine does not start and the ignition system is operating correctly. The engine will not start if the fuel pump fails to supply fuel to the fuel rail (EFI models) or mechanical high pressure fuel pump (HPDI models). The electric fuel pump is located within the vapor separator tank (**Figure 64**).

1. Listen for pump operation while an assistant turns the ignition key switch to the RUN or ON position then back to the OFF position. Do not start the engine. The pump must run for a few seconds each time the switch is cycled from the off to run or on position. Allow 30 seconds between each off cycle to let the computer reset.

2. *The electric pump runs*—Check for fuel in the vapor separator tank and test the electric pump pressure as described in this chapter.
3. *The electric pump does not run*—Test the system relay(s) and fuel pump resistor as described in this chapter. Then, test the wiring and connections for faults. If these components test or check correctly, disassemble the vapor separator tank and replace the electric fuel pump as described in Chapter Five.

Checking for fuel in the vapor separator tank

Perform this procedure if the engine does not start and the electric fuel pump is operating. This test verifies that fuel is entering the reservoir within the vapor separator tank (**Figure 64**).

1. Place a container capable of holding 1 L (1 qt.) of fuel under the vapor separator tank drain plug (**Figure 65**).
2. Carefully remove the drain plug and drain all the fuel from the reservoir. Remove the O-ring from the plug. Discard the O-ring.
3. Observe the plug opening while pumping the primer bulb. Fuel must flow from the drain after pumping the bulb.
   a. *Fuel flows from the opening*—Test the electric fuel pump pressure as described in this chapter.
   b. *Fuel does not flow from the opening*—Test the primer bulb and check the fuel supply hose as described in this chapter. If these components test or check correctly, disassemble and inspect the vapor separator tank components as described in Chapter Five.
4. Fit a new O-ring onto the drain plug. Install the drain plug and securely tighten. Pump the primer bulb to fill the reservoir.
5. Check for fuel leaks from the drain plug. Correct leaks before operating the engine.

Checking for a flooding vapor separator tank

Perform this procedure if the engine smokes excessively and runs rough at lower engine speeds. A flooding vapor separator tank allows fuel to flow into the intake through the vent fitting and hose. This provides excess fuel to the engine. At higher engine speeds, the engine is able to burn the excess fuel and generally will operate properly.

1. Remove the hose from the vent fitting (**Figure 65**). Place a shop towel below the vent fitting to capture any spilled fuel.

2. Slowly pump the primer bulb while observing the fitting for fuel discharge.
   a. *Fuel is exiting the fitting*—The vapor separator tank is flooding. Disassemble and inspect the vapor separator tank components as described in Chapter Five.
   b. *Fuel is not exiting the fitting*—The vapor separator tank is not flooding. Test the electric fuel pump pressure as described in this chapter.
3. Connect the vent hose to the fitting. Route the hose to prevent interference with moving components.

Electric fuel pump pressure test

This test verifies that fuel is supplied to the fuel injectors (EFI models) or mechanical high pressure pump (HPDI models) at the required pressure. This test measures the high fuel pressure on EFI models and the medium fuel pressure on HPDI models. While this test can verify that the electric fuel pump and regulator are operating correctly under controlled conditions, it cannot determine if the fuel pressure is correct at all engine operating ranges. Leaking fuel hoses, restricted passages and blocked filters can inhibit fuel flow at higher throttle settings while allowing adequate flow at lower settings. Engine performance falters as the pressure drops. If this occurs, check all fuel filters, the anti-siphon valve, diaphragm type fuel pump, the pickup in the fuel tank and the fuel supply hose.

Use a suitable fuel pressure gauge or the Yamaha recommended gauge for this test.

EFI models—Use Yamaha part No. YB-06766/90890-06766.

HPDI models—Use Yamaha part No. YB-06766/90890-06786.

65

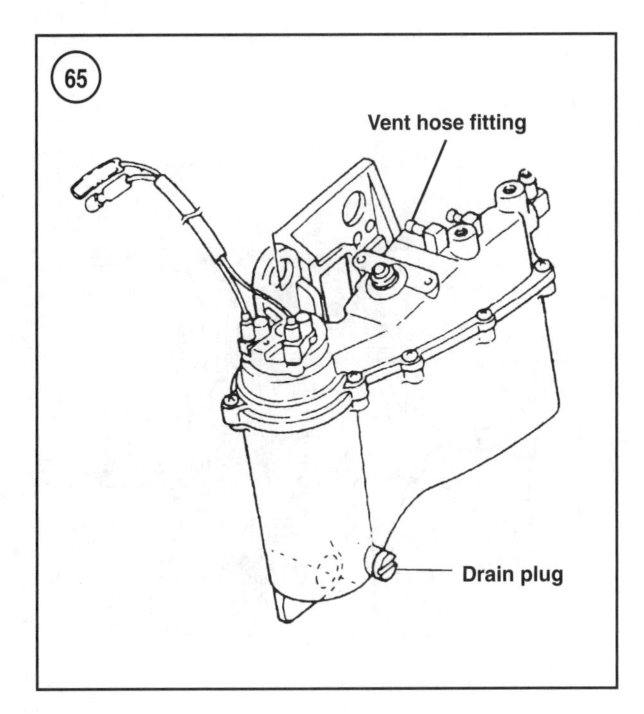

66

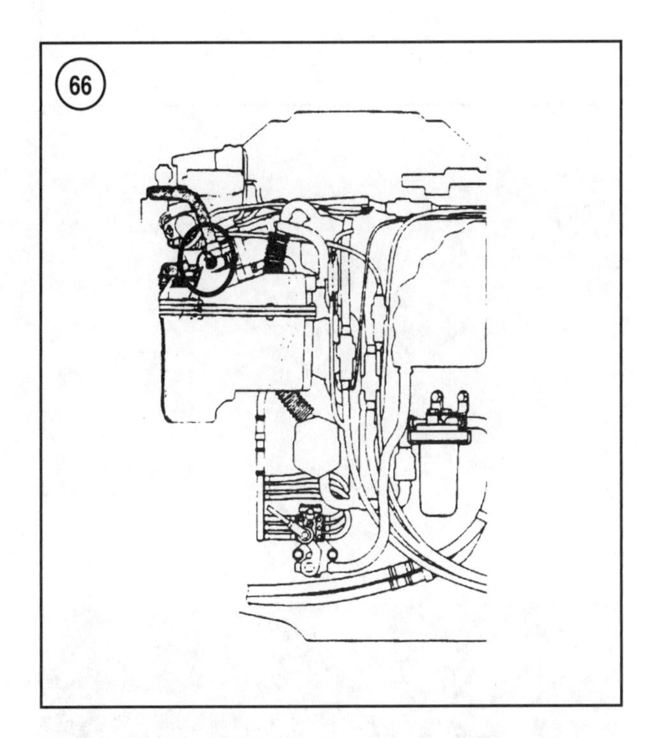

The engine can be run at all recommended throttle settings with the gauge attached to verify fuel pressure.

1. Locate the fuel pressure test port on the vapor separator tank (**Figure 66**). Remove the cap. Then carefully thread the pressure gauge fitting to the test port. Securely tighten the fitting to prevent fuel leaks. Pump the primer bulb to fill the reservoir with fuel.
2. Turn the ignition key switch to the ON or RUN position and immediately check for and correct fuel leakage. Turn the ignition key switch to the OFF position for 30 seconds.
3. Turn the ignition key switch to the ON or RUN position for 5 seconds then turn the ignition key switch to the OFF position. Observe the pressure reading for one minute.
   a. *EFI models*—The pressure should be 280-360 kPa (41-52 psi).
   b. *HPDI models*—The pressure should be 250 kPa (35.6 psi).
   c. *EFI and HPDI models*—The pressure must not drop over 69 kPa (10 psi) over the one minute period.

4A. *Fuel pressure is too low*—Test the fuel pressure regulator as described in this chapter. If the regulator tests correctly, replace the electric fuel pump in the vapor separator tank as described in Chapter Five.

4B. *Fuel pressure is too high*—Test the fuel pressure regulator as described in this chapter. If the pressure regulator tests correctly, remove the pressure regulator tank and inspect the regulator filter as described in Chapter Five.

4C. *Fuel pressure drops excessively*—Test the fuel pressure regulator as described in this chapter. Disassemble the vapor separator tank and inspect the seal between the electric pump and the tank cover if the pressure regulator tests correctly.

5. Route the bleeder hose on the gauge (**Figure 67**) to a suitable container. Open the valve on the gauge to bleed off the fuel pressure.
6. Unthread the gauge fitting from the pressure test port. Thread the cap onto the test port.
7. Check for and correct fuel leakage before putting the engine into service.

Electric fuel pump pressure regulator test

Use the recommended fuel pressure gauge *(see Electric fuel pump pressure test)* and a hand operated pressure/vacuum pump for this test (**Figure 68**, Miti-Vac shown). To ensure accurate test results, the engine should be running for this test.

1. Locate the fuel pressure test port on the vapor separator tank (**Figure 66**). Remove the cap. Then carefully thread the pressure gauge fitting onto the test port. Securely tighten the fitting to prevent fuel leakage. Pump the primer bulb to fill the reservoir with fuel.
2. Locate the reference vacuum hose that is connected onto the pressure regulator fitting (**Figure 69**). Carefully pull the hose from the fitting. Connect the hose on the vacuum pump directly to the pressure regulator fitting.

3. Start the engine and run it at idle speed for one minute to let the fuel pressure stabilize.
4. Note the fuel pressure. Use the hand operated pump to slowly apply a vacuum to the fuel pressure regulator. The fuel pressure should drop as vacuum is applied. The pressure should rise when the vacuum is removed. Replace the fuel pressure regulator as described in Chapter Five if it fails to perform as specified.
5. Route the bleeder hose on the gauge (**Figure 67**) into a suitable container. Open the valve on the gauge to bleed off the fuel pressure.
6. Unthread the gauge fitting from the pressure test port. Thread the cap onto the test port.
7. Check for and correct fuel leakage before putting the engine into service.

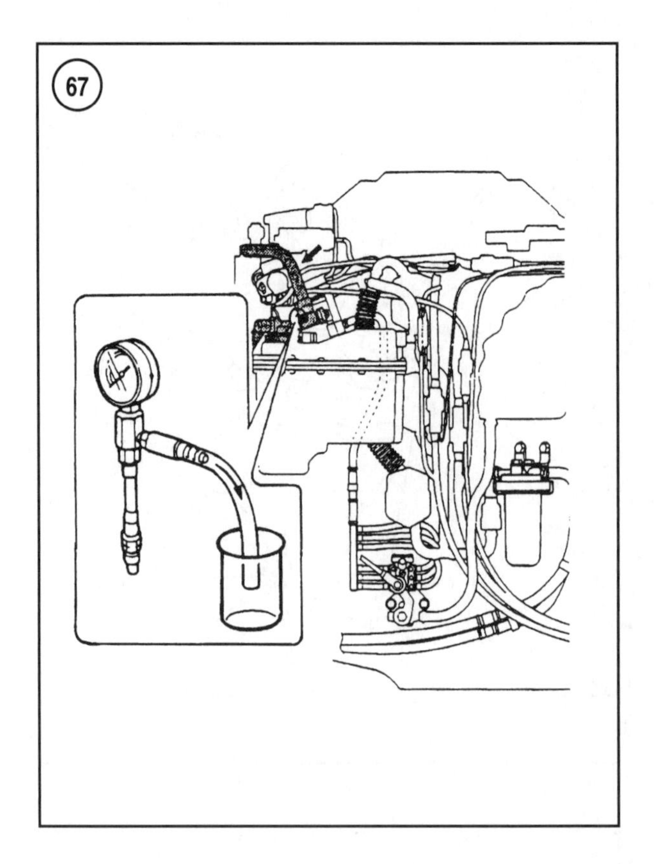

Fuel pressure sensor test

This test applies only to HPDI models. This test determines if the mechanical high pressure fuel pump and the fuel presser sensor are operating properly. The fuel pressure sensor is used to measure the high pressure generated by the mechanical pump. Use a digital multimeter and the three-pin test harness (Yamaha part No. YB-06769/90890-06769) for this procedure. Perform this test only after verifying that the pressure from the electric fuel pump is correct. Perform this procedure with the engine running.
1. Refer to Chapter Six and locate the fuel pressure sensor. Carefully disconnect the engine wire harness connector from the sensor. Connect the test harness to the sensor and the harness connector.
2. Calibrate the multimeter to the 10 or 20 VDC scale.
3. Connect the positive test lead to the pink test harness wire. Connect the negative test lead to the black test harness wire.
4. Route the test harness wires and test leads away from the pump drive belt and pulleys. Secure the test harness wires and test leads with plastic locking type clamps as necessary.
5. Start the engine and run it at idle speed for one minute to let the fuel pressure stabilize.
6. The meter should indicate 2.8-3.2 volts at all operating speeds. The voltage may lower at idle speed and rise at higher speeds.
7A. *Voltage is below the specification at all operating speeds*—Replace the fuel pressure sensor as described in Chapter Six and retest. If the voltage remains low with the replacement sensor, disassemble and repair the mechanical high pressure fuel pump as described in Chapter Five.
7B. *Voltage is above the specification at all operating speeds*—Check for a restriction in the fuel return hose to

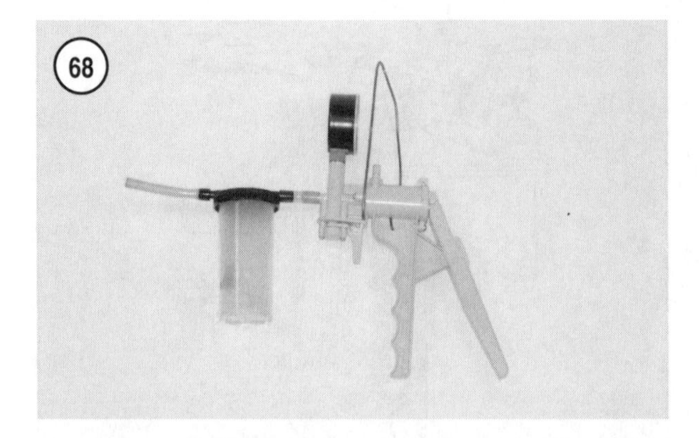

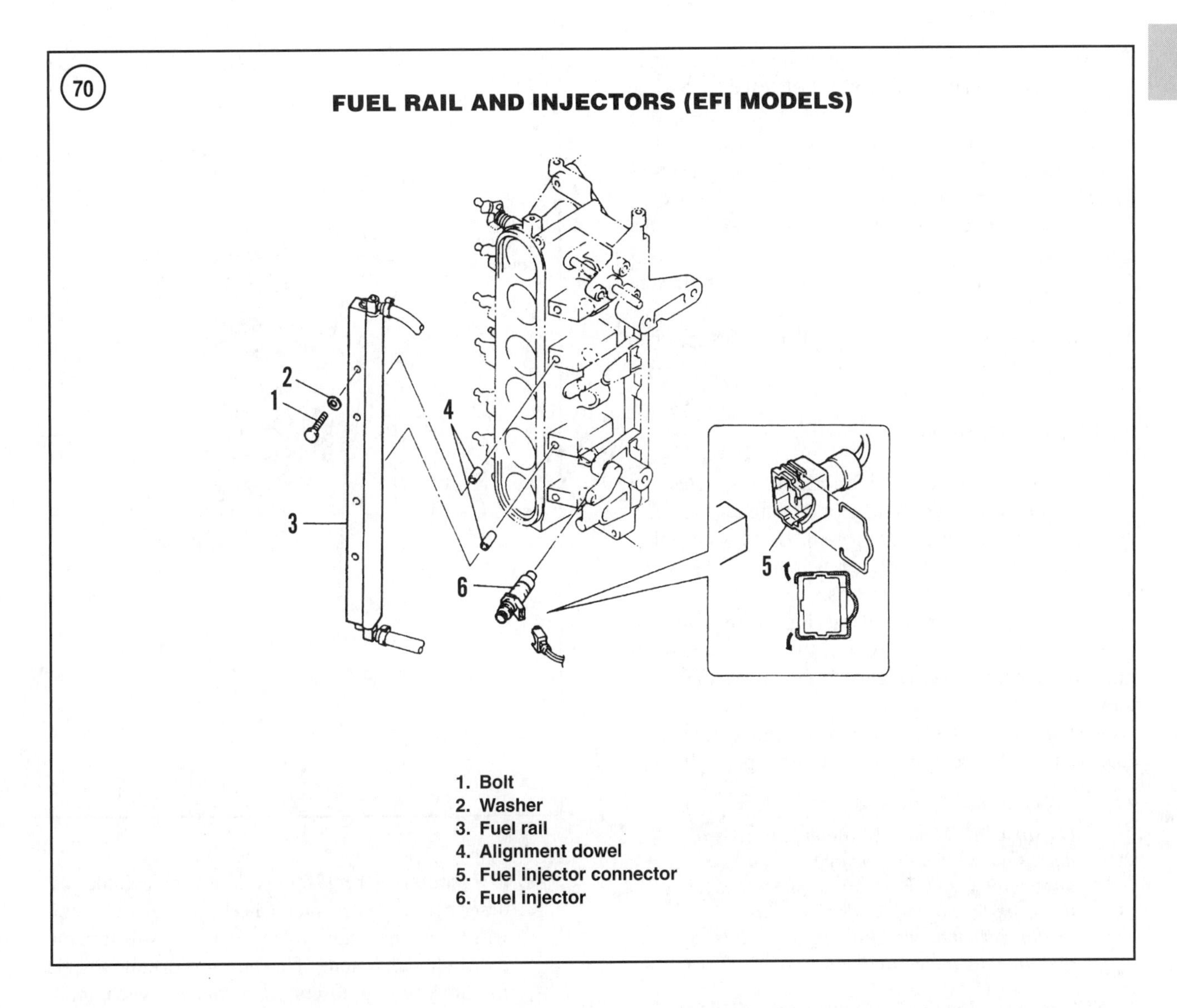

1. Bolt
2. Washer
3. Fuel rail
4. Alignment dowel
5. Fuel injector connector
6. Fuel injector

the vapor separator tank. If the hose, check valve and fitting check correctly, replace the regulator on the mechanical high pressure fuel pump as described in Chapter Five.

7C. *Voltage is below the specification only at higher engine speeds*—Test the electric fuel pump pressure at higher speeds. If the fuel pressure is correct, disassemble and repair the mechanical high pressure fuel pump as described in Chapter Five.

8. Stop the engine, then disconnect the test harness and test leads. Connect the engine wire harness to the fuel pressure sensor. Route the wiring to prevent interference with moving components.

Fuel Rail Leakage Test

The fuel rail provides a means for mounting the fuel injectors as well as supplying fuel pressure to them.

EFI models—The single fuel rail (**Figure 70**) is located on the throttle body at the front of the power head. Remove the silencer cover (**Figure 71**) to check for leakage. Leakage can be detected without running the engine.

HPDI models—The twin fuel rails (**Figure 72**) are located on the cylinder heads at the rear of the power head. The engine must be running to detect leaks.

1. *EFI models*—Remove the silencer cover as described in Chapter Five.

2A. *EFI models*—Pump the primer bulb to fill the vapor separator tank reservoir. Observe each fuel injector to fuel rail mating surface and all fuel hose fittings while an assistant turns the ignition key switch to the RUN or ON position. Repeat this step several times to thoroughly check the system. If there is leakage, disassemble the fuel rail and replace the seals and O-rings as described in Chapter Five.

2B. *HPDI models*—Remove the propeller as described in Chapter Eight. Run the engine on a flush test device while visually inspecting each fuel injector to fuel rail mating surface and all fuel hose fittings. If there is leakage, disassemble the fuel rail and replace the seals and O-rings as described in Chapter Five.
3. Turn the engine OFF and immediately wipe up any spilled fuel.
4. *EFI models*—Install the silencer cover as described in Chapter Five.
5. *HPDI models*—Install the propeller as described in Chapter Eight.

Fuel Injector Operation Test

This test checks for the audible clicking noise generated by the fuel injectors. The injector mounting location varies by the type of fuel system used. Use a mechanic's stethoscope or wooden handle screwdriver to listen to the noise emanating from the injectors.

EFI models—The single fuel rail (**Figure 70**) is located on the throttle body at the front of the power head. Remove the silencer cover (**Figure 71**) to access the injectors.

HPDI models—The twin fuel rails (**Figure 72**) are located on the cylinder heads at the rear of the power head.

NOTE
On some HPDI models, the engine control units shut off two or four cylinders at various engine speeds when the engine is in neutral gear. Do not fail the engine control unit or other components if the injector fails to operate only in neutral gear.

1. *EFI models*—Remove the silencer cover (**Figure 71**) as described in Chapter Five.
2. Remove the propeller as described in Chapter Eight.
3. Disconnect and ground the spark plug leads to prevent accidental starting.
4. Touch the stethoscope or screwdriver to the fuel injector connector (**Figure 73**). Then have an assistant turn the ignition key switch to the START position. A clicking noise should emanate from the injectors. Repeat this test for the remaining five injectors. On some HPDI models, two injectors may not operate at cranking speed.
5. Connect the spark plug leads. Run the engine at idle speed in FORWARD gear using a suitable flush test device. Touch the stethoscope or screwdriver onto the fuel injector connector (**Figure 73**).
6. Each injector should make a clicking noise while running in gear. If not, stop the engine and proceed as follows:

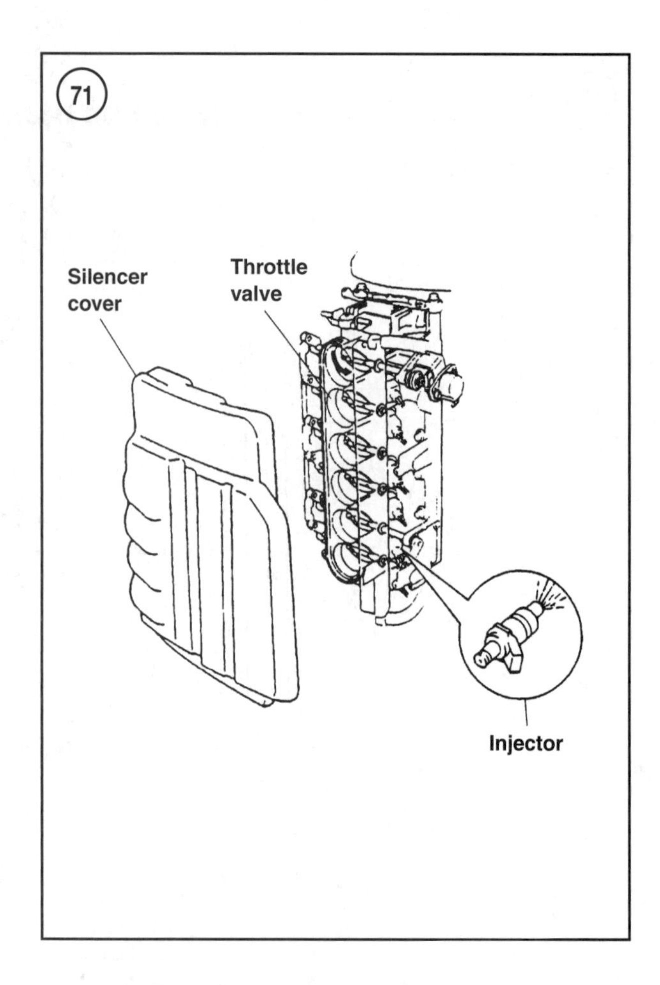

   a. *EFI models*—Check for spark at each cylinder as described in this chapter. Check for faulty injector wiring or connections if the spark is satisfactory on each cylinder. If none of the injectors operate, check for faulty wiring, fuse(s), ignition key switch, ignition system components, injectors or faulty engine control unit. If some injectors operate, check for faulty injector wiring or injectors. Engine control unit failure is unlikely to cause only some of the injectors to operate.
   b. *HPDI models*—Test the injector driver output as described in this chapter.
7. Install the propeller as described in Chapter Eight.
8. *EFI models*—Install the silencer cover as described in Chapter Five.

ELECTRONIC CONTROL SYSTEM (EFI AND HPDI)

Test and verify proper operation of all ignition and fuel system components before testing the electronic control

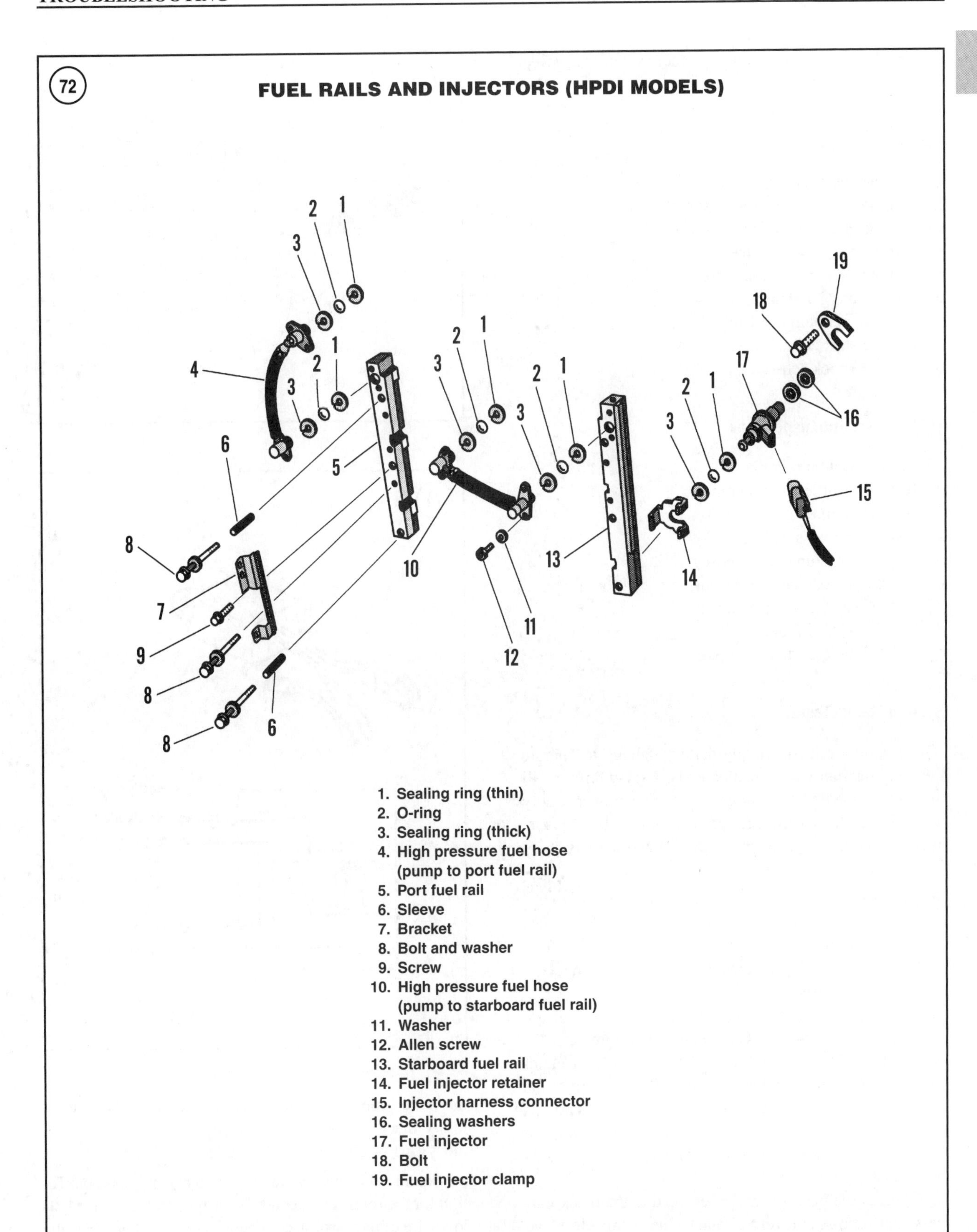
72
FUEL RAILS AND INJECTORS (HPDI MODELS)
1
2
3
4
5
6
7
8
9
10
11
12
13
14
15
16
17
18
19
1. Sealing ring (thin)
2. O-ring
3. Sealing ring (thick)
4. High pressure fuel hose (pump to port fuel rail)
5. Port fuel rail
6. Sleeve
7. Bracket
8. Bolt and washer
9. Screw
10. High pressure fuel hose (pump to starboard fuel rail)
11. Washer
12. Allen screw
13. Starboard fuel rail
14. Fuel injector retainer
15. Injector harness connector
16. Sealing washers
17. Fuel injector
18. Bolt
19. Fuel injector clamp

system. Failure of the ignition or fuel system can affect operation of the electronic control system.

EFI Tests and Inspections

1. Test the system relay(s).
2. Test the throttle position sensor.
3. Test the engine temperature sensor.
4. Test the air pressure sensor.
5. Test the air temperature sensor.
6. Test the oxygen density sensor.
7. Test the fuel pump resistor.
8. Test the shift cutout switch.
9. Test the knock sensor.

HPDI Tests andInspections

1. Test the system relay(s).
2. Test the throttle position sensor.
3. Test the engine temperature sensor.
4. Test the air pressure sensor.
5. Test the air temperature sensor.
6. Test the oxygen density sensor.
7. Test the shift position switch.
8. Test the water in fuel sensor.
9. Test the injector driver output.

System Relay Test

The system relay(s) supply battery voltage to operate the fuel injection system components. A relay failure will prevent engine operation or not remove the power supply when the ignition is switched off. Check for a blown fuse, as described in this chapter, before testing the relay.

EFI models

Use a multimeter for this test.

1. Refer to the wiring diagram and Chapter Six to locate the system relay and connected wiring.
2. Disconnect all wires from the system relay.
3. Select the 20 or 40 VDC scale on the multimeter.
4. Connect the negative test lead to an engine ground. Connect the positive meter test lead to the red wire of the harness connection for the relay. If the meter does not indicate battery voltage, check all connections along the wire up to the battery. Repair wiring or connections as necessary.
5. Connect the negative meter test lead to the black harness connection of the relay. Connect the positive test lead to the red wire of the harness connection for the relay. If the meter does not indicate battery voltage, check for a faulty black harness wire connection to ground. Repair the wiring or connections as needed.
6. Connect the negative lead to the black harness connection for the relay. Connect the positive test lead to the yellow wire of the harness connection for the relay. Place the ignition key switch in the ON or RUN position and ob-

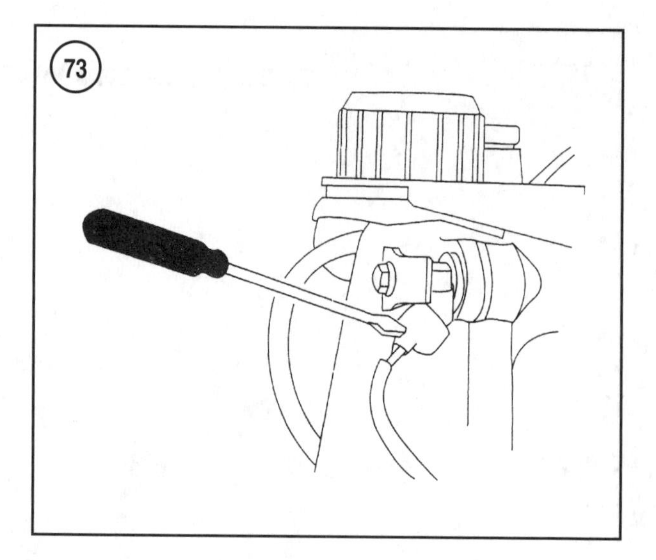

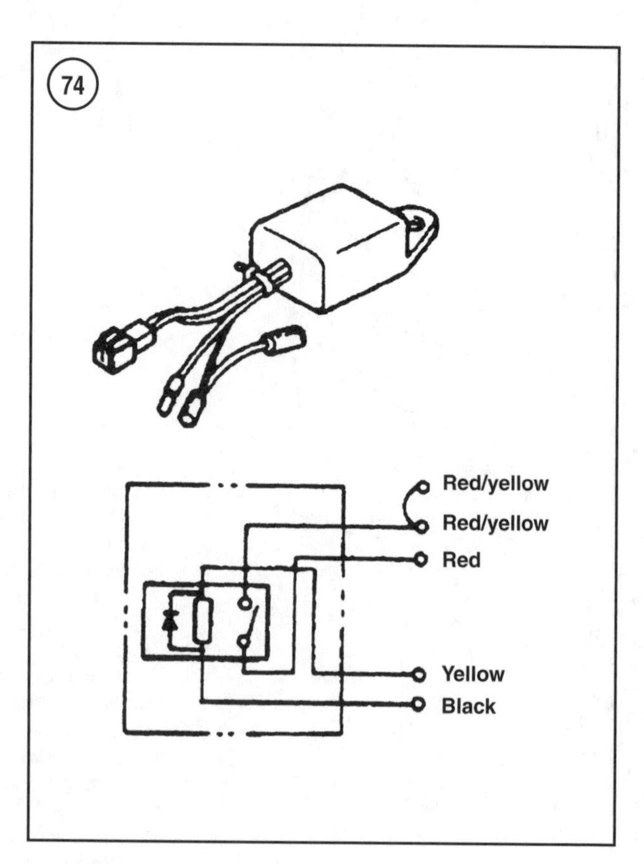

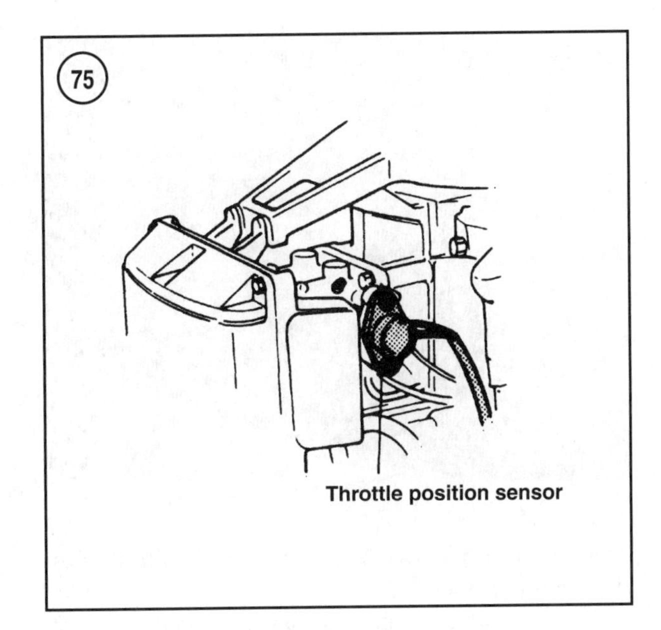

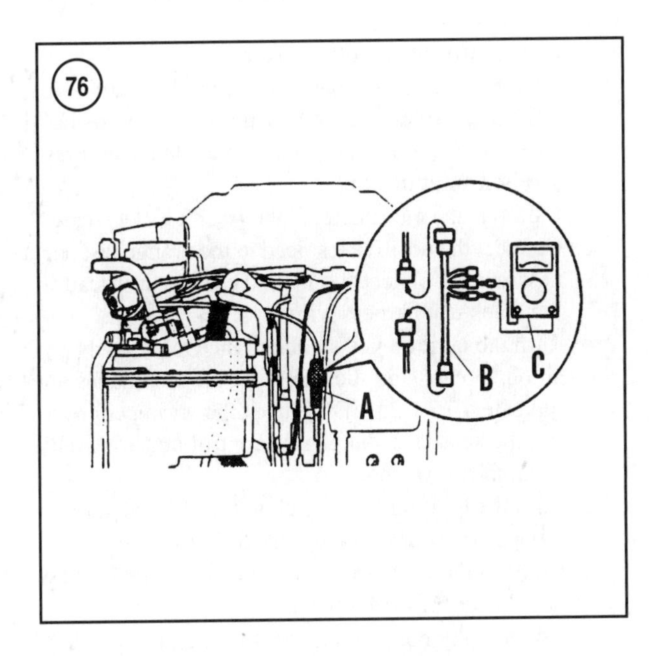

serve the meter. If the meter does not indicate battery voltage, check for faulty wiring or connections along the yellow wire up to the connection to the engine control unit. Test the ignition key switch and related wiring.

7. Connect the yellow and black engine harness wires to the system relay wire. Do not connect the red/yellow or red engine harness wires. Calibrate the multimeter to the 10 ohm scale. Connect the positive test lead to the red system relay wire (**Figure 74**). Connect the negative test lead to the red/yellow system relay wire. Observe the meter while placing the ignition key switch in the ON or RUN position. The meter should indicate continuity. Place the ignition key switch in the OFF position and observe the meter. The meter should indicate no continuity. Replace the relay if it fails to perform as described.

8. Remove the test leads and reconnect the engine harness wires to the system relay. Route the wiring to prevent contact with moving components.

HPDI models

The main relay and injector driver relay plug into the fuse panel on the starboard side of the power head. Hold a finger against the relay while an assistant switches the ignition key switch to the run position. With proper operation, a clicking is felt in the relay and the electric fuel pump in the vapor separator tank will operate to prime the fuel system for starting. Repeat this test several times to check for intermittent faults.

If the relay fails to operate as specified, test the ignition key switch and all fuses as described in this chapter. Replace the relay if the fuses and ignition switch test correctly. Faulty wiring or connections are indicated if the replacement relay fails to operate as described.

Throttle Position Sensor Test

The throttle position sensor (**Figure 75**) is used on all EFI and HPDI models. It is located on the upper port side of the throttle body. This sensor provides a signal to the engine control unit that indicates the amount of throttle opening. This lets the engine control unit provide the optimum amount of fuel and ignition timing advance for the given throttle opening.

This test requires a digital multimeter and a test harness (Yamaha part No. YB-06433/90890-06757).

1. Disconnect the throttle position sensor harness from the engine wire harness (A, **Figure 76**). Connect the test harness (B, **Figure 76**) to the engine and throttle position wire harnesses.

2. Connect the positive test lead to the red test harness wire. Connect the negative test lead to the orange test harness wire.

3. Turn the ignition key switch to the ON or RUN position. Do not start the engine. The meter should indicate 4.76-5.75 VDC.

   a. *Voltage is less than the specification*—Check the battery charge level and check all related wiring and terminal connections for faults, corrosion or damage. Repair as needed. Substitute a known good engine control unit if the wiring tests correctly. Replace the engine control unit if the voltage is within the specification only with the replacement engine control unit.

b. *Voltage is above the specification*—Check the throttle position sensor wiring for damaged insulation or shorted terminals and repair as needed. Substitute a known good engine control unit if the wiring tests correctly. Replace the engine control unit if the voltage is within the specification only with the replacement engine control unit.

4. Connect the positive test lead to the pink test harness wire. Connect the negative test lead to the orange test harness wire. Set the meter to the 1.0 VDC scale.

5. Place the throttle in the fully closed (idle) position, then note the meter reading. The meter should indicate 0.48-0.52 VDC. If not adjust the throttle position sensor as described in Chapter Four. Replace the sensor if it cannot be properly adjusted.

6. Set the meter to the 10 or 20 VDC scale. Observe the meter while slowly moving the throttle roller to the fully open position. If the meter reading does not smoothly increase with corresponding throttle opening and peak at 4.75-5.25 VDC, replace the throttle position sensor as described in Chapter Six.

7. Turn the ignition key switch to the OFF position. Disconnect the test leads and the test harness. Connect the throttle position sensor harness to the engine wire harness.

Engine Temperature Sensor Test

Refer to *Ignition system* in this chapter for engine temperature sensor test procedures. Test both sensors for models using two engine temperature sensors.

Air Pressure Sensor Test

The air pressure sensor (**Figure 77**, typical) is used on all EFI and HPDI models. It provides a varying voltage signal to the engine control unit where it is used to determine intake manifold air pressure. Manifold air pressure is affected by operating altitude and engine load. The engine control unit modifies fuel delivery to provide the optimum amount of fuel for the given operating conditions.

Excessive exhaust smoke, spark plug fouling, poor performance or lean operating conditions can result from operating the engine with a faulty air pressure sensor.

Use a three-pin test harness (Yamaha part No. YB-06769/90890-06769) and a digital multimeter for this procedure.

1. Disconnect and ground the spark plug leads to prevent accidental starting.

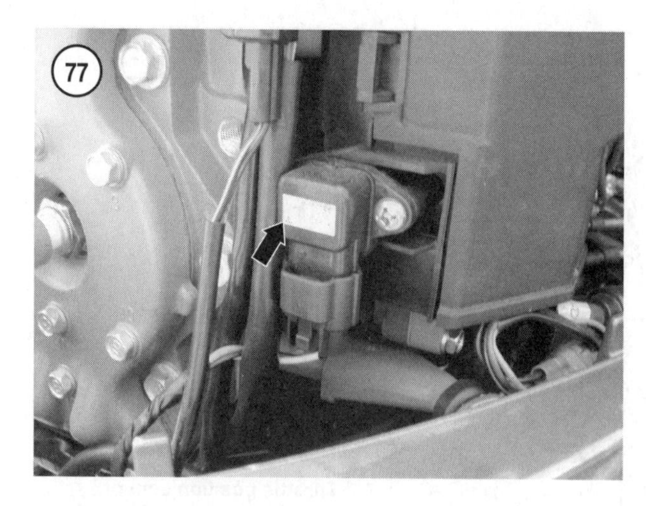

2. Refer to Chapter Six for information and the air pressure sensor location.

3. Disconnect the engine wire harness from the air pressure sensor.

4. Test the sensor input voltage as follows:

a. Connect the test harness (Yamaha part No. YB-06769/90890-06769) to the engine wire harness. Do not connect the test harness to the air pressure sensor at this time.

b. Calibrate the multimeter to the 10 or 20 VDC scale.

c. Connect the positive test lead to the orange test harness wire. Connect the negative meter test lead to the black test harness wire.

d. Turn the ignition key switch to the ON or RUN position. Do not start the engine. If the meter does not indicate 4.75-5.25 VDC, check the wiring connecting the sensor to the engine control unit for faulty terminals or wiring.

e. Turn the ignition key switch to the OFF position.

5. Test the sensor output voltage as follows:

a. Connect the test harness to the engine wire harness and the air pressure sensor.

b. Connect the positive test lead to the pink test harness wire. Connect the negative test lead onto the black test harness wire.

c. Turn the ignition key switch to the ON or RUN position. Do not start the engine. If the meter does not indicate 3.2-4.6 VDC, the air pressure sensor or related wiring is faulty. Check the wiring between the sensor and the engine control unit for faulty terminals or wires. Replace the sensor as described in Chapter Six if the wiring checks satisfactorily.

d. Turn the ignition key switch to the OFF position.

6. Remove the test harness. Connect the engine wire harness to the air pressure sensor. Route the wiring to prevent interference with moving components.

78

79

7. Connect the spark plug leads.

Air Temperature Sensor Test

The air temperature sensor (**Figure 78**) is used on all EFI and HPDI models. It provides the temperature of the air entering the intake manifold to the engine control unit. Cooler incoming air is denser and requires more fuel. Warmer air is less dense and requires less fuel.

Excessive exhaust smoke, poor performance, spark plug fouling or a lean operating condition can occur with a faulty air temperature sensor. Use a digital multimeter and test harness (Yamaha part No. 06768/90890-06768) for this procedure. To ensure accurate test results, perform this test with the engine and the surrounding air at a temperature of 20°C (68°F).

1. Disconnect and ground the spark plug leads to prevent accidental starting.
2. Refer to Chapter Six for information and locate the air temperature sensor.
3. Disconnect the engine wire harness from the air pressure sensor.
4. Test the sensor input voltage as follows:
   a. Connect the test harness (Yamaha part No. YB-06768/90890-06768) to the engine wire harness. Do not connect the test harness to the air pressure sensor at this time.
   b. Calibrate a multimeter to the 10 or 20 VDC scale.
   c. Connect the positive test lead to one of the black/yellow test harness wires. Connect the negative test lead to the other black/yellow test harness wires.
   d. Turn the ignition key switch to the ON or RUN position. Do not start the engine. If the meter does not indicate 4.75-5.25 VDC, check the wiring connecting the sensor to the engine control unit for faulty terminals or wiring.
   e. Turn the ignition key switch to the OFF position.
5. Test the sensor output voltage as follows:
   a. Connect the test harness to the engine wire harness and the air temperature sensor.
   b. Connect the positive test lead to the pink test harness wire. Connect the negative test lead to the black test harness wire.
   c. Turn the ignition key switch to the ON or RUN position. Do not start the engine. The meter should indicate 3.4-5.3 VDC at 20° C (68° F). The reading will be lower at lower temperatures and higher at higher temperatures. If the meter reading is not within the specification, the air temperature sensor or related wiring is faulty. Check the wiring between the sensor and the engine control unit for faulty terminals or wires. Replace the sensor as described in Chapter Six if the wiring checks satisfactorily.
   d. Turn the ignition key switch to the OFF position.
6. Remove the test harness. Connect the engine wire harness to the air temperature sensor. Route the wiring to prevent interference with moving components.
7. Connect the spark plug leads.

Oxygen Density Sensor Test

The oxygen density sensor is used on all EFI and HPDI models. It is located beneath the plastic cover (**Figure 79**) on the starboard side of the power head. Input from the sensor is used by the engine control unit to fine tune the air/fuel mixture delivered to the engine.

Poor performance and spark plug fouling are common symptoms of a faulty oxygen density sensor.

Use a digital multimeter for this procedure.

CAUTION
Never run an outboard without first providing cooling water. Either use a test tank or run the engine under actual conditions. Install a test propeller to run the engine in a test tank.

1. Disconnect and ground the spark plug leads to prevent accidental starting.
2. Trace the sensor wire harnesses from the cover (**Figure 79**) to the respective connections to the engine wire harness. Disconnect both sensor harnesses from the engine wire harness.
3. Test the resistance of the sensor heater coil as follows:
   a. Calibrate the multimeter onto the R × 100 scale.
   b. *EFI models*—Touch the positive test lead to one of the white sensor harness terminals. Touch the negative test lead to the other white sensor harness terminal.
   c. *HPDI models*—Touch the positive test lead to the red/white sensor harness terminals. Touch the negative test lead to the black sensor harness terminal.
   d. If the meter does not indicate approximately 100 ohms for EFI models and 2-100 ohms for HPDI models, replace the oxygen density sensor as described in Chapter Six.
4. Connect the spark plug leads.
5. Test the sensor output as follows:
   a. Connect the sensor heater harness (two white leads or red/white and black) to the engine wire harness.
   b. Calibrate the multimeter to the 1 or 2 VDC scale. Touch the positive test lead to the gray sensor harness terminal. Touch the negative test lead to the black sensor harness terminal. The sensor should not generate voltage with the engine cool and not running.
   c. Remove the propeller if operating the engine on a flush/test device.
   d. Start the engine and observe the meter while operating the engine at idle speed. If the voltage does not increase from 0-0.1 VDC to 0.2-1.0 VDC, replace the oxygen density sensor as described in Chapter Six.
6. Reconnect the sensor harnesses to the engine wire harness. Route the wiring to prevent interference with moving components. Secure the wiring with plastic locking type clamps as necessary.

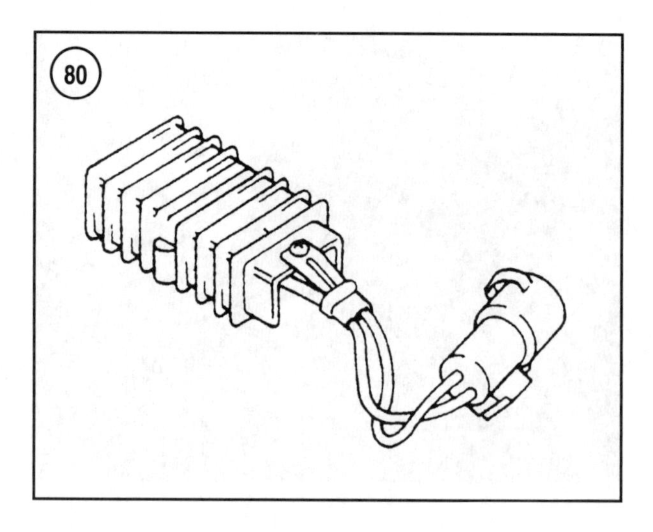

Fuel Pump Resistor Test

The fuel pump resistor is used on all EFI models. The resistor connects in series in the wiring that provides current to the electric fuel pump. The resistor reduces radio frequency interference (RFI) or electrical noise in the system. Use a digital multimeter for this procedure. The resistor (**Figure 80**) is located under the cover on the top and front of the power head. The resistor can be tested without removal from the power head.

1. Disconnect the battery cables and ground the spark plug leads to prevent accidental starting.
2. Refer to Chapter Six locate the resistor. Disconnect the resistor harness from the engine wire harness.
3. Calibrate the multimeter to the 1 ohm scale.
4. Touch the positive test lead to the brown wire terminal in the resistor harness connector. Touch the negative meter test lead to the blue wire terminal in the connector. Do not inadvertently connect the meter test leads to the engine wire harness.
5. If the meter does not indicate 0.53-0.57 ohm, thoroughly clean the terminals and repeat the test. Replace the resistor as described in Chapter Six if it fails to perform as specified.
6. Connect the resistor harness to the engine wire harness. Route the wiring to prevent interference with moving components. Secure the wiring with plastic locking type clamps as necessary.
7. Connect the battery cables and spark plug leads.

Shift Cutout Switch Test

The shift cutout switch is used on 200-250 hp (3.1 liter) EFI models. The switch connects to the engine control unit wiring and is operated by the shift linkages. The

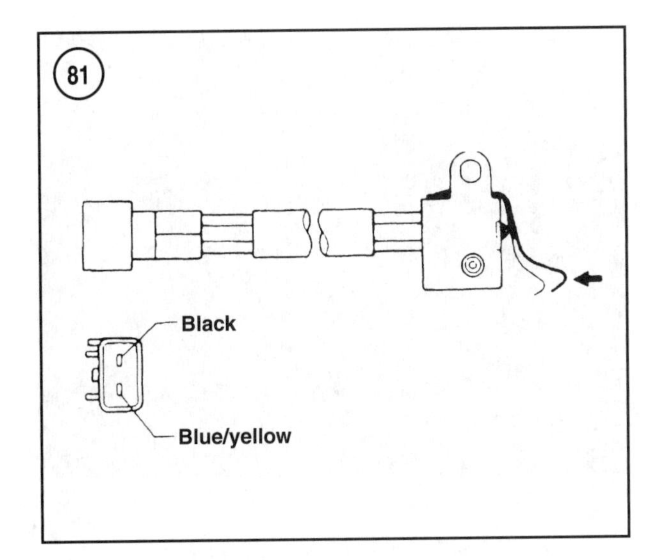

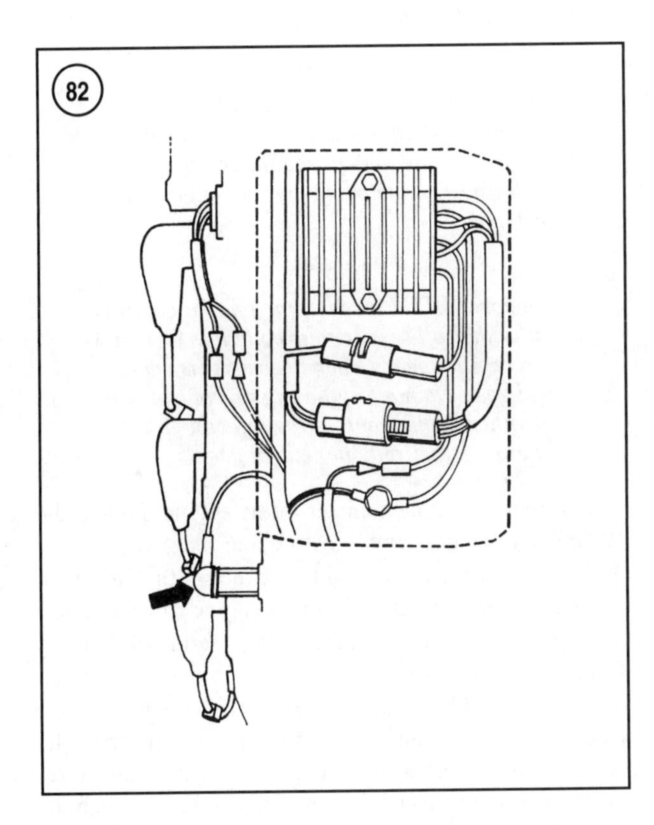

switch provides input to the engine control unit when the gearcase is shifted into neutral gear. The engine control unit interrupts ignition to one or more cylinders to allow easier shifting out of gear. A fault with the switch can cause hard shifting out of gear or an ignition misfire at lower engine speeds. Use a multimeter for this procedure.

1. Remove the shift cutout switch as described in Chapter Six.
2. Calibrate the multimeter onto the R x 1 scale.
3. Connect the positive test lead to the blue/yellow switch harness terminal (**Figure 81**). Connect the negative test lead to the black switch harness terminal.
4. Observe the meter while repeatedly pushing in then releasing the switch lever (**Figure 81**). The meter should indicate no continuity each time the lever is released and continuity each time the switch lever is depressed.
5. Replace the shift cutout switch if it fails to perform as described.
6. Install the shift cutout switch as described in Chapter Six.

Shift Position Switch Test

Perform this procedure on HPDI models. This switch provides a signal to the engine control unit indicating when the engine is shifted into neutral gear. In the engine control unit shuts off ignition to two cylinders at idle speed when in neutral gear. The engine control unit shuts off ignition to four cylinders if engine speed exceeds 1500 rpm in neutral gear.

1. Remove the shift position switch as described in Chapter Six.
2. Calibrate the multimeter onto the R × 1 scale.
3. Connect the positive test lead to the blue/yellow switch harness terminal (**Figure 81**). Connect the negative test lead to the black switch harness terminal.
4. Observe the meter while repeatedly pushing in then releasing the switch lever (**Figure 81**). The meter should indicate no continuity each time the lever is released and continuity each time the switch lever is depressed.
5. Replace the shift position switch if it fails to perform as described.
6. Install the shift position switch as described in Chapter Seven.

Knock Sensor Test

A knock sensor is used on 200-250 hp (3.1 liter) EFI models to communicate with the engine control unit, which in turn alters fuel delivery and/or ignition timing to help prevent spark knock or detonation damage. The knock sensor is located on the lower starboard cylinder head (**Figure 82**). Use a multimeter capable of measuring AC millivolts for this procedure.

1. Disconnect the battery cables and ground the spark plug leads to prevent accidental starting.
2. Disconnect the wire from the knock sensor.
3. Select the AC millivolts scale on the meter.

4. Touch the positive test lead to the wire terminal on the sensor. Touch the negative test lead to unpainted metal portion the body of the sensor.
5. Observe the meter while lightly tapping on the sensor body. If the meter does not display 2-10 mV, replace the sensor as described in Chapter Six.
6. Repeat the test with a replacement sensor to ensure proper operation.
7. Reconnect the wire to the sensor. Route the wiring to prevent interference with moving components. Make sure the sensor wire does not contact the spark plug leads.
8. Connect the battery cables and spark plug leads.

Water in Fuel Sensor Test

The water in fuel sensor is used on HPDI models. The sensor is located in the spin on fuel filter canister (**Figure 83**).

1. Remove the water in fuel sensor as described in Chapter Six.
2. Calibrate the meter on the R × 1 scale.
3. Connect the positive test lead to the blue/white sensor harness terminal. Connect the negative test lead to the black sensor harness terminal.
4. Position the sensor with the open end facing upward. The meter should indicate no continuity.
5. Position the sensor with the open end facing downward. The meter should indicate continuity.
6. Repeat Step 4 and Step 5 several times. Replace the sensor if it fails to perform as specified.
7. Install the water in fuel sensor as described in Chapter Six.

Injector Driver Output Test

Perform this procedure on HPDI models. This test measures the voltage supplied to the injectors by the driver circuits in the engine control unit. Use a multimeter and an accurate shop tachometer for this procedure.

WARNING
The ignition system produces very high voltage current. Never touch the test lead of any engine wiring while the flywheel is rotating. Never stand in water while testing or working around wet wiring or connections. Electrical shock can cause serious injury or death.

WARNING
Use extreme caution when working with the fuel system. Avoid damage to property and/or injury or death. Never smoke around fuel or fuel vapor. Make sure no flame or source of ignition is present.

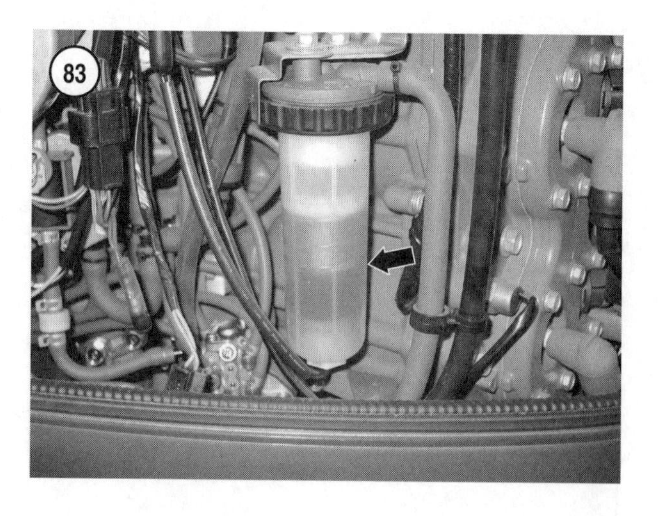

CAUTION
Never run an outboard without first providing cooling water. Either use a test tank or run the engine under actual conditions. Install a test propeller to run the engine in a test tank.

NOTE
On some HPDI models, the engine control units shut off two or four cylinders at various engine speeds when the engine is in neutral gear. Do not fail the engine control unit or other components if the injector fails to operate under only in neutral gear.

1. Connect a shop tachometer to the engine. Follow the tachometer manufacturer's instructions.
2. Refer to Chapter Five and locate the six fuel injectors. Trace the injector wires to the to the engine wire harness.
3. Disconnect the No. 1 injector (top starboard) harness from the engine wire harness.
4. Refer to **Table 14** for the test specifications and test lead connection points. Calibrate the multimeter to the correct peak AC voltage scale. Scale changes may be required when testing at different engine speeds. Touch the meter test leads to the terminals of the engine wire harness connection. Do not inadvertently select the injector harness terminals.
5. Start the engine and run it at fast idle until it reaches normal operating temperature. Record the output voltage while running the engine at the speeds listed in **Table 14**. Change the meter scale as required for the test specification. To check the voltage at cranking speed, disconnect and ground the spark plug leads. Note the meter readings while operating the electric starter.

6. Stop the engine and reconnect the No. 1 injector wire harness to the engine wire harness. Disconnect each injector harness, one at a time and repeat Steps 3-5 for the remaining five fuel injectors. Record all test results.
7. If the injector driver output does not meet or exceed the minimum specification in **Table 14** for each injector and at each designated engine speed, check the wiring and thoroughly clean the terminals. No or low voltage indicates a faulty fuse, relay or engine control unit. Test or replace these components as required.
8. Disconnect the shop tachometer. Reconnect all injector harness connectors. Route the wiring to prevent interference with moving components. Secure the wiring with plastic locking type clamps as necessary.

OIL INJECTION SYSTEM

The primary components of the oil injection system include the oil reservoir, oil level sensor, oil pump, drive and driven gears, hoses, check valves and fittings. An additional electric oil pump is used on HPDI models. Failure of any of these components can cause serious power head damage. Test these components if an oil related failure occurs or the oil injection system is not operating properly. Refer to *Warning System* in this chapter to test the oil level sensor.

WARNING
Stay clear of the propeller shaft while running the engine on a flush/test adapter. To avoid injury or death, remove the propeller before running the engine on a flush/test adapter. Always disconnect the battery and spark plug leads before removing or installing the propeller.

CAUTION
Never run an outboard without first providing cooling water. Use either a test tank or flush/test adapter if the engine cannot be operated under actual conditions. Remove the propeller before running the engine on a flush/test adapter. Install a test propeller to run the engine in a test tank.

Drive/Driven Gear

Oil pump drive or driven gear failure causes insufficient oil delivery to the power head. The oil pump output test described in this section will indicate failure of the gears. The pump can also be removed for visual inspection of the gears. Broken or missing gear teeth, discoloration or debris are sure signs for gear failure. To inspect the gears, remove the oil pump and driven gear as described in Chapter Twelve. Broken gear teeth are caused by interference with debris that has entered the crankcase or by oil pump failure. Replace the oil pump along with the gears to prevent a possible repeat failure.

Oil Pump Output Test

Use an accurate measuring container, shop tachometer and watch or clock for this test. Perform this test only on carburetor equipped and EFI models. Do not perform this test on HPDI models.

CAUTION
The engine must be supplied with a 50:1 fuel/oil mixture when performing the oil pump output test.

NOTE
The measuring container must have 0.1 cc graduations for accurate measurement of the pump output.

1. Run the engine at a fast idle for 10 minutes on a fresh 50:1 fuel/oil mixture. Shut the engine off. Then install a shop tachometer following the manufacturer's instructions.
2. Disconnect the oil pump linkage (**Figure 84**) from the pump lever. Rotate the pump lever to the wide-open position (**Figure 85**). Refer to *Oil Pump Linkage Adjustment* in Chapter Four to determine the wide-open position. The pump lever is rotated in the clockwise direction until it contacts the stop.

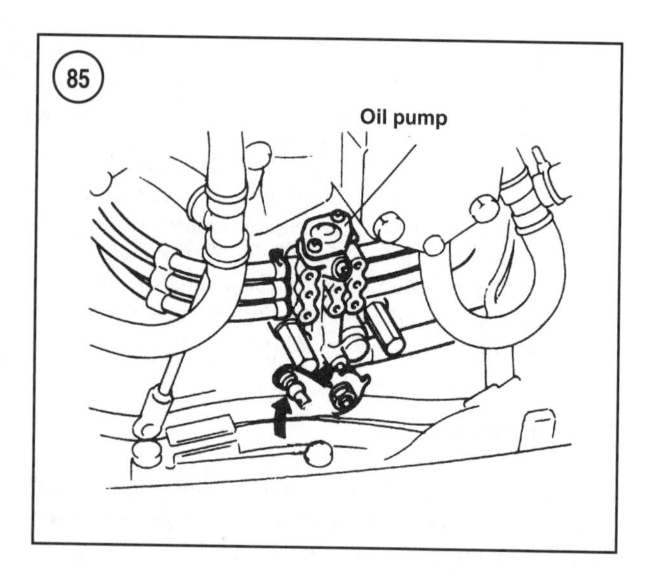

3A. *Carburetor equipped models*—Disconnect one of the oil pump discharge hoses from the intake manifold connection (**Figure 86**).

3B. *EFI models*—Disconnect the oil discharge hose from the connection on the vapor separator tank (**Figure 87**). Refer to the diagrams in Chapter Twelve to assist with oil hose routing and connection points.

4. Direct the hose into the graduated container (**Figure 88**). Make sure the oil discharged from the hose does not contact the side of the container.

5. Start the engine and adjust the throttle to 1500 rpm. Record the oil level in the graduated container. Run the engine for exactly three minutes, then reconnect the oil hose. Shut the engine off.

6. Record the oil level in the container. Subtract the earlier measurement from the last measurement to determine the oil pump output during the three minute run time.

7. *Carburetor equipped models*—Repeat Steps 3-5 for each of the oil pump discharge hoses. Record the output from each hose.

8. The output must be within the specification in **Table 15**.

   a. *Output from each hose is below the specification*—Remove the oil pump and inspect the drive and driven gears. Test the check valves as described in this chapter. Replace the oil pump if the gears and check valves test satisfactorily.
   b. *Output exceeds the specification*—Replace the oil pump. Excessive output is usually caused by installing the wrong oil pump to the engine. Verify the model name and serial number before ordering the replacement pump.
   c. *Uneven output from the hoses (carburetor equipped models)*—Test the check valves as described in this chapter. Replace the oil pump if the valves test correctly and the output is above or below the specification from one or more of the hoses. Check for faulty drive/driven gear if the output is below the specification.

Electronic Oil Pump Test (HPDI Models)

Perform this test with the engine running at idle speed. This test checks for output from the electric oil pump (**Figure 89**). The manufacturer does not list output specifications.

1. Refer to the information in Chapter Twelve and locate the electric oil pump.
2. Start the engine and run it for ten minutes at idle speed.
3. Disconnect the oil discharge hose from the connection on the vapor separator tank (**Figure 87**). Refer to the dia-

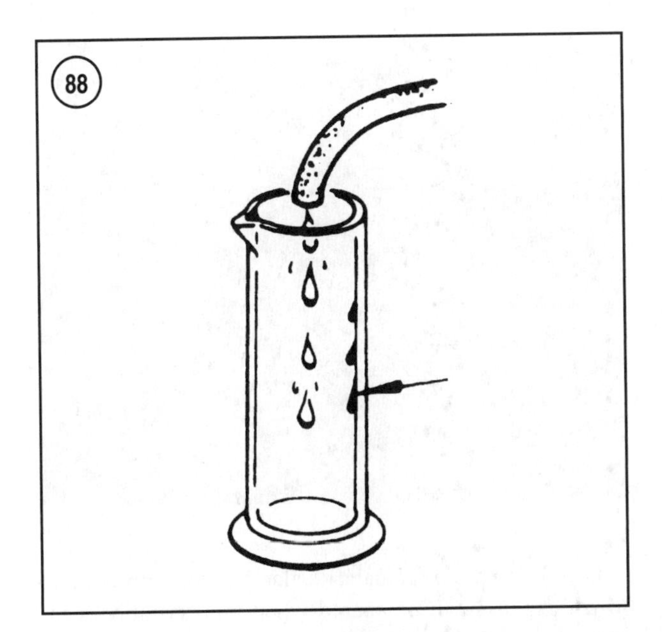

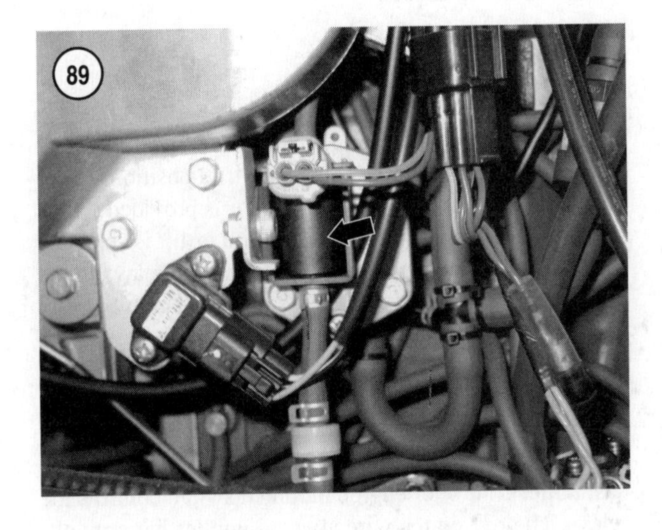

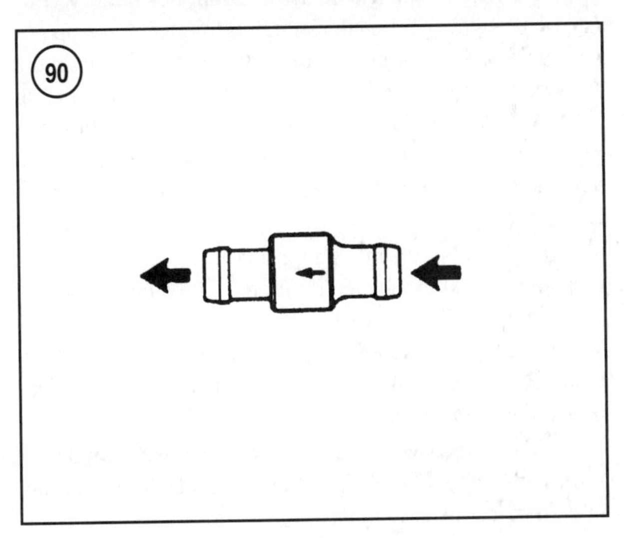

grams in Chapter Twelve to assist with oil hose routing and connection points.

4. Oil should slowly drip from the disconnected hose. Stop the engine and reconnect the oil hose. Wipe up any spilled oil.

5. Replace the electric oil pump as described in Chapter Twelve if it fails to perform as described. After installation, test the replacement pump. Check for faulty wiring or connections if the replacement pump fails to operate.

Oil Hose Check Valves Test

CAUTION

Do not use compressed air to test oil hose check valves. Excessive air pressure or volume may damage or weaken the valve and lead to subsequent failure. Operation with a damaged valve can cause insufficient lubrication and serious power head damage. Use only a hand operated pump and low pressure to test the valve.

An oil hose check valve is used to ensure that the oil flows through the hose in one direction only.

Check valve failure can prevent adequate oil delivery to the power head and lead to serious and expensive power head damage. Use a hand operated pressure/vacuum tester to test the check valve.

1. Remove the oil hose check valves as described in Chapter Twelve.
2. Connect the pressure tester to one of the fittings on the valve (**Figure 90**). Apply light pressure and note if air exits the opposite fitting. Connect the tester to the opposite fitting and repeat the test. Air must flow easily in the direction of the arrow on the valve and not flow opposite the valve.
3. Repeat the test for each check valve on the engine.
4. Replace any check valve that fails to perform as specified.
5. Install the check valves as described in Chapter Twelve.

TILT AND TRIM SYSTEM

A three hydraulic cylinder electrically powered tilt and trim system is used on all models covered in this manual. Although similar in appearance, the system used on 3.1 liter models is different from the system used on V-4 and 2.6 liter V-6 models. The assembly mounts between the clamp brackets (**Figure 91**). It provides the capability to move the engine up or out, against propeller thrust, allowing the operator to change running attitude while underway.

The major components of the system include the bi-rotational electric motor and pump, fluid reservoir, trim cylinders and tilt cylinder. A power head mounted relay unit controls the electric motor/pump rotation. Reversing the motor direction controls fluid movement within the system. When trimming up from a fully down position, fluid is moved into the up side or up cavity of both trim rams and the tilt cylinder. The ends of the trim cylinder rams contact striker plates on the swivel housing. Fluid pressure causes all three hydraulic cylinders to extend and raise the engine. Frequently apply a coating of water resistant grease to the ends of the trim cylinder rams to prevent noisy operations while trimming up.

When the trim cylinders reach the limit of extension, the tilt cylinder remains the only means for raising the engine further. Less power is then developed in the up directions as only one cylinder drives the system. This has an effect of liming the trim range while underway at higher engine speeds. The engine will, however, have a greater tilt speed as all the fluid travels to a single cylinder instead of three.

When trimming down, the electric motor changes direction causing fluid to flow to the down side of the cavity of the cylinders. As the only cylinder connected to the engine at both ends, the tilt cylinder provides all downward movement.

A manual release valve provides manual engine movement without operating the pump. An opening in the starboard side clamp bracket provides access to the valve. If the system should malfunction, check the position of this valve first. Counterclockwise rotation releases the valve to move the engine. Clockwise rotation closes the valve to hold the engine in position.

Typical symptoms of a malfunction include:

1. The engine does not move up.
2. The engine does not move down.
3. The engine leaks down from the tilt position only.
4. The engine tucks under while underway.
5. The engine trails out when in reverse or when slowing down.
6. Hydraulic fluid is leaking from the system.

If the electric motor is not operating, troubleshoot the electric part of the system as described in this chapter. Replacement and repair of the electric motor is described in Chapter Ten. If the motor is operating correctly and the above symptoms are present, check the fluid level and the manual release valve position as described in Chapter Ten.

If both check satisfactorily, have a professional repair the system. Remove the system as described in Chapter Ten and contact a Yamaha dealership for information. Much expense can be spared when the assembly is removed from the engine.

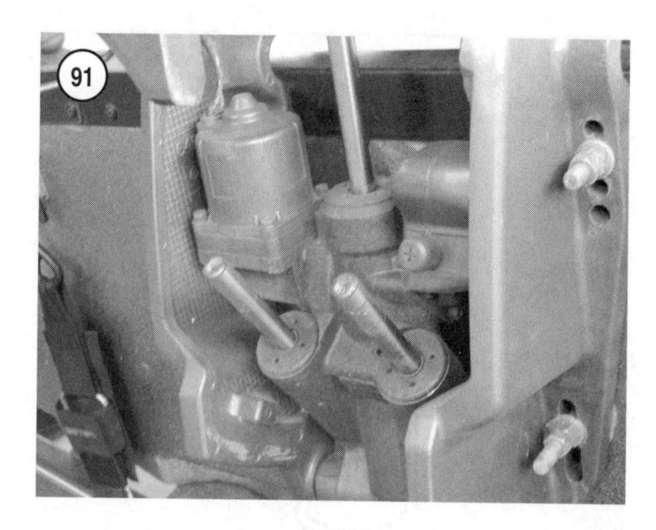

Electrical Tests

The major electrical components of the tilt/trim system include the electric motor, relay unit, trim position sender and switches. The bi-directional motor is provided with a blue and green wire. When the up trim or tilt is selected, the blue wire is connected via wires and the relays to the positive battery terminal. The green wire is connected to ground and the negative terminal by the same means. The electric motor and pump then rotate in the direction that moves fluid to the up side of the cylinders. When the down direction is selected, the relays simply reverse the wire connection and motor rotation. A remote control, dash, tiller control or engine mounted switch controls the relays. The red switch wire always supplies battery voltage to the switch. When the switch is toggled to the up position, the red wire connects to the blue wire contact in the switch. The blue wire directs the battery current to the up relay. When the switch is toggled to the down position, the red wire connects to the green wire contact in the switch. The green wire directs the battery current to the down relay.

When battery current is supplied to either relay, it directs the current to the electric motor. The other relay provides the connection to ground for the electric motor. When energized, the relay opens the ground connection. Both relays must make the proper connection for the electric motor to operate.

All V-6 models covered in this manual use a single assembly that houses both relays (**Figure 92**). If either relay fails, replace the assembly.

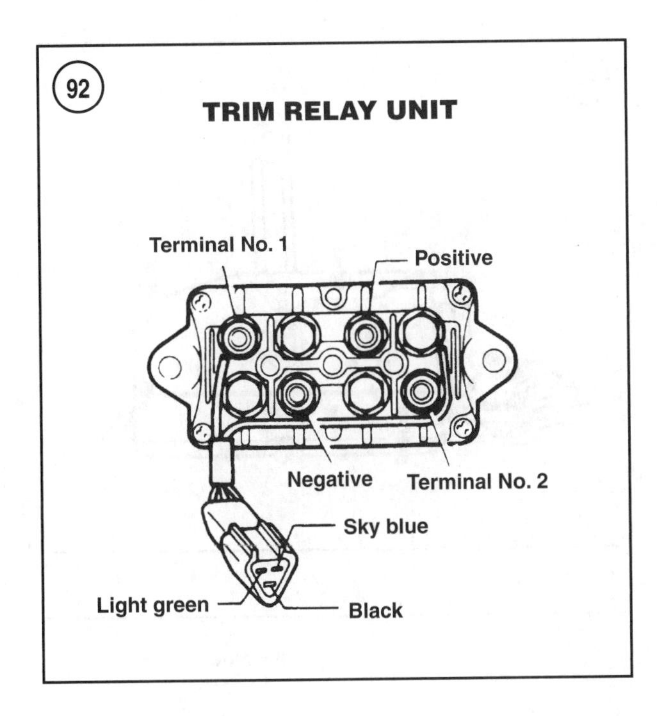

All V-4 models use two independent relays that can be individually replaced if they fail. The two independent relays mount side by side on the port side of the power head. The relays are similar in appearance as the starter relay. To test the relays, perform the *Trim Relay Functional Test* as described in this chapter.

If the trim motor fails to run in either direction, check for a blown fuse and disconnected or faulty wiring. Then, test the trim switch and relay unit as described in this chapter. These tests verify that the trim switch is supplying current to the relays and the relays are supplying current to the electric motor. Repair or replace the electric motor if all other components test correctly and the electric motor does not operate.

If the trim motor fails to run in only one direction, test the trim switches and relay unit as described in this chapter. A fault with the motor is unlikely. If the motor is able to operate in one direction, it can usually operate in the other.

Relay unit continuity test (single relay unit)

Perform this test on models using a single relay unit. Use a common multimeter for this procedure.

1. Remove the relay unit from the engine as described in Chapter Ten.
2. Calibrate the ohmmeter to the R × 1 scale.
3. Connect the positive test lead to the sky blue terminal connection (**Figure 92**). Connect the negative test lead to the black terminal connection. If the meter indicates no continuity, the relay unit has failed open and should be replaced.
4. Connect the positive test lead to the light green terminal connection (**Figure 92**). Connect the negative test lead to the black terminal connection. If the meter indicates no continuity, the relay unit has failed open and should be replaced.
5. Connect the positive test lead to terminal No. 1 on the relay unit (**Figure 92**). Connect the negative test lead to the negative terminal on the relay (**Figure 92**). If the meter indicates no continuity, the relay unit has failed open and should be replaced.
6. Connect the positive test lead to terminal No. 2 on the relay (**Figure 92**). Connect the negative test lead to the negative terminal on the relay (**Figure 92**). If the meter indicates no continuity, the relay unit has failed open and should be replaced.
7. Connect the positive test lead to terminal No. 1 on the relay unit. Connect the negative test lead to the positive terminal on the relay (**Figure 92**). If the meter indicates continuity, the relay unit is shorted internally and should be replaced.
8. Connect the red test lead to terminal No. 2 on the relay unit. Connect the black test lead to the positive terminal on the relay (**Figure 92**). The meter must indicate *no continuity*. If not, the relay unit is shorted internally and must be replaced.
9. Perform an operational test on the relay unit as described in this chapter.

Trim relay functional test (individual relays)

This test requires a fully charged battery, an analog multimeter and suitable jumper wires.

1. Remove the trim relays as described in Chapter Ten. Test the relays individually as described in Steps 2-6.
2. Calibrate the multimeter to the R × 1 scale. Connect the negative test lead to one of the large terminals on the relay. Connect the positive test lead to the other large terminal.
3. Using a jumper wire, connect the black relay lead (small lead terminal) to the negative battery terminal.
4. Connect a jumper wire to the positive battery terminal. Observe the ohmmeter while repeatedly touching the jumper lead to the remaining small lead terminal of the relay. The meter should indicate continuity each time the connection is made and no continuity each time the connection is broken. Also, the relay should make a clicking noise each time the connection is made.
5. Replace the trim relay if it fails to perform as described.
6. Install the trim relays as described in Chapter Ten.

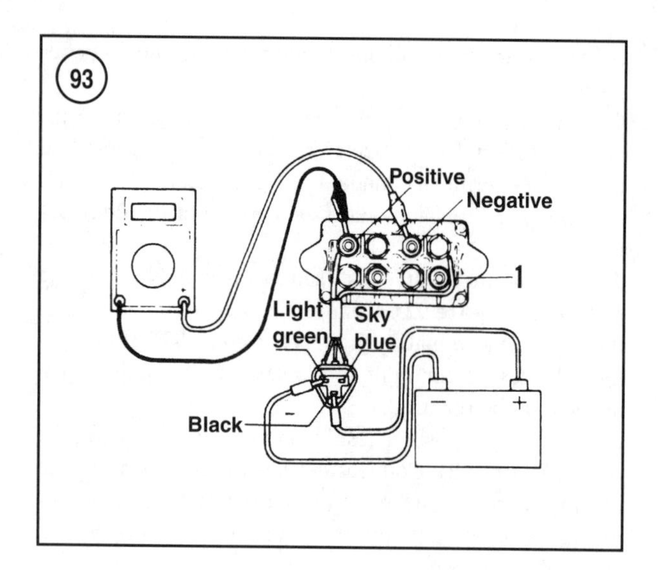

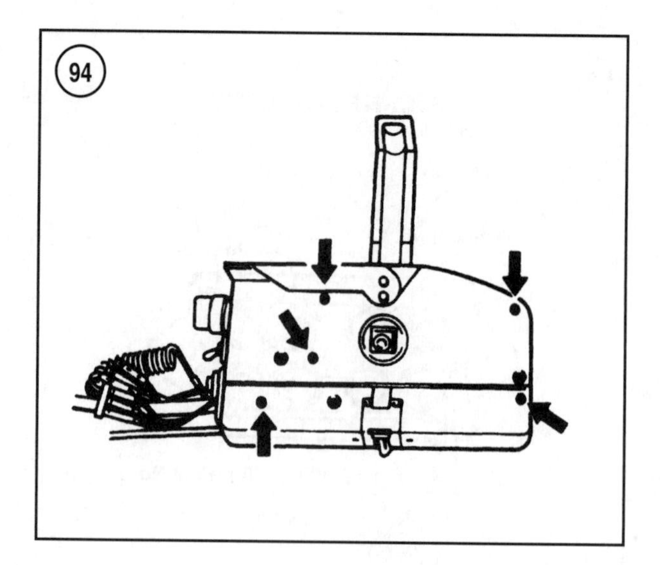

Relay unit operational test (single relay unit)

Perform this test on models using a single relay unit. Use a multimeter, suitable jumper wire and a fully charged battery for this procedure.

1. Remove the relay unit as described in Chapter Ten.
2. Calibrate the meter to the R × 1 scale.
3. Use the jumper wires to connect the positive battery terminal to the light green terminal (**Figure 93**) and the negative battery terminal to the black terminal. Connect the positive lead to the positive terminal on the relay (**Figure 93**). Connect the negative lead to the negative terminal on the relay. If the meter indicates continuity, the relay is not switching internally and must be replaced. Disconnect the jumper wires then the meter test leads.
4. Using the jumper leads, connect the positive battery terminal to the sky blue terminal (**Figure 93**) and the negative battery terminal to the black terminal. Connect the positive test lead to the negative terminal on the relay. Connect the negative test lead to terminal No. 1 on the relay unit (**Figure 93**). The meter indicates no continuity, the relay is not switching internally and must be replaced.
5. Disconnect the jumper wires, then the meter test leads.
6. Install the relay unit as described in Chapter Ten.

Tilt/Trim Switch Test

The tilt and trim system is controlled by a three-position switch mounted on the remote control handle, engine cover or other locations in the boat. Testing procedures are similar for all switches. This rocker type switch is spring loaded to center in the OFF position. The switch activates the up or down direction when toggled to the desired direction. Battery current is supplied to the switch by

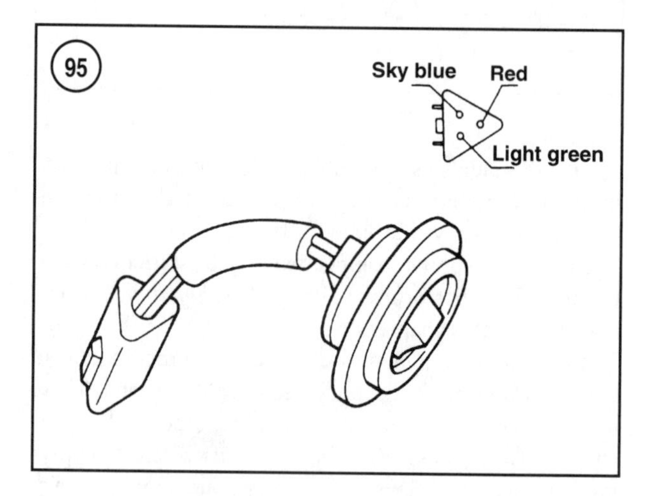

a fused wire. Check the fuse before testing the switch. Refer to *Fuses and Wire Harness* in this chapter for instructions.

If the system fails to operate properly only when operating a single switch, test only that switch. If the system is not functioning when operating multiple switches, check for a blown fuse, faulty wiring, faulty relay or trim motor. Multiple switch failure can prevent the trim system from operating properly by activating both relays simultaneously. Disconnect the trim switches one at a time and check for proper operation. Test the switch if the system operates properly when that switch is disconnected.

Use a multimeter for this procedure.

1. Disconnect the battery cables and ground the spark plug leads to prevent accidental starting.

2A. *Remote control mounted switch*—Remove the remote control and back cover (**Figure 94**) as described in Chap-

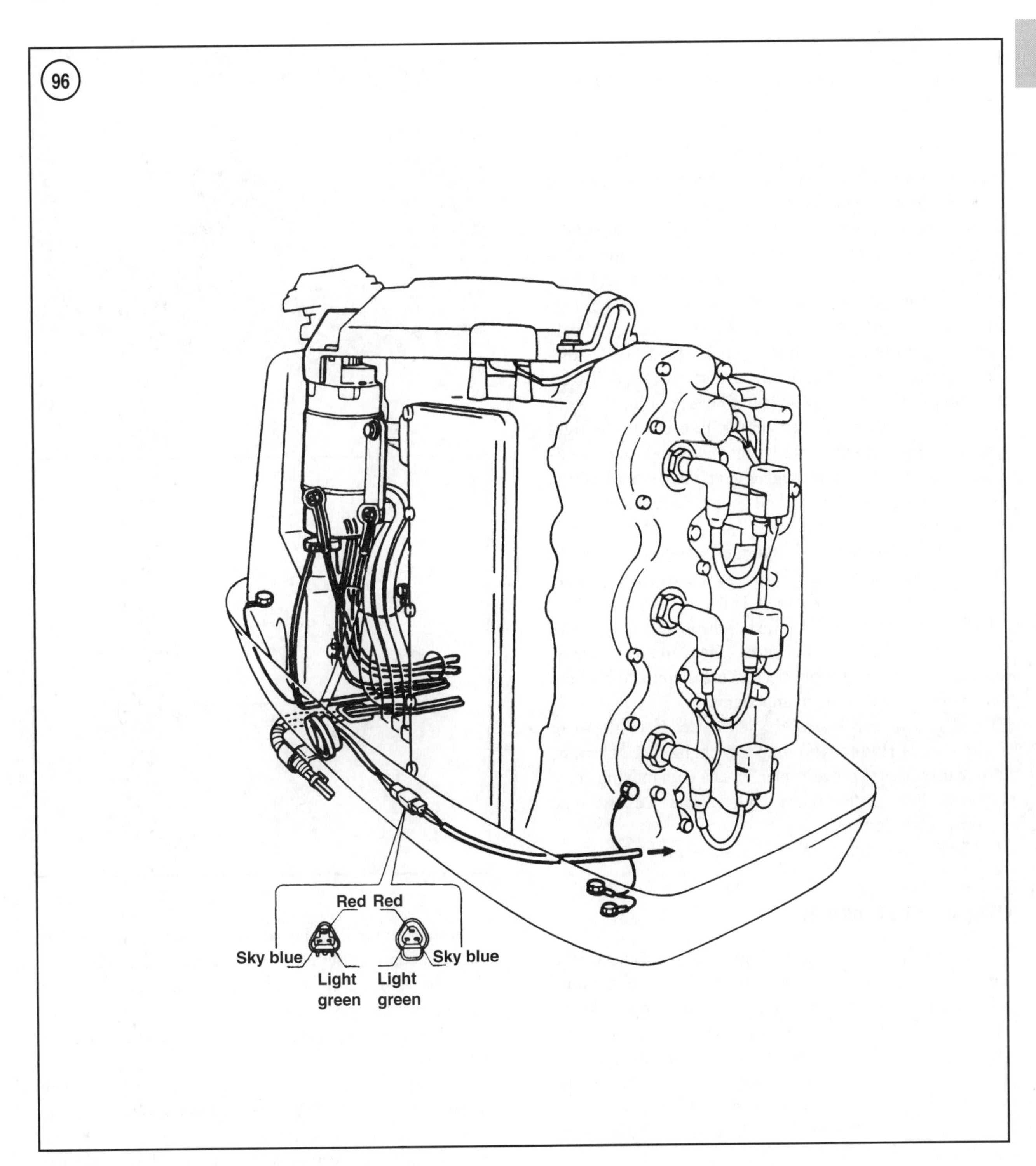

ter Thirteen. Locate and disconnect the trim switch wires from the remote control harness.

2B. *Dash mounted switch*—Disconnect the trim switch wires from the instrument harness. Remove the switch from the dash if necessary for good access to the wire terminals (**Figure 95**).

2C. *Engine cover mounted switch*—Locate and disconnect the trim switch wires (**Figure 96**) from the engine wire harness.

3. Calibrate the meter to the R × 1 scale. Do not inadvertently connect any test leads to the remote control, trim relay or engine wire harness wires.

4. Connect the positive test lead to the red wire pin of the switch. Connect the negative test lead to the sky blue wire pin to the switch. The meter should indicate no continuity with the switch in the OFF position and when toggled to the DOWN position. The meter should indicate continuity when the switch is toggled to the UP position. Replace the switch if it fails to perform as specified.
5. Connect the positive test lead to the red wire pin of the switch. Connect the negative test lead to the light green wire pin of the switch. The meter should indicate no continuity with the switch in the OFF position and when toggled to the UP position. The meter should indicate continuity when the switch is toggled to the DOWN position. Replace the switch if it fails to perform as specified.
6. Connect the positive test lead to the sky blue wire pin to the switch. Connect the negative test lead to the light green wire pin of the switch. If the meter indicates continuity when the switch is toggled to all three positions, the switch is shorted internally and should be replaced.
7A. *Remote control mounted switch*—Reconnect the trim switch wires. Route the wires to prevent interference with moving components. Install the back cover and remote control as described in Chapter Thirteen.
7B. *Dash mounted switch*—Install the switch if removed for access to the wires. Reconnect the trim switch leads. Secure the wires with plastic locking clamps to prevent entanglement with other components.
7C. *Engine cover mounted switch*—Connect the trim switch wires (**Figure 96**) onto the engine wire harness. Route the wires to prevent interference with moving components. Secure the wires with plastic locking type clamps as needed.
8. Connect the battery cables and spark plug leads.

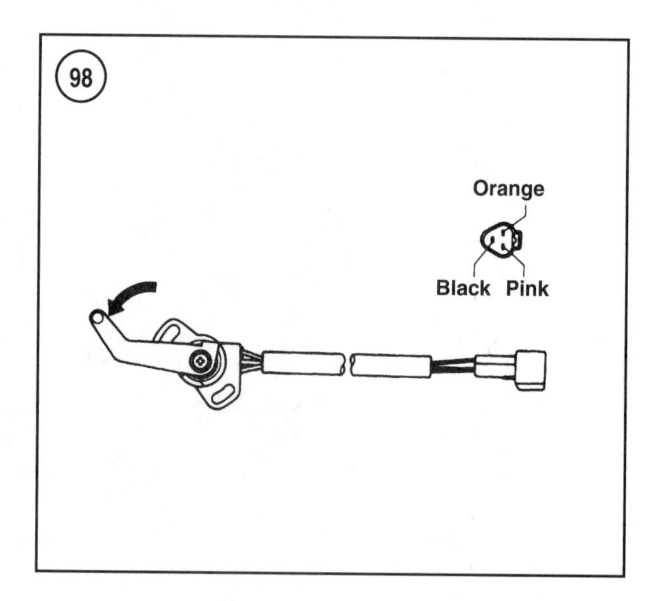

Trim Position Sender Test

The trim position sender (**Figure 97**) mounts to the swivel bracket. The arm of the sender contacts the tilt tube mounted lever and moves as the engine tilts up and down. Movement of the arm changes the sensor resistance and causes a varying voltage to the dash mounted gage. The voltage and gauge reading corresponds to the engine tilt/trim angle. Use a multimeter to test the sender.
1. Remove the trim position sender as described in Chapter Ten.
2. Calibrate an ohmmeter to measure 582-873 ohms of resistance.
3. Connect the positive test lead to the pink sender wire pin (**Figure 98**). Connect the negative test lead to the black sender wire pin. Note the meter reading while slowly moving the sender arm through the full range of travel. The meter should change resistance smoothly. The highest meter reading should not exceed 873 ohms and the lowest meter reading must not fall below 582 ohms. Replace the sender if the meter reading is erratic or the reading exceeds the specification.
4. Calibrate an ohmmeter to measure 800-1200 ohms of resistance.
5. Connect the positive lead to the orange sender wire pin (**Figure 98**). Connect the negative test lead to the black sender wire pin. Note the meter reading while slowly moving the sender arm to the full range of travel. The meter must change resistance smoothly. The highest meter reading should not exceed 1200 ohms and the lowest meter reading should not fall below 800 ohms. Replace the sender if the meter reading is erratic or the reading exceeds the specification.
6. Install the trim position sender as described in Chapter Ten. Adjust the sender as described in Chapter Four.

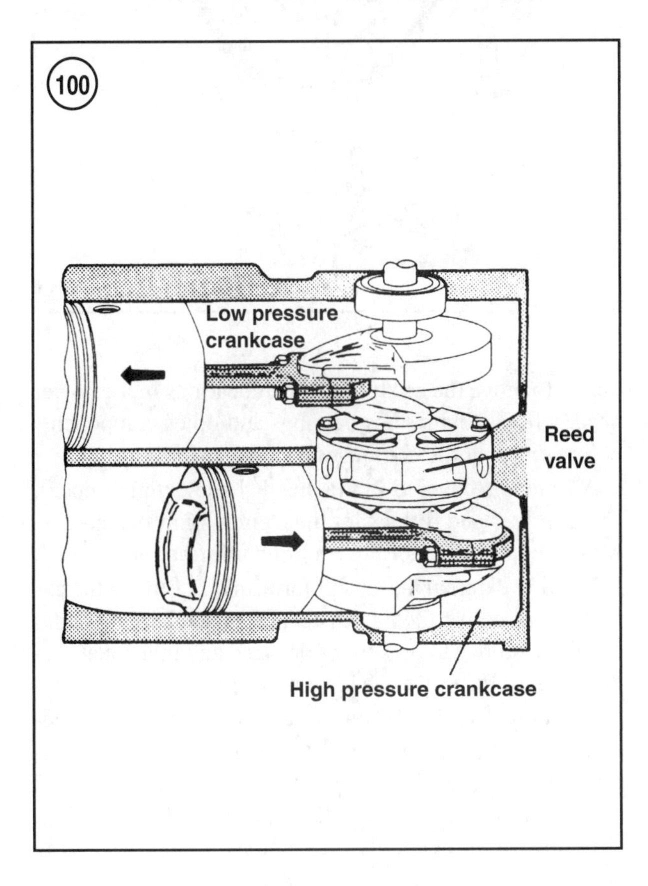

POWER HEAD

This section covers testing and troubleshooting to determine if there is a problem in the power head. Subjects covered include compression testing, engine noises, lubrication failure, detonation, preignition, engine seizure and water entering the cylinder(s).

Compression Test

Older engines or engines with high operating hours often experience hard starting, poor idle quality or poor performance. Perform a compression test if the fuel and ignition system have been ruled out as causing these symptoms. Avoid relying solely on the compression test to indicate the condition of the engine. Sometimes a compression test indicates satisfactory readings when there is still something wrong in the power head. Closer inspection of the piston, rings and cylinder walls may reveal scoring or scuffing not detected by compression testing. A leak-down test is a more reliable indicator of the piston, ring and cylinder wall condition. Equipment for a leakdown test is far more costly and more difficult to use than a typical compression gauge. Allow a reputable marine dealership to perform a leak-down test if needed.

1. Remove the spark plugs. To prevent arcing, ground all the spark plug leads.
2. Remove the propeller as described in Chapter Eight.
3. Install the compression gauge in the No. 1 cylinder spark plug opening (**Figure 99**). Securely tighten the adapter to prevent leakage at the spark plug opening.
4. Manually hold the throttle valve(s) in the wide-open position.
5. Operate the electric until the flywheel rotates at least six revolutions.
6. Record the compression reading.
7. Repeat Steps 3-6 for the remaining cylinders.
8. Install the propeller as described in Chapter Eight.
9. Compare the compression readings noting the highest and lowest readings. Ideally the lowest reading should be within 10% of the highest reading. One or more cylinders with significantly lower readings or a cylinder that reads 517 kPa (75 psi) or less indicates a problem that deserves immediate attention. Remove the cylinder head or exhaust cover and inspect the piston and cylinder wall before troubleshooting or tuning the engine. An engine with inadequate or unbalanced compression cannot be tuned properly or expected to perform correctly. Power head repair is described in Chapter Seven.
10. Install the spark plug(s) and connect the lead(s).

Crankcase Sealing

Carbureted two-stroke engine operate with alternating pressure and vacuum in the crankcase (**Figure 100**, typical). As the piston moves up on the compression stroke, the volume in the crankcase increases and creates a vacuum. Atmospheric pressure forces fuel and air to flow through the carburetor, reed valve and into the crankcase (**Figure 101**) to fill the vacuum. As the piston moves down on the power stroke the volume in the crankcase decreases, forming pressure (**Figure 102**). Pressure formed in the crankcase closes the reed valve (**Figure 102**), trapping the fuel/air mixture in the crankcase. Further downward movement further increases crankcase pressure until the piston moves past the transfer port (**Figure 102**). This allows the pressurized mixture to flow into the combustion chamber, providing a fresh fuel/air mixture for the next cycle.

Crankcase pressure and vacuum are also used to power the fuel pump. Weak crankcase pressure can contribute to poor fuel pump operation during low-speed operation. If the engine runs out of fuel while idling for a few minutes, pump the primer bulb. If the engine operates properly only while pumping the bulb, disassemble and inspect the fuel pump. Suspect weak crankcase pressure if no faults are found with the fuel pump.

Faulty pistons, rings and cylinder walls can affect crankcase pressure by allowing combustion chamber gasses to leak into the crankcase, diluting the fuel/air mixture. Likewise, faulty crankshaft seals, crankcase sealing surfaces, reed valves and cylinder block castings can adversely affect crankcase pressures. Low crankcase pressure prevents the engine from receiving adequate fuel and air into the cylinder, reducing engine efficiency. The engine will produce less power, causing bogging on acceleration and poor idle quality. In extreme cases, the engine can be tuned to idle at the correct out of gear speed, only to have the engine stall when shifted into gear. The reduced efficiency prevents the engine from producing enough power to offset the load of the propeller. Typically, inadequate crankcase sealing affects engine operation only at lower speeds. At higher engine speeds the leakage is a lower percentage of the volume of air entering the engine and the symptoms lessen.

If a crankcase or crankcase gasket leak is suspected, spray a soap and water solution onto the mating surfaces and casting while cranking the engine. Bubbles indicate a leak. On multiple cylinder engines, internal leaks can occur. The symptoms are consistent with external leakage and only disassembly and inspection will reveal the components that are causing the leak(s). Check for leakage at the intake manifold and reed housing before disassembling the power head. Faulty reeds or leaking gaskets cause low crankcase pressure. Power head disassembly, inspection and repair are described in Chapter Seven.

Recirculation System

During low speed operation, a fuel and oil mixture tends to collect in certain areas in the power head. The recirculation system uses crankcase pressure and vacuum

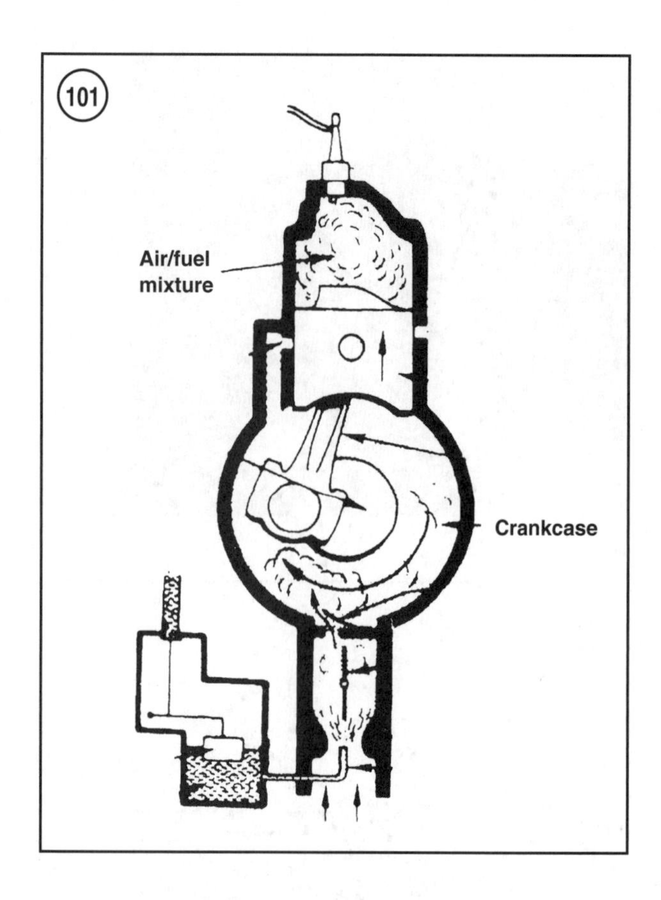

pulses to move the mixture to other locations in the power head, where it lubricates bearings and other components before entering the combustion chamber for burning.

All models covered in this manual use external hoses, check valves and fittings for the recirculation system. Refer to Chapter Five for hose routing information.

Leaking external hoses will form an oily film on the external surfaces of the power head. Replace leaking hoses or faulty fittings to prevent oil leakage and inadequate lubrication of the power head components.

A plugged recirculation system prevents fuel movement, causing fuel to eventually be drawn into the combustion chamber along with the fuel/air mixture. This causes a rich fuel mixture, the engine to run very rough, slow down and eventually stall. The symptoms occur only at lower engine speeds, since fuel generally does not collect at higher engine speeds. With a faulty recirculation system, the engine usually bogs down if accelerated after idling for an extended time. This is due to a rich running condition as the puddled fuel is quickly drawn into the combustion chamber.

To check for a plugged recirculation system, operate the engine at a fast idle until the engine reaches full operating temperature. Return the engine to idle speed and quickly note the engine idle speed and running characteristics.

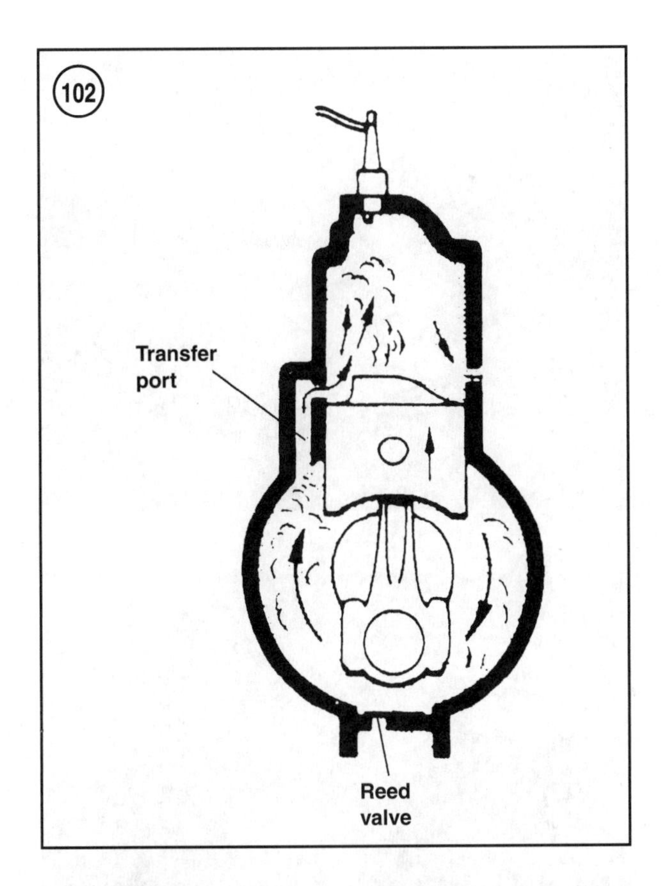

Check for other problems if the symptoms surface fairly quickly after returning to idle speed.

If the recirculation system is suspect, check the condition on external hoses and hose routing as described in Chapter Five. Test the check valve as described in this chapter.

On carbureted models, a faulty recirculation system is not always caused by plugged passages or faulty check valves. A flooding or improperly adjusted carburetor may cause more of the mixture to collect than the system can move. Weak crankcase pressure or faulty reed valves reduce the capability for moving fuel and increases the volume of fuel that collects in the power head. Check these components if the symptoms persist and no faults are found with the hoses or check valves.

Check valve fittings test

Use a hand-operated pressure/vacuum pump or syringe for this test.

1. Disconnect the battery cables and ground the spark plug lead(s) to prevent accidental starting.
2. Remove the hose from the fitting. Connect a hand-operated pressure/vacuum pump or syringe and hose to the fitting. For accurate test results, the hose must fit tightly to the check valve fitting. Clamp the hose to the fitting as needed.
3. Alternately apply pressure and vacuum to the check valve fitting. The check valve must allow air to flow in one direction and not the other (**Figure 90**). Replace the check valve fitting if it does not perform as described.
4. Repeat the test for the remaining check valves on the engine.
5. Connect the battery cables and spark plug leads.

Engine Noises

Some noise does occur in the power head during normal operation. A ticking noise or a heavy knocking noise that intensifies when under heavy load, such as during acceleration, is a reason for concern. A ticking noise commonly occurs when one more of the cylinders has failed. The noise is generally the result of a piece of piston ring that has embedded into the piston dome and is striking the cylinder head. Inspect the spark plug for damage or aluminum deposits (**Figure 103**) and perform a compression test as described in this chapter. Disassemble and inspect the power head if there is a problem with the compression or if there are aluminum deposits or physical damage on the spark plug(s).

CAUTION

Continued running with an unusual noise may cause increased damage that renders a power head repair impossible.

A whirring noise that is most pronounced as the throttle is decreased is usually related to a problem with the crankshaft and rod bearings. The bearing surface on the connecting rod and crankshaft may have a washboard-like surface that causes this noise. Improper storage preparation can lead to corrosion etched surfaces at the contact surfaces for the needle bearings (**Figure 104**). These surfaces may later develop into washboard-like surfaces. Only major repair to the power head corrects this condition.

Sometimes using a mechanics stethoscope can identify the cylinder that is creating the noise. Compare the noise emanating from one area of the power head with noise from the same area but different cylinder. The noise will be more pronounced when the stethoscope probe contacts the problem cylinder. The noise will be more pronounced when the stethoscope probe contacts the crankcase if the problem is related to the crankshaft or connecting rod. Special insulated pliers are available that allows spark plug lead removal while the engine is running. The noise may lessen when the spark plug lead is disconnected from the suspect cylinder. This procedure is difficult to perform successfully and may cause electrical system damage. The stethoscope method is far safer and effective for the amateur technician.

Lubrication Failure

Lack of sufficient lubrication is due to failure to properly add oil to the fuel on a premix model or failure of the automatic oil injection system. In either event, the engine will likely operate fine for a few minutes and then begin to slow down without moving the throttle. The engine may eventually stall and not crank over with the starter. On occasion, the engine will start after cooling down, but will likely slow down and stop again. Upon restarting, the engine will likely run very rough and stall at idle speed. Performance will be lacking as well. Continued operation will cause eventual failure, requiring extensive repairs to the power head.

If the engine is run without sufficient lubrication, perform a compression test as described in this chapter. The top cylinders generally suffer the most damage when oil is lacking or overheating has occurred. If the engine fails the compression test, disassemble and inspect the power head components. A bluish tinge is normally present on the crankshaft and connecting rod components if subjected to insufficient lubrication. The pistons and cylinder wall may develop scuffing (**Figure 105**) and scoring, particularly near the exhaust port areas. Crankshaft seals may also have a burned appearance and the crankshaft components may not have the normal oil film. Test the fuel for oil by taking a sample of the fuel from the carburetor bowl on premix models or vapor separator tank on EFI and HPDI models. Apply a small amount of fuel on a white piece of cardboard and let it dry in a safe area. If oil was present in the fuel, a blue stain of oil will appear after the fuel has evaporated. While oil maybe present in the fuel or in the oil reservoir, improper mixing can be the cause of poor lubrication. Improper mixing can result in excessive oil one minute and insufficient oil the next. Oil injected models should have the entire oil injection system tested and inspected if power head failure occurred due to poor lubrication.

Detonation

Detonation damage (**Figure 106**) is the result of heat and pressure in the combustion chamber becoming too great for the fuel that is used. Fuel normally burns at a controlled rate that produces the expanding gasses to drive the piston down on the power stroke. If conditions in the engine allow heat and pressure to get too high, the fuel may explode violently. These violent explosions cause serious damage to the piston and other power head compo-

106

107

nents. Carbon deposits, overheating, lean fuel conditions, over advanced timing and lugging are some of the conditions that lead to damaging detonation. The octane rating of the fuel used must meet or exceed the octane requirement for the engine. Never use a fuel with lower than the recommended octane rating. It may cause detonation under normal operating conditions.

During detonation, the combustion chamber temperature and pressure rise dramatically, creating a strong shock wave that is often referred to as a spark knock. Due to the normal noise generated by a two-stroke outboard engine, the noise may not be detectable while underway. If detonation occurs, the engine loses power and idles roughly or stalls. The engine may seize if the damage is great enough. A compression test will likely reveal one or more cylinders with significantly low compression. Inspect the spark plug(s) when removed for the compression test. Aluminum deposits (**Figure 103**) are a sure sign of serious piston damage. Complete power head repair is required if detonation has occurred. Refer to Chapter Seven.

Preignition

Preignition is the premature ignition of the air/fuel charge in the combustion chamber. Preignition is caused by hot spots in the combustion chamber. Wrong heat range spark plugs, carbon deposits and inadequate cooling are some of the causes of preignition. Preignition causes loss of power and eventually leads to serious damage to the piston and other power head components. Preignition may lead to detonation as the early ignition causes the heat and pressure in the combustion chamber to rise dramatically. Typically the piston suffers the brunt of the damage. Often times the preignition causes localized heating in the piston dome that eventually heats the material to the melting point, causing a hole to form in the piston dome (**Figure 107**). A compression test and spark plug inspection will likely reveal low compression and aluminum deposits consistent with detonation.

Engine Seizure

Although the power head can seize up at any operating speed, instantaneous seizure at high speed is uncommon. The condition causing the seizure will result in a gradual loss of power and engine speed. The most common causes of engine seizure is power head failure, are detonation or preignition. Always inspect the gearcase or jet pump before removing and disassembling the power head. Gearcase or jet pump failure can prevent the crankshaft from rotating creating the same symptoms as power head seizure. Inspect the gearcase lubricant for metal contamination (see Chapter Three) or check the jet pump for debris lodged in the impeller. If necessary, remove the gearcase (Chapter Eight) or jet pump (Chapter Nine) and check for flywheel rotation. The power head must be disassembled and inspected if the flywheel cannot be rotated with the gearcase or jet pump removed.

Water Entering the Cylinder

Water can enter the cylinder from a number of areas. Water in the fuel, water entering from the carburetor or throttle body, leaking exhaust plate/cover and gaskets, leaking cylinder head gasket and internal cylinder block flaws are some of the common causes. The typical symptom of water in a cylinder is rough running, particularly at idle. Water intrusion is normally discovered when the spark plug(s) are removed for inspection. Remove the cylinder head and exhaust cover if there is water or a white deposit on any of the spark plugs. The wet cylinder usually has significantly less carbon deposits on the cylinder head and piston dome. The steam formed during combustion removes carbon deposits. Continued operating with water entering the cylinder eventually causes power head failure. Inspect the cylinder head gasket, cylinder head, exhaust plate and exhaust cover gaskets. A black or white deposit over a sealing surface or physical damage to the gasket indicates the point of leakage. Internal leakage due to a cylinder flaw can be difficult if not impossible to locate. Small cracks, pin holes and voids in the material may not be visually apparent and may not leak until the engine reaches normal operating temperature. Cylinder block replacement is suggested if water is entering the cylinder and the point of entry is not located. Refer to Chapter Eight for power head repair procedures.

COOLING SYSTEM

A water pump (**Figure 108**, typical) is mounted on the gearcase or jet pump. The pump supplies water to the exhaust area of the power head first, then to the cylinder head and cylinder block. The water exits the power head near the power head mounting surface and travels out through the drive shaft housing. As the water travels through the power head, it absorbs heat and carries it away. If the engine is experiencing overheating, the water is not flowing through the power head with sufficient volume or is not absorbing the heat adequately.

All models covered in this manual are equipped with a thermostat (**Figure 109**) to help maintain a minimum power head temperature. They work by restricting exiting water until a minimum temperature is reached.

A stream of water (**Figure 110**) will appear at the rear of the lower engine cover when the water is exiting the power head. The fitting for the stream will commonly become blocked with debris and cease flowing. Clean the opening with small stiff wire. Inspect the cooling system if the water stream does not appear after cleaning the fitting.

Engine Temperature Verification

If overheating is suspected, verify the engine temperature using Thermomelt sticks (**Figure 111**). They are made of material formulated to melt at a specified temperature. Rub the sticks on the cylinder head near the overheat switch (see Chapter Six). On engines not equipped with an alarm system, make a mark or hold the stick in contact with the cylinder head near the spark plug opening. The marking or stick melts if the surface reaches the temperature listed on the stick. Try to check the temperature during or immediately after the suspected overheating occurs. Use different temperature sticks to determine the approximate engine temperature. Stop the engine if the temperature exceeds 90° C (195° F) to avoid damaging the power head.

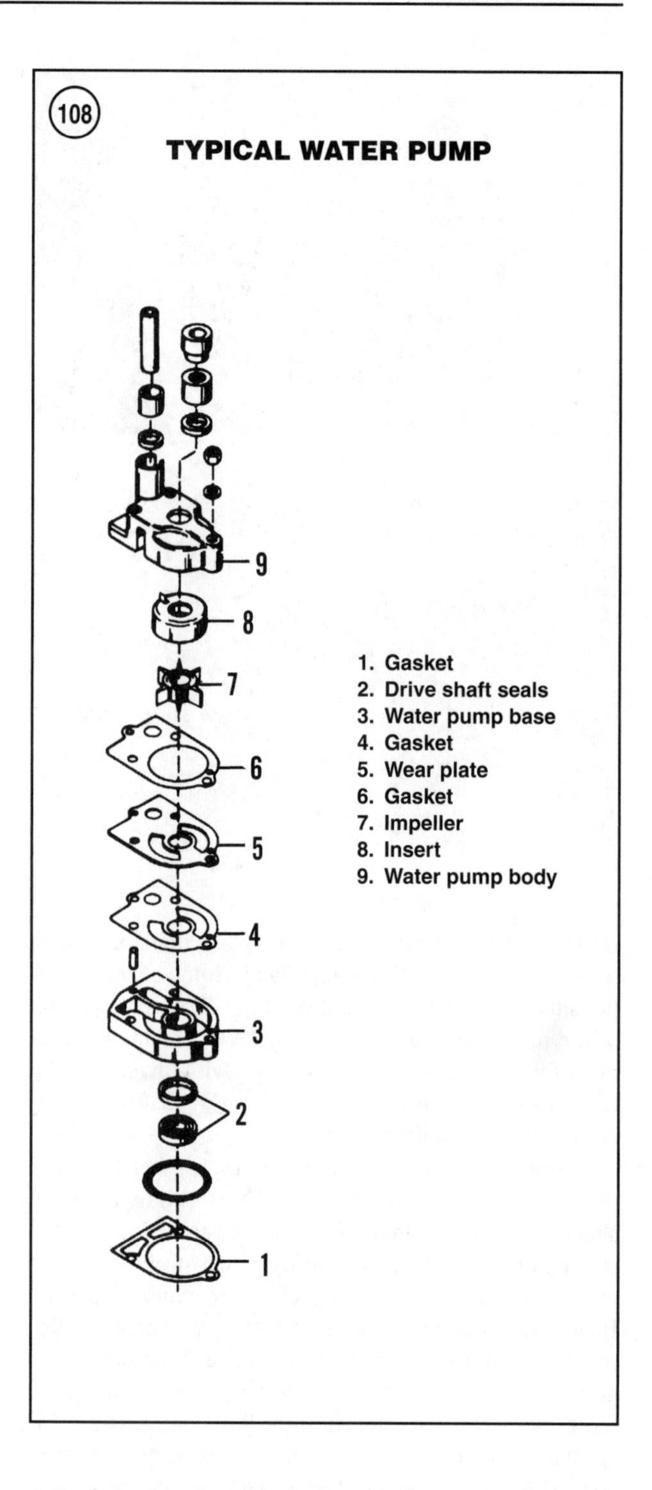

Perform this test with the boat in the water and running under actual conditions. Operating the engine unloaded and in neutral gear may not generate enough heat to test the cooling system. Do not troubleshoot the cooling system while running the engine using a flush/test adapter. The volume of water flowing through the hose does not equate to the water supplied under normal conditions by the water pump. Troubleshooting with a flush/test adapter

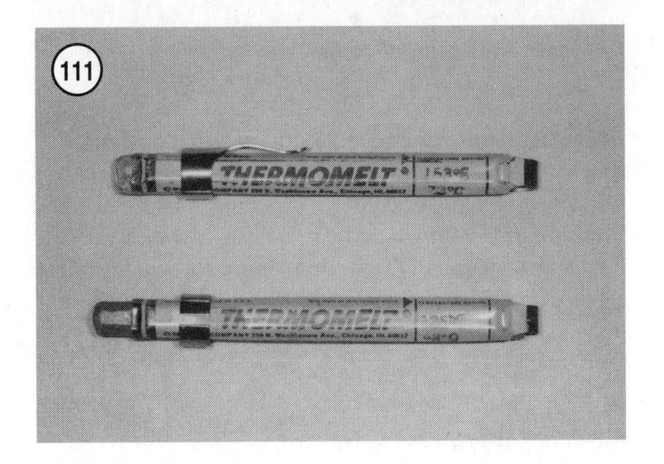

may also mask a problem or cause overheating at high engine speeds.

Cooling System Inspection

If overheating is indicated by the temperature gauge, the water stream is missing or the overheat alarm sounds,

inspect the water pump. Water pump disassembly, inspection and assembly is described in Chapter Eight.

Inspect and test the thermostat if overheating is occurring and the water pump checks satisfactorily. Refer to *Thermostat Test* in this chapter.

Remove the water jacket covers and inspect the cooling system passages for debris or deposits if the engine is overheating and there is no problem with the water pump or thermostat. Rocks, pieces of a failed water pump impeller, sand, shells and other debris may restrict water flow.

Salt, calcium or other deposits can form in the cooling passages, restrict water flow and prevent efficient heat transfer. Use a cleaner specifically designed to remove cooling system deposits. Make sure the cleaner is suitable for use on aluminum material. Always follow the manufacturer's instructions when using one of these products. These cleaners are usually available at marine specialty stores. Water jacket, thermostat and exhaust cover removal and installation are described in Chapter Seven.

Thermostat Test

Test the thermostat if the engine is overheating or is unable to reach normal operating temperature. Use a liquid thermometer, piece of string and a container of water that can be heated for this procedure.

1. Remove the thermostat as described in Chapter Seven.
2. Tie the string to the thermostat and suspend the thermostat in the container of water (**Figure 112**).
3. Place the thermometer in the water and begin heating the water. Observe the thermostat and thermometer.

4. Note the temperature when the thermostat begins to open at the point indicated in **Figure 113**. This should occur at 48-52° C (118-126° F).
5. Observe the thermostat and thermometer while continuing to heat the water. The thermostat should be fully open at approximately 60° C (140° F).
6. Allow the container of water to completely cool. Then remove the thermostat. The thermostat must completely close when cooled to room temperature.
7. Replace the thermostat if it fails to close at room temperature or opens above or below the specified temperature.
8. Install the thermostat as described in Chapter Seven.

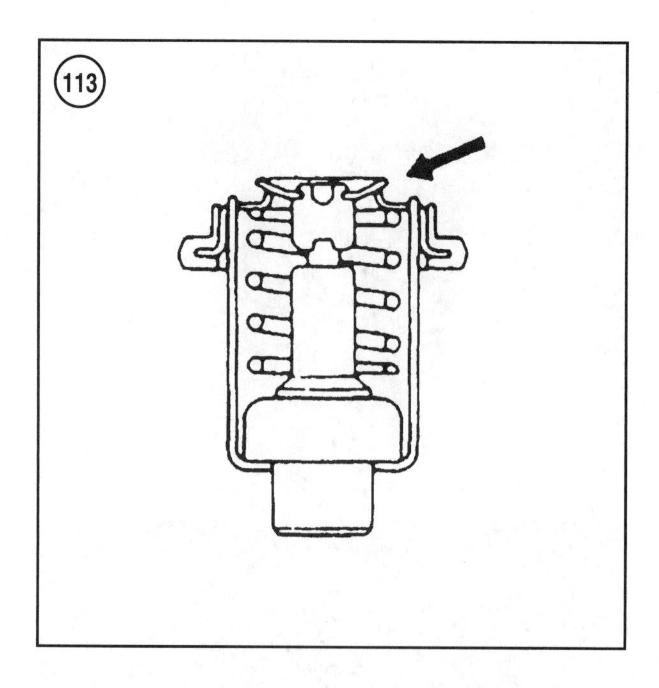

GEARCASE

Problems with the gearcase are generally related to leakage and failed internal components. Signs of trouble include unusual noise, shifting difficulty or slipping under load.

At the first sign of trouble, check the gearcase lubricant level and condition as described in Chapter Three. If the lubricant level is low or there is water in the lubricant, pressure test the gearcase as described in Chapter Eight. If there is an excessive amount or larger metal particles on the gearcase drain plug (**Figure 114**), disassemble and inspect the gearcase as described in Chapter Eight.

Water in the Gearcase

Under certain conditions, a small amount of water may be present in the gearcase lubricant. The conditions would involve a gearcase that has not received normal maintenance for several years and has been stored with the gearcase in the water. Only a pressure test will determine if a faulty seal, gasket, O-ring or other components is allowing water into the gearcase. Gearcase pressure testing is described in Chapter Eight. Failure to correct the leakage eventually leads to extensive damage to internal gearcase components.

Lubricant Leakage

The presence of lubricant on the exterior surface or dripping below the gearcase indicates leakage. Pressure test the gearcase (Chapter Eight). Failure to correct the leakage causes eventual failure of the gears and bearings due to lack of sufficient lubrication. Under certain conditions, the gearcase may vent lubricant through the shift shaft seal. This occurs if the gearcase is overfilled with lubricant and exposed to intense sunlight for an extended period of time. Overfilling prevents the formation of the air pocket at the top of the drive shaft bore. The fluid pressure escapes past the seal due to heating by the sunlight and the lack of the air pocket for expansion. If there is leakage, pressure test the gearcase as described in Chapter Eight and refill the gearcase to the proper level as described in Chapter Three.

Gearcase Vibration or Noise

Gearcase noise does occur during normal operation. A rough growling noise or a high-pitched whine is reason for concern. On occasion, a knocking noise emanates from the gearcase, which may indicate power head failure. An inspection of the gearcase lubricant will in almost all cases reveal metal contamination if the gearcase components have failed.

Knocking or grinding noise

If a knocking or grinding noise is coming from the gearcase, the cause is likely from damaged gears or other components in the gearcase. Inspecting the gearcase lubricant should reveal large metal particles. The gears may have suffered damage from the propeller impacting an underwater object or from shifting into gear at high engine speed.

114

High-pitched whine

A high-pitched whine normally indicates a bearing problem or the gears are running out of alignment. The only way to verify a bearing problem is to disassemble the gearcase and inspect the components. On occasion, a high-pitched whine is created by the propeller during normal operation. Try a different but suitable propeller to check for this condition. On high speed applications, a damaged propeller, gearcase housing or damaged hull can generate a hydro-dynamic vibration that causes a whine noise. Repair damaged components or surfaces to correct this condition.

Vibration

Vibration in the engine can and commonly does originate in the gearcase and is often due to a bent propeller shaft or damaged propeller. A propeller can appear perfect but really be out of balance. The best way to determine this is to have the propeller trued and balanced at a reputable propeller repair shop or simply try a different yet suitable propeller.

Always check for a bent propeller shaft if a vibration occurs. A bent propeller shaft is usually from an impact with an underwater object and other damage in the gearcase. Inspection of the propeller shaft and other components is described in Chapter Eight.

Never run an engine with severe vibration. A large amount of stress is placed on gears, bearings and other gearcase components. Excessive vibration can compromise the durability of the entire engine.

Slipping Under Load

Slipping under load is usually the result of failure of the propeller drive hub. The hub is designed to cushion the shifting action and absorb minor impact damage. If the propeller hub has spun in the bore, the engine rpm will increases, yet the boat will not increase speed. In most cases the boat will not accelerate to planing speed. This may lead the operator to believe a problem exists in the gearcase.

To check for hub slippage, make a reference marking on the propeller shaft that aligns with a marking made on the propeller. Run the engine until the slippage occurs. Stop the engine and inspect the reference marking alignment. Have a new hub installed by a reputable propeller repair shop if the markings are not aligned after running the engine.

Shifting Difficulty

Hard shifting or difficult gear engagement is usually the result of improperly adjusted shift linkages or cables. If shifting difficulty occurs, adjust the linkages and cables as described in Chapter Four. Disassemble and inspect the shifting components in the gearcase if shifting problems cannot be corrected with proper adjustment. Gearcase disassembly and inspection are described in Chapter Eight.

Table 1 STARTING SYSTEM TROUBLESHOOTING

Symptom	Possible causes	Corrective action
Electric starter does not energize	Engine not in neutral gear	Shift into neutral gear
	Weak or discharged battery	Test or charge the battery (Chapter Six)
	Dirty or corroded battery terminals	Thoroughly clean battery terminals
	Faulty neutral safety switch	Test neutral safety switch
	Faulty ignition key switch	Test ignition key switch
	Faulty starter relay	Test starter relay
	Faulty starter solenoid (HPDI)	Test starter solenoid

(continued)

Table 1 STARTING SYSTEM TROUBLESHOOTING (continued)

Symptom	Possible causes	Corrective action
Electric starter does not energize (continued)	Loose or dirty wire connection	Clean and tighten starter wire connections
	Faulty electric starter	Repair the electric starter
Electric starter engages flywheel (flywheel rotates slowly)	Weak or discharged battery	Test or charge the battery (Chapter Six)
	Dirty or corroded battery terminals	Thoroughly clean battery terminals
	Loose or dirty wire connections	Clean and tighten starter wire connections
	Engine is in gear	Correct for improper shift adjustment
	Faulty electric starter	Repair the electric starter
	Internal power head damage	Inspect power head for damage
	Internal gearcase damage	Inspect gearcase lubricant for debris
Electric starter engages (flywheel does not rotate)	Weak or discharged battery	Test or charge the battery (Chapter Six)
	Dirty or corroded battery terminals	Thoroughly clean battery terminals
	Loose or dirty wire connections	Clean and tighten starter wire connections
	Engine is in gear	Correct improper shift adjustment
	Water in the cylinder(s)	Inspect spark plug(s) for water contamination
	Damaged starter drive gear	Inspect starter drive gear
	Damaged flywheel gear teeth	Inspect flywheel gear teeth
	Faulty electric starter	Repair the electric starter
	Seized power head	Inspect power head for damage
	Seized gearcase	Inspect gearcase lubricant for debris
Electric starter (Noisy operation)	Dirty or dry starter drive gear	Clean and lubricate starter drive gear
	Damaged starter drive gear teeth	Inspect starter drive gear
	Damaged or corroded flywheel gear	Inspect flywheel drive gear teeth
	Loose starter mounting bolt(s)	Tighten starter mounting bolt(s)
	Worn or dry starter bushing(s)	Repair electric starter (Chapter Six)

Table 2 FUEL SYSTEM TROUBLESHOOTING

Symptom	Possible causes	Corrective action
Engine does not start	Closed fuel tank vent	Open the fuel tank vent
	Old or contaminated fuel	Provide the engine with fresh fuel
	Disconnected fuel hose	Connect fuel hose
	Faulty primer bulb	Check the primer bulb
	Choke valve not operating	Check choke valve operation
	Faulty electric fuel pump	Check for fuel pump operation
	Faulty belt driven pump (HPDI)	Check the belt driven pump pressure
	Air or fuel leaks in hose fittings	Inspect hose fittings for leakage
	Blocked fuel filter	Inspect the filter for contaminants
	Stuck carburetor inlet needle	Repair the carburetor(s) (Chapter Five)
	Improper float level adjustment	Repair the carburetor(s)
	Blocked carburetor passages	Repair the carburetor(s)
	Faulty diaphragm type fuel pump	Inspect fuel pump (Chapter Five)
	Injectors not operating	Check for fuel injector operation
	Faulty fuel pressure regulator	Test the fuel pressure regulator
	Faulty throttle position sensor	Test the throttle position sensor
	Faulty air pressure sensor	Test the air pressure sensor
	Faulty air temperature sensor	Test the air temperature sensor
	Faulty engine temperature sensor	Test the engine temperature sensor
	Faulty injector driver circuit	Measure the driver circuit output

(continued)

Table 2 FUEL SYSTEM TROUBLESHOOTING (continued)

Symptom	Possible causes	Corrective action
Stalls or runs rough at idle	Old or contaminated fuel	Provide the engine with fresh fuel
	Improper idle speed adjustment	Adjust idle speed
	Closed or blocked fuel tank vent	Open or clear vent
	Blocked carburetor passages	Repair carburetor(s) (Chapter Five)
	Flooding carburetor	Check for carburetor flooding
	Faulty primer bulb	Check the primer bulb
	Air or fuel leaks at hose fittings	Inspect hose fittings for leakage
	Blocked fuel filter	Inspect the filter for contaminants
	Faulty diaphragm type fuel pump	Inspect fuel pump (Chapter Five)
	Faulty electric fuel pump	Check for fuel pump operation
	Faulty belt driven fuel pump (HPDI)	Check the belt driven pump pressure
	Worn or damaged reed valve	Check reed valves
	Faulty throttle position sensor	Test the throttle position sensor
	Faulty air pressure sensor	Test the air pressure sensor
	Faulty air temperature sensor	Test the air temperature sensor
	Faulty engine temperature sensor	Test the engine temperature sensor
	Faulty recirculation system component	Test the check valves and hoses
	Low crankcase pressure	Check for crankcase leak
	Sticking choke valve	Check choke valve operation
	Low crankcase pressure	Check for leaking crankcase
	Faulty shift cutout switch	Test the shift cutout switch
	Faulty shift position switch (HPDI)	Test the shift position switch
Idle speed too high	Improper idle speed adjustment	Adjust idle speed (Chapter Four)
	Improper throttle linkage adjustment	Adjust throttle linkages (Chapter Four)
	Binding throttle linkage(s)	Check linkage(s)
	Faulty throttle position sensor	Test the throttle position sensor
	Faulty air pressure sensor	Test the air pressure sensor
	Faulty air temperature sensor	Test the air temperature sensor
	Faulty engine temperature sensor	Test the engine temperature sensor
Hesitation during acceleration	Old or contaminated fuel	Supply the engine with fresh fuel
	Faulty fuel pump	Inspect the fuel pump
	Sticking choke valve	Check choke valve operation
	Blocked carburetor passages	Repair carburetor(s)
	Blocked fuel filter	Inspect the filter(s) for contaminants
	Air or fuel leaks at hose fittings	Inspect hose fittings for leakage
	Closed or blocked fuel tank vent	Open or clear tank vent
	Flooding carburetor	Check for carburetor flooding
	Faulty throttle position sensor	Test the throttle position sensor
	Faulty air pressure sensor	Test the air pressure sensor
	Faulty air temperature sensor	Test the air temperature sensor
	Faulty engine temperature sensor	Test the engine temperature sensor
	Faulty recirculation system component	Test the check valves and hoses
Misfire or poor high speed performance	Old or contaminated fuel	Supply the engine with fresh fuel
	Faulty diaphragm type fuel pump	Inspect the fuel pump
	Sticking choke valve	Check choke valve operation
	Faulty primer bulb	Test the primer bulb (Chapter Five)
	Misadjusted throttle linkages	Adjust throttle linkages (Chapter Four)
	Blocked carburetor passages	Repair carburetor(s)
	Blocked fuel filter(s)	Inspect the filter(s) for contaminants
	Air or fuel leaks at hose fittings	Inspect hose fittings for leakage

(continued)

Table 2 FUEL SYSTEM TROUBLESHOOTING (continued)

Symptom	Possible causes	Corrective action
Misfire or poor high speed performance (continued)	Closed or blocked fuel tank vent	Open or clear vent
	Faulty electric fuel pump	Check for fuel pump operation
	Faulty belt driven fuel pump (HPDI)	Check the belt driven pump pressure
	Faulty shift cutout switch	Test the shift cutout switch
	Faulty shift position switch (HPDI)	Test the shift position switch
Excessive exhaust smoke	Flooding carburetor	Check for carburetor flooding
	Blocked carburetor passages	Repair carburetor(s)
	Improper float level	Repair carburetor(s)
	Sticking choke valve	Check choke valve operation
	Faulty diaphragm type fuel pump	Check the fuel pump (Chapter Five)
	Faulty throttle position sensor	Test the throttle position sensor
	Faulty air pressure sensor	Test the air pressure sensor
	Faulty air temperature sensor	Test the air temperature sensor
	Faulty engine temperature sensor	Test the engine temperature sensor
	Faulty recirculation system component	Test the check valves and hoses

Table 3 IGNITION SYSTEM TROUBLESHOOTING

Symptom	Possible causes	Corrective action
Engine does not start	Lanyard switch activated	Check lanyard switch
	Fouled spark plug(s)	Check or replace spark plug(s)
	Improper ignition timing adjustment	Adjust the ignition timing (Chapter Four)
	Faulty system relay	Test system relay
	Faulty spark plug lead	Check for arcing spark plug lead
	Faulty spark plug cap	Test ignition coil and cap
	Shorted stop circuit	Test for shorted stop circuit
	Faulty ignition charge coil	Test ignition charge coil
	Faulty pulser coil	Test pulser coil
	Faulty crankshaft position sensor(s)	Test crankshaft position sensor(s)
	Faulty ignition coil	Test ignition coil
	Faulty CDI or engine control unit	Check CDI or engine control unit
	Faulty shift cutout switch	Test the shift cutout switch
Stalls or runs rough at idle	Fouled spark plug(s)	Check or replace spark plug(s)
	Improper ignition timing adjustment	Adjust the ignition timing (Chapter Four)
	Faulty spark plug cap	Test ignition coil and cap
	Faulty spark plug lead	Check for arcing spark plug lead
	Partially shorted stop circuit	Test for shorted stop circuit
	Faulty ignition charge coil	Test ignition charge coil
	Faulty pulser coil	Test pulser coil
	Faulty crankshaft position sensor(s)	Test crankshaft position sensor(s)
	Faulty ignition coil	Test ignition coil
	Faulty CDI or engine control unit	Check CDI or engine control unit
Idle speed too high	Improper ignition timing adjustment	Adjust the ignition timing (Chapter Four)
	Faulty pulser coil	Test pulser coil
	Faulty crankshaft position sensor	Test crankshaft position sensor
	Faulty CDI or engine control unit	Check low speed timing (Chapter Four)

(continued)

Table 3 IGNITION SYSTEM TROUBLESHOOTING (continued)

Misfire or poor high speed performance	Engine reaching rev limit	Check full speed engine rpm
	Fouled spark plug(s)	Check or replace spark plug(s)
	Improper ignition timing adjustment	Adjust the ignition timing (Chapter Four)
	Faulty spark plug lead	Check for arcing spark plug lead
	Faulty spark plug cap	Test ignition coil and cap
	Partially shorted stop circuit	Test for shorted stop circuit
	Faulty ignition charge coil	Test ignition charge coil
	Faulty pulser coil	Test pulser coil
	Faulty crankshaft position sensor(s)	Test crankshaft position sensor(s)
	Faulty ignition coil	Test ignition coil
	Faulty CDI or engine control unit	Check CDI or engine control unit
Engine will not stop	Faulty stop circuit	Test stop circuit
	Faulty system relay	Test the system relay

Table 4 CHARGING SYSTEM CAPACITY

Model	Maximum output (amps)
80 Jet and 115-130 hp (except C115)	20
C115	15
105 Jet and 150-200 hp (except EFI and HPDI models)	25
150-250 hp (EFI models)	35
150-250 hp (HPDI models	45

Table 5 BATTERY CHARGE COIL PEAK OUTPUT SPECIFICATIONS

Model	Cranking speed (volts)	1500 rpm (volts)	3500 rpm (volts)
80 Jet and 115-130 hp (except C115)	7.0	35.0	85.0
C115	6.0	30.0	75.0
105 Jet and 150-200 hp (except EFI and HPDI)			
All models except C150	5.5	35.0	85.0
C150	3.0	20.0	50.0
150-250 hp (EFI models)	*	14.0	14.0
150-250 hp (HPDI models)	7.5-8.0	12.0	12.0

*The battery charge coil output specification is not provided at cranking speed.

Table 6 RECTIFIER/REGULATOR PEAK OUTPUT SPECIFICATIONS

Model	Cranking speed (volts)	1500 rpm (volts)	3500 rpm (volts)
150-250 hp EFI models	*	12.0	12.0
150-250 hp HPDI models	7.5	12.7	12.7

* The rectifier/regulator output specification is not provided at cranking speed.

Table 7 DIAGNOSTIC CODES

Code	Reason	Corrective action
1	Normal condition	None required, no faults detected
12	Ignition charge coil failure	Test ignition charge coil output
13	Pulser coil failure	Test pulser coil output
14	Crankshaft position sensor failure	Test crankshaft position sensor output
15	Engine temperature sensor failure	Test the engine temperature sensor
17	Faulty knock sensor signal	Test the knock sensor
18	Throttle position sensor failure	Test the throttle position sensor
19	Low battery voltage	Test the battery and charging system
22	Faulty air pressure sensor	Test the air pressure sensor
23	Faulty air temperature sensor	Test the air temperature sensor
25	Incorrect high pressure fuel reading	Test the fuel pressure
26	Fuel injector signal faulty	Check for fuel injector operation
27	Water detected in the fuel	Check for water is the filter
28	Shift position switch fault	Test the shift position switch
32	Shift switch is operating	Normal during shifting
33	Ignition timing advanced for warm-up	None required, normal condition
35	Adding fuel to reduce knocking	Test the knock sensor
41	Engine is overspeeding	Reduce engine speed
42	Overheating or low oil level	Correct overheating and oil level
43	Warning horn sounding	Verify horn operation
44	Engine stop switch activated	Test the stop circuit

Table 8 IGNITION CHARGE COIL PEAK OUTPUT SPECIFICATION

Model	Cranking speed (volts)	1500 rpm (volts)	3500 rpm (volts)
80 Jet and 115-130 hp (except C115)			
Red and brown wires	160	165	170
Black/red and blue wires	451	65	170
C115			
Red and brown wires	95	160	160
Black/red and blue wires	30	160	160
105 Jet and 150-200 hp (except EFI and HPDI)			
Oil injected models			
Red and brown wires	160	165	165
Black/red and blue wires	55	165	165
Premix models (C150)			
Red and brown wires	90	165	165
Black/red and blue wires	30	160	170
150-250 hp (EFI models)			
Red and brown wires	85	150	150
Black/red and blue wires	85	150	150

Table 9 PULSER COIL PEAK OUTPUT SPECIFICATION

Model	Cranking speed (volts)	1500 rpm (volts)	3500 rpm (volts)
80 Jet and 115-130 hp			
Oil injected models			
White/red and white/yellow wires	2.5	7.0	11.0
White/black and white/green wires	2.5	7.0	11.0

(continued)

Table 9 PULSER COIL PEAK OUTPUT SPECIFICATION (continued)

Model	Cranking speed (volts)	1500 rpm (volts)	3500 rpm (volts)
80 Jet and 115-130 hp (continued)			
Premix models			
White/red and white/yellow wires	2.5	8.0	12.0
White/black and white/green wires	2.5	8.0	12.0
105 Jet and 150-200 hp (carburetor equipped)			
Oil injected models			
White/red and white/green wires	2.0	8.0	14.0
White/yellow and white/brown wires	2.0	8.0	14.0
Premix models			
White/red and white/green wires	2.0	9.5	16.0
White/black and white/blue wires	2.0	9.5	16.0
White/yellow and white/brown wires	2.0	9.5	16.0
150-250 hp (EFI models)			
White/red and black wires	3.0	16.0	30.0
White/black and black wires	3.0	16.0	30.0
White/yellow and black wires	3.0	16.0	30.0
White/green and black wires	3.0	16.0	30.0
White/blue and black wires	3.0	16.0	30.0
White/brown and black wires	3.0	16.0	30.0
150-250 hp (HPDI models)			
White/red and black wires	5.0	20.0	35.0
White/black and black wires	5.0	20.0	35.0
White/yellow and black wires	5.0	20.0	35.0
White/green and black wires	5.0	20.0	35.0
White/blue and black wires	5.0	20.0	35.0
White/brown and black wires	5.0	20.0	35.0

Table 10 CRANKSHAFT POSITION SENSOR PEAK OUTPUT SPECIFICATIONS

Model	Cranking speed (volts)	1500 rpm (volts)	3500 rpm (volts)
Carburetor equipped models	2.0	5.5	6.0
EFI models	0.5	3.0	4.0
HPDI models	4.0	13.0	20.0

Table 11 FULL THROTTLE ENGINE SPEED RECOMMENDATIONS

Model	Engine speed (rpm)
80 Jet and 115 hp	4500-5500
130 hp	5000-5000
150-250 hp (except P200)	4500-5500
P200	5500-6000

Table 12 CDI UNIT AND ENGINE CONTROL UNIT OUTPUT SPECIFICATIONS

Model	Cranking speed (volts)	1500 rpm (volts)	3500 rpm (volts)
80 Jet and 115-130 hp (except C115)			
Coil black wire and black/white wire	125	140	145
C115			
Coil black wire and black/white wire	85	140	135
105 Jet and 150-200 hp (carburetor equipped)			
Oil injected models			
Coil black wire and black/white wire			
Cylinders No. 1, 3, 4 and 6	130	145	145
Cylinders No. 2 and 5	0	0	145
Premix models			
Coil black wire and black/white wire	65	144	135
150-250 hp (EFI models)			
Coil black wire and black/white wire	100	150	130
Coil black wire and black/green wire	100	150	130
Coil black wire and black/brown wire	100	150	130
Coil black wire and black/orange wire	100	150	130
Coil black wire and black/yellow wire	100	150	130
Coil black wire and black/blue wire	100	150	130
150-250 hp (HPDI models)			
Black/white and red/yellow wire	140	205	220
Black/green and red/yellow wire	140	205	220
Black/brown and red/yellow wire	140	205	220
Black/orange and red/yellow wire	140	205	220
Black/yellow and red/yellow wire	140	205	220
Black/blue and red/yellow wire	140	205	220

Table 13 ENGINE TEMPERATURE SENSOR SPECIFICATIONS

Model	Test specification (ohms)
105 Jet and 150-200 hp (carburetor equipped)	
At 20° C (68° F)	54,200-69,000
At 100° C (212° F)	3100-3500
150-250 hp (EFI and HPDI models)	
At 5° C (41° F)	128,000
At 20° C (68° F)	54,000-69,000
At 100° C (212° F)	3020-3480

Table 14 INJECTOR DRIVER OUTPUT SPECIFICATIONS

Wire color	Cranking speed (volts)	1500 rpm (volts)	3500 rpm (volts)
Orange/red to purple/red	60	65	65
Orange/black to purple/black	60	65	65
Orange/yellow to purple/yellow	60	65	65
Orange/green to purple/green	65	65	65
Orange/blue to purple/blue	60	65	65
Orange/white to purple/white	60	65	65

Table 15 OIL PUMP OUTPUT SPECIFICATIONS

Model	Output at 1500 rpm
80 Jet and 115 hp	2.5-3.9 cc
130 hp	4.7-4.9 cc
105 Jet and 150-175 hp (carburetor equipped)	2.7-4.1 cc
200 hp (carburetor equipped)	3.7-5.9 cc
150-175 hp (EFI models)	16.2-24.6 cc
200 hp (EFI models [2.6 liter])	22.2-35.4 cc
200-250 hp (EFI models [3.1 liter])	31.8-47.4 cc

Chapter Three

Lubrication, Maintenance and Tune-up

Performing regular maintenance and tune-ups is the key to getting the most out of your outboard. This chapter describes lubrication, maintenance and tune-up procedures for the Yamaha outboards covered in this manual. Engine break-in procedures are described at the end of this chapter. All specifications are listed in **Tables 1-5** located at the end of this chapter.

LUBRICATION

Lubrication is the most important maintenance item for any outboard. An outboard simply will not operate for any length of time without lubrication. Lubricant for the power head, gearcase and other components prevents excessive wear, guards against corrosion and provides smooth operation of turning and sliding surfaces (tilt tubes, swivel brackets and control linkages). Power head, gearcase and jet pump lubrication procedures are described in this chapter.

Power Head Lubrication

Two methods are used to lubricate the internal components of a two-stroke Yamaha outboard: premix and oil injection. Both methods introduce the lubricant into the internal power head components. The lubricant provides protection as it passes through the engine, and is eventually burned with the fuel during the combustion process. Either system is only as reliable as the person operating the outboard. Become familiar with proper fuel and oil mixing on premix models. Become familiar with proper usage of Precision Blend oil injection and warning systems. Chapter Two describes warning system operation.

The engine may require additional oil during new engine break-in or after a power head repair. Refer to the following guidelines:

115-150 hp premix models—The engine must be provided with double the normal fuel/oil mixture. Mix the oil into the fuel at double the normal rate during the break-in period.

With Precision Blend oil injection (except HPDI)—The engine must be provided with double the normal fuel/oil mixture. During the break-in period, a 50:1 fuel/oil mixture must be used in the fuel tank in addition to the oil provided by the oil injection system.

HPDI models—Do not add oil into the fuel. On these models, the fuel is injected directly into the combustion chamber, and any oil mixed into the fuel will not lubricate the internal power head components. The Precision Blend oil injection system injects oil directly into the crankcase

1

2

to lubricate the internal components. In addition, a small amount of oil is injected into the fuel in the vapor separator tank to lubricate the high pressure fuel injectors and other fuel system components.

Engine Oil Recommendation

Yamaha recommends using Yamalube Two-Cycle Outboard Oil. This oil meets or exceeds the TCW-3 standards set by the NMMA (National Marine Manufactures Association). If Yamalube is not available, use a major engine manufacturer's oil that meets or exceeds the NMMA TC-W3 specification. Look for the NMMA logo on the container (**Figure 1**). If at all possible, avoid mixing different brands as some oil may form a gel-like deposit in the oil reservoir on oil injected models.

CAUTION

Never use oil that is not specified for two-stroke water cooled engines. Do not substitute automotive motor oil. It does not provide adequate lubrication for the internal engine components. Operating the engine without adequate lubrication causes serious power head damage or engine seizure.

Premix Models

The oil must be mixed into the fuel for models with a premix system. The fuel/oil passes through the entire fuel system and into the engine for burning. Some of the oil collects on internal components and provides protection. Residual oil collects in certain areas of the power head where it is moved by the recirculation system to other areas of the power head. The oil lubricates bearings and other components as it moves through the various passages and eventually passes into the combustion chamber for burning.

Using too little oil can result in insufficient lubrication and possible power head damage. Using too much oil does not decrease wear in the engine but it may cause spark plug fouling, excessive smoke, lean fuel conditions and increased combustion chamber deposits. Lean fuel conditions and increased combustion chamber deposits can lead to preignition or detonation (**Figure 2**). All 115-150 hp premix models use a 50:1 fuel/oil ratio for all conditions after break-in.

Thorough mixing of the oil and fuel is crucial for correct engine operation. If the oil is simply poured into the fuel, it will likely settle onto the bottom of the tank. The engine may operate with a very rich oil mixture one-minute and a lean oil mixture the next. Erratic engine running will persist until the oil mixes with the fuel from wave action or movement of the vessel.

Mixing the fuel and oil

First determine the amount of fuel that is needed. Then determine the amount of oil for the fuel. **Table 4** lists the amount of oil needed for in both the 50:1 and 25:1 ratios.

NOTE

Mix fuel and oil only when it will be used within a few weeks. Fuel has a relatively short shelf life. Fuel begins to deteriorate in as little as 14 days. If let to sit for an extended period of time, the oil may settle onto the bottom of the tank causing inconsistent fuel/oil delivery to the engine.

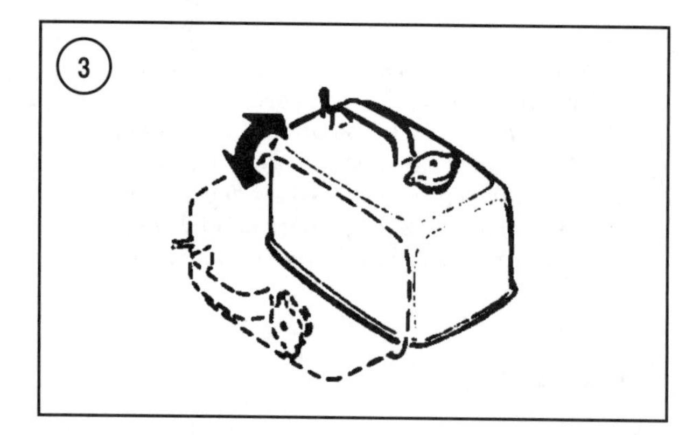

If using a portable fuel tank, pour in the correct amount of oil for the fuel needed and half of the fuel. Install the fill cap and shut the vent or valve. Disconnect the fuel line if the tank is equipped with an automatic fuel tank vent. Disconnecting the fuel line closes the vent on most tanks with this feature. Carefully tip the tank to one side and back to the upright position (**Figure 3**) several times to mix the fuel. Remove the fill cap, then add the other half of the fuel.

If the boat is equipped with a built-in fuel tank (**Figure 4**, typical), mix the required amount of oil for the fuel into a suitable container with 3.8 L (1 gal.) of fuel. Carefully mix the oil into the fuel. Insert a large metal funnel into the filler neck and pour the mixture in as the fuel is added to the tank (**Figure 5**).

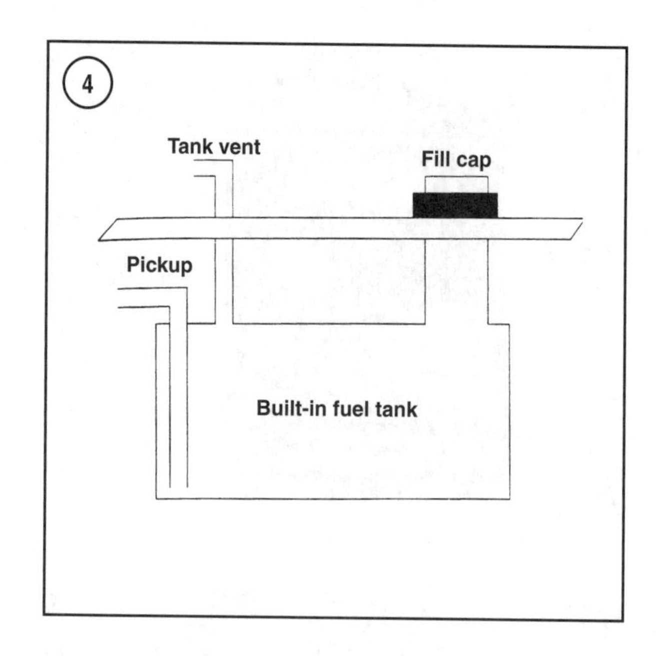

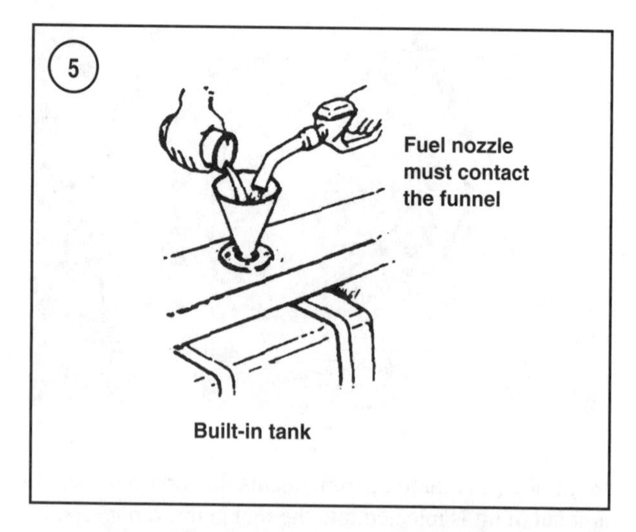

Oil Injected Models

Oil injection eliminates the need to mix fuel and oil after break-in. A gear-driven variable rate oil pump (**Figure 6**) delivers the correct amount of oil to the engine at all speeds. **Figure 7** shows a diagram of the typical oil injection system. The oil is injected directly into the engine at the intake manifold. Air and fuel flowing into the engine disperse the oil throughout the power head. Oil is never introduced into the carburetors to reduce the formation of deposits in the carburetors during extended storage. However, oil is injected into the vapor separating tank on EFI and HPDI models to lubricate the fuel injectors and electric fuel pump.

A warning system continuously monitors the oil level in the engine mounted oil reservoir (10, **Figure 7**) using the oil level sensor (**Figure 8**). As oil is consumed, the sensor mounted float drops until it activates the oil reservoir refill switch in the sensor assembly. The switch signals the microcomputer to switch on the electric oil pump in the onboard mounted oil reservoir (11, **Figure 7**). Oil is then pumped from the onboard reservoir until the engine

7

OIL INJECTION SYSTEM (TYPICAL)

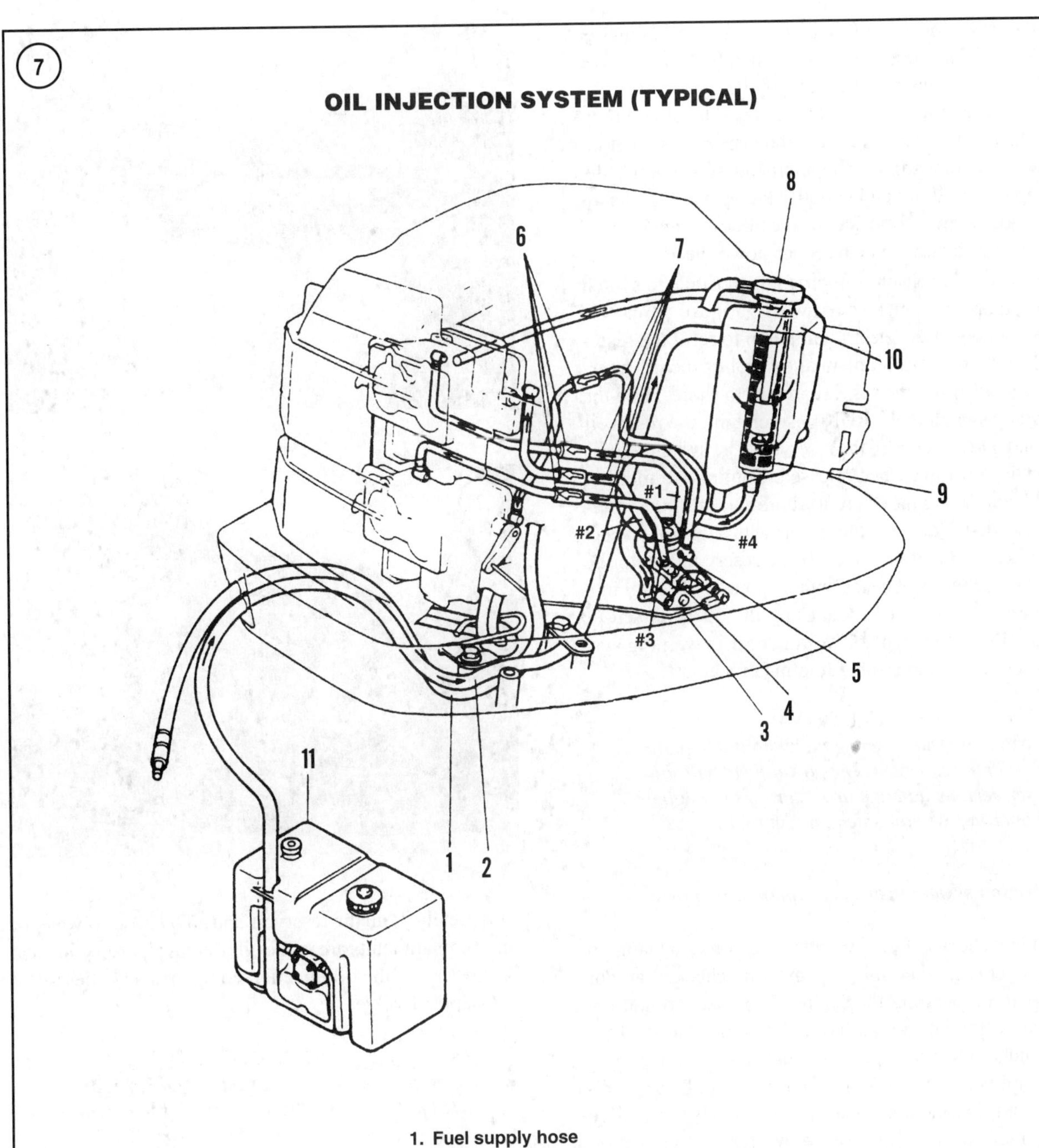

1. Fuel supply hose
2. Oil hose
3. Pump control linkage (gear driven pump)
4. Gear driven oil pump
5. Oil outlet fitting
6. Oil line check valves
7. Oil outlet hoses
8. Oil level sensor
9. Oil strainer
10. Oil reservoir (engine mounted)
11. Oil reservoir (onboard mounted)

mounted reservoir is refilled. Then the oil pump is switched off. This process repeats to maintain an adequate oil level in the engine reservoir. If the system fails to maintain the engine reservoir level, the oil level will drop below the normal level, causing the sensor to switch on the low oil level warning. The microcomputer sounds the warning horn and/or lamp to notify the operator. If the oil level drops below a critical level, the microcomputer initiates power reduction to help reduce power head damage.

If the oil level sensor or microcomputer fails to switch on the pump, the emergency switch (**Figure 9**) can be used to switch on the electric oil pump to refill the reservoir. This feature allows manual control of the oil reservoir level, allowing the vessel to return to shore. Hold the emergency switch in the RUN position until the reservoir is refilled. Do not overfill the reservoir. The switch returns to the OFF position when released. Continue to monitor the oil level during the return to shore and refill the reservoir as needed. Never run the engine with a low oil level in the engine mounted reservoir. If the reservoir cannot be refilled using the emergency switch, pour oil directly into the engine mounted reservoir using the oil level sensor opening. Refer to Chapter Two for specific warning system descriptions and troubleshooting procedures.

CAUTION
Never operate the engine without adequate oil in the reservoir. The power head will suffer serious damage or seizure if the engine operates without adequate lubrication.

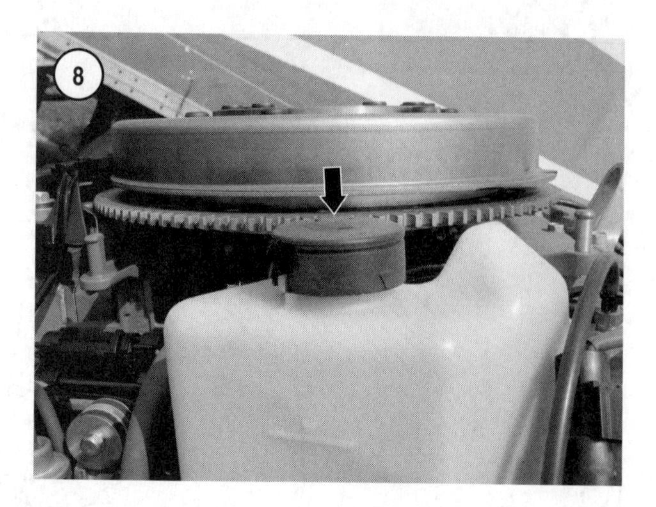

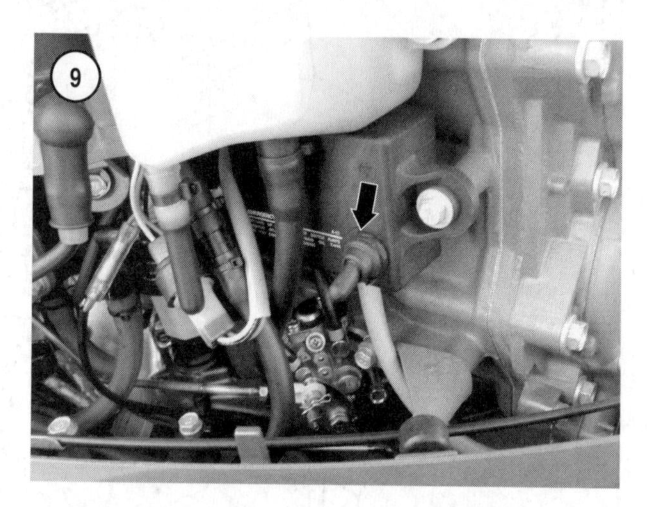

Oil injection system inspection and maintenance

Frequent checks of this system are required to maintain proper operation. The oil level must be checked and/or filled before operating the engine. The areas around the oil hoses, oil pump and reservoir should be inspected for oily residue. Oily residue may indicate oil leakage or it may simply be residue from the lubricant applied to linkages during routine maintenance. Corrosion preventative sprays also leave an oily residue that tends to collect on the bottoms of hoses and other components. Oil leakage from the oil injection system must be corrected before operating the engine. Check for leakage when performing routine maintenance and anytime the engine cover is removed.

Check the oil pump control linkage (3, **Figure 7**) for free movement and proper adjustment. Oil pump adjustment is described in Chapter Four.

Inspect the water drain tube (1, **Figure 10**) for water by removing the hose over a suitable container. Drain a few ounces from the reservoir and quickly reconnect the hose. Completely drain the reservoir and all oil hoses if water or other contaminants are in the oil. Removal, inspection and installation of the oil injection components are described in Chapter Twelve.

CAUTION
*The oil strainer and gasket (2 and 3, **Figure 10**) must be installed correctly for the oil injection system to operate properly. An improperly installed gasket or strainer can block oil flow to the oil pump, causing inadequate lubrication for the power head. Follow the procedures described in Chapter Twelve when removing or installing any oil injection system components.*

Remove the oil level sensor (4, **Figure 10**) to gain access to the oil strainer (3). Remove the strainer and clean it with a soft brush and mild solvent. Dry the components with compressed air before installing them into the reservoir. Removal and installation instructions for these and

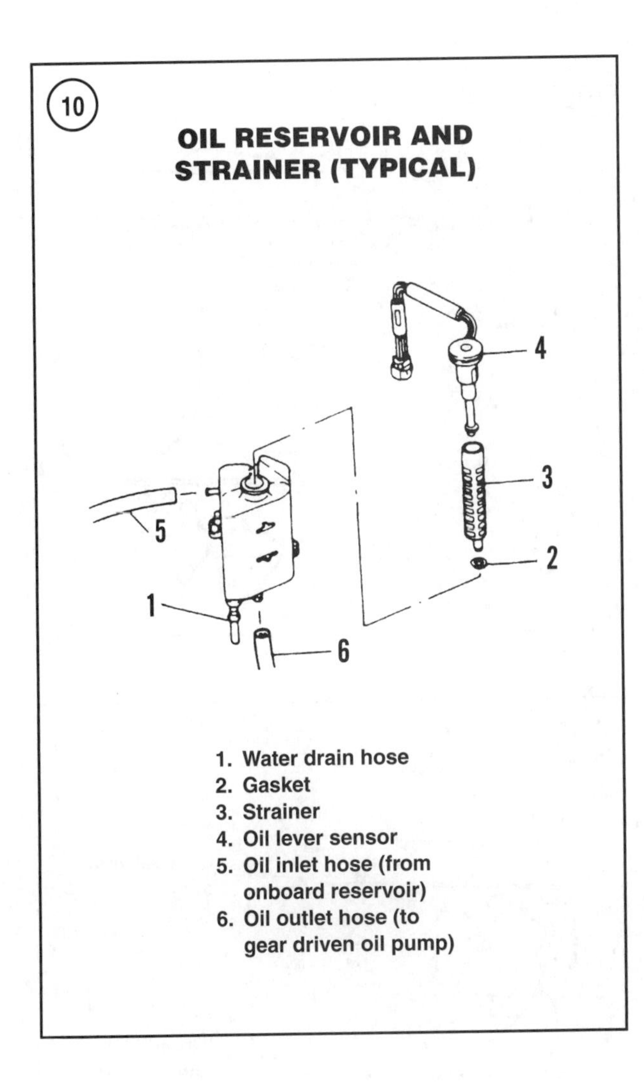

other oil injection components are described in Chapter Twelve.

Oil reservoir filling procedure

CAUTION
Never overfill the oil reservoir. The oil may seep past the fill cap and leak from the oil fill opening. Leakage occurs when the oil expands from heat. When properly filled, an air space is present to allow for heat expansion.

To add oil, wipe debris or contaminants from the oil fill cap (**Figure 11**) and surrounding surfaces. Remove the cap and use a clean funnel to pour oil into the onboard reservoir. Check the fill cap seal for deteriorated or damaged surfaces. Replace the seal as needed. Install and securely tighten the fill cap. If the oil level in the engine mounted reservoir is low, run the engine or use the emergency

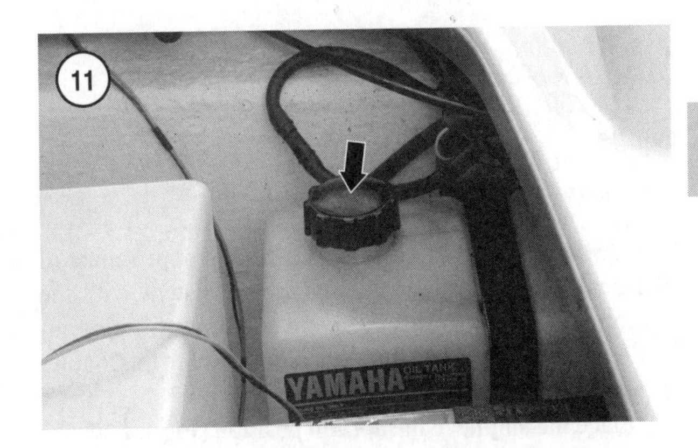

switch to activate the electric oil pump. Refer to Chapter Two for troubleshooting procedures if the system fails to fill the reservoir.

Gearcase Lubrication

Change the gearcase lubricant after the first 10 hours of use on a new engine, new gearcase or after major gearcase repair. After the first 10 hours, change the gearcase lubricant at 50 hour intervals or once a season (whichever occurs first). Use Yamaha gearcase lubricant or a suitable gearcase lubricant that meets or exceeds marine GL5 specifications. Refer to **Table 3** for gearcase lubricant capacities.

Refer to the recommendations on the container for proper application. Check the gearcase lubricant level and condition at regular intervals (see **Table 2**). Correct problems before they lead to gearcase component failure.

CAUTION
Never use automotive gear lubricant in an outboard gearcase. These types of lubricants are not suitable for marine applications. Using them can lead to increased wear and corrosion of internal components.

CAUTION
Operating the engine with low lubricant level or water contaminated lubricant causes in serious damage to internal gearcase components. When removed, inspect the sealing washer on the drain/fill and level/vent plugs. Replace missing or damaged plugs to prevent water or lubricant from seeping past the threads.

NOTE
Some gearcases use two level/vent plugs. Refer to the illustrations in Chapter Eight to determine the quantity and location(s) of the

plug(s). If so equipped, remove both level/vent plugs when checking the lubricant level or filling the gearcase.

Lubricant level and condition check

Significant differences exist in size and appearance of the gearcases. These procedures are consistent with the majority of the models. Level/vent and drain/fill plug locations may vary from the illustration. Refer to Chapter Eight for specific information.

1. Place the engine in the upright position (**Figure 12**) for at least one hour before checking the lubricant level.
2. *Electric start models*–Disconnect the battery cables.
3. Disconnect and ground the spark plug lead(s) to prevent accidental starting.
4. Slowly loosen and remove the drain/fill plug (**Figure 13**). Drain a small sample (about a thimble full) of lubricant from the gearcase. Quickly replace and securely tighten the drain plug.
5. Inspect the lubricant for water or a milky appearance. Pressure test the gearcase (Chapter Eight) to determine the source of leakage if either of these conditions exist.

NOTE

*A small amount of very fine metal particles is usually present in the gearcase lubricant and on the magnetic tip of the drain plug. These fine particles form during normal operation and they do not indicate a problem. Large particles (**Figure 14**), however, indicate a potential problem within the gearcase.*

6. If the lubricant feels contaminated or gritty, disassemble and inspect the gearcase for damaged components. Refer to Chapter Eight for disassembly and inspection procedures.
7. Remove the level/vent plug (**Figure 13**). The lubricant level must be even with the bottom of the threaded level/vent plug opening. Add lubricant through the drain plug opening (**Figure 15**) until full. If over an ounce is required to fill the gearcase, pressure test the gearcase as described in Chapter Eight. If lubricant is added, allow the gearcase to sit in a shaded area for at least one hour and check the lubricant level again. Top off again if necessary.
8. *Electric start models*—Connect the battery cables.
9. Connect the spark plug lead(s).

Gearcase lubricant change

Table 3 lists gearcase fluid capacities. Although differences exist in size and appearance of the gearcases, these procedures are consistent with the majority of the models. Level/vent and drain/fill plug locations may vary from the illustration. Refer to Chapter Eight for specific information.

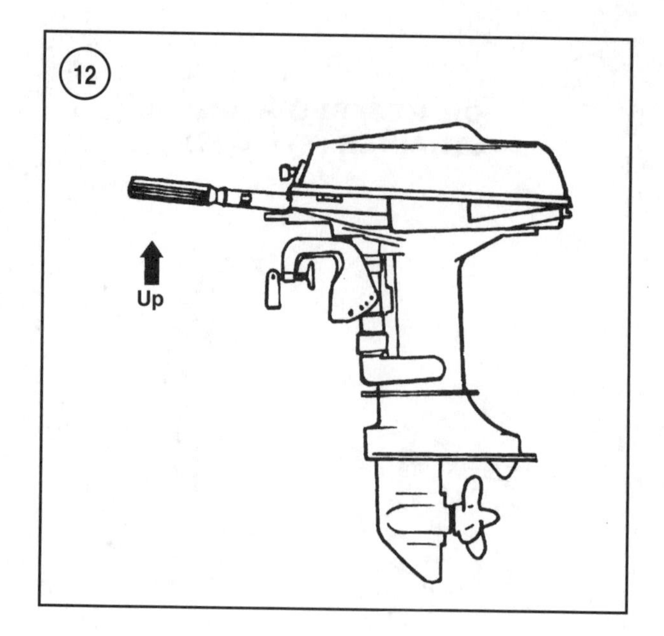

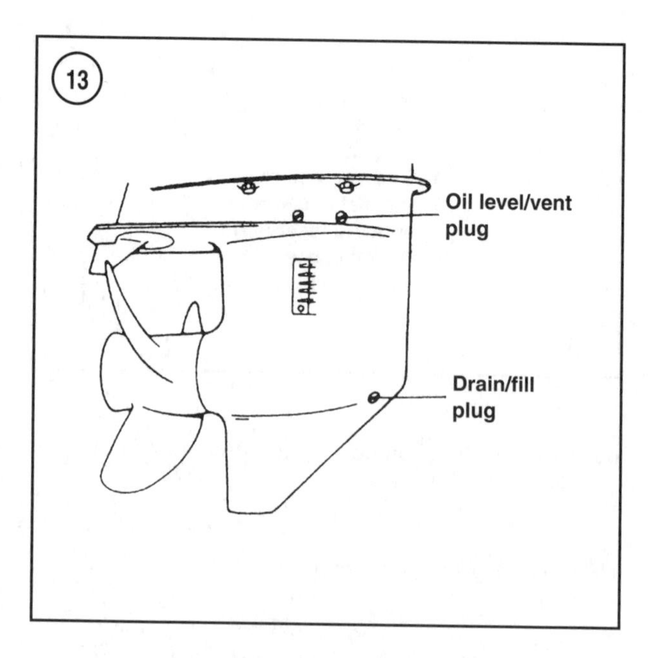

1. Place the engine in the upright position (**Figure 12**).
2. Disconnect the battery cables.
3. Disconnect and ground the spark plug lead(s) to prevent accidental starting.
4. Place a suitable container under the gearcase. Remove the drain/fill plug and level/vent plug(s) (**Figure 13**).
5. Drain the gearcase completely, then tilt the engine to ensure the drain/fill plug is at the lowest point.

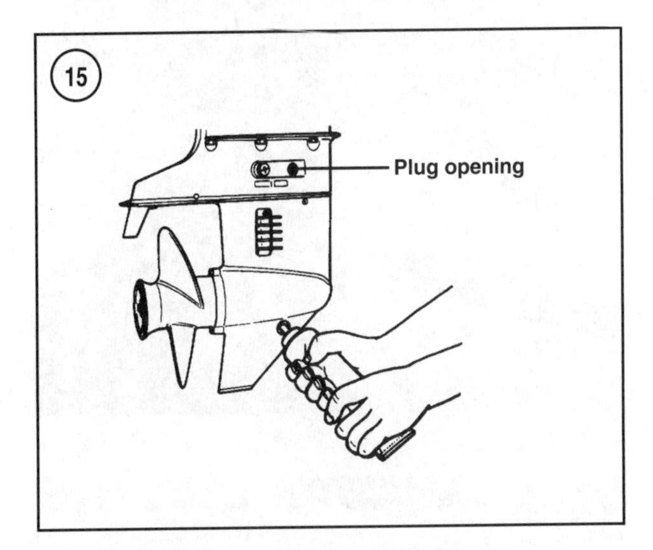

6. Inspect the lubricant for water or a milky appearance. Pressure test the gearcase (Chapter Eight) to determine the source of leakage if either of these conditions exist.
7. Feel for particles in the lubricant. Disassemble and inspect the gearcase (Chapter Eight) for damaged components if the lubricant feels contaminated or gritty.
8. Return the engine to the upright position (**Figure 12**).
9. Use a pump type dispenser or squeeze tube and slowly pump the gearcase lubricant into the fill/drain opening (**Figure 15**) until it flows out of the level/vent opening(s). On models with two level/vent openings, continue until lubricant flows from each opening.
10. Without removing the pump from the drain/fill opening, install and securely tighten the level/vent plug(s). Remove the pump. Quickly install and tighten the drain/fill plug securely.
11. Allow the gearcase to sit in a shaded area for at least one hour and check the lubricant level again. Top off again if necessary.
12. Connect the battery cables.
13. Connect the spark plug lead(s).
14. Dispose of the drained gearcase lubricant in a responsible manner. Many automotive parts retailers accept used gearcase lubricant for recycling.

Jet Drive Lubrication

A jet drive unit attaches to the drive shaft housing in place of a propeller drive gearcase. Jet drive units require maintenance at frequent intervals. The drive shaft bearings must be lubricated after each operating period. Lubrication is also required after each 10 hour period of operation and when preparing the engine for extended storage. Use Yamaha All-Purpose Grease or a grease that meets or exceeds NLGI No. 1 rating.

NOTE
Slight discoloration of the expelled jet drive grease is normal during the break-in period.

1. Locate the capped vent hose (1, **Figure 16**) on the port side of the jet drive.
2. Disconnect the vent hose from the fitting.
3. Connect a grease gun onto the fitting on the jet drive. Pump grease into the fitting until grease exits the vent hose (3, **Figure 16**).
4. Inspect the exiting grease. Disassemble the jet drive and inspect the internal components (Chapter Nine) if metal particles, dark gray coloration or a significant amount of water is present.
5. Wipe the expelled grease from the hose and connect the hose to the jet drive grease fitting.

Other Lubrication Points

Other lubrication points include the throttle and shift linkages, steering cable, swivel shafts, tilt tube, tilt lock mechanism and reverse lock mechanism. In certain applications, lubrication is required for the gearcase bearing carrier. Refer to **Table 2** for maintenance frequency and a description of the lubricating points.

Throttle and shift linkages

Apply a light film of Yamaha All-Purpose Grease or equivalent to pivot points and sliding surfaces of the throttle and shift linkages (**Figure 17**, typical). Have an assistant move the throttle and shift controls while observing the linkages to identify the lubrication points. Use just enough grease to leave a film on the surfaces.

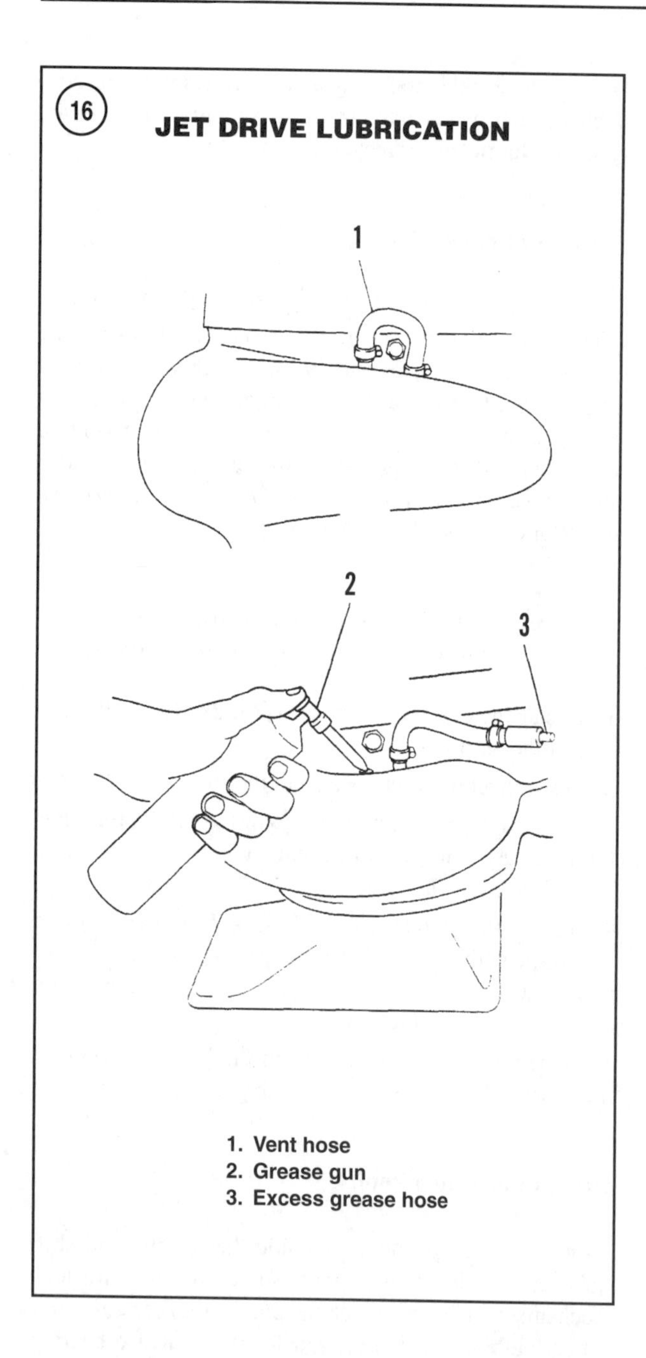

JET DRIVE LUBRICATION

1. Vent hose
2. Grease gun
3. Excess grease hose

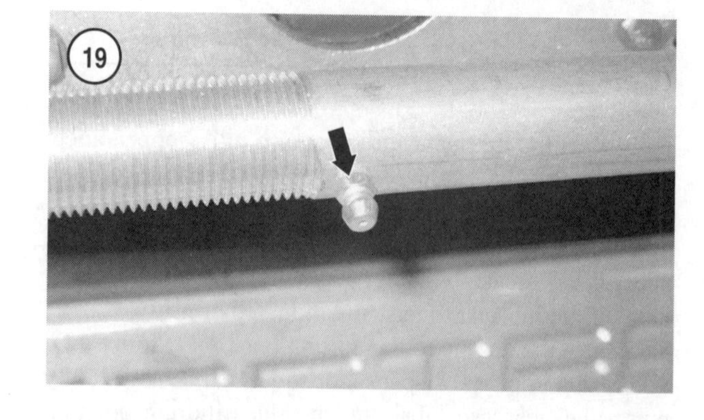

Steering cable

CAUTION
The steering cable must be in the retracted position when injecting grease into the fitting. The cable can become hydraulically locked if grease is injected with the cable extended. Refer to the cable manufacturer's recommendation for the type and frequency of lubrication.

Apply Yamaha All-Purpose Grease or equivalent to the pivot point and sliding surfaces of the steering cable (**Figure 18**, typical). Some steering cables are equipped with a grease fitting (**Figure 19**). Regular lubrication of the steering cable will dramatically increase the life of the cable.

Swivel shaft and tilt tube

Locate the grease fittings on the swivel housing (**Figure 20** and **Figure 21**) and lubricate the swivel shaft and tilt tube. Inject Yamaha All-Purpose Grease or equivalent into the fittings until grease comes out of the joints.

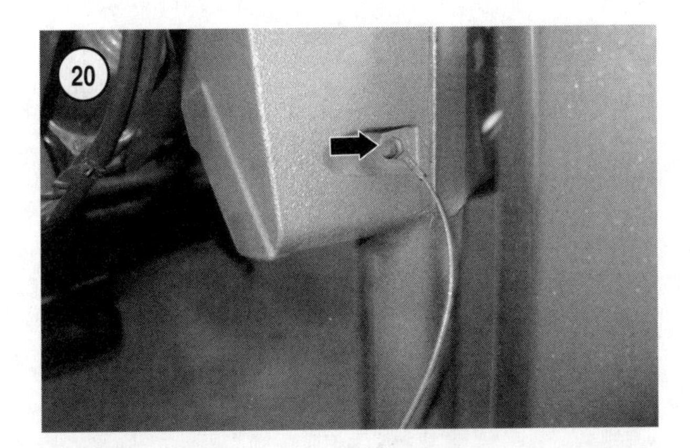

Tilt lock mechanism

Raise the engine to the full tilt position to access the reverse lock mechanism pivot points (**Figure 22**, typical). Engage the tilt lock or secure the engine with an overhead support during this procedure. Apply a light coating or Yamaha All-Purpose Grease or equivalent onto the pivot points and sliding surfaces of the tilt lock mechanism. Use only enough grease to leave a film on the surfaces. Refer

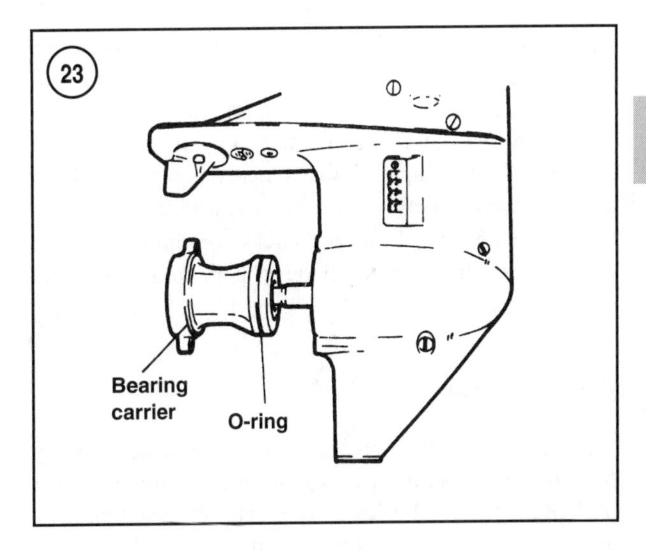

to the illustrations in Chapter Eleven to identify the tilt lock components and pivot points.

Bearing carrier

If the engine is operated frequently in saltwater, remove the bearing carrier (**Figure 23**) and clean all salt deposits from the carrier and gearcase housing. Salt deposits build up at the carrier-to-housing contact area and eventually damage the housing. Install a new O-ring (**Figure 23**) during assembly. Refer to Chapter Eight for bearing carrier removal and installation procedures.

MAINTENANCE

This section describes maintenance procedures for the propeller shaft and propeller, cooling system, electrical system, fuel system and corrosion prevention. Preparation for storage, fitting out and special instructions for submerged engines.

Propeller Shaft and Propeller

Remove the propeller nut, attaching hardware and the propeller as described in Chapter Eight. Clean deposits and old grease from the propeller shaft and propeller splined sections.

Inspect the propeller for bent blades (**Figure 24**), cracked surfaces, nicks or missing blade sections. Dress small nicks in the propeller blade with a file. Have the propeller replaced or have it repaired if cracks are found, blade sections are missing or significant bending is noted.

Inspect the propeller shaft for twisted splines (**Figure 25**) or a bent shaft. Inspect the tapered surface where the thrust washer contacts the propeller shaft (**Figure 26**) for rough, deeply grooved or excessively worn surfaces. Operating the engine with an insufficiently tightened propeller nut causes rapid wear on this surface. Replace the propeller shaft (Chapter Eight) if bent, has twisted splines or is worn excessively on the tapered surface.

Apply Yamaha All-Purpose Grease or equivalent onto the propeller shaft and install the propeller as described in Chapter Eight.

24

Cooling System

Inspect the cooling system for proper operation every time the engine is run. A stream of water exiting the lower back of the engine (**Figure 27**) indicates the water pump is operating. Never run the engine if it is overheating or the water indicator stream is weak. Inspect the water pump after every 50 hours of operation, once a year or whenever the engine is running warmer than normal. Water pump inspection and repair is described in Chapter Eight.

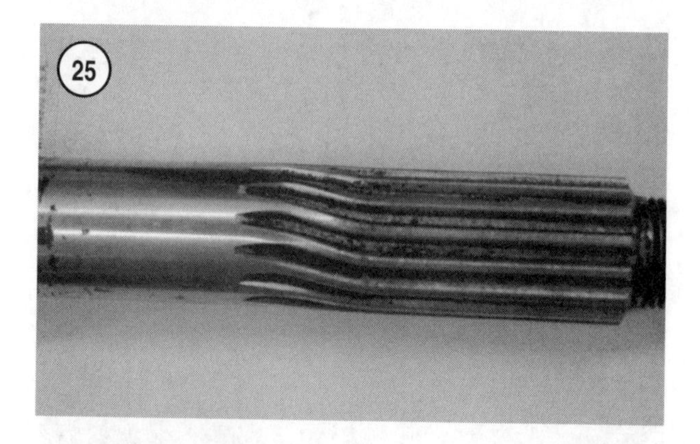

25

Flush the cooling system at regular intervals to help prevent corrosion and deposit buildup. Flush the system after each operation in salt or brackish water, polluted water or water that is laden with silt or sand. More frequent water pump inspection is required if the engine is operated in these conditions. The type of gearcase determines the type of flush/test adapter required.

Standard single propeller gearcase—Use the type of adapter shown in **Figure 28**.

Twin counter rotating propeller gearcase—Use the type of adapter shown in **Figure 29**. The adapter must completely cover the water inlet screens on the lower front of the gearcase.

Jet drive models—Use the thread-in type adapter (Yamaha part No. 6EO-28193-00-94). See **Figure 30**.

26

Flushing the cooling system

CAUTION

Never run the engine without first providing cooling water. Use either a test tank or flush/test adapter. Remove the propeller before running the engine.

WARNING

To avoid serious injury or death, stay clear of the propeller shaft while running an engine on a flush/test adapter. Disconnect the battery cables and spark plug lead(s) before removing or installing the propeller.

1. Remove the propeller as described in Chapter Eight.

27

28

29

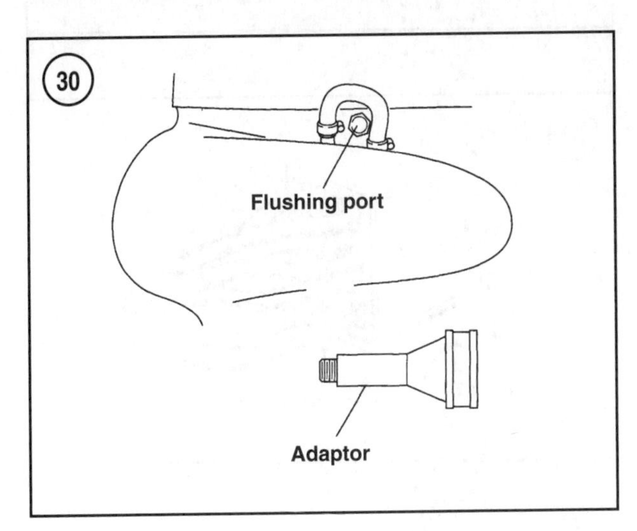

30

2A. Install the adapter as follows (If using a test tank go to Step 2B):

a. *Jet drive models*—Remove the plug from the flushing port (**Figure 30**). Thread the adapter fully into the port.
b. *Propeller drive gearcase*—Slip the flush test adapter over the antiventilation plate (**Figure 28**) or sides of the gearcase (**Figure 29**).
c. Thread a garden hose onto the adapter. Connect the hose to a freshwater supply.
d. Turn the water on. Make sure the cup(s) on the adapter completely cover the water screen(s).

2B. *Using a test tank*—Place the engine in the full tilt position. Move the boat or tank to position the engine over the tank. Tilt the engine downward until in the upright position. Make sure the water screen is well below the water line. Add water as needed.

3. Run the engine at a fast idle in NEUTRAL gear until the engine reaches full operating temperature.
4. Run the engine until the water exiting the lower cover (**Figure 27**) is clear and the engine has run a minimum of 10 minutes. Monitor the engine temperature. Stop the engine if it begins to overheat.
5. Bring the engine back to idle speed for a few minutes and stop the engine.
6. Remove the flush/test adapter or remove the engine from the test tank. Place the engine in the upright position for several minutes to drain the cooling system passages.
7. Install the propeller as described in Chapter Eight.

Electrical System

Many problems with a modern outboard engine are directly related to a lack of sufficient maintenance to the electrical system. Components requiring maintenance include:

1. Cranking battery.
2. Wiring and electrical connections.
3. Starter motor.

Cranking battery

The cranking battery requires more maintenance than any other single component. Unlike automobiles, boats may set idle for several weeks or longer. Without proper maintenance the battery loses charge and begins to deteriorate. Clean the battery terminals and charge the battery at no more than 30-day intervals when subjected to long-term storage. Battery testing, changing and maintenance are described in Chapter Six.

Wiring and electrical connections

Periodically loosen the clamp screw (**Figure 31**), and then disconnect the instrument harness from the engine wire harness. Inspect the main harness connector (**Figure 32**) for corrosion or faulty pin connections. Carefully scrape corrosion from the contacts. Apply a light coating of Yamaha All-Purpose Grease or equivalent onto the

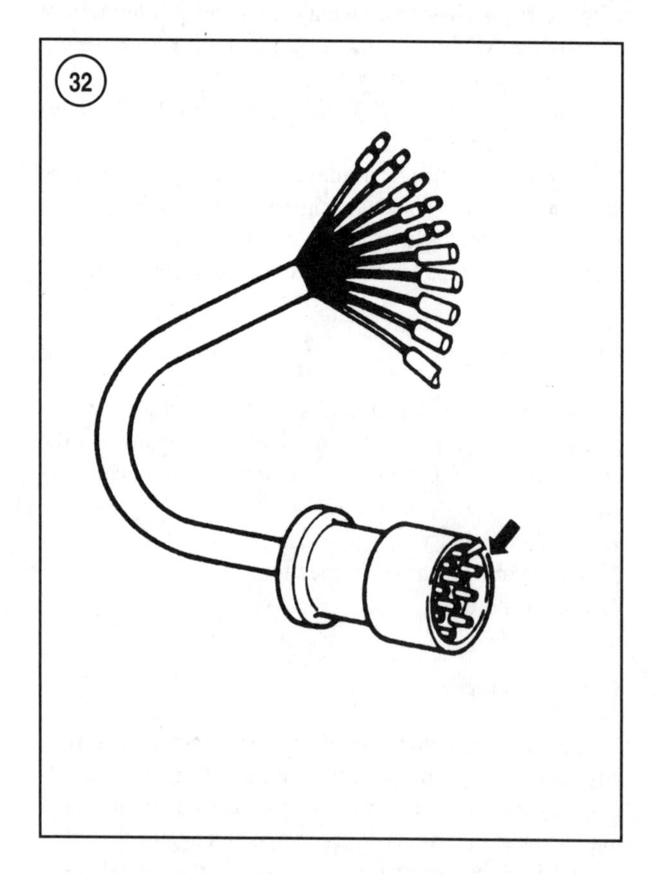

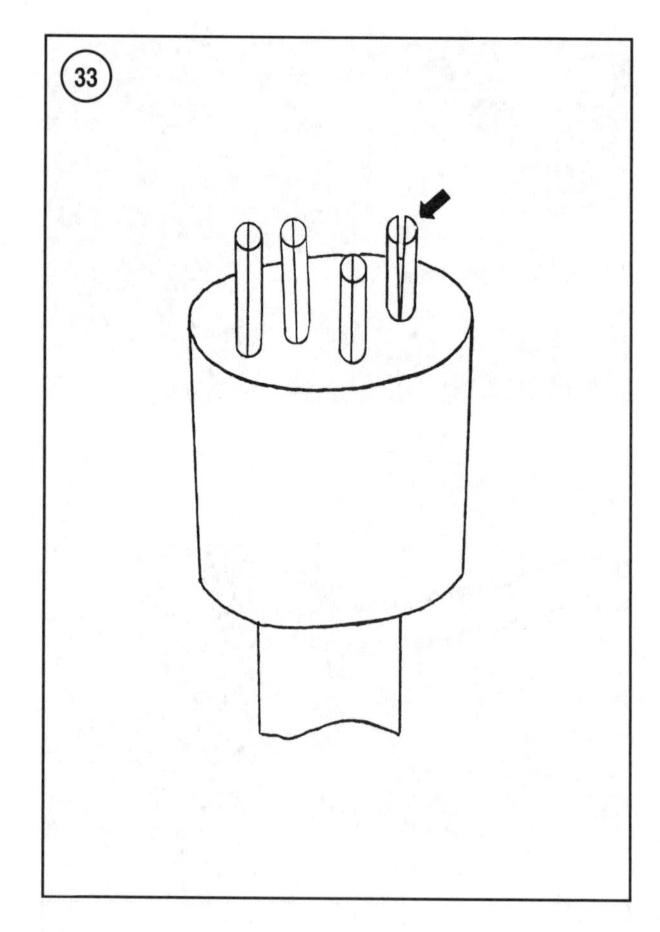

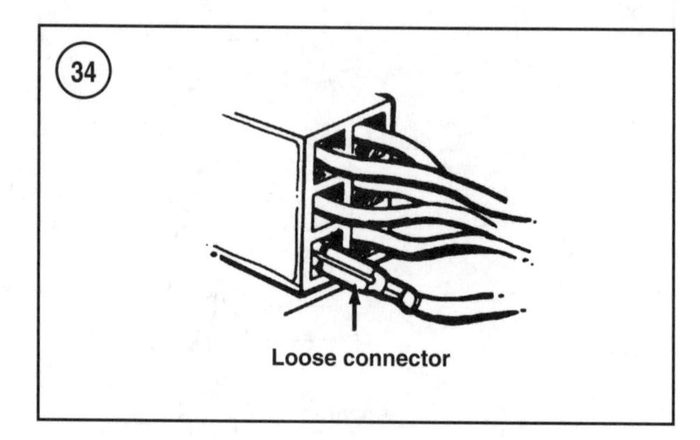

main harness plug to seal out moisture and prevent corrosion.

CAUTION

Apply only enough grease to provide a light film on electrical contacts. Excessive grease may seep into some electrical components such as relays and cause a malfunction.

The male pins of the main harness connector tend to squeeze together and may not make good contact with the female end. Use a very small blade screwdriver and slightly spread the pins (**Figure 33**) to maintain a reliable connection. Disconnect and inspect the wire connectors for corrosion, loose connectors (**Figure 34**) and/or bent terminal pins (**Figure 35**).

Starter motor

Use only a light coating of grease where required. Excessive amounts of grease may attract dirt or debris and

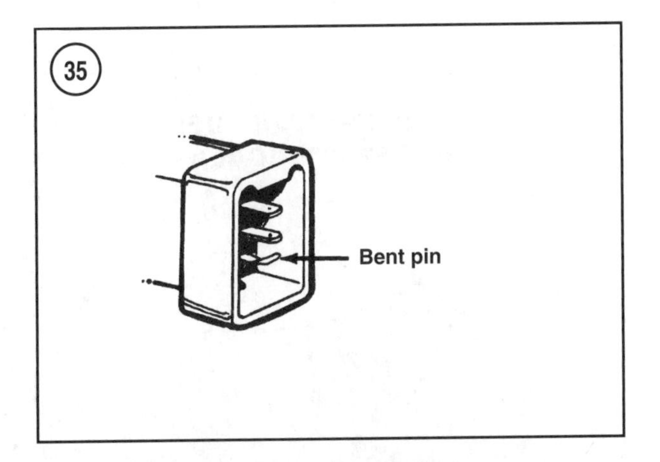

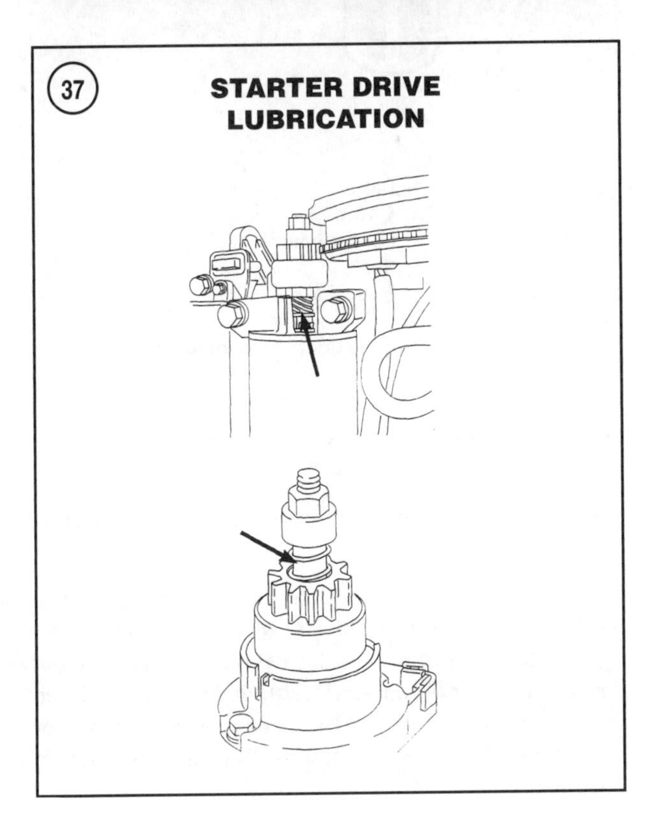

lead to starter malfunctions. Refer to Chapter Six if removal or disassembly is required to access the pinion drive and armature shaft.

1. Disconnect the battery cables. Disconnect and ground the spark plug leads to prevent accidental starting.
2. Disconnect the starter cable terminal (**Figure 36**) and clean any corrosion from the cable terminal and terminal post. Reconnect the cable and securely tighten the nut. To prevent corrosion, apply a coating of Liquid Neoprene onto the terminal. Liquid Neoprene is available from automotive parts retailers.
3. Rotate the pinion drive gear clockwise to expose the helical splines on the armature shaft (**Figure 37**). Apply a light coating of Yamaha All-Purpose Grease or equivalent onto the splines. Release the pinion drive gear and apply the same grease onto the armature shaft and the teeth on the pinion drive gear.
4. Reconnect the battery cables and check for proper electric starter operation.

Fuel System

Topics covered in this section include fuel requirements, fuel filter servicing and inspecting the hoses, clamps and shut-off valve.

CAUTION

Never run an outboard on old or stale fuel. Serious power head damage could result from using fuel that has deteriorated. Varnish-like deposits form as fuel deteriorates and block fuel flow. Lean fuel conditions lead to preignition and detonation damage. Dispose of fuel that has been stored for a considerable length of time without proper treatment. If at all possible, fuel should not be stored over 60 days even with proper treatment. Contact a local marine dealership or an automotive repair facility for information on disposal of old or stale fuel.

Fuel requirements

Use fuel with a minimum average octane rating of 86. This fuel should meet the requirement for the engine if operated under normal conditions. Use a higher octane rating if the engine is used for commercial or heavy-duty service.

Using premium grade fuel from a major supplier offers advantages. The higher octane rating provides added protection against damaging detonation. Major brands generally have a high turnover of fuel, helping to ensure fresh fuel for your engine. Premium grade fuel normally con-

tains a high quality detergent that helps keep the internal components of the power head and fuel system clean.

Fuel additives

Use fuel stabilizer such as Sta-Bil to help prevent the formation of gum or varnish-like deposits in the fuel system during extended storage.

Add Yamaha Ring Free or another outboard manufacturer's similar additive into the fuel on a continual basis. Using this type of additive combined with a good quality oil have shown to be effective in reducing deposits on the piston rings and in the combustion chamber.

Mix the additives into the fuel in the concentrations listed on the container.

Fuel filter

Clean, inspect or replace the fuel filter at the intervals specified in **Table 2**. The models covered in this manual use the serviceable canister-type fuel filter (**Figure 38**). Water collects from the fuel system into the bottom of the canister. Inspect the fuel system components, including the fuel tank, for contamination if there is water in the canister.

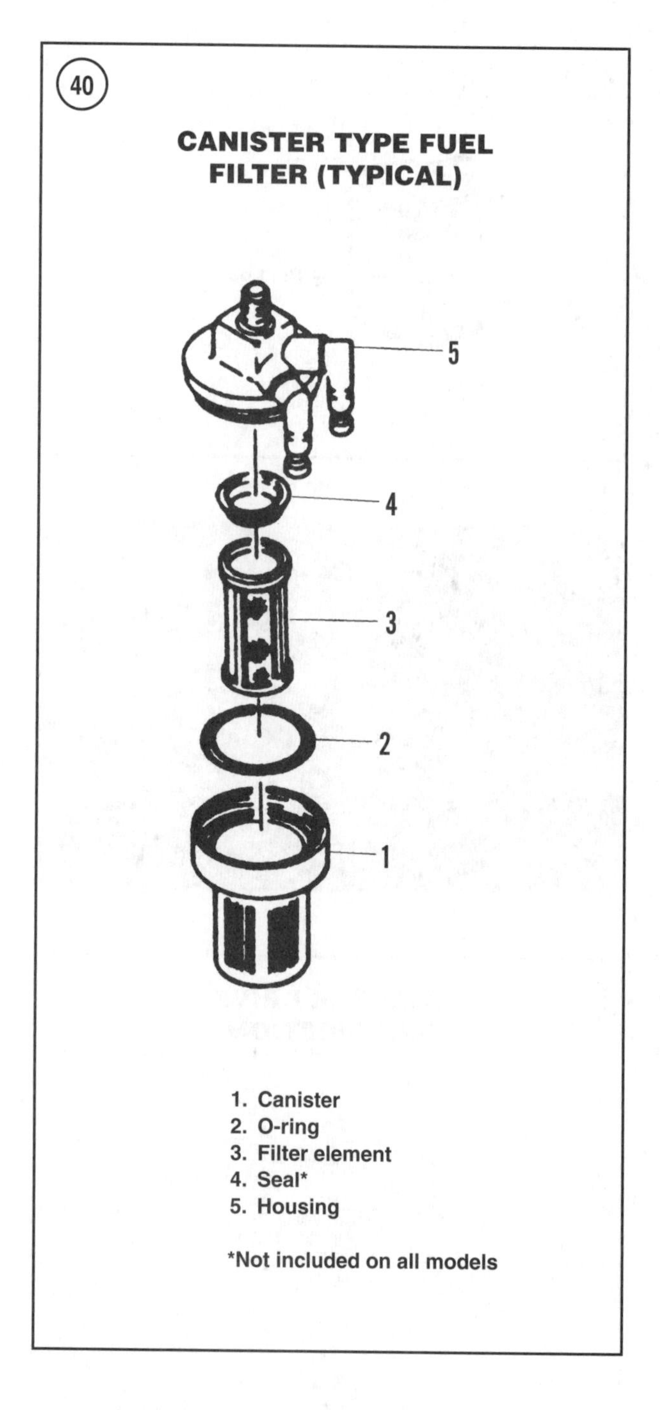

HPDI models also use in-line fuel filters (**Figure 39**). This type of filter is made of a translucent material that allows easy visual inspection for debris or contamination. The in-line filter is non-serviceable and must be replaced if contaminated. If a visual inspection reveals debris or discoloration, replace the in-line filter as described in Chapter Five.

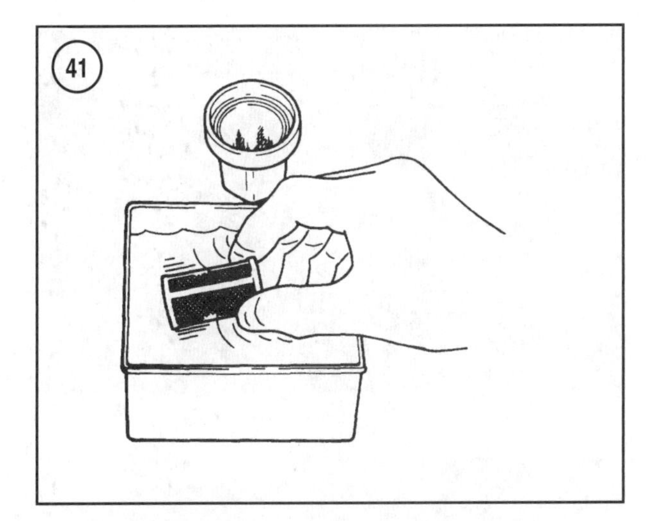

Servicing the canister fuel filter

1. Disconnect the battery cables.
2. Disconnect and ground the spark plug lead to prevent accidental starting.
3. Place a suitable container or shop towels under the filter to capture spilled fuel.
4. While supporting the filter housing (5, **Figure 40**), carefully twist the canister (1), counterclockwise as viewed from the bottom to remove the canister. Wipe up any spilled fuel.
5. Empty the canister into a see-through container and check for contaminants. Inspect the fuel tank if there is water or a significant amount of debris in the fuel.
6. Remove and inspect the O-ring (2, **Figure 40**) and replace if pinched or damaged in any way.
7. Carefully pull the filter element (3, **Figure 40**) and upper seal (4) from the filter housing. Some models do not use a seal in this location.
8. Clean the canister and filter element using a mild solvent (**Figure 41**). Do not scrub the screen. Use a toothpick to remove the gummy material that tends to collect in the bottom of the canister.
9. Inspect the filter element for deposits, debris or damage to the screen. Replace the screen if the screen is damaged or the element cannot be adequately cleaned.
10. Inspect the canister for cracks or damaged threads. Replace the canister if not found in excellent condition.
11. Install the filter element into the canister with the open end facing upward. Position the seal (4, **Figure 40**) between the element and the filter housing. Some models do not use a seal in this location.

WARNING
Use caution when installing the canister onto the filter housing. Make certain the canister threaded section is not cross-threaded during installation. A fuel leak may cause a fire or an explosion. Always check the entire fuel system for leakage after servicing the fuel system. Correct fuel leakage before putting the engine into service.

12. Install the O-ring into the recess in the canister. Carefully thread the canister onto the filter housing and hand-tighten. Using other means for tightening may crack or damage the canister.
13. Squeeze the primer bulb to fill the fuel system while checking for leaks. Correct any fuel leaks before putting the engine into service.
14. Connect the spark plug leads.
15. Connect the battery cables.

High-Pressure Fuel Pump

A belt driven high pressure fuel pump (**Figure 42**) is used on all 150-250 hp HPDI models.

Drive belt inspection

Inspect the drive belt (**Figure 43**) at the intervals specified in **Table 2**.

1. Disconnect the battery cables. Disconnect and ground the spark plug leads to prevent accidental starting.
2. Remove the flywheel cover.
3. Use compressed air to blow debris from the flywheel and fuel pump pulleys.
4. Inspect the belt surfaces for worn, cracked or oil soaked surfaces.
5. Manually rotate the flywheel while inspecting all the drive belt cogs.
6. Inspect the flywheel and fuel pump pulleys for worn, corroded or damaged surfaces. Replace the drive belt and

related components as described in Chapter Five if there are any defects.
7. Install the flywheel cover.
8. Install the spark plugs and leads. Connect the battery cables.

Pump oil level check

Check the oil level in the high pressure fuel pump at the intervals specified in **Table 2**.
1. Disconnect the battery cables. Disconnect and ground the spark plug leads to prevent accidental starting.
2. Remove the flywheel cover.
3. Remove the oil level plug (A, **Figure 44**) from the pump. Inspect the gasket on the oil level plug for torn or damaged surfaces and replace as needed.
4. Insert a thin screwdriver or other suitable object into the opening to check the oil level. The oil must be even with the plug opening.
5. If the oil level is low, slowly pour Yamaha gearcase lubricant or a good quality SAE 90 hypoid gear oil into the opening as needed.
6. Install and securely tighten the oil level plug.
7. Install the flywheel cover. Connect the battery cables and spark plug leads.
8. Run the engine for a few minutes at idle speed. Stop the engine and allow the pump to cool for approximately 15 minutes. Recheck the oil level as described previously. Correct the oil level as needed.

Changing the pump oil

Change the oil in the high pressure fuel pump at the interval specified in **Table 2**. Use Yamaha gearcase lubricant or a good quality SAE 90 hypoid gear oil when filling the pump reservoir. **Table 3** lists the pump oil capacity.
1. Disconnect the battery cables. Disconnect and ground the spark plug leads to prevent accidental starting.

2. Remove the flywheel cover (Chapter Seven).
3. Place a suitable container under the drain/fill plug (B, **Figure 44**). Remove the drain/fill and oil level plug (A, **Figure 44**). Then allow the gear oil to fully drain. Inspect the gasket on the oil level plug for torn or damaged surfaces and replace as needed.
4. Inspect the plugs and a sample of the gear oil for large metal particles or other contaminants. Small metal particles form during normal operation and do not necessarily indicate a problem. Disassemble and inspect the high pressure pump (Chapter Five) if larger metal particles or other contaminants are found in the oil.
5. To fill the pump reservoir, insert the tip of the gear oil container firmly into the drain/fill plug opening (B, **Figure 44**). Squeeze the container until gear oil flows from the oil level plug (A, **Figure 44**). Quickly thread the oil level plug into the opening. Pull the tip of the container from the drain/fill opening and quickly thread the drain/fill plug into the opening. Securely tighten both plugs.
6. Install the flywheel cover. Connect the battery cables and spark plug leads.
7. Run the engine for a few minutes at idle speed. Stop the engine and allow the pump to cool for approximately 15

minutes. Recheck the oil level. Correct the oil level as needed.

Fuel Hoses and Clamps

At the recommended intervals (**Table 2**), check the entire fuel system for leaking hoses or connections. Check the condition of all fuel hose clamps. Remove and replace any plastic locking type clamps (**Figure 45**) that appear old or have become brittle. Carefully tug on the fuel hoses to ensure a tight fit at the connections. Inspect spring type hose clamps (**Figure 46**). Replace clamps that are corroded, appear distorted, damaged or have weak spring tension. Always replace damaged or questionable hose clamps.

Replace fuel hoses that have become hard or brittle, have a cracked or weathered appearance, have a spongy feel or are leaking. Use only the recommended hose available from a Yamaha dealership. Fuel hoses available from automotive parts stores may not meet the demands placed upon the hose or may not meet coast guard requirements.

Corrosion Prevention

Corrosion prevention is an excellent way to increase the life and reliability of the engine.

Decrease corrosion in the power head cooling passages by flushing the cooling system as soon as possible after running the engine in salt, brackish or polluted water.

Clean and inspect external mounted anodes (**Figure 47**) at the recommended intervals (**Table 2**). Frequently test the electrical continuity of the anodes. Clean, inspect and test the anodes more frequently if the engine is run or stored in salt, brackish or polluted water. Use a stiff brush to remove vegetation or powdery deposits. Replace the anode if it has lost 40% or more of its original material. Never paint or cover the anode with a protective coating. Doing so will dramatically decrease its ability to protect underwater components from corrosion. Always clean the mounting area before installing the anode to ensure a proper connection. Inspect the anode mounting area when there is corrosion on the engine components, but the anode is not experiencing corrosion.

Preparation for Storage and Fitting Out

The objective when preparing the engine for long term storage is to prevent unnecessary corrosion or deterioration during the storage period. If done properly the engine should operate as well as before it was stored. All major systems require some preparation during storage. Perform any maintenance that will come due during the storage period.

Preparing the engine for storage

1. Check the gearcase lubricant level and condition as described in this chapter. Check and top off the tilt/trim system fluid as described in Chapter Eleven.
2. *HPDI models*—Check the oil level in the high pressure fuel pump as described in this chapter.
3. Lubricate the propeller shaft as described in this chapter. Remove the battery from the boat and maintain it as described in Chapter Six.
4. Check all electrical harnesses for corrosion or faulty connections and repair as needed. Apply grease to the terminals as described in this chapter.
5. Lubricate all steering, throttle and shift linkages.
6. Lubricate all pivot and swivel shafts.
7. Lubricate the jet drive if so equipped.
8. Drain fuel from the engine mounted fuel tank. Treat the fuel in portable or built in tanks with an additive such as Sta-Bil.

9. Clean the exterior of the gearcase and mid-section to remove vegetation, dirt or deposit buildup.
10. Wipe down the components under the engine cover and apply a good corrosion preventative spray such as CR66 or its equivalent.
11. Inspect, clean or change the fuel filter(s). Clean the screen for the oil injection system.
12. Flush the cooling system and treat the internal components of the power head with a storage sealing agent to prevent corrosion during the storage period. Refer to the procedures described in this chapter.
13. Drain each carburetor float bowl or vapor separator tank (EFI and HPDI models). Refer to the diagrams in Chapter Five to locate the drain fitting. Disconnect the fuel supply hose at the engine and fuel tank. Direct the outlet end of the hose into the fuel tank and pump the primer bulb to remove residual fuel from the hose. Reconnect the fuel hose to prevent contamination during the storage period.
14. Check all water drain openings to ensure they are clear.
15. *Oil injected models*—Fill the on board and engine mounted oil reservoirs to reduce the air space in the reservoir. This important step can dramatically reduce condensation and the formation of water in the reservoir.
16. Store the engine in a protected area. If at all possible, store the engine in the upright position. Place a piece of cardboard or other suitable material under to engine to capture any fluids that may leak from the engine.

Applying the storage sealing agent

Apply a storage sealing agent such as Yamaha Store-Rite Fogging Oil to help prevent corrosion of internal power head components during the storage period.
1. Remove the silencer cover from the carburetors or throttle body as described in Chapter Five.
2. Run the engine at idle speed in a test tank or on a flush/test adapter for 10 minutes or until the engine reaches normal operating temperature.
3. Raise the engine speed to 1500 rpm. Spray the sealing agent into the carburetors or throttle body openings until the engine stalls out. Try to spray evenly into all carburetors or throttle body openings.
4. Remove the engine from the test tank or remove the flush/test adapter. Place the engine in the upright position for several minutes to drain water from the cooling passages.
5. Remove each spark plug and pour in approximately one ounce of Yamalube Two-Stroke Outboard Oil into each cylinder. Crank the engine over a few times to distribute the oil.
6. Install the spark plugs.

Fitting out

When the time comes to remove the engine from storage, a few items need attention. Perform all required maintenance on jet drive models, service the water pump and replace the impeller as described in Chapter Eight. This vital component deteriorates during extended storage. Prepare and run the engine as follows:
1. Change or correct all lubricant levels.
2. Supply the engine with fresh fuel.
3. *Carburetor and EFI models*—If the engine has been in storage for one year or longer, operate the engine on a 50:1 fuel/oil mixture for the first few hours. This applies to oil injected engines as well.
4. Pump the primer bulb and check for carburetor or vapor separator tank flooding as described in Chapter Two.
5. Install the battery as described in Chapter Six.
6. Supply the engine with cooling water and start the engine.
7. Run the engine at idle speed until it reaches full operating temperature. Check for proper operation of the cooling, fuel, electrical and oil injection systems. Stop the engine immediately if there is fuel or oil leakage, overheating or the warning system activates. Correct any problems before continuing.
8. Avoid running at wide-open throttle settings for an extended period of time for the first hour of operation.
9. Avoid continued operation if the engine is not operating properly. Continued operation may damage the power head or other components. Refer to Chapter Two for troubleshooting and testing procedures.

Submersion

If the engine has been submersed, three factors need to be considered. Was the engine running when the submersion occurred? Was the engine submerged in salt, brackish or heavily polluted water? How long has it been since the engine was retrieved from the water?

Submerged while running

Completely disassemble and inspect the power head if the engine was submerged while running. Internal damage to the power head, such as a bent connecting rod, is likely when this occurs. Refer to Chapter Seven for power head repair procedures.

Submerged in salt, brackish or heavily polluted water

Many components of the engine will suffer the corrosive effects of submersion in salt, brackish or heavily polluted water. The wire harness and connectors are the most commonly affected components. Replace the wire harness and all connectors to ensure a reliable repair. The starter motor, relays and any switch or sensor on the engine will likely fail if not thoroughly cleaned or replaced.

Retrieve and service the engine as soon as possible. Vigorously wash all debris from the engine with freshwater. If there is sand silt or other gritty material within the engine cover, disassemble and inspect the power head as described in Chapter Seven.

Service the engine to the point that it is running within a few hours of retrieval. Have a qualified assistant help if it is not possible to do the required service alone within a few hours.

Thoroughly clean the engine and submerge it in a barrel or tank of clean freshwater if it cannot be serviced within a few hours of retrieval. This protective submersion prevents exposure to air and decreases the potential for corrosion. This does not preserve the engine indefinitely. Service the engine within a few days of protective submersion.

Submerged engine servicing

1. Remove the engine cover and wash all material from the engine with freshwater. Completely disassemble and inspect the internal power head components if there is sand, silt or other gritty material inside the engine cover.
2. Dry the exterior of the engine with compressed air or other means.
3. Remove the spark plugs and ground the spark plug leads.
4. Remove the propeller as described in Chapter Eight.
5. Drain all water and residual fuel from the carburetor float bowls or vapor separator tank (EFI and HPDI models), filters and fuel hoses.
6. Position the engine with the spark plug openings facing downward. Slowly rotate the flywheel clockwise to force water from the cylinders. Manually turn the flywheel.
7. Rotate the flywheel several times to note if the crankshaft is turning over freely. Disassemble and inspect the internal power head components if there is interference or binding.

WARNING
Rubbing alcohol is extremely flammable. Never smoke when using rubbing alcohol. Never use rubbing alcohol around any flame sparks or other sources of ignition. Fire or explosions can occur and cause serious injury or death and damage to property.

8. *Carburetor models*—Position the engine with the spark plug openings facing downward and pour rubbing alcohol into the carburetor opening(s). Allow the engine to set for a few minutes and slowly rotate the flywheel one complete revolution.
9. Position the engine with the spark plug opening facing upward and pour rubbing alcohol into the spark plug opening(s).
10. Slowly rotate the flywheel one complete revolution. Position the engine with the spark plug opening(s) facing downward and drain the alcohol from the cylinders. Rotate the flywheel to remove residual alcohol from the power head.
11. Place the engine in the normal upright position. Pour one teaspoon of Yamalube Two-Stroke Outboard Oil into the cylinder(s) and carburetor opening(s).
12. Rotate the flywheel several revolutions to distribute the oil.
13. *Oil injected models*—Remove, clean and reinstall the oil reservoir, screen and oil hoses as described in Chapter Twelve.
14. Disconnect and air-dry all electrical connections.
15. Remove, disassemble and inspect the starter motor as described in Chapter Six.

16A. *Carburetor and EFI models*—Provide the engine with a fresh supply of fuel and oil mixed at a 50:1 ratio.

16B. *HPDI models*—Provide the engine with a fresh supply of fuel.

17. Provide the engine with cooling water, then start the engine and run at idle speed for at least one hour to remove any residual water. If the engine cannot be started, refer to Chapter Two for troubleshooting procedures.
18. Perform all routine maintenance (**Table 2**) before putting the engine into service. Install the propeller as described in Chapter Eight.

TUNE-UP

This section describes required adjustments, inspections and spark plug replacement.

Compression Test

Perform a compression test (**Figure 48**) before beginning a tune-up. An engine with weak or unbalanced compression cannot be tuned properly. Refer to Chapter Two for compression testing procedures and recommendations.

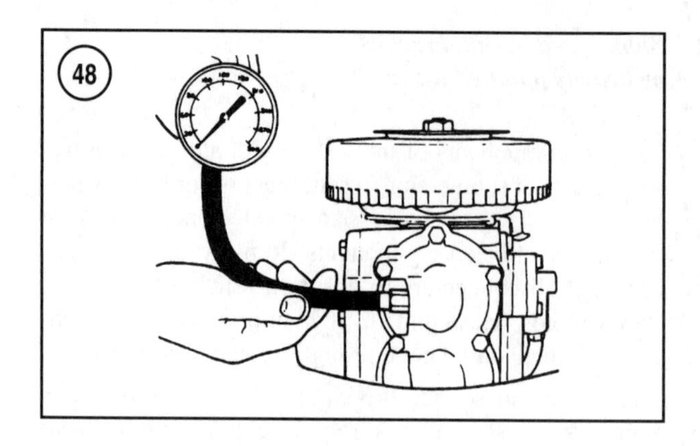

Spark Plugs

No tune-up is complete without servicing the spark plug(s). Although spark plugs can be cleaned and gapped to restore performance, they do not last as long as new plugs. Replace spark plugs that are not in like-new condition. Both standard type and surface gap plugs are used on Yamaha outboards. Remove the plugs and compare them to the plugs shown in **Figure 49** for standard type plugs and **Figure 50** for surface gap plugs. Spark plugs can give a clear indication of problems with the engine, before symptoms occur. Refer to the information provided in **Figure 49** or **Figure 50** for spark plug diagnosis.

Correct spark plug

NOTE
*Earlier production 2001 and prior models use spark plugs with a single ground electrode. All later production 2001 and all 2002 HPDI engines use spark plugs with two ground electrodes. The engine will not perform properly if operated with the wrong type of plugs as different ignition coils and engine control units are used for each type of plug. Refer to the color of the coil harness connector to determine which plugs to use. Earlier production 2001-prior models use black color connectors (**Figure 51**) on each ignition coil and must use the single ground electrode plugs. Later production 2001 and 2002-on models use gray color connectors and must use the spark plugs with two ground electrodes.*

Using the correct spark plug is vital to the performance and durability of the outboard. Refer to **Table 5** for the correct spark plug. Use the NGK part No. or, if necessary, the equivalent Champion spark plug. Refer to the spark plug manufacturer's application guide if using a different brand of spark plug. On HPDI models, note the color of the ignition coil to the engine wire harness connector prior to referring to **Table 5**.

Spark plug removal

CAUTION
Dirt or other foreign material surrounding the spark plug can fall into the spark plug opening when the spark plug is removed causing serious power head damage when the engine is started.

1. Clean the area around the spark plug(s) using compressed air or a suitable brush.
2. Disconnect the spark plug lead(s) by carefully twisting the cap back and forth on the plug while pulling outward. Pulling on the lead instead of the cap can damage the lead.
3. If there is corrosion at the spark plug threaded opening, apply a penetrating oil on the threaded opening and allow it to soak for several hours.
4. Remove the spark plugs using the correct size spark plug wrench. Metric size is required on many spark plugs. Arrange the plugs in the order of the cylinder from which they were removed.
5. Closely examine each spark plug for the conditions shown in **Figure 49** for standard type plugs or **Figure 50** for surface gap plugs.
6. Check each plug for brand and part No. All plugs must be identical.
7. Discard the spark plugs. Although they can be cleaned and reused if in good condition, they will not last as long as new plugs. New plugs are relatively inexpensive and are far more reliable.

Gapping standard type spark plugs

New spark plugs must be carefully gapped to ensure a reliable consistent spark. Use a special spark plug gapping tool (**Figure 52**) and feeler gauges (**Figure 53**) or a combination tool (**Figure 54**) to adjust the gap.

49

SPARK PLUG CONDITION

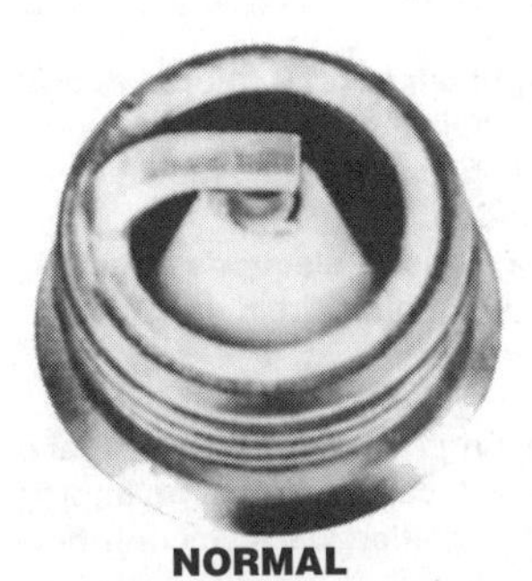

NORMAL
- Identified by light tan or gray deposits on the firing tip.
- Can be cleaned.

GAP BRIDGED
- Identified by deposit buildup closing gap between electrodes.

OIL FOULED
- Identified by wet black deposits on the insulator shell bore and electrodes.
- Caused by excessive oil entering combustion chamber through worn rings and pistons, excessive clearance between valve guides and stems or worn or loose bearings. Can be cleaned. If engine is not repaired, use a hotter plug.

CARBON FOULED
- Identified by black, dry fluffy carbon deposits on insulator tips, exposed shell surfaces and electrodes.
- Caused by too cold a plug, weak ignition, dirty air cleaner, too rich fuel mixture or excessive idling. Can be cleaned.

ADDITIVE FOULED
- Identified by dark gray, black, yellow or tan deposits or a fused glazed coating on the insulator tip.
- Caused by using gasoline additives. Can be cleaned.

WORN
- Identified by severely eroded or worn electrodes.

FUSED SPOT DEPOSIT
- Identified by melted or spotty deposits resembling bubbles or blisters.
- Caused by sudden acceleration. Can be cleaned.

OVERHEATING
- Identified by a white or light gray insulator with small black or gray brown spots with bluish-burnt appearance of electrodes.
- Caused by engine overheating, wrong type of fuel, loose spark plugs, too hot a plug or incorrect ignition timing. Replace the plug.

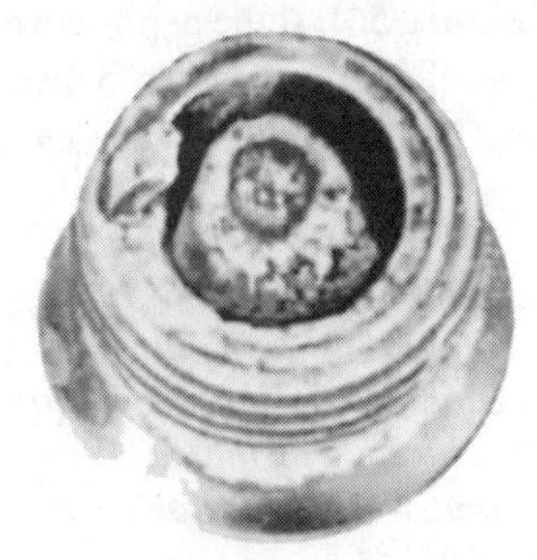

PREIGNITION
- Identified by melted electrodes and possibly blistered insulator. Metallic deposits on insulator indicate engine damage.
- Caused by wrong type of fuel, incorrect ignition timing or advance, too hot a plug, burned valves or engine overheating. Replace the plug.

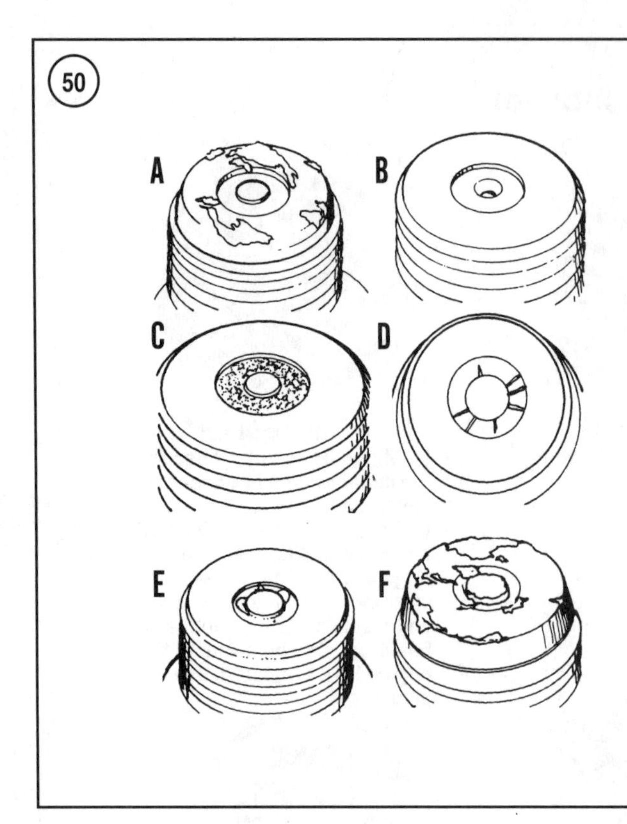

SPARK PLUG ANALYSIS (SURFACE GAP SPARK PLUGS)

A. Normal—Light tan or gray colored deposits indicate that the engine/ignition system condition is good. Electrode wear indicates normal spark rotation.

B. Worn out—Excessive electrode wear can cause hard starting or a misfire during acceleration.

C. Cold fouled—Wet oil or fuel deposits are caused by "drowning" the plug with raw fuel mix during cranking, rich carburetion or an improper fuel:oil ratio. Weak ignition will also contribute to this condition.

D. Carbon tracking—Electrically conductive deposits on the firing end provide a low-resistance path for the voltage. Carbon tracks form and can cause misfires.

E. Concentrated arc—Multi-colored appearance is normal. It is caused by electricity consistently following the same firing path. Arc path changes with deposit conductivity and gap erosion.

F. Aluminum throw off—Caused by preignition. This is not a plug problem but the result of engine damage. Check engine to determine cause and extent of damage.

1A. *All models except later production 2001 HPDI models*—Insert the appropriate size gauge (**Table 5**) between the electrodes (**Figure 55**). If the gap is correct, there will be a slight drag as the gauge passes between the electrodes.

1B. *Later production 2001 and 2002-on HPDI models (gray coil connector)*—Insert the appropriate size gauge (**Table 5**) between each ground electrode and the center electrode (**Figure 56**). If the gap is correct, there will be a slight drag as the gauge passes between the electrodes. The gap must be the same for each electrode.

CAUTION

Never attempt to narrow the spark plug gap by tapping the spark plug on a hard object. This can damage the spark plug and possibly fracture the insulator. Always use a gapping tool to increase or narrow the spark plug gap.

2. Carefully bend the side electrode with a gapping tool (**Figure 52**) to adjust the gap. Always measure the gap after the adjustment.

3. Inspect the spark plug threads in the power head and clean with a thread chaser (**Figure 57**) as needed. Occasionally the aluminum material from the spark plug opening will come out with the spark plug. A threaded insert

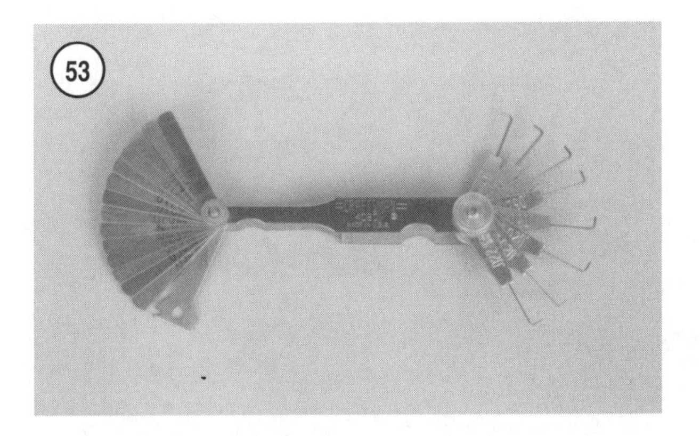

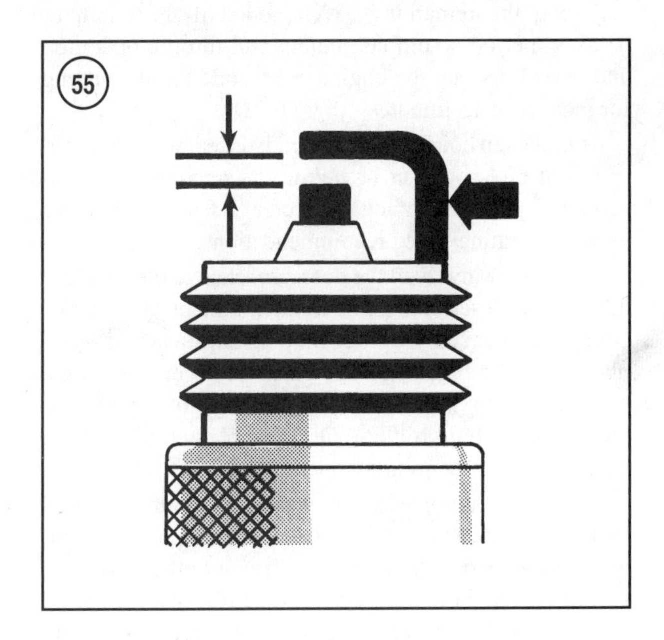

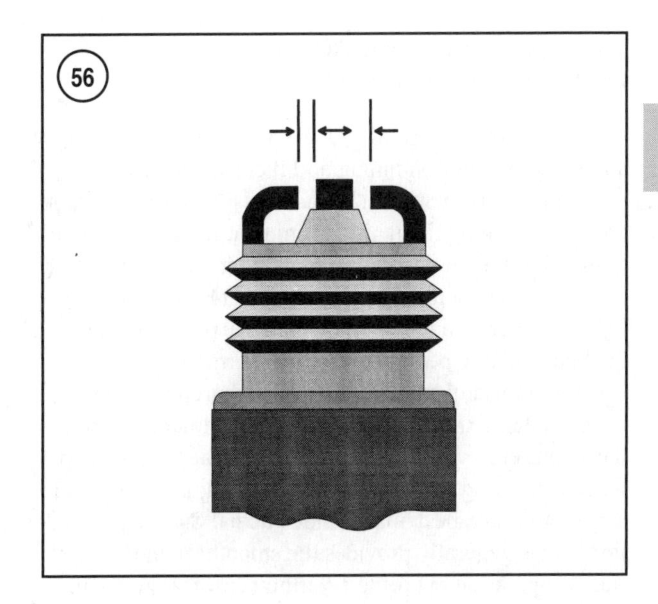

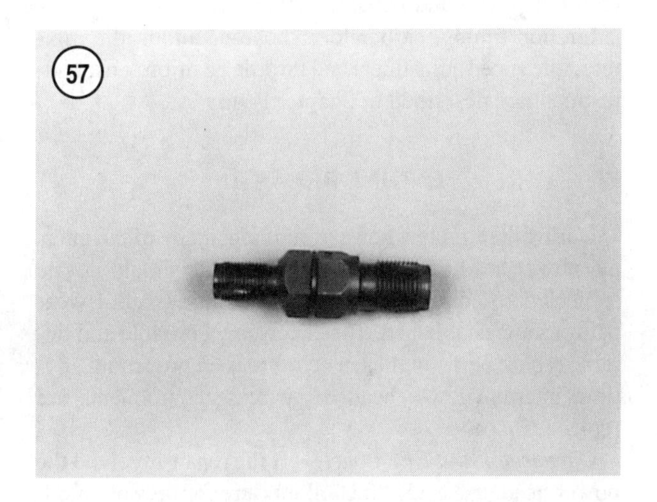

can be installed into the cylinder head to repair this condition. Have a reputable marine dealership or machine shop perform this operation unless you are familiar with this operation and have access to the required tools.

4. Make sure the gasket is installed onto the spark plug.

5. Wipe the opening clean before installing the new plug. Screw the plug into the power head by hand until it seats. Very little force is required. If force is necessary, the threads may be dirty or the plug may be cross-threaded. Unscrew the plug and clean the threads before installing the plug.

6. Tighten the spark plug to the specification in **Table 1**. If a torque wrench is not available, seat the plug finger-tight, then tighten the plug an additional quarter turn of the wrench.

7. Inspect the spark plug lead and cap before connecting it onto the spark plug. Replace the lead if the insulation is damaged or deteriorated. Replace the cap if the contacts are corroded or the cap is cracked.

8. Install the cap onto the plug and push until the spark plug terminal fully seats into the cap.

Timing, Carburetor and Throttle Position Sensor Adjustment

The remaining steps in a tune-up include checking and adjusting the ignition timing and fuel system.

Timing adjustments can be made easily at lower engine speeds. Checking the ignition timing at higher speeds requires that the engine be test ran under actual operating conditions with a timing light installed (**Figure 58**). Timing adjustments require an assistant, one person to operate the boat and one person to check the timing.

Carburetor adjustments involve carburetor synchronization, idle mixture and idle speed adjustments. Perform the carburetor synchronization and linkage adjustments before adjusting the carburetors. Final adjustments to the carburetors are best made under actual running conditions. This generally provides the smoothest and most efficient operation. Throttle position sensor adjustment is required on EFI and HPDI models.

Ignition timing, carburetor synchronization, idle mixture, idle speed adjustment and throttle position sensor adjustment are described in Chapter Four.

ENGINE BREAK-IN

During the first few hours of running, many of the internal power head and gearcase components should not be subjected to continuous full-load conditions until wear patterns are established. To help ensure a reliable and durable repair, perform the ten hour break-in procedure anytime internal power head or gearcase components are replaced.

Carburetor and EFI models—If the repair involved the power head, use a 25:1 fuel/oil mixture for pre-mix models during the first 10 hours of operation. On oil injected models, use a 50:1 fuel/oil mixture in the fuel tank during this time period in addition to the oil provided by the oil injection system.

HPDI models—Do not add oil into the fuel during the break-in period. On these models, the fuel is injected directly into the combustion chamber and any oil added to the fuel will not provide any benefit to the internal components.

Operate the engine at fast idle, approximately 1500 rpm, in NEUTRAL for the first 10 minutes of operation. During the next 50 minutes of operation, avoid full throttle operation, except to quickly plane the boat. Change the

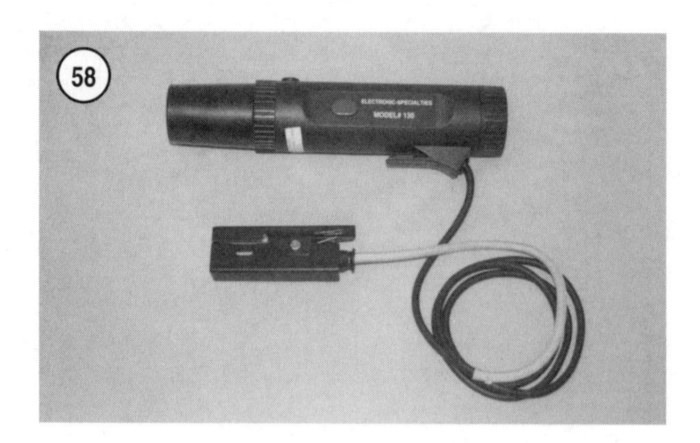

throttle setting frequently. Do not exceed 3000 rpm during the 50 minute time period.

During the second hour of operation, use full throttle only as needed to plane the boat, then reduce the throttle setting to a maximum of three-fourths open. Vary the throttle setting frequently during the second hour of operation.

During the third hour of operation, run the engine at varying throttle openings. Occasionally advance the throttle to full open for a short period, up to five minutes. Then reduce the throttle setting to three-fourths open or lower to cool the engine.

During the remaining seven hours of the ten hour break-in period, avoid continuous full-throttle operation and do not operate the engine at any one throttle setting for more than 15 minutes.

After the ten hour break-in period, the engine can be operated at any speed at or below the maximum recommended by the manufacturer. Refer to Chapter Two for engine operating speed recommendations.

If the repair involved the gearcase, change the gearcase lubricant as described in this chapter. Fine metal particles commonly form on the drain/fill plug during new gear replacement and break-in. Fine particles do not indicate a problem within the gearcase. Larger or gritty particles indicate a potential problem. Disassemble and inspect the gearcase before continuing operation.

If the repair involved the power head, use the standard fuel/oil mixture as recommended in this chapter. If the engine is equipped with oil injection, monitor the amount of fuel burned and the oil used from the reservoir. If the oil usage is not consistent with the fuel burned, test the oil pump output as described in Chapter Two.

Table 1 MAINTENANCE AND TUNE-UP TORQUE SPECIFICATIONS

Fastener	N•m	in.-lb.	ft.-lb.
Gearcase plugs	7	62	–
Spark plug	25	–	18

Table 2 MAINTENANCE SCHEDULE

After each use	Check for loose nuts, bolts, or spark plugs
	Check the propeller
	Check the oil reservoir level*
	Flush the cooling system
	Lubricate the jet drive bearings*
	Check for and correct leaking fluids
	Wash the exterior of gearcase and drive shaft housing
	Touch up paint damage on external surfaces
Before each use	Check for and correct fuel leakage
	Check the steering and controls for proper operation
	Check the oil reservoir level*
	Check for a proper cooling system operation (water stream)
	Check for proper operation of the neutral only start system
Initial 10 hours or one month	Lubricate the swivel tube, tilt tube and steering system
	Check throttle operation
	Check shift linkages for proper operation
	Check tightness of all accessible nuts and bolts
	Check power tilt and trim operation*
	Check choke lever operation*
	Inspect fuel filter for contamination
	Inspect fuel hoses and connections
	Adjust the idle speed* (Chapter Four)
	Inspect mid-section components (Chapter Eleven)
	Inspect the spark plug(s)
	Adjust the oil pump linkage* (Chapter Four)
	Inspect oil reservoir for water or contamination*
	Check electrical wiring and connections
	Check power head for water and exhaust leakage
	Check gearcase lubricant level and condition
	Check condition and charge level of battery
	Check carburetor synchronization and adjustments
	Check cylinder compression (Chapter Two)
Initial 50 hours or 90 days	Lubricate the swivel tube, tilt tube and steering system
	Adjust the carburetor(s)
	Inspect fuel filter for contamination
	Check spark plug condition and gap
	Check and adjust the ignition timing
	Check the oil injection system*
	Check electrical wiring and connections
	Check power head for water and exhaust leakage
	Check gearcase lubricant level and condition
	Inspect the water pump impeller (Chapter Eight)
	Check the propeller
	Check propeller nut for tightness
	Clean and inspect sacrificial anodes
	Check all accessible nuts and bolts for tightness
	Check cylinder compression (Chapter Two)
Each 100 hours of usage or 180 days	Lubricate the swivel tube, tilt tube and steering system
	Check carburetor synchronization and adjustments
	Inspect fuel filter for contamination
	Check fuel hoses and clamps for leakage
	Check the spark plug condition and gap

(continued)

Table 2 MAINTENANCE SCHEDULE (continued)

Each 100 hours of usage or 180 days (continued)	Check the power tilt/trim fluid level (Chapter Ten) Inspect the mid-section components. Check the oil injection system* Check electrical wiring and connections Check power head for water and exhaust leakage Check gearcase lubricant level and condition Check the condition and charge level of the battery* Clean and inspect sacrificial anodes Check all accessible nuts and bolts for tightness Check the propeller nut for tightness Check cylinder compression
Each 200 hours of usage or one year	Inspect fuel tank, hoses and clamps Clean or rep0lace the fuel filter Replace the water pump impeller Check the fuel pump oil level* (HPDI models) Adjust the throttle position sensor (EFI and HPDI models) Inspect the fuel pump drive belt* (HPDI models)
Each 1000 hours or five years	Change the fuel pump oil* (HPDI models) Replace the fuel pump drive belt* (HPDI models)

*This maintenance item does not apply to all models.

Table 3 FLUID CAPACITIES

Model	Capacity (approximate)
Gearcase	
115 and 130 hp	
Standard RH rotation	760 ml (25.7 oz.)
Optional LH rotation	715 ml (24.2 oz.)
150-200 hp (2.6 liter models)	
Standard RH rotation	980 ml (33.1 oz.)
Optional LH rotation	870 ml (29.4 oz.)
Twin counter rotating propellers	900 ml (30.4 oz.)
200-250 hp (3.1 liter models)	
Standard RH rotation	1.15 L (38.9 oz.)
Optional LH rotation	1.0 L (33.8 oz.)
High pressure fuel pump	50 ml (1.7 oz.)
Oil Reservoir	
115 and 130 hp	
On board reservoir	10.5 L (11.1 qt.)
Engine mounted reservoir	0.9 L (0.95 qt.)
150-200 hp (2.6 liter models)	
On board reservoir	10.5 L (11.1 qt.)
Engine mounted reservoir	0.9 L (0.95 qt.)
200-250 hp (3.1 liter models)	
On board reservoir	10.5 L (11.1 qt.)
Engine mounted reservoir	1.2 L (1.27 qt.)

Table 4 OIL AND FUEL MIXING RATES

Quantity of fuel	Oil for 50:1 ratio	Oil for 25:1 ratio
3.8 L (1 gal.)	76 cc (2.6 oz.)	152 cc (5.2 oz)
7.6 L (2 gal.)	152 cc (5.2 oz)	304 cc (10.4 oz.)
11.4 L (3 gal.)	228 cc (7.8 oz.)	456 cc (15.6 oz.)
15.4 L (4 gal.)	304 cc (10.4 oz.)	608 cc (20.8 oz.)

(continued)

Table 4 OIL AND FUEL MIXING RATES (continued)

Quantity of fuel	Oil for 50:1 ratio	Oil for 25:1 ratio
18.9 L (5 gal.)	380 cc (12.8 oz.)	760 cc (25.6 oz.)
22.8 L (6 gal.)	456 cc (15.6 oz.)	912 cc (31.2 oz.)
26.6 L (7 gal.)	530 cc (18.2 oz.)	1060 cc (36.4 oz.)
30.8 L (8 gal.)	608 cc (20.8 oz.)	1216 cc (41.6 oz.)
34.2 L (9 gal.)	684 cc (23.4 oz.)	1368 cc (46.8 oz.)
37.8 L (10 gal.)	760 cc (25.6 oz.)	1520 cc (51.2 oz.)
41.6 L (11 gal.)	832 cc (28.2 oz.)	1664 cc (56.4 oz.)
45.6 L (12 gal.)	912 cc (31.2 oz.)	1824 cc (62.4 oz.)

Table 5 SPARK PLUG SPECIFICATIONS

Model	NGK plug	Champion plug	Gap
115 hp	BR8HS-10	QL78C	1.0 mm (0.039 in.)
130 hp	BR9HS-10	QL77CJ4	1.0 mm (0.039 in.)
150 hp			
C150	B8HS-10	L82C	1.0 mm (0.039 in.)
D150	BR8HS-10	QL82C	1.0 mm (0.039 in.)
P150, S150, L150, DX150, LX150, PX150 SX150, VX150	BR7HS-10	RL82C	1.0 mm (0.039 in.)
HPDI			
With black coil connector	BKR6E-S1-10	*	1.0-1.1 mm (0.039-0.043 in.)
With gray coil connector	BKR6E-KU-10	*	0.6 mm (0.024 in.)
175 hp			
Carburetor and EFI models	B8HS-10	QL78C	1.0 mm (0.039 in.)
HPDI models			
With black coil connector	BKR7E-S1-10	*	1.0-1.1 mm (0.039-0.043 in.)
With gray coil connector	BKR7E-KU-10	*	0.6 mm (0.024 in.)
200 hp			
2.6 Liter models with carburetors	B8HS-10	QL78C	1.0 mm (0.039 in.)
2.6 Liter models with EFI	BR8HS-10	QL78C	1.0 mm (0.039 in.)
3.1 Liter models	BR8HS-10	QL78C	1.0 mm (0.039 in.)
HPDI model			
With black coil connector	BKR7E-S1-10	*	1.0-1.1 mm (0.039-0.043 in.)
With gray coil connector	BKR7E-KU-10	*	0.6 mm (0.024 in.)
225 and 250 hp (except HPDI)	BR9HS-10	*	1.0 mm (0.039 in.)

*The manufacturer does not provide a Champion spark plug part No. for this model.

Chapter Four

Timing, Synchronization and Adjustment

Synch and link refers to the correct adjustment of both the ignition and fuel systems. Refer to **Tables 1-4** for synch and link specifications at the end of this chapter. Adjustments to the shift linkages, jet drive, and trim position sender are in *Miscellaneous Adjustments* at the end of this chapter. Read *Safety Precautions* before beginning any procedures.

TIMING, SYNCHRONIZATION AND ADJUSTMENT

If an outboard engine is to deliver maximum efficiency, performance and reliability, the engine should have both the ignition and fuel systems correctly adjusted. Failure to properly synch and link an engine may lead to power head failure. Perform all synch and link procedures during a tune-up or when replacing, servicing, or adjusting any ignition or fuel system components. Synch and link procedures differ according to engine model, fuel system, and lubrication system.

To make the most accurate adjustments possible, perform synch and link procedures with the engine running under actual operating conditions.

CAUTION
Do not run the engine without an adequate water supply and do not exceed 3000 rpm without an adequate load. Refer to ***Safety Precautions*** *at the beginning of this chapter.*

Safety Precautions

Wear approved eye protection at all times, especially when machinery is in operation. Wear approved ear protection during all running tests and in the presence of noisy machinery. Keep loose clothing tucked in and long hair tied back and secured.

When making or breaking any electrical connection, always disconnect the negative battery cable. When performing tests that require cranking the engine without starting, disconnect and ground the spark plug leads to prevent accidental starts and sparks.

Securely cap or plug all disconnected fuel lines to prevent fuel discharge when cranking the engine or squeezing the primer bulb.

Thoroughly read all manufacturers' instructions and safety sheets when using test equipment and special tools.

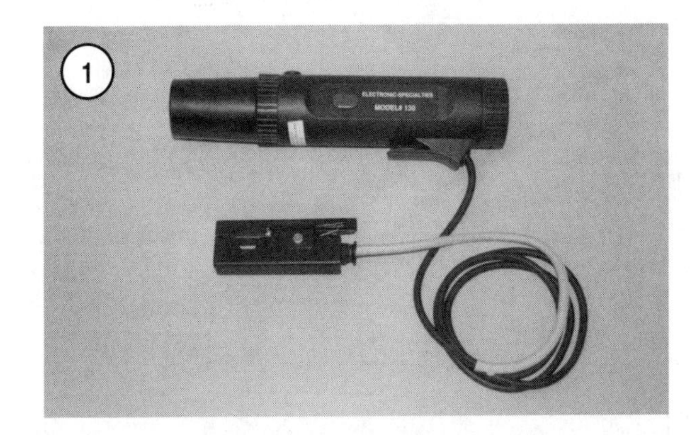

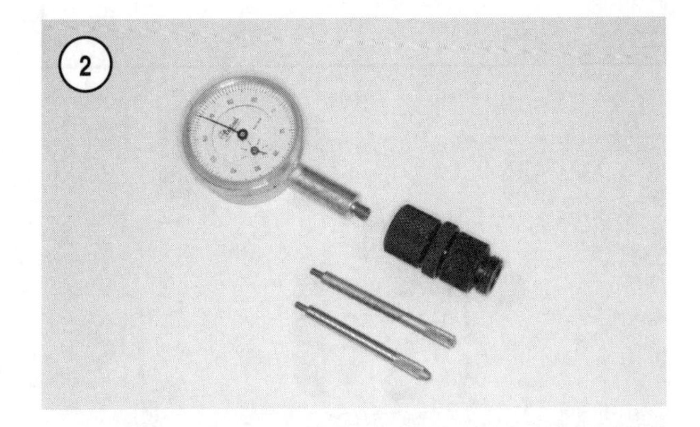

Do not substitute parts unless they meet or exceed the original manufacturer's specifications.

Never run an outboard engine without an adequate water supply. Never run an outboard engine at wide-open throttle without an adequate load. Never exceed 3000 rpm in neutral (no load).

Always perform on water tests with at least two people; one person to operate the boat, the other to monitor the gauges or test instruments. All personnel must remain seated inside the boat at all times. Do not lean over the transom while the boat is under way. Use extensions to allow all gauges and meters to be located in the normal seating area.

A test propeller is an economical alternative to the dynamometer. A test propeller is also a convenient alternative to on-water testing. A test propeller can be made by modifying (turning down) the diameter of a standard low pitch aluminum propeller until the recommended wide-open throttle speed can be obtained with the engine mounted on the boat and the trailer backed into the water. Propeller repair stations can provide the modification service. Normally, approximately 1/3 to 1/2 of the outer blade surface is removed. However, it is far better to remove too little than too much. It may take several tries to achieve the correct full throttle speed, but once achieved, no further modifications are required. Many propeller repair stations have experience with this type of modification and may be able to recommend a starting point.

Be careful if the boat is tied to the dock. The test propeller develops considerable thrust. Some docks may not be able to withstand the load.

Test propellers also allow simple tracking of engine performance. The full-throttle test speed of an engine fitted with a correctly modified test propeller can be tracked from season to season. It is not unusual for a new or rebuilt engine to show a slight increase in test propeller speed as complete break-in is achieved and then to hold that speed over the normal service life of the engine. As the engine begins to wear out, the test propeller speed shows a gradual decrease. Tracking the engine performance is a good point of reference to keep the engine running efficiently and safely.

Ignition Timing

All models use some form of timing marks that allow the ignition timing to be checked using a suitable stroboscopic timing light (**Figure 1**). On models with adjustable timing, a linkage adjustment is made to bring the timing into specification. If the timing is not within specification on models with nonadjustable timing, either a mechanical or electrical defect is present in the system. Chapter Two covers ignition troubleshooting for all models.

The full advance timing is best checked at wide-open throttle. This method is not always practical, however, as the outboard must be operated at full throttle in forward gear (under load) to verify maximum timing advance. If at all possible, use a test propeller, as timing an engine while speeding across open water can be hazardous. Refer to *Safety Precautions* in the previous section.

The maximum timing advance on some models is set by holding the ignition linkage in the full-throttle position while the engine is being cranked. This is an acceptable (where noted). This procedure is not considered as accurate as the full advance throttle check. Whenever possible, check the full advance timing under actual operating conditions.

Timing equipment

Static adjustment of the ignition timing and/or verification of the timing pointer requires the use of a suitable dial indicator (**Figure 2**) to position the No. 1 piston at top dead center (TDC) accurately before making timing adjustments. This procedure required removal of the No. 1 spark plug. The dial indicator threads into the spark plug opening. Adjustments to the timing pointer synchronize the pointer to the actual piston position relative to TDC. See **Figure 3**.

All ignition timing checks and adjustments require the use of a stroboscopic timing light connected to the No. 1 spark plug lead. As the engine is cranked or operated, the light flashes each time the spark plug fires. When the light is pointed at the moving flywheel, the mark on the flywheel appears to stand still (**Figure 4**). The specified timing marks will align if the timing is correctly adjusted.

CAUTION

*Factory timing specifications provided by Yamaha Marine are listed in **Table 4**. However, Yamaha Marine has occasionally found it necessary to modify their specifications during production. If the engine has a decal attached to the power head silencer cover, always follow the specification listed on the decal instead of the specification in **Table 4**.*

NOTE

Do not use timing lights with built-in features (such as a timing advance function) on outboard engines. Instead, use a basic high-speed timing light with an inductive pickup. Yamaha part No. YM-33277 fulfills these requirements.

Use an accurate shop tachometer to determine engine speed during timing adjustment. Do not rely on the tachometer installed in a boat to provide accurate engine speed readings. Use Yamaha part No. YU-8036-A or other suitable tachometer.

SYNCH AND LINK (80 JET, 105 JET, AND 115-200 HP [EXCEPT EFI AND HPDI MODELS])

These models use mechanical timing advancement. Perform the following adjustments and checks:

1. Static timing adjustment.
2. Carburetor synchronization.
3. Pilot screw adjustment.
4. Idle speed adjustment.
5. Pickup timing adjustment.
6. Oil pump adjustment.
7. Throttle cable adjustment.
8. Ignition timing check.

WARNING

Use extreme caution when working around a running engine. It is easy to get entangled in the flywheel or propeller and suffer serious injury or death. Stay clear of the flywheel, pulleys and the propeller.

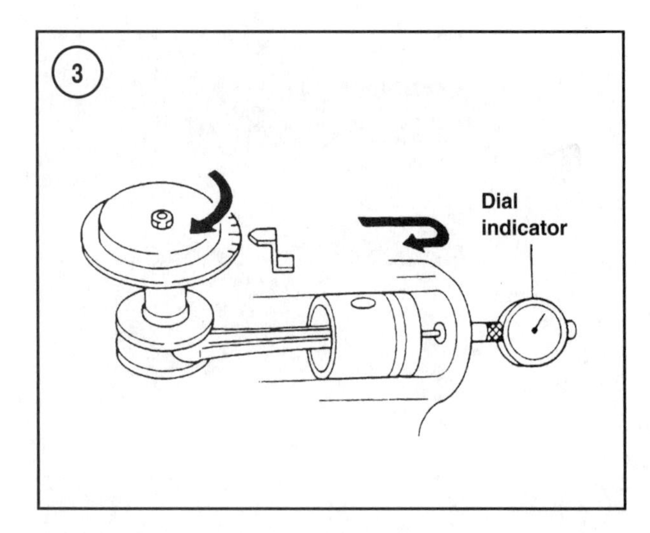

CAUTION

Never run an outboard without first providing cooling water. The water pump will suffer damage within a few seconds if ran dry. A damaged water pump may supply insufficient cooling water to the power head and cause overheating and eventual power head failure.

Static Timing Adjustment

1. Disconnect the battery cables and ground the spark plug leads. Remove all four spark plugs.
2. Disconnect the throttle cable (**Figure 5**) from the throttle arm.
3. Remove the flywheel cover.
4. Observe the timing pointer and markings on the flywheel (**Figure 6**) while slowly rotating the flywheel in the

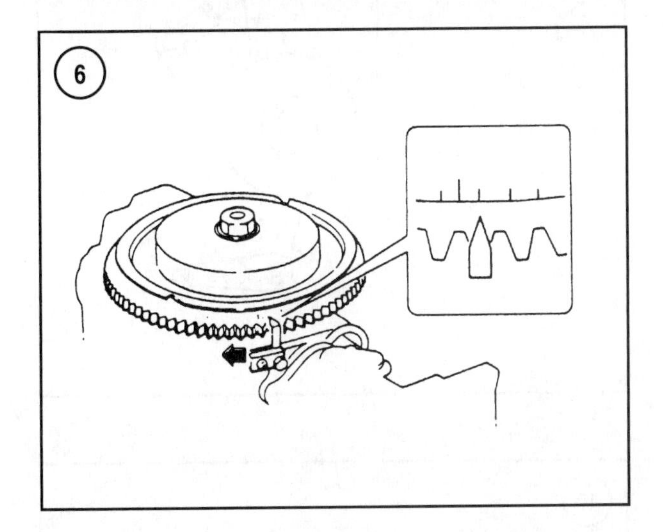

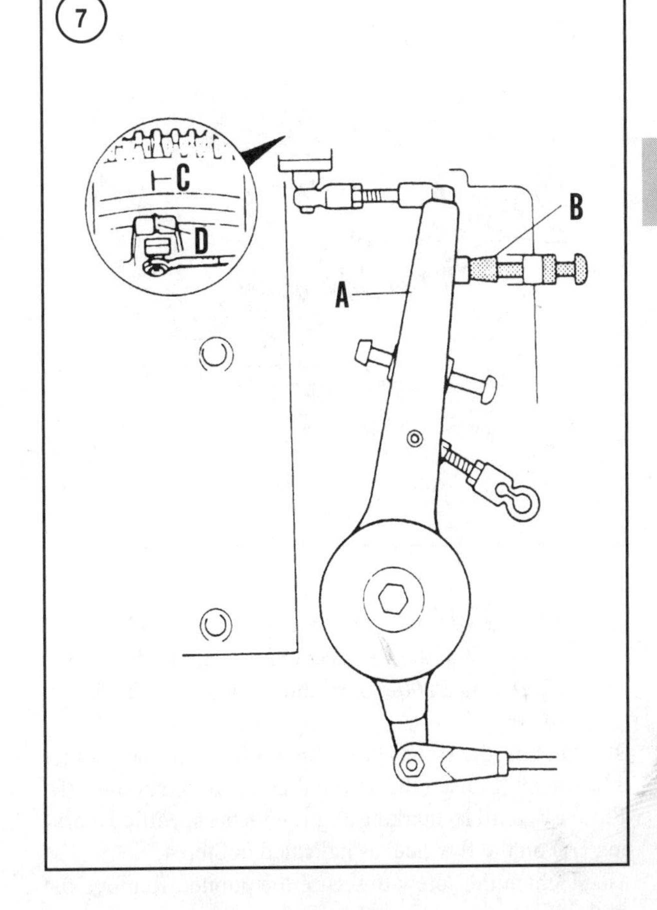

clockwise direction. Stop when the timing pointer aligns with the specified marking.

a. *80 Jet and 115 hp*—Stop at the 25° BTDC marking.

b. *130 hp, 105 Jet, 150, S150, L150 and S175*—Stop at the 22° BTDC marking.

c. *C150*—Stop at the 19° BTDC marking.

d. *D150, P150, P175, V175 and P200*—Stop at the 23° BTDC marking.

e. *V150 and 200 hp (except L200, S200 and P200)*—Stop at the 20° BTDC marking.

f. *S200 and L200*—Stop at the 21° BTDC marking.

5. Maintain the flywheel in this position. Move the throttle arm (A, **Figure 7**) until it contacts the plastic cap on the full timing advance adjusting screw (B). Look under the flywheel to check the alignment of the marks on the pulser coil housing (D, **Figure 7**) and the flywheel (C). The marks must align. If the marks align go to Step 13. If not, perform Steps 6-14.

6. Thread the dial indicator mount into the No. 1 spark plug opening. The No. 1 spark plug opening is the top cylinder on the starboard cylinder bank. Install the dial indicator into the mount with the indicator stem inserted into the No. 1 spark plug opening. Adjust the mount to position the stem in direct contact with the piston as it nears the top of its compression stroke (**Figure 3**).

7. Slowly rotate the flywheel in the clockwise direction while observing the dial indicator. Stop when the flywheel reaches the top of its stroke. This is the point of rotation when the needle on the dial indicator just reverses its direction as the flywheel is rotated.

8. Rotate the dial on the dial indicator until the needle aligns with the *0* marking on the dial. Observe the dial indicator while slowly rotating the flywheel in the counterclockwise direction. Stop rotating the flywheel when the dial indicates the piston has moved down the specified distance BTDC.

a. *For 80 Jet and 115 hp models*—Stop the flywheel at 3.91 mm (0.154 in.).

b. *130 hp, 105 Jet, 150, S150, L150 and S175*—Stop the flywheel at 3.05 mm (0.120 in.).

c. *C150*—Stop the flywheel at 2.28 mm (0.089 in.).

d. *D150, P150, P175, V175 and P200*—Stop the flywheel at 3.33 mm (0.131 in.).

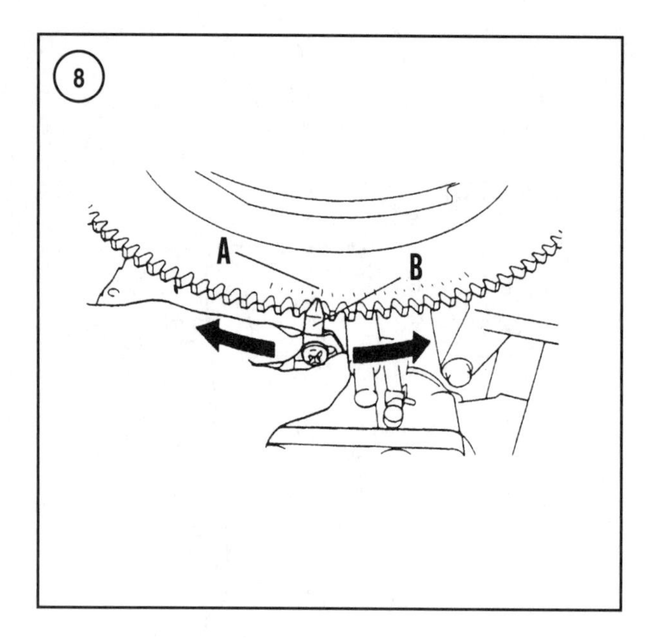

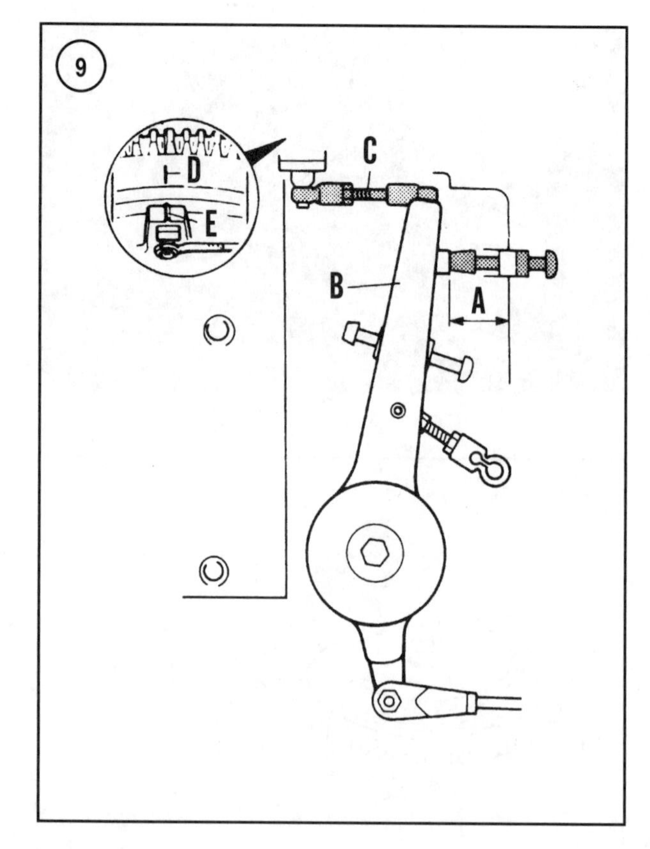

e. *V150 and 200 hp (except L200, S200 and P200*—Stop the flywheel at 2.53 mm (0.099 in.).
f. *S200 and L200*—Stop the flywheel at 2.78 mm (0.109 in.).

9. Maintain the flywheel in this position during pointer adjustment. Loosen the screw, then move the pointer (B, **Figure 8**) until its marking aligns with the specified marking (A) on the flywheel as indicated in Step 4.
10. Tighten the screw to secure the pointer. Remove the dial indicator from the spark plug opening.
11. Turn the full advance adjusting screw to achieve the indicated length (A, **Figure 9**).
   a. *80 Jet and 115 hp models (except C115)*—Adjust the length to 26.0 mm (1.02 in.).
   b. *C115 models*—Adjust the length to 22.0 mm (0.87 in.).
   c. *130 hp models*—Adjust the length to 20.9 mm (.82 in.).
12. Carefully pry the plastic ball socket connector on the linkage (C, **Figure 9**) from the throttle arm. Loosen the jam nut on the linkage. Move the linkage until the marking on the pulser coil housing (E, **Figure 9**) aligns with the marking on the flywheel (D). Verify that the timing pointer remains aligned as specified in Step 4. Rotate the linkage into or out of the connector until the linkage can be connected without disturbing the marking alignment (D and E, **Figure 9**). Securely tighten the jam nut. Then snap the linkage onto the throttle arm.
13. Observe the timing pointer and markings on the flywheel (**Figure 6**) while rotating the flywheel in the clockwise direction. Stop when the timing pointer aligns with the specified marking on the flywheel. For 80 Jet and

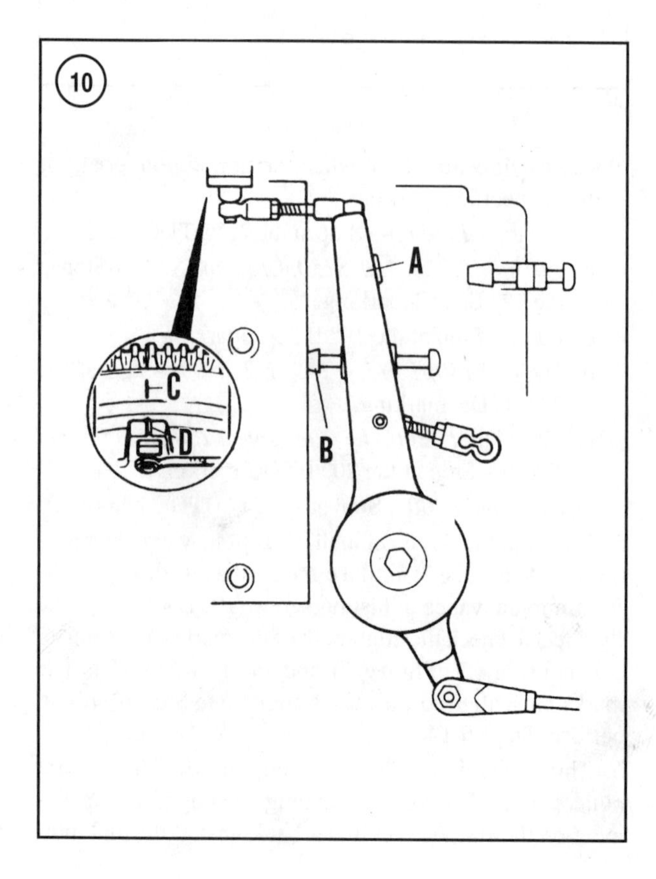

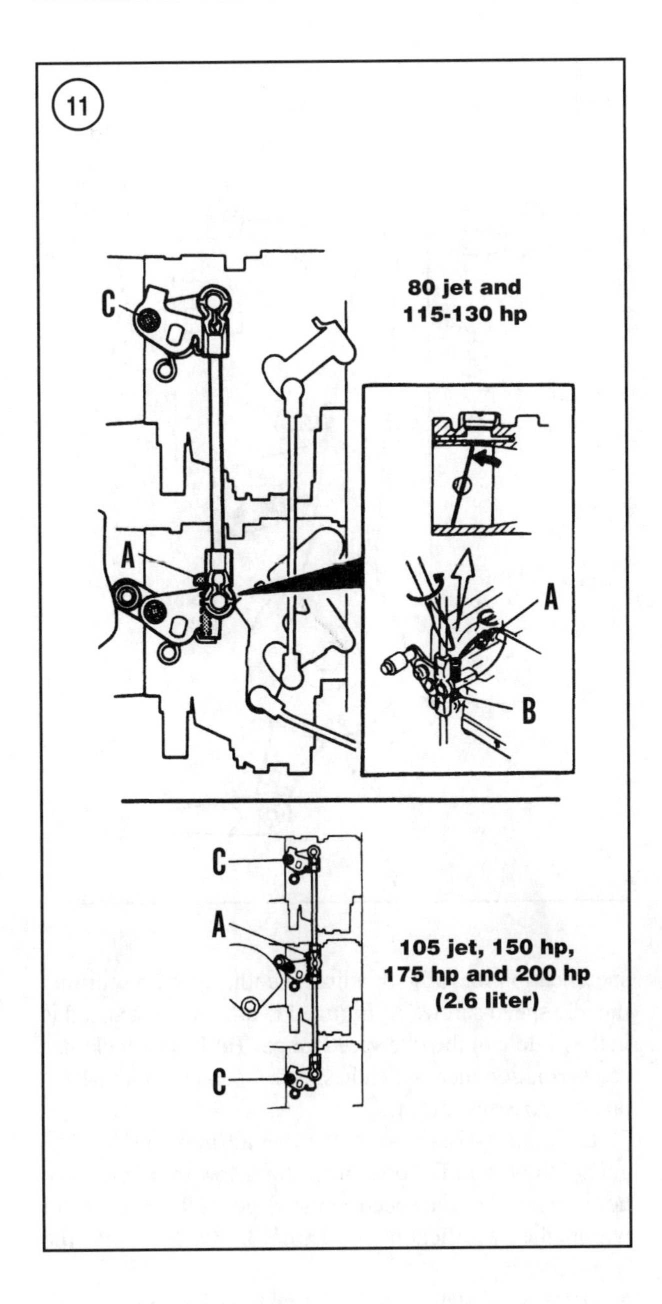

115-130 hp models, align the pointer with the 5° ATDC marking.

14. Move the throttle arm until the cap on the full retard adjusting screw (B, **Figure 10**) just contacts the stop on the cylinder block. Maintain the throttle arm in this position during the adjustment. Loosen the jam nut, then turn the full retard adjusting screw until the marking on the pulser coil housing (D, **Figure 10**) aligns with the marking on the flywheel (C). Securely tighten the jam nut.

15. Install the spark plugs. Do not connect the spark plug leads at this time.

16. Reconnect the throttle cable onto the throttle arm (**Figure 5**).

17. Synchronize the carburetors as described in the following chapter.

Carburetor Synchronization

1. Remove the silencer cover as described in Chapter Five. Do not remove the carburetors.
2. Move the throttle arm until the cap on the full retard adjusting screw (B, **Figure 10**) contacts the stop on the cylinder block. Maintain the arm in this position during the adjustment.
3. Look into the No. 2 carburetor, then loosen the idle speed screw (A, **Figure 11**) until the throttle shutter just reaches the full closed position. Loosen the screw on the throttle roller lever.
4. Loosen the throttle lever screws (C, **Figure 11**). Inspect the throttle shutters to ensure all are fully closed.
5. Lightly pull up on the throttle linkage to remove the slack. Support the linkage while securely tightening the throttle lever screws (C, **Figure 11**). Move the throttle roller lever until the roller just contacts the throttle cam (**Figure 12**). Hold the roller in light contact with the cam. Then securely tighten the screw in the throttle roller lever.
6. Look into the carburetor openings while slowly turning the idle speed screw (A, **Figure 11**). Stop turning when the throttle shutters just begin to move.
7. While looking into the carburetor openings, move the throttle roller to the full throttle and back to the idle position. All throttle shutters must move at the same time and return to the same partially open position when closed. If not, repeat the synchronization procedure.
8. Install the silencer cover as described in Chapter Five.

Pilot Screw Adjustments

CAUTION
Use extreme caution when seating the pilot screws prior to adjustment. The tapered seat

will be permanently damaged if using excessive force. Turn the screw with very little force and stop the instant resistance is felt.

1. Disconnect the battery cables and ground the spark plug leads.
2. Locate the two pilot screws (**Figure 13**) on each carburetor.
3. One at a time, turn the pilot screws clockwise until they lightly seat. Then, turn each screw counterclockwise the number of turns specified in **Table 2**. The specification may vary by screw location on the power head.
4. Repeat Step 3 for all pilot screws.
5. Adjust the idle speed as described in this chapter.

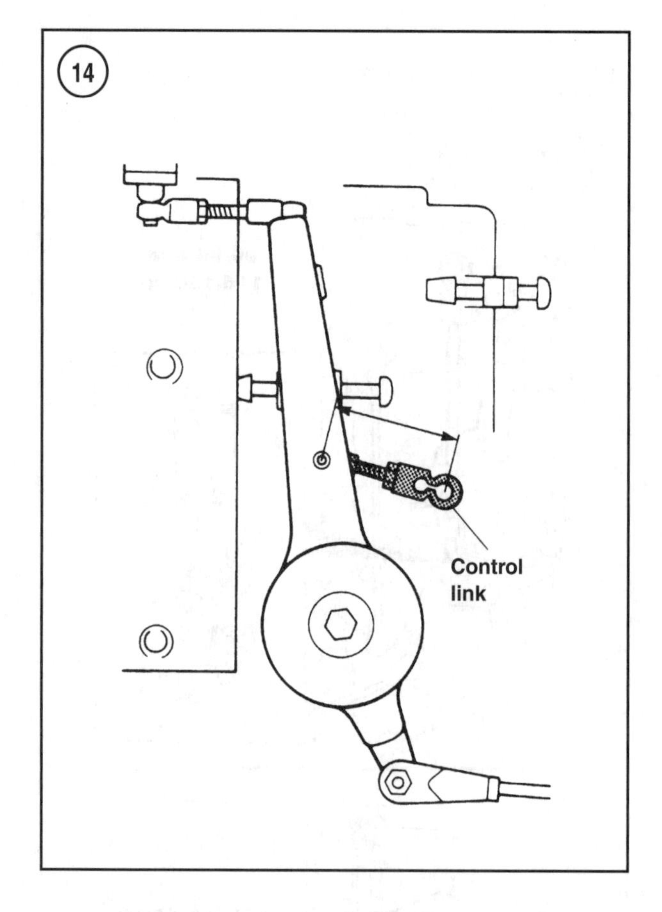

Idle Speed Adjustment

Perform this adjustment under actual running conditions. Adjust the idle speed while a qualified assistant operates the controls. Operate the boat in a safe area away from boat traffic.

1. Turn the idle speed screw (A, **Figure 11**) counterclockwise until a gap exists between the screw tip and the carburetor throttle lever. Slowly rotate the screw clockwise until the tip just contacts the lever. Turn the screw one complete turn clockwise to slightly open the throttle.
2. Place the remote control in the NEUTRAL gear idle position. Check the throttle to ensure that the cap on the full retard adjusting screw (B, **Figure 10**) contacts the stop on the cylinder block. If not, adjust the throttle cable as described in this chapter.
3. Connect the battery cables and spark plug leads.
4. Connect a suitable shop tachometer onto the spark plug lead.
5. Start the engine and advance the throttle to approximately 1500 rpm in NEUTRAL gear for ten minutes or until the engine reaches full operating temperature.
6. Place the throttle control in the idle position and note the idle speed. If the in-neutral idle speed is not within the specification in **Table 3**, adjust the idle speed by turning the idle speed screw (A, **Figure 11**) until the idle speed is in the middle of the idle speed range (**Table 3**). Clockwise screw rotation increases idle speed. Counterclockwise rotation decreases idle speed.
7. Have an assistant shift the engine into FORWARD gear. Allow the idle to stabilize for a few minutes. Then note the in-gear idle speed. If the in-gear idle speed is not within the specification in **Table 3**. Readjust the idle speed screw.
8. Have an assistant shift the engine into NEUTRAL gear. Allow the idle to stabilize for a few minutes. Then note the idle speed. If the neutral gear idle speed does not remain within the specification in **Table 3**. Readjust the idle speed screw until both the in-gear and in-neutral idle speeds are within specifications. Several adjustments may be required.
9. Shift the engine into FORWARD gear and allow a few minutes for the idle to stabilize. Slowly advance the throttle. Then bring it down to the idle position. Check for binding linkages or readjust the idle speed if the engine does not return to the specified range within 30-40 seconds.
10. Adjust the pickup timing as described in this chapter.

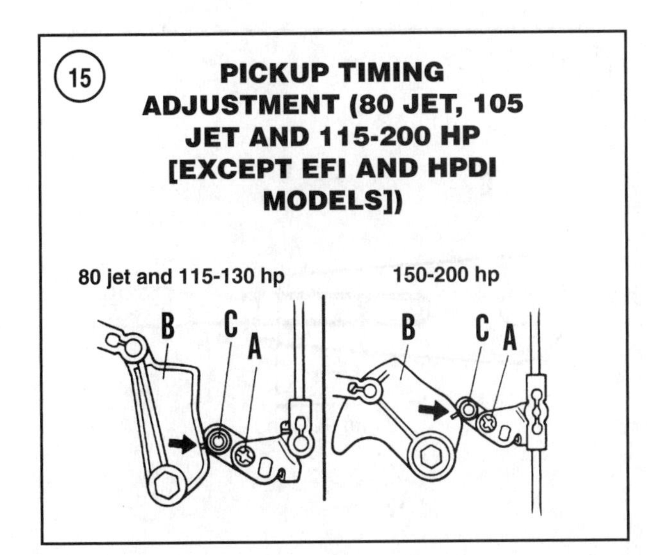

15 PICKUP TIMING ADJUSTMENT (80 JET, 105 JET AND 115-200 HP [EXCEPT EFI AND HPDI MODELS])

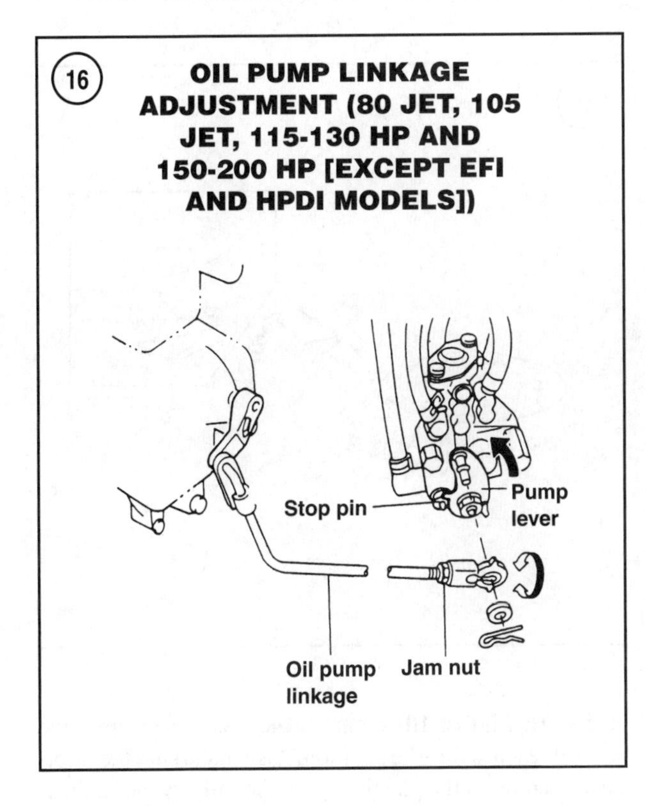

16 OIL PUMP LINKAGE ADJUSTMENT (80 JET, 105 JET, 115-130 HP AND 150-200 HP [EXCEPT EFI AND HPDI MODELS])

Pickup Timing Adjustment

Throttle pickup is the point in throttle advancement in which the throttle arm and cam just start to open the carburetors. A control link (**Figure 14**) connects the throttle arm to the throttle cam. The throttle cam contacts the throttle roller (**Figure 12**) that opens the carburetors. Proper adjustment allows the throttle cam to contact the roller at the proper ignition timing. Changing the length of the control link moves the cam in relation to the roller.

Perform this adjustment under actual running conditions. Adjust the idle speed while a qualified assistant operates the controls. Operate the boat in a safe area away from boat traffic.

1. Adjust the idle speed prior to adjusting the pickup timing.
2. Shift the engine into FORWARD gear.
3. Connect a timing light onto the No. 1 spark plug lead. The No. 1 cylinder is the top cylinder on the starboard side.
4. Direct the timing light onto the timing pointer (**Figure 6**).
5. Observe the throttle roller (C, **Figure 15**) while slowly advancing the throttle. Stop when the throttle cam (B, **Figure 15**) just contacts the roller and note the ignition timing. Also note the point of contact on the throttle cam. The marking on the throttle cam should align with the center of the roller upon contact and the pickup timing at this point should be within the specification in **Table 4**. If not, adjust as follows:
   a. Loosen the screw (A, **Figure 15**) to allow the roller to pivot.
   b. Adjust the length of the control link (**Figure 14**) until the cam contacts the roller at the specified point. Securely tighten the screw (A, **Figure 15**).
   c. Check the pickup timing and readjust as necessary. Several adjustments may be required.
6. Stop the engine. Adjust the oil pump as described in this chapter.

Oil Pump Linkage Adjustment

1. Disconnect the battery cables and ground the spark plug leads.
2. Locate the oil pump on the lower port side of the power head.
3. Place the throttle in the fully closed position. If the tab on the pump lever does not contact the stop pin (**Figure 16**) on the oil pump body, adjust as follows:
   a. Remove the spring clip and washer. Carefully pry the oil pump linkage connector from the pump lever (**Figure 16**).
   b. Rotate the pump lever in the counterclockwise direction until the tab on the pump lever just contacts the stop pin (**Figure 16**). Hold the pump lever in this position during adjustment.
   c. Loosen the jam nut. Rotate the connector until it aligns with the ball pin on the pump lever.
   d. Carefully snap the linkage connector onto the pump lever ball pin. Install the washer and spring clip.

4

e. Support the linkage connector while securely tightening the jam nut.
4. Place the throttle in the fully open position. If the tab on the pump lever does not lightly contact the stop pin on the pump body, readjust the linkage as described in Step 3.
5. Move the throttle to fully closed and fully open positions several times while checking for binding or linkage interference with other components. Correct any binding or interference before operating the engine.
6. Connect the battery cables and spark plug leads.

Throttle Cable Adjustment

1. Perform the static timing adjustments and carburetor synchronization prior to adjusting the throttle cable.
2. Disconnect the battery cables and ground the spark plug leads to prevent accidental starting.
3. Locate the cable connection on the throttle arm (**Figure 5**).
4. Remove the locking clip and washer. Carefully pull the cable from the throttle arm pin.
5. Move the throttle arm (A, **Figure 10**) until the plastic cap (B) on the full retard adjusting screw contacts the stop on the cylinder block. Hold the arm in this position during the adjustment.
6. Loosen the jam nut (A, **Figure 17**) to allow rotation of the cable connector (B). Rotate the cable connector until it can be slid over the throttle arm pin without moving the throttle arm away from the stop.
7. Check the cable connector for adequate thread engagement. If the threaded end of the cable does not thread into the connector a minimum of 8 mm (0.31 in.), check the timing and synchronization adjustments. Install the connector onto the throttle arm pin. Then install the washer and locking clip. Securely tighten the jam nut (A, **Figure 17**) against the cable connector (B).
8. With the throttle in the idle position, use light pressure while moving the screw cap (B, **Figure 10**) away from the full retard stop. If the arm can be moved with light pressure, perform the following steps:
   a. Disconnect the cable connector from the throttle arm.
   b. Adjust the connector to remove the free movement. Do not excessively preload the cable. Make sure the threaded end of the cable threads into the connector a minimum of 8 mm (0.31 in.).
   c. Install the connector onto the throttle arm pin. Secure the connector with the washer and locking clip.
   d. Securely tighten the jam nut (A, **Figure 17**).
9. Move the remote control lever from idle to the full throttle several times while checking for binding or looseness. Make sure the plastic cap on the full retard adjusting screw (B, **Figure 10**) contacts the block each time the control reaches the idle position. Also make sure the throttle arm contacts the plastic cap on the full advance adjusting screw each time the remote control reaches the full throttle position. Readjust the cable if the arm fails to reach either stop.
10. Connect the battery cables and spark plug leads.

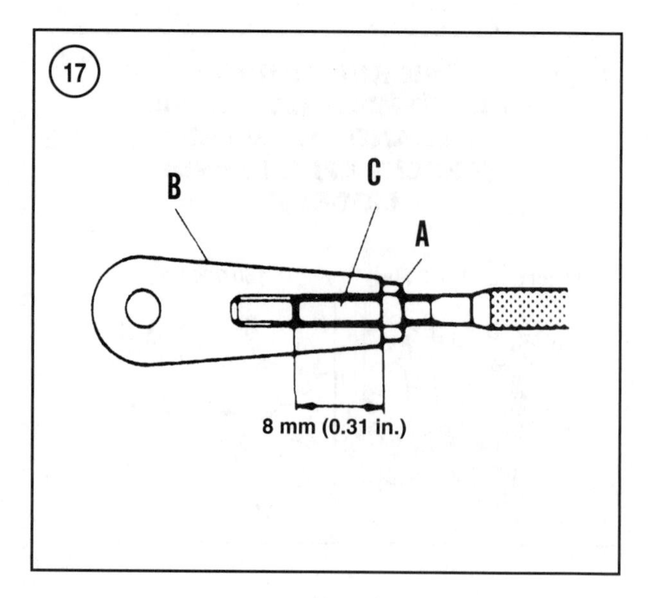

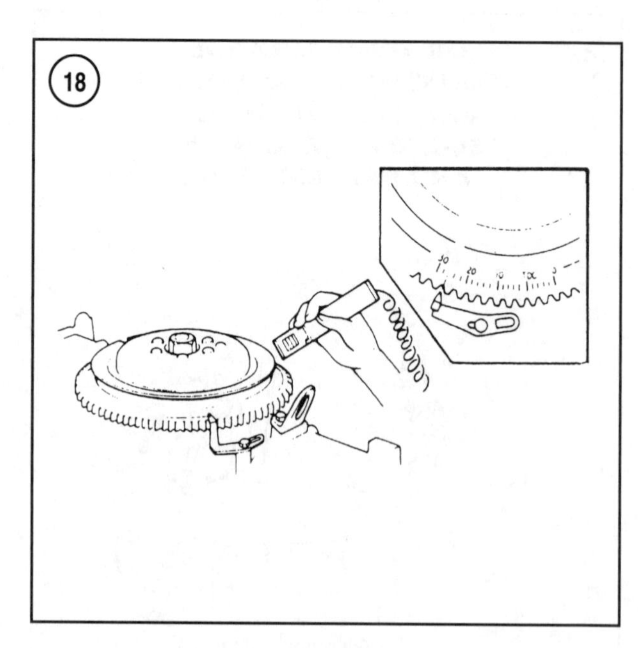

Ignition Timing Check

The ignition timing must be checked under actual running conditions. Check the timing while a qualified assis-

tant operates the controls. Operate the boat in a safe area away from boat traffic.

1. Connect a suitable timing light onto the No. 1 (top) spark plug lead. The No. 1 cylinder is the top cylinder on the starboard bank.

2. Start the engine and advance the throttle to approximately 1500 rpm in NEUTRAL gear for ten minutes or until the engine reaches full operating temperature.

3. Place the throttle in the idle position and allow the idle speed to stabilize for a few minutes. Direct the timing light onto the timing pointer and flywheel markings (**Figure 18**). When the light flashes, the timing pointer should align with the idle speed timing mark(s) specified in **Table 4**.

4. Shift the engine into FORWARD gear. Observe the tachometer and timing pointer while having an assistant advance the throttle to full open. The timing pointer (**Figure 18**) should align with the full timing advance marking specified in **Table 4**.

5. Return the throttle to idle. Shift the engine into NEUTRAL, then stop the engine.

6. If the markings do not align as specified, readjust the static timing.

SYNCH AND LINK (150-250 HP [EFI AND HPDI MODELS])

These models use electronic timing advancement, so timing adjustment is not required. Synchronization and linkage adjustment and an ignition timing check are required. Perform the following adjustments and checks:

1. Throttle valve synchronization.
2. Idle speed adjustment.
3. Throttle position sensor adjustment.
4. Throttle cable adjustment.
5. Oil pump adjustment.
6. Ignition timing check.

WARNING
Use extreme caution when working around a running engine. It is easy to get entangled in the flywheel or propeller and suffer serious injury or death. Stay clear of the flywheel, pulleys and the propeller.

CAUTION
Never run an outboard without first providing cooling water. The water pump will suffer damage within a few seconds if ran dry. A damaged water pump may supply insufficient cooling water to the power head and cause overheating and eventual power head failure.

Throttle Valve Synchronization

1. Disconnect the battery cables and ground the spark plug leads to prevent accidental starting.

2. Remove the silencer cover as described in Chapter Five.

3. Carefully disconnect the throttle linkage from the ball post on the throttle cam (**Figure 19**). The throttle cam is located on the starboard side of the power head. If necessary, pry the plastic connector from the ball post. Work carefully to avoid breaking the plastic linkage connector.

4. Carefully disconnect the oil pump linkage from the ball post on the throttle lever (**Figure 20**). The throttle lever is located on the port side of the throttle body. If necessary, pry the plastic connector from the ball post. Work carefully to avoid breaking the plastic linkage connector.

5. Locate the idle speed adjusting screw (**Figure 21**) on the starboard side of the throttle body. Rotate the screw

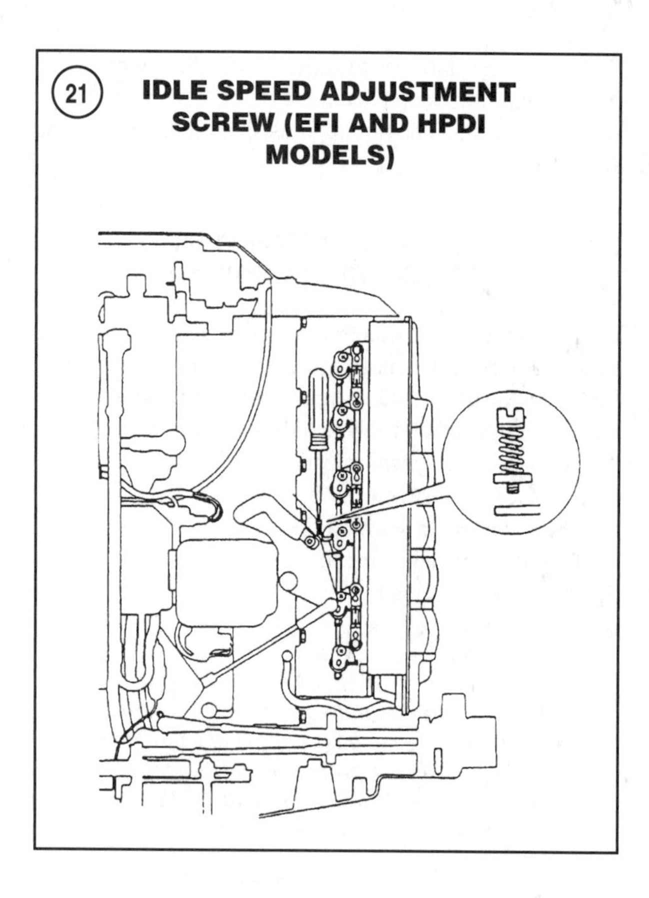

counterclockwise until a gap exists between the screw tip and the throttle lever

NOTE
The synchronization screws on the throttle levers use left-hand threads. Turn the screws clockwise to loosen and counterclockwise to tighten.

6. Loosen the synchronization screws (A, **Figure 22**) for throttle valves No. 1, 2, 3, 5 and 6 by turning clockwise. Do not loosen the screw for the No. 4 throttle valve (B, **Figure 22**).

7. Press lightly on the shutter to close the No. 1 throttle valve. Hold the valve closed while tightening the No. 1 synchronization screw. Repeat this step, one at a time for throttle valves 2, 3, 5 and 6. Make sure all throttle valves fully close. Repeat the synchronization if any of the throttle valves remain open.

8. Observe the throttle valves while slowly rotating the idle speed adjusting screw (**Figure 21**) clockwise. Stop when the throttle valve just begins to open. Note the idle speed screw position. Then rotate the screw exactly one turn clockwise.

9. Observe the throttle valves while opening and closing the throttle with the throttle roller. All valves must open

and close simultaneously. If each throttle shutter does not open the same amount with the throttle in the idle position, repeat the synchronization procedure.

10. Carefully snap the oil pump linkage connector (**Figure 20**) onto the throttle lever ball post.

11. Carefully snap the throttle linkage connector (**Figure 19**) onto the throttle cam post.

12. Install the silencer cover as described in Chapter Five.

13. Connect the battery cables and spark plug leads.

14. Adjust the idle speed as described in this chapter.

Idle Speed Adjustment

Perform this adjustment under actual running conditions. Adjust the idle speed while a qualified assistant operates the controls. Operate the boat in a safe area away from boat traffic.

1. Connect a suitable shop tachometer onto the spark plug lead.

2. Start the engine and advance the throttle to approximately 1500 rpm in NEUTRAL gear for ten minutes or until the engine reaches full operating temperature.
3. Place the remote control in the NEUTRAL gear idle position, then stop the engine.
4. Loosen the screw (B, **Figure 22**) on the No. 4 throttle lever by turning the screw clockwise. The throttle roller lever must pivot freely during idle speed adjustment.
5. Start the engine and allow the idle speed to stabilize. If the in-neutral gear idle speed is not steady and within the specification in **Table 3**, slowly turn the idle speed screw (**Figure 21**) until the idle speed is within the middle of the speed range (**Table 3**). Clockwise screw rotation increases idle speed. Counterclockwise rotation decreases idle speed.
6. Stop the engine, then pivot the throttle roller lever to align the center of the throttle lever with the marking on the throttle cam (**Figure 23**). Hold the roller in contact with the cam. Securely tighten the screw (B, **Figure 22**).
7. Start the engine and allow the idle speed to stabilize. If the idle speed does not remain within the specification in **Table 3**, make sure the throttle cam is not pushing on the throttle roller. Readjust the roller if necessary. If the roller is adjusted correctly, readjust the idle speed.
8. Stop the engine and remove the tachometer.

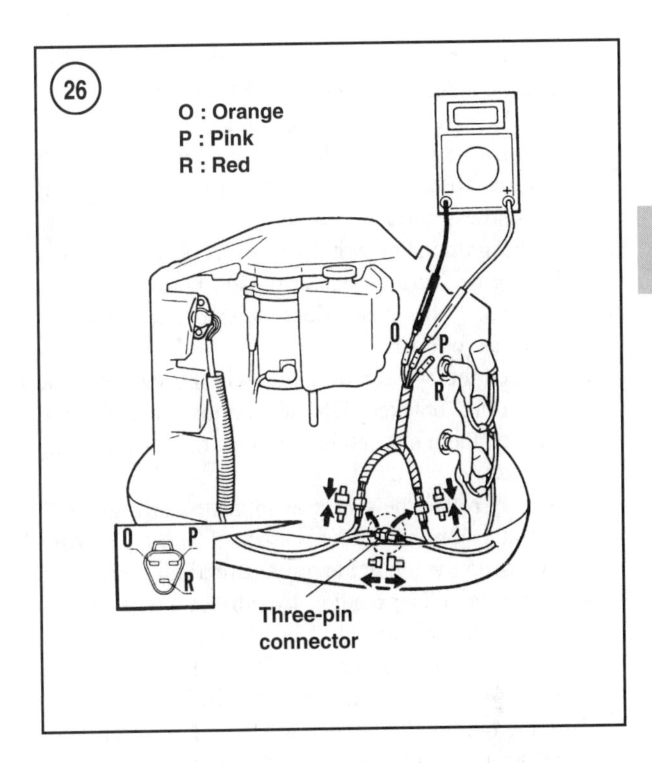

9. Adjust the throttle position sensor as described in this chapter.

Throttle Position Sensor Adjustment

Use a digital multimeter and test harness (Yamaha part No. YB-06443/90890-06757) to adjust the throttle position sensor.

NOTE

An inaccurate meter reading may occur if an unused lead of the test harness contacts any part of the engine. Isolate or tape over any unused leads to prevent accidental contact.

1. Place the remote control lever in the NEUTRAL gear idle position.
2. Carefully disconnect the throttle linkage from the ball post on the No. 1 (top) throttle lever (**Figure 24**). If necessary, pry the connector from the ball post. Work carefully to avoid breaking the plastic connector. This step places the No. 1 throttle shutter shaft in the closed position as required for sensor adjustment.
3. Locate the throttle position sensor (**Figure 25**) on the upper port side of the throttle body. Trace the sensor wire harness to the connection to the engine wire harness (see **Figure 26**). Disconnect the sensor harness connector from the engine harness connector.

4. Connect the test harness to the engine harness and sensor harness connectors as shown in **Figure 26**.
5. Select the 1 VDC scale on the digital multimeter. Connect the positive meter test lead to the pink test harness wire (**Figure 26**) and the negative test lead onto the orange test harness wire.
6. Turn the ignition key switch to the ON position. Do not start the engine. Turn the meter on and note the voltage measurement. If the meter does not indicate 0.49-0.51 volt, adjust as follows:
   a. Loosen the two screws that retain the sensor to the mounting bracket. Do not remove the screws. Loosen them just enough to rotate the sensor on the bracket.
   b. Slowly rotate the sensor as indicated in **Figure 27** until the correct reading is attained. Hold the sensor in position while tightening the retaining screws.
   c. Check the meter reading for correct sensor adjustment. The adjustment often changes after tightening the retaining screws. Readjust the sensor if necessary. Several adjustments may be required.
7. Turn the ignition key switch to the OFF position.
8. Turn the meter off. Disconnect the meter test leads from the test harness wires.
9. Disconnect the test harness wire from the sensor and engine harness connectors. Connect the sensor harness connector onto the respective engine harness connector. Route the wiring to prevent interference with moving components. Secure the wiring with plastic locking type clamps as needed.
10. Carefully snap the throttle linkage onto the ball post on the No. 1 (top) throttle lever (**Figure 24**). Work carefully to avoid breaking the plastic linkage connector.

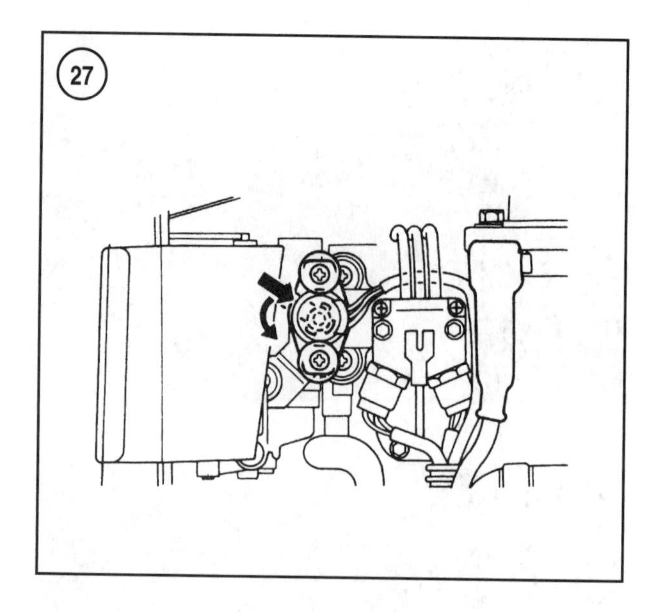

Throttle Cable Adjustment

1. Disconnect the battery cables and ground the spark plug leads to prevent accidental starting.
2. Locate the throttle linkages on the lower starboard side of the power head.
3. Remove the locking pin (3, **Figure 28**). Pull the washer and cable connector (6) from the pin on the throttle lever (4).
4. *HPDI models*—Loosen the jam nut (1, **Figure 28**), then rotate the closest throttle stop screw counterclockwise to achieve a 9.5 mm (3/8 in.) or greater gap between the screw tip and the contact surface on the throttle lever (4).
5. Move the throttle lever (4, **Figure 28**) until the center of the throttle roller (7) aligns with the marking (8) on the throttle cam. Hold the throttle arm in this position during throttle cable adjustment.
6. Loosen the jam nut (5, **Figure 28**). Rotate the cable connector (6) until it can be installed over the pin on the throttle arm without moving the arm.
7. Check the cable connector for adequate thread engagement. If the threaded end of the cable does not thread into the connector a minimum of 8 mm (0.31 in.), perform the following:
   a. Carefully disconnect the throttle link connector (10, **Figure 28**), loosen the jam nut (9) and rotate the connector (10) to lengthen the rod enough to achieve adequate thread engagement.
   b. Snap the throttle link connector onto the throttle cam ball post. Work carefully to avoid breaking the plastic linkage connector.
   c. Securely tighten the jam nut (9, **Figure 28**).
8. Fit the cable connector (6, **Figure 28**) onto the pin on the throttle lever (4). Check for proper alignment of the roller and cam marking as described in Step 5. Remove the connector from the pin and readjust as necessary.
9. Install the washer and locking pin (3, **Figure 28**) onto the throttle arm pin. Hold the cable connector. Then securely tighten the jam nut (5).
10. Move the remote control lever to the full throttle then the idle position several times. Check for proper alignment of the roller and cam marking each time the throttle reaches the idle position. Readjust the cable if necessary.
11. *HPDI models*—Move the remote control to the NEUTRAL idle position. Turn the closed throttle stop screw (2, **Figure 28**) until the plastic cap just contacts the throttle lever (4). Hold the screw while securely tightening the jam nut (1, **Figure 28**).
12. Connect the battery cables and spark plug leads. Adjust the oil pump as described in this chapter.

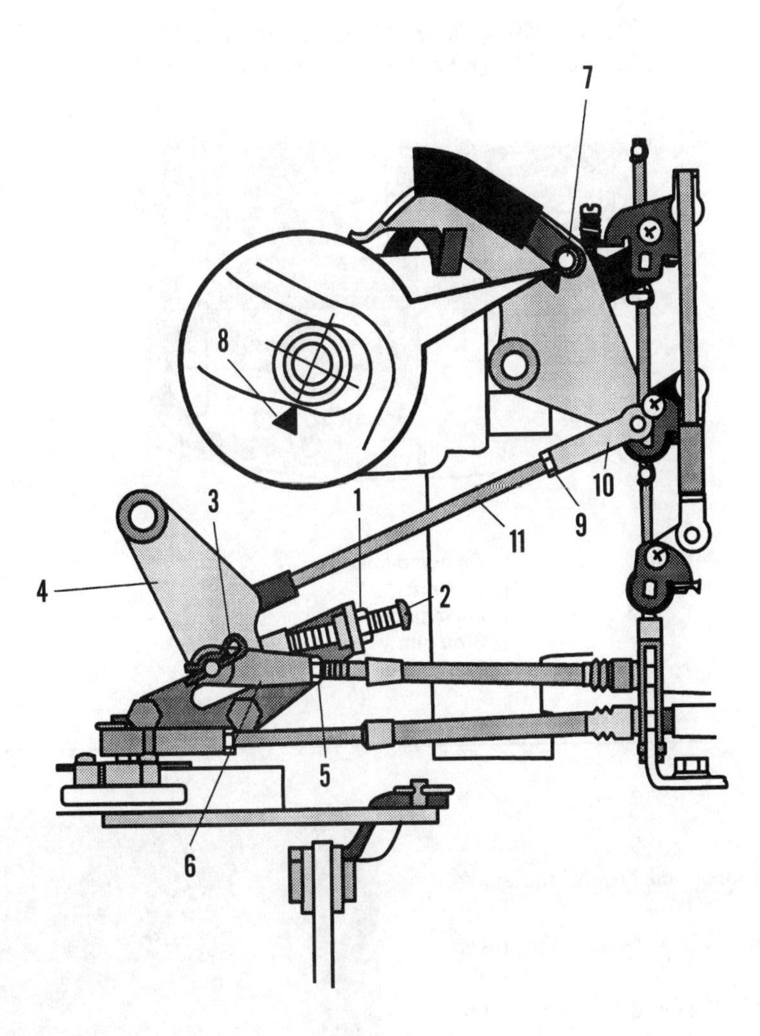

1. Jam nut (HPDI models)
2. Closed throttle stop screw (HPDI models)
3. Locking pin
4. Throttle lever
5. Jam nut
6. Cable connector
7. Throttle roller
8. Throttle cam marking
9. Jam nut
10. Connector
11. Throttle link

OIL PUMP LINKAGE ADJUSTMENT (150-250 HP [EFI AND HPDI MODELS])

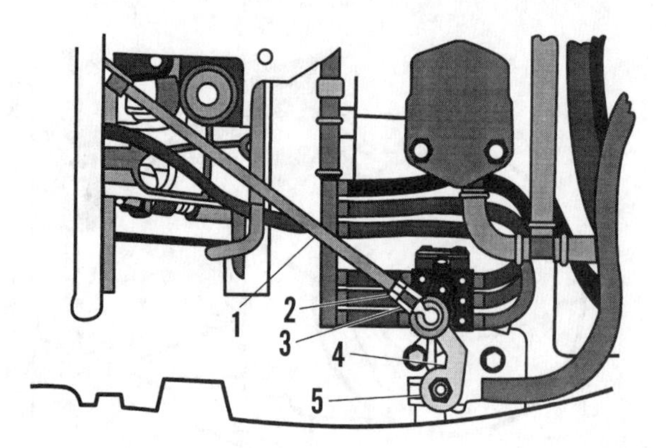

1. Oil pump linkage
2. Jam nut
3. Linkage connector
4. Stop pin
5. Pump lever

Oil Pump Adjustment

1. Disconnect the battery cables and ground the spark plug leads.
2. Locate the oil pump on the lower port side of the power head.
3. Place the throttle in the full closed position. If the pump lever (5, **Figure 29**) does not contact the stop pin (4) on the oil pump body, adjust as follows:
   a. Carefully pry the oil pump linkage connector (3, **Figure 29**) from the ball pin on the pump lever (5). Work carefully to avoid breaking the plastic linkage connector.
   b. Rotate the pump lever counterclockwise until the pump lever just contacts the stop pin. Hold the pump lever in this position during adjustment.
   c. Loosen the jam nut (2, **Figure 29**). Rotate the connector (3) until it aligns with the ball pin on the pump lever.
   d. Carefully snap the linkage connector onto the pump lever ball pin.
   e. Support the linkage connector while securely tightening the jam nut.
4. Place the throttle in the full open position. If the tab on the pump lever does not lightly contact the stop pin on the pump body, readjust the linkage as described in Step 3.

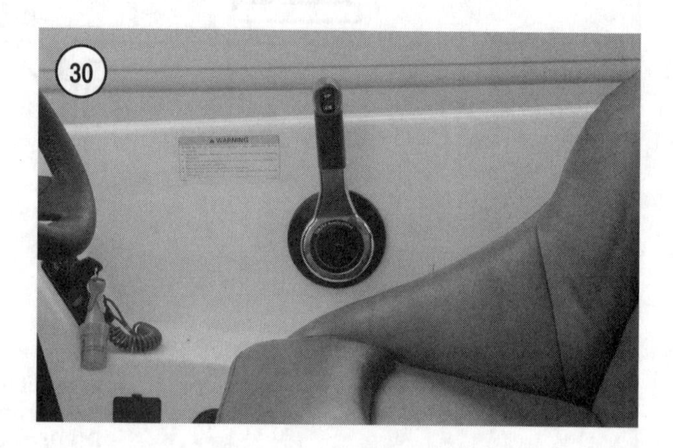

5. Move the throttle to full closed and full open positions several times while checking for binding or linkage interference with other components. Correct any binding or interference before operating the engine.
6. Connect the battery cables and spark plug leads.

Ignition Timing Check

Check the ignition timing under actual running conditions. Check the timing while a qualified assistant operates the controls. Operate the boat in a safe area away from boat traffic.

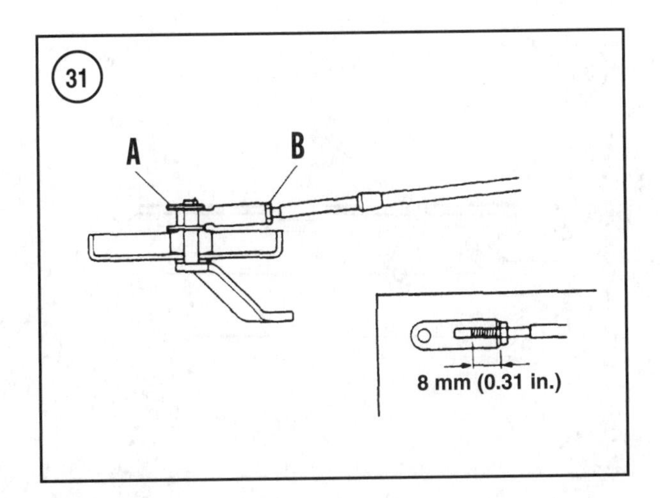

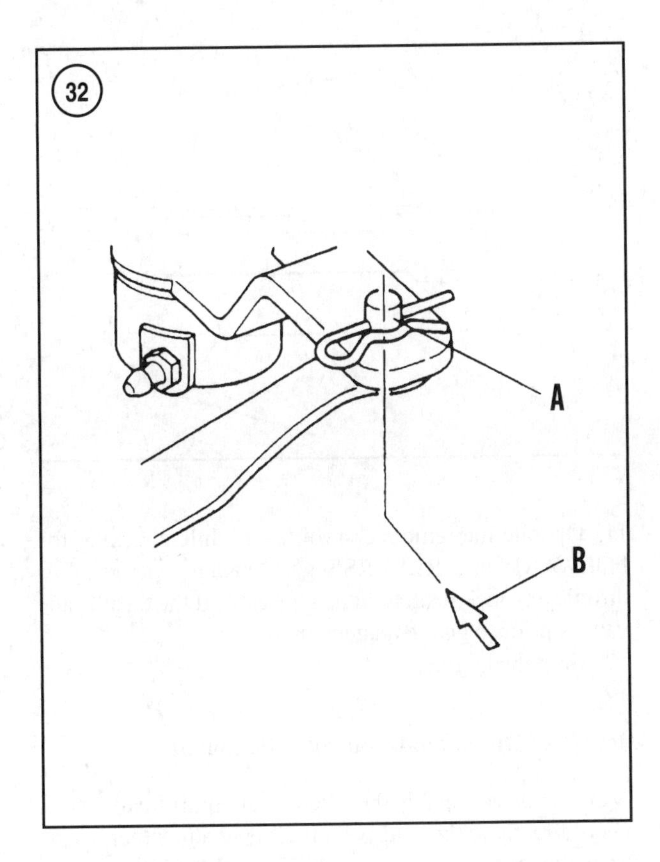

1. Connect a suitable timing light to the No. 1 (top) spark plug lead. The No. 1 cylinder is the top cylinder on the starboard bank.
2. Start the engine and advance the throttle to approximately 1500 rpm in NEUTRAL gear for ten minutes or until the engine reaches full operating temperature.
3. Place the throttle in the idle position and allow the idle speed to stabilize for a few minutes. Direct the timing light onto the timing pointer and flywheel markings (**Figure 18**). When the light flashes, the timing pointer should align with the idle speed timing mark(s) specified in **Table 4**.
4. Shift the engine into FORWARD gear. Observe the tachometer and timing pointer while having an assistant advance the throttle to full open. The timing pointer (**Figure 18**) must align with the full timing advance marking specified in **Table 4**.
5. Return the throttle to idle. Shift the engine into NEUTRAL. Stop the engine.
6. The timing is electronically advanced on these models. If the markings do not align as specified, test the ignition system components as described in Chapter Two.

MISCELLANEOUS ADJUSTMENTS

WARNING
Use extreme caution when working around a running engine. It is easy to get entangled in the flywheel or propeller and suffer serious injury or death. Stay clear of the flywheel, high pressure fuel pump drive belt and pulleys and the propeller.

CAUTION
Never run an outboard without first providing cooling water. The water pump will suffer damage within a few seconds if ran dry. A damaged water pump may supply insufficient cooling water to the power head and cause overheating and eventual power head failure.

Shift Cable Adjustment

1. Disconnect the battery cables and ground the spark plug leads to prevent accidental starting.
2. Place the remote control in the NEUTRAL gear position (**Figure 30**).
3. Locate the shift cable attaching points on the lower starboard side of the power head.
4. Remove the locking pin and washer (A, **Figure 31**). Lift the shift cable from the shift lever attaching pin.
5. Move the shift lever until the center of the pivot pin (A, **Figure 32**) aligns with the arrow (B) on the lower engine cover.
6. Loosen the jam nut (B, **Figure 31**). Rotate the cable connector until it can be installed onto the shift lever attaching pin without moving the lever. If the threaded end of the cable does not thread into the cable connector a minimum of 8.0 mm (0.31 in.), check for improper shift lever positioning or improper upper to lower shift shaft alignment. See *Gearcase Installation* in Chapter Eight.

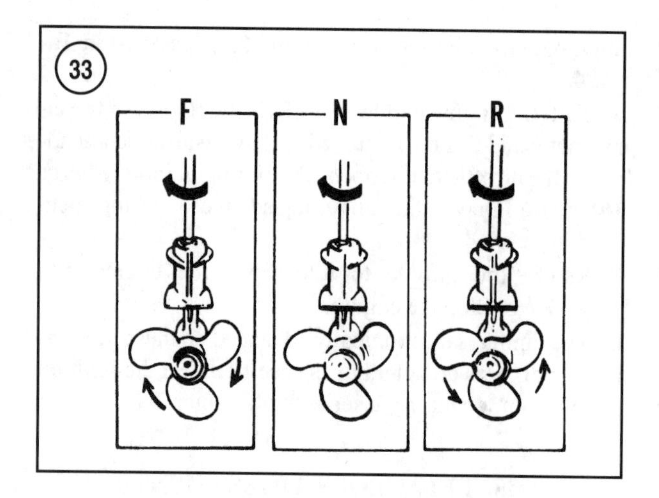

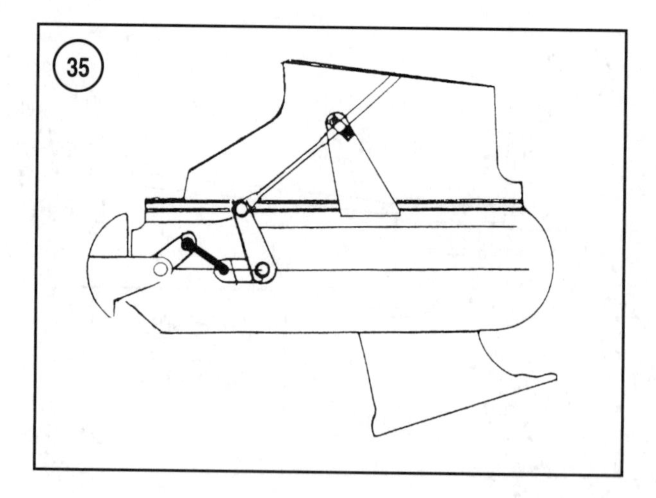

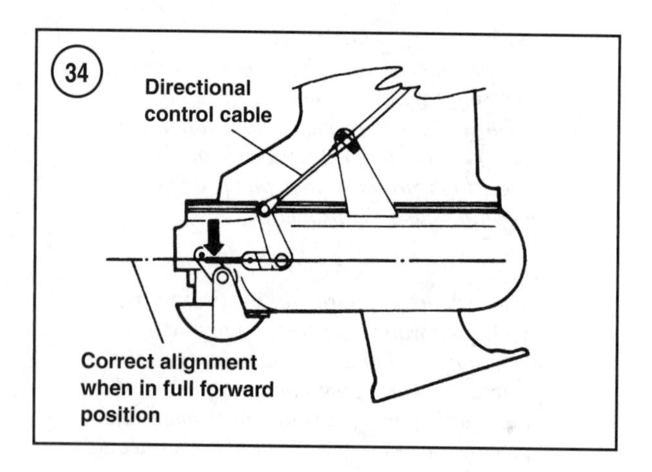

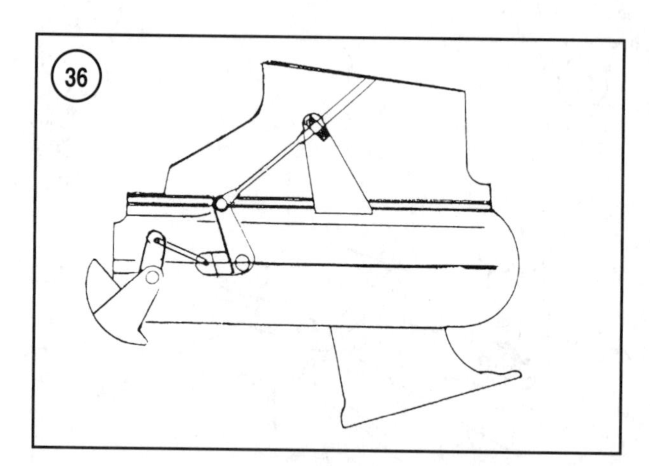

7. Fit the cable connector over the shift lever attaching pin. Install the washer (A, **Figure 31**). Install the locking clip through the hole in the pin. Hold the cable connector and securely tighten the jam nut (B, **Figure 31**).
8. Connect the battery cables and spark plug leads.
9. Prepare the engine for operation on a flush test adapter as described in *Flushing the cooling system* in Chapter Three.
10. Start the engine. View the propeller shaft rotational direction from a safe distance, while shifting to FORWARD, NEUTRAL and REVERSE gears.
   a. *Standard rotation (RH) gearcase*—The propeller shaft must rotate in the directions indicated in **Figure 33**.
   b. *Counter rotation (LH) gearcase*—The propeller shaft must rotate opposite the directions indicated in **Figure 33**.
   c. *Twin propeller gearcase*—Both propeller shafts must rotate in forward gear. One propeller shaft should only rotate in reverse gear. Neither shaft should rotate in neutral gear.

11. Operate the remote control lever while checking for FORWARD and REVERSE gear engagement prior to throttle advancement. Readjust the cable if the throttle advances prior to gear engagement.
12. Stop the engine.

Jet Drive Directional Control Adjustment

On jet drive models the directional control cable connects directly to the jet drive unit. Proper adjustment positions the thrust gate completely clear of the outlet nozzle when in forward gear (**Figure 34**). The thrust gate must completely cover the outlet nozzle when in reverse gear (**Figure 35**). The thrust gate should be midway between the forward and reverse points when in neutral (**Figure 36**).

1. Disconnect and ground the spark plug leads to prevent accidental starting.
2. Place the control in FORWARD gear position.
3. Loosen the cable retainer or jam nut and adjust the directional control cable until the pivot bracket aligns with

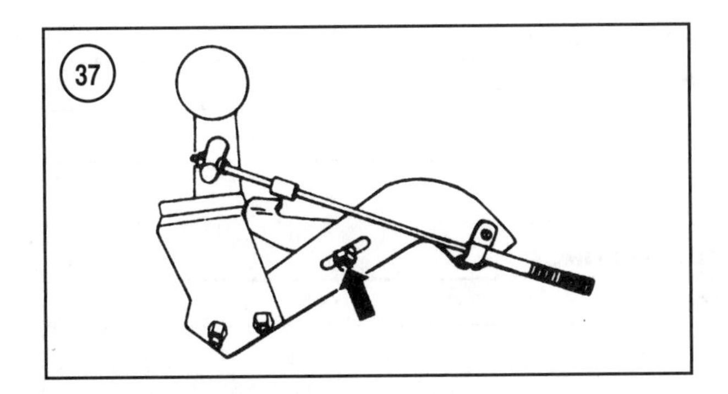

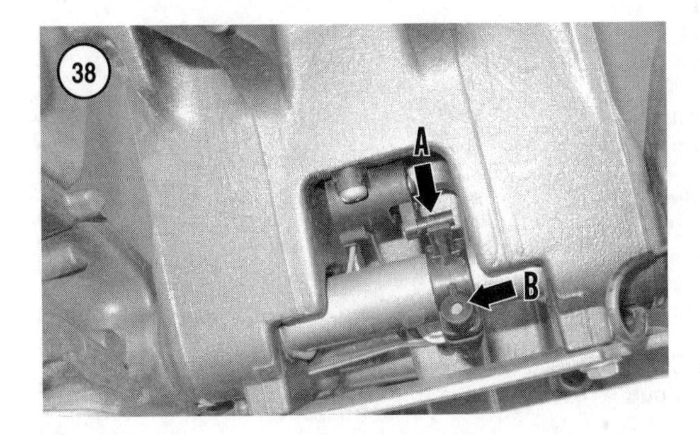

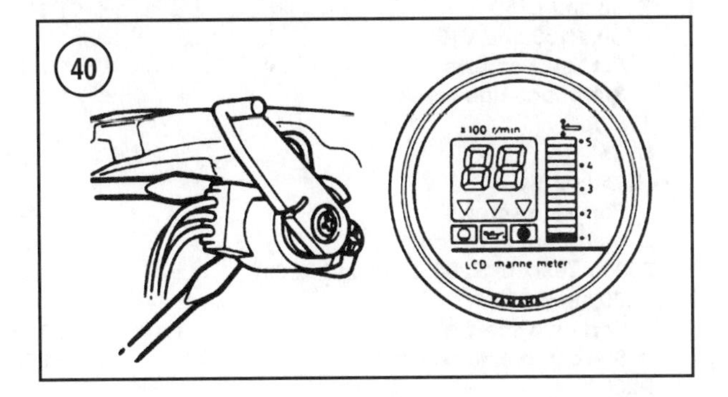

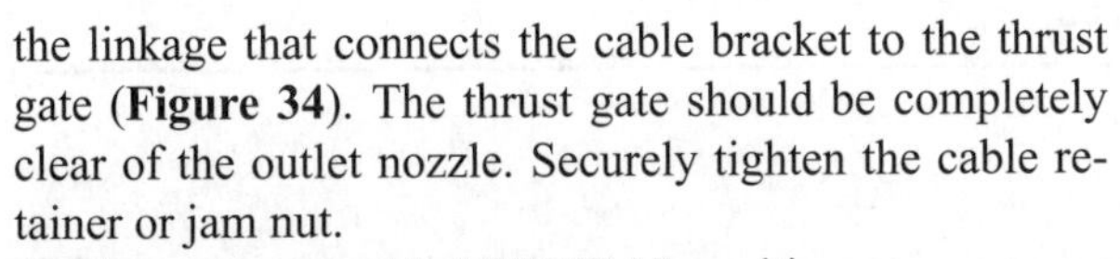

the linkage that connects the cable bracket to the thrust gate (**Figure 34**). The thrust gate should be completely clear of the outlet nozzle. Securely tighten the cable retainer or jam nut.

4. Place the control in NEUTRAL position.
5. Loosen the nut (**Figure 37**) and move the stop bracket until it just contacts the thrust gate lever.
6. Connect the battery cables and spark plug leads.
7. Tie the vessel securely onto a sturdy dock or other appropriate structure. Start the engine and run it at idle speed only.
8. Place the control in the FORWARD, REVERSE and NEUTRAL positions while checking for proper directional control. Stop the engine and perform additional cable and stop bracket adjustments if further adjustments are needed.

Trim Position Sender Adjustment

Adjustment to the trim position sender involves moving the tilt tube mounted lever (A, **Figure 38**) that contacts the trim sender lever until the gauge reads correctly.

1. Ground the spark plug leads to prevent accidental starting.
2. Turn the ignition key switch to the RUN position. Do not engage the starter.

3A. *Models using the analog gauge*—Check the reading on the gauge (**Figure 39**) when the engine reaches the full down position. The gauge should reach the down marking at the same time the engine reaches its down limit.

3B. *Models using the digital gauge*—Check the gauge (**Figure 40**) when the engine reaches the full down position. One segment on the gauge should be displayed with the engine in the full down position.

4. If necessary, adjust the trim position sender as follows:
   a. Place the engine in the full down position.
   b. Loosen the clamping screw (B, **Figure 38**) on the tilt tube mounted lever (A).
   c. Rotate the lever to attain a correct gauge reading.
   d. Hold the lever in position, while tightening the clamping screw (B, **Figure 38**) to the specification in **Table 1**.
   e. Monitor the gauge for proper operation while cycling the engine to the full up and down positions. Readjust the sender as needed.
5. Turn the ignition key switch OFF. Connect the spark plug leads.

Table 1 TORQUE SPECIFICATIONS

Fastener	N•m	in.-lb.	ft.-lb.
Trim sender lever	2	18	–

Table 2 PILOT SCREW SETTINGS

Model	Specification
80 Jet and 115 hp	5/8 turns out
130 hp	7/8 turns out
105 Jet and 150 hp	
(except C150, D150, P150 and V150)	1 turn out
C150	1 turn out
S150 and L150	1 1/4 turns out
D150, P150 and V150	
Port side screws	1 1/16 turns out
Starboard side screw	1 9/16 turns out
175 hp	
S175	1 1/8 turns out
P175 and V175	
Port side screws	1 1/8 turns out
Starboard side screw	1 5/8 turns out
200 hp (except S200, L200 and P200)	1 1/8 turns out
S200 and L200	
Port side screws	5/8 turns out
Starboard side screw	1 1/8 turns out
P200	
Port side screws	3/4 turns out
Starboard side screw	1 1/4 turns out

Table 3 IDLE SPEED SPECIFICATIONS

Model	Neutral gear (rpm)	Forward gear (rpm)
80 Jet and 115-130 hp	700-800	600-700
105 Jet, 150 hp and 175 hp		
(except EFI and HPDI models)	675-725	550-600
150-250 hp (EFI models)	700-760	*
150-250 hp (HPDI models)	670-730	*
P200 (carburetor equipped)	700	600
S200 and L200 (carburetor equipped)	700	575

* In gear idle speed adjustment is not required on these models.

Table 4 IGNITION TIMING SPECIFICATIONS

Model	Specification
80 Jet and 115 hp	
Idle speed timing	4°-6° ATDC
Pickup timing	3°-5° ATDC
Full timing advance	24°-26° BTDC

(continued)

Table 4 IGNITION TIMING SPECIFICATIONS (continued)

Model	Specification
130 hp	
Idle speed timing	4°-6° ATDC
Pickup timing	3°-5° ATDC
Full timing advance	21°-23° BTDC
C150 (carburetor equipped)	
Idle speed timing	6°-8° ATDC
Pickup timing	6°-8° ATDC
Full timing advance	18°-20° BTDC
105 Jet, 150, L150 and S150 (carburetor equipped)	
Idle speed timing	5°-9° ATDC
Pickup timing	6°-8° ATDC
Full timing advance	20°-24° BTDC
D150 and P150 (carburetor equipped)	
Idle speed timing	5°-9° ATDC
Pickup timing	6°-8° ATDC
Full timing advance	21°-25° BTDC
V150 (carburetor equipped)	
Idle speed timing	5°-9° ATDC
Pickup timing	6°-8° ATDC
Full timing advance	19°-21° BTDC
150 hp (EFI models)	
Idle speed timing	1°-5° ATDC
Full timing advance	19°-23° BTDC
150 hp (HPDI models)	
Idle speed timing	
Long shaft models	6° BTDC-0° TDC
Extra long shaft models	1°-7° BTDC
Full timing advance	15°-19° BTDC
S175 (carburetor equipped)	
Idle speed timing	5°-9° ATDC
Pickup timing	5°-9° ATDC
Full timing advance	20°-24° BTDC
P175 and V175 (carburetor equipped)	
Idle speed timing	5°-9° ATDC
Pickup timing	5°-9° ATDC
Full timing advance	21°-25° BTDC
175 hp (HPDI models)	
Idle speed timing	6° BTDC-0° TDC
Full timing advance	15°-19° BTDC
200 hp (carburetor equipped [except P200, S200 and L200])	
Idle speed timing	5°-9° ATDC
Pickup timing	5°-9° ATDC
Full timing advance	19°-21° BTDC
P200 (carburetor equipped)	
Idle speed timing	5°-9° ATDC
Pickup timing	5°-9° ATDC
Full timing advance	21°-24° BTDC
S200 and L200 (carburetor equipped)	
Idle speed timing	5°-9° ATDC
Pickup timing	5°-9° ATDC
Full timing advance	19°-23° BTDC
200 hp (EFI [2.6 liter])	
Idle speed timing	3°-7° ATDC
Full timing advance	21°-25° BTDC
200 hp (EFI [3.1 liter])	
Idle speed timing	1°-5° ATDC
Full timing advance	18°-22° BTDC

(continued)

Table 4 IGNITION TIMING SPECIFICATIONS (continued)

Model	Specification
200 hp (HPDI)	
Idle speed timing	1°-7° BTDC
Full timing advance	15°-19° BTDC
L225, LX225, S225 and SX225 (EFI)	
Idle speed timing	1°-5° ATDC
Full timing advance	15°-19° BTDC
V225 and VX225	
Idle speed timing	1°-5° ATDC
Full timing advance	16°-20° BTDC
L250, LX250, S250 and SX250 (EFI)	
Idle speed timing	1°-5° ATDC
Full timing advance	14°-18° BTDC
VX250	
Idle speed timing	4° ATDC
Full timing advance	20° BTDC

Chapter Five

Fuel System

This chapter describes removal, repair and installation of fuel system components. Topics covered include:

1. Fuel tank.
2. Primer bulb.
3. Fuel hoses, check valves and connectors.
4. Low pressure mechanical fuel pump.
5. Vapor separator tank.
6. Fuel rail(s), injectors and silencer cover.
7. Throttle body.
8. High pressure mechanical fuel pump.
9. Carburetor and silencer cover.
10. Choke solenoid.
11. Reed housing/intake manifold.
12. Recirculation system.

Table 1-3 lists fuel system specifications. **Tables 1-3** are at the end of this chapter.

FUEL SYSTEM SAFETY AND GENERAL INFORMATION

WARNING

Use caution when working with the fuel system to avoid damage to property and potential injury or death. Never smoke around fuel or fuel vapors. Make sure no flame or source of ignition is present in the work area. Flame or sparks can ignite fuel or fuel vapor resulting in a fire or an explosion.

WARNING

Wear protective eyewear and work in a well-ventilated area when repairing the fuel system. Take all necessary precautions against fire or explosions. Always disconnect the battery before servicing any outboard.

Pay close attention when removing and installing components, especially carburetors, to avoid installing them in the wrong location. Mark them if necessary.

To avoid fuel or air leaks, replace all gaskets, seals or O-rings anytime a fuel system component is disassembled.

The most important step in a fuel system repair is the cleaning process. Use a good quality solvent to remove varnish-like deposits that commonly form in fuel systems. Spray-type carburetor cleaners available at auto parts stores are effective in removing most stubborn deposits. Never use any solvent that is not suitable for aluminum material. Some solvents may irreparably damage the carburetor, vapor separator tank, fuel rails and fuel pump castings. Remove all plastic or rubber components from the fuel pump, fuel filter or carburetor before cleaning them with solvent. Carefully scrape gasket material from the components with a razor scraper. Work carefully to avoid removing material or scratching sealing surfaces. Use a stiff bristle brush and solvent to remove deposits from the

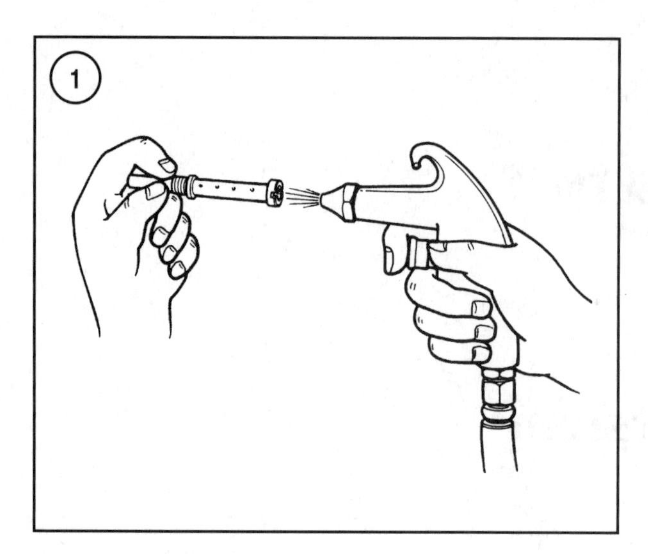

carburetor fuel bowl. Never use a wire brush as delicate sealing surfaces can be quickly damaged. Use compressed air to blow out all passages and orifices (**Figure 1**). A piece of straw from a broom works well to clean out small passages. Never use stiff wire for this purpose. The wire may enlarge the size of the passage and alter the fuel calibration. Soak the component in the solvent if the deposits are difficult to remove. Never compromise the cleaning process. Continue to clean until all deposits and debris are removed.

FUEL TANKS

Two types of fuel tanks are used with the Yamaha outboards covered in this manual: a portable remote tank (**Figure 2**) and a vessel mounted fuel tank (**Figure 3**).

Portable Remote Fuel Tank Servicing

Portable remote tanks (**Figure 2**, typical) can be used on any of the engines. They are used primarily on the lower horsepower engines. Higher horsepower engines use considerably more fuel than lower horsepower engines; the smaller capacity of a portable tank requires frequent refills if used on a higher horsepower engine. Portable tanks are manufactured by several companies. Go to a marine dealership or repair shop when purchasing replacement parts for the tank.

Portable remote fuel tanks require periodic cleaning and inspection. If there is water in the tank, make sure to inspect the remainder of the fuel system for potential contamination.

1. Remove the fuel cap (11, **Figure 2**), then pour the fuel into a suitable container.

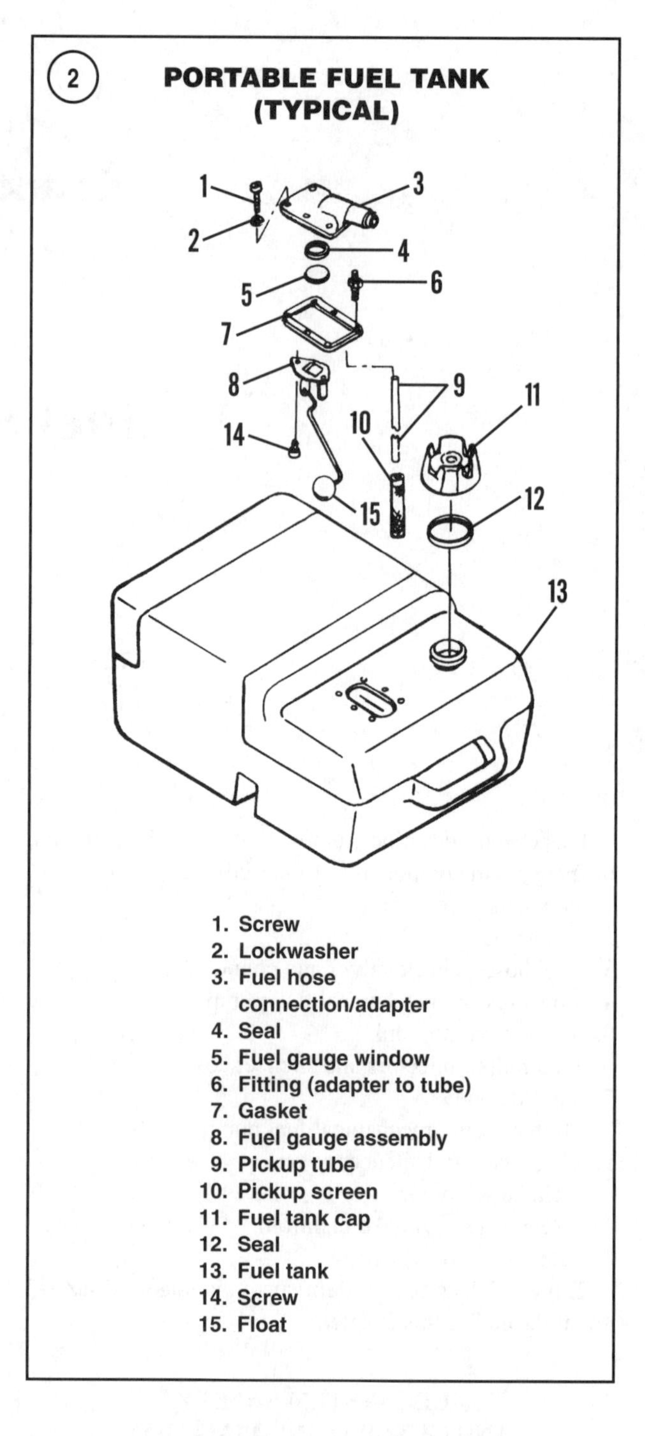

1. Screw
2. Lockwasher
3. Fuel hose connection/adapter
4. Seal
5. Fuel gauge window
6. Fitting (adapter to tube)
7. Gasket
8. Fuel gauge assembly
9. Pickup tube
10. Pickup screen
11. Fuel tank cap
12. Seal
13. Fuel tank
14. Screw
15. Float

2. Remove the screws (1, **Figure 2**). Then carefully pull the connector/adapter (3) from the tank. Remove and discard the gasket (7, **Figure 2**) between the adapter and the tank.

3. Check for free movement of the float arm on the gauge assembly (**Figure 4**). Replace the assembly if binding cannot be corrected by bending the float arm into the correct position. Inspect the float for deteriorated or physi-

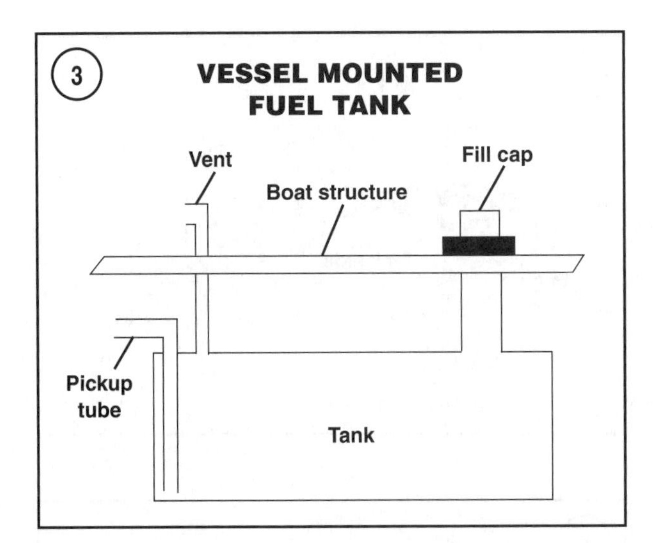

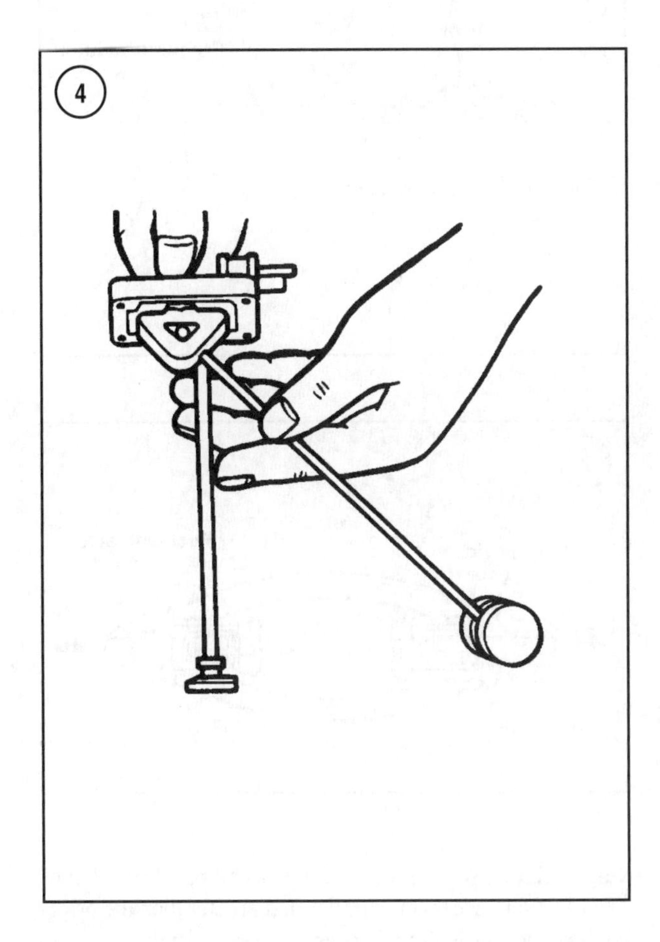

cally damaged surfaces. Replace the float if damaged or if it appears to be saturated with fuel.

4. Add a small amount of solvent into the fuel tank. Block the fuel gauge opening with a shop towel and install the fill cap. Shake the tank for a few minutes. Drain the solvent and blow dry with compressed air.

5. Replace the tank if internal or external rusting is present or if physical damage is evident. Replace the tank if there are fuel leaks or the tank is leaking. Repeat Step 4 if residual debris or deposits are found in the tank.

6. Install the fuel gauge assembly into the tank with a new gasket. Install and securely tighten the screws.

7. Check for and correct fuel leaks before using the tank.

Vessel Mounted Fuel Tank Servicing

Vessel mounted tanks (**Figure 3**) can be used on any of the engines covered in this manual. However, vessel mounted tanks are sometimes difficult to access for repair or service. Removable panels in some boats provide reasonable access to the fitting and sender assembly.

The only components that can be serviced without major disassembly of the boat include the fuel pickup tube, fuel fill fitting, fuel level sender and antisiphon device. These components are available from many different suppliers. Removal, inspection and installation procedures vary by the brand and model of the tank. Contact the tank manufacturer or boat manufacturer for specific instructions.

PRIMER BULB

The primer bulb (**Figure 5**) is located in the fuel supply hose between the fuel tank and the engine. Use a hand operated pressure pump (**Figure 6**) to test the primer bulb. Purchase the pump from an auto parts store, tool supplier or Yamaha dealership (Yamaha part No. YB-35956/90890-06756).

1. Disconnect the fuel supply hose from the engine. Drain the fuel from the hose into a suitable container. Remove the hose clamps from both connections to the primer bulb. Pull the hoses from the primer bulb fittings.

2. Place the primer bulb over a container suitable for holding fuel. Direct the outlet side toward the container,

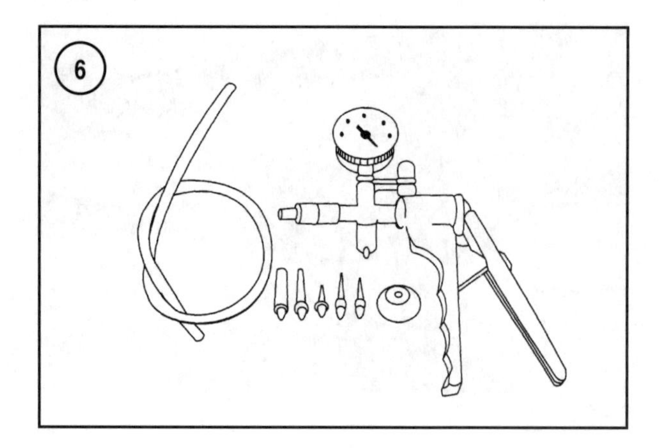

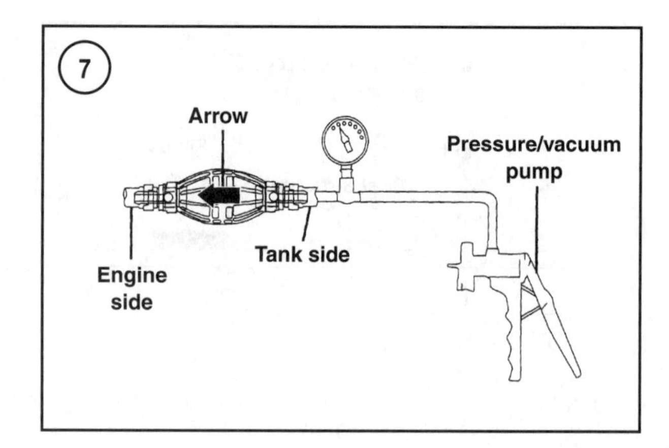

and squeeze the primer bulb until fully collapsed. The outlet side connects to the hose leading to the engine. Replace the primer bulb if it does not freely expand when released or sticks together. Replace the primer bulb if it appears weathered, has surface cracks or is hard to squeeze.

3. Connect the hand operated air pump to the check valve fitting on the fuel tank side of the primer bulb (**Figure 7**). The arrow molded into the bulb points toward the engine side check valve fitting. If air does not exit the check valve fitting on the engine side of the primer bulb while the pump is operated, replace the primer bulb.
4. Connect the pressure pump to the check valve fitting on the engine side of the primer bulb (**Figure 8**). The arrow molded into the bulb points toward the engine side check valve fitting. Air should not exit the fuel tank side check valve fitting while operating the pump. Replace the primer bulb if air exits the fitting.
5. Submerge the primer bulb, with the air pump hose attached to the engine side, into clear water. Block the fuel tank fitting with a finger. Operate the pump and check for bubbles on the primer bulb surface and fittings. Replace the primer bulb if leakage is indicated from the surfaces or leakage from the fittings cannot be corrected by installing new clamps. Thoroughly dry the primer bulb before installation.
6. Connect the fuel hoses to the primer bulb fittings. Note the direction of fuel flow before connecting the hoses (**Figure 9**). Use the arrow molded into the primer bulb surface for correct orientation.

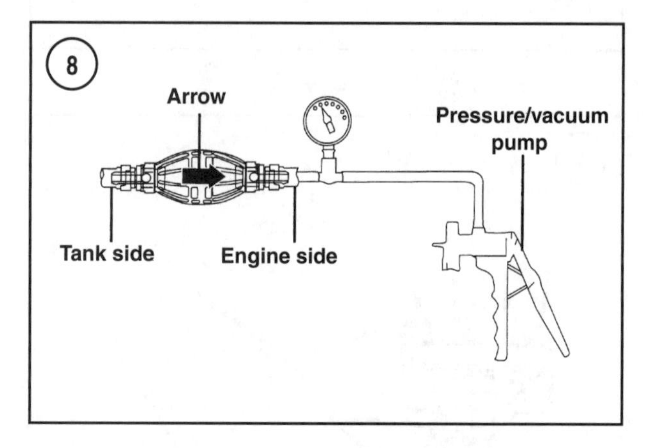

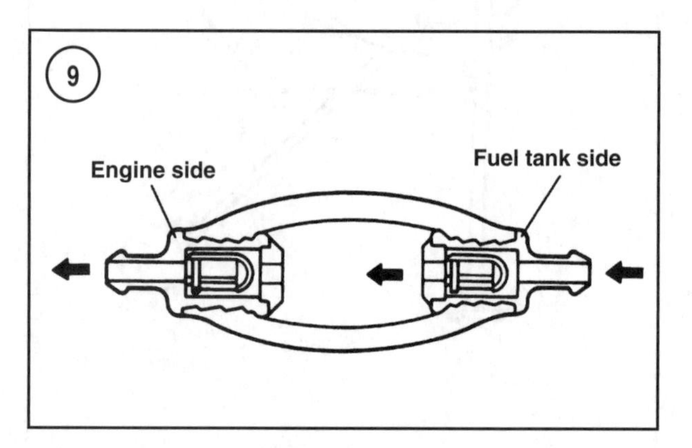

FUEL HOSES

Fuel hose sizes and routing vary by model. Refer to **Figures 10-14** for the appropriate fuel system diagram when routing and connecting hoses.

Only use Yamaha replacement fuel hoses. This is especially important with EFI and HPDI models that create very high pressure in the fuel hoses and passages. Never install a fuel hose that is smaller in diameter than the original hose. Replace all fuel hoses at the same time unless the situation calls to replace only one fuel hose. If one hose fails, other hoses are suspect.

Replace hoses that feel sticky spongy, are hard and brittle or have surface cracks. Always replace hoses that split on the end instead of cutting off the end and reattaching the hose. The hose will likely split again. To avoid hoses

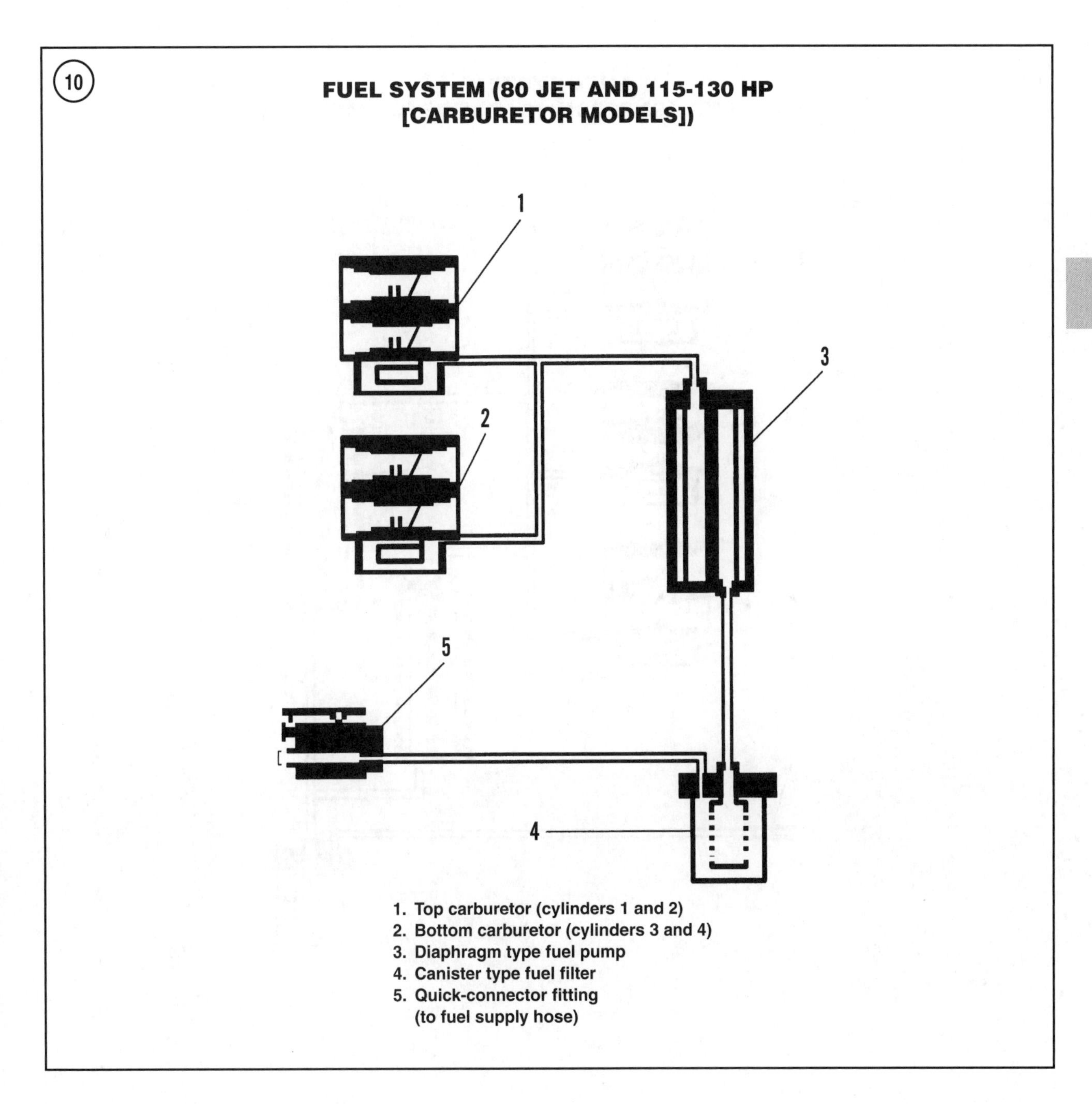

1. Top carburetor (cylinders 1 and 2)
2. Bottom carburetor (cylinders 3 and 4)
3. Diaphragm type fuel pump
4. Canister type fuel filter
5. Quick-connector fitting (to fuel supply hose)

kinking or interference with other components, never cut replacement hoses shorter or longer than the original.

FUEL HOSE CONNECTORS

Connectors used on the fuel hoses include the plastic locking type hose clamps, spring type hose clamps, quick-connector fittings and crimp type hose clamps. Plastic locking type and crimp type hose clamps must be replaced if loosened or removed. Never substitute a different type of clamp than was originally used on the connections. Worm gear type hose clamps generally work loose over time and are not suitable for use on any fuel system component.

Plastic Locking Clamp

The plastic locking type clamps (**Figure 15**) must be cut to remove. Replace them with the correct Yamaha part. Some plastic locking type clamps are not suitable for the

(11)

FUEL SYSTEM
(105 JET AND 150-200 HP [CARBURETOR MODELS])

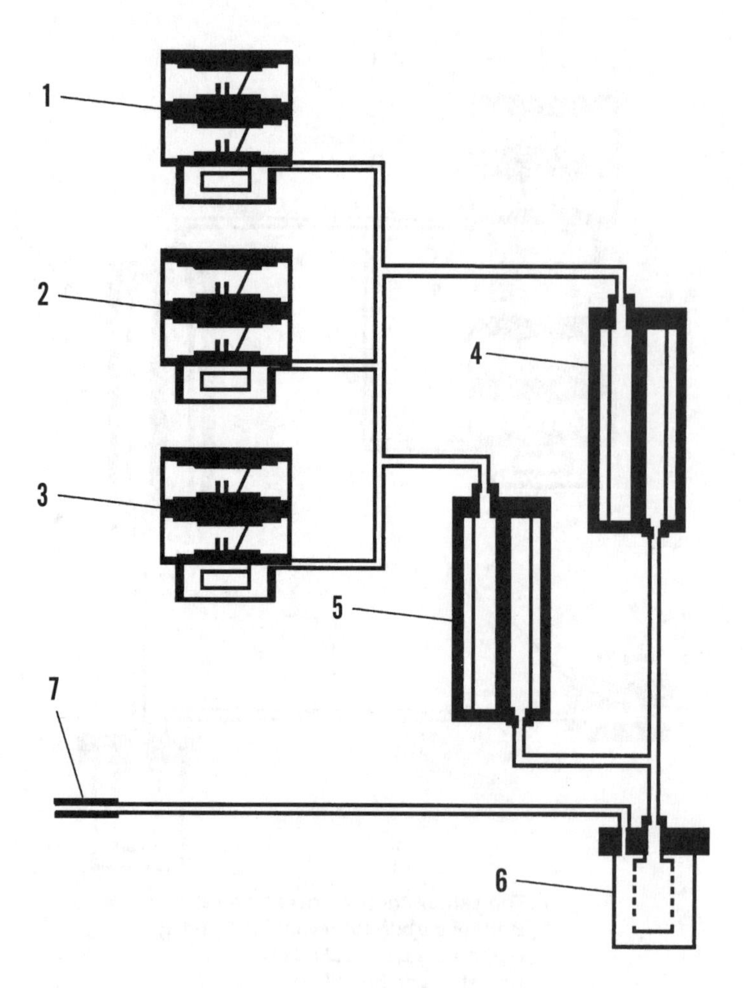

1. Top carburetor (cylinders 1 and 2)
2. Middle carburetor (cylinders 3 and 4)
3. Bottom carburetor (cylinders 5 and 6)
4. Diaphragm type fuel pump (upper)
5. Diaphragm type fuel pump (lower)
6. Canister type fuel filter
7. Barb type fuel hose connector
   (to fuel supply hose)

(12)

FUEL SYSTEM
(150-200 HP EFI [2.6 LITER MODELS])

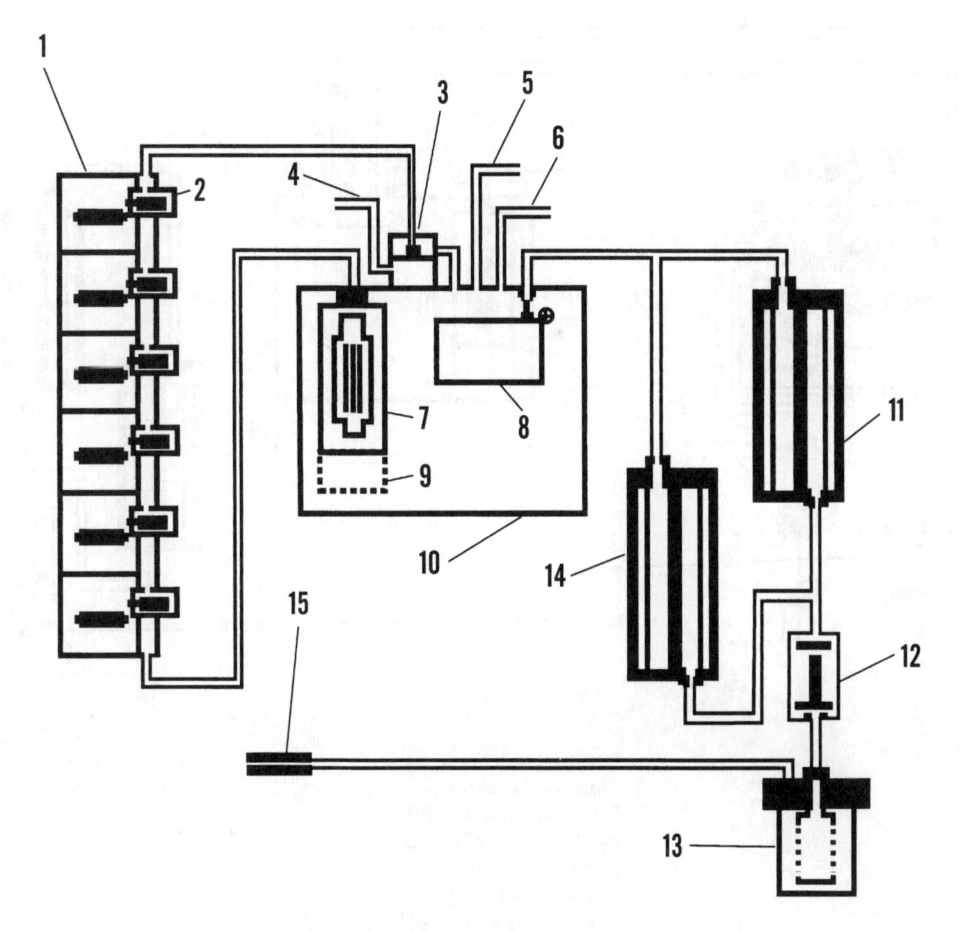

1. Throttle body
2. Fuel injector (6)
3. Fuel pressure regulator
4. Vacuum hose to intake manifold
5. Vent hose to intake manifold
6. Oil hose (from oil pump)
7. Electric high pressure fuel pump
8. Float
9. Filter
10. Vapor separator tank assembly
11. Diaphragm type fuel pump (upper)
12. Check valve
13. Canister type fuel filter
14. Diaphragm type fuel pump (lower)
15. Barb type fuel hose connector (to fuel supply hose)

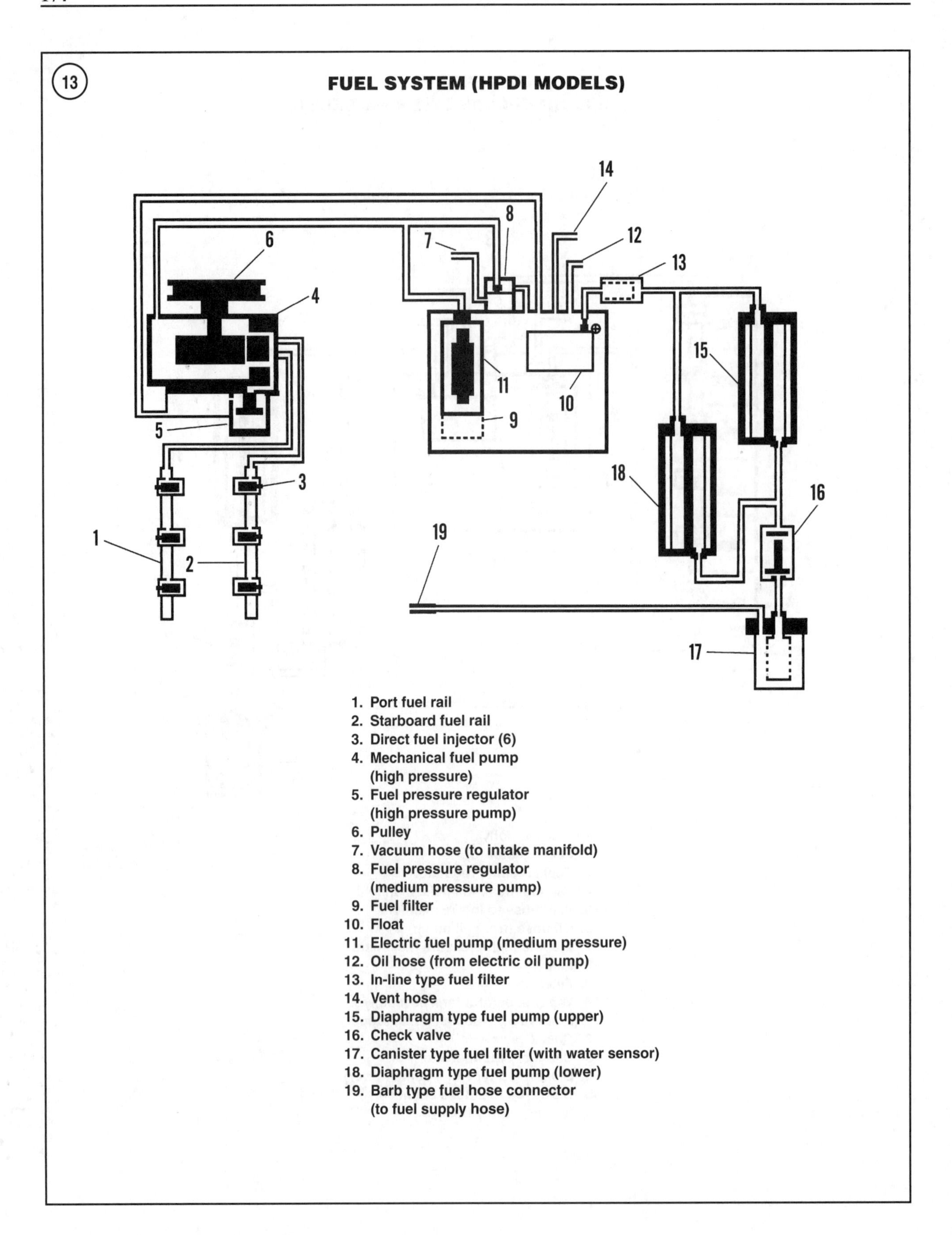
13
FUEL SYSTEM (HPDI MODELS)
14
8
12
7
6
13
4
15
11
10
9
5
18
16
3
1
19
2
17
1. Port fuel rail
2. Starboard fuel rail
3. Direct fuel injector (6)
4. Mechanical fuel pump (high pressure)
5. Fuel pressure regulator (high pressure pump)
6. Pulley
7. Vacuum hose (to intake manifold)
8. Fuel pressure regulator (medium pressure pump)
9. Fuel filter
10. Float
11. Electric fuel pump (medium pressure)
12. Oil hose (from electric oil pump)
13. In-line type fuel filter
14. Vent hose
15. Diaphragm type fuel pump (upper)
16. Check valve
17. Canister type fuel filter (with water sensor)
18. Diaphragm type fuel pump (lower)
19. Barb type fuel hose connector (to fuel supply hose)

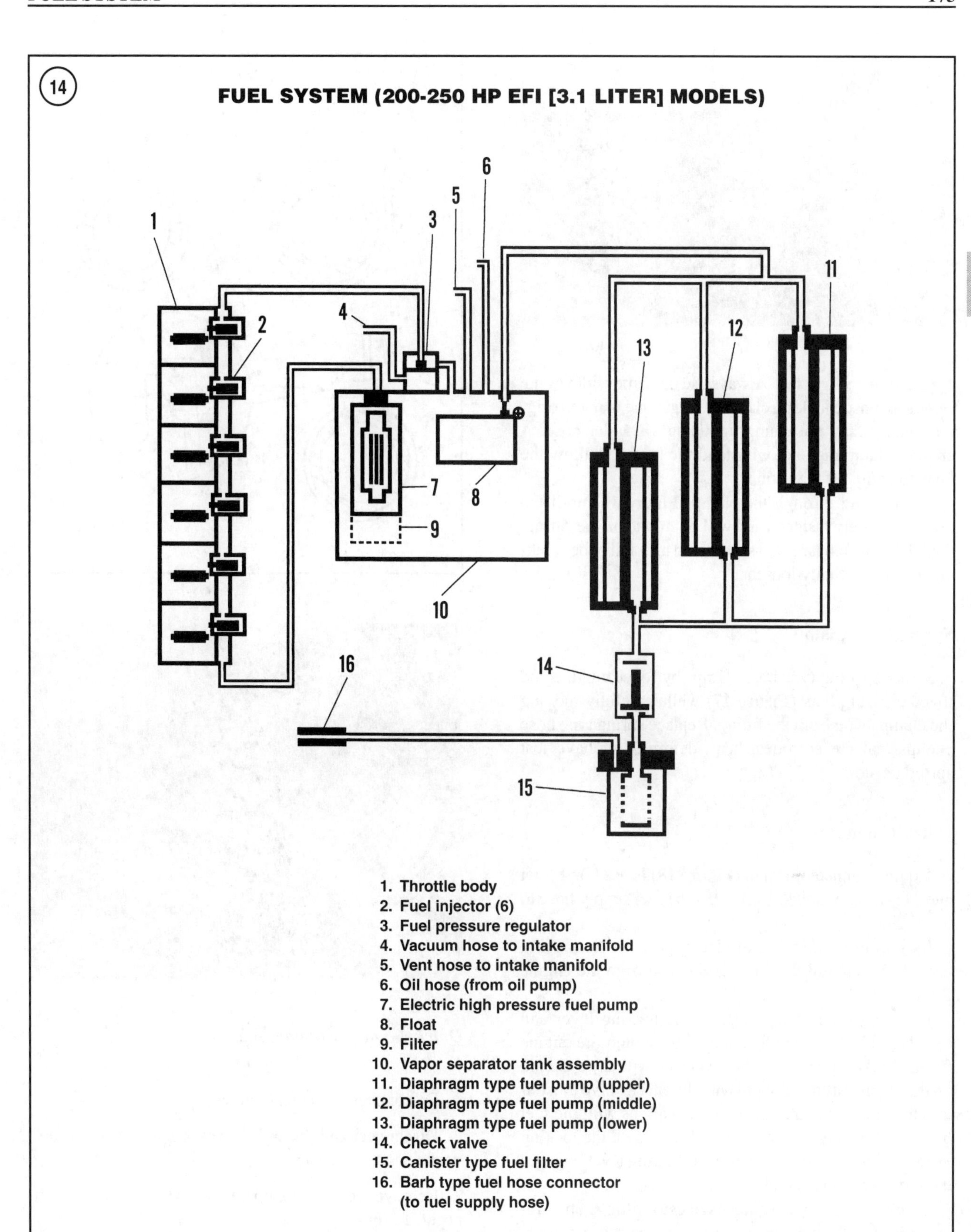

1. Throttle body
2. Fuel injector (6)
3. Fuel pressure regulator
4. Vacuum hose to intake manifold
5. Vent hose to intake manifold
6. Oil hose (from oil pump)
7. Electric high pressure fuel pump
8. Float
9. Filter
10. Vapor separator tank assembly
11. Diaphragm type fuel pump (upper)
12. Diaphragm type fuel pump (middle)
13. Diaphragm type fuel pump (lower)
14. Check valve
15. Canister type fuel filter
16. Barb type fuel hose connector (to fuel supply hose)

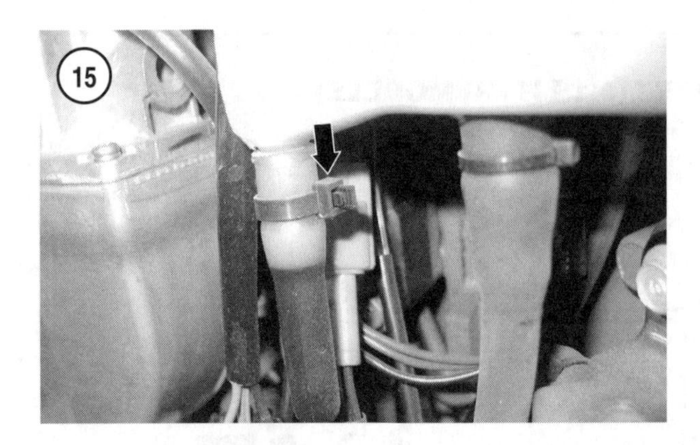

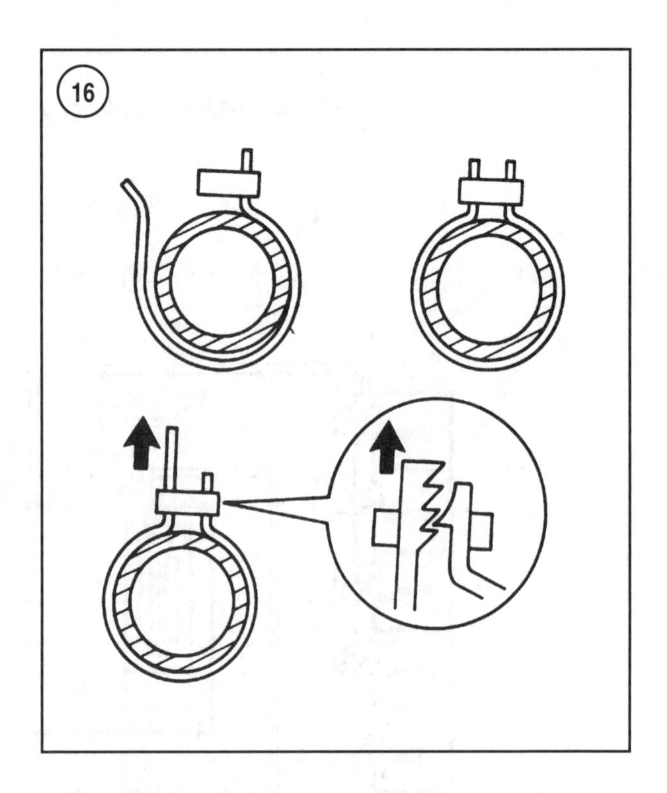

application and will fail. Always use the same width as the removed plastic locking clamp. A larger one than the original clamp may not clamp tightly to a smaller hose. A smaller clamp may not withstand the load and allow the hose to come off the fitting.

Pull the end through the clamp (**Figure 16**) until the hose is securely fastened and will not rotate on the fitting. Avoid pulling too harshly as the clamp may fail or be weakened and eventually loosen.

Spring Hose Clamp

Remove spring type hose clamps by squeezing the end together with pliers (**Figure 17**) while carefully moving the clamp away from the fitting. Replace spring type hose clamps that are corroded, bent, deformed or have lost spring tension.

Quick-Connector

A quick-connector clamp (**Figure 18**) is used on 80 Jet and 115-130 hp models to connect the fuel supply hose to the engine.

To disconnect this type of clamp, push on the locking lever and then pull the fuel supply hose from the engine fitting.

To connect the clamp, depress the locking lever and then carefully push the fuel supply hose onto the engine fitting. Make sure to align the lever side with the solid pin on the engine fitting. The pin with the check valve must fit into the opening in the connector that aligns with the fuel hose. Push firmly on the fitting. Then release the locking lever. Pull on the hose to ensure the locking lever engages the groove in the solid pin.

Check for leaks at the quick connector fittings on a frequent basis. Observe the connection while squeezing the primer bulb. Replace both fittings if leakage is detected at the connection.

Quick-connector replacement

1. Disconnect the battery cables.

2. Disconnect and ground the spark plug lead(s) to prevent accidental starting.

3. Remove the hose clamps. Pull the hose from the connector fittings.

4. Remove the mounting bolt and pull the engine fitting from the lower engine cover.

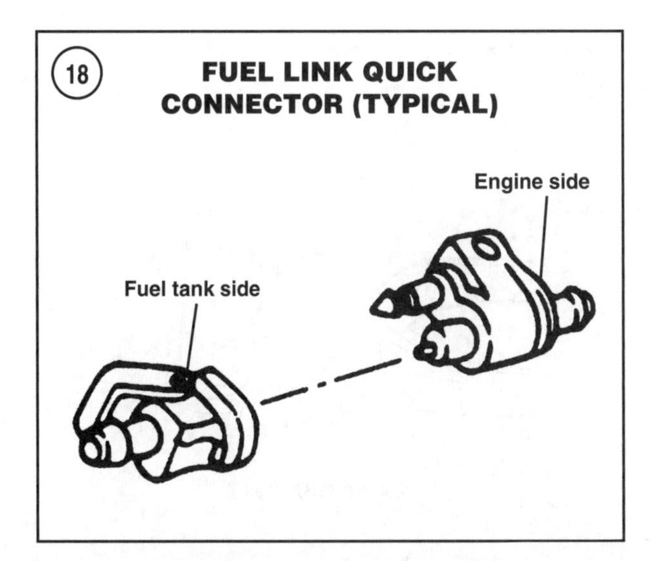

5. Fit the replacement quick connector onto the lower engine cover. Align the fitting with the opening and install the mounting bolt. Securely tighten the bolt.
6. Connect the fuel supply quick connector onto the engine quick connector as described.
7. Push the engine and fuel supply hoses onto the respective quick connector fitting. Secure the hose with appropriate clamps. Route the hoses to avoid interference with moving components.
8. Squeeze the primer bulb while checking for leakage. Correct leakage as necessary.
9. Connect the spark plug lead(s).
10. Connect the battery cables.

Crimp Type Hose Clamp

A crimp type hose clamp (**Figure 19**) is used to connect the high pressure fuel hoses used on EFI and HPDI models. This type of clamp is damaged during removal and must be replaced anytime it is loosened or removed. Use only the correct Yamaha part when replacing crimp type hose clamps. Never substitute other types of hose clamps where this type of clamp is used. Other types of clamps may loosen or fail and cause dangerous fuel leaks.

CAUTION
Always follow the proper procedure when removing crimp hose clamps. The hoses may be damaged if the clamps are removed when twisting the clamp. Always relieve the pressure in the hose prior to removing the hose clamp.

Crimp type hose clamp replacement

1. Relieve the fuel pressure in the hose. See *Vapor Separator Tank* in this chapter.
2. Use sharp cutters to cut the clamp at the point indicated in **Figure 20**. Do not twist the clamp while cutting.
3. Place a suitable container or shop towel under the hose and fitting. Residual fuel will likely spill from the hose and fitting upon removal.
4. Carefully pull the hose from the fitting. Avoid twisting the hose. If necessary, use a blunt tip screwdriver to push the hose off the fitting.
5. Drain the fuel from the hose. Wipe up any spilled fuel.
6. Fit the new hose clamp over the fitting and the end of the hose as shown in **Figure 20**.
7. Carefully push the hose fully onto the fitting. Position the hose clamp onto the section of hose that aligns with the fitting. Do not position the clamp over the barb at the end of the fitting.
8. Use cutters to squeeze the squared end of the clamp as shown in **Figure 20**. Do not use excessive force; otherwise the clamp or hose may become damaged.
9. Tug on the hose to verify a secure fit. Inspect the hose for areas damaged during clamp installation. Replace the clamp if the hose or clamp appears damaged.

FUEL PUMPS

Three types of fuel pumps are used on the Yamaha outboards covered in this manual.

One or more diaphragm type fuel pumps (**Figure 21**) are used on all models. This type of pump is commonly referred to as the low-pressure pump.

All EFI and HPDI models use an electric fuel pump. The pump mounts inside the vapor separator tank (**Figure 22**). On EFI models, this pump is referred to as the high pressure pump. On HPDI models, this pump is referred to the medium pressure pump.

A belt driven mechanical fuel pump (**Figure 23**) is used on all HPDI models. This pump provides fuel under high pressure to the direct fuel injectors. On HPDI models, this pump is referred to as the high pressure pump.

This section describes replacement and repair for the diaphragm type fuel pump. Refer to *Vapor Separator Tank* in this chapter for replacement instructions for the electric fuel pump. Refer to *High Pressure Mechanical Fuel Pump* in this chapter for replacement and repair instructions.

Refer to the appropriate diagram (**Figures 10-14**) to assist with fuel hose routing and connections. Replace all gaskets, diaphragms, check valves and seals when servicing the fuel pump. Upon completion of the repair, check for proper engine operation and fuel leakage. Correct any fuel leakage before putting the engine into service.

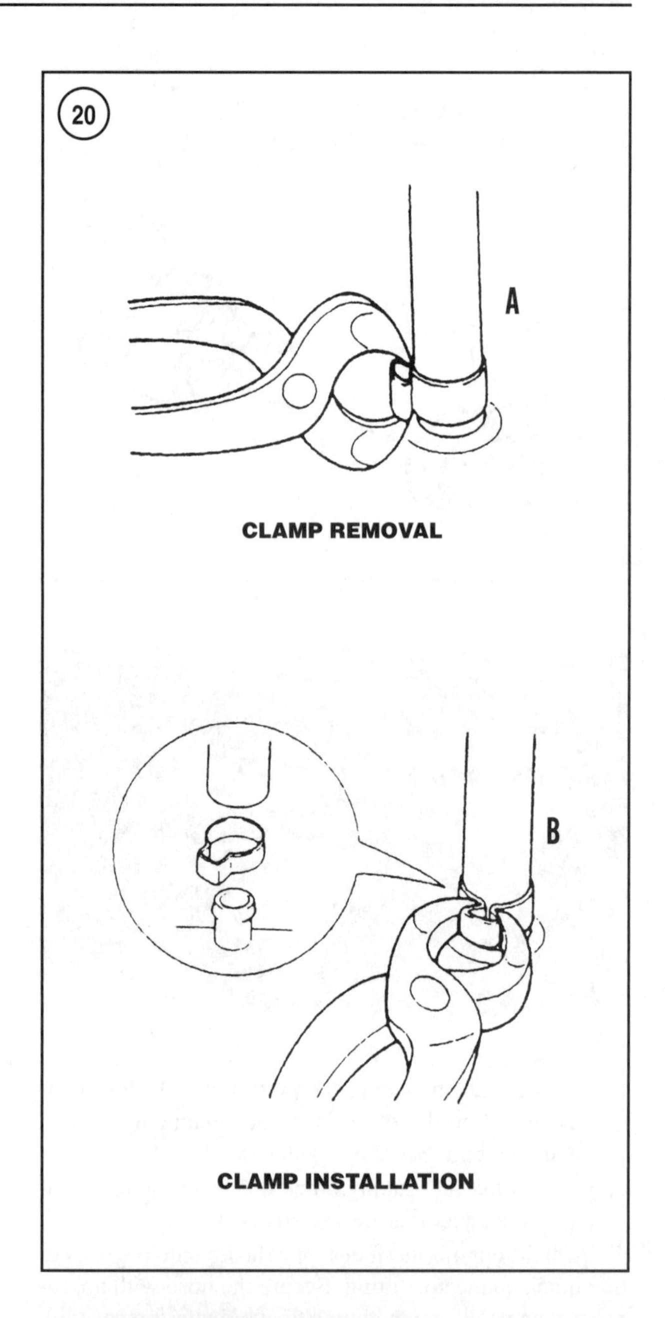

Diaphragm Fuel Pump

Removal

Deformed or damaged diaphragms, brittle gaskets and faulty check valves are some common causes of fuel pump failure. Mark all components during disassembly to ensure proper orientation during assembly.

A pressure/vacuum tester (Miti-Vac or Yamaha part No. YB-35956/90890-06756) is required to pressure test the fuel pump after assembly. Use a shop towel or suitable container to capture any residual fuel that spills from disconnected hoses.

1. Disconnect the battery cables and ground the spark plug leads to prevent accidental starting.
2. Locate the inlet and outlet fittings on the fuel pump body (**Figure 24**). Cut and dispose of the plastic locking type clamps used on some models. Remove spring type hose clamps by squeezing the ends together.
3. Position a container or shop towel under the fuel pump hoses, then carefully push the hoses from the fittings. Work carefully and do not tug on the hoses or exert side force against the fittings. The fittings will break if excessive force is used. Gently twist difficult hoses to free them from the fittings.
4. Drain residual fuel from the hoses. Remove the two fuel pump mounting screws (**Figure 25**); the mounting screws are closest to the inlet fitting.
5. Carefully pull the pump from the power head. Remove the gasket (1, **Figure 26**) from the power head or fuel pump surface. Discard the gasket.

Disassembly/assembly

Refer to **Figure 26** for this procedure.

22

23

24

25

1. Remove the three screws (**Figure 27**) that hold the assembly together.
2. Working carefully to avoid damaging gasket surfaces, carefully pry the fuel pump cover off the fuel pump body (**Figure 28**).
3. Remove the outer gasket and diaphragm from the body (**Figure 29**). Be extremely careful if the gaskets must be scraped for removal. Avoid damaging the gasket mating surfaces. Discard all gaskets and diaphragms.
4. Use the same procedures to remove the back cover, diaphragm and gasket from the pump body (**Figure 30**).
5. Remove the boost spring and cap (10 and 11, **Figure 26**). Inspect the spring for bending or corrosion. Replace as needed.
6. Remove the screws and nuts. Lift the check valves from the body (**Figure 31**). Inspect the check valve for bent or corroded surfaces. Inspect the valve contact surfaces on the fuel pump body for wear or deterioration. Replace the fuel pump body and/or check valves unless they are in excellent condition.
7. Use a straightedge to check the fuel pump body, outer cover and inner cover for warped surfaces. Replace warped components.
8. Inspect gasket and diaphragm contact surfaces on the body and covers for scratches, nicks or deteriorated surfaces. Replace components with damaged surfaces.
9. Assembly is the reverse of disassembly. Note the following:
   a. Install new gaskets and diaphragms during assembly.
   b. Use the screw openings in the diaphragms and gaskets to assist with proper orientation.
   c. Fit the boost spring and cap into the fuel pump body as shown in **Figure 32**.
   d. Carefully position the diaphragm over the boost spring and cap. The spring and cap are easily dislodged.
   e. Align the screw openings while holding the components together.
   f. Evenly and securely tighten the three screws to hold the assembly together.
10. Pressure test the fuel pump as described in this chapter.

Installation

1. Install a new gasket (1, **Figure 26**) onto the back cover of the fuel pump. Slip the two mounting screws through the pump and gasket to retain the gasket.
2. Install the fuel pump onto the power head. Make sure the mounting gasket is not dislodged during installation. Thread the mounting screws (**Figure 25**) into the power

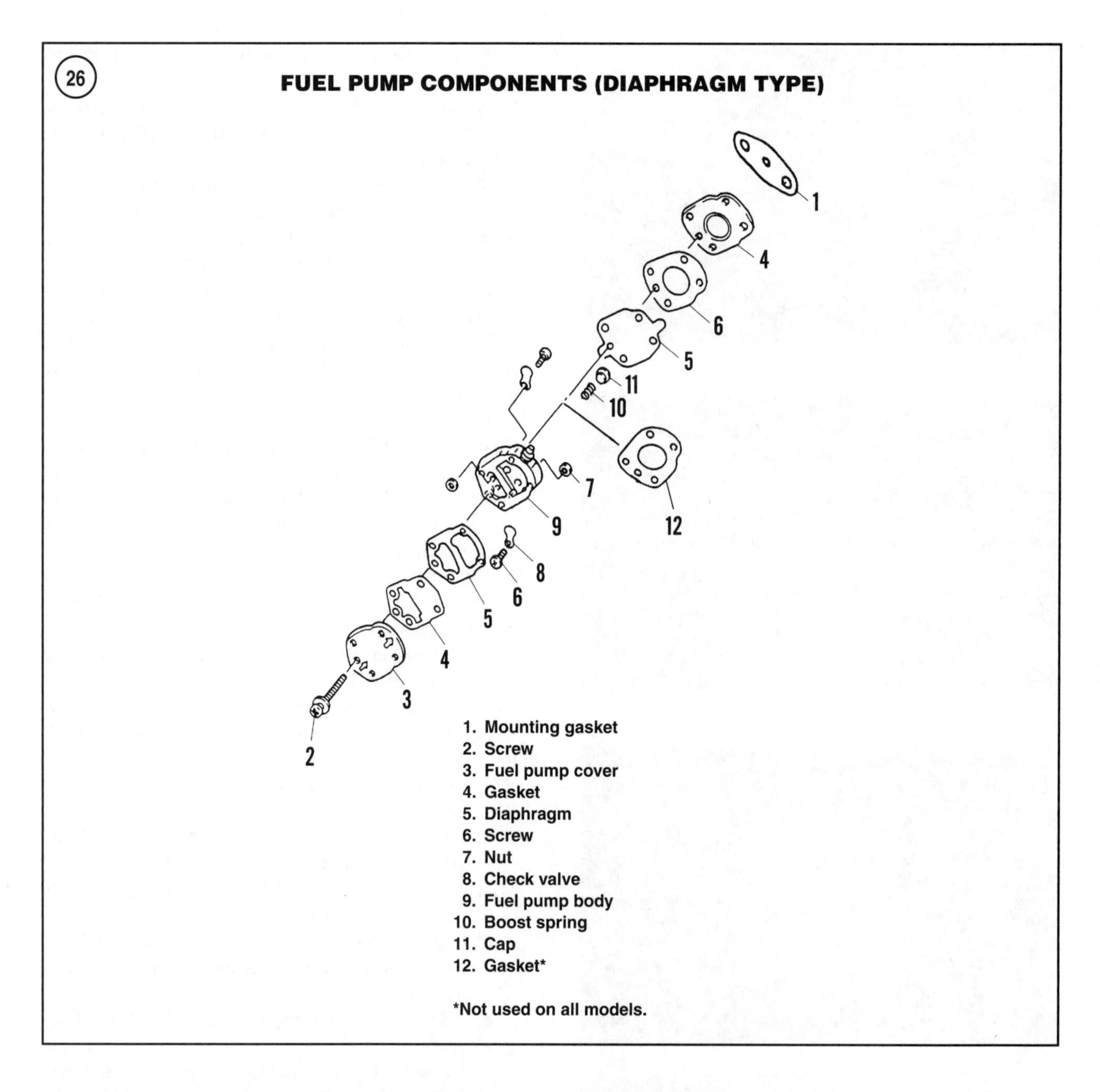
26
FUEL PUMP COMPONENTS (DIAPHRAGM TYPE)
1
4
6
5
11
10
7
9
12
8
6
5
4
3
2
1. Mounting gasket
2. Screw
3. Fuel pump cover
4. Gasket
5. Diaphragm
6. Screw
7. Nut
8. Check valve
9. Fuel pump body
10. Boost spring
11. Cap
12. Gasket*
*Not used on all models.

27
YAMAHA

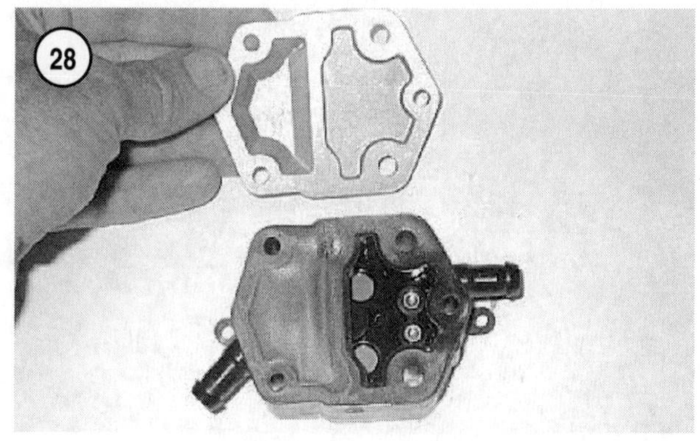
28

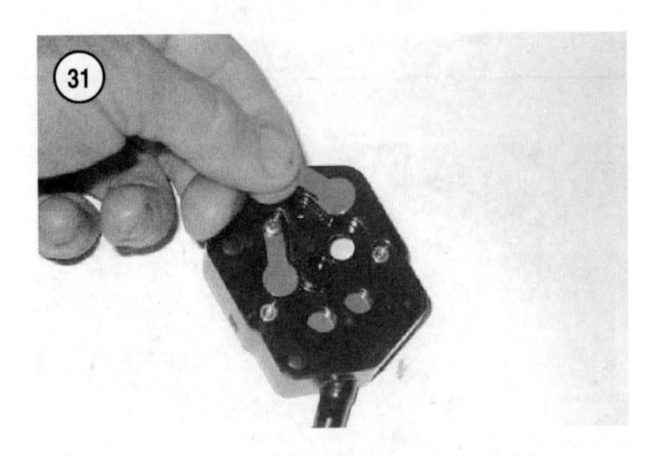

head openings. Evenly and securely tighten the mounting screws.

3. Connect the fuel hoses onto the fuel pump fittings (**Figure 24**). The hose connecting to the outlet side should lead to the carburetors or vapor separator tank. The hose connecting to the inlet side should lead to the canister type fuel filter.

4. Observe the fuel pump for leaks while squeezing the primer bulb. Correct any leaks before putting the engine into service.

5. Connect the battery cables and spark plug leads.

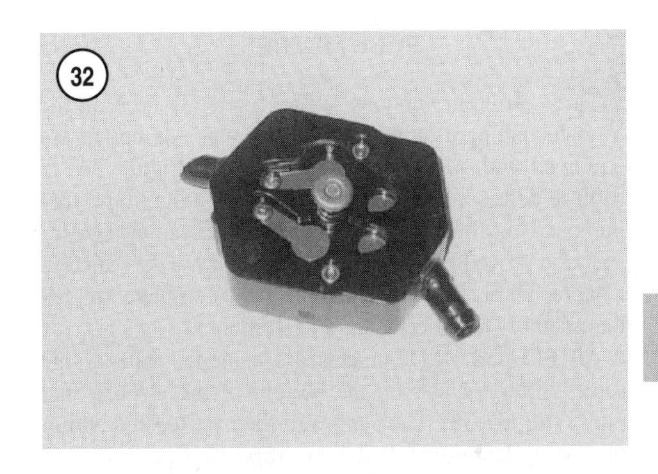

Fuel pump pressure test

Use a pressure/vacuum tester (Miti-Vac or Yamaha part No. YB-35956/90890-06756) for this procedure. Purchase the tester from an auto parts store or Yamaha dealership.

NOTE
Put a small amount of fuel in the fuel pump fittings before pressure testing the fuel pump. The fuel is necessary to simulate the fuel present on the sealing surfaces during operation. Use only enough fuel to wet the inner components and check valve surfaces.

1. Connect a hand operated vacuum/pressure pump onto the inlet fitting of the fuel pump. Block the outlet fitting with a finger (A, **Figure 33**). Slowly apply pressure until reaching 50 kPa (7.2 psi.). Faulty gaskets or incorrect assembly is indicated if the test pressure cannot be attained.

2. Connect the vacuum/pressure pump onto the inlet fitting of the fuel pump. Do not block the outlet fitting (B, **Figure 33**). Apply a vacuum until reaching 30 kPa (8.9 Hg). The check valve is faulty if the test vacuum cannot be attained.

3. Connect the vacuum/pressure pump onto the outlet fitting of the fuel pump (C, **Figure 33**). Do not block the inlet fitting. Slowly apply pressure until reaching 50 kPa (7.2 psi.). The check valve is faulty if the test pressure cannot be attained.

4. Disassemble and inspect the fuel pump if failure is noted in Step 1. Disassemble and inspect the check valves and pump body if failure is noted in Step 2 or Step 3. Replace the check valves if there are no faults with the pump body. Assemble the pump and retest. Replace the pump body if the pump again fails the pressure test.

FUEL FILTERS

Three different versions of fuel filters are used on the Yamaha outboards covered in this manual. All models are equipped with a canister type fuel filter (**Figure 34**) to capture debris before it reaches the diaphragm type fuel pumps. This filter is fully serviceable. Cleaning and inspection procedures for this type of filter are described in Chapter Three. If necessary, replace the complete fuel filter assembly as described in this section.

All EFI and HPDI models are equipped with a fine screen filter located on the bottom of the electric fuel pump (**Figure 35**). The pump and filter are located within the vapor separator tank. Refer to *Vapor Separator Tank* in this chapter for filter inspection and replacement procedures.

All HPDI models are equipped with an in-line type fuel filter (**Figure 36**) to capture debris before it reaches the vapor separator tank. Replace the in-line type filter as described in this section.

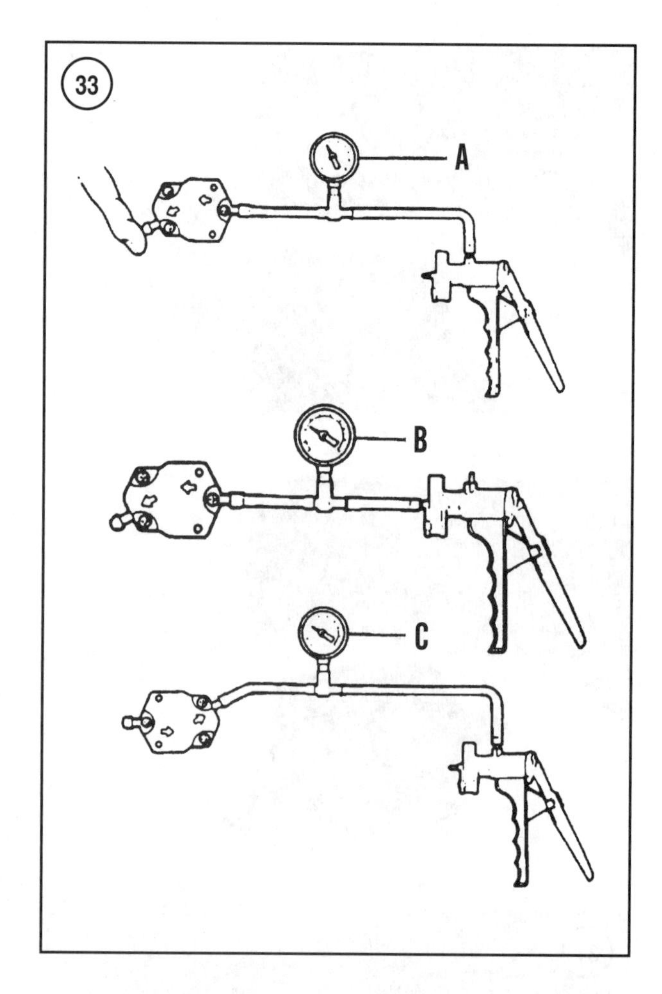

Canister Type Fuel Filter Replacement

1. Disconnect the battery cables and ground the spark plug leads to prevent accidental starting.
2. Trace the fuel hose from the quick-connector or barb type fitting to the fuel filter assembly (**Figure 34**).
3. Place a shop towel under the fuel filter to capture spilled fuel. Cut and remove plastic locking type clamps at the fuel filter fittings. Squeeze the ends of spring type clamps and move them away from the fittings. Replace weak or corroded clamps.
4. Carefully twist and pull to remove the hoses from the filter fittings. Cut stubborn hoses to remove them and replace them with the correct type of fuel hose. Drain residual fuel from the hoses.
5. Remove the fasteners and pull the filter assembly from the power head or filter mounting bracket or power head.
6. Fit the replacement filter assembly on the bracket or power head mount. Install the filter assembly mounting bolts and nuts and securely tighten them.
7. Refer to the fuel system diagrams (**Figures 10-14**) to determine the correct fuel flow and connection points for the specified model.
8. Note the molded in arrows near the filter housing fittings to determine the direction of fuel flow through the filter. Push the hoses onto the fuel filter fittings. The arrow pointing toward the hose fitting is the outlet fitting. The hose connected to this fitting must lead to the diaphragm type fuel pump(s).
9. Install new plastic locking type hose clamps onto the fuel hose. Tighten the clamps until the hoses fit snug on the fittings. Squeeze the ends together to open spring type clamps. Slide spring type clamps over the fittings. Then release the ends. Tug on the hose to verify a secure connection.

10. Observe the fuel filter, fittings and hoses while squeezing the primer bulb. Correct any fuel leaks before putting the engine into service.
11. Connect the battery cables and spark plug leads.

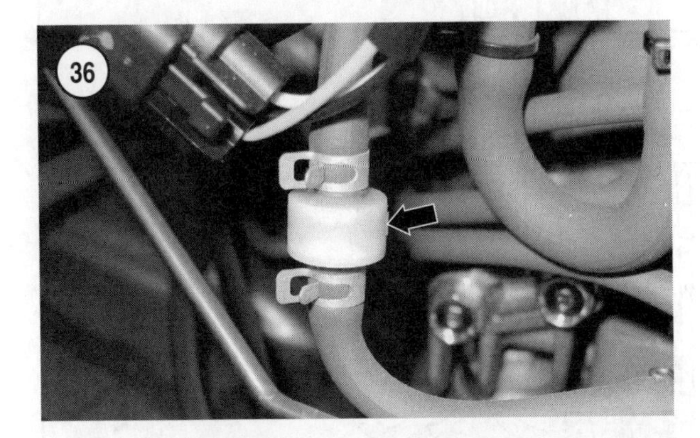

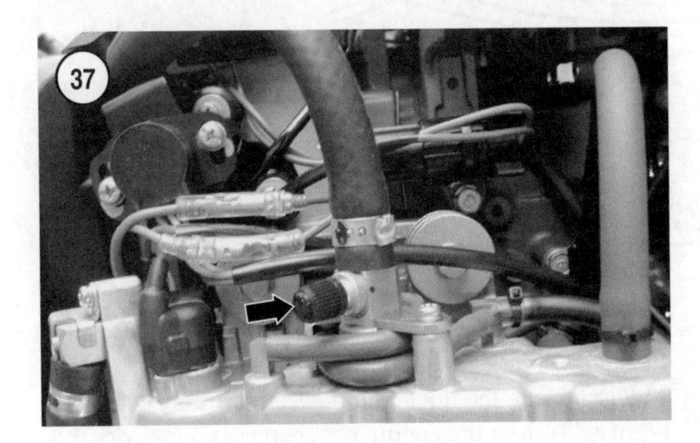

12. Clean and inspect the filter element at regular intervals as described in Chapter Three.

Inline Type Fuel Filter Replacement

1. Disconnect the battery cables and ground the spark plug lead(s) to prevent accidental starting.
2. Trace the fuel hose from the vapor separator tank to the inline fuel filter (**Figure 36**).
3. Place a shop towel under the fuel filter to capture spilled fuel. Remove any clamps at the fuel filter fittings. Refer to *Fuel Hose Connectors* in this chapter for removal procedures.
4. Carefully twist and pull to remove the hoses from the filter fittings. Cut stubborn hoses to remove them and replace with the correct type of fuel hose. Drain residual fuel from the hoses. Discard the fuel filter.
5. Push the hoses onto the fuel filter fittings. The arrow molded into the filter body should point toward the hose leading to the vapor separator tank.
6. Install new plastic locking type hose clamps onto the fuel hose. Tighten the clamps until the hoses fit snugly on the fittings. Tug on the hoses to verify a secure connection.
7. Route the fuel hoses and position the filter to prevent contact with any moving components.
8. Observe the fuel filter, fittings and hoses while squeezing the primer bulb. Correct any fuel leaks before putting the engine into service.
9. Connect the battery cables and spark plug leads.

VAPOR SEPARATOR TANK

This section describes removal, disassembly, inspection, assembly and installation procedures for the vapor separator tank (**Figure 22**). The vapor separator tank is used on all EFI and HPDI models. Although similar in appearance, the tank used on EFI models differs from the tank used on HPDI models.

Differences include the hose connection points and the electrical connections for the electric fuel pump. However, removal and installation of the complete assembly is similar.

Relieving Fuel System Pressure

Perform this procedure prior to disconnecting any fuel hoses from the vapor separator tank, fuel rails or high pressure mechanical fuel pump. The procedure requires a suitable fuel pressure gauge with a bleed off hose or Yamaha part No. YB-06766/90890-06766.

1. Disconnect the battery cables and ground the spark plug leads to prevent accidental starting.
2. Remove the cap from the fuel pressure test port (**Figure 37**).
3. Wrap a shop towel around the grip on the pressure gauge fitting to capture spilled fuel. Screw the fitting onto the test port as shown in **Figure 38**. Wipe up any spilled fuel.
4. Direct the bleed off hose into a suitable container. Press the valve until fuel stops flowing from the hose.

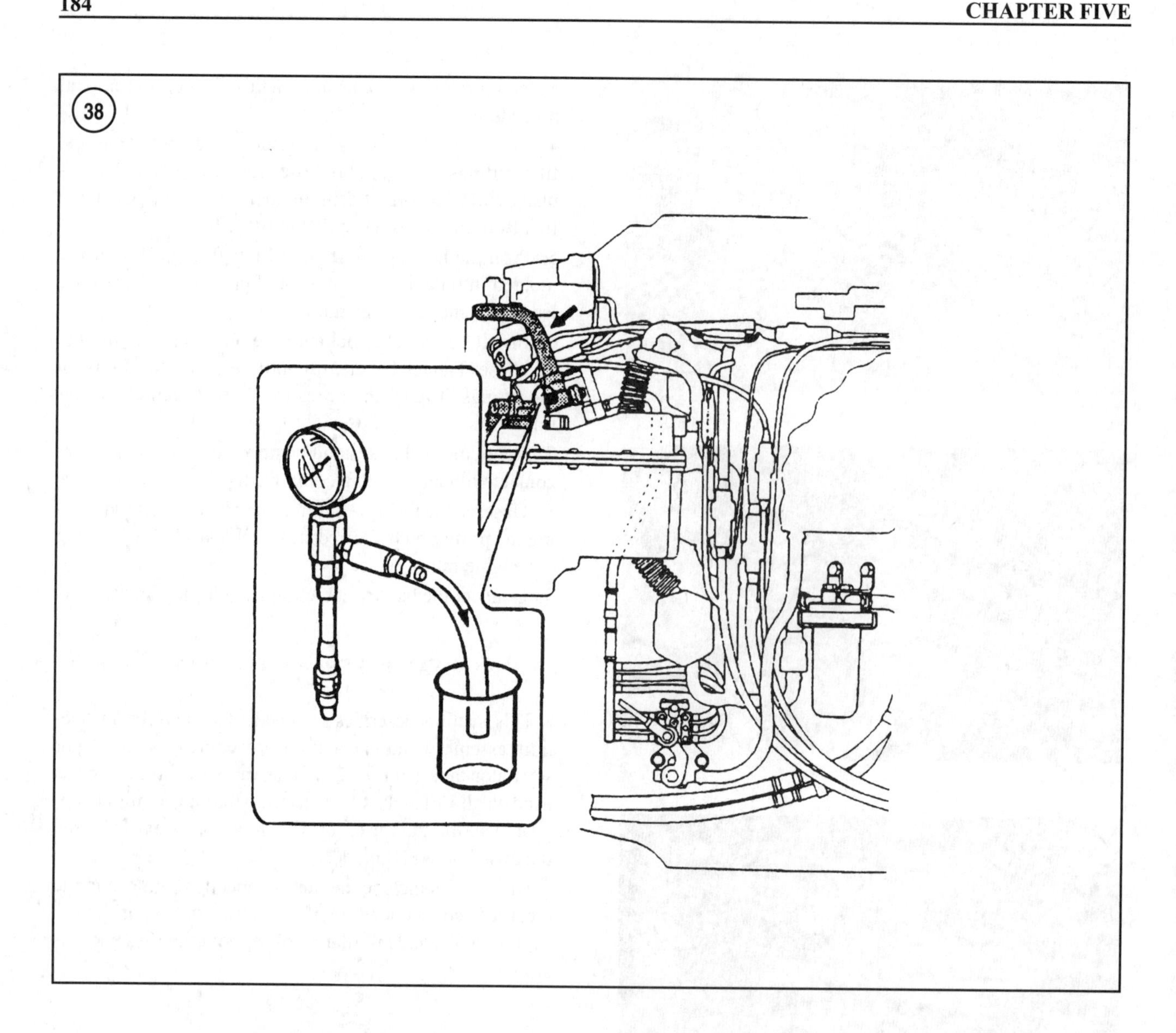

5. Wrap the towel around the grip on the pressure gauge fitting to capture spilled fuel, then unscrew the fitting. Wipe up any spilled fuel.
6. Screw the cap onto the pressure test port.

Removal

Mark all fuel hose connections to the vapor separator tank prior to removing any of the hoses. Many of the hoses use the same size fitting and can be easily connected to the wrong fitting after installation. Refer to **Figure 39** for this procedure.

1. Relieve the fuel system pressure as described in this section.
2. Mark all hoses to identify the connection points. Remove the clamps then disconnect all hoses from the vapor separator tank. Refer to *Fuel Hose Connectors* in this chapter for fuel clamp removal procedures.
3. Disconnect the fuel pump electrical connectors (3, **Figure 39**) from the engine wire harness connectors.
4. *150-250 hp (HPDI models)*—Remove the air pressure sensor from the vapor separator tank as described in Chapter Six. Remove the electric oil pump as described in Chapter Twelve.
5. Place a container suitable for holding fuel under the fuel drain plug (**Figure 40**). Remove the plug and drain all fuel from the tank. Disassemble and clean the vapor separator tank if debris or deposits are on the tip of the drain plug. Inspect the O-ring on the plug for worn, flattened or damaged surfaces. Replace the O-ring as needed. Install and securely tighten the plug.
6. Support the vapor separator tank, then remove the three mounting bolts and washer.

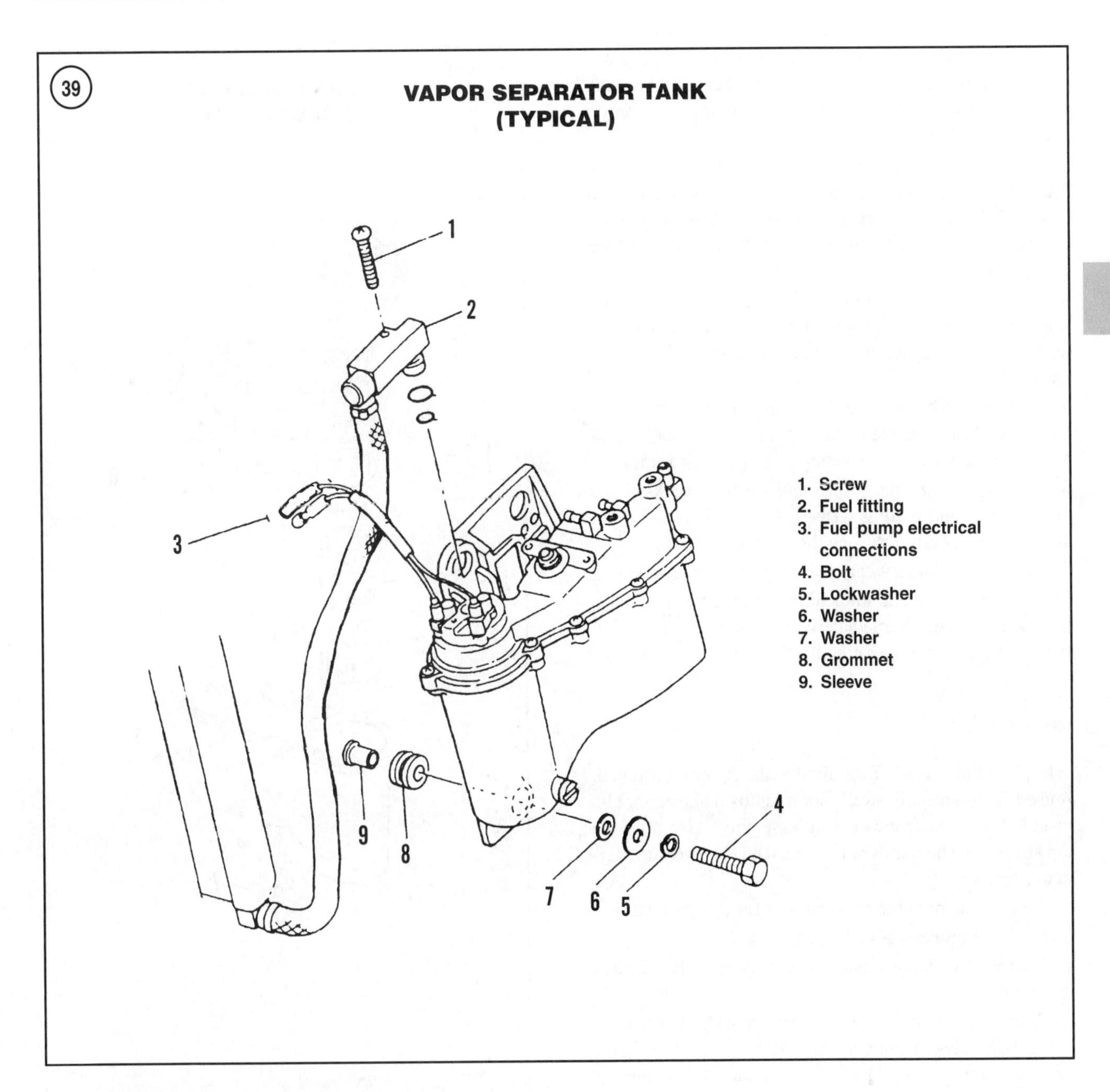

7. Retrieve the grommets (8, **Figure 39**) and sleeves (9) while pulling the tank from the power head.

Installation

Refer to **Figure 39** for this procedure.

1. Fit the three grommets (8, **Figure 39**) and sleeves (9) into the openings in the vapor separator tank.
2. Align the three mounting bolt openings while fitting the tank onto the power head. Thread the three bolts and washer into the vapor separator tank and power head.

Make sure that no hoses, wiring cables or linkages are pinched between the tank and the power head. Tighten the three mounting bolts to the general tightening torque specification in Chapter One.

3. Connect the fuel pump electrical connectors (3, **Figure 39**) onto the respective engine wire harness connectors. Route the wiring to prevent interference with moving components. Secure the wiring with plastic locking type clamps as necessary.

4. *150-250 hp (HPDI models)*—Install the air pressure sensor onto the vapor separator tank as described in Chapter Six. Install the electric oil pump as described in Chapter Twelve.

5. Connect all hoses onto the respective fittings on the tank. Install all hose clamps. Refer to *Fuel Hose Connectors* in this chapter for fuel clamp installation procedures.

6. Slowly squeeze the primer bulb while checking the hose fittings and mating surfaces for fuel leaks. Correct any fuel leaks before starting the engine.

7. Connect the battery cables and spark plug leads.

8. Start the engine and immediately check for fuel leaks. If leakage occurs, immediately stop the engine. Correct fuel leakage before operating the engine.

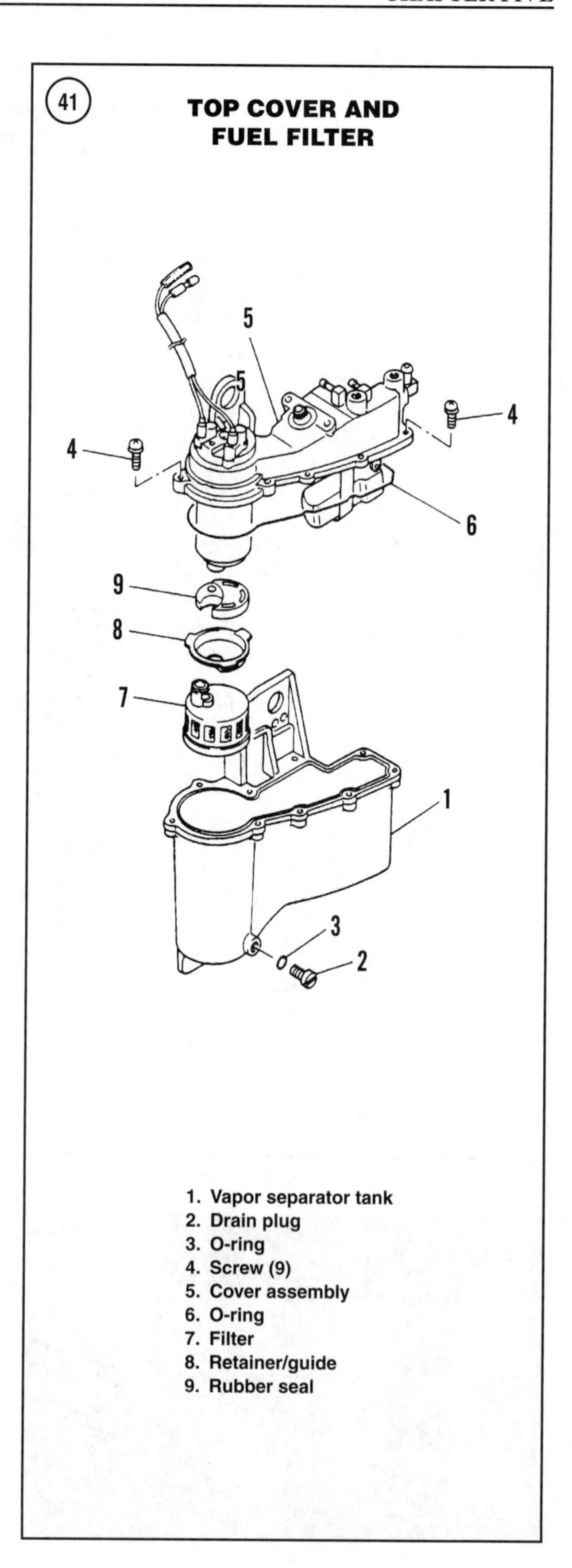

1. Vapor separator tank
2. Drain plug
3. O-ring
4. Screw (9)
5. Cover assembly
6. O-ring
7. Filter
8. Retainer/guide
9. Rubber seal

Disassembly

Replace the cover O-ring anytime the cover is removed. A used O-ring usually swells upon removal and cannot be installed into the groove during assembly. Also note the top cover must be removed to access the fuel pressure regulator screws.

Use a tire stem valve removal tool for this procedure.

Refer to **Figures 41-44** for this procedure.

1. Remove the vapor separator tank as described in this section.

2. Remove the nine screws (4, **Figure 41**), then lift the cover assembly (5) from the tank. Remove the O-ring (6, **Figure 41**) from the tank groove . Discard the O-ring.

3. Remove the fuel pressure regulator as follows:

   a. Remove the two screws and washers (2, **Figure 42**), then carefully pull the regulator away from the cover.

   b. Retrieve the screen (8, **Figure 42**) and O-ring (7) from the opening in the cover.

   c. Remove the O-ring (4, **Figure 42**) from the bottom of the regulator or opening in the cover.

   d. Pull the sleeve (5, **Figure 42**) and O-ring (6) from the opening in the cover.

4. Remove the fuel filter and related components as follows:

FUEL PRESSURE REGULATOR

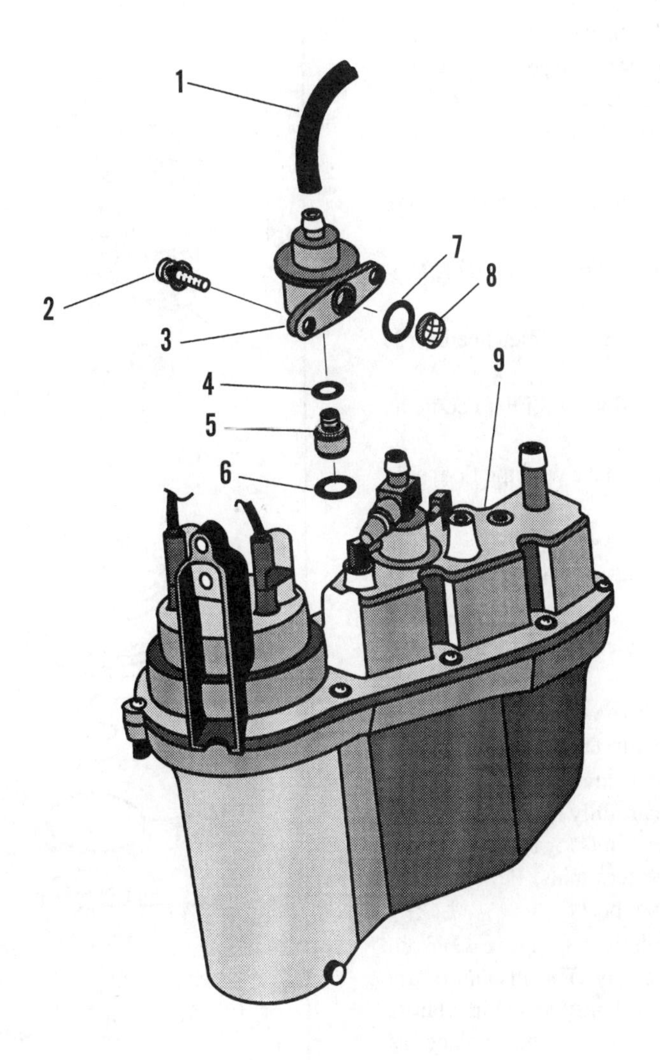

1. Hose (regulator to intake manifold)
2. Screw and washer
3. Fuel pressure regulator
4. O-ring
5. Sleeve
6. O-ring
7. O-ring
8. Screen
9. Vapor separator tank

a. Grasp the fuel filter (7, **Figure 41**), then carefully rotate the filter clockwise as viewed from the bottom of the electric fuel pump.
b. Pull the filter from the pump when the hook clears the slot in the retainer (8, **Figure 41**).
c. Pull the retainer (8, **Figure 41**) and seal (9) from the bottom of the pump.

5A. *EFI models*—Remove the float and inlet needle as follows:

a. Place the cover on a work surface with the float side facing up.
b. Remove the screw that retains the float pin (1, **Figure 43**) to the top cover.
c. Lift the float (2, **Figure 43**) and inlet needle (3) from the cover.

5B. *HPDI models*—Remove the float and inlet needle as follows:

a. Place the cover on a work surface with the float side facing up.
b. Remove the screw (4, **Figure 44**).
c. Lift the float (2, **Figure 44**) and inlet needle (3) from the cover.

6A. *EFI models*—Remove the electric fuel pump as follows:

a. Place the cover assembly on a work surface with either side of the electric pump facing down.
b. Carefully pry the terminal covers (4, **Figure 43**) from the terminal. Work carefully to avoid breaking the covers. Remove the two nuts (5, **Figure 43**) and lockwashers (6) from the terminals. Pull the wire terminals from the terminal posts.
c. Carefully pry the two insulators (8, **Figure 43**) from the top cover. Work carefully. The insulators are easily broken. Upon removal, inspect the insulators for broken or cracked surfaces and replace as needed. Remove the O-rings (9, **Figure 43**) from the insulators. Discard the O-rings.
d. Pull the electric fuel pump (10, **Figure 43**) from the cover. Pull the O-ring (16, **Figure 43**) from the outlet nozzle of the pump. Then lift the aluminum spacer (15) from the pump.
e. Pull the retainer (13, **Figure 43**) from the top of the pump. Remove the O-ring (14, **Figure 43**) from the retainer or the top cover.
f. Remove the O-rings (12, **Figure 43**) and plastic washers (11) from the terminal posts of the pump.

6B. *HPDI models*—Remove the electric fuel pump as follows:

a. Place the cover assembly on a work surface with either side of the electric pump facing down.

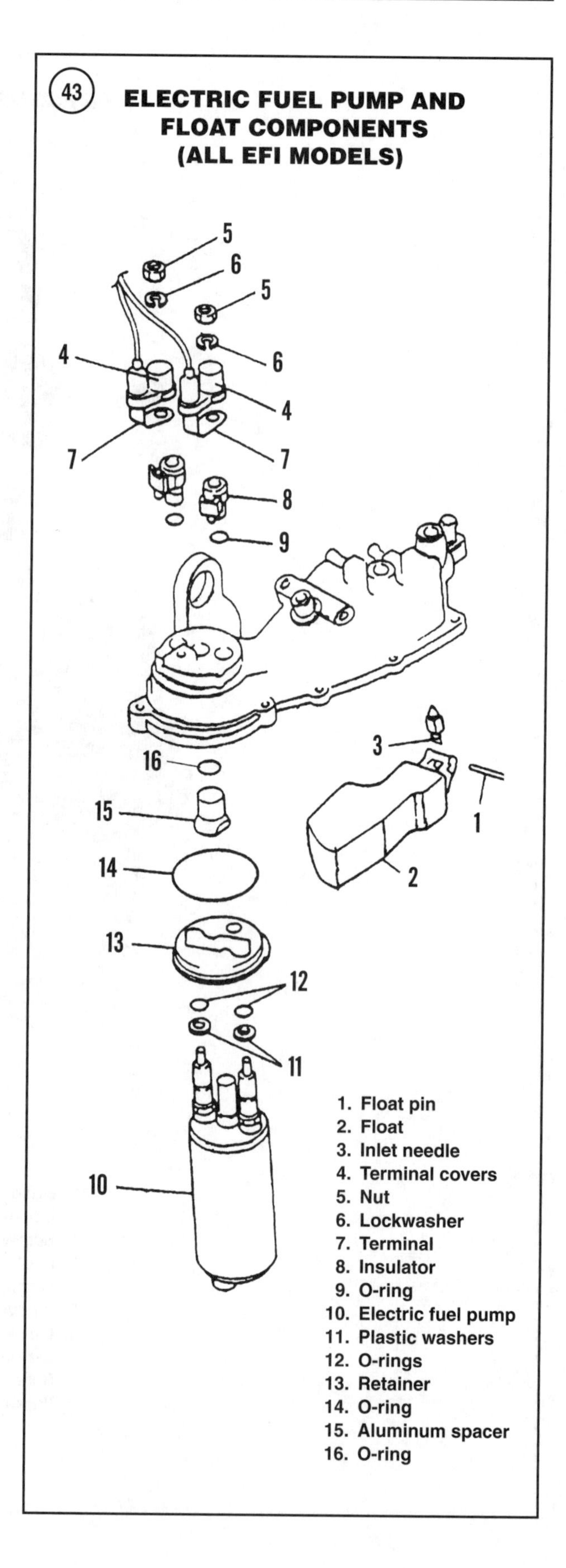

44

ELECTRIC FUEL PUMP AND FLOAT COMPONENTS (ALL HPDI MODELS)

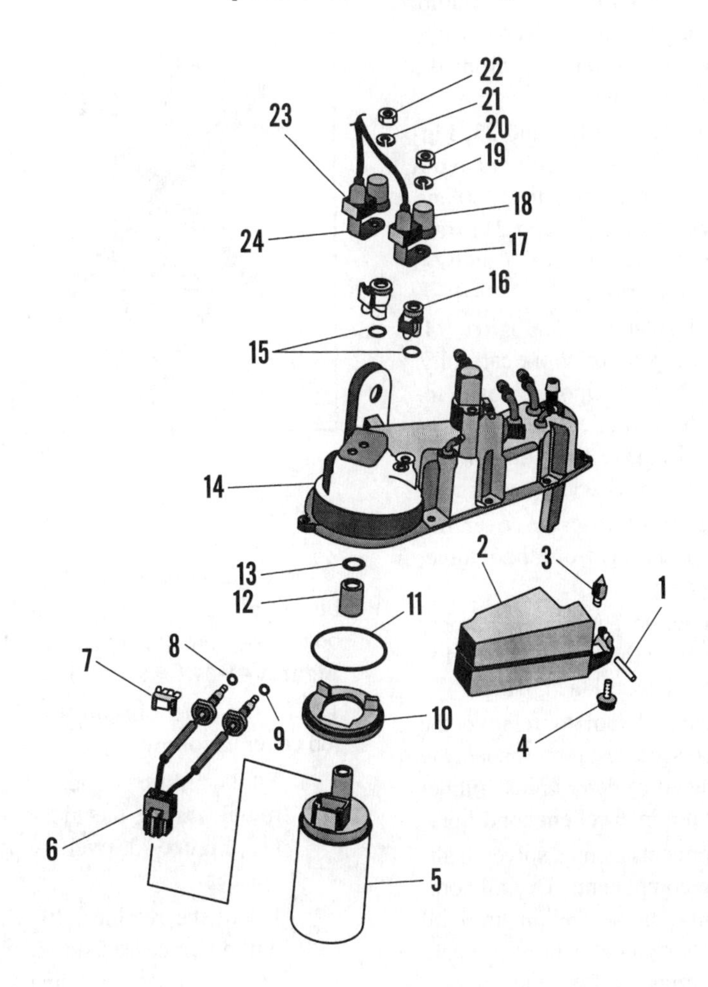

1. Float pin
2. Float
3. Inlet needle
4. Screw
5. Electric fuel pump
6. Electric pump harness connector
7. Retainer
8. O-ring
9. O-ring
10. Retainer
11. O-ring
12. Spacer
13. O-ring
14. Top cover
15. O-rings
16. Insulators
17. Wire terminal
18. Terminal cover
19. Lockwasher
20. Nut
21. Lockwasher
22. Nut
23. Terminal cover
24. Wire terminal

b. Carefully pull the electric fuel pump (5, **Figure 44**) out of the cover enough to access the electrical pump harness connector (6).
c. Disconnect the connector from the electric pump, then pull the pump from the cover. Remove the spacer (12, **Figure 44**) and O-ring (13) from the pump nozzle. Discard the O-ring.
d. Carefully pry the terminal covers (18 and 23, **Figure 44**) from the terminal. Work carefully to avoid breaking the covers. Remove the two nuts (20 and 22, **Figure 44**) and lockwashers (19 and 21) from the terminals. Pull the wire terminals (17 and 24, **Figure 44**) from the terminal posts.
e. Carefully pry the two insulators (16, **Figure 44**) from the openings in the top cover. Work carefully. The insulators are easily broken. Upon removal, inspect the insulators for broken or cracked surfaces and replace as needed. Remove the O-rings (15, **Figure 44**) from the insulators. Discard the O-rings.
f. Pull the retainer (10, **Figure 44**) from the cover. Remove the O-ring (11, **Figure 44**) from the retainer or cover. Discard the O-ring.
g. Carefully pull the terminal post and pump harness from the top cover. Remove the O-rings (8 and 9, **Figure 44**) from the post. Discard the O-rings.

7. Use a tire stem valve removal tool to remove the schrader valve from the fuel pressure test port. Inspect the valve for corrosion and damaged or deteriorated rubber surfaces. Replace the valve if not in excellent condition.
8. Thoroughly clean all components using a solvent suitable for aluminum and plastic components. Dry all components with compressed air. Direct the air into all passages and orifices to remove residual debris. Use care when cleaning debris or contamination from the filter (7, **Figure 41**). The screen is easily damaged. A damaged screen does not provide adequate protection for the fuel injectors.
9. Inspect all components for cracked, corroded or damaged surfaces and replace as needed.
10. Inspect the filter (7, **Figure 41**) for a torn or damaged screen. Replace the filter if the screen is damaged or if contamination cannot be completely removed by cleaning.
11. Inspect the inlet needle for a worn or damaged tip (**Figure 45**). Replace the needle if its condition is questionable.
12. Inspect the float for cracked, worn or deteriorated surfaces. Push a thumbnail against the float to check for fuel saturation. The float is saturated if fuel appears in the thumbnail contact area. Replace the float if the surfaces are damaged or if saturated with fuel.

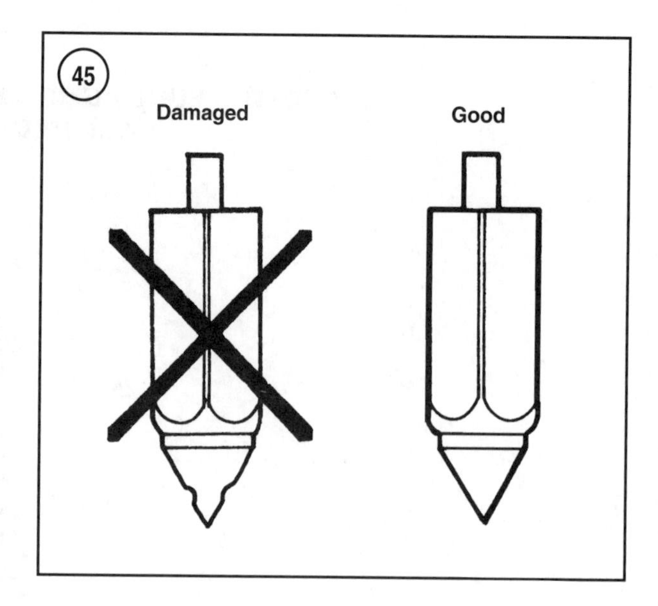

Assembly

Install new O-rings in all locations during tank assembly. Apply a light coating of Yamalube two-cycle outboard oil onto the O-rings before installing. Refer to **Figures 41-44** for this procedure.

1A. *EFI models*—Install the electric fuel pump into the top cover as follows:

a. Fit the plastic washers (11, **Figure 43**) onto the terminal post of the fuel pump. Install new O-rings (12, **Figure 43**) over the posts and rest them on the washers.
b. Install the retainer (13, **Figure 43**) over the post with the recessed side facing the pump. Rotate the retainer until the retainer fits over the terminals and drops down onto the pump.
c. Slip the aluminum spacer (15, **Figure 43**) over the outlet nozzle of the pump. The flat sides of the spacer must align with the terminal post. The larger end of the spacer must face toward the retainer as shown in **Figure 43**. Fit the O-ring (16, **Figure 43**) over the pump nozzle. Then seat the O-ring against the spacer.
d. Install a new O-ring (14, **Figure 43**) onto the retainer (13). Make sure the O-ring rests on the step of the retainer.
e. Carefully guide the terminal posts into the respective openings while installing the pump into the top cover. Rotate the pump assembly to align the retainer with the recess in the cover. The retainer seats against the cover when the pump is oriented correctly.

f. Install new O-rings (9, **Figure 43**) onto the insulators (8). Fit the O-rings into the grooves in the insulators. Carefully manipulate the pump to center the terminal post in the openings. Then slip the insulators over the post. Align the protrusion on the bottom of the insulators with the opening in the cover. Support the pump while seating the insulators into the cover.

g. Fit the wire terminals (7, **Figure 43**) onto the respective posts. The terminals have different size openings. Install the lockwashers (6, **Figure 43**) and nuts (5). Securely tighten the nuts without overtightening. Do not over-tighten the nuts. Otherwise the insulators may break. Slip the terminal covers (4, **Figure 43**) onto the terminals. Gently push down on the covers until they lock onto the terminals.

1B. *HPDI models*—Install the electric fuel pump into the top cover as follows:

a. Fit new O-rings (8 and 9, **Figure 44**) into the groove on each terminal post. Carefully insert the terminal posts into the correct openings in the top cover.

b. Install new O-rings (15, **Figure 44**) onto the insulators (16). Fit the O-rings into the grooves in the insulators. Carefully manipulate the terminal posts to center the them in the openings, then slip the insulators over the posts. Align the protrusion on the bottom of the insulators with the opening in the cover. Support the pump while seating the insulators into the cover.

c. Fit the wire terminals (17 and 24, **Figure 44**) onto the correct posts. The terminals have different size openings. Install the lockwashers (19 and 21, **Figure 42**) and nuts (20 and 22). Securely tighten the nuts without overtightening. Do not over-tighten the nuts. Otherwise the insulators may break. Slip the terminal covers (18 and 23, **Figure 44**) on the terminals. Gently push down on the covers until they lock to the terminals.

d. Place a new O-ring (11, **Figure 44**) onto the retainer (10). The O-ring must rest on the step on the retainer. Guide the pump wire harness through the opening while installing the retainer into the cover. The tabs on the retainer must face toward the cover. The O-ring must fit on the step and between the retainer and the cover.

e. Fit the spacer (12, **Figure 44**) over the outlet nozzle on the pump. Install a new O-ring (13, **Figure 44**) over the nozzle and seat against the spacer.

f. Connect the fuel pump wire harness connector (6, **Figure 44**) onto the electric fuel pump. Make sure the locking tab of the retainer (7, **Figure 44**) engages the pump. The wires must fit into the two slots in the retainer. Tug on the connector to verify a secure connection.

g. Rotate the pump to align the connector with the offset opening in the retainer while inserting the pump into the cover. Guide the nozzle into the cover opening while seating the pump into the cover.

2A. *EFI models*—Install the inlet needle and float as follows:

a. Place the cover on a work surface with the float side facing up.

b. Slip the pull-off wire on the inlet needle over the tab on the float arm.

c. Insert the float pin (1, **Figure 43**) through the opening in the float arm.

d. Guide the inlet needle into the seal while lowering the float onto the cover. Fit the float pin into the slot in the cover.

e. Lightly coat the threads with Loctite 271. Then install the screw that retains the float pin into the cover. Securely tighten the screw. Wipe excess Locktite from the screw and cover. Allow a few hours for the Loctite to cure before proceeding.

2B. *HPDI models*—Install the inlet needle and float as follows:

a. Place the cover onto a work surface with the float side facing up.

b. Slip the pull-off wire on the inlet needle over the tab on the float arm.

c. Insert the float pin (1, **Figure 44**) through the opening in the float arm.

d. Guide the inlet needle into the seal while lowering the float onto the cover. Fit the float pin into the slot in the cover.

e. Lightly coat the threads with Loctite 271. Then install the screw (4, **Figure 44**) into the cover. Securely tighten the screw. Wipe excess Loctite from the screw and cover. Allow a few hours for the Loctite to cure before proceeding.

3. Install the fuel filter and related components as follows:

a. Install the rubber seal (9, **Figure 41**) to the bottom of the electric fuel pump. The rounded notch in the seal must fit around the inlet nozzle on the pump.

b. Fit the retainer (8, **Figure 41**) to the bottom of the pump with the recess facing the pump. Rotate the retainer to align the opening with the inlet nozzle on the pump. When aligned, seat the retainer against the rubber seal.

c. Carefully insert the filter opening into the opening in the retainer. This aligns the opening with the pump inlet nozzle. Rotate the filter to position the

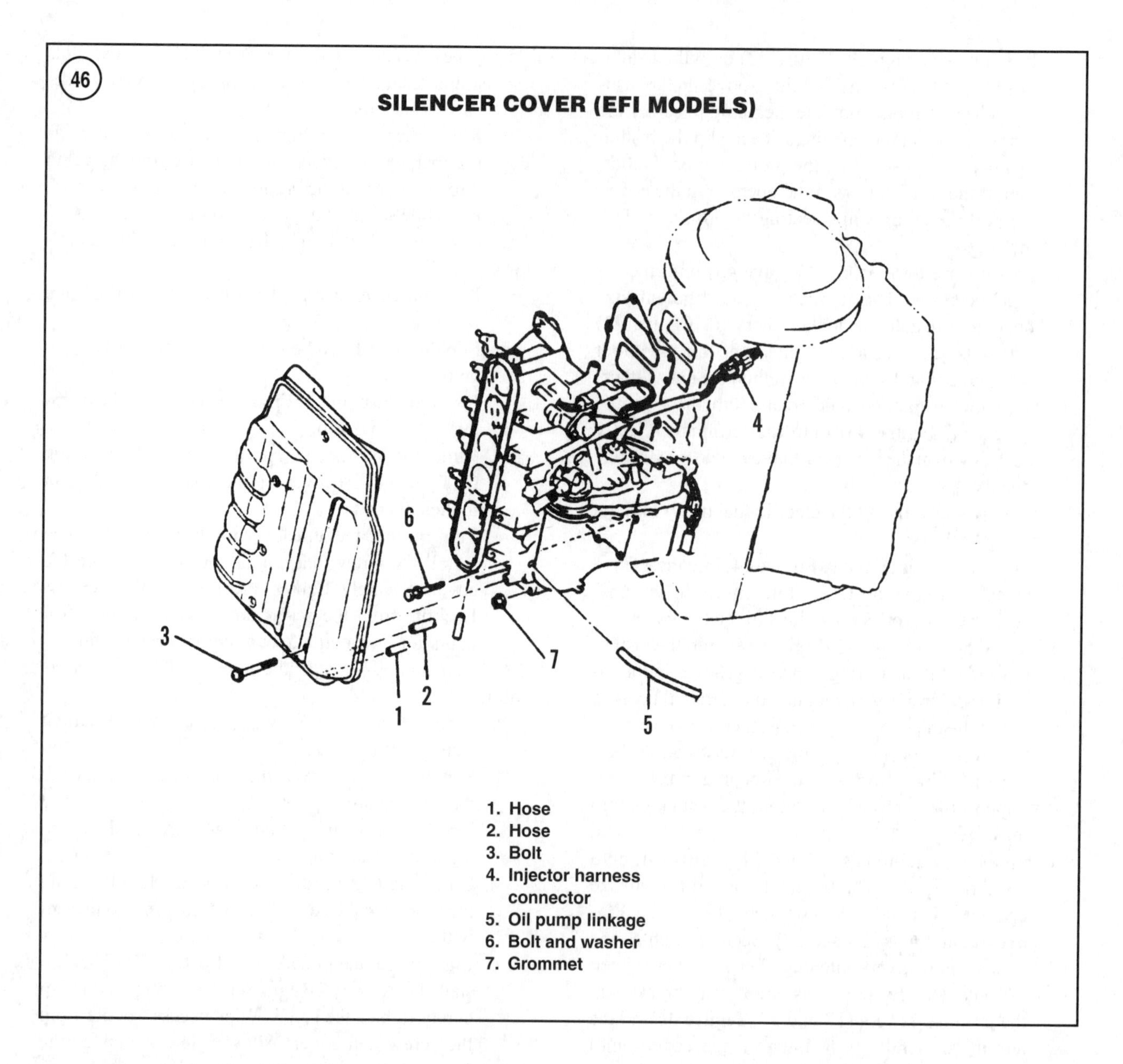

filter offset from the center of the pump. Gently push the filter onto the pump nozzle until the hook on the filter aligns with the slot in the retainer. Rotate the filter counterclockwise as viewed from the bottom of the pump until the hook enters the slot. Continue rotating the filter until the filter is centered with the pump.

4. Install the fuel pressure regulator as follows:
   a. Fit a new O-ring (6, **Figure 42**) onto the larger diameter side of the sleeve (5). Carefully insert the sleeve into the vapor separater tank (9, **Figure 42**) with the larger diameter side facing down or into the opening.
   b. Fit a new O-ring (4, **Figure 42**) onto the smaller diameter side of the sleeve. Seat the O-ring onto the step on the sleeve.
   c. Install a new O-ring (7, **Figure 42**) into the recess in the fuel pressure regulator (3). Install the screen (8, **Figure 42**) into the regulator so the cup side of the screen faces the regulator.
   d. Guide the lower end of the regulator over the sleeve (5, **Figure 42**) while installing the regulator onto the cover. Make sure the screen (8, **Figure 42**) fits into the cover opening. Work carefully to avoid damaging the O-ring and screen.

(47) **FUEL RAIL AND HOSE CONNECTIONS (EFI MODELS)**

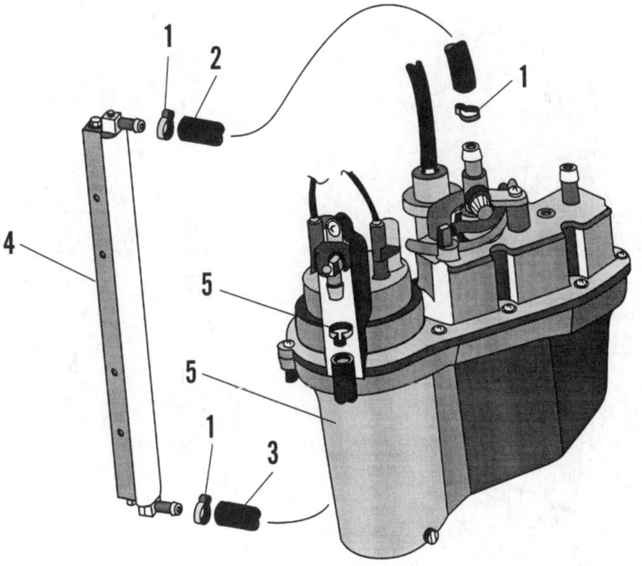

1. Crimp type hose clamps
2. Fuel return hose
3. Fuel outlet hose
4. Fuel rail
5. Vapor separator tank

e. Seat the regulator against the cover, then install the two screws and washers (2, **Figure 42**). Evenly tighten the screws to the general torque specification in Chapter One.

5. Install a new O-ring (6, **Figure 41**) into the groove in the tank. Guide the fuel filter and retainer into the opening while installing the cover assembly onto the tank. Check for improper alignment of the retainer (8, **Figure 41**) to the bosses in the filter opening if the cover does not easily seat against the tank. Lightly coat Loctite 271 on the threads of the screws (4, **Figure 41**). Install the screws. Then tighten them securely using a crossing pattern.

6. Use a tire valve stem tool to install the schrader valve into the fuel pressure test port opening.

7. Install the vapor separator tank as described in this section.

SILENCER COVER, FUEL RAIL AND INJECTORS (EFI AND HPDI MODELS)

On all EFI models, the fuel rail and injectors mount onto the throttle body at the front of the engine. On HPDI models, the fuel rails and injectors mount onto each cylinder head. To prevent dangerous fuel leaks, replace all seals, grommets and O-rings anytime the fuel rail and injectors are removed. Lubricate the seals, grommets and O-rings with a light coating of Yamalube two-cycle outboard oil prior to assembly.

Removal/Disassembly (EFI Models)

Refer to **Figures 46-49** for this procedure.

1. Relieve the fuel system pressure as described in this chapter.

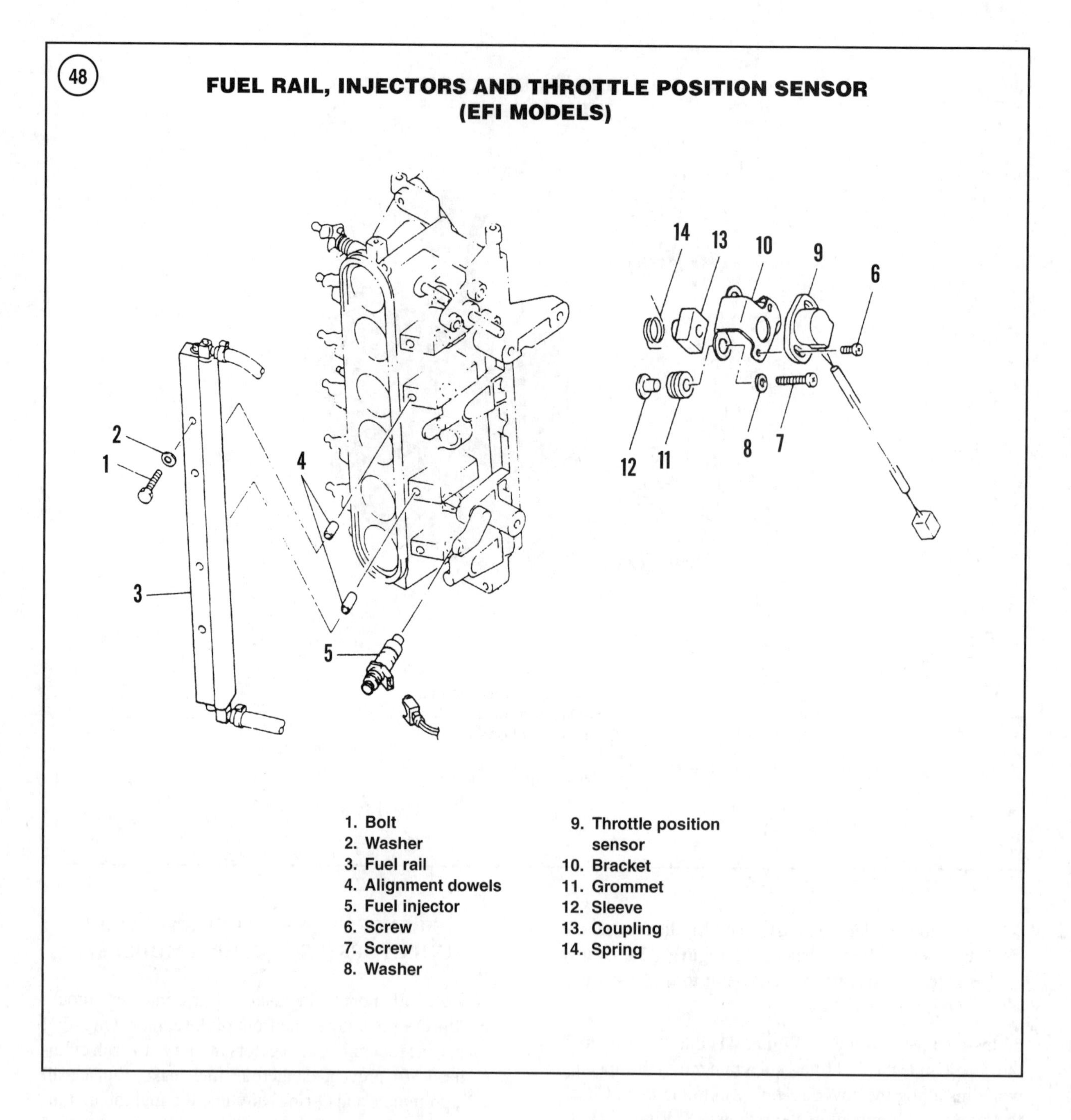

FUEL RAIL, INJECTORS AND THROTTLE POSITION SENSOR (EFI MODELS)

1. Bolt
2. Washer
3. Fuel rail
4. Alignment dowels
5. Fuel injector
6. Screw
7. Screw
8. Washer
9. Throttle position sensor
10. Bracket
11. Grommet
12. Sleeve
13. Coupling
14. Spring

2. Remove the six bolts (3, **Figure 46**). Then pull the silencer cover away from the throttle body.
3. Note the routing and connection points, then disconnect the hoses (1 and 2, **Figure 46**) from the silencer cover. Remove the cover.
4. Remove the four hose clamps (1, **Figure 47**) as described in this chapter. See *Fuel Hose Connectors*.
5. Place a suitable container or shop towel under the fittings to capture spilled fuel. Then carefully pull the hoses (2 and 3, **Figure 47**) from the fuel rail (4) fittings. Drain residual fuel from the hoses. Then pull the hoses from the vapor separator tank fittings (5, **Figure 47**).
6. Disconnect the injector harness connector (4, **Figure 46**) from the engine wire harness connector.
7. Note the wire routing and connection points then disconnect the connector from each fuel injector. Push in on the rounded end of the wire clip to spread the ends of the clip, then carefully pull the connector from the injector.

49

FUEL INJECTOR SEALS (EFI MODELS)

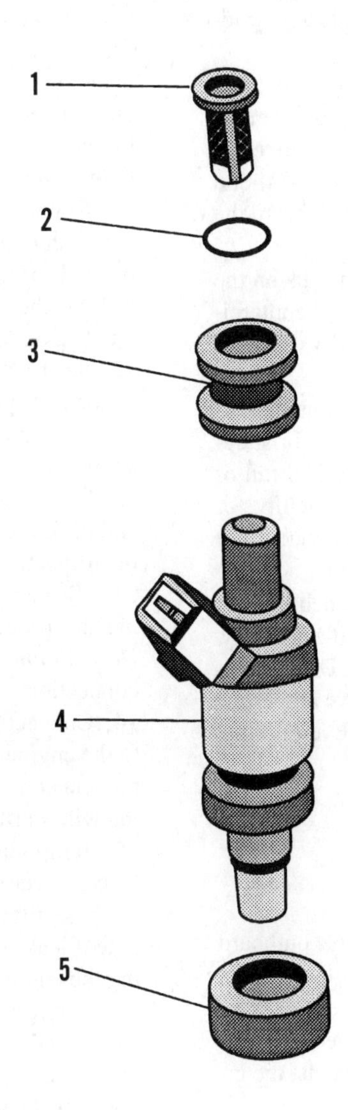

1. Filter
2. O-ring
3. Grommet
4. Fuel injector
5. Seal

Remove the plastic locking type clamp that secures the injector wire harness to the fuel rail.
8. Remove the four bolts (1, **Figure 48**) and washers (2). Carefully pull the fuel rail (3, **Figure 48**) slightly away from the throttle body. Pull each injector from the throttle body while removing the fuel rail. Twist the injectors slightly to break any bond in the opening. Drain any residual fuel from the hose fittings on the rail.
9. Mark each injector to identify the mounting location on the rail. Then carefully pull each fuel injector from the fuel rail openings. Carefully pull the filter (1, **Figure 49**) from the injector opening. Remove the O-ring, seal and grommet from each fuel injector (**Figure 49**). Discard the seals, grommets and O-rings.
10. Inspect the injector filters for debris, deposits on the screen or physical damage. Replace the filters if contaminated or damaged. The cleaning process usually damages the filters.
11. Thoroughly clean the fuel rail using a suitable solvent. Have injectors professionally cleaned if debris, deposits or other contaminants are found in the fuel rail or injector openings. Ask an automotive repair facility for sources that can perform this service. If this service is not available, replace the contaminated injector(s).
12. Take all necessary precaution to prevent debris from entering the fuel rail or injector openings. If necessary, store the fuel injectors in airtight plastic bags. Direct pressurized air into the fuel rail opening to remove all debris. Cover all fuel rail openings after cleaning. Even minute particles can foul a fuel injector.

Assembly/Installation (EFI Models)

Refer to **Figures 46-49** for this procedure.
1. Apply a light coating of Yamalube two-cycle outboard oil on the O-rings, seals and grommets and the contact surfaces on the injector.
2. Insert the filter (1, **Figure 49**) into the injector opening. Make sure the flange on the filter seats against the injector.
3. Fit the grommet (3, **Figure 49**) onto the injector body. Seat the grommet against the step on the injector.
4. Install the new O-ring (2, **Figure 49**) onto the injector. Make sure the O-ring fits into the groove on the injector.
5. Fit the seal (5, **Figure 49**) onto the injector. Make sure the seal seats against the step on the injector.
6. Lightly coat Yamalube two-cycle outboard oil into the injector opening in the fuel rail. Carefully insert each injector, filter end first, into the openings. Work carefully to avoid pinching or damaging the O-ring (2, **Figure 49**) or dislodging the filter (1).
7. Rotate the injectors to position the connector end facing toward the port side of the engine, as if the rail was installed.
8. Lightly coat Yamalube two-cycle outboard oil onto the injector openings in the throttle body.
9. Guide the injectors into the openings while installing the fuel rail onto the throttle body (see **Figure 48**). Carefully rock the fuel rail to help seat the injectors into the openings.
10. Thread the four bolts (1, **Figure 48**) and washers (2) through the fuel rail and into the throttle body. Evenly tighten the bolts to the specification in **Table 1**.
11. Connect each injector harness connector onto injector. Push in on the rounded end of the wire clip to expand the ends of the clip. Push the connector onto the injector. Release the wire clip to lock the connector onto the injector. Tug on the connector to verify a secure connection.
12. Secure the injector harness to the fuel rail with a plastic locking type clamp. Make sure to trim the clamp to prevent it from entering the throttle opening during engine operation.
13. Slip new crimp type clamps (1, **Figure 47**) over each end of the fuel hoses (2 and 3). Slide the fuel hoses onto the respective fittings on the fuel rail and vapor separator tank. Secure the hoses with the crimp type clamps following the procedures described in this chapter (See *Fuel Hose Connectors*). Tug on the hoses to verify a secure connection.
14. Connect the injector harness connector (4, **Figure 46**) to the engine wire harness connector. Route the wiring to prevent contact with moving components. Secure the wiring with plastic locking type clamps as needed.
15. Temporarily run the engine at idle speed with the silencer cover removed while checking for fuel leaks from the hose fittings and injectors. Immediately stop the engine if leaks occur. Remedy any fuel leaks before operating the engine.
16. Connect the hoses (1 and 2, **Figure 46**) to fittings on the silencer cover. Route all hoses to prevent contact with moving components. Secure the hoses with plastic locking type clamps as necessary. Do not overtighten the clamps. Otherwise the hoses may collapse.
17. Fit the silencer cover onto the throttle body. Thread the six bolts (3, **Figure 46**) through the cover and into the throttle body. Evenly tighten the bolts to the specification in **Table 1**.

Removal/Disassembly (HPDI Models)

The silencer cover and throttle body used on HPDI models are very similar to the components used on EFI models. On EFI models, the fuel rail and injectors are mounted on the throttle body. On HPDI models, the fuel rails and injectors are mounted on the cylinder heads.

Refer to **Figure 46** and **Figure 50** for this procedure.

1. Relieve the fuel system pressure as described in this chapter.
2. Remove the six bolts (3, **Figure 46**). Then pull the silencer cover away from the throttle body.
3. Note the routing and connection points, then disconnect the hoses (1 and 2, **Figure 46**) from the silencer cover. Remove the cover.
4. Disconnect the harness connector (15, **Figure 50**) from each fuel injector.
5. Place a shop towel under the hose connection on the starboard fuel rail. Slowly loosen the two screws (12, **Figure 50**) until fuel begins flowing from the connection. Allow the fuel pressure to dissipate. Then remove the screws and washers. Pull the hose flange away from the starboard fuel rail. Drain residual fuel into a suitable container. Remove the sealing rings and O-ring (1-3, **Figure 50**) from the hose flange or the fuel rail opening. Discard the sealing rings and O-ring.
6. Place a shop towel under the hose connection to the port fuel rail to capture spilled fuel. Carefully remove the two screws and washers that retain the hose flange onto the port fuel rail. Pull the hose flange away from the fuel rail. Then drain residual fuel from the hose. Remove the sealing rings and O-ring (1-3, **Figure 50**) from the hose flange or the fuel rail opening. Discard the sealing rings and O-ring.
7. Remove the six bolts (18, **Figure 50**). Then pull the injector clamps (19) from the cylinder head bosses.
8. Remove the bolt (9, **Figure 50**), then pull the bracket (7) from each fuel rail.
9. Remove the three bolts (8, **Figure 50**). Then carefully pull the port fuel rail assembly from the cylinder head. Remove the sleeves (6, **Figure 50**) from the top and bottom bolts or the bolt openings in the fuel rail. Repeat this step to remove the starboard fuel rail assembly. Drain residual fuel from the rails using the fuel hose opening. Place the fuel rail assemblies onto a clean work surface.
10. Remove the two sealing washers (16, **Figure 50**) from each fuel injector. Remove the washers from the cylinder head openings if they are not found on the injectors. Discard the sealing washers.
11. Carefully pry the hooked end of the fuel injector retainers (14, **Figure 50**) from the fuel rail. Then slide the retainer off the injectors.
12. Pull the fuel injectors (17, **Figure 50**) from the fuel rail openings. Remove the sealing rings and O-ring (1-3, **Figure 50**) from the fuel injectors or the openings in the fuel rail. Discard the sealing rings and O-rings.
13. Clean all debris, oily residue or other contamination from the injector opening in the cylinder heads. The opening must be completely clean for proper sealing for assembly.
14. Thoroughly clean the fuel rail using a suitable solvent. Have injectors professional cleaned if debris, deposits or other contaminants are found in the fuel rail or injector openings. Contact an automotive repair facility for sources in the area that can perform this service. If this service is not available, replace the contaminated injector(s).
15. Take all necessary precautions to prevent debris from entering the fuel rail or injector openings. Direct pressurized air into the fuel rail opening to remove all debris. Cover all openings in the fuel rail after cleaning. Even minute particles can foul a fuel injector.

Assembly/Installation (HPDI Models)

Refer to **Figure 46** and **Figure 50** for this procedure.

1. Lightly coat Yamalube two-cycle outboard oil onto the O-rings, seals and grommets and the contact surfaces on the injector.
2. Install the fuel injectors onto the fuel rails as follows:
   a. Insert a new thick sealing ring (3, **Figure 50**), new O-ring (2), then a thin sealing ring (1) onto the fuel injector (17). Seat these components onto the step on the inlet side of the fuel injector.
   b. Carefully insert the fuel injector into the opening in the fuel rail. Work carefully to avoid pinching or damaging the O-rings. Carefully twist the injector for easy installation.
   c. Rotate the injector to position the harness connection side of the injector facing outward with the rail installed. Carefully slide the slot in the retainer (14, **Figure 50**) over the fuel injector body. Then snap the hooked end of the retainer onto the fuel rail. Carefully tug on the injector to verify a secure connection.
   d. Repeat this step for the remaining injectors and fuel rail.
3. Install the fuel rails onto the cylinder head as follows:
   a. Lightly coat Yamalube two-cycle outboard oil onto the injector openings in the cylinder heads. Do not use an excessive amount of oil. The oil may enter the combustion chamber and foul the spark plug.
   b. Fit two new sealing washers (16, **Figure 50**) onto the discharge end of each fuel injector. Seat the washers against the step on the injector body.
   c. Guide the injectors into the openings while installing the port fuel rail onto the cylinder head. Care-

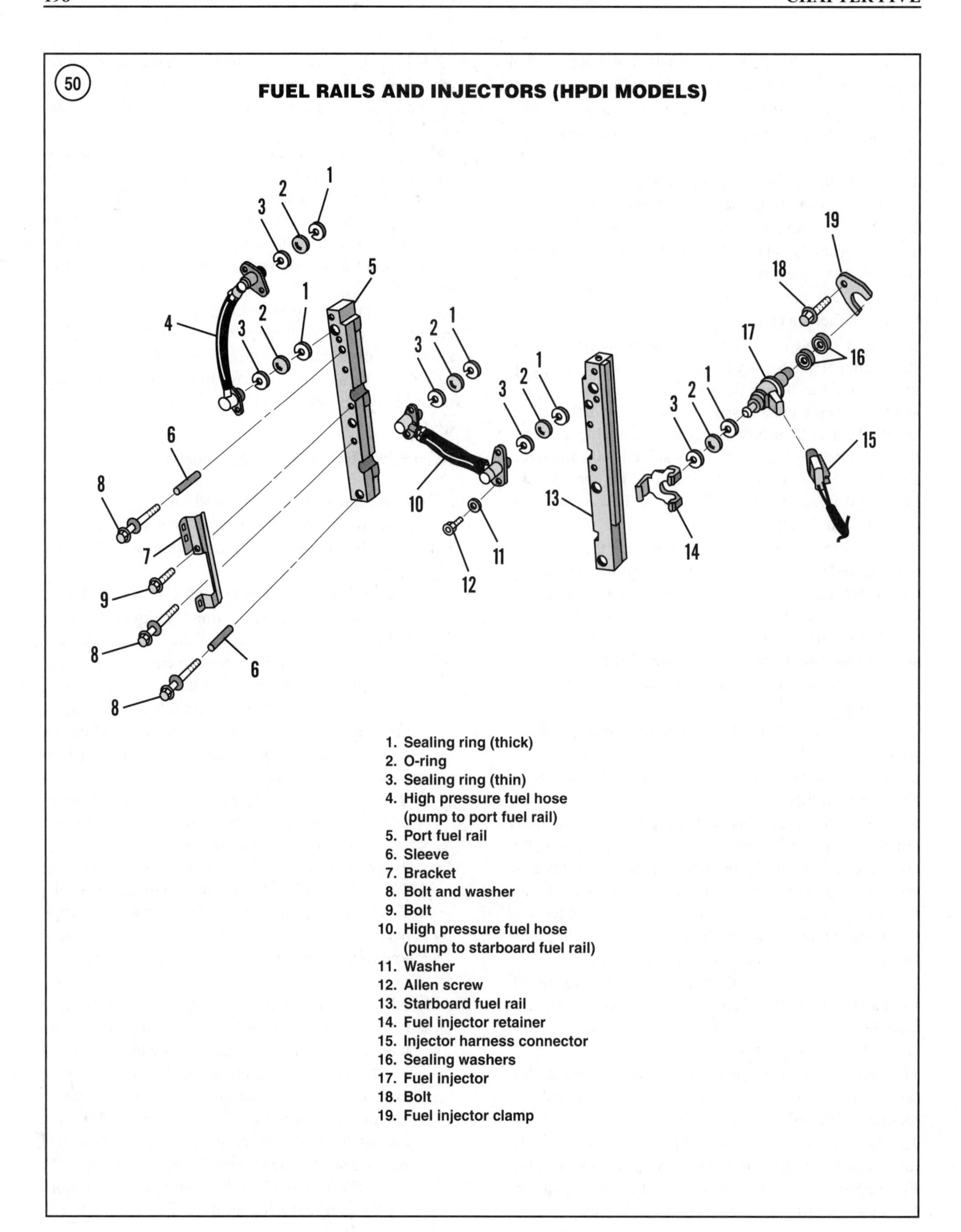
50
FUEL RAILS AND INJECTORS (HPDI MODELS)
1
2
3
4
5
6
7
8
9
10
11
12
13
14
15
16
17
18
19
1. Sealing ring (thick)
2. O-ring
3. Sealing ring (thin)
4. High pressure fuel hose (pump to port fuel rail)
5. Port fuel rail
6. Sleeve
7. Bracket
8. Bolt and washer
9. Bolt
10. High pressure fuel hose (pump to starboard fuel rail)
11. Washer
12. Allen screw
13. Starboard fuel rail
14. Fuel injector retainer
15. Injector harness connector
16. Sealing washers
17. Fuel injector
18. Bolt
19. Fuel injector clamp

fully rock the fuel rail to assist with seating the injectors into the openings.
   d. Fit the three injector clamps (19, **Figure 50**) onto the fuel injectors and cylinder head bosses. Thread the bolts (18, **Figure 50**) into the clamps and cylinder head. Verify proper engagement of the clamp onto the injector. Then tighten each clamp bolt to the specification in **Table 1**.
   e. Install the sleeves (6, **Figure 50**) into the upper and lower bolt opening in the fuel rail. Install the three bolts and washers (8, **Figure 50**) into the fuel rail and cylinder head. Evenly tighten the three bolts to the specification in **Table 1**.
   f. Repeat this step for the starboard fuel rail.
4. Connect each injector harness connector onto the respective injector. Push in on the connector tab. Then push the connector fully onto the injector. Release the tab. Then tug on the connector to verify a secure connection.
5. Lightly coat Yamalube two-cycle outboard oil onto the surfaces. Then install a new thick sealing washer (3, **Figure 50**), new O-ring (2) and then a thin sealing washer (1) onto the flange of the high pressure hose (10). Seat the washers and O-rings against the flange. Carefully insert the flange into the opening in the starboard fuel rail (13, **Figure 50**), then seat the flange. Align the openings. Then thread the screw (12, **Figure 50**) and washer (11) into the flange and fuel rail. Securely tighten the screws. Repeat this step for the hose connecting to the port side fuel rail.
6. Install the brackets (7, **Figure 50**) onto each fuel rail and secure with the screws.
7. Run the engine at idle while checking for fuel leaks from the hose fittings and injectors. Immediately stop the engine if leaks occur. Remedy any fuel leaks before operating the engine.
8. Connect the hoses (1 and 2, **Figure 46**) onto the respective fittings on the silencer cover. Route all hoses to prevent contact with moving components. Secure the hoses with plastic locking type clamps as necessary. Do not over-tighten the clamps. Otherwise the hoses may collapse.
9. Fit the silencer cover onto the throttle body. Thread the six bolts (3, **Figure 46**) through the cover and into the throttle body. Evenly tighten the bolts to the specification in **Table 1**.

THROTTLE BODY

The silencer cover and throttle body used on HPDI models is very similar to the components used on EFI models. On EFI models, the fuel rail and injectors are mounted on the throttle body. On HPDI models, the fuel rails and injectors are mounted onto the cylinder heads.

Throttle Body Removal

Refer to **Figure 46** for this procedure.

1. Relieve the fuel system pressure as described in this chapter.
2. Remove the six bolts (3, **Figure 46**). Then pull the silencer cover away from the throttle body.
3. Note the routing and connection points, then disconnect the hoses (1 and 2, **Figure 46**) from the silencer cover. Remove the cover.
4. *150-250 hp (EFI models)*—Remove the fuel rail and fuel injectors as described in this chapter.
5. Remove the throttle position sensor as described in Chapter Six.
6. *150-250 hp (HPDI models)*—Remove the air temperature sensor from the throttle body as described in Chapter Six.
7. Carefully disconnect the throttle linkage arm from the pivot on the throttle cam. The throttle cam is located on the starboard side of the throttle body.
8. Remove the vapor separator tank as described in this chapter.
9. *150-250 hp (EFI models)*—Remove the fuel pump resistor as described in Chapter Six.
10. Disconnect the oil pump linkage from the throttle shaft arm. The throttle shaft arm is located on the lower port side of the throttle body.
11. Note the routing and connection points, then disconnect all hoses from the throttle body fittings.
12. Remove the bolts and nuts that secure the throttle body onto the intake manifold/reed housing.
   a. *150-200 hp models (2.6 liter)*—11 bolts with washers and two nuts are used.
   b. *200-250 hp models (3.1 liter)*—12 bolts and two nuts are used.
13. Remove the gasket from the throttle body or intake manifold/reed housing surfaces. Discard the gasket. Thoroughly clean all gasket mating surfaces. Improper gasket sealing may cause lean fuel/air mixture and possible power head damage.
14. Clean all corrosion, debris, oily deposits and other contaminants from the throttle body. Do not remove the screws that secure the throttle plates onto the throttle shaft. Damage to the threaded openings occurs if the screws are removed.

Installation

Refer to **Figure 46** for this procedure.

1. Install a new gasket onto the intake manifold/reed housing. Align the gasket with the two studs on the intake manifold. Seat the gasket onto the housing.

2. Install the throttle body to the intake manifold/reed housing. Thread the bolts into the throttle body and housing. Thread the two nuts onto the studs.
   a. *150-200 hp models (2.6 liter)*—11 bolts with washers and two nuts are used.
   b. *200-250 hp models (3.1 liter)*—12 bolts and two nuts are used.
3. Tighten the bolts and nuts and three even steps to the specification in **Table 1**. Use a crossing pattern working from the middle toward the upper and lower end of the throttle body.
4. Connect all hoses to the fittings on the throttle body. Route the hoses to prevent contact with moving components. Secure the hoses with plastic locking type clamps as needed. Do not overtighten the clamp. Otherwise the hoses may collapse.
5. Carefully connect the oil pump linkage onto the throttle shaft arm. Work carefully to avoid breaking the plastic linkage connector.
6. Install the vapor separator tank as described in this chapter.
7. *150-250 hp (EFI models)*—Install the fuel rail and fuel injectors as described in this section.
8. *150-250 hp (EFI models)*—Install the fuel pump resistor as described in Chapter Six.
9. Install the throttle position sensor as described in Chapter Six.
10. *150-250 hp (HPDI models)*—Install the air temperature sensor from the throttle body as described in Chapter Six.
11. Carefully connect the throttle linkage arm to the throttle cam pivot. Work carefully to avoid breaking the plastic linkage connector.
12. Connect the hoses (1 and 2, **Figure 46**) to the respective fittings on the silencer cover. Route all hoses to prevent contact with moving components. Secure the hoses with plastic locking type clamps as necessary. Do not overtighten the clamps. Otherwise the hoses may collapse.
13. Fit the silencer cover onto the throttle body. Thread the six bolts (3, **Figure 46**) through the cover and into the throttle body. Evenly tighten the bolts to the specification in **Table 1**.
14. Perform all applicable adjustments as described in Chapter Four.

HIGH PRESSURE MECHANICAL FUEL PUMP

The high pressure mechanical fuel pump (**Figure 51**) is used on all HPDI models. This section describes drive belt removal and installation of the belt tensioner removal and installation, pulley removal and installation and pump

disassembly and assembly. If a torque specification is not given when tightening a fastener, refer to Chapter One for general torque specifications.

NOTE

If disassembling the fuel pump or replacing the pulley, remove the pulley before removing the pump from the power head. The power head provides a sturdy mount required when removing the pulley.

Drive Belt Removal and Installation

WARNING

Wear gloves and eye protection when working with the belt tensioner spring. The spring may unexpectedly release from the tensioner and cause serious injury.

1. Disconnect the battery cables and ground the spark plug leads to prevent accidental starting.
2. Remove the flywheel cover.
3. Use locking grip pliers to grasp the hook of the spring (A, **Figure 52**). Carefully extend the spring and remove the spring hook from the tensioner bracket.

4. Pivot the tensioner pulley (B, **Figure 52**) away from the belt, then lift the belt from the pulleys.
5. Use compressed air to remove belt material and loose debris from the pulleys and the top of the power head. Use a shop towel and mild solvent to remove all oily deposits form the pulleys. Oily deposits can dramatically reduce belt life. Inspect the drive belt as described in Chapter Three.
6. Pull the tensioner pulley toward the starboard side of the power head, then fit the belt onto the pulleys. The tensioner pulley must contact the flat side of the belt. Make sure the numbers on the side of the belt (**Figure 53**) face up and the cogs on the belt fit into the cogs on the pulleys.
7. Pivot the tensioner pulley toward the belt. Use locking grip pliers to hook the spring (A, **Figure 52**) into the slot in the tensioner.
8. Install the flywheel cover.
9. Connect the battery cables and spark plug leads.

Belt Tensioner Removal and Installation

Refer to **Figure 54** for this procedure.
1. Remove the drive belt as described in this section.
2. Remove the bolt and washer (2, **Figure 54**). Lift the tensioner pulley from the bracket.
3. Remove the pivot bolt (15, **Figure 54**), then lift the bracket (4) from the power head.
4. Remove the screw (13, **Figure 54**), then lift the spring bracket (12) from the tensioner bracket.
5. Except for the pulley clean all components using a suitable solvent. Use compressed air to dry the components and remove debris from the tensioner pulley.
6. Rotate the pulley and check for roughness, looseness or wobbling. Roughness indicates bearing failure or debris within the pulley bearing. A loose feel or wobbling indicates an excessively worn pulley bearing. Replace the tensioner pulley if these or other defects occur.
7. Clean corrosion and other contaminants from the tension mounting boss on the power head.
8. Install the spring bracket onto the tensioner bracket. Orient the spring slot in the bracket as indicated in **Figure 54**. Thread the screw (13, **Figure 54**) into the spring and tensioner bracket. Then tighten to the general torque specification in Chapter One.
9. Fit the tensioner bracket (4, **Figure 54**) onto the power head mounting boss. Lightly coat Yamaha All-Purpose grease or equivalent onto the shank of the pivot bolt (15, **Figure 54**). Then thread the bolt into the tensioner bracket and power head boss.
10. Insert the bolt and washer (2, **Figure 54**) into the tensioner pulley (3). Fit the pulley onto the tensioner bracket, then hand-thread the bolt into the tensioner bracket.
11. Hook the spring (14, **Figure 54**) into the slot in the pulley. Do not hook the spring onto the spring bracket (12, **Figure 54**) before installing the drive belt.
12. Install the drive belt as described in this section.
13. Tighten the tensioner pulley bolt (2, **Figure 54**) to the specification in **Table 1**.
14. Install the flywheel cover. Connect the battery cables and spark plugs leads.

Pulley Removal and Installation

Use a pulley bracket holding tool (Yamaha part No. YU-01235/90890-01235) to remove and install the flywheel pulley. Use a strap wrench and a common two jaw gear puller to remove and install the fuel pump pulley.
1. Disconnect the battery cables and ground the spark plug leads to prevent accidental starting.
2. Remove the drive belt as described in this section.

NOTE
Remove the pulley from the flywheel only if it must be replaced or if the flywheel must be removed. Corrosion damage may seize the pulley to the pulley bracket. If seized, the pulley bracket and pulley are usually damaged during the removal process.

3. Remove the pulley from the flywheel as follows:
   a. Engage the pins of the holding tool into the pulley bracket openings as shown in **Figure 55**.
   b. Hold the tool to prevent flywheel rotation. Then remove the bolt (5, **Figure 54**). Remove the pulley cover (6, **Figure 54**) from the pulley.
   c. Carefully pry the pulley from the bracket. If the pulley is seized onto the bracket and must be replaced or removed to access the flywheel nut, break the hub of the bracket by striking the side of the pulley

(54)

DRIVE BELT, TENSIONER AND PULLEYS (HPDI MODELS)

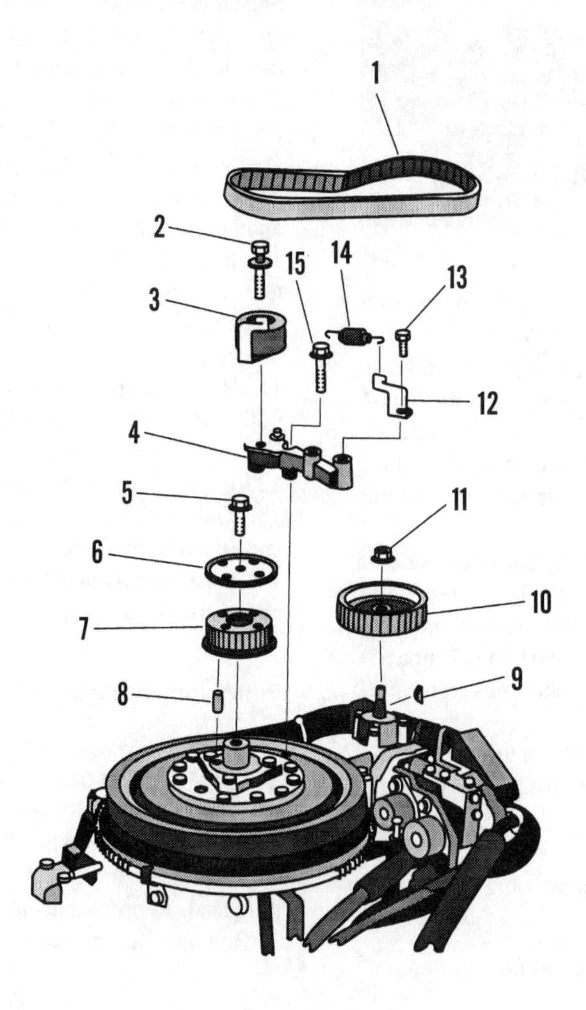

1. Drive belt
2. Bolt and washer
3. Tensioner pulley
4. Tensioner bracket
5. Bolt
6. Pulley cover
7. Flywheel pulley
8. Pin
9. Drive key
10. Fuel pump pulley
11. Nut
12. Spring bracket
13. Screw
14. Spring
15. Pivot bolt

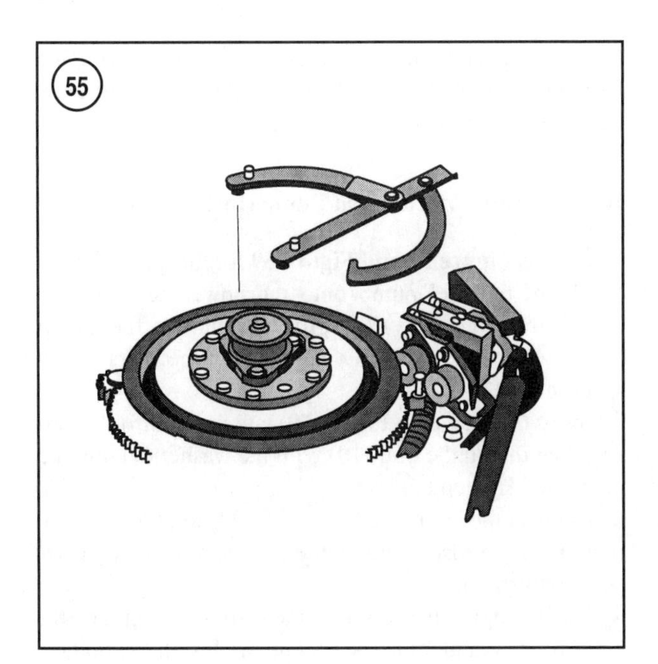

55

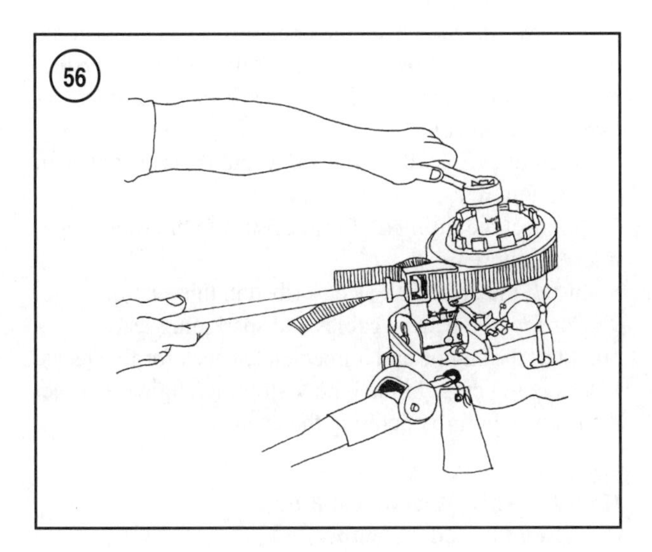

56

with a suitable mallet. The pulley must be replaced along with the bracket if removed by this method. Refer to Chapter Seven under *Flywheel Removal* for bracket replacement procedures.

d. Pull the pin (8, **Figure 54**) from the bracket or pulley.

4. Remove the pulley from the fuel pump as follows:
   a. Engage a strap wrench (**Figure 56**, typical) onto the pump pulley.
   b. Hold the strap wrench to prevent pulley rotation. Then remove the nut (11, **Figure 54**).
   c. Fit the jaws of the puller onto the bottom of the pulley (**Figure 57**, typical) and align the puller bolt

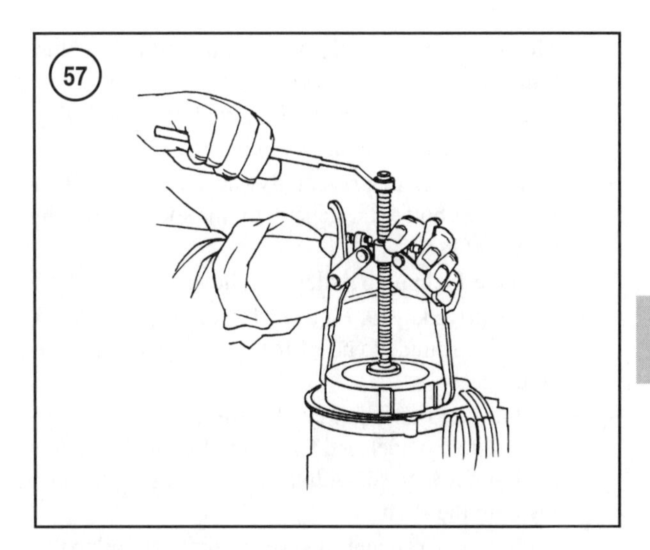

57

with the pulley shaft. Turn the puller bolt to free the pulley.

   d. Remove the pulley (10, **Figure 54**) from the pump. Remove the drive key (9) from the shaft.

5. Use an aerosol type carburetor cleaner to clean all corrosion, debris and oily deposits from the pulley surfaces.

6. Use an oil soaked shop towel to wipe all corrosion deposits and other contaminants from the fuel pump pulley shaft. Inspect the shaft for worn, damaged or rough surfaces. Replace the pulley shaft if not in excellent condition.

7. Inspect the drive key (9, **Figure 54**) for bent or deformed surfaces. Replace the key if not in excellent condition. If the key is bent or deformed, disassemble and inspect the internal components of the fuel pump.

8. Inspect the pulleys for worn cogs and cracked or damaged surfaces. Replace the pulley(s) if not in excellent condition. Inspect the hub of the pulley bracket for rough or damaged surfaces and replace as needed. Pulley bracket removal and installation is described in Chapter Seven under *Flywheel Removal.*

9. Install the flywheel pulley as follows:
   a. Fit the pin (8, **Figure 54**) into the opening in the pulley bracket.
   b. Install the flywheel pulley (7, **Figure 54**) onto the hub of the bracket with the wider diameter side facing the power head. Rotate the pulley to align the pin with the opening in the pulley. Then seat the pulley against the bracket.
   c. Fit the pulley cover onto the pulley with the cupped side facing up. Align the four bosses on the pulley with the openings in the pulley cover. Then seat the cover onto the pulley.

d. Thread the bolt (5, **Figure 54**) into the pulley and pulley bracket.
e. Engage the pins of the holding tool into the pulley bracket openings as shown in **Figure 55**.
f. Hold the tool to prevent flywheel rotation. Then tighten the bolt to the general torque specification in Chapter One.

10. Install the fuel pump pulley as follows:
a. Fit the drive key (9, **Figure 54**) into the pulley shaft slot. The rounded end of the key must fit into the slot.
b. Fit the fuel pump pulley (10, **Figure 54**) onto the shaft with the open end facing up. Rotate the pulley to align the key slot with the key. Then seat the pulley onto the shaft.
c. Lightly coat Yamalube two-cycle outboard oil to the threads. Then thread the nut (11, **Figure 54**) onto the pulley shaft.
d. Engage a strap wrench (**Figure 56**, typical) onto the pump pulley.
e. Hold the strap wrench to prevent pulley rotation. Then tighten the nut.

11. Install the drive belt as described in this section.

High Pressure Mechanical Pump (Removal)

Refer to **Figure 58** and **Figure 59** for this procedure.

1. Disconnect the battery cables and ground the spark plug leads to prevent accidental starting.
2. Remove the drive belt as described in this section.
3. *Pump will be disassembled*—Remove the pulley as described in this section.
4. Disconnect the high pressure fuel hoses (2 and 3, **Figure 58**) from the fuel rails as described in this chapter. See *Fuel rails, injectors and silencer cover.* If the fuel pump will not be disassembled, cover the fuel hose openings with tape to prevent contaminants from entering the fuel system.
5. Disconnect the connector (1, **Figure 58**) from the fuel pressure sensor.
6. Remove the bolt (5, **Figure 58**), then pull the clamp (4) from the pump.
7. Disconnect the fitting (2, **Figure 59**) from the vapor separator tank hose. Drain residual fuel from the hose. Then route the hose through any passages. Remove the hose and fuel pump at the same time.
8. Remove the bolt (10, **Figure 59**) and washer (9), then pull the sleeve (8) from the manifold opening.
9. Remove the three bolts (7, **Figure 58**). Then lift the fuel pump (6) from the power head.
10. Clean all corrosion, debris and contaminants from the fuel pump mounting bosses. Use compressed air to blow debris from the fuel pump surfaces.

High Pressure Mechanical Pump (Installation)

Refer to **Figure 58** and **Figure 59** for this procedure.

1. Mount the fuel pump onto the power head bosses. Thread the three bolts (7, **Figure 58**) into the fuel pump and power head bosses. Evenly tighten the bolts to the specification in **Table 1**.
2. Insert the sleeve (8, **Figure 59**) into the manifold opening, then thread the bolt (10) with the washer (9) into the manifold. Tighten the bolt.
3. Connect the harness connector (1, **Figure 58**) onto the fuel pressure sensor. Tug on the connector to verify a secure connection.
4. Fit the engine harness into the clamp (4, **Figure 58**). Then fit the clamp onto the fuel pump. Install the bolt (5, **Figure 58**) and tighten.
5. Remove the tape from the openings. Then connect the high pressure fuel hoses (2 and 3, **Figure 58**) onto the fuel rails as described in this chapter. See *Fuel rails, injectors and silencer cover.*
6. If removed, install the fuel pump pulley as described in this section.
7. Connect the fitting (2, **Figure 59**) onto the vapor separator tank hose.
8. Install the drive belt as described in this section.
9. Connect the battery cables and spark plug leads.
10. Start the engine and immediately check for fuel leaks. If leakage occurs, immediately stop the engine. Correct the leakage before operating the engine.

High Pressure Mechanical Pump (Disassembly and Assembly)

Refer to **Figures 59-61** for this procedure.

1. Remove the fuel pump pulley as described in this section.
2. Remove the high pressure mechanical pump as described in this section.
3. Drain the lubricating oil from the pump as described in Chapter Three.
4. Remove the screw (22, **Figure 59**), then pull the hose guide from the manifold.
5. Remove the fuel pressure sensor as described in Chapter Six.
6. Remove the manifold from the fuel pump as follows:
a. Support the manifold (18, **Figure 59**) while removing the bolt (16) and washer (17).

HIGH PRESSURE FUEL PUMP

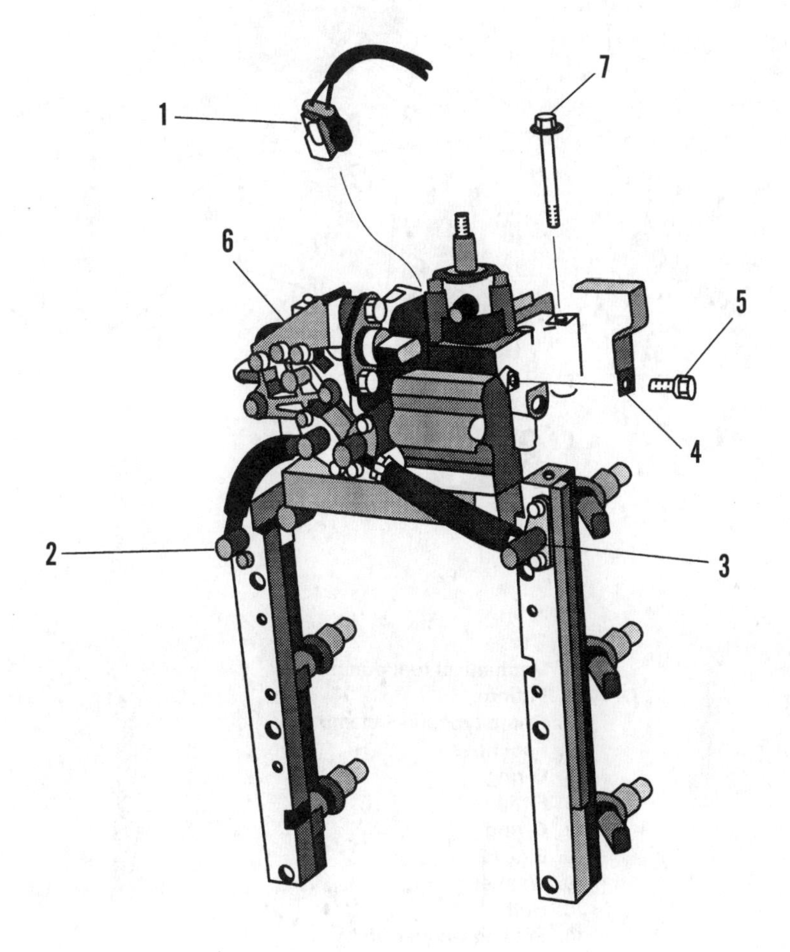

1. Fuel pressure sensor connector
2. Port fuel rail hose connection
3. Starboard fuel rail hose connection
4. Harness clamp
5. Bolt
6. High pressure fuel pump
7. Bolt

(59)

HIGH PRESSURE FUEL PUMP MANIFOLD AND FUEL HOSES

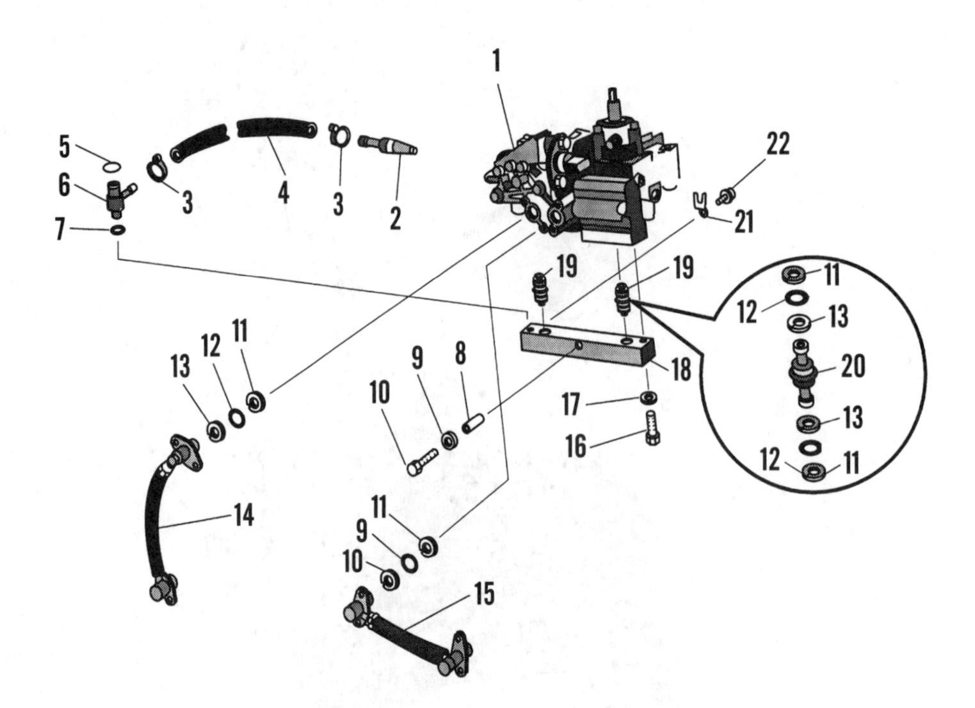

1. Mechanical fuel pump
2. Fitting
3. Crimp type hose clamp
4. Fuel hose
5. O-ring
6. Fitting
7. O-ring
8. Sleeve
9. Washer
10. Bolt
11. Sealing washer (thin)
12. O-ring
13. Sealing washer (thick)
14. High pressure fuel hose (port)
15. High pressure fuel hose (starboard)
16. Bolt
17. Washer
18. Manifold
19. Sleeve assembly
20. Sleeve
21. Hose guide
22. Screw

(60)

HIGH PRESSURE FUEL PUMP COMPONENTS

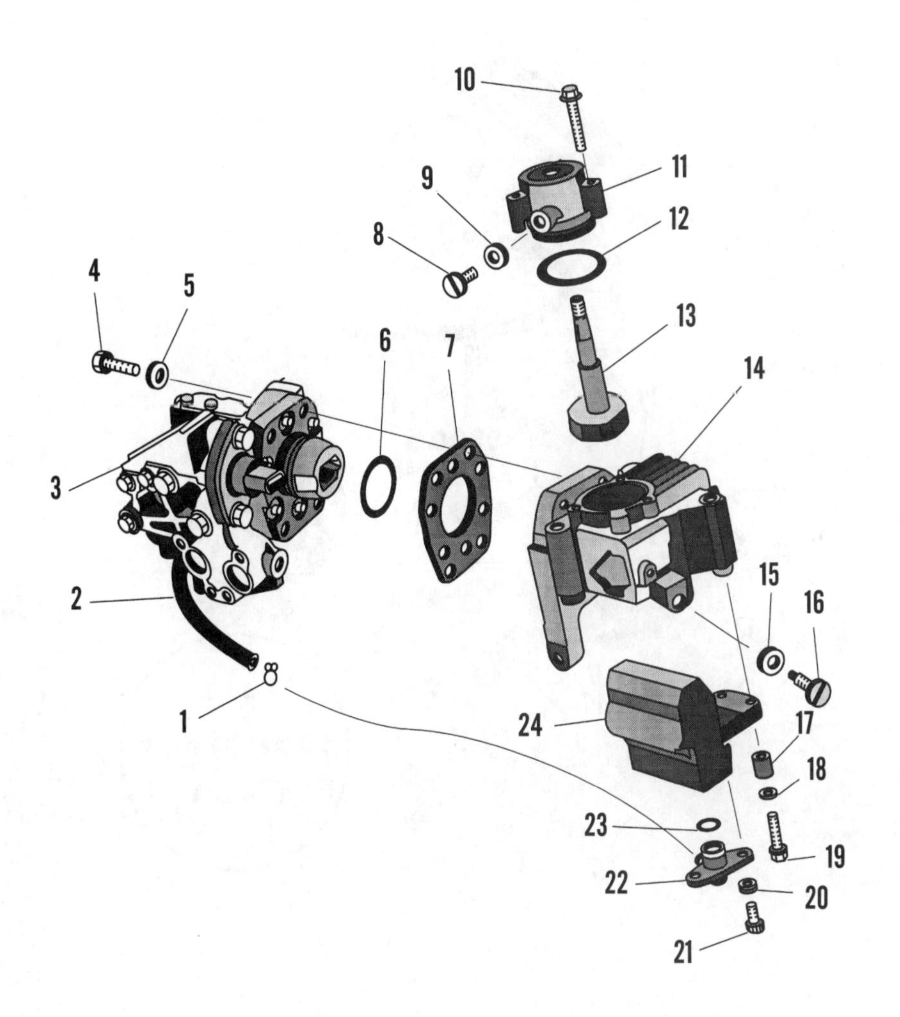

1. Hose clamp
2. Fuel return hose
3. Fuel pump
4. Bolt
5. Washer
6. O-ring
7. Plate
8. Level/vent plug
9. Sealing washer
10. Bolt
11. Pump cover
12. O-ring
13. Camshaft
14. Camshaft housing
15. Sealing washer
16. Drain/fill plug
17. Sleeve
18. Washer
19. Bolt
20. Washer
21. Screw
22. Hose fitting (flange)
23. O-ring
24. High pressure fuel regulator

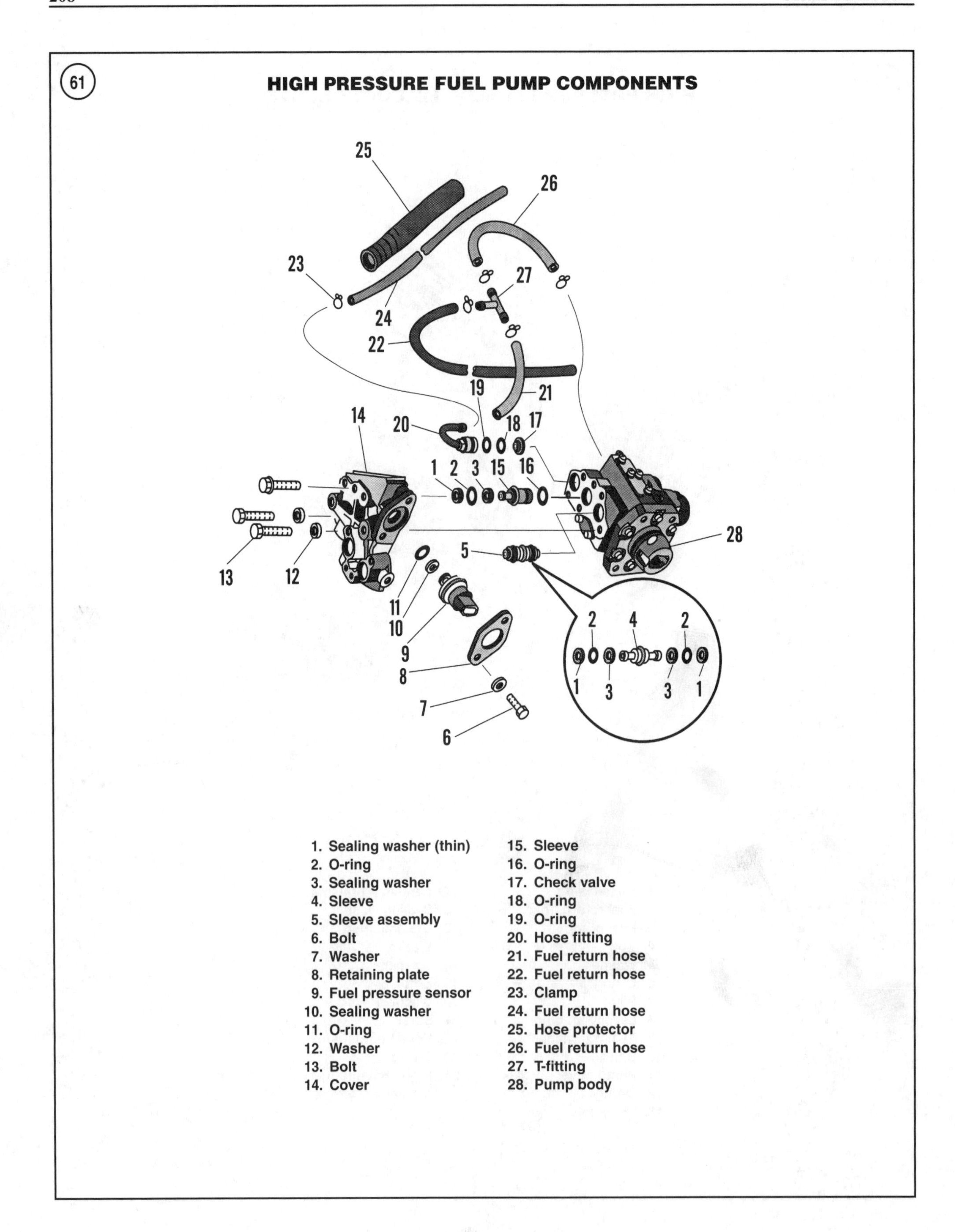
61
HIGH PRESSURE FUEL PUMP COMPONENTS
1. Sealing washer (thin)
2. O-ring
3. Sealing washer
4. Sleeve
5. Sleeve assembly
6. Bolt
7. Washer
8. Retaining plate
9. Fuel pressure sensor
10. Sealing washer
11. O-ring
12. Washer
13. Bolt
14. Cover
15. Sleeve
16. O-ring
17. Check valve
18. O-ring
19. O-ring
20. Hose fitting
21. Fuel return hose
22. Fuel return hose
23. Clamp
24. Fuel return hose
25. Hose protector
26. Fuel return hose
27. T-fitting
28. Pump body

b. Position the pump assembly over a container to hold the fuel, then carefully separate the manifold (18, **Figure 59**) from the pump. Drain residual fuel from the openings.
c. Carefully pull the sleeves (19, **Figure 59**) from the manifold or fuel pump openings. Remove the sealing washers (11 and 13, **Figure 59**) and O-rings (12) from the sleeves. Discard the sealing washers and O-rings.
d. Pull the fuel fitting (6, **Figure 59**) from the manifold or fuel pump openings. Remove the O-rings (5 and 7, **Figure 59**) from the fittings or openings in the fuel pump and manifold. Discard the O-rings.

7. Remove the screws from the flange. Pull the port fuel hose (14, **Figure 59**) from the fuel pump. Remove the sealing washers (11 and 13) and O-ring (12) from the flange or fuel pump. Repeat this step for the starboard fuel hose (15, **Figure 59**).
8. Remove the high pressure fuel regulator as follows:
   a. Remove the hose clamp (1, **Figure 60**), then remove the return hose (2) from the fitting (22). Drain residual fuel from the hose.
   b. Remove the two bolts (19, **Figure 60**), washers (18) and sleeves (17), then remove the regulator (24) from the camshaft housing.
   c. Remove the two screws (21, **Figure 60**) and washers (20), then remove the hose fitting (22) from the regulator.
   d. Remove the O-ring (23, **Figure 60**) from the fitting or the opening in the regulator. Discard the O-ring.
9. Remove the four bolts (4, **Figure 60**) and washers (5). Separate the pump (3) from the camshaft housing (14). Remove the plate (7, **Figure 60**) and O-ring (6) from the pump or camshaft housing. Discard the O-ring.
10. Remove the three bolts (10, **Figure 60**). Use a blunt tip instrument to carefully pry the cover (11) and camshaft (13) from the housing. Work carefully to avoid damaging the mating surfaces. Remove the O-ring (12, **Figure 60**) from the cover or housing. Discard the O-ring. Remove the camshaft from the cover.
11. Remove the clamp (23, **Figure 61**). Disconnect the fuel return hose (21) from the fitting (20). Drain residual fuel from the return hoses.
12. Remove the three bolts (13, **Figure 61**) and washers (12). Use a blunt tip instrument to carefully pry the cover (14) from the pump body (28). Work carefully to avoid damaging the mating surfaces.
13. Remove the sleeve assembly (5, **Figure 61**) from the cover or pump body. Remove the sealing washers (1 and 3, **Figure 61**) and the O-ring (2) from the sleeve. Discard the sealing washers and the O-ring.
14. Remove the sleeve (15, **Figure 61**) from the pump cover or pump body. Remove the O-rings (2 and 16, **Figure 61**) and sealing washers (1 and 3) from the sleeve, cover or pump body. Discard the O-rings and sealing washers.
15. Remove the fitting (20, **Figure 61**) from the pump body. Remove the O-rings (18 and 19, **Figure 61**) and check valve (17) from the fitting or pump body opening. Discard the O-rings.
16. Clean all components in a suitable solvent and dry with compressed air. Take all precautions necessary to prevent debris or contaminants from entering the pump during assembly. Even minute particles can foul the fuel injectors.
17. Inspect all components for worn or damaged surfaces. Replace any damaged or questionable components.
18. Pump assembly is the reverse of disassembly. Note the following:
    a. Install new O-rings and sealing washers in all locations.
    b. Refer to **Figures 59-61** to assist with component location and orientation.
    c. Fill the pump with the recommended lubricant as described in Chapter Three.
19. Install the fuel pump and the pulley as described in this section.

5

CARBURETOR AND SILENCER COVER

Refer to the appropriate illustration during the removal and installation procedures. Use a suitable container to capture fuel from the hoses and other components while removing them. After installing the carburetors and silencer cover, refer to Chapter Four for carburetor synchronization and adjustments. If a torque specification is not given in **Table 1**, refer to general torque specification in Chapter One.

Removal

Refer to **Figure 62** or **Figure 63** for this procedure.

1. Disconnect the battery cables and ground the spark plug leads to prevent accidental starting.
2. Disconnect the two choke solenoid wires from the engine wire harness. The choke solenoid is located on the upper starboard side of the silencer cover. Remove the O-ring that retains the solenoid linkage onto the choke lever. The choke lever is located on the starboard side of the upper carburetor.
3. Note the routing and connection points. Disconnect all hoses from the fittings on the bottom of the silencer cover.

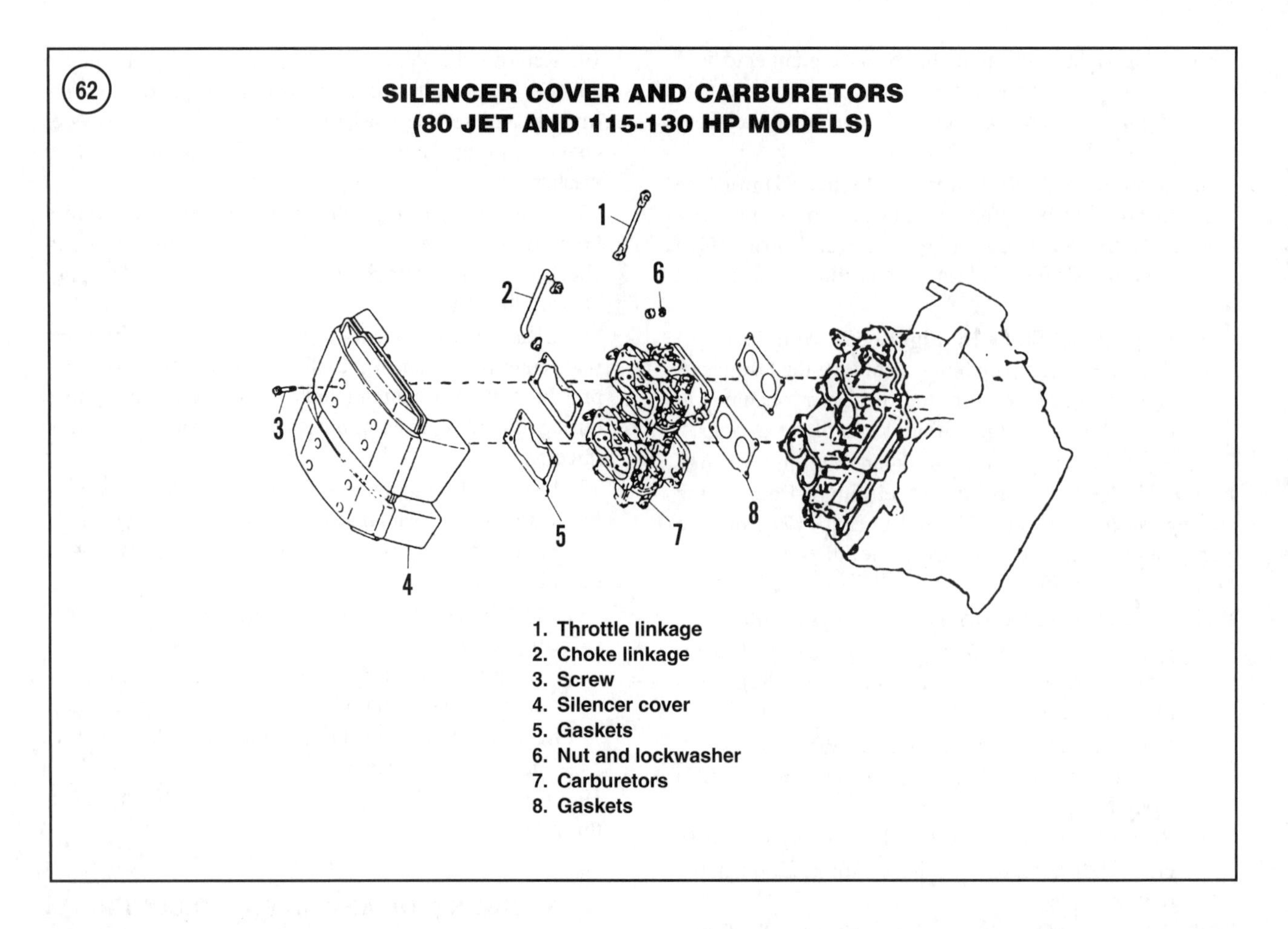

Plug the hoses to prevent contaminants from entering the fuel system.

4A. *80 Jet and 115-150 hp models*—Remove the eight screws (3, **Figure 62**) while supporting the silencer cover, the pull the cover (4) from the carburetors.

4B. *105 Jet and 150-200 hp models (2.6 liter)*—Remove the twelve screws (4, **Figure 63**) while supporting the cover. Then pull the cover (5) from the carburetors.

5. Remove the gaskets (5, **Figure 62** or 6, **Figure 63**) from the silencer cover or carburetor flanges. If necessary, scrape the gaskets from the surface with a razor scraper. Avoid damaging the mating surfaces.

6. Pry the throttle and choke linkages from the carburetor levers. Avoid damaging the linkage connectors.

7. Note the routing and connections points, then disconnect the fuel hoses from the carburetor fittings. If spring type hose clamps are used, inspect them for corrosion, distortion and weak spring tension. Replace the clamps if not in excellent condition.

8. Use a permanent ink marker to mark the mounting location on each carburetor. Make sure the ink marker is able to remain legible after soaking in the cleaning solvent.

9. Remove the four nuts and lockwashers (6, **Figure 62** or 7, **Figure 63**) from each carburetor, then remove the individual carburetors from the intake manifold/reed housing. If necessary, use a blunt tip instrument to remove the carburetors from the intake manifold. Avoid damaging the mating surfaces.

10. Remove the gaskets (8, **Figure 62** or 9, **Figure 63**) from the intake manifold or carburetor flanges. If necessary, scrape the gaskets from the surface with a razor scraper. Avoid damaging the mating surfaces.

11. Place the carburetors on a clean work surface. If the carburetors will not require repair, place them in a covered container to prevent contaminants from entering the fuel system.

12. Use an aerosol carburetor cleaner and clean shop towels to remove all corrosion deposits, oil residue and other contaminants from the gasket mating surfaces.

13. If the silencer cover requires cleaning, remove the choke solenoid from the cover as described in this chapter. Clean the silencer cover surfaces with soap and water.

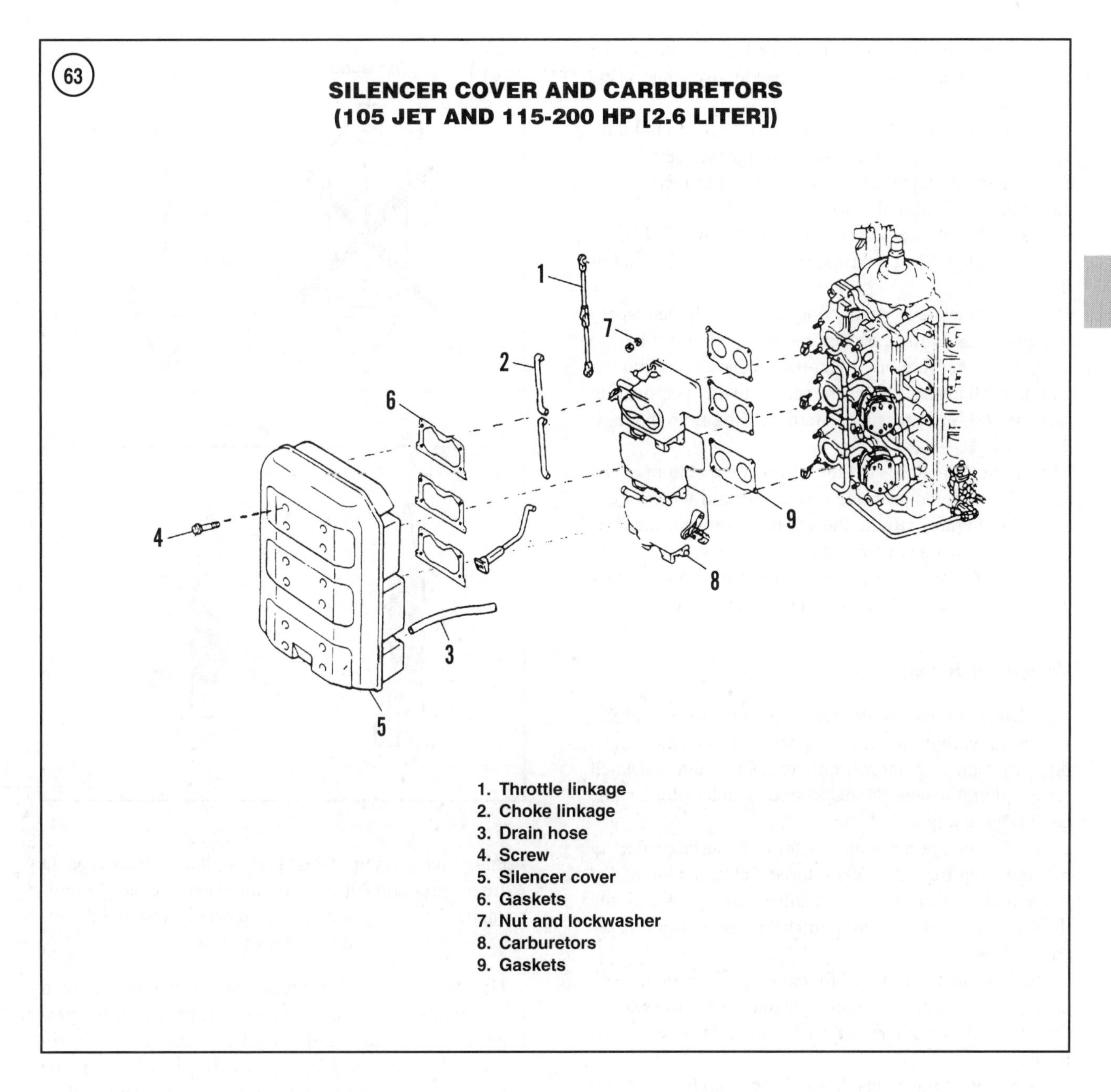

1. Throttle linkage
2. Choke linkage
3. Drain hose
4. Screw
5. Silencer cover
6. Gaskets
7. Nut and lockwasher
8. Carburetors
9. Gaskets

When cleaned, rinse the cover with clean water, then thoroughly dry the cover with compressed air.

14. If necessary, repair the carburetors, one at a time, as described in this section.

Installation

1. Fit new the carburetor mounting gaskets (8, **Figure 62** or 9, **Figure 63**) over the carburetor studs, then seat them against the intake manifold. Do not apply sealant to the gaskets.
2. Fit each carburetor onto the mounting studs. Make sure the carburetors are installed in the original location. Note the markings made prior to removing the carburetors.
3. Thread the four lockwashers and nut onto each carburetor mounting stud.
4. Tighten the four mounting nuts. Use a crossing pattern and tighten the nuts in two or more steps.
5. Connect the fuel hoses to the carburetor fittings. Install new plastic locking type clamps or spring type hose clamps to secure the hoses. Refer to *Fuel hose connectors* in this chapter for instructions.

6. Snap the throttle linkage onto the ball sockets on the carburetor levers. The connects break easily with excessive force.
7. Install the choke linkages to the carburetor choke levers. Avoid damaging the plastic linkage retainers.
8. If removed, install the choke solenoid to the silencer cover as described in this section.
9. Insert the silencer cover mounting screws (3, **Figure 62** or 4, **Figure 63**). Fit the gaskets on the threaded end of the bolts.
10. Align the silencer mounting screws with the respective openings in the carburetor flanges while fitting the cover to the carburetors. Hand-thread the screws into the carburetor flanges. Tighten the screws to the specification in **Table 1**. Use a crossing pattern working from the middle outward.
11. Connect the hoses to the fittings at the bottom of the silencer cover. Connect the choke solenoid wires to engine wire harness. Route the wiring to prevent interference with moving components.
12. Perform the carburetor synchronization and adjustment procedures as described in Chapter Four.

Carburetor Repair

A clean working environment is essential to performing successful carburetor repairs. Mark all hose connections prior to removing them from the carburetor(s). Match each carburetor with the respective cylinder. Repair one carburetor at a time.

Place all components on a clean work surface after removing them from the carburetor and cleaning them. Arrange these components in a manner consistent with the illustrations. This saves time and helps ensure a proper repair.

Using a solvent designed for carburetor cleaning, clean all deposits from the carburetor passages, jets, orifices and float bowl. Use compressed air to help remove contaminants.

Carefully remove jets from the carburetor. Use the proper size screwdriver and maintain downward force while removing jets. If a jet cannot be easily removed, soak the carburetor in solvent for several hours before attempting to remove them. Clean the carburetor without removing the jets if they are especially difficult to remove.

When disassembling carburetors, pay very close attention to location and orientation of internal components. Jets normally have a jet number stamped on the side or opening end. Purchase replacement jets from a Yamaha dealership or carburetor specialty shop. Replacement jets should have the same size and shape of opening as the original jets. Engines used at higher altitudes or in extreme environments may require alternate jets for optimum performance. Contact a Yamaha dealership if the engine will be used in these environments.

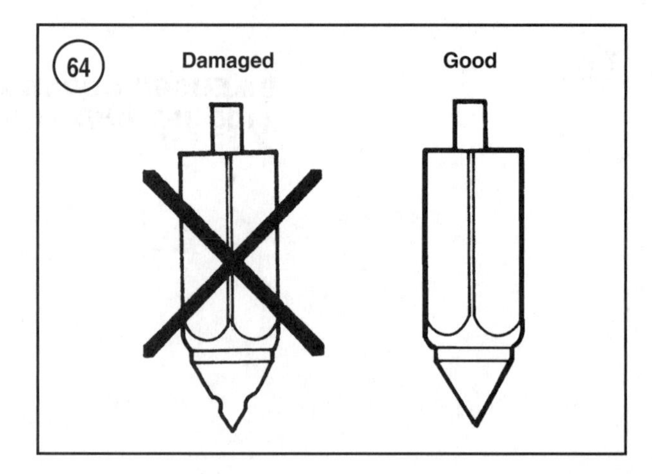

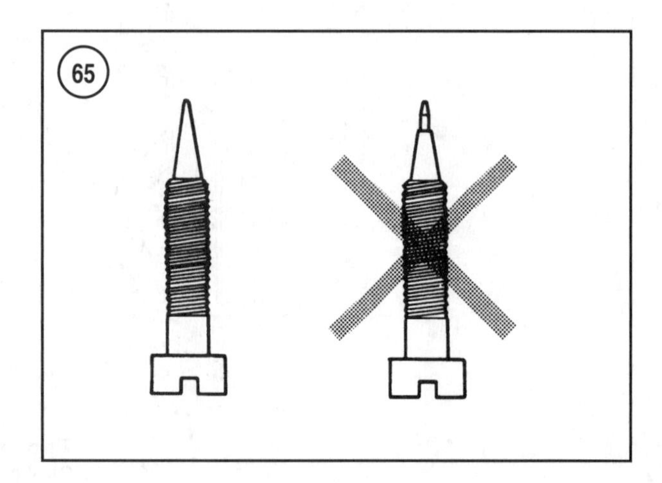

Use great care and patience when removing jets and other threaded or pressed-in components. Clean the passage without removing the jet if it cannot be easily removed. Carburetor jets are easily damaged if the screwdriver slips in the slot. Never install a damaged jet into the carburetor as it may altar the fuel or airflow. Altering the fuel or airflow can cause performance problems or potentially serious engine damage.

CAUTION

Never use stiff steel wire to clean carburetor passages. Material can be inadvertently removed from the inner diameter of the orifice, disturbing the carburetor fuel calibration. Carburetor replacement may be required if passages are damaged during the cleaning process.

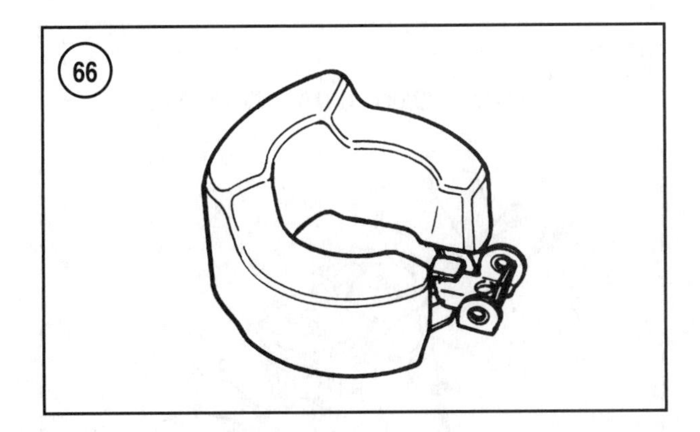

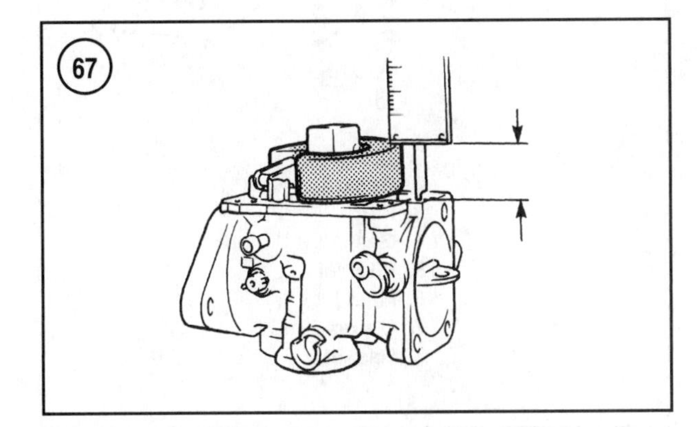

CAUTION
Always replace damaged carburetor jets. Seemingly insignificant damage to a jet can have a profound effect on fuel delivery and adversely affect engine performance and durability.

NOTE
Never compromise the cleaning process of the carburetor(s). Residual debris or contaminants may break free at a later time and block passages.

Upon removal, inspect the fuel inlet valve for damaged sealing surfaces (**Figure 64**). Inspect the valve seat for pitting, worn or irregular surfaces. Replace worn or damaged components. The seat is not available separately on these carburetors. Replace the carburetor if the seat is faulty.

Inspect the pilot screw for worn or damaged surfaces (**Figure 65**). Damage in this area is usually the result of using excessive effort when seating the screw. If the screw is damaged, inspect the seat in the screw opening for damage. Replace the screw if damaged. Replace the carburetor if the seat is damaged. The engine will not operate properly at lower speeds with a damaged screw or seat.

Inspect the float (**Figure 66**) for fuel saturation and worn or damaged surfaces. Some floats are constructed with a translucent material that allows visual inspection for saturation. On models using non-translucent floats, gently press a thumbnail against the float surface. Remove the thumbnail and inspect the thumbnail indentation in the float. Replace the float if fuel appears in the indentation.

Use an accurate depth gauge or ruler to check the float height setting (**Figure 67**, typical). Set the float exactly as specified to help ensure maximum performance and durability. Float settings have a profound effect on fuel calibration.

The float height can be adjusted without completely disassembling the carburetor. Refer to the instructions in the carburetor disassembly/assembly procedures.

Carburetor Disassembly

Note the original location of all air and fuel jets prior to removing them from the carburetor.

Refer to **Figures 68-70** during this procedure.

1. Remove the carburetor(s) as described in this section.
2. Position the carburetor over a container suitable for holding fuel. Remove the drain screws (**Figure 71**). Drain all residual fuel from the bowl(s). Remove the gaskets from the plugs. Discard the gaskets.
3. Remove the screws (**Figure 72**). Remove the float bowl (**Figure 73**). If necessary, use a blunt tip instrument and carefully pry the bowl from the carburetor body. Avoid damaging the mating surfaces.
4. Remove the gasket from the float bowl or carburetor body. Discard the gasket.
5. Remove each float pin. Remove the float from the carburetor (**Figure 74**). Remove the inlet valves from the float tabs or the inlet valve seats.
6. Remove the main jets from the drain plug openings (**Figure 75**). Work carefully to avoid damaging the jets or float bowl.
7. Remove the carburetor cover screw, then remove the cover (**Figure 76**).
8. Remove the pilot jet plugs (**Figure 77**). Remove the pilot jets. Remove and discard the gaskets from the pilot jet plug.
9. Remove the pilot screws and springs (**Figure 78**).
10. Remove the two plugs from the top of the carburetor (**Figure 79**). Remove and discard the gaskets from the plugs. If so equipped, remove the pilot air jet (**Figure 80**) from the throttle bore openings.
11. Insert a flat blade screwdriver into the carburetor throttle bore, then remove the main nozzles (**Figure 81**) from the carburetor. Work carefully to avoid damaging the

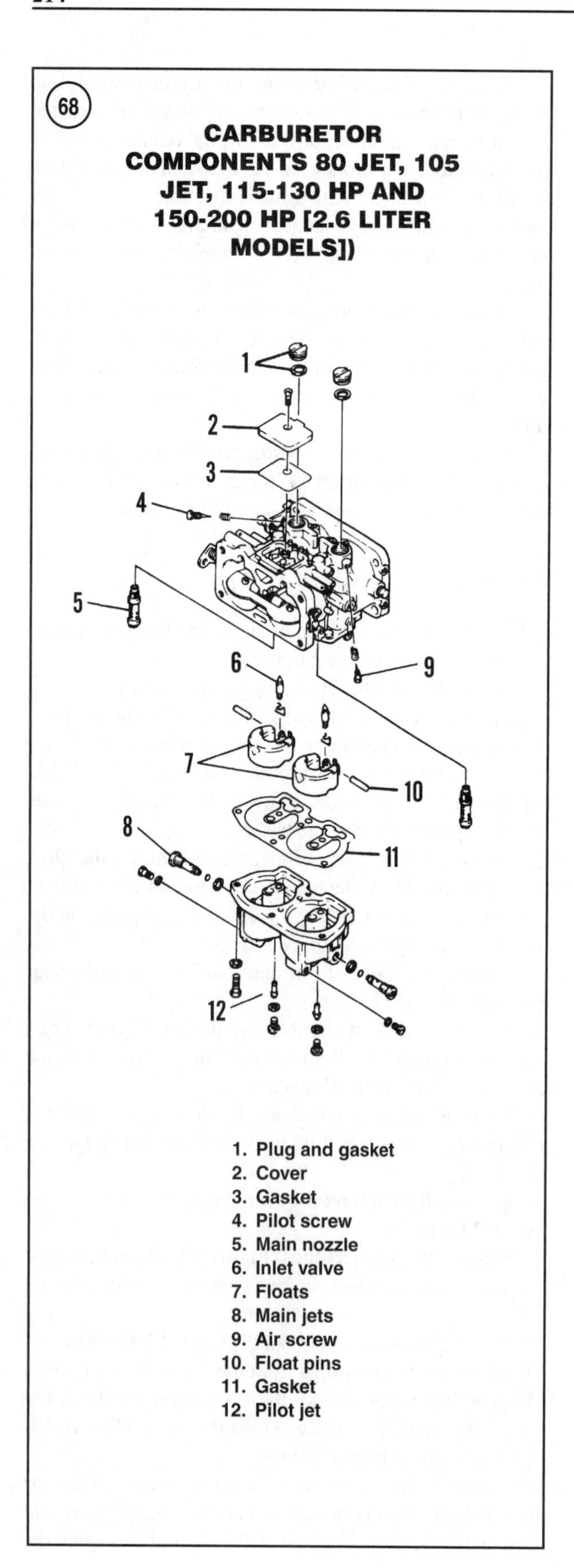

68

CARBURETOR COMPONENTS 80 JET, 105 JET, 115-130 HP AND 150-200 HP [2.6 LITER MODELS])

1. Plug and gasket
2. Cover
3. Gasket
4. Pilot screw
5. Main nozzle
6. Inlet valve
7. Floats
8. Main jets
9. Air screw
10. Float pins
11. Gasket
12. Pilot jet

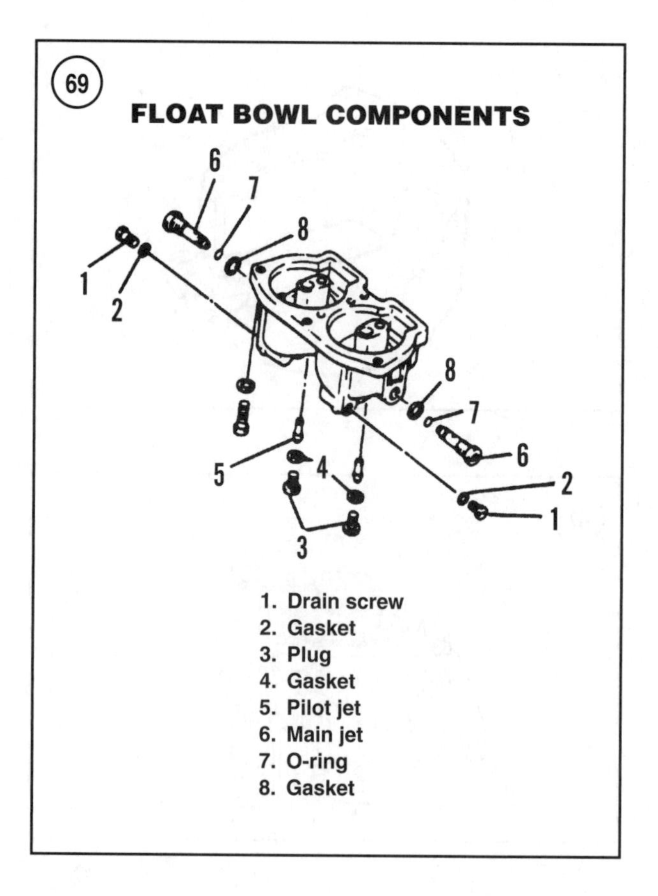

69

FLOAT BOWL COMPONENTS

1. Drain screw
2. Gasket
3. Plug
4. Gasket
5. Pilot jet
6. Main jet
7. O-ring
8. Gasket

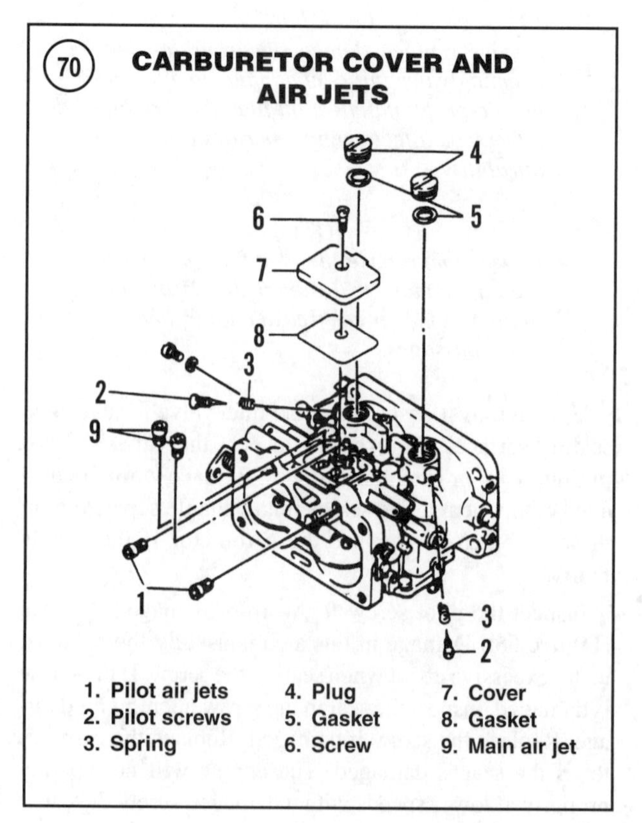

70

CARBURETOR COVER AND AIR JETS

1. Pilot air jets
2. Pilot screws
3. Spring
4. Plug
5. Gasket
6. Screw
7. Cover
8. Gasket
9. Main air jet

71

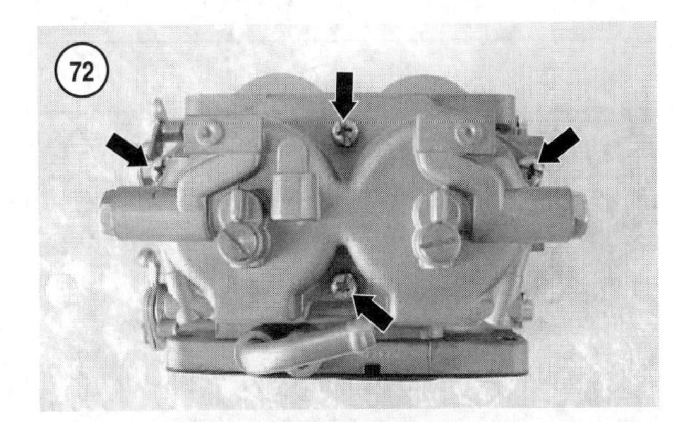
72

73

74

75

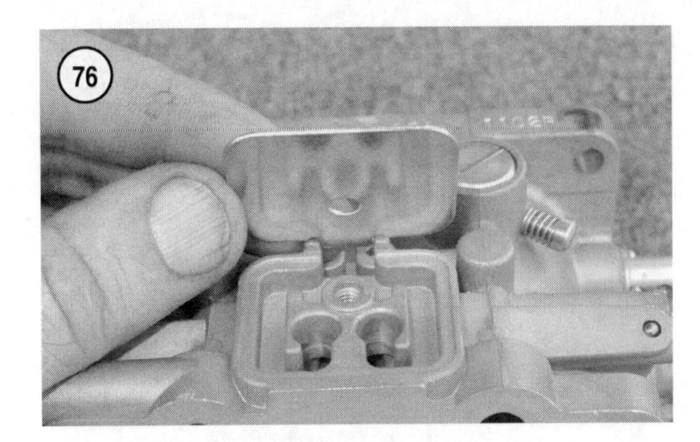
76

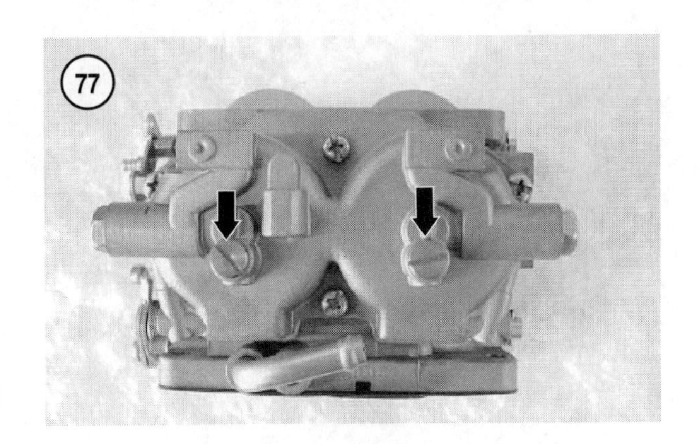
77

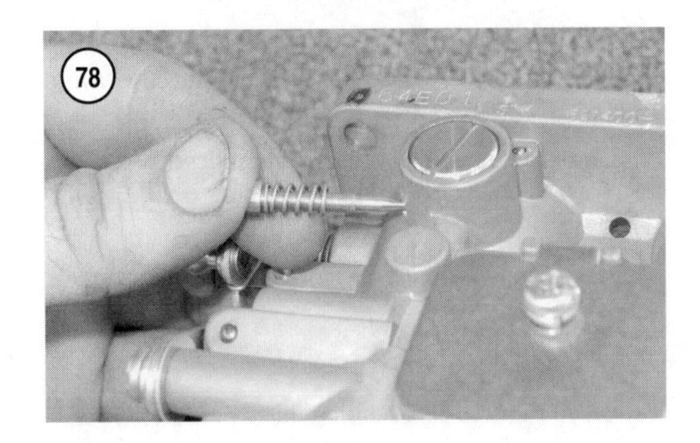
78

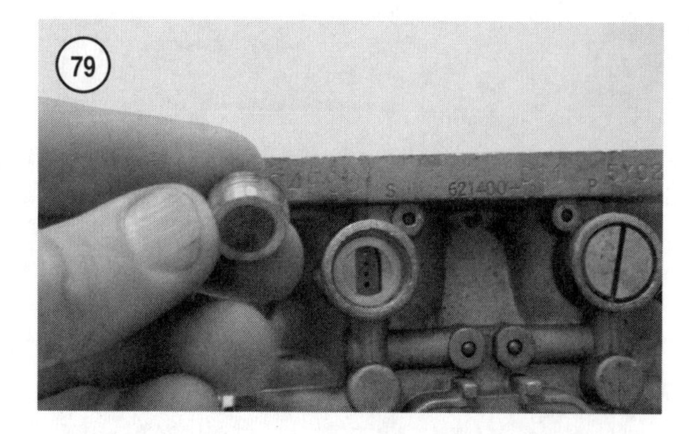

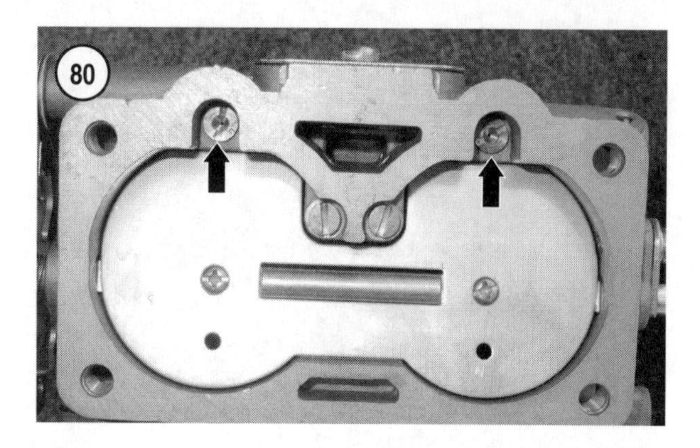

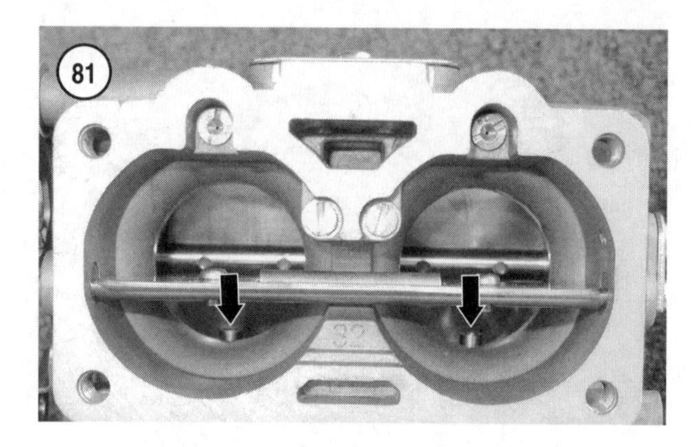

throttle bores or nozzles. If the nozzles cannot be easily removed, clean the passages with them intact to avoid damage.

12. Clean and inspect all components as described in this section. Install new gaskets in all locations during assembly.

Assembly

Refer to **Figures 68-70** during this procedure.

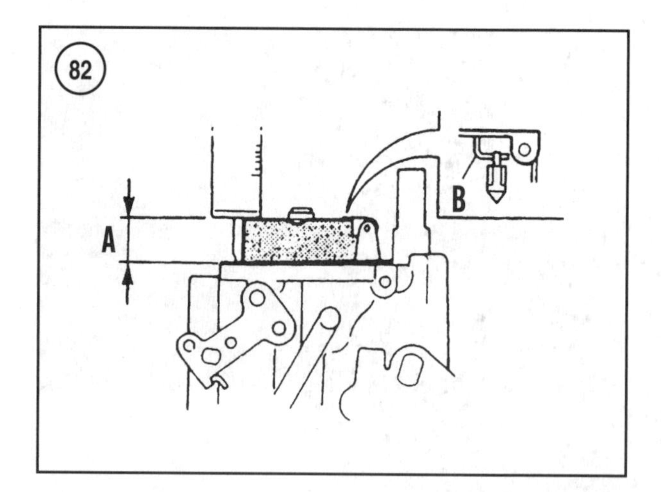

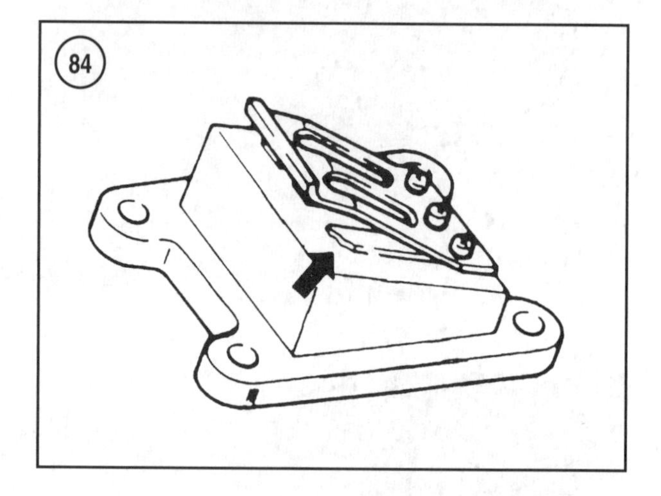

1. If so equipped, install the pilot air jets into the original throttle bore openings. The jet size may not be the same on each side.

2. Install the pilot screw and springs. Turn the screw in until lightly seated. Turn the screws out the number of turns specified in **Table 2**.

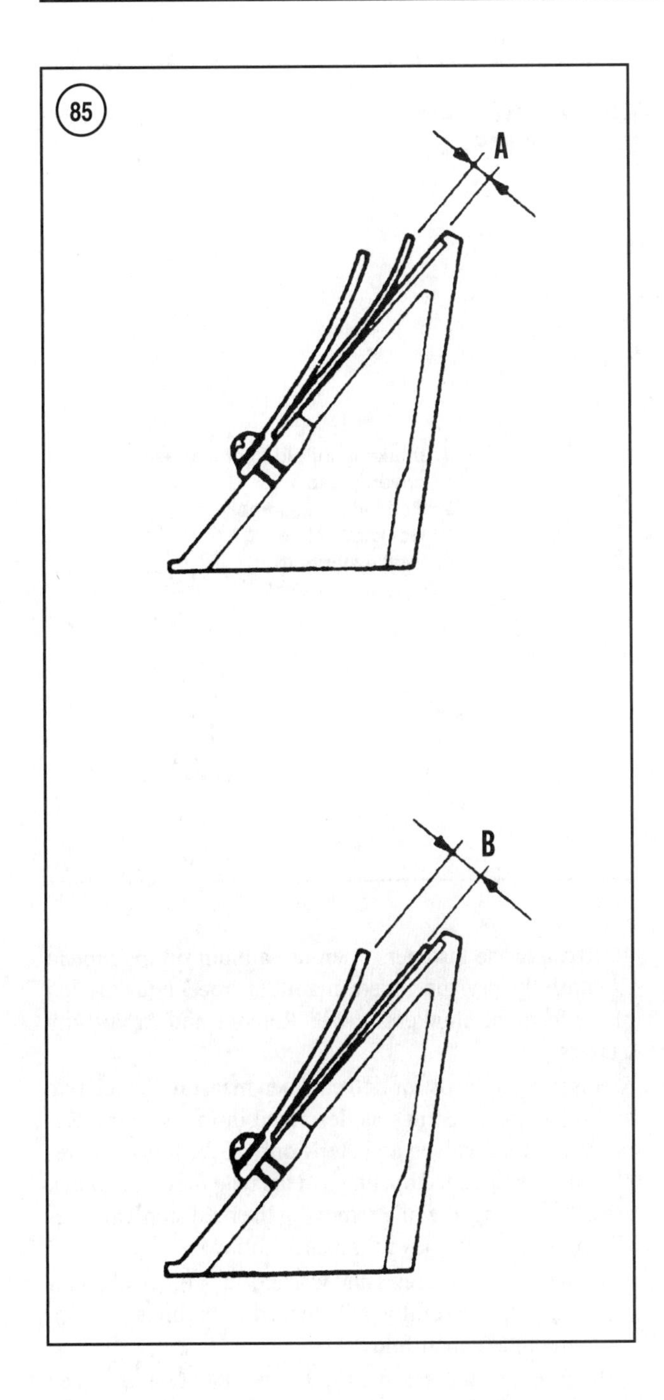

3. Install new gaskets onto the plugs, then install both plugs into the openings in the top of the carburetor. Tighten the plugs to the specification in **Table 1**.

4. Thread the pilot jets into the carburetor. Securely tighten the screws. Install the pilot jet cover/plugs and tighten to the specification in **Table 1**.

5. Install the main jets into the float bowls. Securely tighten the main jets. Install a new gasket onto the drain plugs. Then thread the drain plugs into the main jet. Tighten the bowl drain screws to the specification in **Table 1**.

6. Insert the main nozzles into the carburetor body openings. Make sure the nozzles extend into the throttle bore and fully seat in the opening.

7. Fit the pull-off clip onto the groove on the flat end of the inlet needle. Slip the pull-off clip over the float tab, then install the float into the carburetor body. Repeat this step for the other float and needle. Insert the float pin through the pin openings in the carburetor and float. The float pin must protrude the same amount on each side of the pin bosses.

8. Place the carburetor on a work surface with the topside facing down. With the floats resting on the inlet needle, measure the distance from the bottom of the float to the carburetor casting (A, **Figure 82**). If the measurement is not within the float height specification in **Table 2**, remove the float and carefully bend the tab (B, **Figure 82**) until the float height is correct. Several adjustments may be required.

9. Install a new gasket onto the top cover, then install the cover and gasket onto the carburetor. Install and securely tighten the screw to the cover.

10. Install a new gasket to the float bowl, then fit the bowl to the carburetor body. Secure the bowl with the four screws. Securely tighten the screws using a crossing pattern.

11. Install the carburetor(s) as described in this chapter.

Choke Solenoid

Two screws secure the choke solenoid (**Figure 83**) to the silencer cover or intake manifold (V-6 models).

1. *V-4 models*—Disconnect the solenoid wires. Remove the silencer cover as described in this section.
2. Remove the two screws, then lift the solenoid from the silencer.
3. Fit the replacement solenoid into the recess in the silencer cover.
4. Insert the mounting screw through the ground wire terminal, then thread both screws through the solenoid and silencer cover or intake manifold (V-6 models). Securely tighten the screws.
5. *V-4 models*—Install the silencer cover as described in this section.

REED HOUSING/INTAKE MANIFOLD

Upon removal, inspect the reed valves for bent, cracked or missing sections (**Figure 84**, typical). Measure the reed tip opening (A, **Figure 85**) and reed stop opening (B) dur-

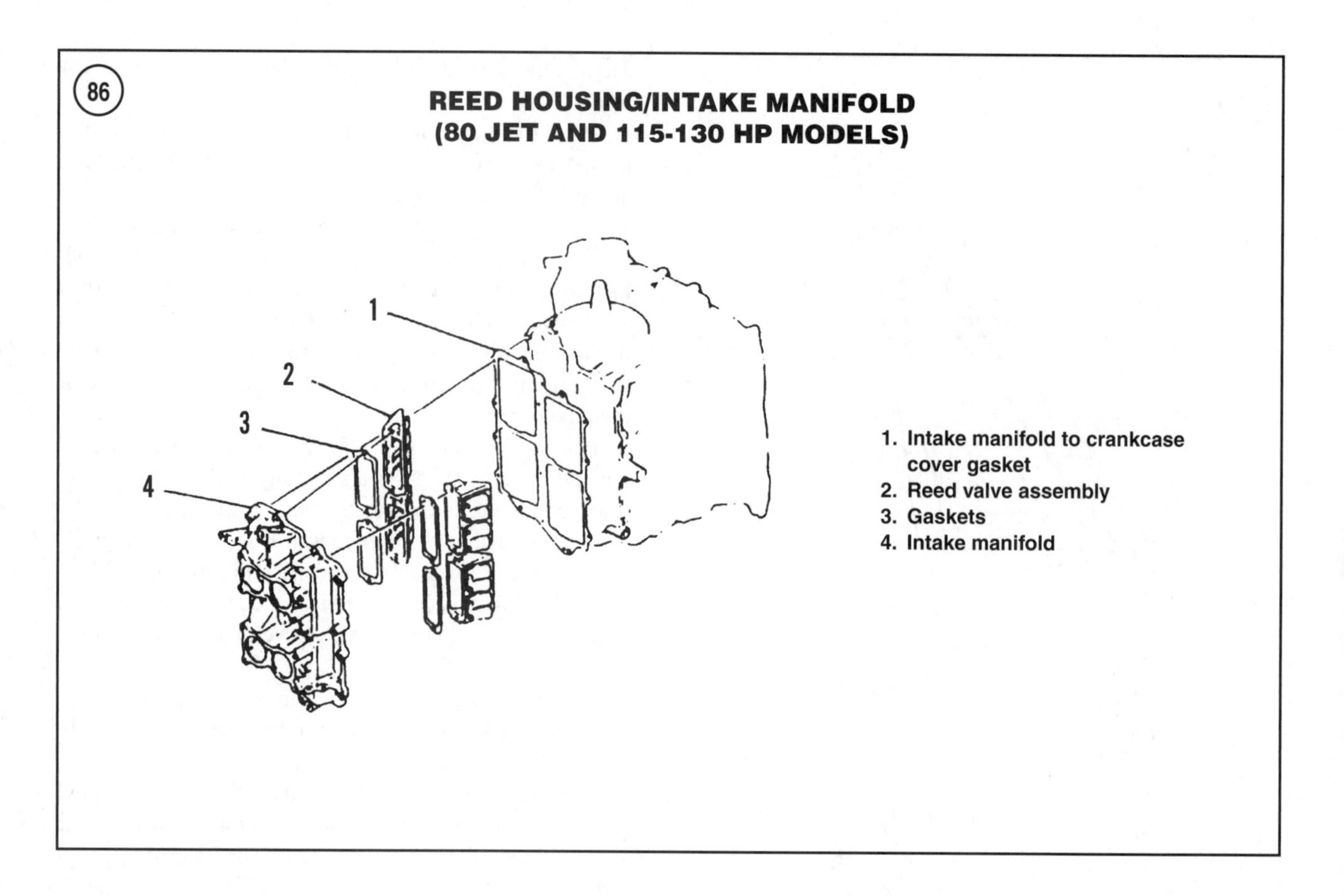

ing inspection and after installing new components. If the measurement(s) are incorrect, replace the reed valve and/or reed stop. Reed valve service specifications are listed in **Table 3**.

Reed housing and intake manifold designs vary by model. Refer to **Figures 86-88** for the appropriate illustration during the removal, disassembly and installation procedures.

Removal

1A. *80 Jet, 105 Jet and 115-200 hp (except EFI and HPDI models)*—Remove the carburetors as described in this chapter.

1B. *150-250 hp (EFI models)*—Remove the fuel rail, fuel injectors and throttle body as described in this chapter.

1C. *150-250 hp (HPDI models)*—Remove the throttle body as described in this chapter.

2. *Oil injected models*—Note the hose routing and connection points, then disconnect the oil lines from the intake manifold fittings.

3. Note the hose routings and connection points, then disconnect the recirculation hoses from the intake manifold or reed housing fittings.

4. Remove the fasteners, then use a blunt tip instrument to carefully pry the intake manifold, reed housing and plates from the crankcase cover. Remove and discard any gaskets.

5. Inspect the reeds for defects, then measure the reed tip and reed stop openings as described in this section. Remove the reed valves and stops only if they must be replaced. The threaded openings in the reed mounting block are often damaged while removing the reed stop/valve retaining screws. Remove threads as follows:

   a. Remove the screws and washers (5, **Figure 88**, typical), then carefully pull the reed assemblies (7) from the intake manifold.
   b. Remove the screws (8, **Figure 88**, typical), reed stops (9) and reed valves (10) from the reed block.

6. Use a suitable solvent to thoroughly clean all components. Dry the components with compressed air. Clean all residual Loctite from the reed screw and screw openings.

Reed Valve Inspection

1. Inspect the reed valves for cracked, chipped or missing petals. Replace all reeds in the engine if any are faulty. The conditions contributing to reed failure may have af-

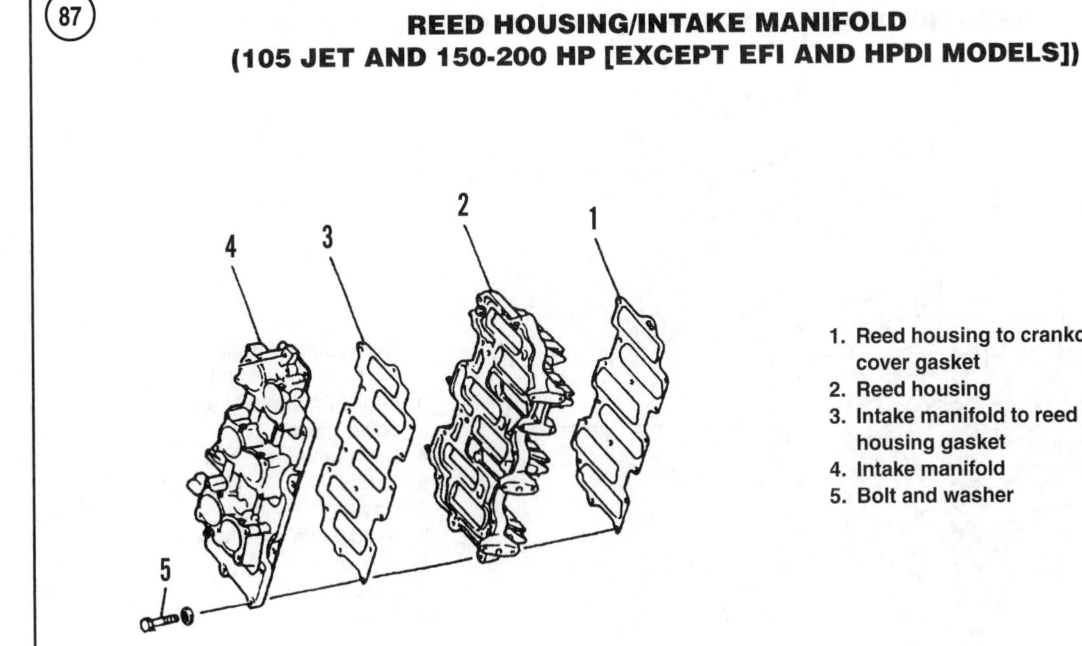

fected other reed valves. Remove the cylinder head and inspect the piston and cylinder head if any petals are missing. Refer to Chapter Eight for cylinder head removal and installation instructions.

2. Inspect the reed contact surface for worn or irregular wear on the surfaces. Replace the reed body, intake manifold or reed mounting plate if these or other defects are noted.

3. Inspect the screw opening for damaged or missing threads. The threads may be damaged during screw removal. Replace the reed block if the threads are damaged.

4. Inspect the reed stop for cracked or damaged surfaces and replace as needed.

5. Use feeler gauges to measure the reed tip opening at the points indicated (A, **Figure 85**). If the reed tip opening exceeds the specification in **Table 3**, replace the reed valves.

6. Use a depth micrometer to measure the distance from the tip of the reed valve to the reed stop surface (B, **Figure 85**). If the measurement is not within the reed stop opening specification in **Table 3**, replace the reed stop.

Installation

Refer to the appropriate illustration (**Figures 86-88**) during the assembly and installation instructions.

1. Install the reed valve (10, **Figure 88**, typical) and reed stop (9) to the reed block. Center the reed petals over the openings.
2. Apply Loctite 242 to the threads of the reed and reed stop mounting screws (8, **Figure 88**, typical). Thread the screws into the intake manifold, reed plate or reed body. Check for properly centered reed petals. Then tighten the screws to the specification in **Table 1**.
3. Install new gaskets (6, **Figure 88**, typical), then fit the reed valve assemblies to the reed housing or intake manifold. Apply Loctite 242 onto the threads, then hand-thread the reed assembly screws into the reed assemblies, plate and intake manifold. Tighten the screws evenly to the specification in **Table 1**.

4A. *80 Jet and 115-130 hp models*—Install a new gasket (1, **Figure 86**), then install the intake manifold assembly (4, **Figure 86**) onto the crankcase cover. Thread the intake manifold mounting bolts into the manifold and crankcase cover. Refer to the tightening sequence indicated in **Figure 89**. Then tighten the bolts to the specification in **Table 1**.

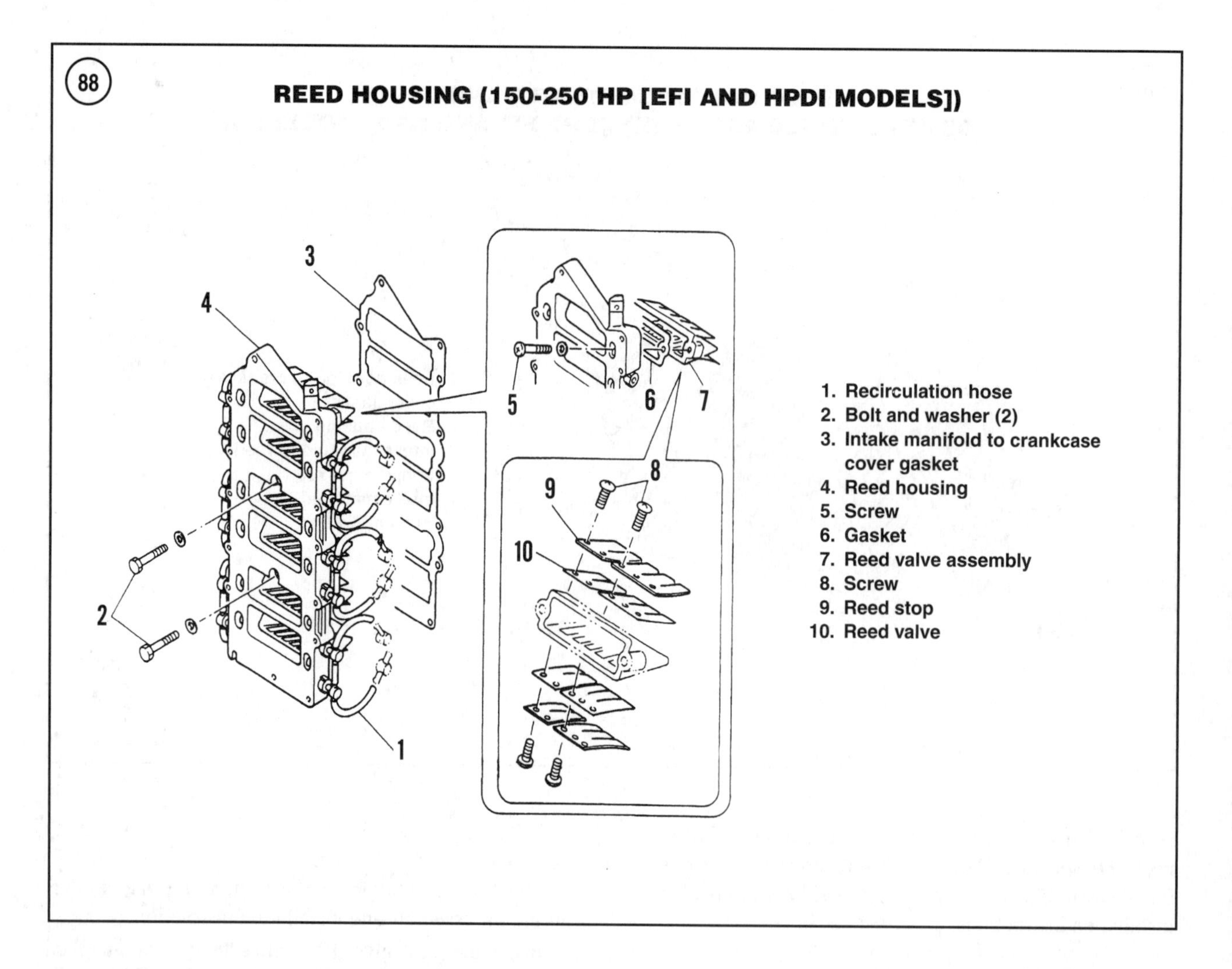

4B. *105 Jet and 150-200 hp models (except EFI and HPDI models)*—Install a new intake manifold gasket (1, **Figure 87**). Install the reed housing assembly (2) onto the crankcase cover. Install a new gasket (3, **Figure 87**). Then install the intake manifold (4, **Figure 87**) onto the reed housing assembly. Thread the intake manifold mounting bolts into the manifold, reed housing and crankcase cover. Refer to the tightening sequence indicated in **Figure 90**. Then tighten the bolts to the specification in **Table 1**.

4C. *150-250 hp models (EFI and HPDI models)*—Install a new reed housing gasket (3, **Figure 88**), then install the reed housing assembly (4) onto the crankcase cover. Thread the two bolts and washers into the reed housing and crankcase cover. Tighten the bolts to the specification in **Table 1**.

5. Connect the recirculation hoses to the fittings on the intake manifold or reed housing.

6. *Oil injected models*—Connect the oil lines to the intake manifold fittings.

7A. *80 Jet, 105 Jet and 115-200 hp (except EFI and HPDI models)*—Install the carburetors as described in this chapter.

7B. *150-250 hp (EFI models)*—Install the throttle body, fuel injectors and fuel rail as described in this chapter.

7C. *150-250 hp (HPDI models)*—Install the throttle body as described in this chapter.

8. Perform all applicable fuel system synchronization and adjustment as described in Chapter Four.

9. *Oil injected models*—Bleed air from the oil injection system as described in Chapter Twelve.

Recirculation System

Refer to **Figures 91-96**.

Faulty hoses in the recirculation system cause oil and fuel leakage, hard starting, and rough idling. Inspect the recirculation hoses for tears, leaks or deterioration. Replace all of the recirculation hoses if any are faulty. The

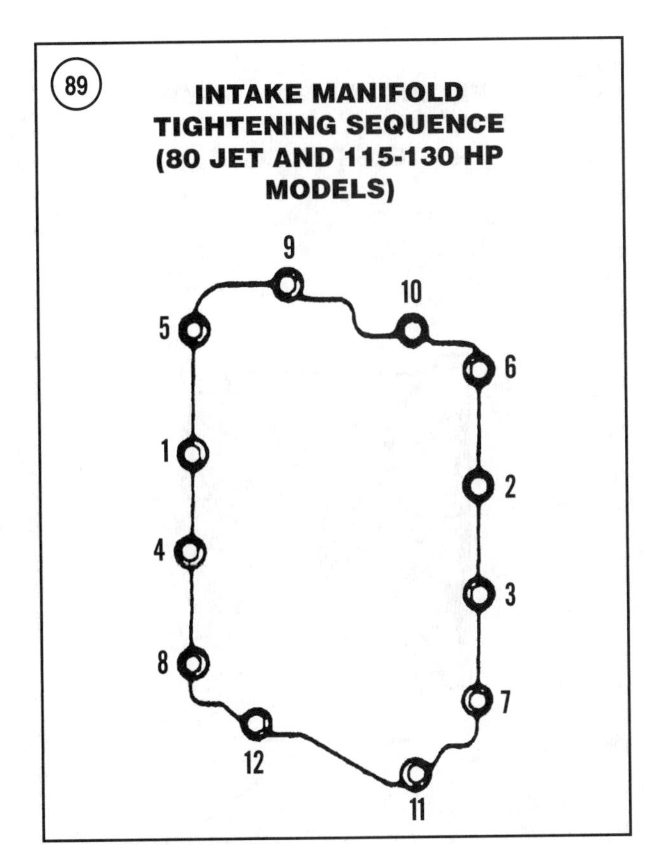
89
INTAKE MANIFOLD TIGHTENING SEQUENCE (80 JET AND 115-130 HP MODELS)
9
10
5
6
1
2
4
3
8
7
12
11

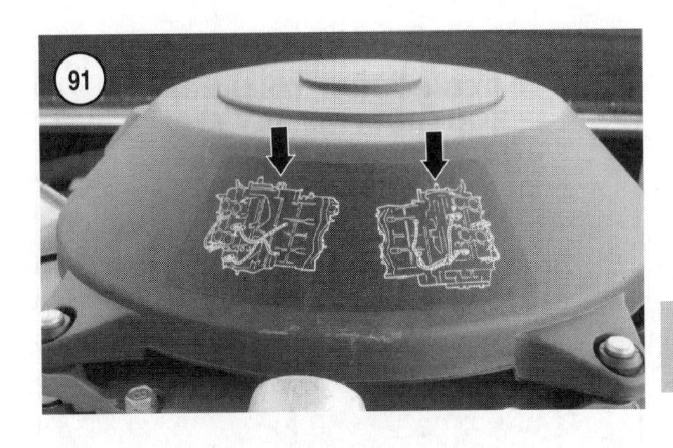
91

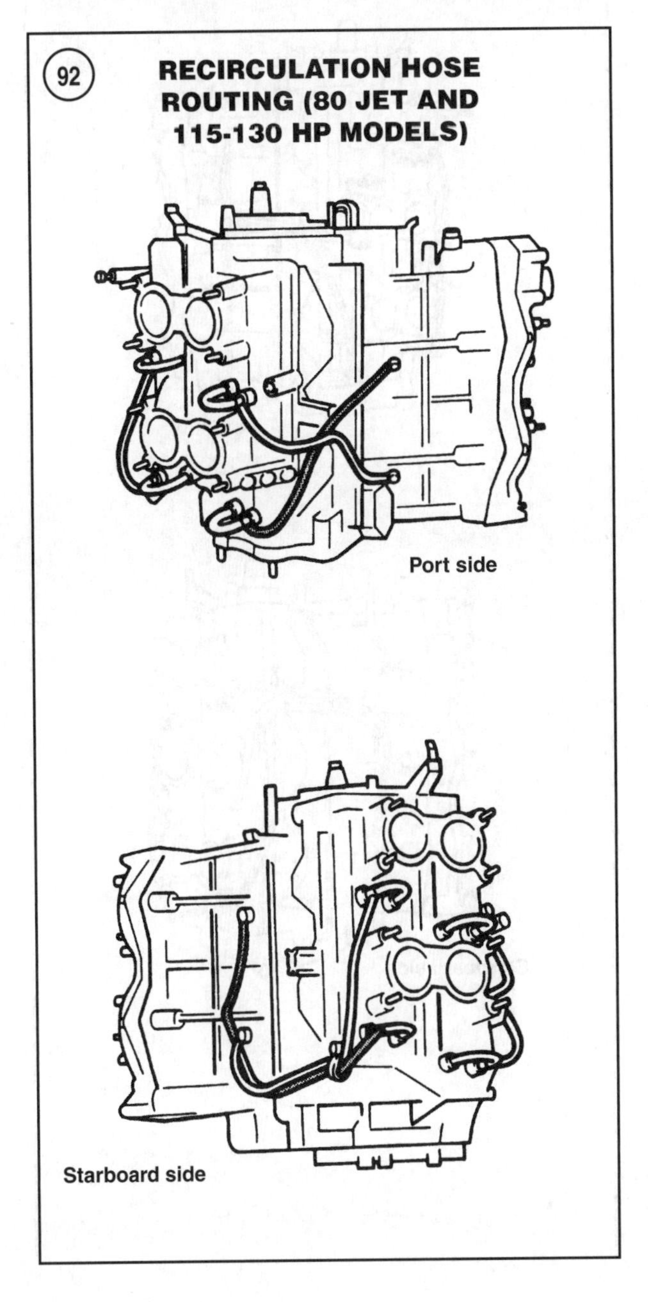
92
RECIRCULATION HOSE ROUTING (80 JET AND 115-130 HP MODELS)
Port side
Starboard side

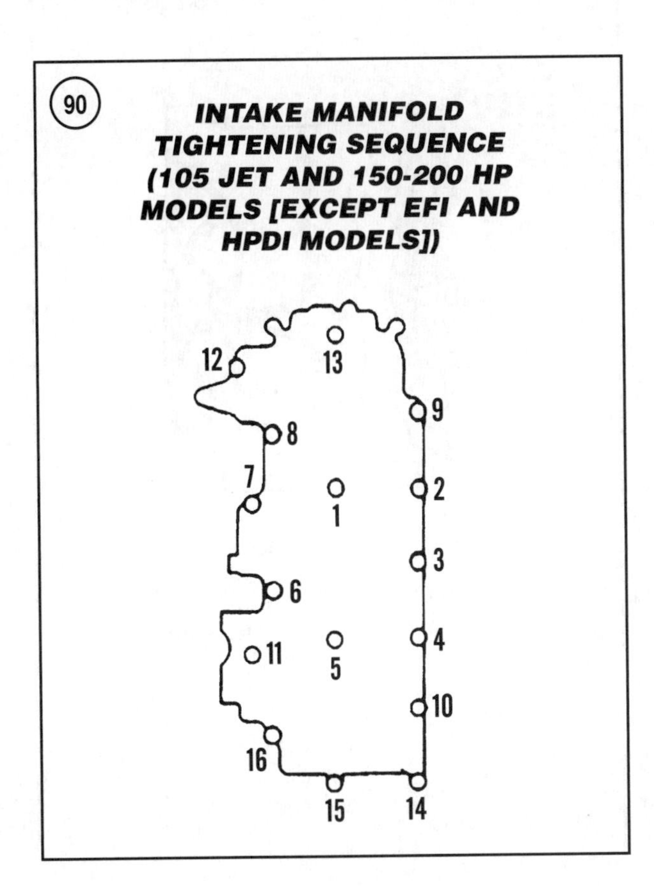
90
INTAKE MANIFOLD TIGHTENING SEQUENCE (105 JET AND 150-200 HP MODELS [EXCEPT EFI AND HPDI MODELS])
12
13
9
8
7
1
2
3
6
11
5
4
10
16
15
14

93

RECIRCULATION HOSE ROUTING (105 JET AND 150-200 HP [EXCEPT EFI AND HPDI MODELS])

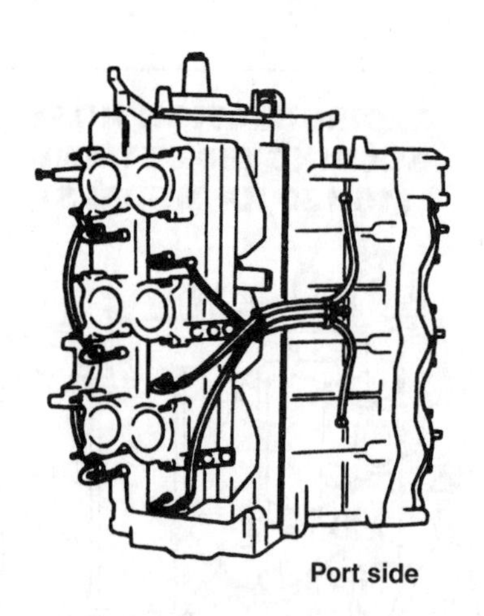

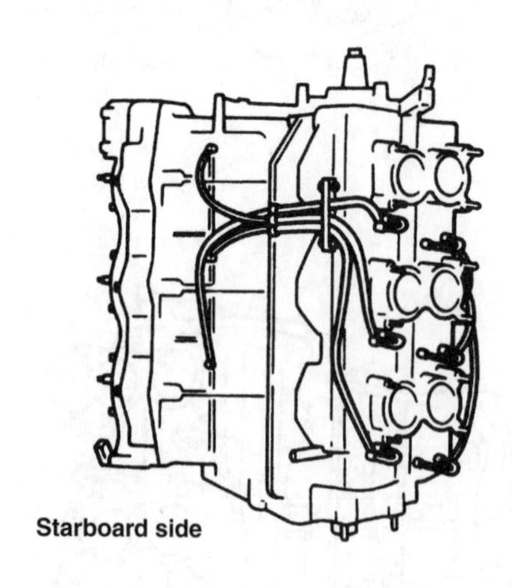

94

RECIRCULATION HOSE ROUTING (150-200 HP EFI MODELS [2.6 LITER])

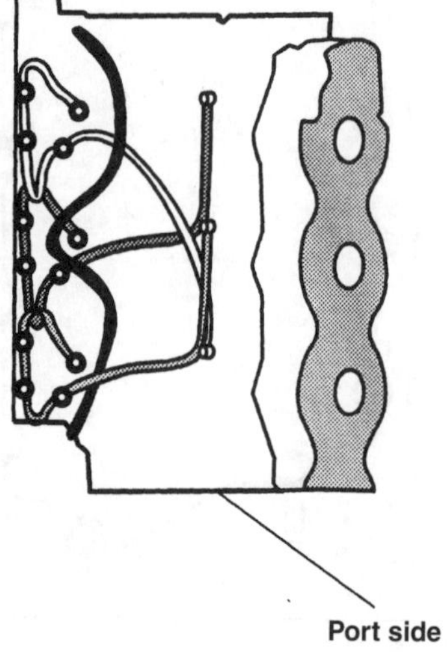

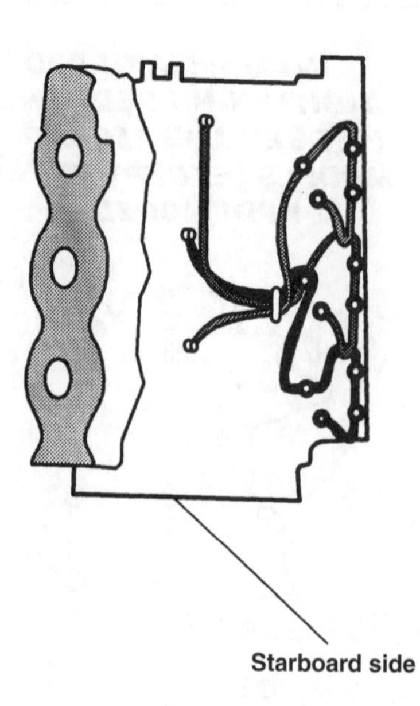

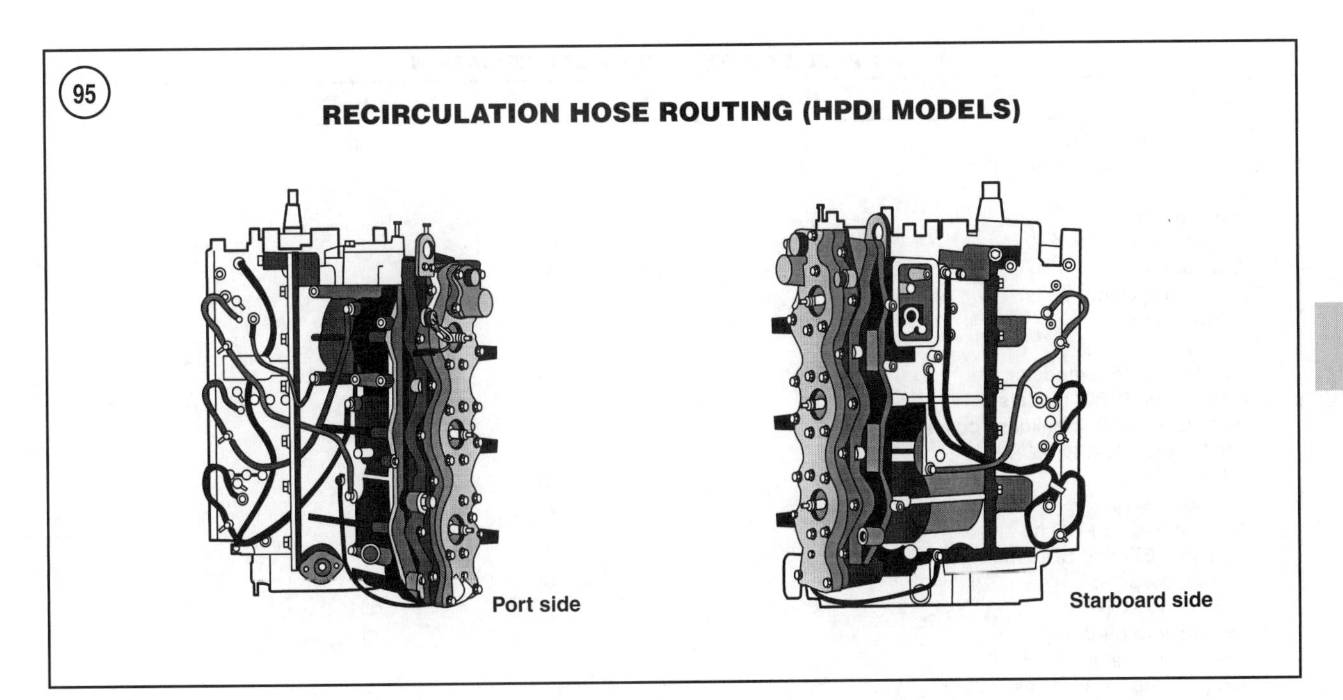

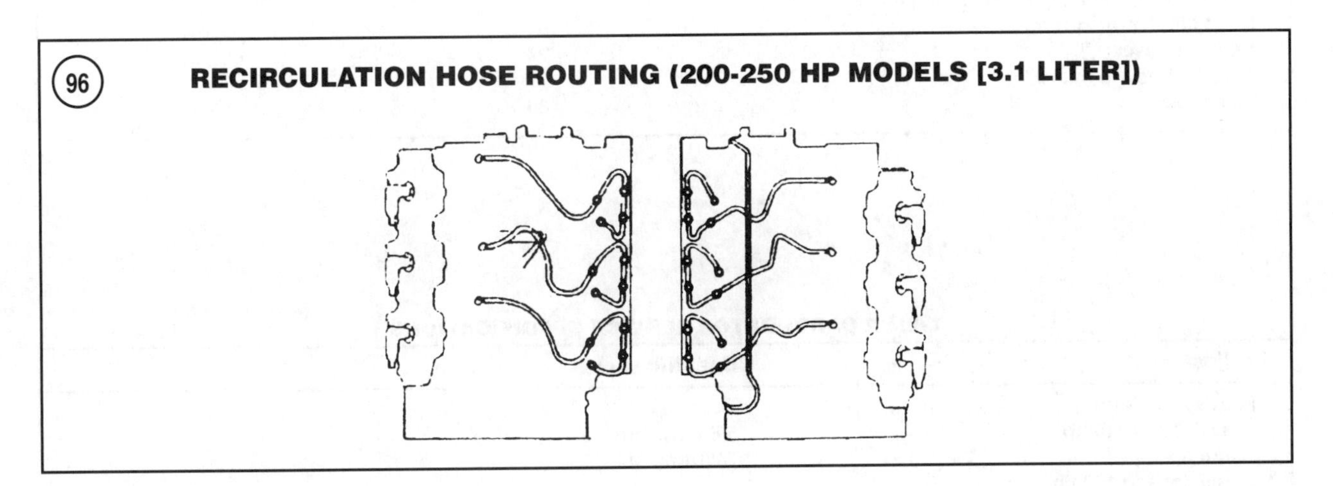

condition failed one hose is almost certainly affecting the others. Use only the hose available from a Yamaha outboard dealership. An automotive vacuum hose is not suitable for marine applications.

To prevent hose kinking and blocked passages, never cut hoses longer or shorter than the original. Note the hose routing and connection points prior to removal. For proper recirculation system operation, the hoses must be routed correctly and be connected to the proper fittings. Always route the hose to the fitting in which it was removed. On most models, a hose routing diagram is located on the flywheel housing (**Figure 91**). If the decal is not present or illegible, refer to **Figures 92-96**.

Tables 1-3 are on the following pages.

5

Table 1 FUEL SYSTEM TORQUE SPECIFICATIONS

Fastener	N·m	in.-lb.	ft.-lb.
Belt tensioner pulley	40	–	30
Belt-driven fuel pump mounting	23	–	17
Carburetor plugs			
Bowl drain	5	44	–
Pilot jet cover	3	27	–
Top plugs	6	53	–
Fuel injector clamp			
HPDI models	26	–	19
Fuel rail			
150-250 hp (EFI models)	10	89	–
150-250 hp (HPDI models)	23	–	17
Intake manifold to crankcase cover			
80 Jet and 115-130 hp			
First step	4	35	–
Second step	8	71	–
105 Jet and 150-200 hp (except EFI and HPDI models)			
First step	4	35	–
Second step	8	71	–
Reed block to reed housing/intake manifold	3	27	–
Reed housing to crankcase cover 150-250 hp (EFI and HPDI models)	10	89	–
Reed stop/valve screws	1	9	–
Silencer cover	3	27	–
Throttle body to intake manifold	10	89	–

Table 2 CARBURETOR SERVICE SPECIFICATIONS

Adjustment	Specification
Pilot screw setting	
80 Jet and 115 hp	5/8 turns out
130 hp	7/8 turns out
105 Jet and 150 hp (except C150, D150, P150 and V150)	1 turn out
C150	1 turn out
S150 and L150	1 1/4 turns out
D150, P150 and V150	
Port side screws	1 1/16 turns out
Starboard side screw	1 9/16 turns out
175 hp	
S175	1 1/8 turns out
P175 and V175	
Port side screws	1 1/8 turns out
Starboard side screw	1 5/8 turns out
200 hp (except S200, L200 and P200)	1 1/8 turns out
S200 and L200	
Port side screws	5/8 turns out
Starboard side screw	1 1/8 turns out
P200	
Port side screws	3/4 turns out
Starboard side screw	1 1/4 turns out
Float height	15.5-16.5 mm (0.61-0.65 in.)

Table 3 REED VALVE SERVICE SPECIFICATIONS

Item/model	Specifications
Maximum reed tip opening	0.2 mm (0.008 in.)
Reed stop opening	
80 Jet and 115-130 hp	6.2-6.8 mm (0.244-0.268 in.)
105 Jet and 150-200 hp (except EFI and HPDI models)	6.2-6.8 mm (0.244-0.268 in.)
150-200 hp (EFI and HPDI models [2.6 liter])	8.7-9.3 mm (0.342-0.366 in.)
200 hp (3.1 liter)	8.7-9.3 mm (0.342-0.366 in.)
225 hp	
S225 and L225	7.6-8.2 mm (0.299-0.323 in.)
V225	8.7-9.3 mm (0.342-0.366 in.)
250 hp	8.7-9.3 mm (0.342-0.366 in.)

Chapter Six

Electrical and Ignition Systems

This chapter provides service procedures for the battery, starting system, charging system and ignition system used on outboard engines covered in this manual. Refer to the wiring diagrams at the end of the manual during electrical system testing. Due to changes made by the manufacturer, in some cases the wire colors called out in the text may not match a particular model. If there is a difference, make sure the correct component(s) is identified.

Electrical and ignition system specifications are listed in **Tables 1-4** are located at the end of the chapter.

BATTERY

Batteries used in marine applications endure far more rigorous treatment than those used in automotive applications. Marine batteries (**Figure 1**) generally have a thicker exterior case to cushion the plates during tight turns and rough water operation. Thicker plates are also used, with each one individually fastened within the case to help prevent premature failure. Spill-resistant caps on the battery cells help prevent electrolyte from spilling into the bilge. Automotive batteries should be used in a boat *only* during an emergency situation when a suitable marine battery is not available.

CAUTION

Sealed or maintenance-free batteries are not recommended for unregulated charging systems. Excessive charging during continued high-speed operation will cause the electrolyte to boil, resulting in battery failure. Since water cannot be added to sealed batteries, prolonged overcharging destroys the battery.

Battery Rating Methods

The battery industry developed specifications and performance standards to evaluate batteries and battery energy potential. Several rating methods are available to provide information on battery selection.

Cold cranking amps (CCA)

This figure represents in amps the current flow the battery can deliver for 30 seconds at -17.6° C (0° F) without dropping below 1.2 volts per cell (7.2 volts on a standard 12 volt battery). The higher the number, the more amps are delivered to crank the engine. CCA times 1.3 equals MCA.

Marine cranking amps (MCA)

This figure is similar to the CCA test figure except that the test is run at 0° C (32° F) instead of -17.6° C (0° F).

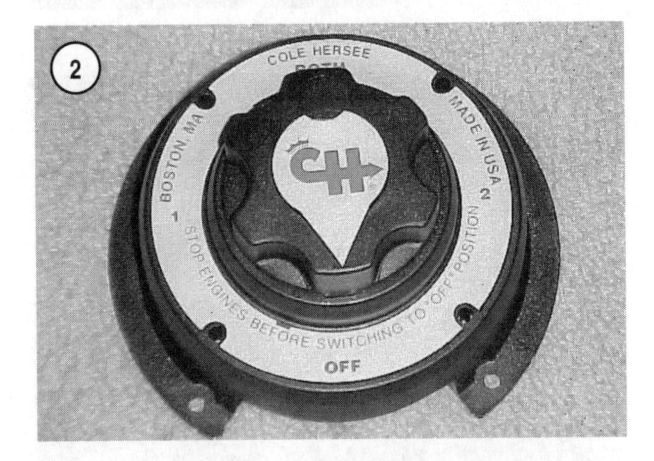

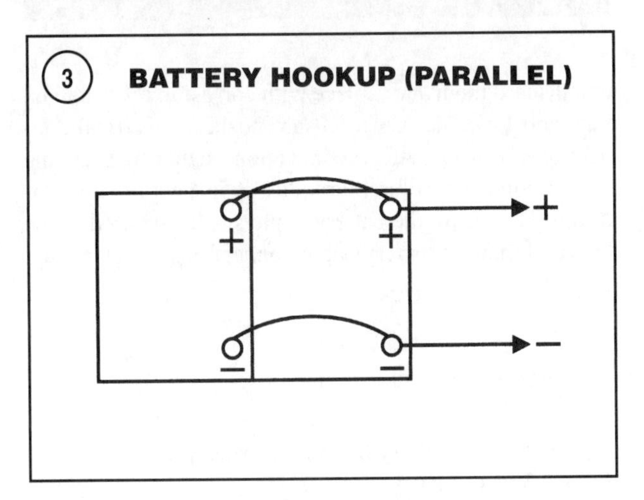

This is more aligned with actual boat operating environments. MCA times 0.77 equals CCA.

Reserve capacity

This figure represents the time (in minutes) that a fully charged battery at 26.7° C (80° F) can deliver 25 amps, without dropping below 1.75 volts per cell (10.5 volts on a standard 12 volt battery). The reserve capacity rating defines the length of time that a typical vehicle can be ran after the charging system fails. The 25 amp figure takes into account the power required by the ignition, lighting and other accessories. The higher the reserve capacity rating, the longer the vehicle could be operated after a charging system failure.

Amp-hour rating

The ampere hour rating method is also called the 20 hour rating method. This rating represents the steady current flow that the battery will deliver for 20 hours while at 26.7° C (80° F) without dropping below 1.75 volts per cell (10.5 volts on a standard 12 volt battery). The rating is actually the steady current flow times 20 hours. Example: A 60 amp-hour battery will deliver 3 amps continuously for 20 hours. This rating method has been largely discontinued by the battery industry. Cold cranking amps (or MCA) and reserve capacity ratings are now the most common battery rating methods.

Battery Recommendations

Using a battery with an inadequate capacity causes hard starting or an inability to start the engine. All models covered in this manual require a battery that meets or exceeds the following minimum requirements.

1. 100 amp hour.
2. 512 cold cranking amps.
3. 182 minutes of reserve capacity.

Use a battery with a capacity exceeding the minimum requirement whenever possible and, if the boat is equipped with numerous electrical accessories (radios, depth finders, pumps). Consider adding an additional battery and installing a battery switch (**Figure 2**) on such applications. The switch allows starting and charging using one or both batteries. The switch can be turned off with the boat at rest or during storage to prevent discharge that occurs from some on-board accessories.

Separate batteries are used to provide power for accessories such as lighting, fish finders and depth finders. To determine the required capacity of such batteries, calculate the accessory current (amperage) draw rate of the accessory and refer to **Table 4**.

Two batteries may be connected in parallel to double the ampere-hour capacity while maintaining the required 12 volts. See **Figure 3**. For accessories that require 24 volts, batteries may be connected in series (**Figure 4**), but only accessories specifically requiring 24 volts should be

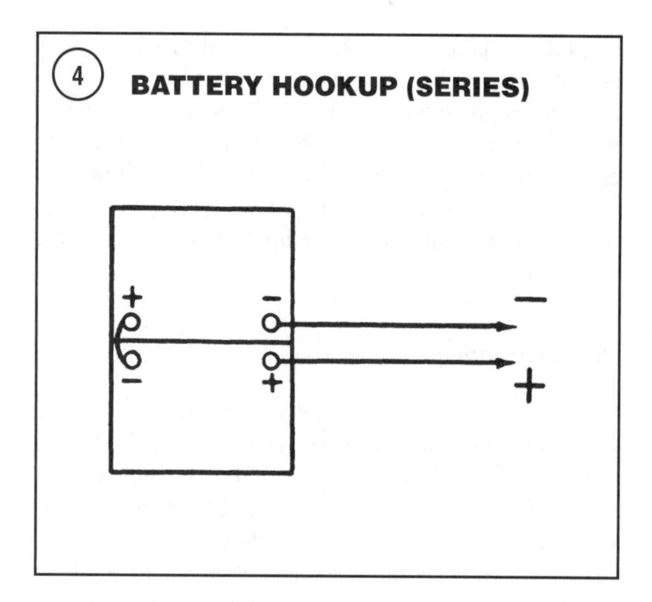

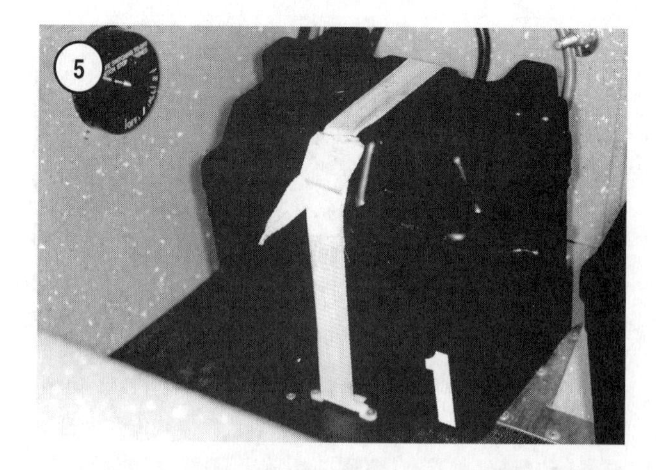

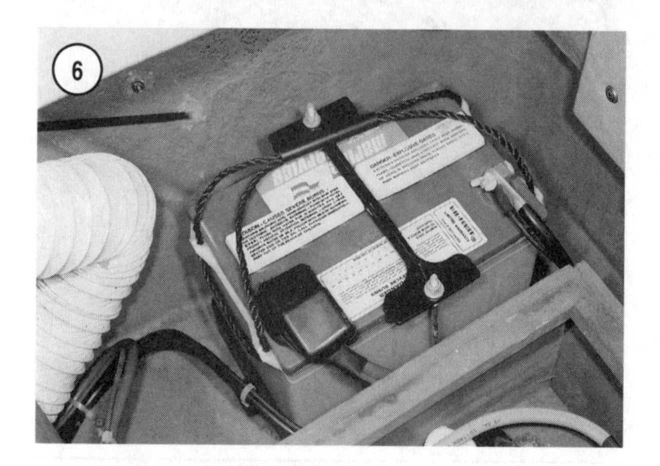

connected to the system. If charging becomes necessary, batteries connected in a parallel or series circuit should be disconnected and charged individually.

Safety Concerns

The battery must be securely fastened in the boat to prevent the battery from shifting or moving in the bilge area. The positive battery terminal (or the entire top of the battery) must also be covered with a nonconductive shield or boot.

If the battery is not properly secured it may contact the hull (or metal fuel tank) in rough water or while being transported. If the battery shorts against the metal hull or fuel tank, the resulting short circuit will cause sparks and an electrical fire. An explosion could follow if the fuel tank or battery case is compromised.

If the battery is not properly grounded and the battery contacts the metal hull, the battery will try to ground through the control cables or the wiring harness, causing sparks and possibly an electrical fire.

Observe the following preventive steps when installing a battery in any boat, especially a metal boat or a boat with a metal fuel tank.

1. Choose a location as far as practical from the fuel tank while still providing access for maintenance.
2. Secure the battery to the hull with a plastic battery box and tie-down strap (**Figure 5**) or a battery tray (**Figure 6**) with a nonconductive shield or boot covering the positive battery terminal.
3. Make sure all battery cable connections (two at the battery, two at the engine) are clean and tight. Do not use wing nuts to secure battery cables. If wing nuts are present, discard them and replace with corrosion resistant hex nuts and lock washers to ensure positive electrical connections. Loose battery connections can cause engine malfunction and failure. Periodically inspect the installation to make sure the battery is physically secured to the hull and that the battery cable connections are clean and tight.

Care and Inspection

1. Remove the battery tray top or battery box cover. See **Figure 5** or **Figure 6**.
2. Disconnect the negative battery cable, then the positive battery cable.

NOTE
*Some batteries have a built-in carry strap (**Figure 1**) for use in Step 3.*

3. Attach a battery carry strap to the terminal posts. Remove the battery from the boat.

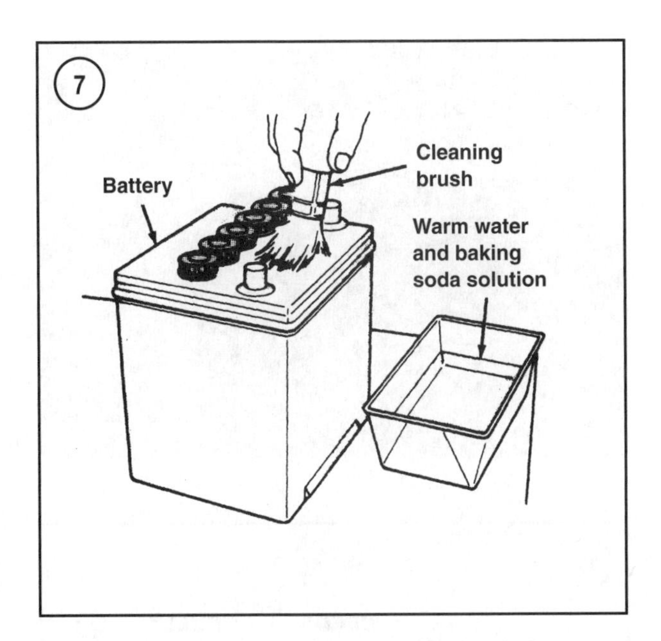

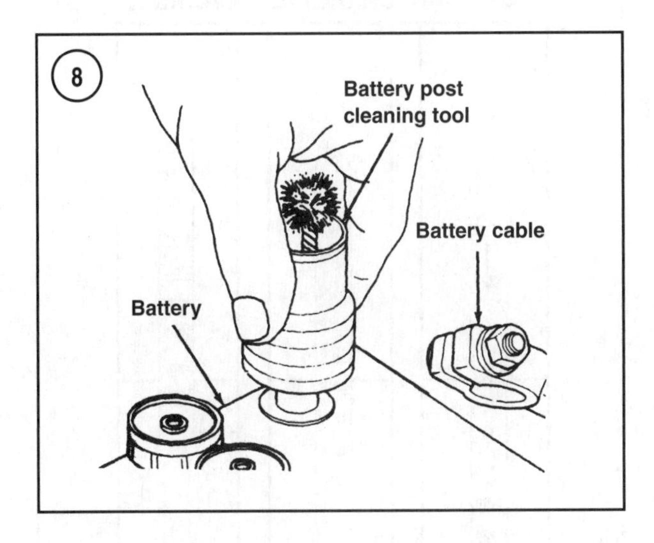

4. Inspect the entire battery case for cracks, holes or other damage.

5. Inspect the battery tray or battery box for corrosion or deterioration. Clean as necessary with a solution of baking soda and water.

NOTE
Do not allow the baking soda cleaning solution to enter the battery cells in Step 6 or the electrolyte will be severely weakened.

6. Clean the top of the battery with a stiff bristle brush using the baking soda and water solution (**Figure 7**). Rinse the battery case with clear water and wipe dry with a clean cloth or paper towel.

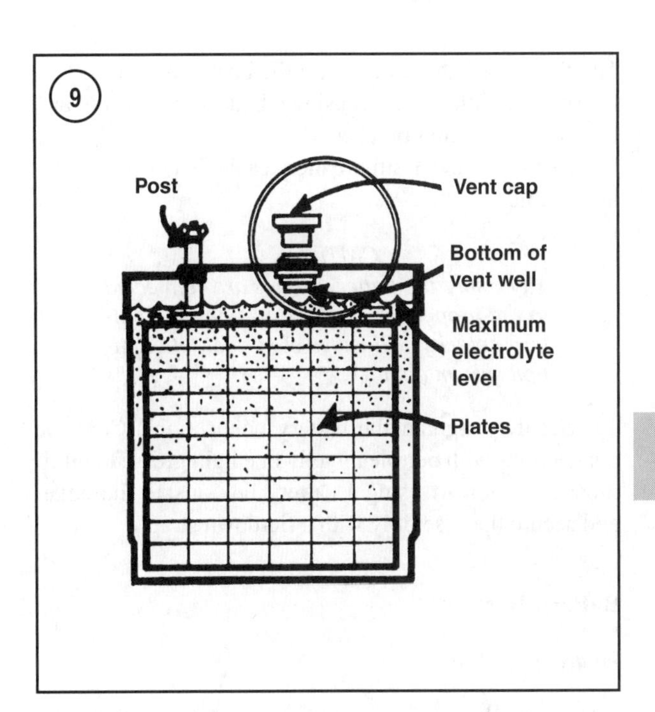

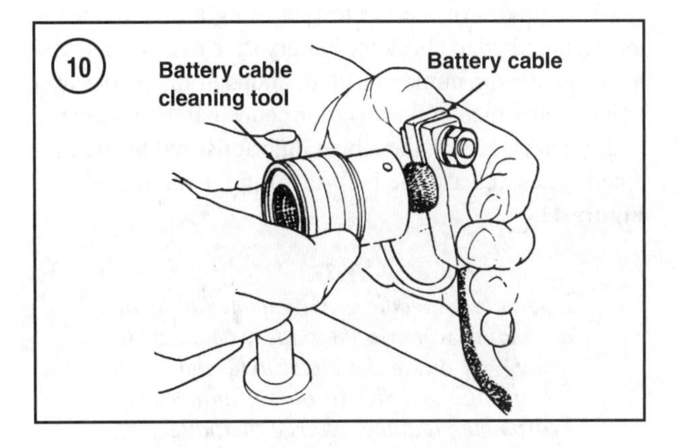

7. Clean the battery terminal posts with a stiff wire brush or battery terminal cleaning tool (**Figure 8**).

NOTE
Do not overfill the battery cells in Step 8. The electrolyte expands due to heat from the charging system and will overflow if the level is more than 4.8 mm (3/16 in.) above the battery plates.

8. Remove the filler caps and check the electrolyte level. Add distilled water to bring the level up to 4.8 mm (3/16 in.) above the plates in the battery case. See **Figure 9**.

9. Clean the battery cable clamps with a stiff wire brush (**Figure 10**).

6

10. Place the battery back into the boat and into the battery tray or battery box. If using a battery tray, install and secure the retaining bracket.
11. Reconnect the positive battery cable first, then the negative cable.

CAUTION
Make sure the battery cables are connected to the proper terminals. Reversing the battery polarity will cause electrical and ignition system damage.

12. Securely tighten the battery connections. Coat the connections with petroleum jelly or a light grease to minimize corrosion. If using a battery box, install the cover and secure the assembly with a tie-down strap.

Battery Tests

Hydrometer test

On batteries with removable vent caps, checking the specific gravity of the electrolyte using a hydrometer is the best method to check the battery state of charge. Use a hydrometer with numbered graduations from 1.100-1.300 points rather than one with color-coded bands. To use the hydrometer, squeeze the rubber bulb and insert the tip into a cell. Then release the bulb to fill the hydrometer. See **Figure 11**.

NOTE
Do not test specific gravity immediately after adding water to the battery cells, as the water will dilute the electrolyte and lower the specific gravity. To obtain an accurate hydrometer reading, charge the battery after adding water.

Draw a sufficient amount of electrolyte to raise the float inside the hydrometer. When using a temperature-compensated hydrometer, discharge the electrolyte back into the battery cell and repeat the process several times to adjust the temperature of the hydrometer to that of the electrolyte.

Hold the hydrometer upright and note the number on the float that is even with the surface of the electrolyte (**Figure 12**). This number is the specific gravity for the cell. Discharge the electrolyte into the cell from which it came.

A fully charged cell will read 1.260 or more at 80° F (26.7° C). A cell that is 75 percent charged will read from 1.220-1.230 while a cell with a 50 percent charge will read from 1.170-1.180. Any cell reading 1.120 or less is discharged. All cells should be within 30 points specific

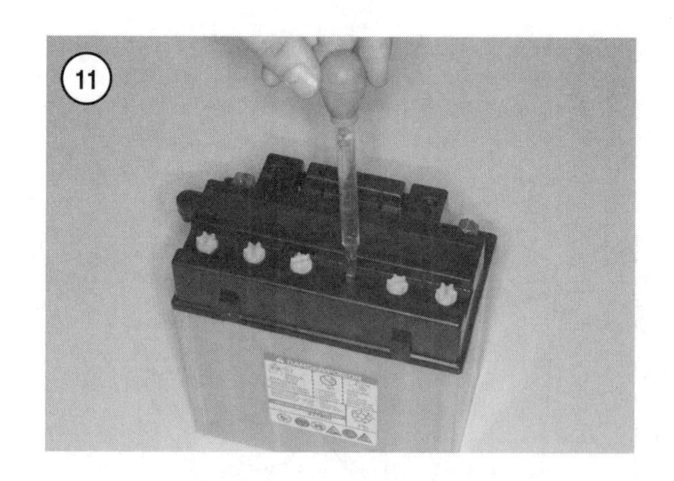

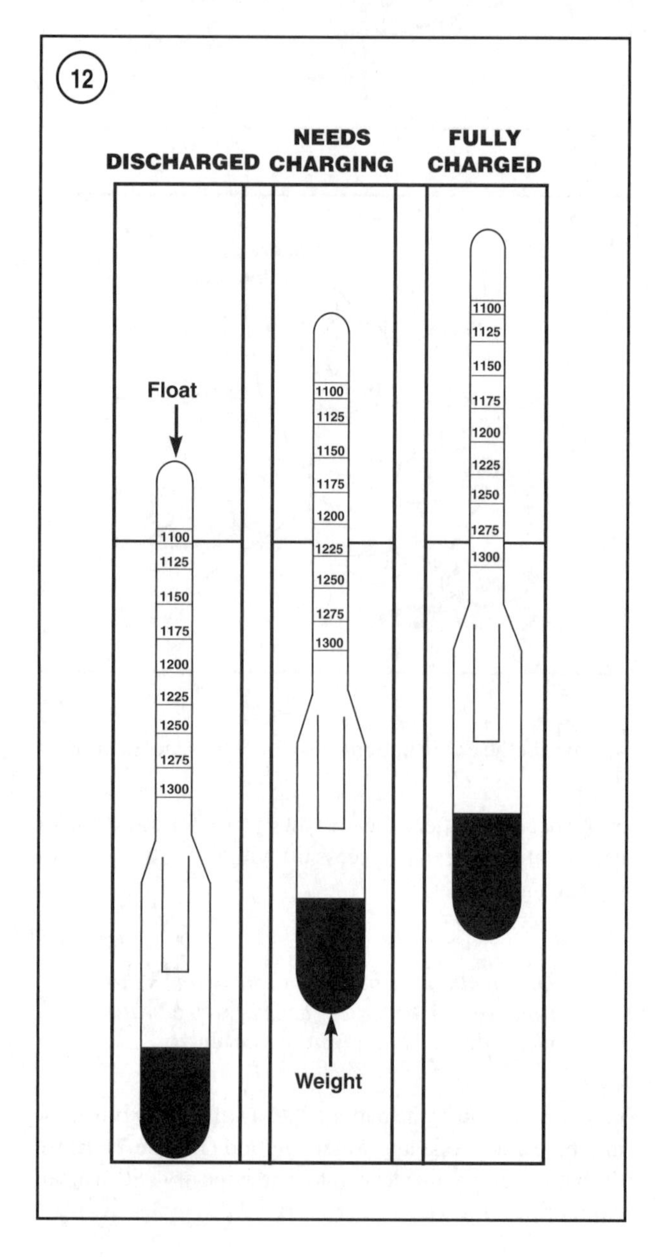

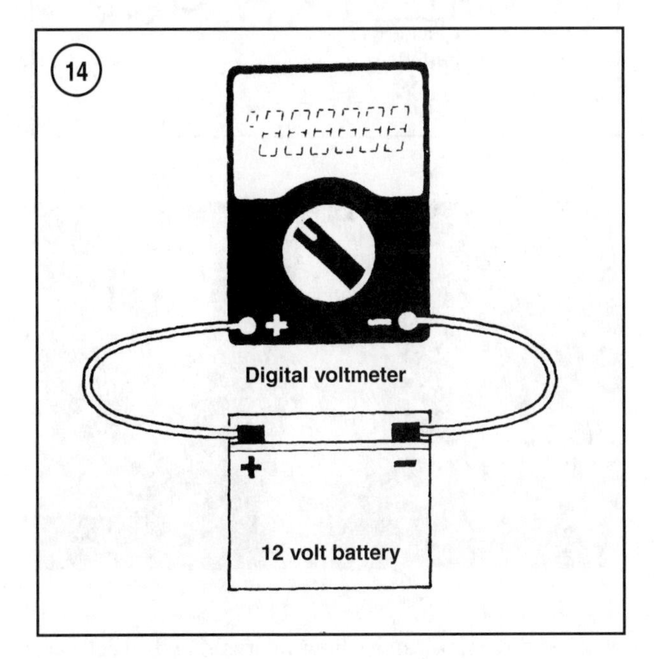

gravity of each other. If there is over 30 points variation, the battery condition is questionable. Charge the battery and recheck the specific gravity. If 30 points or more variation remains between cells after charging, the battery has failed and should be replaced. Refer to **Table 3** for battery charge level based on specific gravity readings.

NOTE
If a temperature-compensated hydrometer is not used, add 4 points specific gravity to the actual reading for every 10° above 80° F (26.7° C). Subtract 4 points specific gravity for every 10° below 80° F (26.7° C).

Open-circuit voltage test

On sealed or maintenance free batteries (vent caps not removable), check the state of charge by measuring the open-circuit (no load) voltage of the battery. Use a digital voltmeter for best results. For the most accurate results, set the battery at rest for at least 30 minutes to stabilize the battery. Then, observing the correct polarity, connect the voltmeter to the battery and note the meter reading. If the open-circuit voltage is 12.7 volts or higher, the battery is fully charged. A reading of 12.4 volts means the battery is approximately 75% charged, a reading of 12.2 means the battery is approximately 50% charged and a reading of 12.1 volts means that the battery is approximately 25% charged.

Load test

Two common methods are used to load test batteries. A commercially available load tester (**Figure 13**) measures the battery voltage as it applies a load across the terminal. Measure the cranking voltage by following the instructions in this section if a load tester is not available.

1. Attach a voltmeter across the battery as shown in **Figure 14**.
2. Remove and ground the spark plug leads to the power head to prevent accidental starting.
3. Crank the engine for approximately 15 seconds while noting the voltmeter reading. Note the voltage at the end of the 15 second period.

4A. *Voltage is 9.5 volts or higher*—The battery is sufficiently charged and of sufficient capacity.

4B. *Voltage is below 9.5 volts*—One of the following conditions is present:

   a. The battery is discharged or defective. Charge the battery and retest.
   b. The battery is of too small capacity. Refer to *Battery Recommendations* in this chapter.
   c. The starting system is drawing excessive current causing the battery voltage to drop. Refer to Chapter Two for starting system troubleshooting procedures.
   d. A mechanical defect is present in the power head or gearcase creating excessive load (and current draw) on the starting system. Inspect the power head and gearcase for mechanical defects.

Battery Storage

Wet cell batteries slowly discharge when stored. They discharge faster in a warm climate. Before storing a battery, clean the case with a baking soda and water solution. Rinse with clear water and wipe dry. The battery should be fully charged and then stored in a cool, dry location. Check the electrolyte level and state of charge frequently during storage. If specific gravity falls to 40 points or more below full charge (1.260), or the open circuit voltage falls below 12.4 volts, recharge the battery.

Battery Charging

A good state of charge must be maintained in batteries used for starting. Check the state of charge with a hydrometer or digital voltmeter as described in the previous section.

Remove the battery from the boat. Charging a battery releases highly explosive hydrogen gas. In many boats, the area around the battery is not well ventilated and the gas may remain in the area for hours after the charging process has been completed. Sparks or flames occurring near the battery can cause it to explode spraying battery acid over a wide area.

If the battery cannot be removed for charging, make sure the bilge access hatches, doors or vents are open to allow adequate ventilation. Observe the following precautions when charging batteries:

1. Never smoke in close proximity to any battery.
2. Make sure all accessories are turned off before disconnecting the battery cables. Disconnecting a circuit that is electrically active will create a spark that can ignite explosive gas that may be present.
3. Always disconnect the negative battery cable first, then the positive cable.
4. On batteries with removable vent caps, always check the electrolyte level before charging the battery. Maintain the correct electrolyte level throughout the charging process.
5. Never attempt to charge a battery that is frozen.

WARNING
Be extremely careful not to create any sparks around the battery when connecting the battery charger.

6. Connect the charger to the battery, negative charger lead to the negative battery terminal and positive charger lead to the positive battery terminal. If the charger output is variable, select a setting of approximately 4 amps. Charge the battery slowly at low amp settings, rather than quickly at high amp settings.
7. If the charger has a dual voltage setting, set the voltage switch to 12 volts. Then switch the charger on.
8. If the battery is severely discharged, allow it to charge for at least 8 hours. Check the charging process with a hydrometer. Consider the battery fully charged when the specific gravity of all cells does not increase when checked three times at one hour intervals, and all cells are gassing freely.

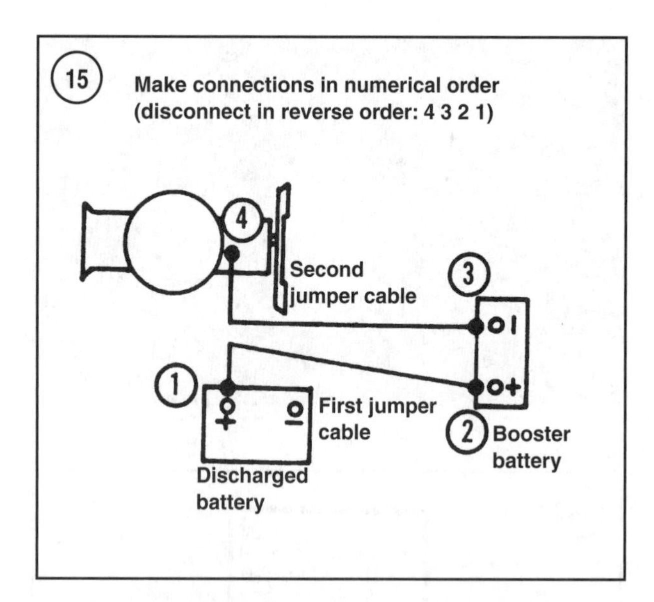

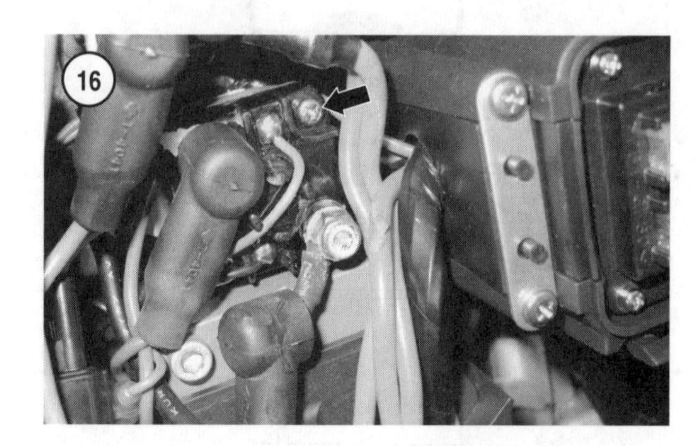

Jump Starting

If the battery becomes severely discharged, it is possible to jump start the engine from another battery (in or out of a vehicle). Jump starting can be dangerous if the proper procedure is not followed. Always use caution when jump starting.

Check the electrolyte level of the discharged battery before attempting the jump start. If the electrolyte is not visible or if it is frozen, do not jump start the discharged battery.

WARNING
Use extreme caution when connecting the booster battery to the discharged battery to avoid injury or damage to the system. Be

certain the jumper cables are connected in the correct polarity.

1. Connect the jumper cables in the order and sequence shown in **Figure 15**.

WARNING
An electrical arc may occur when the final connection is made. This could cause an explosion if it occurs near the battery. For this reason, the final connection should be made to a good engine ground, away from the battery and not to the battery itself.

2. Check that all jumper cables are out of the way of moving engine parts.

CAUTION
*Do not run the engine without an adequate water supply and do not exceed 3000 rpm without an adequate load. Refer to **Safety** in Chapter Two.*

3. Start the engine. Once it starts, run it at a moderate speed (fast idle).

CAUTION
Running the engine at high speed with a discharged battery can damage the charging system.

4. Remove the jumper cables in the exact reverse of the order shown in **Figure 15**. Remove the cable at point 4, then 3, then 2 and finally 1.

STARTING SYSTEM COMPONENTS

This section describes removal, inspection and installation for all electric starting system components.

Starter Relay Removal/Installation

Starter relay mounting locations vary by model. Use the wiring diagrams at the back of the manual to identify the starter relay wire colors. Refer to the following information and trace the wires to the relay (**Figure 16**, typical).

80 Jet and 115-130 hp models—The relay is located forward of the starter motor on the port side of the power head.

105 Jet and 150-250 hp models—The relay is on the starboard side of the power head. The plastic electrical component cover (**Figure 17**) must be removed to access the relay.

1. Disconnect the cables from the battery. Disconnect and ground the spark plug leads to prevent accidental starting.
2. *105 Jet and 150-250 hp models*—Remove the plastic electrical component cover (**Figure 17**).
3. Mark all relay terminals with the wire color and note the wire routing prior to disconnecting the relay wires.
4. Carefully pull the rubber boots (**Figure 18**) away from the wire terminals. Remove the hex nuts and washers. Then pull the two wire terminals from the relay terminals.
5. Locate the brown and black wires leading to the relay. Unplug the brown and black wire bullet connectors from the relay. If the black wire does not have a bullet connector, trace the wire from the relay to the connection to the power head ground. Remove the screw and lockwasher (5 and 6, **Figure 19**) to disconnect the wire.
6. Carefully pry the relay (4, **Figure 19**) from the rubber mount (3). Work carefully to avoid damaging the relay. If poor access prevents removing the relay easily, remove the bolt and washer (2, **Figure 19**), then lift the relay and mounting bracket from the engine.
7. Inspect the rubber mount for tears or deterioration. If there are defects, carefully pull the mount from the arms of the mounting bracket (1, **Figure 19**).
8. If removed, fit the arms of the mounting bracket through the slots on each side of the rubber mount. Push

the mount onto the bracket until the tabs on each arm protrude from the slots.

9. Align the large terminals with the arms of the relay mount, then insert the replacement relay into the mount. Push the relay in until the tabs in the mount opening capture the relay. Apply a light coating of window cleaning solution into the rubber mount for easier installation.

10. If the relay bracket (1, **Figure 19**) was removed, install the bracket onto the power head, electrical component plate or lower engine cover. Secure the bracket with the bolt and washer (2, **Figure 19**). Tighten the mounting bolt to the general torque specification in Chapter One.

11. Reconnect the brown and black wire bullet connectors. If the black wire does not use a bullet connector, connect the black wire to the engine ground using the screw and lockwasher (5 and 6, **Figure 19**).

12. Connect the large diameter wire terminals to the large terminals of the relay. Secure the terminals with the hex nuts and tighten securely.

13. Fit the rubber boots (**Figure 18**) onto the terminals. The rubber boots must completely cover the terminals. Route the relay wiring as noted prior to removal.

14. *105 Jet and 150-250 hp models*—Install the plastic electrical component cover (**Figure 17**).

15. Connect the battery cables and spark plug leads.

Key Switch Removal/Installation

Key switch mounting locations vary by the type of control box used.

1. *Surface mount remote control*—The switch mounts into the remote control (**Figure 20**).

2. *Panel or binacle mount control*—The key switch mounts into the dash (**Figure 21**) or other suitable surface near the control station.

Remote control mounted switch

1. Disconnect the battery cables and ground the spark plug leads to prevent accidental starting.

2. *Switch mounted in the remote control*—Disassemble the remote control (Chapter Thirteen) to the point necessary to access the ignition key switch wires and retaining nut (**Figure 22**).

3. Note the wire routing and connection points, then carefully disconnect each ignition key switch wire form the remote control harness.

4. Loosen the retaining nut, then carefully pull the switch from the control.

5. Connect each of the ignition key switch wires onto the respective remote control harness wires.

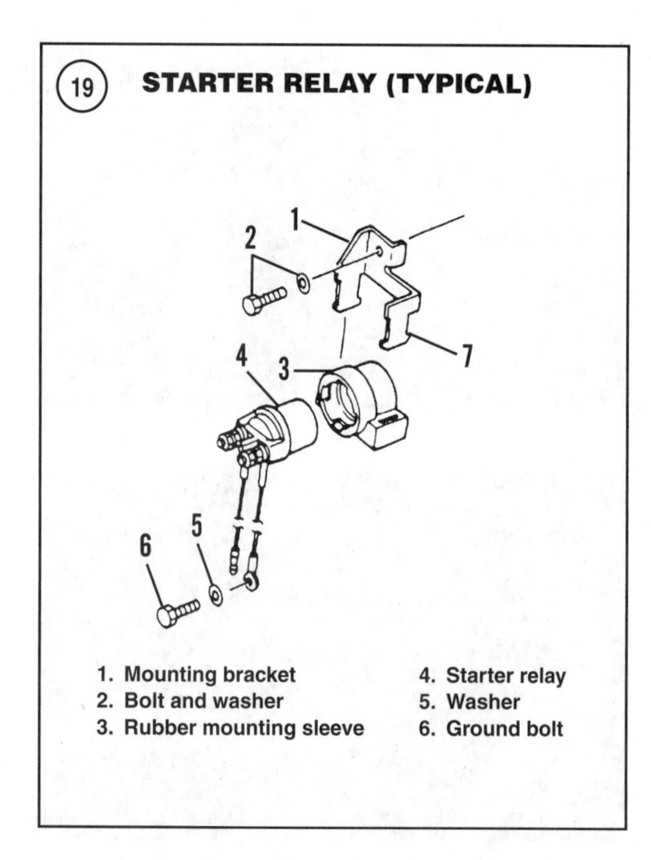

19 **STARTER RELAY (TYPICAL)**

1. Mounting bracket
2. Bolt and washer
3. Rubber mounting sleeve
4. Starter relay
5. Washer
6. Ground bolt

20

6. Fit the ignition key switch into the control housing and secure with the nut. Carefully press the ignition key switch wires into the remote control cavity. Route the wiring to prevent any contact with moving components.

7. Assemble the remote control as described in Chapter Thirteen.

8. Reconnect the spark plug leads and battery cables. Check for proper operation of the ignition key switch and neutral safety switch before putting the engine into service.

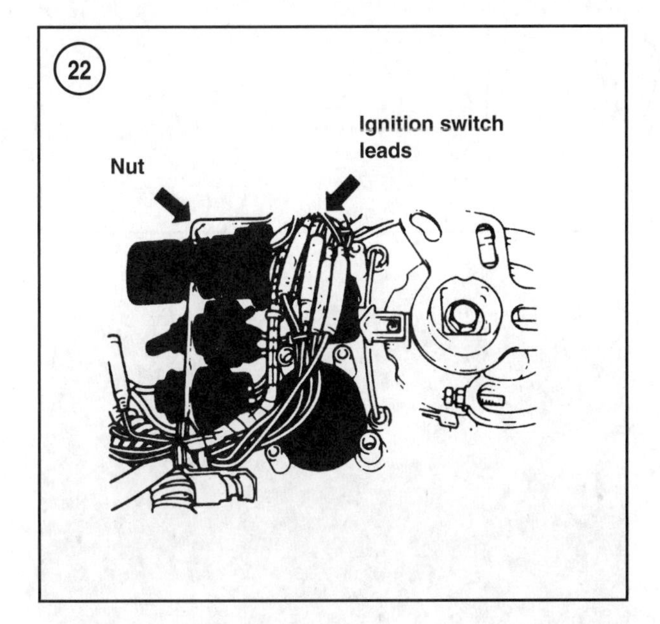

Dash mounted switch

1. Disconnect the battery cables and ground the spark plug leads to prevent accidental starting.
2. Note the color of each wire connection onto the ignition key switch. The instrument harness wire colors may not match the ignition key switch wires. Carefully disconnect each of the switch wires.
3. Pull the rubber seal from the outer nut. Remove the outer nut (**Figure 21**) and pull the switch from the dash.
4. Thread the inner nut onto the shaft of the replacement switch. Install the lockwasher over the shaft.
5. Insert the key switch into the dash opening. The drain holes in the switch body (**Figure 23**) must face downward.
6. Thread the outer nut (**Figure 21**) onto the key switch. If the switch fits loosely in the opening, loosen the outer nut slightly, then thread the inner nut toward the dash until the switch is securely mounted.

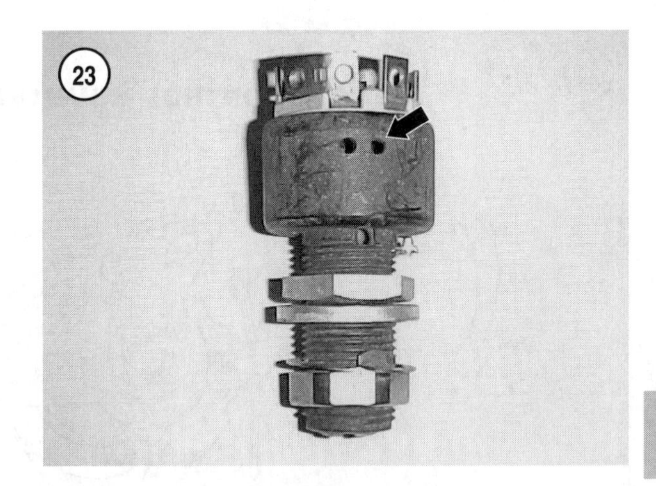

7. Install the rubber seal into the outer nut. Use the ignition key to align the slot in the seal with the key slot in the switch.
8. Connect all wires to the corresponding wires of the switch as noted before removal. If replacing the switch, refer to the switch manufacturer's instructions for wire connecting points.
9. Connect the battery cables and spark plug leads.

Neutral Safety Switch

The neutral safety switch is located within the remote control housing on all models covered in this manual. The neutral safety switch prevents operation of the electric starter if the engine is shifted into forward or reverse gears. The shift position switch used on HPDI models is used by the ECU to control the fuel and ignition system and does not prevent the engine from starting in gear.

NOTE
Some Yamaha outboards are not equipped with a Yamaha control. Contact the manufacturer of the control or a reputable marine dealership for repair information and replacement parts for other types of controls.

Removal/installation

Refer to **Figure 24**, typical for this procedure.

1. Disconnect the battery cables and ground the spark plug leads to prevent accidental starting.
2. Refer to Chapter Thirteen and disassemble the remote control to the point that the switch is accessible.
3. Locate then disconnect the two neutral switch wires from the remote control harness.
4. Remove the screw and retainer (2, **Figure 24**), then lift the switch (3) from the control.

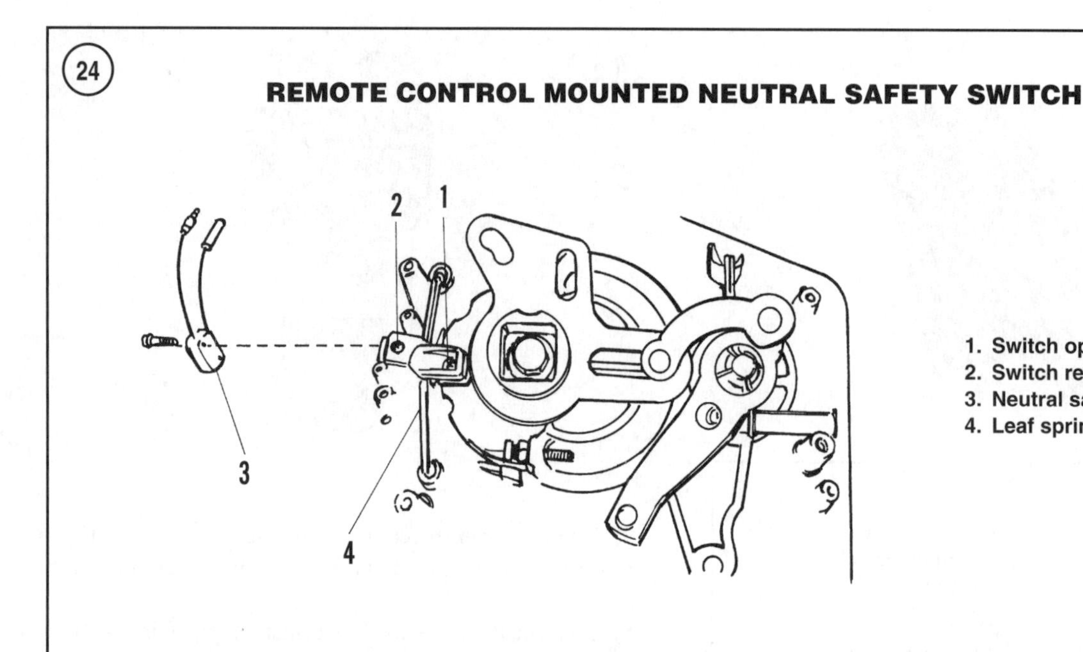

5. Inspect the switch operating arm (1, **Figure 24**) and leaf spring (4) for cracks and worn surfaces. Replace if necessary.

6. Position the replacement switch onto the mounts and secure it with the retainers screw.

7. Connect the neutral switch wires to the remote control harness wires. Route the switch wires to prevent interference with moving components. Secure the wires with plastic locking type clamps as needed.

8. Reassemble and install the remote control as described in Chapter Thirteen.

9. Connect the battery cables and spark plug leads.

Starter Motor

Starter motor appearance and mounting locations vary by model:

1. *80 Jet and 115-130 hp models*—The standard type starter (without a solenoid) is mounted to the port side of the power head (**Figure 25**).

2. *105 Jet and 150-200 hp (2.6 liter) models (except HPDI models)*—The standard type starter (without a solenoid) is mounted to the starboard side of the power head (**Figure 26**).

3. *150-250 hp (HPDI models)*—The solenoid type starter is mounted to the starboard side of the power head (**Figure 27**).

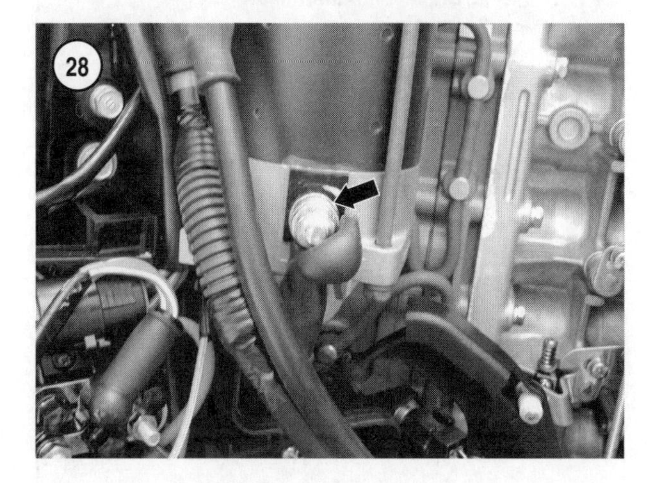

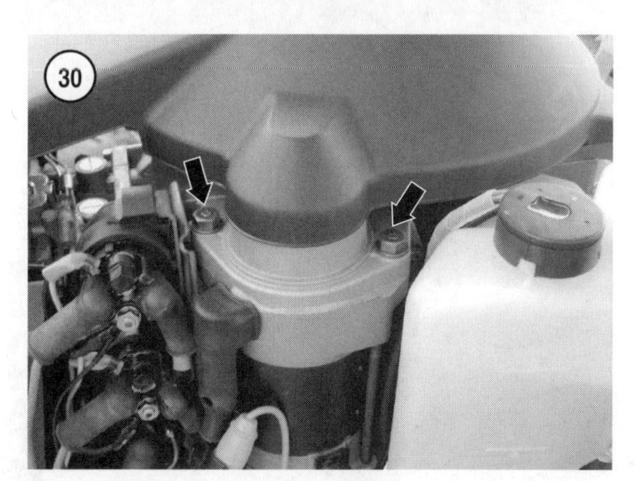

4. *200-250 hp (3.1 liter) models*—The standard type starter (without a solenoid) is mounted to the starboard side of the power head (**Figure 26**).

Removal/installation

1. Disconnect the battery cables and ground the spark plug leads to prevent accidental starting.
2. Locate the starter motor on the side of the power head.
3A. *Except HPDI models*—Pull the insulating boot from the large starter wire terminal, then remove the terminal nut (**Figure 28**). If access to the terminal is difficult, remove the terminal nut after removing the starter mounting bolts and pulling the starter away from the power head.
3B. *HPDI models*—Pull the insulating boots away, then remove the terminal nuts from the positive battery wire and red starter relay wires (**Figure 29**) from the starter solenoid. Pull the red wires away from the terminals. Carefully disconnect the black solenoid ground wire from the engine harness wire.
4. Support the starter, then locate and remove the starter mounting bolts (**Figure 30**, typical). Note any battery grounding cables connected to the mounting bolts during removal. Refer to the following information and illustrations to locate the mounting bolts.
   a. *80 Jet and 115-130 hp models*—Refer to **Figure 31**. Two bolts secure the upper starter mounting bracket onto the power head. Two top mounted bolts secure the starter to the bracket. The single sided mounted bolt, located at the bottom of the starter motor secures the lower end of the starter to the power head. Remove the upper mounting bracket by removing the two top mounted bolts and the single lower bolt.
   b. *105 Jet and 150-250 hp models (except HPDI)*—Refer to **Figure 32**. Two bolts secure the upper starter mounting bracket and battery cable terminal to the power head. Two top mounted bolts secure the starter motor to the bracket. The single bolt, located at the bottom of the starter, secures the lower end of the starter to the power head. The starter can be removed, without removing the upper

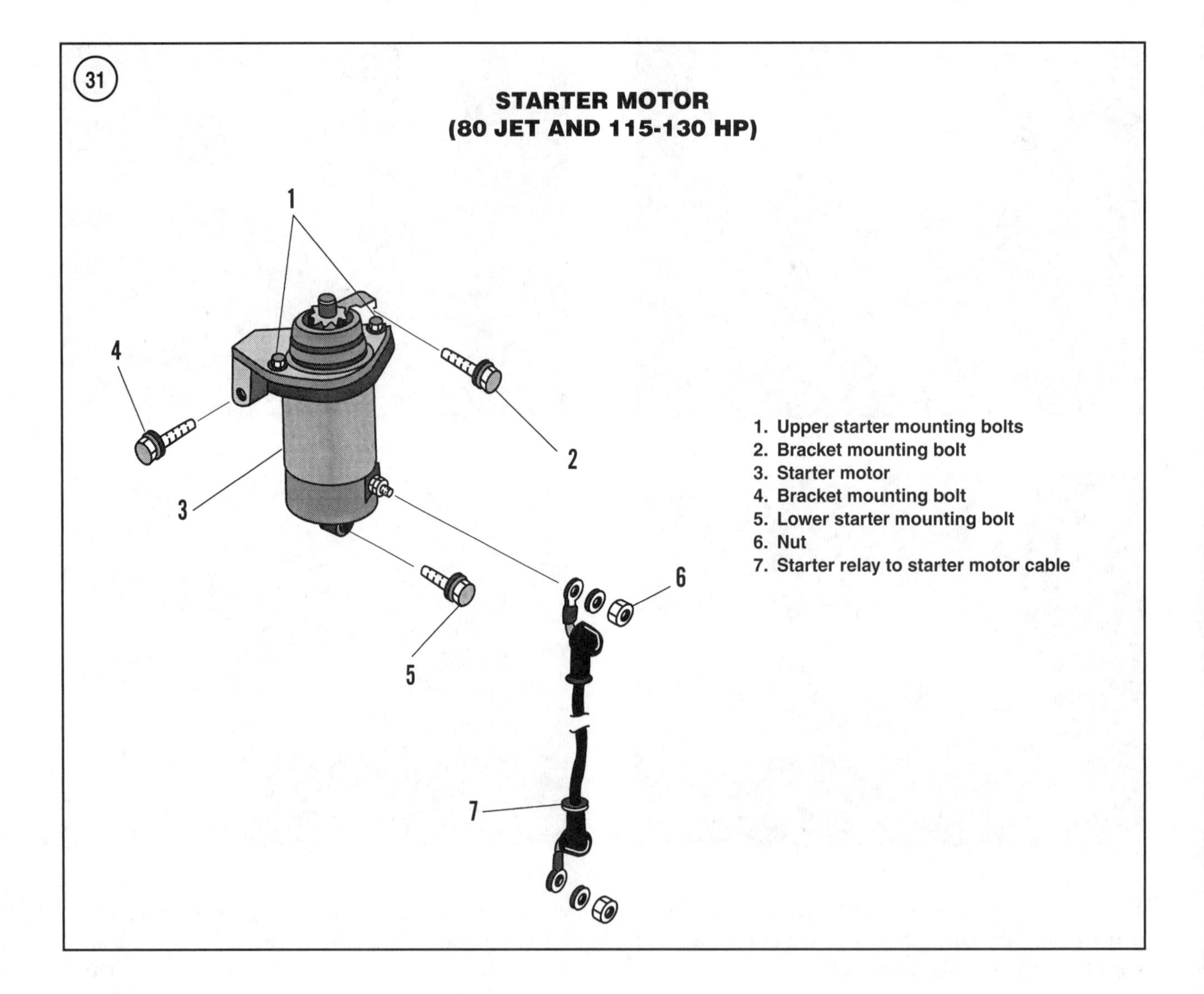

mounting bracket, by removing the two top mounted bolts, the single lower bolt and the spacer.

c. *150-250 hp (HPDI models)*—Refer to **Figure 33**. Three sided mounting bolts and washers secure the starter motor to the power head. Remove the three bolts and washers to remove the starter.

5. If accessibility to the starter terminal is difficult after starter installation, connect the cable and tighten the nut prior to installing the starter to the bracket.

6. Install the replacement starter motor to the mounting bracket. Verify that the starter is pinching no wire or hoses. Then thread the mounting bolts into the starter and mounting bracket. If so equipped, make sure the mounting bolts pass through any battery ground cable terminals before passing through the starter. Route the ground wire to prevent interference before tightening the mounting bolts.

7. Tighten the starter mounting bolts to the specification in **Table 1** or general torque specificatons in Chapter One.

8A. *Except HPDI models*—Connect the starter relay-to-starter motor cable to the terminal (**Figure 18**). Install and securely tighten the terminal nut (general torque specifications in Chapter One). Slide the insulating boot onto the terminal. Make sure the boot completely covers the terminal.

8B. *HPDI models*—Connect the battery positive and starter relay cables to the respective terminals of the starter solenoid. Install the terminal nuts. Slide the insulating boots onto the terminals. Make sure the boots completely covers the terminal. Connect the black solenoid ground wire to the respective engine harness wire.

9. Connect the battery cables and spark plug leads. Check for proper starter and neutral safety switch operation before putting the engine into service.

32

STARTER MOTOR
(105 JET AND 150-250 HP MODELS [EXCEPT HPDI])

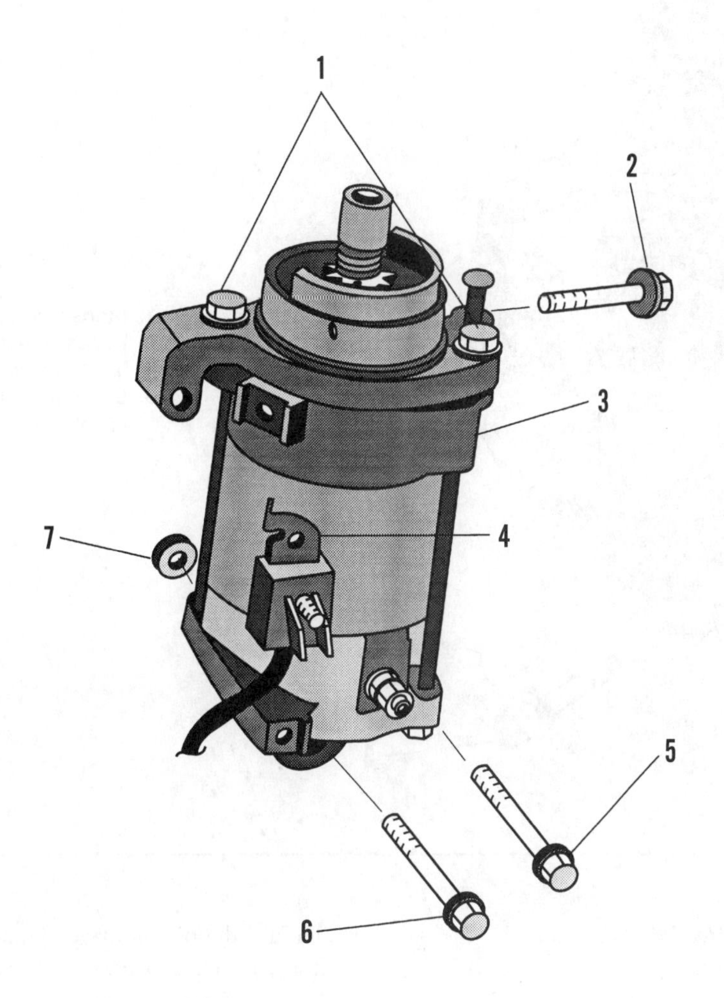

1. Upper starter mounting bolts
2. Bracket mounting bolt
3. Starter motor
4. Terminal block
5. Bracket mounting bolt
6. Lower starter mounting bolt
7. Spacer

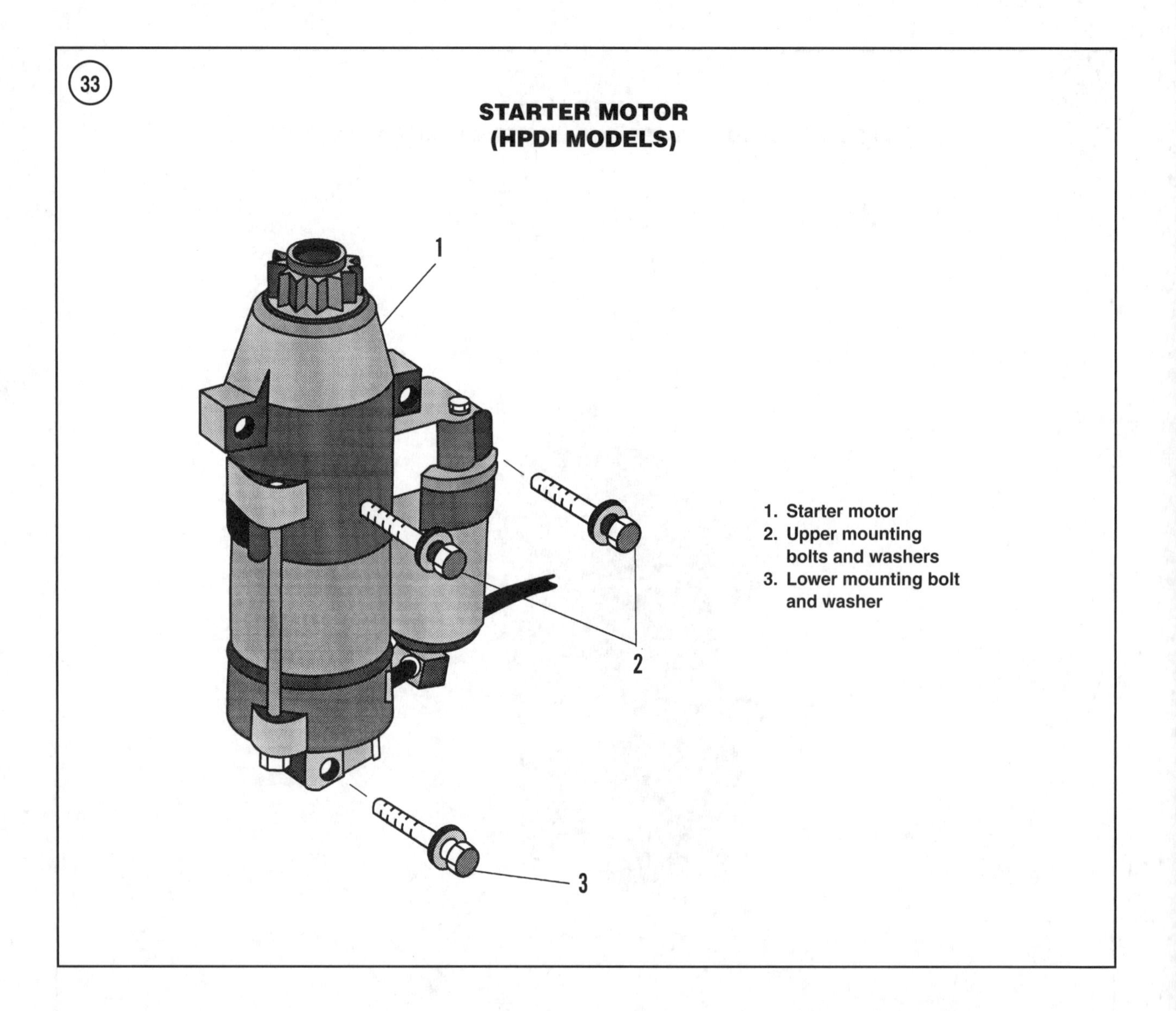

Disassembly (80 Jet and 115-250 hp [except HPDI models])

CAUTION
Never strike the frame of the starter motor. The permanent magnets in the frame may crack or break and cause starter motor failure.

Refer to **Figure 34** for this procedure.

1. Remove the starter as described in this chapter.
2. Clamp the starter lightly into a vise with protective jaws. Grasp the pinion drive and stopper (**Figure 35**), then pull them toward the starter to expose the locking clip. Carefully pry the locking clip from the armature shaft. Remove the stopper and spring from the armature shaft.
3. Turn the pinion drive counterclockwise to remove it from the helical splines of the armature.
4. Make reference markings (**Figure 36**) on the top and bottom cover of the frame. Support both ends of the starter motor and remove the throughbolts (**Figure 37**).
5. Carefully pull the bottom cover from the armature (**Figure 38**). If necessary, carefully tap on the bottom cover with a plastic mallet. Remove and discard the O-ring, if so equipped.
6. Lay the frame assembly on its side. Grasp the frame and lightly tap on the end of the armature to free the top cover from the frame. Do not tap on the commutator part of the armature.
7. Pull the armature from the frame assembly. Note the orientation of the composite and metal washers (8, **Figure 34**) on the armature shaft. Remove them from the arma-

(34)

STARTER MOTOR COMPONENTS (80 JET, 105 JET AND 150-250 HP [EXCEPT HPDI MODELS])

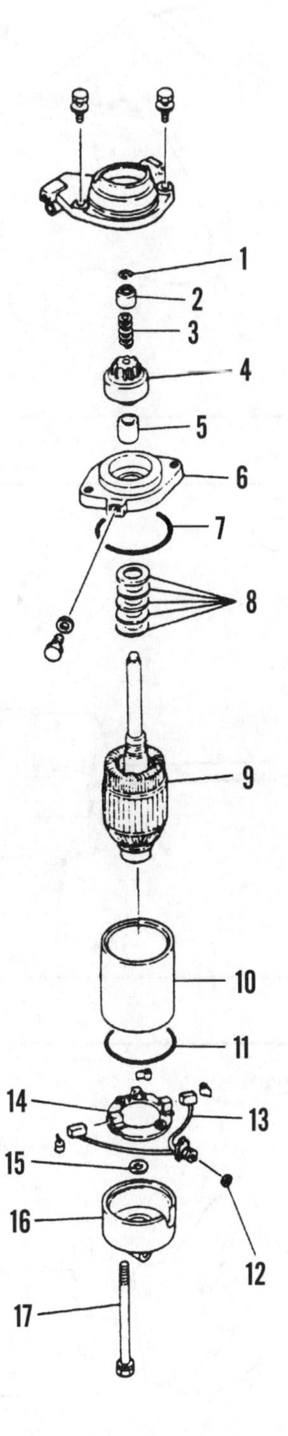

1. Locking clip
2. Pinion stopper
3. Spring
4. Pinion drive
5. Collar
6. Top cover
7. O-ring
8. Washers
9. Armature
10. Frame
11. O-ring
12. Nut
13. Brushes
14. Brush plate
15. Washer
16. Bottom cover
17. Throughbolt

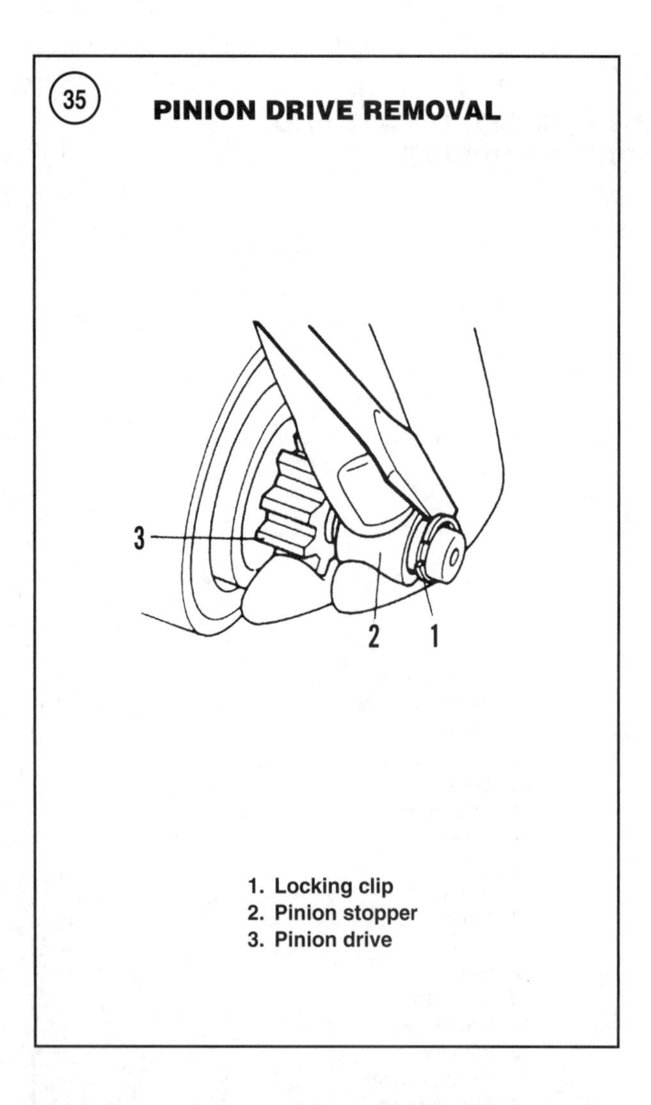

35

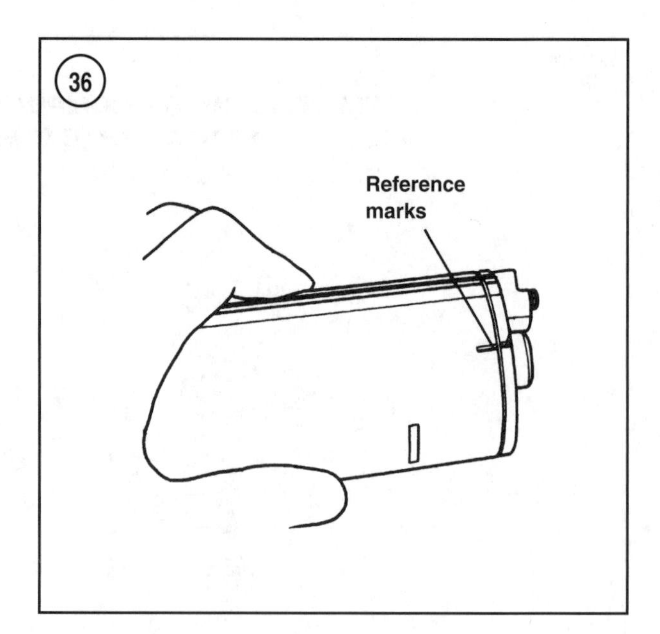

36

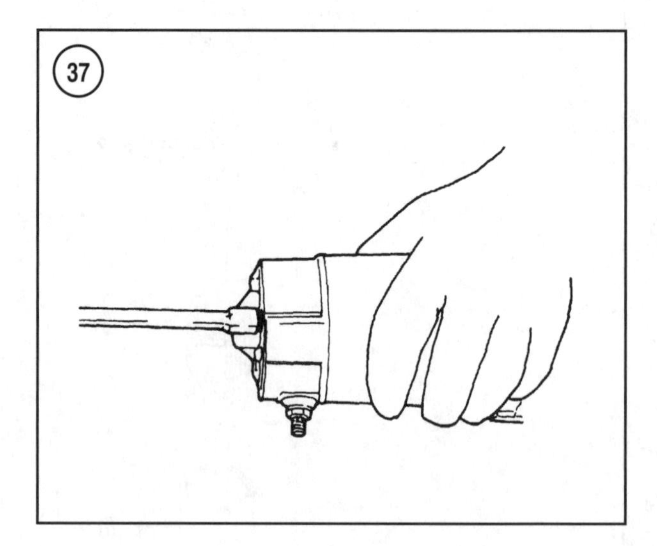

37

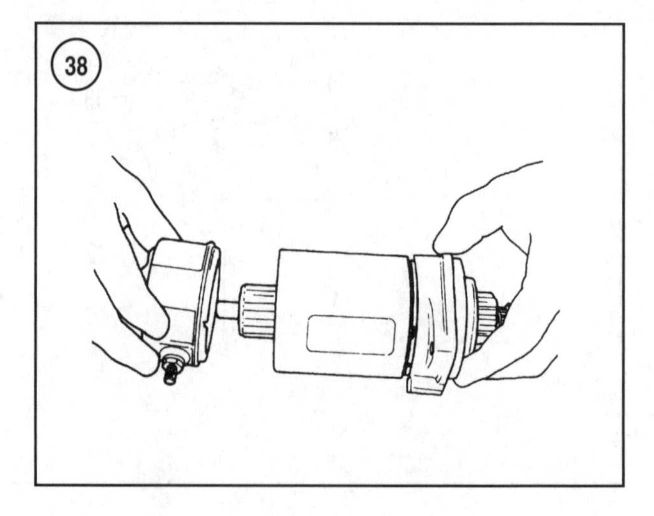

38

ture and wire them together to ensure they are installed in the same order and position during assembly.

8. Remove the nut (**Figure 39**). Then pull the washers and bushing or insulator from the starter terminal.

9. Mark the brush plate and bottom cover to ensure correct orientation to the bottom cover during assembly. Remove the screws (**Figure 40**). Then lift the brush plate from the bottom cover.

10. Use compressed air to remove debris from the starter motor components. Use a mild solvent to remove oily material from the armature, covers, frame assembly and pinion drive. Do not clean the brushes in solvent.

Solenoid removal (150-250 hp [HPDI models])

Refer to **Figure 41** for this procedure.

1. Remove the starter motor following the instructions in this chapter. Remove the larger nut retaining the short ca-

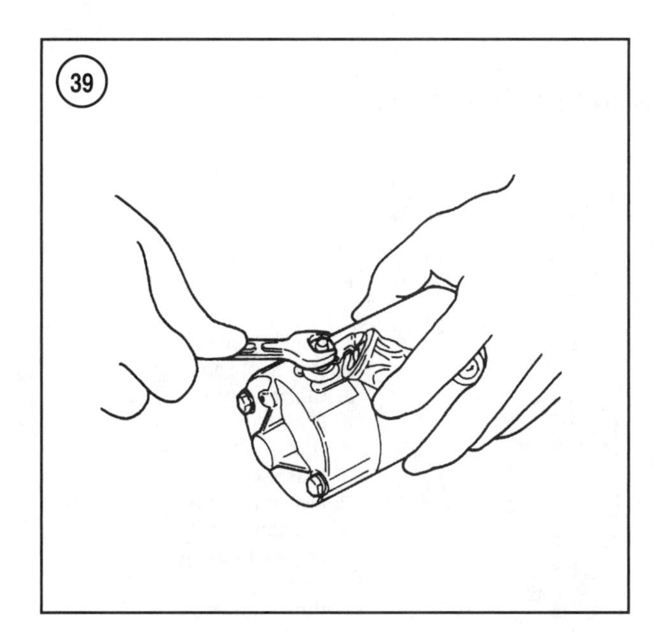

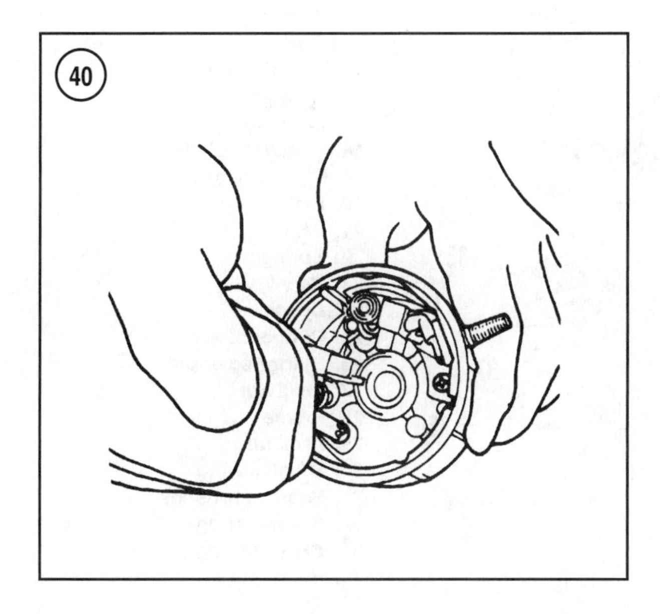

ble to the starter solenoid. Disconnect the cable from the starter solenoid.

2. Remove both screws (1, **Figure 41**) and lift the starter solenoid (9) from the starter motor housing (2).

3. Pull the starter lever (4, **Figure 41**) from the starter motor housing (2). Pull the lever spring (5, **Figure 41**) from the mounting holes in the starter solenoid (9).

4. Remove both screws (6, **Figure 41**), then lift the cover (7) from the starter solenoid (9).

5. Test the solenoid as described in this chapter. See *Component inspection and testing.*

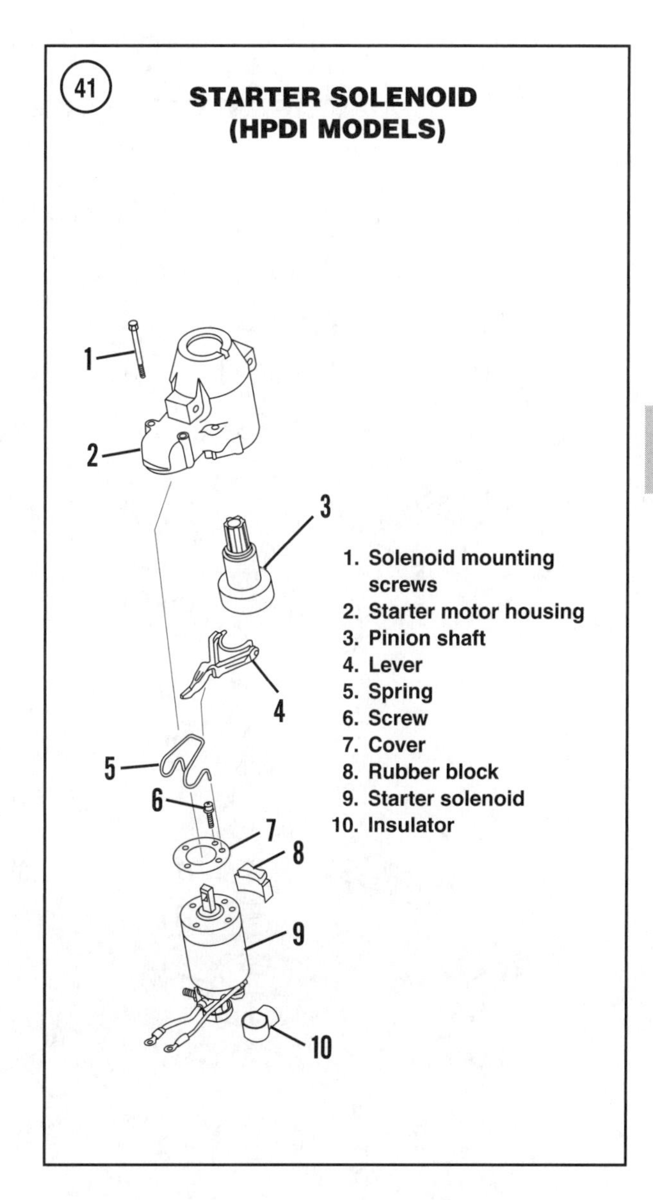

Disassembly (150-250 hp [HPDI models])

CAUTION
Never strike the frame of the starter motor. The permanent magnets in the frame may crack or break and cause starter motor failure.

Refer to **Figure 42** for this procedure.

1. Remove the starter solenoid (26, **Figure 42**) as described in this chapter.
2. For reference during assembly, place reference marks on the starter frame, upper and bottom covers (**Figure 36**).
3. Secure the electric starter motor into a vice with soft jaws. Do not overtighten the vice. Pry the starter pinion (3, **Figure 42**) downward. Tap the edge of the pinion stopper

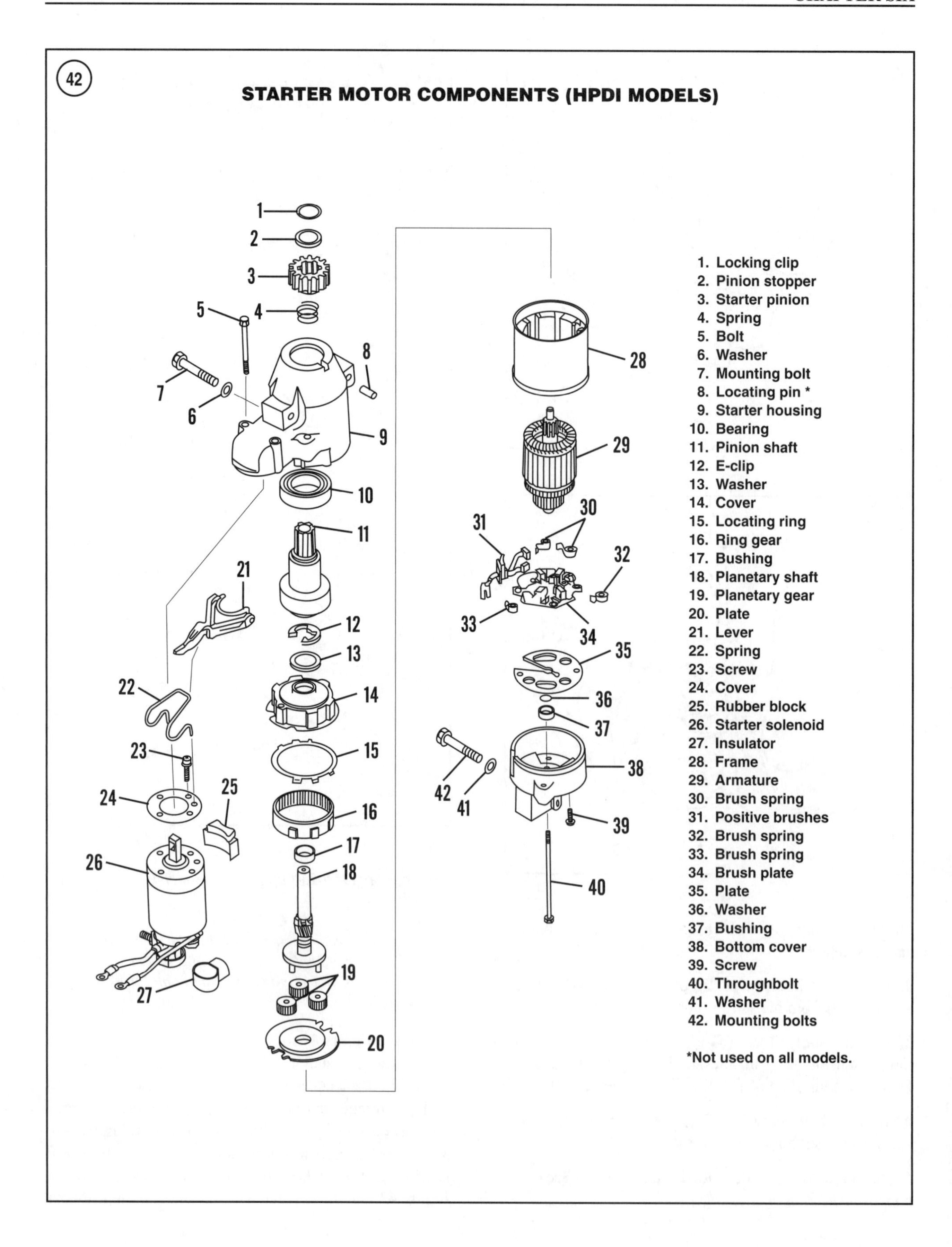

STARTER MOTOR COMPONENTS (HPDI MODELS)

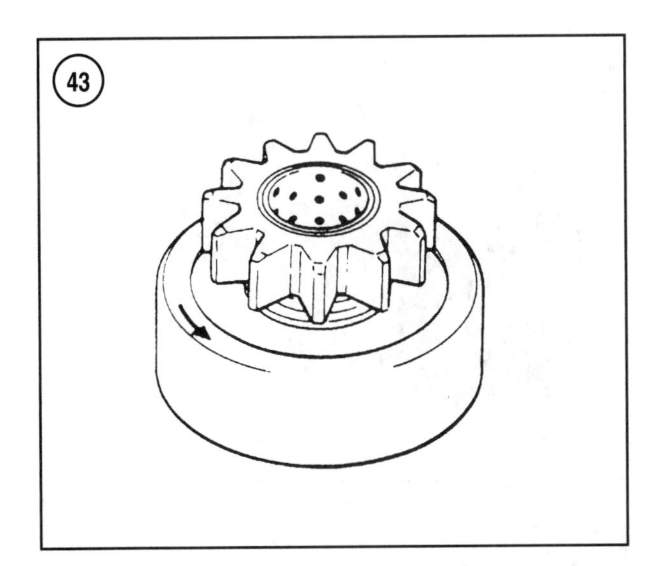
43

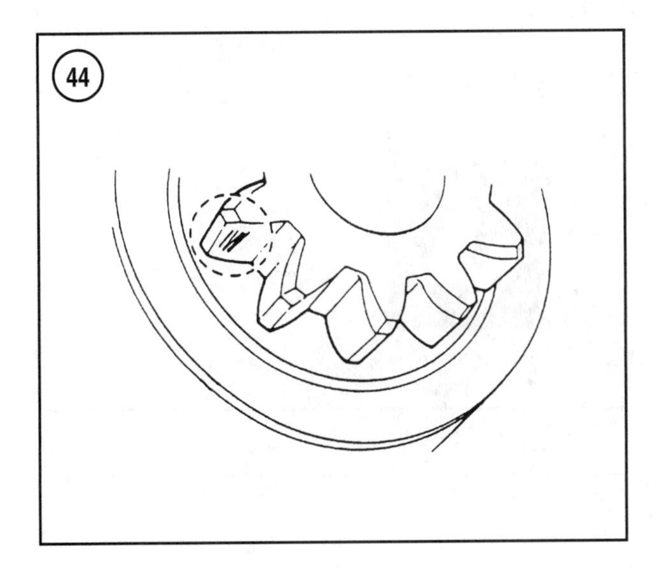
44

(2, **Figure 42**) down with a small hammer until the locking clip (1) is exposed. Carefully pry the locking clip from the pinion stopper.

4. Rotate the starter pinion counterclockwise to remove it from the pinion shaft (11, **Figure 42**). Pull the spring (4, **Figure 42**) from the shaft.
5. Remove both screws (39, **Figure 42**) from the bottom cover (38). Remove both throughbolts (40, **Figure 42**). Then carefully tap the bottom cover (38) free from the frame (28). Pull the bottom cover (38, **Figure 42**) away from the starter.
6. Using a small punch, tap the plate (35, **Figure 42**) from the groove on the lower end of the armature (29). Pull the brush plate (34, **Figure 42**) from the commutator.
7. Pull the frame and armature (28 and 29, **Figure 42**) from the starter housing (9). Mark the armature side and remove the plate (20, **Figure 42**) from the starter housing (9) or frame (28). Remove the three planetary gears (19, **Figure 42**) from the planetary shaft (18).
8. Remove the planetary assembly from the starter housing by tapping lightly on the pinion shaft (11, **Figure 42**) with a plastic hammer. Pull the pinion shaft from the planetary assembly.
9. Carefully pry the E-clip (12, **Figure 42**) from the planetary shaft (18). Lift the washer and cover (13 and 14, **Figure 42**) from the planetary shaft. Pull the locating ring and gear (15 and 16, **Figure 42**) from the planetary shaft. Slide the bushing off the planetary shaft.
10. Clean the starter housing, bottom cover, armature and frame assembly using a mild solvent. Dry all components with compressed air.
11. Inspect all components for worn, damaged or shorted components as described in this section.

Component inspection and testing

1. Place the pinion drive on a flat work surface. Rotate the pinion drive in the clockwise and counterclockwise directions (**Figure 43**). Replace the pinion drive if it does not turn freely when rotating it in the clockwise direction and lock to the center hub when rotating it in the counterclockwise direction.
2. Inspect the pinion drive teeth (**Figure 44**) for cracked, chipped or excessively worn teeth. Replace the pinion drive if any of the teeth are damaged or if a step has worn on the contact surfaces. Inspect the helical splines at the pinion drive end of the armature. Replace the armature if there are chipped areas or the pinion drive does not thread smoothly onto the shaft.
3. *150-250 hp (HPDI models)*—Inspect the planetary gears (19, **Figure 42**) and ring gear (16) for worn, cracked or broken gear teeth. Replace the gears if they are not found in excellent condition.
4. Carefully secure the armature into a vise with protective jaws (**Figure 45**). Tighten the vise only enough to secure the armature. Excessive force will damage the armature. Use 600 grit carburundum (wet or dry polishing cloth) to remove corrosion deposits and glazed surfaces from the commutator. Work the area enough to just clean the surfaces. Avoid removing too much material. Rotate the armature to polish the surfaces evenly.
5. Calibrate a multimeter to the 1 ohm scale. Connect the negative meter test lead to one of the commutator contacts and the positive test lead to the laminated section of the armature. If the meter indicates continuity, the armature is shorted and must be replaced.
6. Connect the negative meter test lead to one of the commutator contacts and positive test lead to the armature

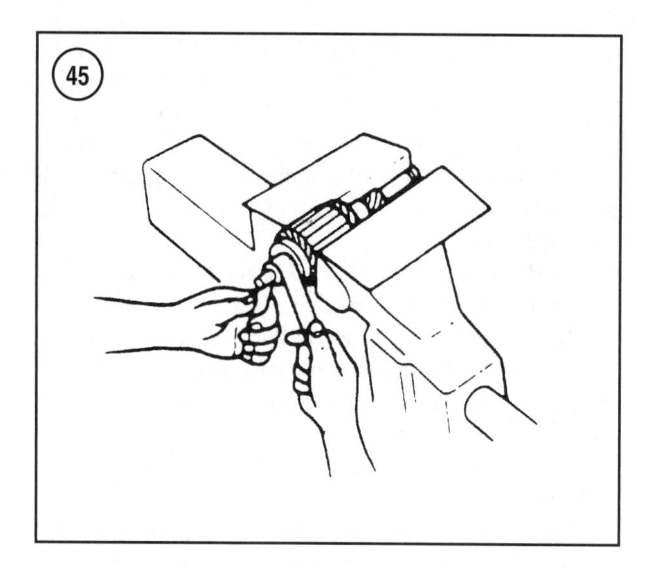

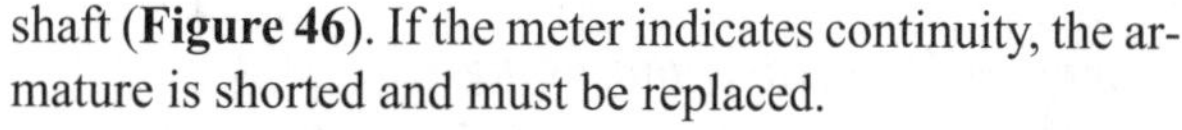

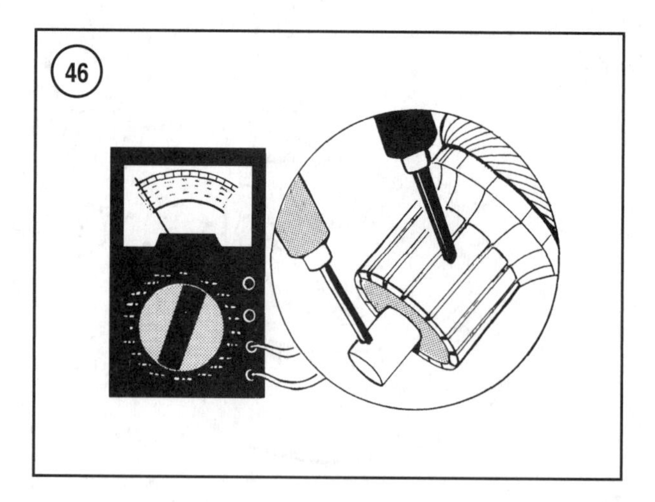

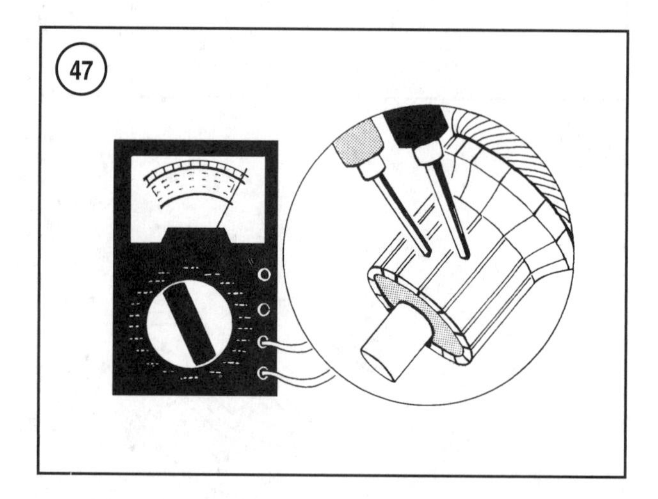

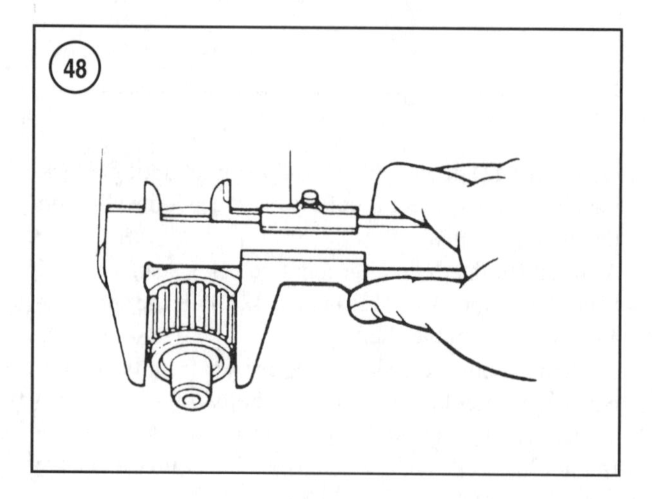

shaft (**Figure 46**). If the meter indicates continuity, the armature is shorted and must be replaced.

7. Connect the negative meter test lead to one of the commutator contacts. Connect the positive test lead to each of the remaining contacts (**Figure 47**). If the meter does not indicate continuity for each connection, the armature has failed open and must be replaced.

8. Use an accurate caliper or micrometer to measure the diameter of the commutator (**Figure 48**). Replace the armature if the measurement is less than the minimum specification in **Table 2**.

9. Use a disposable nail file or a suitable small file (**Figure 49**) to remove metal and mica particles from the undercut, or area between and below the commutator contacts.

10. Use compressed air to blow away any loose particles and use a depth micrometer to measure the depth of the undercut (**Figure 50**). Replace the armature if the measurement is less then minimum specification in **Table 2**.

11. Support the end of the armature shaft on V-blocks or other suitable supports. Position a dial indicator with the plunger in direct contact with the upper bushing contact surface (**Figure 51**). Slowly rotate the armature while reading the deflection on the dial indicator. Replace the armature if the measurement exceeds the specification in **Table 2**.

12. Inspect the brushes for corroded, chipped or oil contaminated surfaces. Replace the brushes if any of these conditions are present. Use an accurate caliper to measure the length of each of the brushes (**Figure 52**). Replace both brushes if either of them has worn to less than the minimum specification in **Table 2**.

13. Inspect the brush springs for corroded or distorted loops and replace as needed. Corroded or damaged springs will not apply adequate pressure for the brushes, causing poor starter motor performance.

14. Inspect the magnets in the frame assembly for corrosion or other contaminants. Inspect the frame assembly

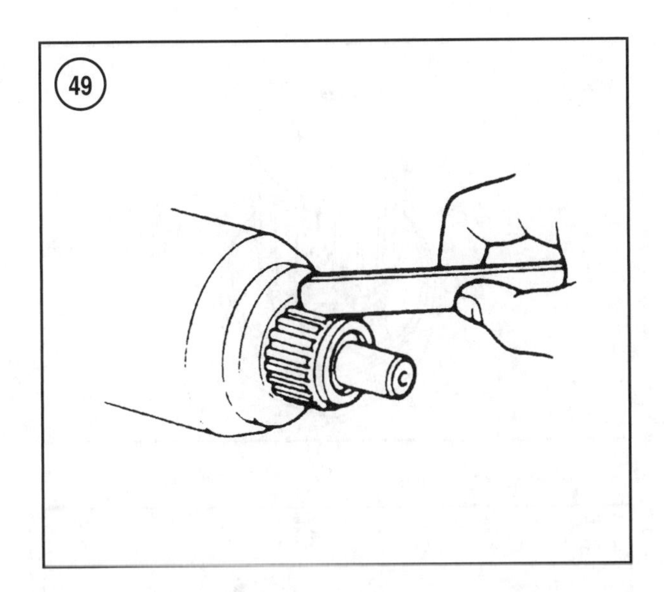

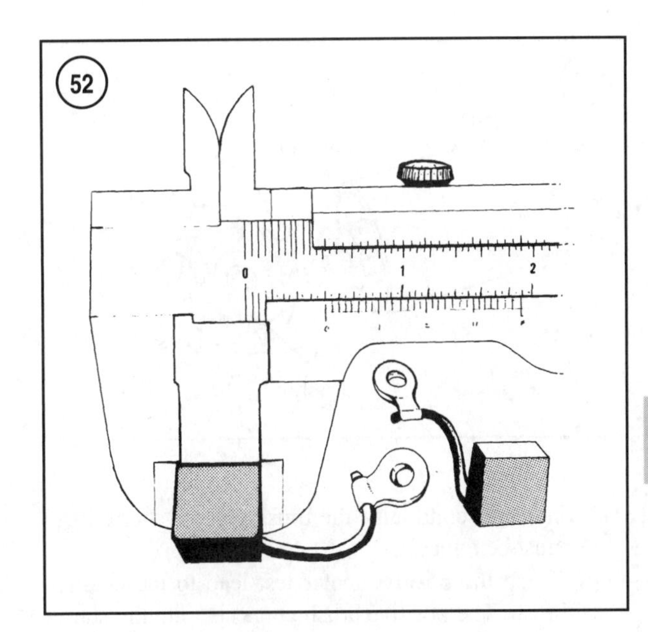

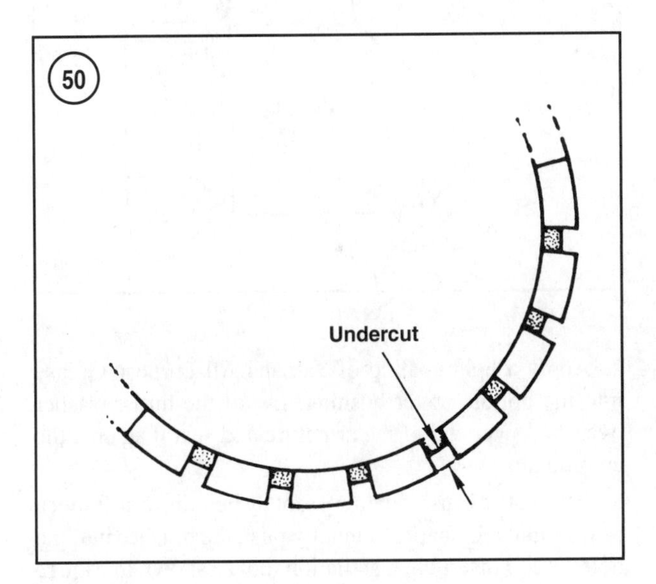

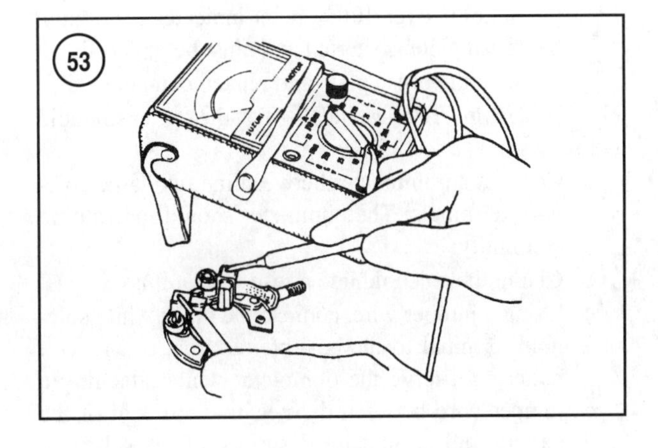

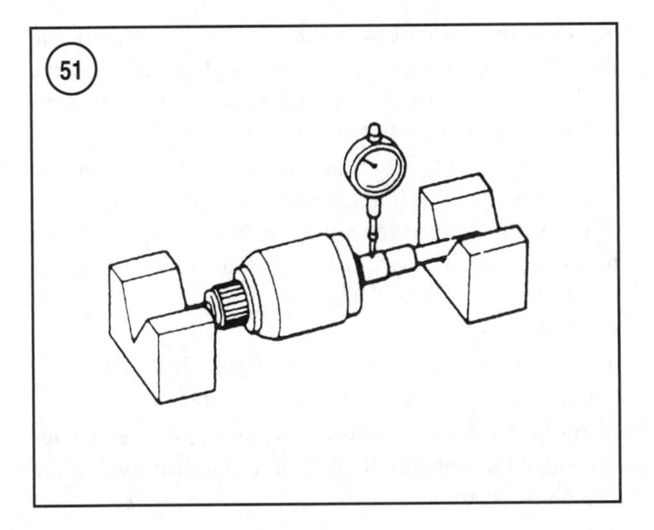

for loose or cracked magnets. Replace the frame assembly if it cannot be adequately cleaned or if the magnets are damaged.

15. Inspect the bushings in the upper and lower covers and the bushing contact surfaces on the armature for discoloration and excessive or uneven wear. Replace the armature if rough or unevenly worn bushing surfaces are present. Replace the cover(s) if the bushing is worn. The bushing is not available separately.

16. Temporarily install the brush plate, brushes, bushing and washers into the bottom cover. Do not install the brush springs. Test the brush plate and bottom cover for a shorted condition as follows:

a. Calibrate a multimeter to the 1 ohm scale.

b. Touch the positive and negative test leads to the brush holders as shown in **Figure 53**. If the meter

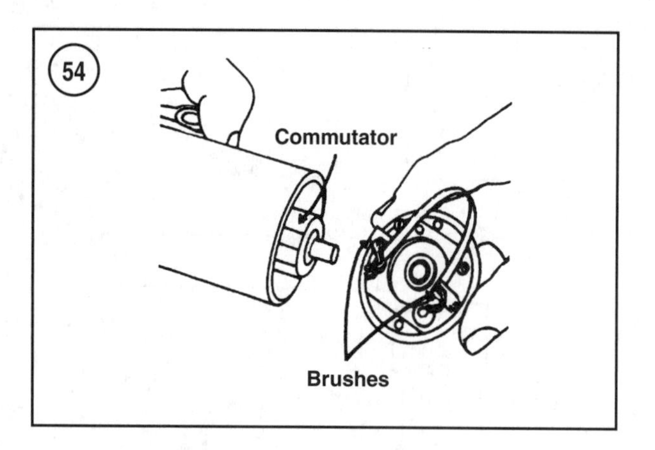

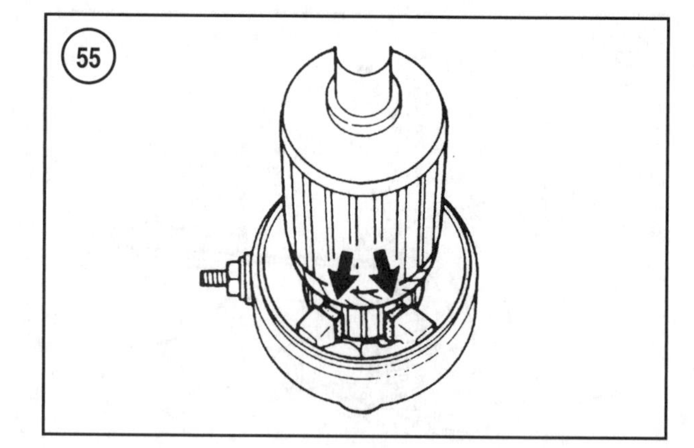

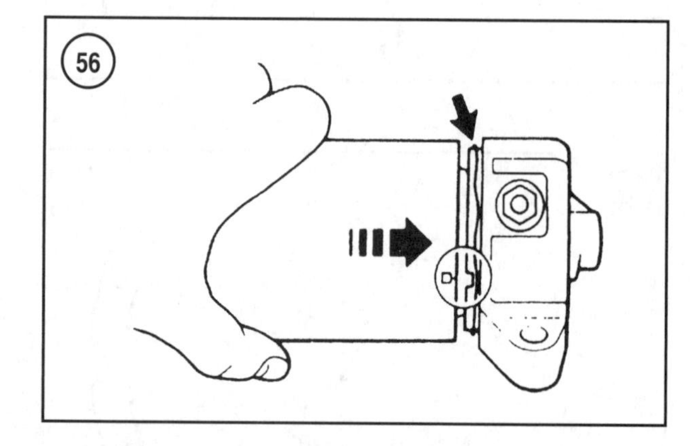

indicates continuity, the brush plate is shorted and must be replaced.

c. Touch the positive meter test lead to the positive brush. The positive brush connects onto the starter terminal post. Touch the negative meter test lead to the bottom cover. If the meter indicates continuity, the brush plate is shorted and must be replaced.

d. Remove the brush plate and related components.

17. *150-250 hp (HPDI models)*—Test the starter solenoid as follows:

a. Connect an ohmmeter between the two large solenoid terminals. The ohmmeter should indicate no continuity.

b. Clamp the solenoid into a vise with soft jaws.

c. Using a jumper wire, connect the black/white solenoid terminal to the positive terminal of a12-volt battery. Observe the ohmmeter while attaching a jumper wire between the negative terminal of the battery and an unpainted surface of the solenoid. Continuity should be present and the solenoid shaft should extend rapidly when the jumper wire is attached. If the solenoid does not perform as described, replace the solenoid.

Assembly (80 Jet and 115-250 hp [except HPDI models])

Apply a light coating of Yamaha All-Purpose Grease to the bushing in the covers and the helical splines on the armature during assembly. Do not allow grease to contact the brushes or the commutator surfaces. Replace grease contaminated brushes.

Refer to **Figure 34** for this procedure.

1. Align the markings and install the brush plate and brushes into the bottom cover. Securely tighten the brush plate screws. Insert the springs then brushes into the brush holders.

2. Apply a light coating of Yamaha All-Purpose Grease into the bottom cover bushing. Install the thrust washer over the lower end of the armature and seat it against the commutator.

3. Fabricate a brush retaining tool by bending a stiff piece of wire into a U shape. Manually push the brushes into the holders and insert the installation tool as shown in **Figure 54**.

4. Guide the armature shaft into the bushing while fitting the bottom cover over the commutator. Carefully remove the brush retaining tool. Inspect the brushes to ensure contact with the commutator contacts (**Figure 55**).

5. Install a new O-ring onto the bottom cover. Seat the O-ring into the step in the cover, then carefully slide the frame assembly over the armature. Maintain pressure on the opposite end of the armature shaft to prevent the armature from pulling from the bottom cover. The magnets are quite strong.

6. Align the reference markings (**Figure 36**) and anti-rotation structures (**Figure 56**) used on some models.

7. Install the stack of washers (8, **Figure 34**) over the upper end of the armature shaft. Seat the washers against the step on the shaft.

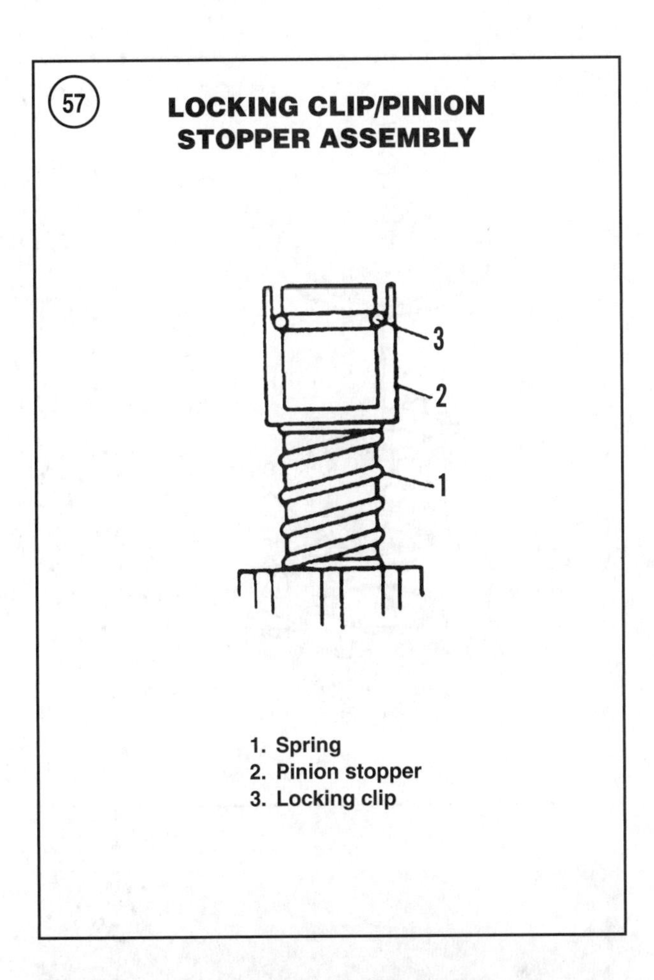

1. Spring
2. Pinion stopper
3. Locking clip

8. Apply a light coating of Yamaha All-Purpose Grease into the bushing in the upper cover. Install a new O-ring onto the upper cover. Seat the O-ring into the step in the cover.

9. Slide the upper cover over the armature shaft. Align the reference markings (**Figure 36**) and seat the cover into the frame assembly.

10. Guide the throughbolts through the bottom cover and frame, then hand-thread them into the upper cover. Tighten the throughbolts evenly to the tightening torque specification in **Table 1**.

11. Apply a light coating of Yamaha All-Purpose Grease to the helical splines in the armature shaft. Thread the pinion drive onto the armature shaft.

12. Install the spring (1, **Figure 57**) and pinion stopper (2) onto the shaft. Push the pinion stopper toward the starter to expose the locking clip groove. Snap the locking clip (3, **Figure 57**) into the groove. Use pliers to form the locking clip if it was distorted during installation.

13. Slowly release the pinion stopper. Inspect the stopper and locking clip for proper installation. The locking clip must be positioned into the pinion stopper opening as shown in **Figure 57**.

14. Install the starter motor as described in this chapter.

Assembly (150-250 hp [HPDI models])

Refer to **Figure 42** for this procedure.

1. Lubricate the entire planetary shaft (18, **Figure 42**) with a light coat of Yamaha All-Purpose Grease. Slide the bushing (17, **Figure 42**) over the planetary shaft (18). Place the gear (16, **Figure 42**), locating ring (15) and cover over the planetary shaft as indicated in **Figure 42**.
2. Install the washer (13, **Figure 42**) over the planetary shaft (18). Carefully snap the E-clip (12, **Figure 42**) into the groove on the planetary shaft (18).
3. Apply a light coat of Yamaha All-Purpose Grease to the upper end of the planetary shaft (18, **Figure 42**). Place the pinion shaft (11, **Figure 42**) onto the planetary shaft.
4. Lubricate the bearing surfaces of the pinion shaft (11, **Figure 42**) then slide the pinion assembly into the starter housing. Align the locating tabs on the locating ring (15, **Figure 42**), cover (14) and gear (16) with the notch in the starter housing (9).
5. Lubricate the three planetary gears (19, **Figure 42**) with Yamaha All-Purpose Grease and mesh them with the larger gear (16). Install the plate (20, **Figure 42**) onto the starter housing (9).
6. Make sure all bushes and springs are in position prior to installing the brush plate (34, **Figure 42**). Hold the brushes fully retracted in the brush plate while sliding the brush plate over the commutator (29, **Figure 42**). Release the brushes only after the brush plate slides into position.
7. Carefully tap the plate (35, **Figure 42**) onto the groove at the lower end of the armature (29). Apply a light coating of Yamaha All-Purpose Grease to the gear teeth and bearing surfaces of the armature.
8. Install the frame (28, **Figure 42**). Hold the armature in the planetary assembly while cautiously sliding the frame assembly over the armature.
9. Rotate the armature (29, **Figure 42**) while inserting the upper end into the planetary assembly. The armature will drop into the opening as the gear teeth on the armature meshes with the planetary gears (19, **Figure 42**).
10. Apply a drop or two of engine oil to the bushing in the bottom cover. Align the screw opening in the brush plate (34, **Figure 42**), plate (35) and bottom cover (38) while carefully sliding the bottom cover over the brush plate. Fit the insulator portion of the positive brush (31, **Figure 42**) into the groove in the bottom cover (38).
11. Install both screws (39, **Figure 42**). Securely tighten the screws. Align all reference marks (**Figure 36**) and aligning structures (**Figure 56**). Then install both

throughbolts (40, **Figure 42**). Tighten the throughbolts to the specification in **Table 1**.

12. Slide the spring (4, **Figure 42**) and starter pinion (3) onto the planetary shaft. Place the stopper (2, **Figure 42**) onto the planetary shaft with the open end facing upward.
13. Push the stopper down until the locking clip (1, **Figure 42**) groove is exposed. Insert the clip into the groove. Then pull the stopper up and over the locking clip. The stopper must cover the sides of the clip.
14. Install the starter solenoid as described in this section.

Solenoid installation (150-250 hp [HPDI models])

Refer to **Figure 41** for this procedure.

1. Align the screw and spring holes in the cover (7, **Figure 41**) with the matching holes in the solenoid (9). Install and securely tighten both screws (6, **Figure 41**). Install both ends of the spring into the cover and solenoid.
2. Install the lever (4, **Figure 41**) into the electric starter motor housing (2). Engage the lever and the pinion shaft (3, **Figure 41**) and orient the lever with the spring notch facing outward.
3. Position the rubber block (8, **Figure 41**) within the starter solenoid opening. Carefully insert the solenoid into the opening in the electric starter motor housing. Make sure the spring (5, **Figure 41**) contacts the notch and the tip of the lever passes through the opening in the solenoid shaft.
4. Hold the solenoid in position and install both screws (1, **Figure 41**). Tighten both bolts evenly to the standard tightening torque specification in Chapter One. Connect the short cable attached to the starter motor to the starter solenoid terminal. Securely tighten the terminal nut.
5. Install the starter motor following the instructions in this chapter.

CHARGING SYSTEM

This section describes removal and installation instructions for the battery charge/ignition charge coil and rectifier/regulator.

Battery Charge Coil (Stator)

The battery charge and ignition charge coils are integrated into a single component (**Figure 58**) that is commonly referred to as the stator. The stator mounts under the flywheel along with the pulser coil assembly. If either coil fails replace the stator. The flywheel must be removed to access the stator. Refer to Chapter Seven for flywheel removal and installation procedures.

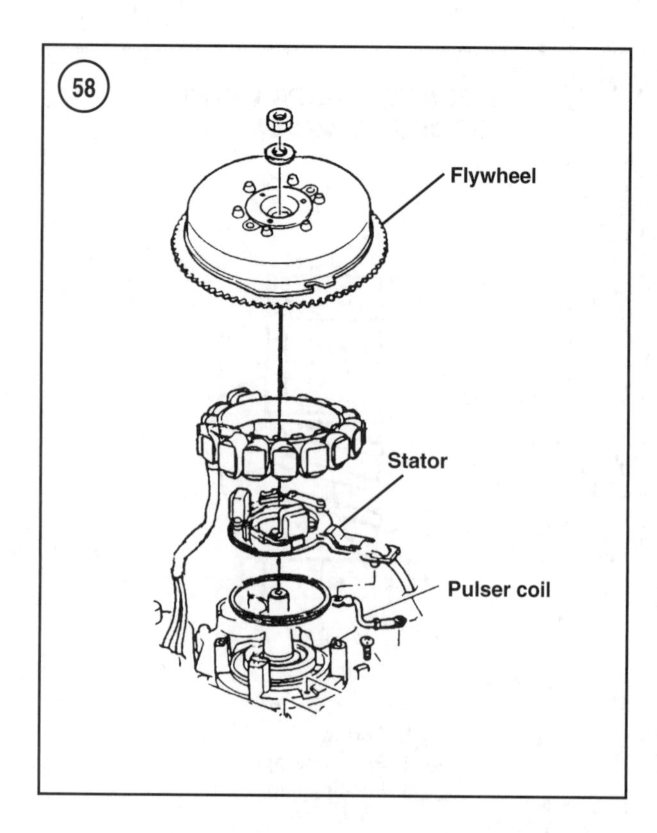

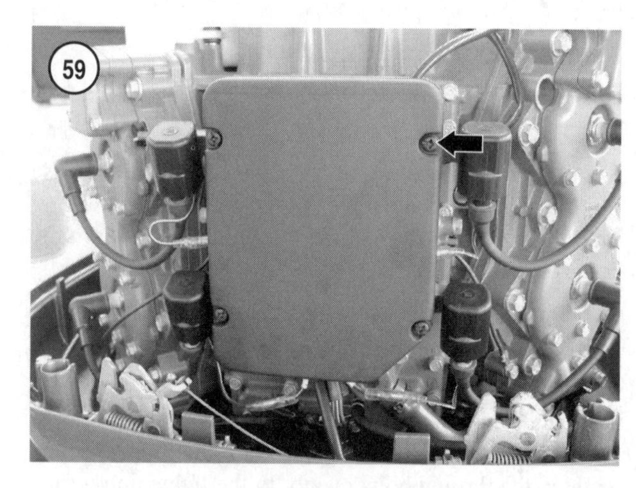

Prior to removing the stator, make a drawing or take a photograph of the wire routing and stator mounting location. This step helps ensure proper wire routing and charge coil positioning during installation.

CAUTION

On many models, the stator mounting bolt openings allow stator installation with the stator oriented in several directions. However, this can cause ignition system malfunction, as the stator may be out of phase with ignition system charging cycles. Al-

ways make reference markings on the stator to a fixed surface on the cylinder block, and note the wire harness orientation prior to removing the stator. Align the reference markings and align the wire harness as noted prior to removal during installation.

Removal

Refer to the wiring diagrams at the end of the manual to identify the wire colors of the battery charge/lighting coil. Trace the wires from to the coil mounting location under the flywheel.

1. Disconnect the battery cables and ground the spark plug leads to prevent accidental starting.
2. Remove the flywheel as described in Chapter Seven.
3. *80 Jet, 105 Jet, 115 hp, 130 hp and C150 models*—Remove the CDI unit cover (**Figure 59**) from the rear of the power head.
4. Remove the electrical component cover (**Figure 60**) from the starboard side of the power head.
5. Mark the wire terminals and take note of the wire routing and connection points, then disconnect the battery charge/ignition charge coil wire harness from the rectifier/regulator and CDI unit or engine control unit.
6. Make reference markings on the battery charge/lighting coil(s) and the mounting base or power head. This step helps ensure proper orientation of the coil on assembly.

CAUTION
It may be necessary to use an impact driver to loosen the stator mounting bolts. Work carefully and do not use excessive force. The stator mounting base on the power head can sustain considerable damage if excessive force is used.

7A. *150-250 hp models*—Remove the four bolts, then lift the stator retaining plate from the stator.
7B. *80 Jet and 115-130 hp models*—Remove the three bolts (**Figure 61**) from the stator.
8. Mark each bolt to indicate the mounting location upon removal. On some models the mounting bolts are different lengths. Route the wire harnesses through any openings while lifting the stator from the mounting base. If necessary, carefully tap on the stator laminations with a plastic hammer to free the coil. Do not tap on the coil windings or pry the stator loose. The coils are easily damaged if the pry bar contacts the windings.
9. Clean the coil mounting location of corrosion, and other debris or contaminants. Clean, inspect and repair, if necessary, the threads for the coil mounting screws.

Installation

CAUTION
The stator must be installed with the correct orientation for proper phasing of the ignition system. Always align the reference markings on the stator and power head during coil installation. Poor ignition system performance can result if the engine operates with an out of phase ignition charge coil.

1. If replacing the stator, transfer the reference marking onto the replacement stator.
2. Carefully route the wire harness through any openings while installing the stator onto the power head.
3. Align the reference marks and seat the stator onto the mounting bosses on the power head.
4. *150-250 hp models*—Fit the stator retaining plate onto the stator. The flat side of the retaining plate must contact the stator. Rotate the retaining plate to align the mounting bolt opening.

5. Apply a light coating of Loctite 271 onto the threads, then thread the mounting bolt into the retaining plate (if so equipped), the stator and the mounting base openings. Verify proper seating and reference marking alignment. Then tighten the screws to the standard tightening torque specification in Chapter One.
6. Connect the battery charge coil wire harness onto the rectifier/regulator connections.
7. Connect the ignition charge coil wire harness onto the CDI unit or engine control unit connections.
8. Route the wires to avoid interference with moving components. Secure the wiring with plastic locking type clamps as necessary.
9. Install the flywheel as described in Chapter Eight. Manually rotate the flywheel while checking for interference with the stator coils. Remove the flywheel and check the installation if there is binding or rubbing. Improper seating or debris on the stator mounting base can tilt the stator and allow contact with the rotating flywheel.
10. *80 Jet, 105 Jet, 115 hp, 130 hp and C150 models*—Install the CDI unit cover (**Figure 59**) onto the rear of the power head.
11. Install the electrical component cover (**Figure 60**) onto the starboard side of the power head.
12. Connect the battery cables and spark plug leads.

Rectifier/Regulator

All of the Yamaha outboards covered in this manual have a rectifier/regulator (**Figure 62**, typical). The rectifier/regulator contains diodes that convert the alternating current (AC) produced by the battery charge coil to direct current (DC) for charging the battery. The rectifier/regulator contains additional circuits that limit current flowing into the battery when it reaches full charge. Refer to the wiring diagrams located at the end of the manual to identify the wire colors connected to the rectifier/regulator prior to removal.

Removal and installation instructions for either components are similar. Refer to the following information and the wiring diagrams to help locate the rectifier regulator.

80 Jet, 105 Jet, 115 hp, 130 hp and C150 models—The air cooled rectifier/regulator (**Figure 62**) mounts onto the electrical component mounting plate on the rear starboard side of the power head.

150-250 hp models (except C150)—The water cooled rectifier/regulator (2, **Figure 63**) mounts onto the inner side of the electrical component mounting plate. The mounting plate is located on the starboard side of the power head. Clamps secure the cooling water hoses onto the cooling block (9, **Figure 63**). The cooling block transfers heat produced by the rectifier/regulator to the cooling water.

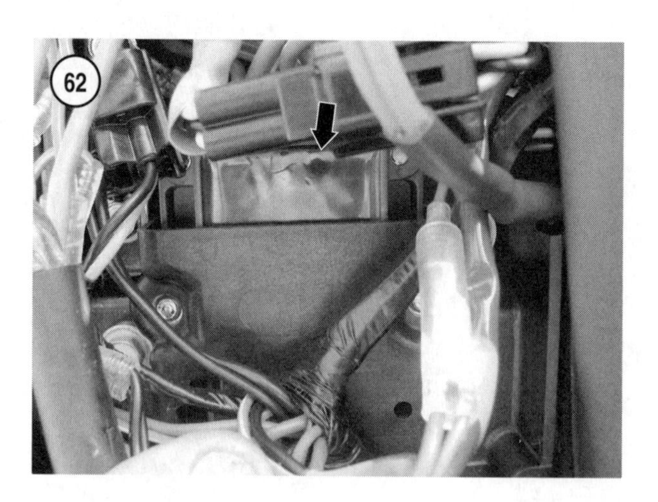

Removal (80 Jet, 105 Jet, 115 hp, 130 hp and C150 models)

1. Disconnect the battery cables and ground the spark plug leads to prevent accidental starting.
2. Remove the electrical component cover (**Figure 60**) from the starboard side of the power head.
3. Disconnect the rectifier/regulator harness connectors from the battery charge coil wires of the stator. Disconnect the red wire connector from the engine wire harness connector or starter relay terminal.
4. Remove the mounting screw(s), then pull the rectifier/regulator from the plate. If so equipped, remove the screw to disconnect the ground wire from the component. On some models, the mounting screw secures the ground wire terminal.
5. Clean corrosion and contamination from the rectifier mounting and ground screw openings.
6. Installation is the reverse of removal. Note the following:

Route the wiring to avoid interference with moving components. Clamp the wiring with plastic locking type clamps as necessary.

Removal (150-250 hp [except C150 models])

Refer to **Figure 63** for this procedure.
1. Disconnect the battery cables and ground the spark plug leads to prevent accidental starting.
2. Remove the electrical component cover (**Figure 60**) from the starboard side of the power head.
3. Remove the bolts that retain the electrical component mounting plate onto the power head. Pull the plate away

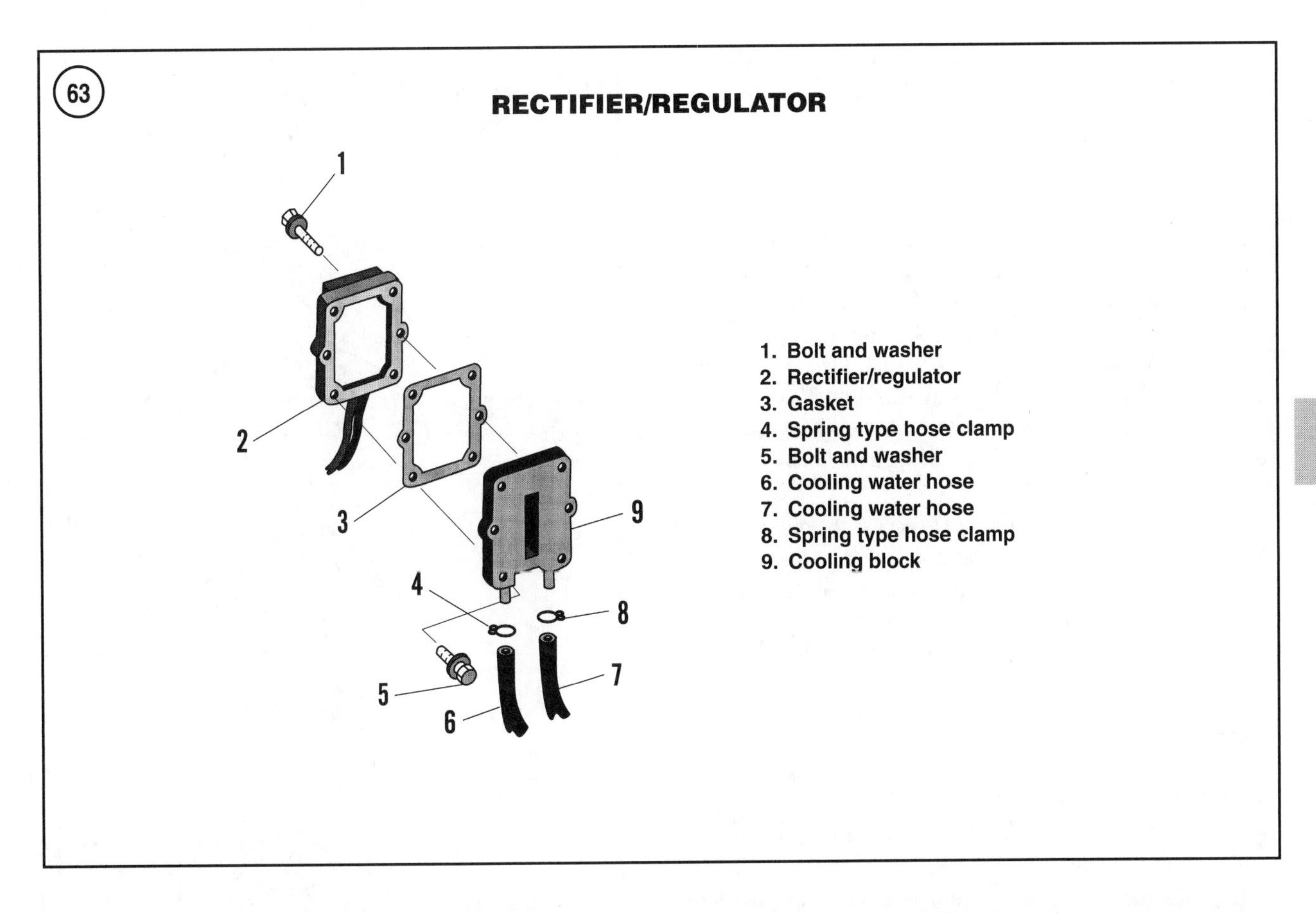

from the power head to access the rectifier/regulator mounting bolts (1, **Figure 63**). Retain the grommets and washers from the bolt openings.

4. Disconnect the rectifier/regulator harness connectors from the battery charge coil wires of the stator. Disconnect the red wire connector from the engine wire harness connector or starter relay terminal.
5. Remove the clamps (4 and 8, **Figure 63**). Then pull the cooling hoses (6 and 7) from the cooling block fittings.
6. Remove the two bolts (1, **Figure 63**), then pull the rectifier/regulator (2) and cooling block (9) from the mounting plate.
7. Remove the four bolts and washers (5, **Figure 63**) then, pull the cooling block (9) from the rectifier/regulator. Remove and discard the gasket (3, **Figure 63**).
8. Clean corrosion and contamination from the rectifier mounting and ground screw openings.
9. Installation is the reverse of removal. Note the following:

Refer to **Figure 63** for this procedure.

a. Fit a new gasket (3, **Figure 63**) onto the rectifier/regulator (2). Do not apply sealant onto the gasket or mating surfaces. Install the cooling block (9, **Figure 63**) onto the rectifier/regulator. Install the four bolts and washers (5, **Figure 63**). Tighten the bolts to the general torque specification in Chapter One.
b. Fit the electrical component mounting plate onto the power head bosses. Make sure all grommets and washers are in position before installing the plate. Check for pinched hoses or wiring, then install the mounting bolts. Make sure that any ground terminals are reconnected onto the mounting bolts. Tighten the mounting bolts to the general tightening torque specification in Chapter One.

IGNITION SYSTEM

This section describes removal and installation procedures for the ignition charge coil, pulser coil, ignition coil, crankshaft position sensor and CDI unit or engine control unit.

Ignition Charge Coil

The ignition charge coil and battery charge coil are integrated into a single component commonly referred to as the stator. If either the battery charge or ignition charge coil should fail, replace the stator. Refer to *Battery Charge Coil* in this chapter for removal and installation instructions.

Pulser Coil

The pulser coil assembly (2, **Figure 64**) is located under the flywheel along with the stator. The assembly consists of coil windings incorporated into the ring shaped assembly. Two coil windings are used on 80 Jet, 115 hp and 130 hp models. Three coil windings are used on 105 Jet and 150-250 hp models. A bearing (1, **Figure 64**) is used to support the pulser coil on models using a moveable pulser coil.

The flywheel must be removed to access the pulser coil. The pulser coil appearance and connected wiring varies by model. Refer to the wiring diagrams at the end of the manual to identify the wire colors for the pulser coil. Trace the wire colors to identify the pulser coil.

Removal

Refer to **Figure 64** for this procedure.

1. Remove the battery charge/ignition charge coil as described in this chapter.
2. Remove the electrical component covers (**Figure 59** and **Figure 60**) to access the pulser coil harness connectors. Carefully disconnect the pulser coil harness connectors from the CDI unit or engine control unit connectors.
3. Note the wire harness routing and make reference markings on the pulser coil and corresponding power head surfaces prior to removal. If replacing the pulser coil, transfer the reference markings onto the replacement component.

4A. *80 Jet, 115 hp, 130 hp and C150 models*—Carefully unsnap the throttle arm linkage from the pulser coil connection point (4, **Figure 64**).

4B. *150-250 hp (except C150 models)*—Remove the three screws that retain the pulser coil to the power head.

5. Carefully guide the wire harness through any openings while pulling the pulser coil from the power head.
6. *80 Jet, 115 hp, 130 hp and C150 models*—Remove the retainers, then lift the bearing (1, **Figure 64**) from the power head. Inspect the bearing for worn, corroded or damaged surfaces. Replace the bearing if not found in excellent condition.
7. Clean any corrosion, oily residue or other contamination from the pulser coil mounting surface.
8. Clean and inspect the mounting screw openings. Repair corroded or damaged threads as necessary.

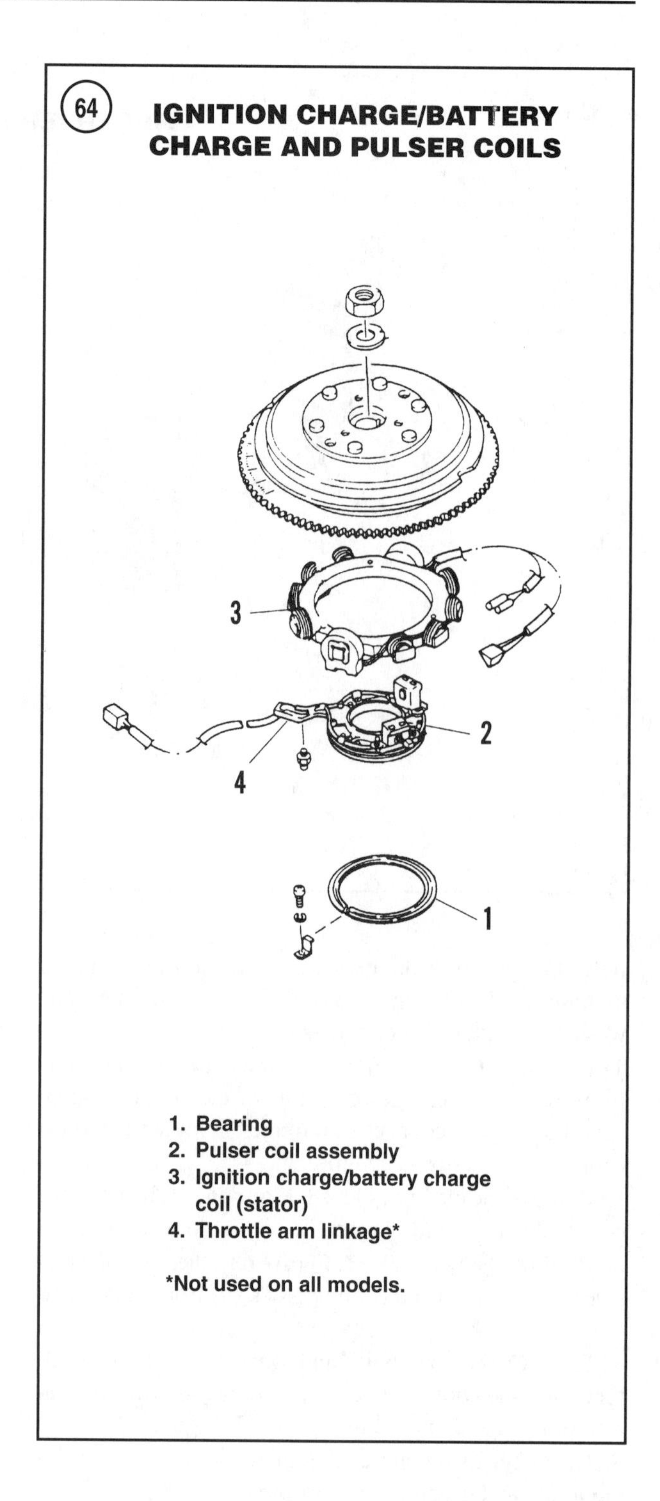

1. Bearing
2. Pulser coil assembly
3. Ignition charge/battery charge coil (stator)
4. Throttle arm linkage*

*Not used on all models.

Installation

Refer to **Figure 64** for this procedure.

1. *80 Jet, 115 hp, 130 hp and C150 models*—Install the bearing (1, **Figure 64**) onto the power head. Center the bearing opening with the crankshaft. Install the retainers. Then securely tighten the retainer screws. Apply a light

coating of Yamaha All-Purpose Grease onto the bearing surfaces that contact the pulser coil.

2. Guide the wire harness through any openings while installing the pulser coil onto the power head.

3A. *80 Jet, 115 hp, 130 hp and C150 models*—Rest the pulser coil onto the bearing (1, **Figure 64**). Rotate the pulser coil on the mount to align the linkage connection point with the throttle arm linkage (4, **Figure 64**). Carefully snap the throttle arm linkage onto the pulser coil.

3B. *150-250 hp (except C150 models)*—Rest the pulser coil on the power head. Rotate the pulser coil to align the mounting screw openings and the reference markings made prior to removal. Install and securely tighten the three mounting screws.

4. Carefully connect the pulser coil harness connection onto the CDI unit or engine control unit connectors. Route the wiring to prevent interference with moving components. Secure the wiring with plastic locking type clamps as necessary. On models with a moveable pulser coil, do not clamp the pulser coil harness in a manner that prevents the coil from rotating on the bearing.

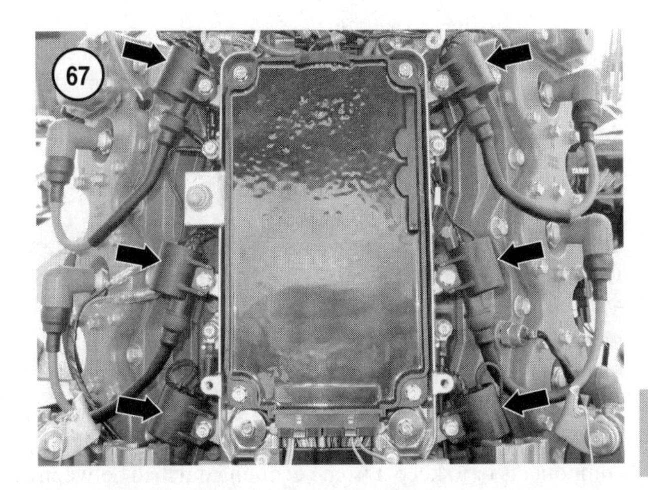

5. Install any electrical component covers (**Figure 59** and **Figure 60**) that were removed to access the pulser coil harness.

6. Install the battery charge/ignition charge coil as described in this chapter.

7. Perform all applicable ignition timing adjustments as described in Chapter Four.

Ignition Coil Removal/Installation

Ignition coil appearances and mounting locations vary by model. On all models covered in this manual, the ignition coils are mounted at the rear of the power head. Refer to the following to locate the coils.

80 Jet, 115 hp, 130 hp and 150-200 hp models (2.6 liter [except HPDI models])—The ignition coils (**Figure 65**) are mounted on each side of the CDI unit mounting plate.

150-250 hp (HPDI models)—Two ignition coils are mounted on each side of the engine control unit mounting plate. Two additional coils are mounted on top of the mounting plate. See **Figure 66**.

200-250 hp (3.1 liter) models—Three ignition coils are mounted on each side of the engine control unit mounting plate. The electrical component cover on the rear of the power head (**Figure 59**) must be removed to access the coils. See **Figure 67**.

Refer to the wiring diagrams at the end of the manual to identify the wire colors connected to the ignition coil(s). The wire color indicates the cylinder number for the ignition coil.

1. Disconnect the battery cables and ground the spark plug leads to prevent accidental starting.

2. Remove the electrical component cover (**Figure 59**) from the rear of the power head.

3. Disconnect the ignition coil harness from the engine wire harness connection and the spark plug lead from the coil.
4. Remove the coil mounting bolt and lift the coil from the mounting plate. Clean any corrosion, oily residue or other contaminants from the coil mount.
5. Install the replacement coil onto the mounting plate. Install and securely tighten the coil mounting fasteners. Make sure the mounting bolt passes through the coil ground lead before installing the bolt.
6. Connect the ignition coil wire harness connector onto the engine wire harness connector.
7. Route the wiring to prevent interference with moving components or to keep it from becoming pinched between the cover (**Figure 59**) and the mounting plate. Secure the wiring with plastic locking type clamps as necessary.
8. Install the electrical component cover onto the rear of the power head. Do not pinch the coil wires between the cover and the mounting plate.
9. Connect the battery cables and spark plug leads.

68

69

Crankshaft Position Sensor Removal/Installation

The crankshaft position sensor (**Figure 68**) is located in line with and next to the flywheel ring gear teeth. Air gap adjustment is not required. However, the sensor must not contact the flywheel at any point. Check for an incorrectly installed flywheel or sensor if contact occurs.

Refer to the wiring diagrams at the back of the manual to identify the wire color for the crankshaft position sensor. Trace the wires to the sensor.

1. Disconnect the battery cables and ground the spark plug leads to prevent accidental starting.
2. Remove the flywheel cover (**Figure 69**, typical) to access the crankshaft position sensor mounting screws.
3. Disconnect the crankshaft position sensor harness from the engine wire harness connector. Remove any plastic locking type clamps from the sensor harness.
4. Remove the mounting screws and washers, then lift the sensor from the mounting bosses. Carefully guide the sensor wire harness through any opening while removing the sensor.
5. Clean corrosion, oily residue or other contaminants from the mounting bosses and mounting screw openings.
6. Install the replacement sensor to the mounting bosses as shown in **Figure 68**. Install and securely tighten the two mounting screws and washers. Manually rotate the flywheel while checking for adequate clearance between the sensor and the flywheel ring gear teeth. Check for incorrect sensor installation if contact occurs. Check for improper flywheel installation if contact occurs and the sensor is mounted correctly.
7. Guide the sensor wire harness through any openings while routing the wire to the engine harness connector. Plug the sensor harness into the engine harness. Route the wiring to prevent interference with moving components. Secure the wiring with plastic locking type clamps as necessary.
8. Install the flywheel cover.
9. Connect the battery cables and spark plug leads.

CDI Unit and Engine Control Unit Removal/Installation

The CDI unit or engine control unit (**Figure 70**, typical) is located on the mounting plate at the rear of the power head.

1. Disconnect the battery cables and ground the spark plug leads to prevent accidental starting.
2. Remove the electrical component cover (**Figure 59**) to access the component.

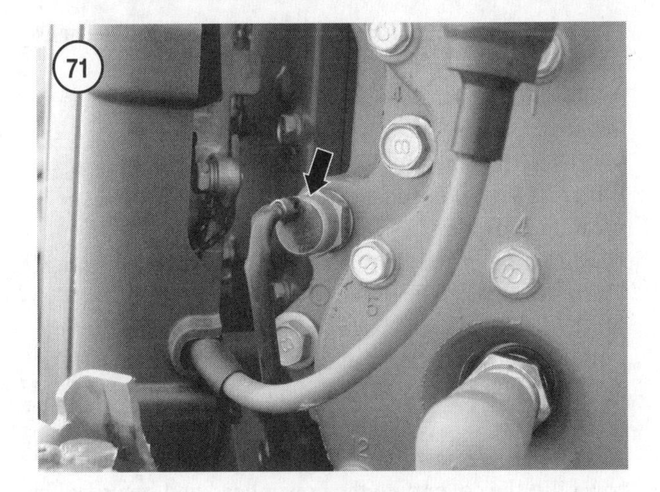

3. Mark or make notes indicating each wire routing and connection point, then carefully disconnect all wire connectors from the CDI unit or engine control unit.

4. Inspect all connectors for bent, corroded or damaged contacts. Correct faulty connections before considering replacing the unit. Most faults are the result of poor connections rather than a fault with the unit.

5. Remove the mounting fasteners, then lift the CDI unit or engine control unit from the mounting plate. Note any ground wires connected to the mounting bolts. They must be connected onto the mounting bolts during installation.

6. Clean corrosion, oily residue or other contaminants from the mounting plate and threaded openings for the mounting bolts.

7. Fit the replacement unit onto the mounting plate. Connect any ground wires to the mounting bolts. Then thread them into the unit and mounting plate. Securely tighten the mounting bolts (general torque specification in Chapter One).

8. Carefully connect all wire connectors onto the CDI unit or engine control unit wire connectors. Route the wiring to prevent interference with moving components or becoming pinched between the cover and the mounting plate. Secure the wiring with plastic locking type clamps as necessary.
9. Connect the battery cables and spark plug leads.
10. Perform all applicable ignition timing adjustments as described in Chapter Four.

SENSORS, SWITCHES AND RELAYS

Engine Temperature Sensor Removal/Installation

This component is used on all EFI and HPDI models. The appearance and mounting locations of the engine temperature sensor (**Figure 71**) varies by model. Proper identification is important as the engine temperature sensor appearance is almost identical to the overheat switch. Refer to the following information to find the sensor.

150-200 hp (2.6 liter [except HPDI models])—The sensor (**Figure 71**) mounts into the starboard side cylinder head near the No. 5 spark plug opening.

150-250 hp (HPDI models)—The sensor (**Figure 71**) mounts into the port side cylinder head near the No. 6 spark plug opening.

200-250 hp models (3.1 liter)—The sensor mounts into the top of the cylinder block (**Figure 72**) and in line with the starboard cylinder bank.

Refer to the wire diagrams at the end of the manual to identify the engine temperature sensor wires. Trace the wires to the correct component on the engine.

NOTE
Two black/yellow wires lead into the engine temperature sensor. A pink and black wire lead into the overheat switch. Also note that the engine temperature sensor threads into the mount opening. A clamp plate and screw

secure the overheat switch into the mount opening.

1. Disconnect the battery cables and ground the spark plug leads to prevent accidental starting.
2. Disconnect the sensor wire harness connector from the engine harness or CDI unit connection.
3. Carefully unthread the sensor from the cylinder head opening. Then remove the sensor and sealing washer. Discard the washer.
4. Use the proper thread chase or tap to remove all corrosion, sealant or other contaminants from the threaded opening.
5. Install a new sealing washer onto the sensor. Use a crowfoot adapter to tighten the sensor to the specification in **Table 1**.
6. Carefully connect the engine temperature sensor harness connector onto the engine wire harness or CDI unit connector. Route the wiring to prevent interference with moving components. Secure the wiring with plastic locking type clamps as necessary.
7. Connect the battery cables and spark plug leads.

Oxygen Density Sensor Removal/Installation

This component is used on all EFI and HPDI models. The oxygen density sensor (**Figure 73**) mounts under the sensor cover on the upper starboard side of the cylinder block. A fault with the sensor will cause poor high speed performance and a rough running condition.

Refer to **Figure 74** for this procedure.

1. Disconnect the battery cables and ground the spark plug leads to prevent accidental starting.
2. Remove the plastic locking type clamp that retains the sensor wire harness onto the rubber boot (1, **Figure 74**). Pull the rubber boot from the cover (3, **Figure 74**).
3. Trace the sensor wire harness to the connection in the engine wire harness. Remove the plastic locking type clamp that retains the sensor harness to the engine temperature sensor harness.
4. Remove the two screws (2, **Figure 74**), then pull the cover (3) away from the power head. Pull the sensor harness through the slot provided in the cover.
5. Remove the three bolts (4, **Figure 74**), then pull the mounting block from the power head. Remove and discard the gasket (6, **Figure 74**).
6. Pull the metal draw tube from the cylinder block opening. If at all possible, replace the draw tube. The replacement draw tube has been changed to a design that increases the service life of the sensor. If the draw tube must be reused, use an aerosol type carburetor to clean all carbon deposits and oily residue from the tube.

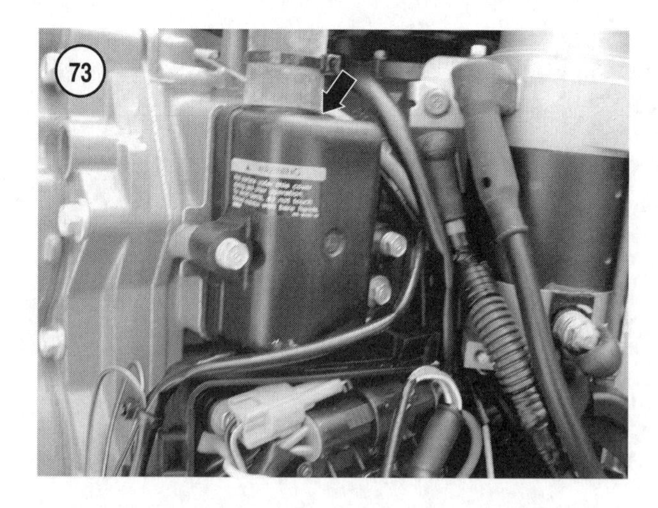

7. Loosen the nut (5, **Figure 74**). Then unthread the sensor from the mounting block.
8. Thread the nut onto the replacement sensor, then thread the sensor fully into the mounting block.
9. Clamp the mounting block into a vise with protective jaws. Use a crowfoot adapter to tighten the nut to the specification in **Table 1**.
10. Insert the metal draw tube into the exhaust opening. The draw tube can be installed in either direction.
11. Install a new gasket onto the mounting block, then install the mounting block onto the power head. Install the three bolts (4, **Figure 74**) and tighten to the specification in **Table 1**.
12. Fit the sensor wire through the slot provided in the cover. Fit the cover onto the power head and secure with the two bolts (2, **Figure 74**). Tighten the cover bolts to the specification in **Table 1**.
13. Carefully connect the sensor harness connector onto the engine wire harness. Secure the harness onto the engine temperature sensor with a plastic locking type clamp.
14. Push the boot (1, **Figure 74**) over the cover. Secure the wire to the boot and the boot to the cover with a plastic locking type clamp.
15. Connect the battery cables and spark plug leads.

Water In Fuel Sensor Removal/Installation

This component is used on 150-250 hp HPDI models. The sensor (**Figure 75**) is integrated into the fuel filter canister. Replace the canister if the sensor is faulty.

WARNING

Use caution when working with the fuel system to avoid damage to property, potential injury or death. Never smoke around fuel or fuel vapor. Make sure no flame or ignition

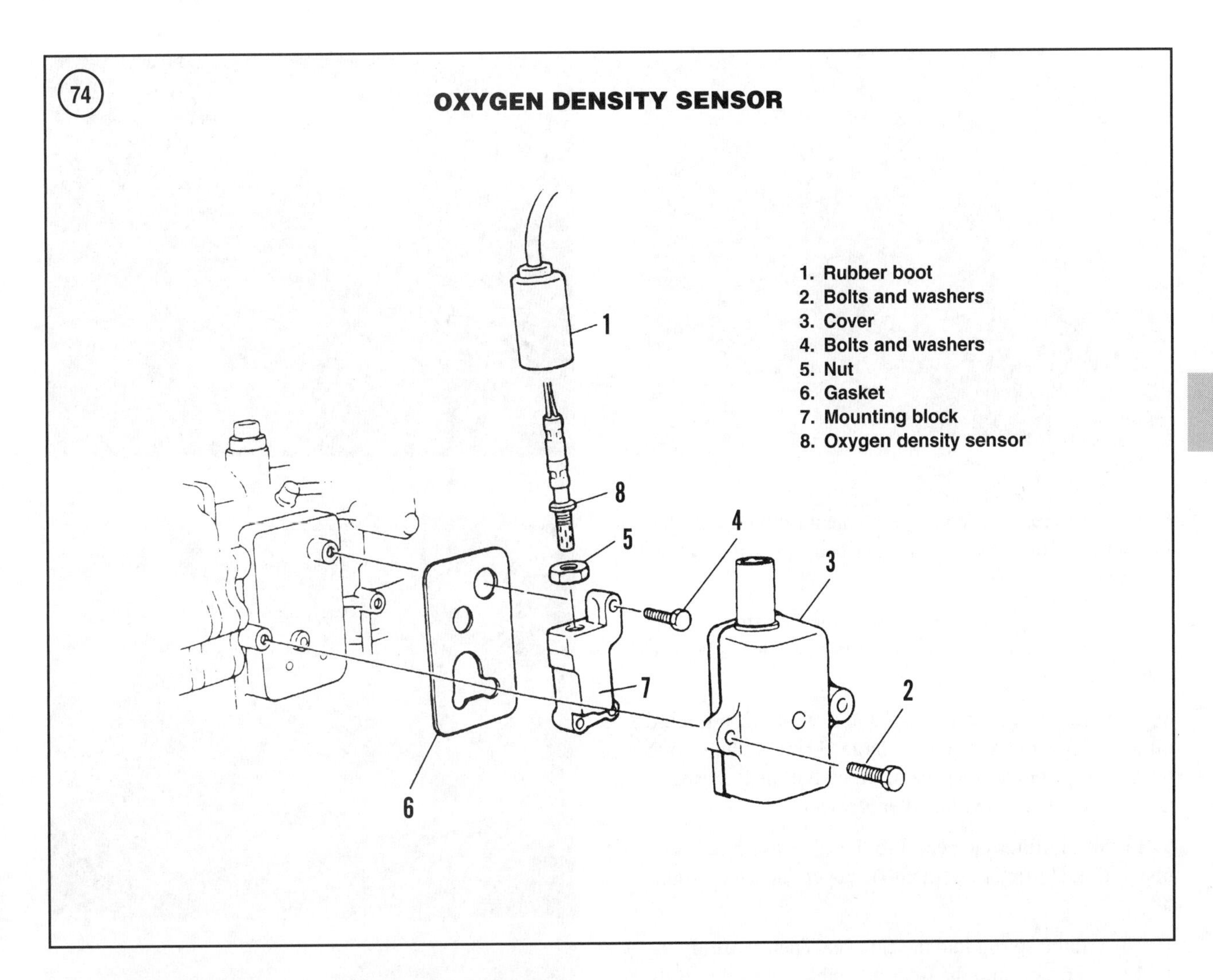

source is present in the work area. Flame or sparks can ignite fuel or vapor causing a fire or an explosion.

WARNING

Fuel leakage can lead to a fire or an explosion causing injury, death or destruction of property. Always check for and correct fuel leaks after any repair made to the fuel system.

1. Disconnect the battery cables and ground the spark plug leads to prevent accidental starting.

2. Trace the sensor harness to connection in the engine wire harness. Note the wire routing and connection points, then disconnect the sensor harness from the engine wire harness.

3. Route the sensor wire through any openings and remove any clamps that interfere with removal of the canister.

4. Place shop towels and a suitable container under the canister to capture spilled fuel.

5. Loosen the ring nut located at the top of the canister, then pull the sensor from the fuel filter assembly.

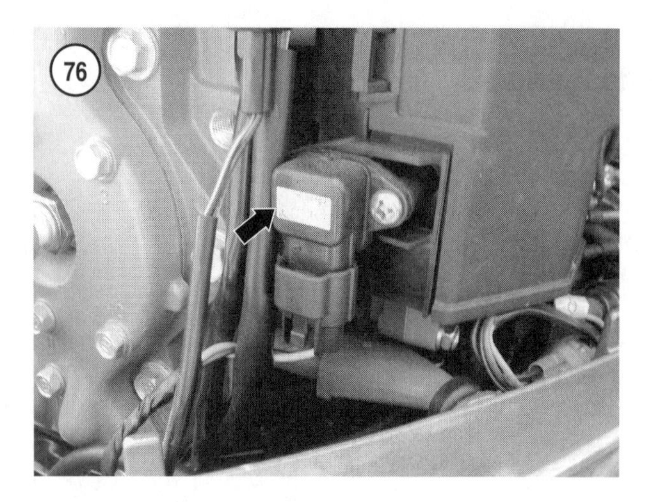

6. Remove the spring and filter element from the canister. Inspect the filter element as described in Chapter Three and replace it as necessary.

7. Remove the large O-ring from the canister and the small O-ring from the top of the filter or filter assembly. Discard the O-rings.

8. Apply a light coating of Yamalube two-cycle outboard oil to the new O-rings. Fit the large O-ring in the step at the top of the replacement canister. Fit the small O-ring in the recess in the top of the filter element.

9. Install the filter element into the filter assembly. The opening and O-ring must fit into the opening in the assembly.

10. Install the spring into the canister. Then carefully install the canister onto the filter assembly. Work carefully to avoid dislodging the filter element and O-ring.

11. Verify proper seating of the canister into the filter assembly and proper seating of the large O-ring. Slip the ring nut over the canister and thread it in to the filter assembly. Do not cross thread the ring nut. Hand-tighten the ring nut.

12. Squeeze the primer bulb while inspecting the filter assembly for fuel leakage. Correct any fuel leakage before operating the engine.

13. Route the sensor wire harness to the connection in the engine wire harness. Route the wiring to avoid interference with moving components. Secure the wiring with plastic locking type clamps as necessary. Connect the sensor harness to the engine wire harness.

14. Connect the battery cables and spark plug leads. Start the engine and immediately check for fuel leakage. Stop the engine and correct any leakage before putting the engine into service.

Air Pressure Sensor Removal/Installation

This component is used on all EFI and HPDI models. The sensor (**Figure 76**, typical) mounting location varies by model. Refer to the following to find the sensor.

150-200 hp (2.6 liter [except HPDI]) models—The sensor (**Figure 76**) mounts to the side of the electrical component mounting plate on the starboard side of the power head.

150-250 hp (HPDI) models—The sensor (**Figure 77**) mounts to a bracket secured to the bottom of the vapor separator tank. The vapor separator tank is located on the port side of the power head.

200-250 hp (3.1 liter) models—The sensor (**Figure 78**) mounts to the junction box top. The junction box is located just forward of the electrical component cover on the starboard side of the power head.

Replace the air pressure sensor as follows:

1. Disconnect the battery cables and ground the spark plug leads to prevent accidental starting.

2. Trace the sensor wire harness to the connection in the engine wire harness. Remove the electrical component cover on the starboard side (**Figure 60**) and rear (**Figure 59**) of the power head as necessary. Note the wire routing and connection points, then disconnect the sensor harness from the engine wire harness.
3. Route the sensor wire through any openings and remove any clamps that may interfere with removing of the sensor and harness.
4. Remove the two mounting screws, then lift the sensor from the mount. Clean corrosion, oily deposits or other deposits from the sensor and mounting surface.
5. Install the replacement sensor to the mounting surface. Install the two mounting screws and tighten to the specification in **Table 1**.
6. Route the sensor wire harness to the connection in the engine wire harness. Route the wiring to avoid interference with moving components. Secure the wiring with plastic locking type clamps as necessary. Connect the sensor harness to the engine wire harness. If removed, install

the starboard side and rear electrical component covers (**Figure 59** and **Figure 60**).
7. Connect the battery cables and spark plug leads.

Air Temperature Sensor Removal/Installation

This component is used on all EFI and HPDI models. The sensor (**Figure 79**, typical) mounting location varies by model. Refer to the following to assist with locating the sensor.

150-200 hp (2.6 liter) models [except HPDI]—The sensor (**Figure 79**) mounts into a boss on front edge of the electrical component mounting plate. The mounting plate is located on the starboard side of the power head.

150-250 hp (HPDI) models—The sensor (**Figure 80**) mounts into a bracket secured to the top of the throttle body assembly. The throttle body assembly is located on the front of the power head.

200-250 hp (3.1 liter) models—The sensor (**Figure 81**) mounts into the junction box. The junction box is located forward of the electrical component cover on the starboard side of the power head.

Replace the air temperature sensor as follows:

1. Disconnect the battery cables and ground the spark plug leads to prevent accidental starting.
2. Carefully disconnect the engine wire harness from the sensor connection.
3. Hold the hex section of the sensor with a suitable wrench, then unthread the sensor mounting nut from the back side of the sensor. Carefully pull the sensor from the mount.
   a. *150-250 hp (HPDI) models*—Remove the washer from the sensor. Inspect the grommet in the sensor mount for torn or deteriorated surfaces and replace as needed.

b. *200-250 hp (3.1 liter) models*—Remove the washer from the retaining nut or mount surface.

4. Install the replacement sensor into the mount opening.

a. *150-250 hp (HPDI) models*—Install the washer onto the sensor prior to fitting the washer through the grommet.

b. *200-250 hp (3.1 liter) models*—Fit the washer onto the threaded end of the sensor after inserting the sensor through the mount opening.

5. Thread the mounting nut onto the sensor and hand-tighten.

6. Hold the hex section of the sensor with a suitable wrench. Then tighten the mounting nut to the specification in **Table 1**.

7. Carefully connect the engine wire harness to the sensor connector.

8. Connect the battery cables and spark plug leads.

Knock Sensor Removal/Installation

This component is used on 200-250 hp (3.1 liter) EFI models. The sensor (**Figure 82**) threads into the side of the starboard cylinder head.

CAUTION

Handle the knock sensor carefully. The sensor may suffer damage that is difficult to visually detect if dropped or handled roughly. A damaged sensor may not detect spark knock that leads to serious power head damage.

1. Disconnect the battery cables and ground the spark plug leads to prevent accidental starting.

2. Trace the green sensor wire to the connection in the engine wire harness. Carefully disconnect the sensor wire.

3. Use a suitable wrench to unthread the sensor from the cylinder head. Clean all corrosion, oily residue or other contaminants from the threaded opening. The sensor mounting opening must be clean for proper sensor operation.

4. Thread the replacement sensor into the opening. Do not use sealant on the threads. Use a crowfoot adapter to tighten the knock sensor to the specification in **Table 1**.

5. Connect the green sensor wire onto the engine wire harness connections. Route all wiring to prevent contact with moving components. Secure wiring with plastic locking clamps as necessary.

6. Connect the battery cables and spark plug leads.

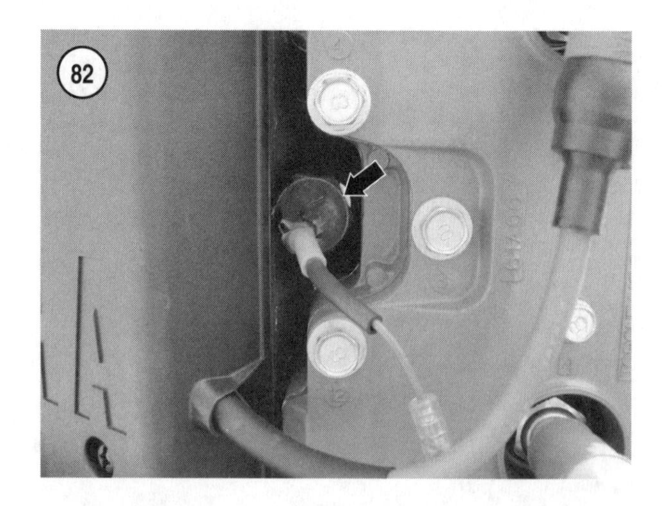

82

83

Throttle Position Sensor Removal/Installaion

This component is used on all EFI and HPDI models. The throttle position sensor (**Figure 83**) mounts to the upper port side of the throttle body. The throttle body is located on the front of the power head.

Refer to **Figure 84** for this procedure.

1. Disconnect the battery cables and ground the spark plug leads to prevent accidental starting.

2. Trace the sensor wire harness to the connection in the engine wire harness. Carefully disconnect the sensor wire.

3. Remove the two sensor mounting screws (6, **Figure 84**). Then carefully pull the sensor from the mounting bracket (10). Wipe all corrosion, oily residue or other contaminants from the mounting bracket and sensor.

4. Align the flat surfaces on the sensor rotor shaft with the slotted opening in the coupling (13, **Figure 84**) while installing the sensor onto the bracket (10). Do not force the sensor into the bracket. Check for proper shaft and cou-

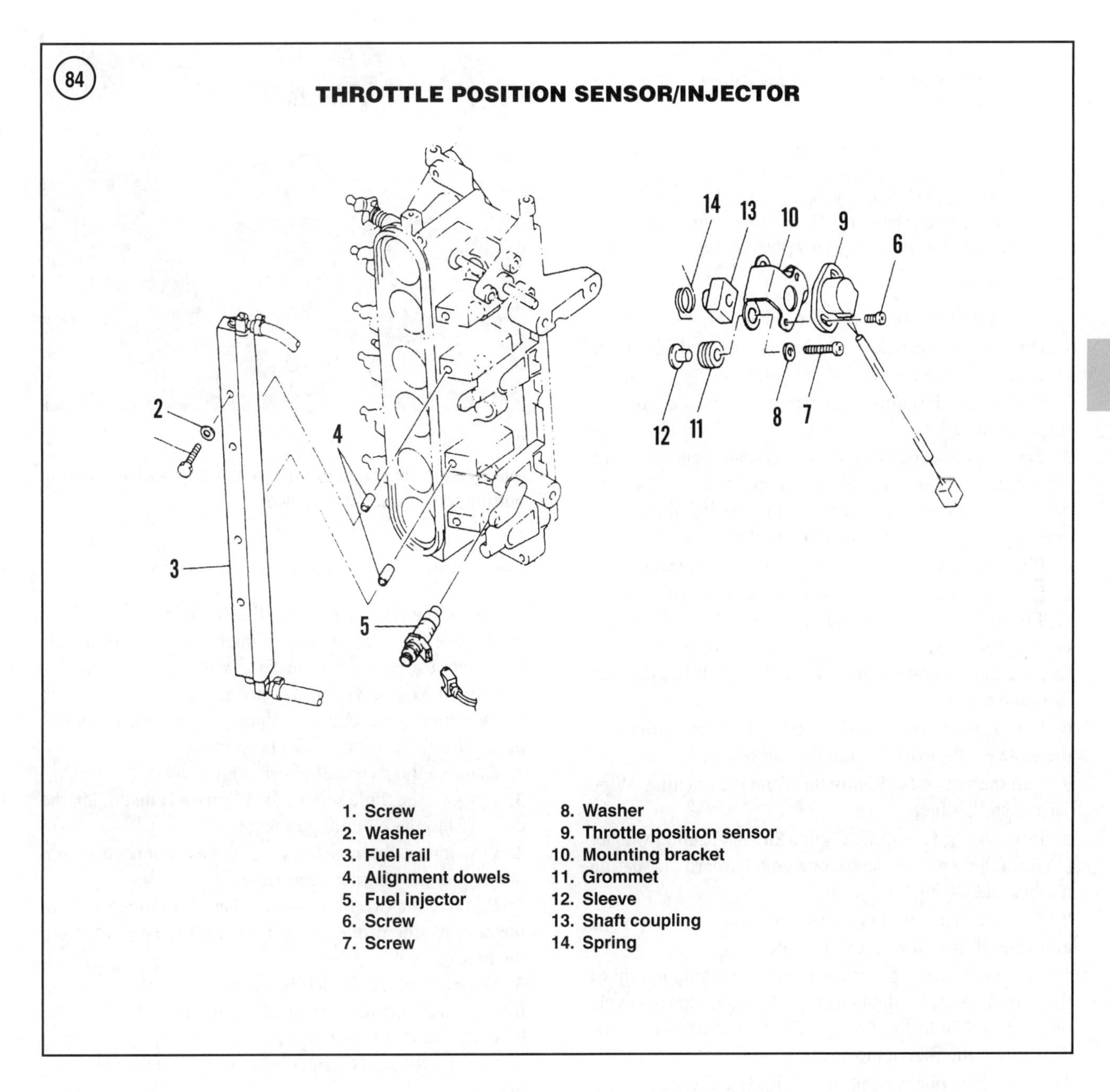

pling alignment if the sensor does not easily contact the bracket.

5. Rotate the sensor enough to align the mounting screw opening. Install the two mounting screws (6, **Figure 84**) and tighten to the specification in **Table 1**.

6. Connect the sensor wire harness connection onto the engine wire harness connection. Route all wiring to prevent contact with moving components. Secure wiring with plastic locking type clamps as necessary.

7. Connect the battery cables and spark plug leads.

8. Adjust the throttle position sensor as described in Chapter Four.

Fuel Pressure Sensor Removal/Installation

This component is used on 150-250 hp HPDI models. A plate and two bolts secure the fuel pressure sensor (**Figure 85**) into an opening in the belt driven high pressure fuel pump.

WARNING

Use caution when working with the fuel system to avoid damage to property, potential injury or death. Never smoke around fuel or fuel vapor. Make sure no flame or source of ignition is present in the work area. Flame or sparks can ignite fuel or fuel vapor causing a fire or an explosion.

WARNING
Fuel leakage can lead to fire or explosion with potential bodily injury, death or destruction of property. Always check for and correct fuel leaks after making any repair to the fuel system.

Refer to **Figure 86** for this procedure.

1. Disconnect the battery cables and ground the spark plug leads to prevent accidental starting.
2. Disconnect the engine wire harness connector from the sensor connection.
3. Thoroughly wipe all debris and contaminants from the sensor mounting area of the fuel pump. Take all necessary precautions to prevent debris from entering the sensor opening. Even small particles can foul the injectors.
4. Place shop towels under the sensor to capture fuel.
5. Slowly and evenly loosen the two retaining plate bolts (1, **Figure 86**). Stop loosening the bolts when fuel begins to leak from the sensor opening. Resume loosening when the internal pressure is relieved and all fuel has drained from the opening.
6. Remove the screws and washers. Then carefully pull the plate (3, **Figure 86**) from the sensor.
7. Pull the sensor (4, **Figure 86**) from the opening. Wipe up any spilled fuel.
8. Remove the O-ring (6, **Figure 86**) and sealing washer (5) from the sensor or sensor opening. Discard the sealing washer and O-ring.
9. Using a clean and lint-free shop towel, carefully wipe all material from the sensor opening.
10. Fit a new sealing washer, then O-ring onto the tip of the sensor. Apply a light coating of Yamalube two-cycle outboard oil onto the O-ring. Then carefully insert the sensor into the fuel pump opening.
11. Install the retaining plate (3, **Figure 86**) onto the sensor. Make sure the opening in the plate seats against the step on the sensor body.
12. Thread the screws and washers (1 and 2, **Figure 86**) through the retaining plate and into the fuel pump. Evenly tighten the bolts to the general tightening torque specification in Chapter One.
13. Connect the engine wire harness connector to the sensor connector. Route the wiring to prevent interference with moving components. Secure the wiring with plastic locking type clamps as needed.
14. Connect the battery cables and spark plug leads. Start the engine and immediately check for fuel leakage. Immediately stop the engine and correct any leakage before putting the engine into service.

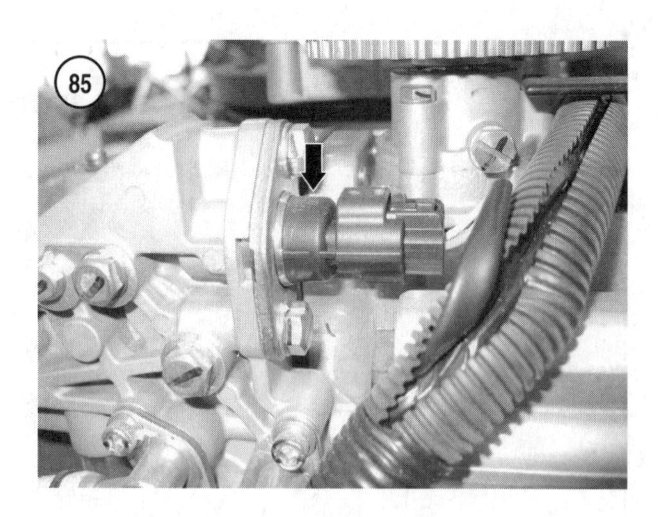

Fuel Pump Resistor Removal/Installation

This component is used on all EFI models. The resistor (5, **Figure 87**) mounts into a bracket above the throttle body and just forward of the flywheel.

Refer to **Figure 87** for this procedure.

1. Disconnect the battery cables and ground the spark plug leads to prevent accidental starting.
2. Remove the flywheel cover (**Figure 69**).
3. Remove the three screws (1, **Figure 87**) that retain the cover (2) to the mounting bracket.
4. Disconnect the resistor wire harness connector from the engine wire harness connector.
5. Remove the two bolts and washer (4, **Figure 87**) from the bottom side of the bracket, then lift the resistor from the bracket.
6. Wipe all corrosion, debris, oil residue or other contaminants from the resistor and mounting surface on the bracket. Debris or contaminants on the mounting surface may hinder the heat dissipating capability of the mounting bracket.
7. Install the replacement resistor into the mounting bracket. Apply a light coating of Loctite 271 onto the threads. Then thread the bolts and washers (4, **Figure 87**) through the mounting bracket and into the resistor. Evenly tighten the two bolts to the general torque specification in Chapter One.
8. Connect the resistor wire harness connector onto the engine wire harness connector. Route the wiring to prevent interference with moving components. Secure the wiring with plastic locking clamps as needed.
9. Install the cover (2, **Figure 87**) onto the mounting bracket. Thread the three screws (1, **Figure 87**) into the

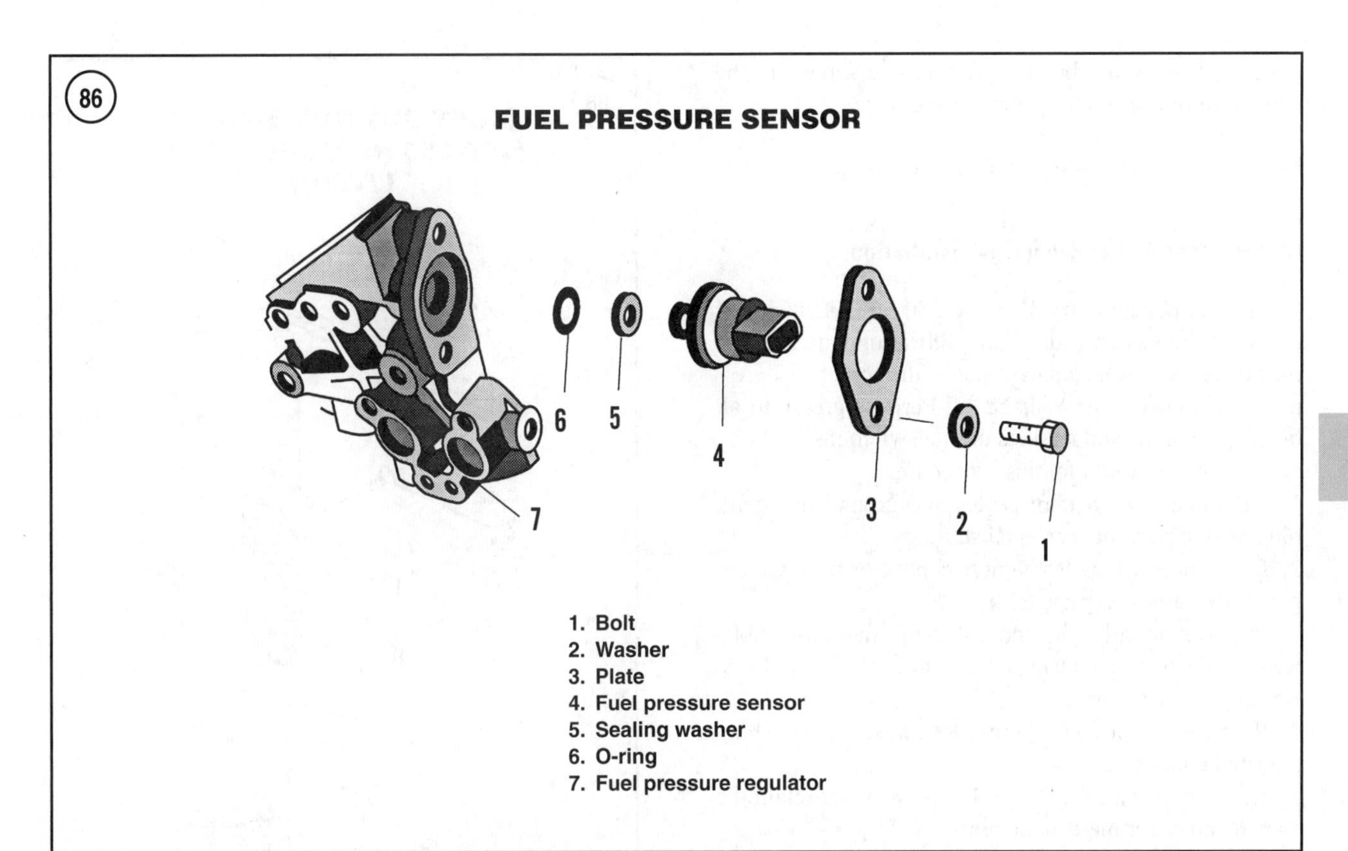
86
FUEL PRESSURE SENSOR
6
5
4
7
3
2
1
1. Bolt
2. Washer
3. Plate
4. Fuel pressure sensor
5. Sealing washer
6. O-ring
7. Fuel pressure regulator

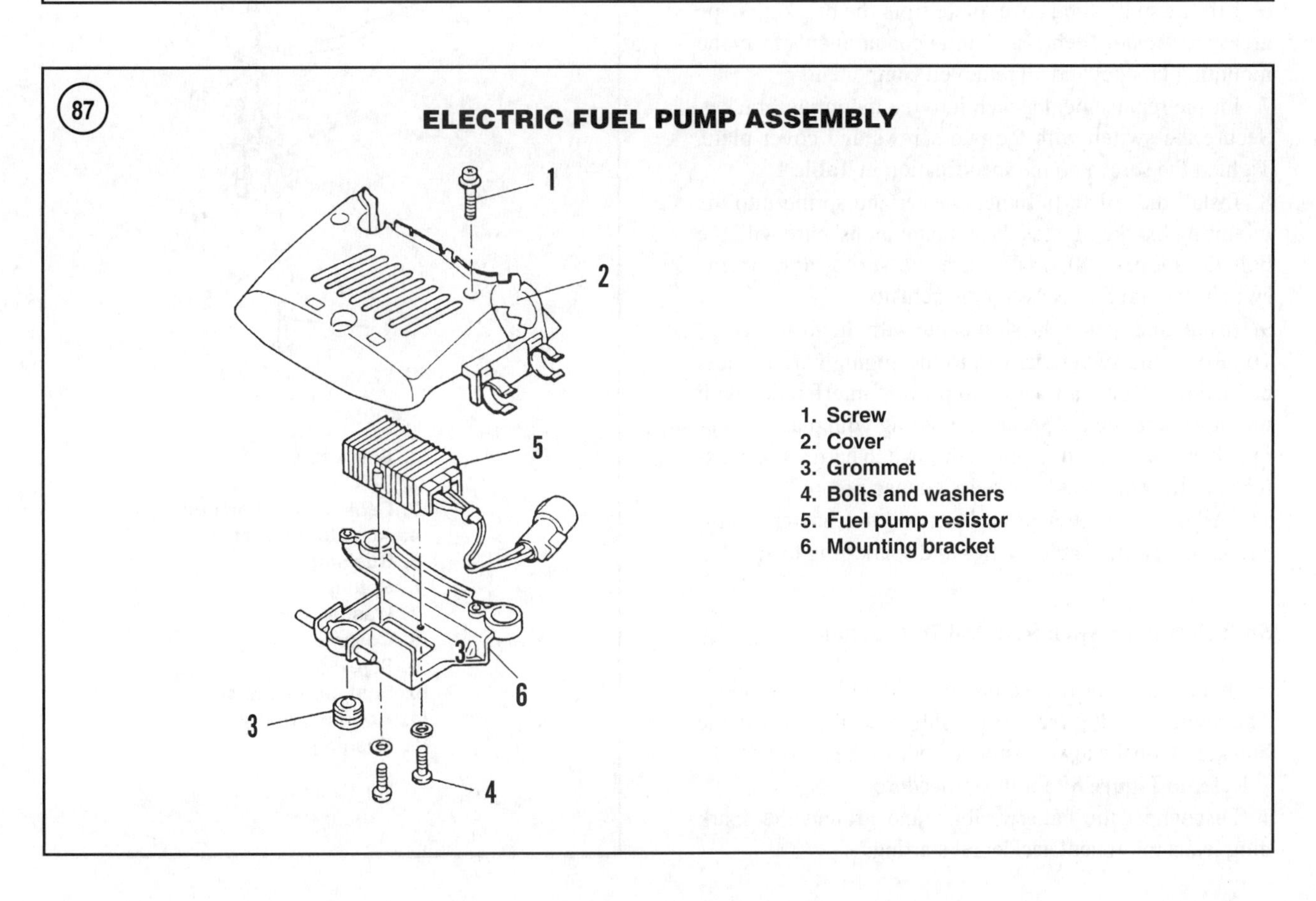
87
ELECTRIC FUEL PUMP ASSEMBLY
1
2
5
3
6
3
4
1. Screw
2. Cover
3. Grommet
4. Bolts and washers
5. Fuel pump resistor
6. Mounting bracket

cover and mounting bracket. Tighten the screws to the general torque specification in Chapter One.
10. Install the flywheel cover.
11. Connect the battery cables and spark plug leads.

Shift Cutout Switch Removal/Installation

This component is used on 200-250 hp (3.1 liter) EFI models. The switch (10, **Figure 88**) is mounted onto a bracket on the lower starboard side of the power head. Apply a light coating of Yamaha All-Purpose grease to all bushings, collars and washers during switch installation.

Refer to **Figure 88** for this procedure.

1. Disconnect the battery cables and ground the spark plug lead to prevent accidental starting.
2. Disconnect the switch harness connector from the engine wire harness connector.
3. Remove the shift cable and fastener from the assembly. Remove the bolts that retain the actuator to the power head. Remove the actuator.
4. Remove the spring (6, **Figure 88**), washer (7) bushing (12) and collar (8).
5. Remove the two screws (9, **Figure 88**) that retain the switch and cover plate to the bracket (11).
6. Lift the switch and cover plate from the bracket. Wipe grease, corrosion, debris and other contaminants from the mounting bracket and all removed components.
7. Fit the replacement switch into the mounting bracket. Secure the switch with the two screws and cover plate. Tighten the screws to the specification in **Table 1**.
8. Install the collar, bushing, washer and spring into the mounting bracket. Install the actuator and secure with the bolt (2, **Figure 88**). Make sure the spring arm on the switch fits into the recess on the actuator.
9. Install and secure the shift cable with the fasteners.
10. Route the switch harness to the engine wire harness connection. Route all wiring to prevent interference with moving components. Secure the wiring with plastic locking clamps as needed. Connect the switch harness connector onto the engine wire harness connector.
11. Adjust the shift cable as described in Chapter Four.
12. Connect the battery cables and spark plug leads.

Shift Position Switch Removal/Installation

This component is used on 150-250 hp HPDI models. The switch (3, **Figure 89**) mounts onto the shift cable bracket (4) on the lower starboard side of the power head.

Refer to **Figure 89** for this procedure.

1. Disconnect the battery cables and ground the spark plug leads to prevent accidental starting.

SHIFT CUT OUT SWITCH (200-250 HP MODELS [3.1 LITER])

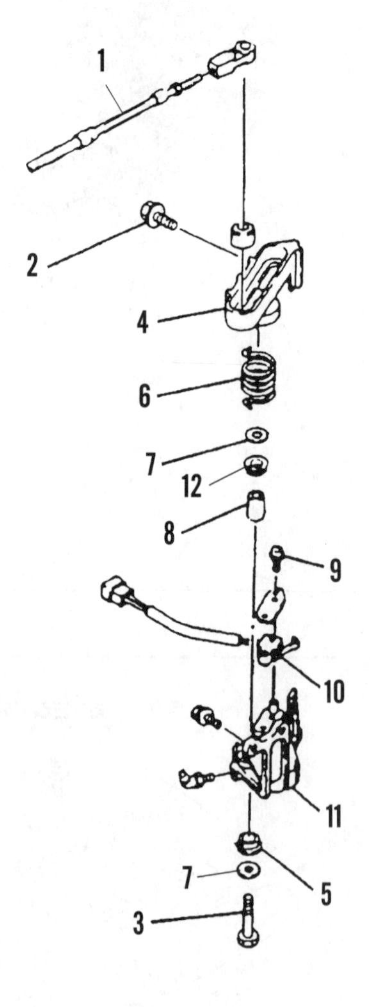

1. Shift cable
2. Bolt
3. Cable connection bolt
4. Actuator bracket
5. Bushing
6. Spring
7. Washer
8. Collar
9. Screws
10. Shift cutout switch
11. Mounting bracket
12. Bushing

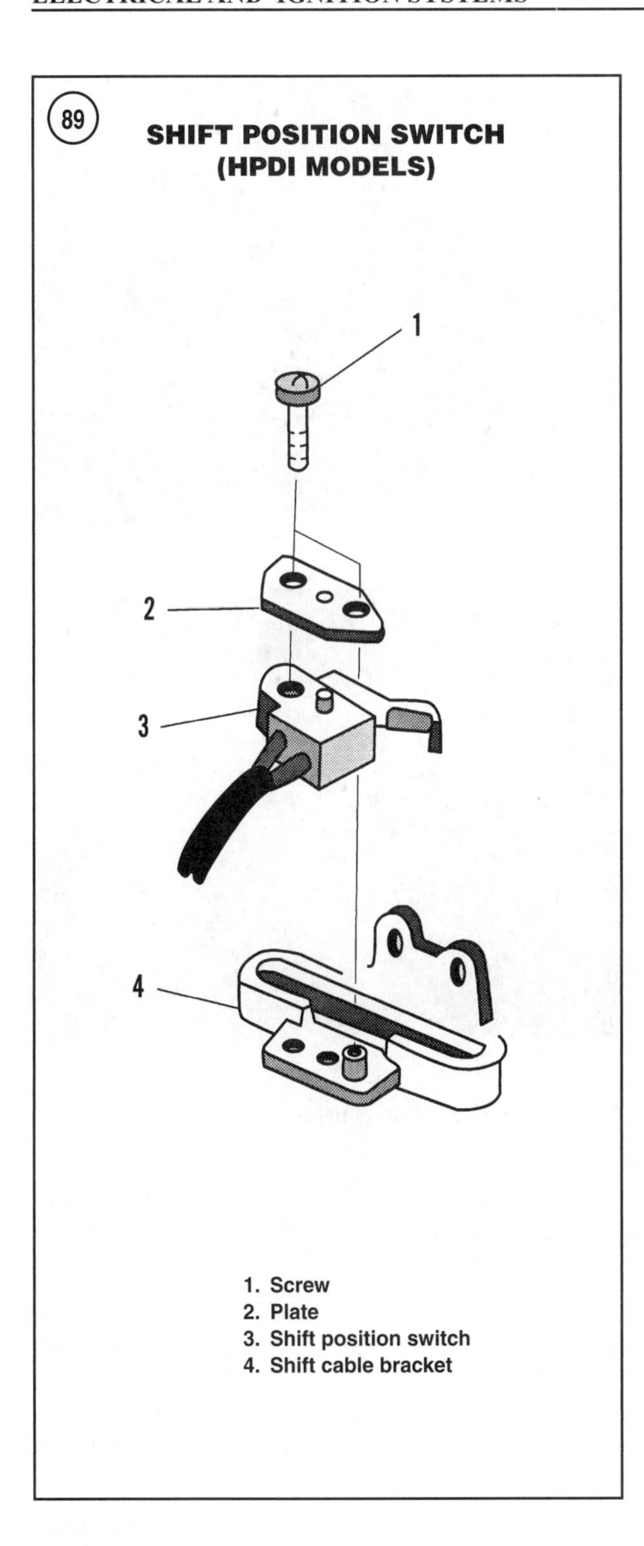

2. Disconnect the switch harness connector from the engine wire harness connector.

3. Shift the engine into NEUTRAL gear.

4. Remove the two screws (1, **Figure 89**), then lift the plate (2) and switch (3) from the bracket (4). Wipe grease, corrosion, debris and other contaminants from the mounting bracket and all removed components.

5. Fit the replacement switch onto the bracket as shown in **Figure 89**. Install the plate. Then thread the screws into the plate, switch and bracket. Tighten the screws to the specification in **Table 1**.

6. Route the switch harness to the engine wire harness connection. Route all wiring to prevent interference with moving components. Secure the wiring with plastic locking type clamps as needed. Connect the switch harness connector onto the engine wire harness connector.

7. Connect the battery cables and spark plug leads.

Ignition System and Fuel Pump Relays Removal/Installation

Ignition and fuel pump relays are used on all EFI and HPDI models. Relay usage and mounting locations vary by model. Refer to the following to find the relay(s).

6

150-250 hp (except HPDI) models—A screw secures the system relay (**Figure 90**) into the junction box on the starboard side of the power head.

150-250 hp (HPDI) models—Two relays are used. Both relays are located in the fuse terminal block on the starboard side of the engine and below the starter motor. Remove the plastic cover to access the relays. The system relay (A, **Figure 91**) plugs into the terminals closest to the front of the power head. The electric fuel pump relay (B, **Figure 91**) plugs into the terminals closest to the rear of the power head.

Refer to the wiring diagrams at the end of the manual to identify the wire colors connected to the relay. Trace the wires to the correct component on the engine.

1. Disconnect the battery cables and ground the spark plug leads to prevent accidental starting.
2. Remove the cover (**Figure 92**) to access the relay(s).

3A. *150-250 hp (except HPDI models)*—Disconnect the relay harness connector from the engine wire harness connector, then disconnect the red and red/yellow relay wires from the engine harness. Remove the mounting screw, then guide the wire harness through an opening while removing the relay.

3B. *150-250 hp (HPDI) models*—Carefully pull the suspect relay from the terminals. Gently rock the relay on the mount for easier removal.

4A. *150-250 hp (except HPDI models)*—Guide the relay harness through any openings while routing the relay wiring to connection points to the engine wire harness. Connect the relay harness connector to the engine wire harness connector, then connect the red and red/yellow wires to the engine harness terminals. Install the relay to the mounting surface in the junction box. Thread the mounting screw through the relay and into the threaded opening. Tighten the mounting screw to the specification in **Table 1**. Route the relay wiring to prevent interference with moving components. Secure the wiring with plastic locking clamps as necessary.

4B. *150-250 hp (HPDI) models*—Align the relay contacts with the respective openings in the fuse terminal block, then carefully push the relay until it seats on the block.

5. Install the cover.
6. Connect the battery cables and spark plug leads.

WARNING SYSTEM COMPONENTS

This section describes replacing the overheat switch and warning horn. Refer to Chapter Twelve for replacement procedures for the oil level sensor and emergency switch.

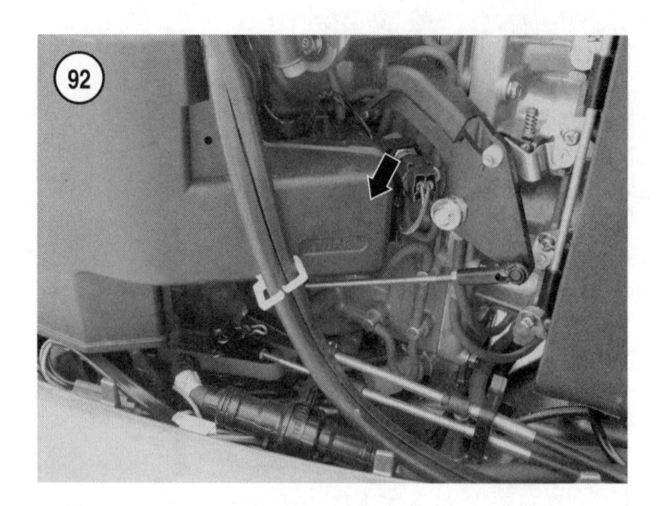

Overheat Switch Removal/Installation

The mounting locations for the overheat switch (**Figure 93**) varies by model. Refer to the following information to finding the switch.

80 Jet, 105 Jet, 115 hp, 130 hp and 150-200 hp (2.6 liter) models—Two overheat switches are used. Each switch fits into an opening at the top of each cylinder head water jacket. A press fit secures the switch into the opening. A retaining clamp is not used to secure the switch on these models.

200-250 hp (3.1 liter) models—Two overheat switches are used. Each switch fits into an opening at the top of each cylinder bank. A bolt and clamp plate secures the switches into the opening.

Refer to the wire diagrams at the end of the manual to identify the wire colors for the switch. Trace the wires to the correct component on the engine.

CAUTION

The engine temperature sensor and overheat switch look similar and mount in similar locations on the power head. Verify the correct component by the wire colors leading into the sensor or switch. Two black/yellow wires lead into the engine temperature sensor. A pink and black wire lead into the overheat switch. Also note that the engine temperature sensor threads into the mount opening. A clamp plate and screw secure the overheat switch into the mount opening.

1. Disconnect the battery cables and ground the spark plug leads to prevent accidental starting.
2. Disconnect the pink and black switch wires from the engine harness connectors.
3. *200-250 hp (3.1 liter) models*—Remove the bolt, then carefully pry the retaining plate from the switch.
4. Grasp the boss on the exposed end of the switch with needlenose pliers. Do not pull on the wires to remove the switch. Gently rock the switch from side to side while pulling it from the opening. If necessary, rotate the switch to break the bond that often forms between the switch and the opening.
5. Thoroughly clean all corrosion, debris, oily deposits or other contaminants from the switch and switch opening.
6. Carefully press the replacement switch into the opening. Apply a soap and water solution to the switch to ease installation. Do not apply grease or other material to the switch. If necessary, gently rock the switch while pushing it into the opening. The switch must seat in the bore for proper operation.
7. *200-250 hp (3.1 liter) models*—Install the retaining plate onto the switch. Do not pinch the switch wires between the plate and switch or cylinder block. Install the bolt into the retaining plate and tighten to the specification in **Table 1**.
8. Connect the black and pink switch wires onto the respective engine wire harness connections. Route the wiring to prevent contact with moving components. Secure the wiring with plastic locking clamps as needed.
9. If necessary, repeat Steps 2-8 for the remaining overheat switch.
10. Connect the battery cables and spark plug leads.

6

Warning Horn

Refer to Chapter Thirteen for replacement procedures if the warning horn is located within the remote control. If the warning horn is located beneath the instrument panel, remove the horn by simply disconnecting the two wires from the horn and removing the mounting clamps. Secure the replacement horn onto a secure mount with a suitable clamp. Then connect the leads. Make sure the horn is located in an area that does not excessively muffle the tone. Relocate the horn if it cannot be heard while underway.

Table 1 TIGHTENING TORQUE

Fastener	N•m	in.-lb.	ft.-lb.
Air pressure sensor			
150-250 hp (except HPDI)	3	27	–
150-250 hp (HPDI)	4	35	–
Air temperature sensor	8	70	–
Engine temperature sensor	15	–	11
Knock sensor	28	–	21
Overheat switch retainer	15	–	11
Oxygen density sensor			
Sensor nut	49	–	36
Mounting block bolts	14	–	10
Sensor cover	9	80	–
Shift cutout switch	3	27	–
Shift position switch	3	27	–

(continued)

Table 1 TIGHTENING TORQUE (continued)

Fastener	N·m	in.-lb.	ft.-lb.
Starter motor			
80 Jet and 150-250			
(except HPDI)			
Mounting bracket	30	–	22
Cable terminal nut	9	80	–
Upper mounting bolts			
(top mounted)	30	–	22
Lower mounting bolt	30	–	22
Through bolts	5	44	–
150-250 hp (HPDI)			
Starter mounting	18	–	13
Starter solenoid	8	70	–
Through bolts	8	70	–
System relay	3	27	–
Throttle position sensor	4	35	–

Table 2 STARTER MOTOR SERVICE SPECIFICATIONS

Model	Specification
Brush length	
80 Jet and 115-130 hp	10.0-17.0 mm (0.394-0.669 in.)
105 Jet and 150-200 hp (2.6 liter	
[except HPDI])	12.0-16.0 mm (0.472-0.630 in.)
150-250 hp (HPDI models)	9.5-15.5 mm (0.374-0.610 in.)
Commutator diameter (minimum)	
80 Jet and 115-130 hp	32.0-33 mm (1.259-1.299 in.)
105 Jet and 150-250 hp	
(except HPDI models)	31.0-33 mm (1.220-1.299 in.)
150-250 hp (HPDI models)	28.0-29.0 mm (1.102-1.141 in.)
Commutator undercut (minimum)	0.2-0.8 mm (0.008-0.031 in.)
Maximum armature deflection	0.05 mm (0.002 in.)

Table 3 BATTERY CHARGE PERCENTAGE

Specific gravity reading	Percentage of Charge Remaining
1.120-1.140	0
1.135-1.155	10
1.150-1.170	20
1.160-1.180	30
1.175-1.195	40
1.190-1.210	50
1.205-1.225	60
1.215-1.235	70
1.230-1.250	80
1.245-1.265	90
1.260-1.280	100

Table 4 BATTERY CAPACITY

Accessory draw	Provides continuous power for:	Approximate recharge time
80 amp-hour battery		
5 amps	13.5 hours	16 hours
15 amps	3.5 hours	13 hours
25 amps	1.6 hours	12 hours
105 amp-hour battery		
5 amps	15.8 hours	16 hours
15 amps	4.2 hours	13 hours
25 amps	2.4 hours	12 hours

Chapter Seven

Power Head

POWER HEAD PRELIMINARY INFORMATION

This chapter provides power head removal/installation, disassembly/assembly, cleaning and inspection procedures for models covered in this manual. Sacrificial anode replacement procedures are at the end of the chapter. The power head can be removed from the outboard without removing the entire outboard from the boat.

This chapter is arranged in a normal disassembly/assembly sequence. When only a partial repair is required, follow the procedure(s) to the point where the faulty parts can be replaced, then jump ahead to reassemble the unit.

Many procedures require the use of manufacturer recommended special tools, which can be purchased from a Yamaha outboard dealership.

Power head work stands and holding fixtures (**Figure 1**) are available from specialty shops or marine and industrial product distributors.

Make sure that the workbench, work station, engine stand or holding fixture is able to support the size and weight of the power head. This is especially important when working on larger engines.

Table 1 lists specific torque specifications for most power head fasteners. Use the general torque specification in Chapter One for fasteners not listed in **Table 1**. **Tables 2-5** list powerhead specifications at the end of the chapter.

SERVICE CONSIDERATIONS

Performing internal service procedures on the power head requires considerable mechanical ability. Carefully consider your capabilities before attempting any operation involving major disassembly of the engine.

If, after studying the text and illustrations in this chapter, you decide not to attempt a major power head disassembly or repair, it may be financially beneficial to perform certain preliminary operations yourself. Consider separating the power head from the outboard and removing the fuel, ignition and electrical systems and all accessories, taking only the basic power head to the dealership for the actual overhaul or major repair.

Repair is quicker and easier if the engine is clean before starting any service procedure. There are many special cleaners (degreasers) available from any automotive supply store. Most of these cleaners are simply sprayed on, then rinsed off with a garden hose after the recommended time period. Always follow all instructions provided by the manufacturer. Never apply cleaning solvent to electri-

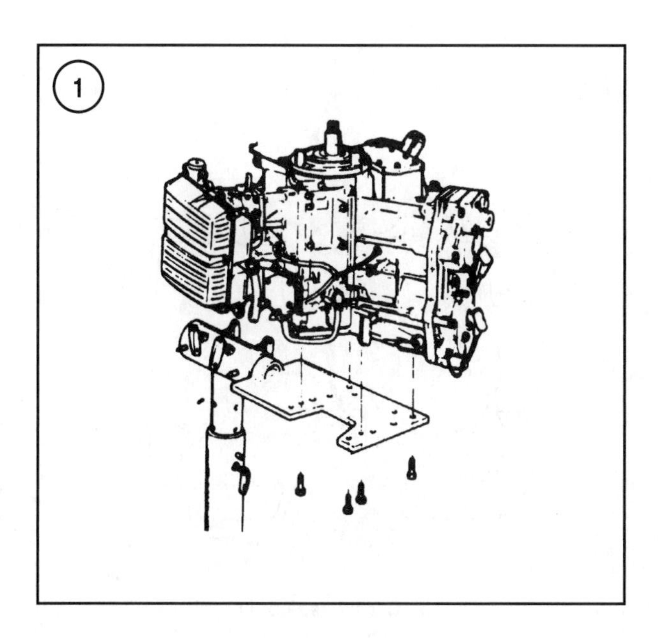

cal and ignition components or spray it into the induction or exhaust system.

WARNING
Never use gasoline as a cleaning agent. Gasoline presents an extreme fire and explosion hazard. Work in a well-ventilated area when using cleaning solvent. Keep a large fire extinguisher rated for gasoline and oil fires nearby in case of an emergency.

After deciding to do the job yourself, thoroughly read this chapter until you understand what is involved to complete the repair. Make arrangements to buy or rent the necessary special tools and obtain a source for replacement parts *before* starting.

NOTE
A series of at least five photographs, taken from the front, rear, top and both sides of the power head (before removal) is helpful during assembly and installation.

Before beginning the job, review Chapter One.

POWER HEAD BREAK-IN

Whenever a power head is rebuilt or replaced, or if *any* new internal parts are installed, the engine must be run on the specified fuel/oil mixture (except HPDI models) and operated in accordance with the recommended break-in procedure as described in Chapter Three.

CAUTION
Failure to follow the recommended break-in procedure causes premature power head failure.

SERVICE RECOMMENDATIONS

If the engine has experienced a power head failure, refer to the *Engine* section in Chapter Two for troubleshooting procedures.

Many failures are caused from using the incorrect or stale fuel and lubricating oil. Refer to Chapter Three for all fuel and oil recommendations.

When rebuilding or performing a major repair on the power head, consider performing the following steps to prevent the failure from reoccurring.

1. Service the water pump. Replace the impeller and all seals and gaskets. See Chapter Eight.
2. Replace the thermostat(s) and remove and inspect the water pressure relief valve as described in this chapter. Replace any suspect components.
3. Drain the fuel tank(s) and dispose of the old fuel in an approved manner.
4. Fill the fuel tank with fresh fuel and add the recommended oil to the fuel tank (except HPDI models) during *break-in* as described in Chapter Three.
5. Replace (or clean) all fuel filters. See Chapter Three.

6A. *Carburetor equipped models*—Clean and adjust the carburetors. See Chapter Five.

6B. *EFI and HPDI models*—Clean or replace the vapor separator tank filter.

7. *Oil injected models*—Drain and clean the oil reservoir(s). Dispose of the old oil in an approved manner. Refill the oil system with the specified oil (Chapter Three) and bleed the oil system as described in Chapter Twelve.
8. Install new spark plugs. Use only the recommended spark plugs listed in Chapter Three. Make sure the spark plugs are correctly tightened. Incorrect torque leads to spark plug overheating.
9. Perform *all* the synchronization and linkage adjustments as described in Chapter Four before returning the engine into service.

LUBRICANTS, SEALANTS AND ADHESIVES

Recommended lubricants, sealants and adhesives are listed in the repair instructions. Equivalent (after-market) products are acceptable to use, as long as they meet or exceed the original specifications.

During power head assembly, all internal engine components must be lubricated with two-cycle (TCW-3) outboard engine oil. Lubricate all seal lips and O-rings with Yamaha All-Purpose Grease.

To efficiently remove the carbon from the pistons and combustion chambers use Yamaha Combustion Chamber Cleaner. Allow ample time for the cleaner to soak into and soften carbon deposits.

Unless specified, do not apply gasket sealing compound onto gasket surfaces. Many of the gaskets used on Yamaha outboards are designed to cure and provide sealing after reaching operating temperature. Gasket sealing compound may prevent proper curing of the gasket. Coat the threads of all external fasteners (when no other sealant or adhesive is specified) with a light coating of two-cycle outboard engine oil to help prevent corrosion and ease future removal.

When sealing the crankcase cover/cylinder block, both mating surfaces must be free of all sealant residue, dirt, oil or other contamination. Locquic Primer, lacquer thinner, acetone or similar solvents work well when used in conjunction with a plastic scraper. Never use solvents with an oil, wax or petroleum base.

CAUTION

Clean all mating surfaces carefully to avoid nicks and gouges. A plastic scraper can be improvised from a common household electrical outlet cover or a piece of Lucite with one edge ground to a 45° angle. Use extreme caution if using a metal scraper, such as a putty knife. Nicks and gouges may prevent the sealant from curing. The crankcase cover-to-cylinder block surface must ***not*** *be lapped or machined.*

Loctite Gasket Maker is the only recommended sealant used to seal the crankcase cover-to-cylinder block mating surface. The sealant bead must be applied to the inside (crankshaft side) of all crankcase cover boltholes. See **Figure 2**.

Apply Loctite 271 threadlocking adhesive to the outer diameter of all seals before pressing the seals into place. Also apply this adhesive to the threads of all internal fasteners (when no other adhesive is specified).

Whenever a Loctite product is called for, always clean the surface to be sealed (or threads to be secured) with Locquic Primer. Locquic Primer cleans and primes the surface and ensures a quick secure bond by leaving a thin film of catalyst on the mating surface or threads. Allow the primer to air dry. Blow-drying disables the catalyst.

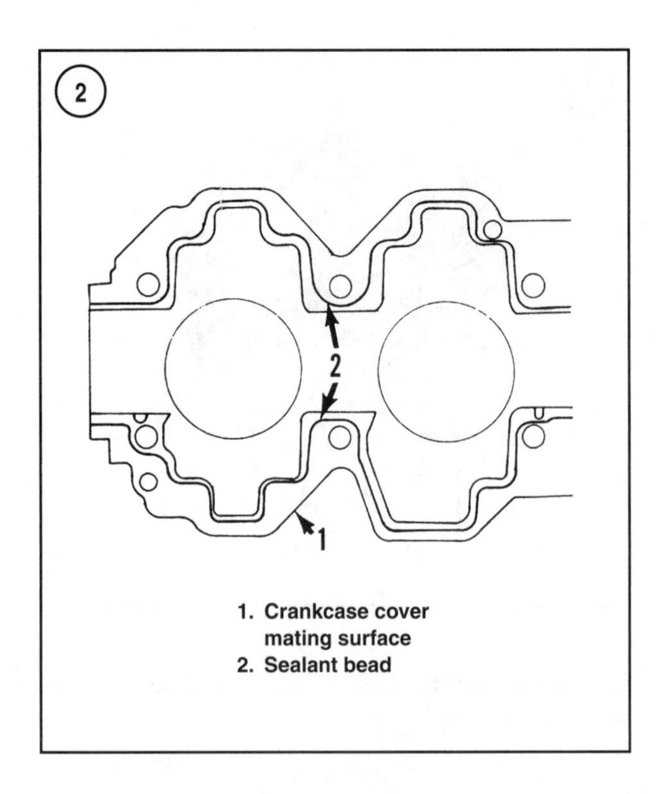

1. Crankcase cover mating surface
2. Sealant bead

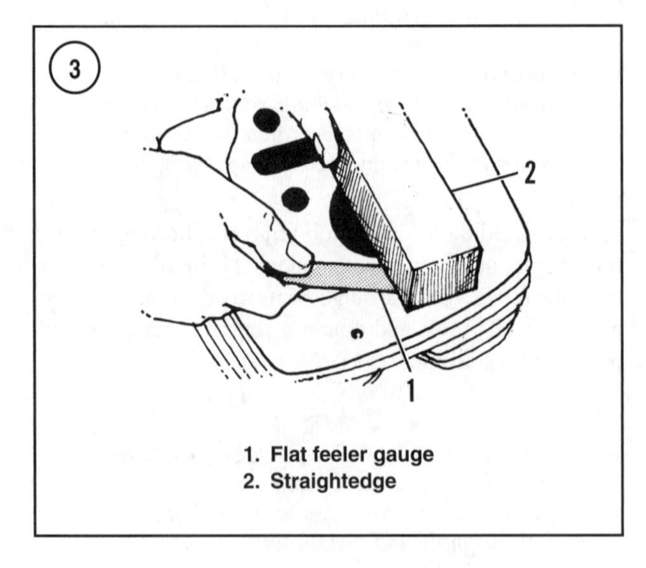

1. Flat feeler gauge
2. Straightedge

SEALING SURFACES

Clean all sealing surfaces carefully to prevent nicks and gouges. Often a shop towel soaked in solvent is used to rub gasket material and/or sealant from a mating surface. If scrapers is used, try using a plastic scraper (such as a household electrical outlet cover) or a piece of Lucite with one edge ground to a 45° angle to prevent damage to the sealing surfaces.

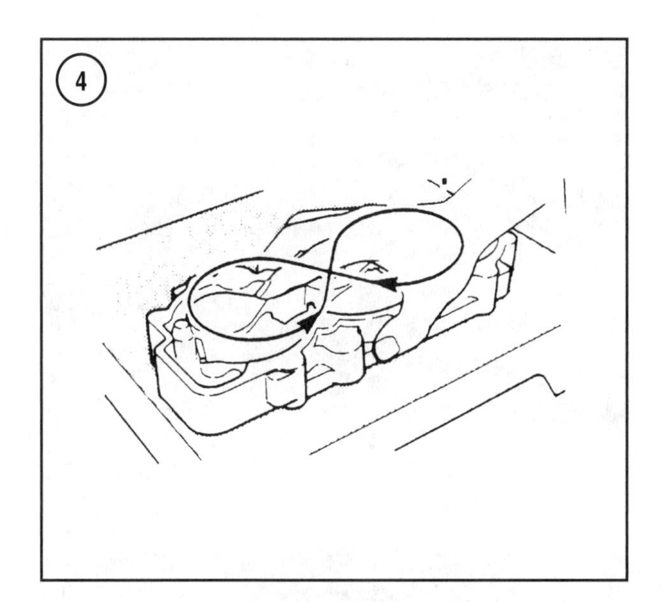

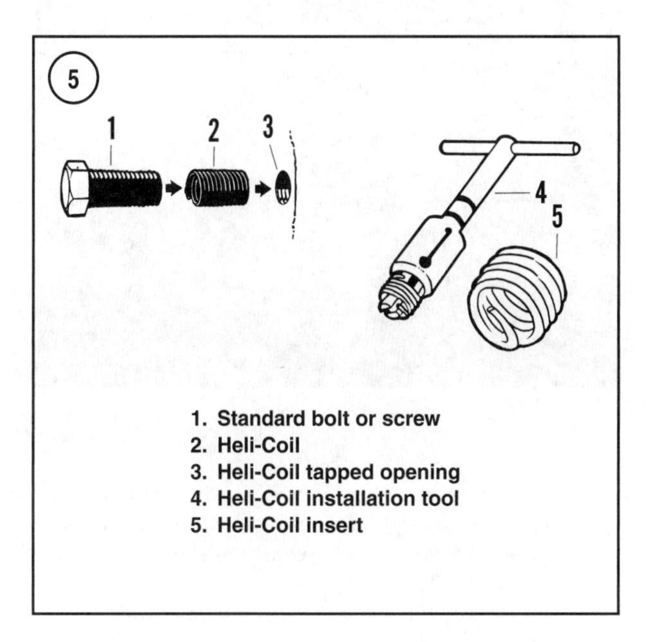

1. Standard bolt or screw
2. Heli-Coil
3. Heli-Coil tapped opening
4. Heli-Coil installation tool
5. Heli-Coil insert

Once the surfaces are clean, check the component for warpage by placing the component onto a piece of plate glass or a machinist's surface plate. Apply uniform downward pressure and try to insert a selection of feeler gauges between the plate and the component.

Use a machinist's straightedge to check areas that cannot be accessed using the glass or surface plate. See **Figure 3**. Warpage exceeding 0.1 mm (0.004 in.) is cause for component replacement, unless otherwise specified.

CAUTION
The cylinder block-to-crankcase cover must not be lapped (on all models).

To remove minor warp, minor nicks or scratches, or traces of sealant or gasket material, place a large sheet of 320-400 grit wet sandpaper onto the plate glass or surface plate. Apply light downward pressure and move the component in a figure-8 pattern as shown in **Figure 4**. Use a light oil (such as WD-40) to keep the sandpaper from loading up. Remove the component from the sandpaper and recheck the sealing surface.

It may be necessary to repeat the lapping process several times to achieve the desired results. Never remove any more material than is absolutely necessary. Make sure the component is thoroughly washed to remove all grit before assembly.

FASTENERS AND TORQUE

Always replace a worn or damaged fastener with one of equal size, type and torque requirement. Power head torque values are listed in **Table 1**. If a specification is not provided for a given fastener, use the general torque values listed in Chapter One according to fastener size. Fastener measurement is described in Chapter One.

Damaged threads in components and castings may be repaired using a Heli-Coil (or equivalent) stainless steel threaded insert (**Figure 5**, typical). Heli-coil kits are available at automotive or marine and industrial supply stores. Never run a thread tap or thread chaser into a hole equipped with a Heli-coil. Damaged Heli-coils may be replaced (if damaged) by gripping the outermost coil with needlenose pliers and unthreading the coil from the hole. Pulling the coil straight out damages the opening.

CAUTION
Metric fasteners are used on these engines. Always match a replacement fastener to the original. Do not run a tap or thread chaser into a hole (or over a bolt) without first verifying the thread size and pitch.

Unless otherwise specified, components secured by more than one fastener should be tightened in a *minimum* of three steps. First, evenly tighten all fasteners hand-tight (snug). Then evenly tighten all fasteners to 50% of the specified torque value. Finally, evenly tighten all fasteners to 100% of the specified torque value as the third step.

Make sure to follow torque patterns (sequences) as directed. If no pattern is specified, start at the center of the component and tighten in a circular pattern, working outward. Torque sequences are listed in the appropriate sections of this chapter.

POWER HEAD REMOVAL/INSTALLATION

If complete disassembly is not necessary, stop disassembly at the appropriate point, then begin assembly where disassembly stopped. Remove the power head as an assembly if major repair must be performed. Power head removal is not required for certain service procedures such as cylinder head removal, intake and exhaust cover removal, ignition component replacement, fuel system component replacement and reed block/intake manifold removal.

Preparation for Removal

Visually inspect the engine to locate the fuel supply hose, control cables and battery connections. Disconnect only the hoses and wires necessary to remove the power head from the engine. Many of the hoses and wires are much more accessible after removing the power head.

Diagrams of the fuel and electrical systems are provided in Chapter Five and Chapter Six. Use them to assist with hose and wire routing.

To speed up installation and ensure correct connections, take several pictures of each side of the power head *before* removing the power head. Make notes and sketches of wire and hose routing, connection points and linkage orientation. This step saves a great deal of time and frustration during assembly.

Secure the proper lifting equipment before removing the power head. A complete power head may weigh 200 lb. (91 kg) or more.

Use assistance when lifting or moving the power head. Attach the hoist into the lifting hook located on top of the power head (**Figure 6**).

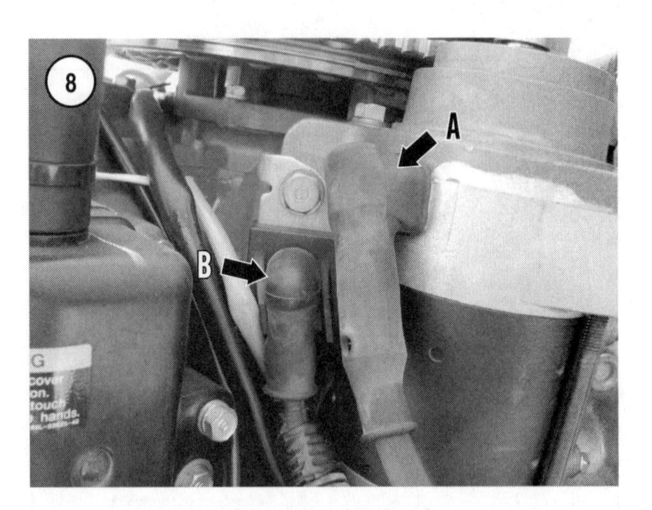

WARNING

The power head may abruptly separate from the mid-section during removal with an overhead hoist causing injury or damaging property. Avoid using excessive force when lifting the power head. Use pry bars to carefully separate the power head from the mount before lifting with a hoist.

NOTE

Corrosion may prevent easy removal of the power head. Apply a penetrating oil onto the mating surface and allow it to soak into the gasket before removing the power head.

CAUTION

Power head mounting fastener location and quantity varies with the models. Refer to the illustrations for assistance in locating the fasteners. Variations also exist as to which components must be removed prior to removing the power head. Make sure all re-

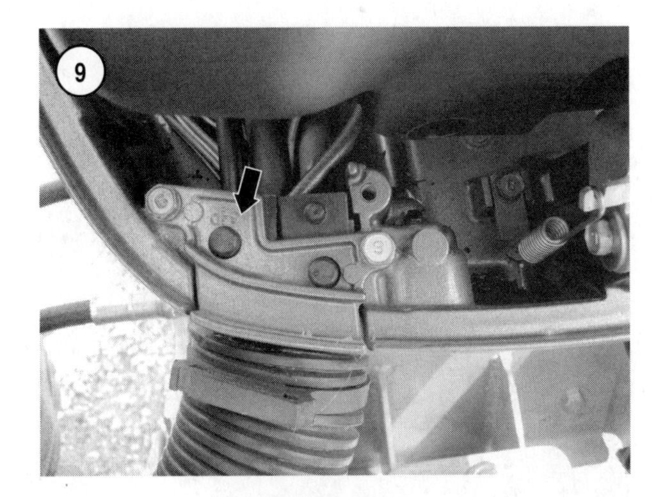

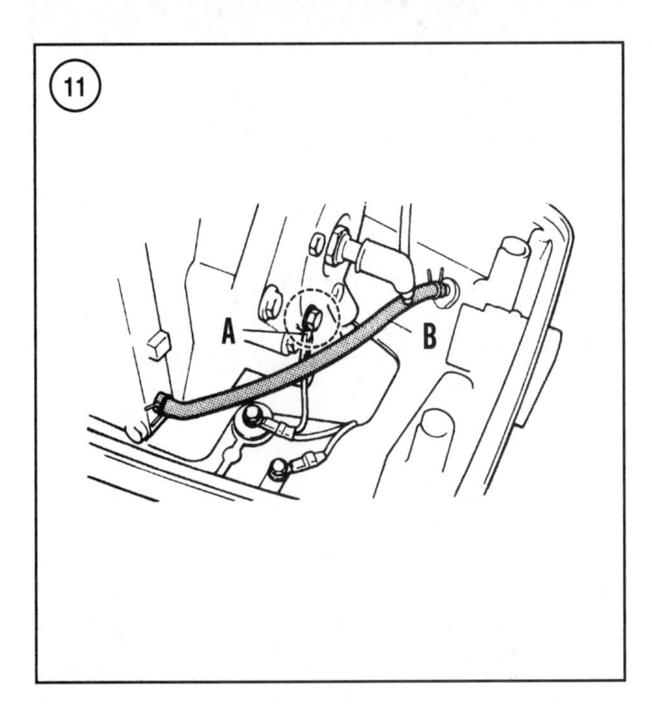

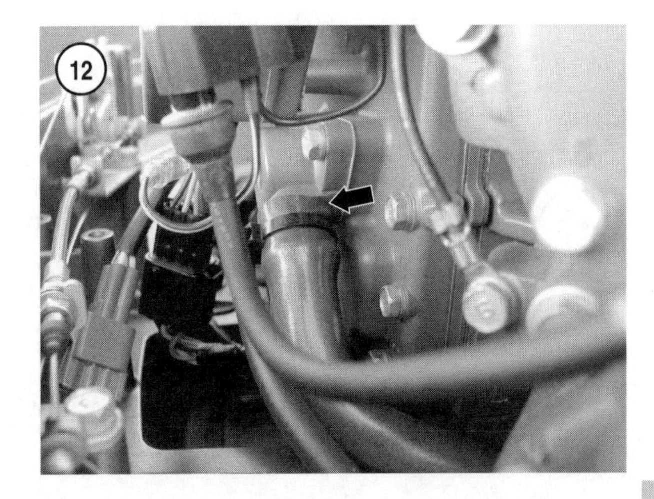

quired fasteners and components are removed before lifting the power head from the mid-section.

Removal

1. Disconnect both cables from the battery.
2. Unscrew the plastic clamp (**Figure 7**). Then unplug the main engine harness from the remote control instrument harness.
3. Disconnect the battery cable from the power head terminals (**Figure 8**).
4. *Oil injected models*—Disconnect the oil supply hose from the engine mounted reservoir. Refer to the diagrams in Chapter Twelve to identify the proper hose. Plug the disconnected hose and reservoir to prevent oil leakage.
5. Remove the retainer (**Figure 9**) to free the hoses and cables from the lower engine cover.
6. Disconnect the remote control shift and throttle cables from the power head (**Figure 10**).
7. Disconnect the trim position sender wires from the engine harness wiring. Disconnect the engine cover mounted trim switch from the engine harness wiring. Disconnect the electric trim motor wires from the trim relay connections.
8. Disconnect the water stream hose (B, **Figure 11**) from the lower engine cover fitting. Disconnect the ground wire (A, **Figure 11**) from the power head.
9. Loosen the clamp, then pull the water hose from the water pressure relief valve fitting (**Figure 12**).
10. Remove the two bolts. Then pull the front apron (B, **Figure 13**) from the lower engine cover. Remove the two bolts. Then pull the rear apron (C, **Figure 13**) from the lower engine cover. Remove the flywheel cover (A, **Figure 13**).

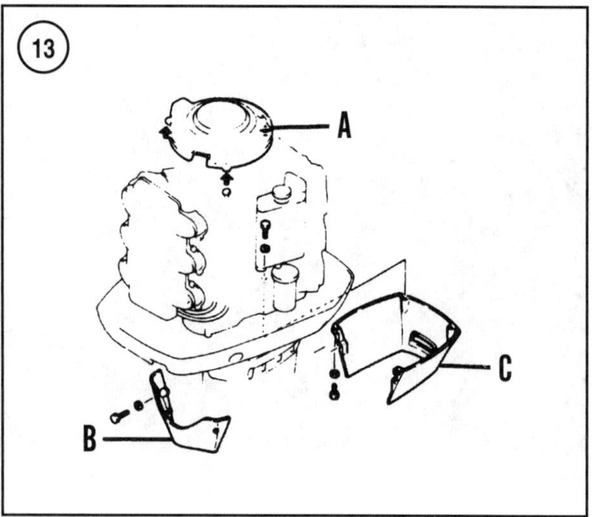

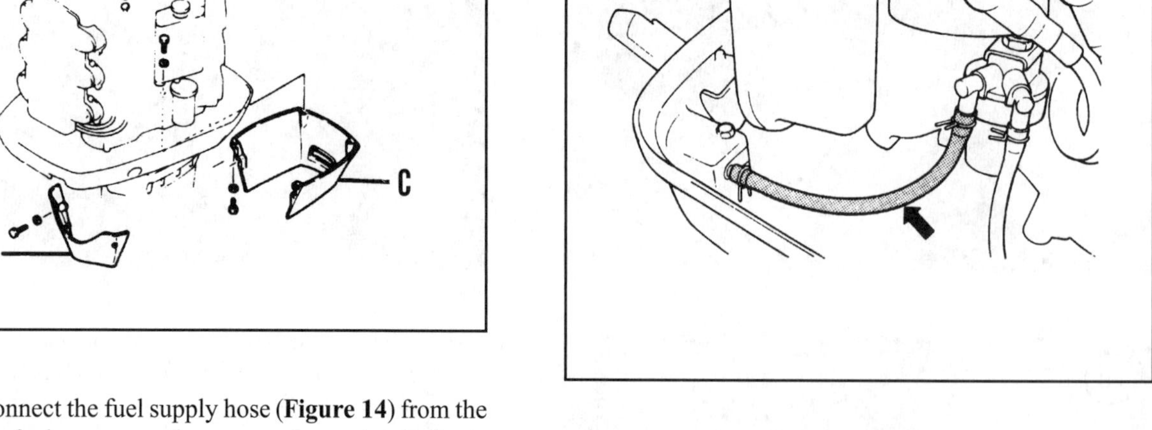

11. Disconnect the fuel supply hose (**Figure 14**) from the fuel filter, fuel pump or other connection point. Refer to Chapter Five for instructions.
12. *HPDI models*—Disconnect the shift position switch wires from the engine wire harness. See Chapter Six.

CAUTION
At this point, there should be no hoses, wires or linkage connecting the power head to the drive shaft housing. Make sure that nothing will interfere with power head removal before continuing.

13. Thoroughly inspect the power head for attached wires, linkages or hoses (**Figure 15**). Remove or disconnect them if necessary for power head removal.
14. *200-250 hp (3.1 liter) models*—Thread the lifting hook (Yamaha part No. YB-6202) fully onto the crankshaft threads (**Figure 16**). The threaded end of the crankshaft must pass completely through the hook casting to ensure adequate thread engagement.
15. Attach an overhead hoist to the lifting hook (**Figure 17**) and provide enough force to just take the slack from the lifting cable or chain.
16. Locate and remove the power head mounting bolts or nuts (**Figure 18**, typical).
   a. *80 Jet, 105 Jet, 115-130 hp and 150-200 hp (2.6 liter) models*—Twelve fasteners attach the power head to the adapter. Two nuts and washers are located at the front power head mounting surface. Two bolts are located at the rear power head mounting surface. Four bolts are located on each side of the power head mounting surface.
   b. *200-250 hp (3.1 liter) models*—Fourteen bolts attach the power head to the adapter. Two bolts are lo-

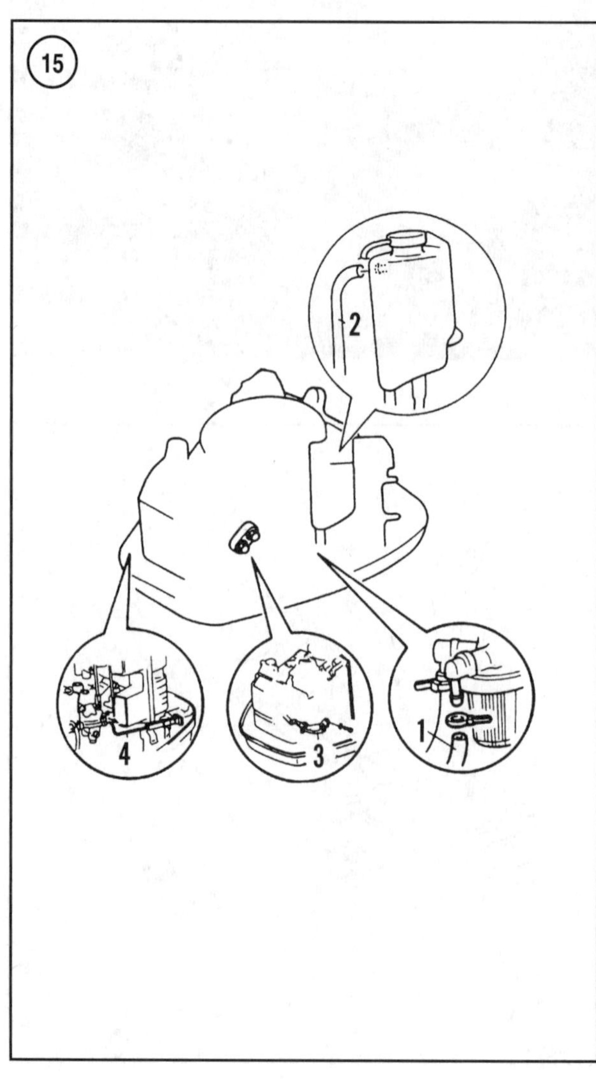

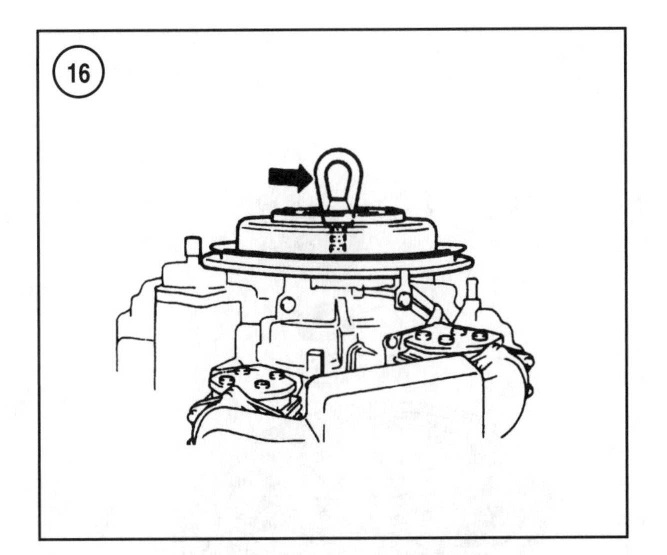

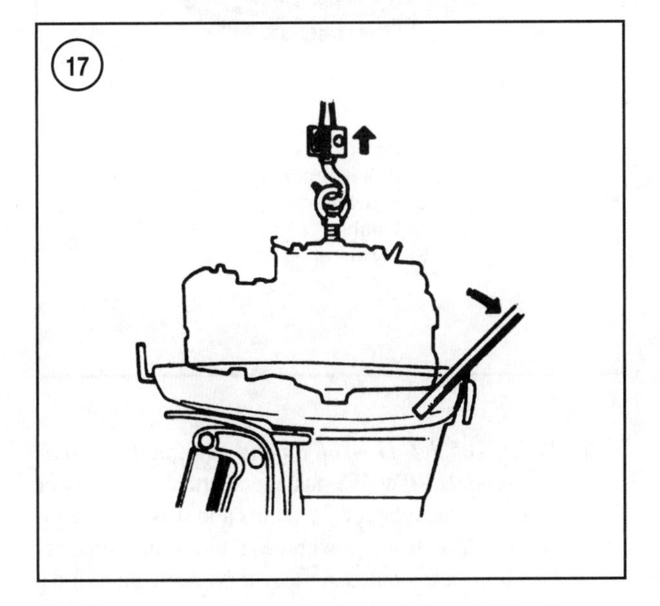

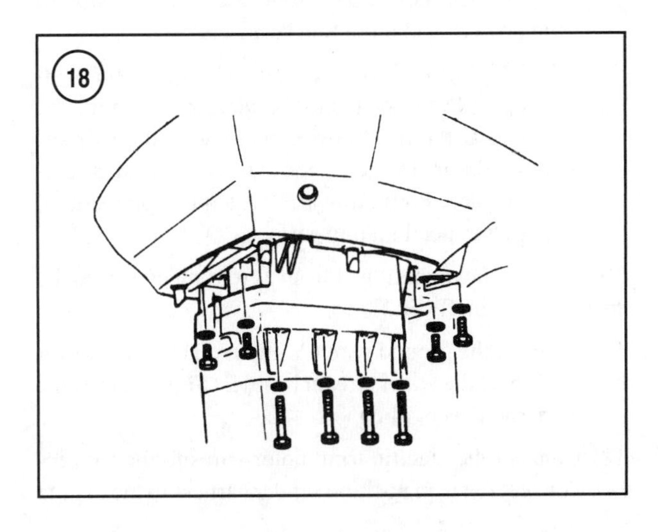

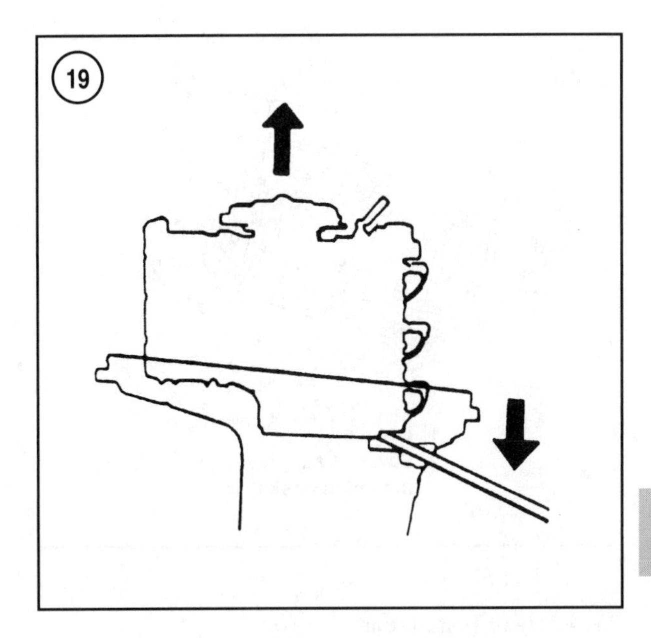

cated at the front power head mounting surface. Four bolts are located at the rear power head mounting surface. Four bolts are located on each side of the power head mating surface.

17. Locate an area that allows a pry bar to be inserted between the power head and mid-section. Notches are provided on most models. Carefully pry and rock the power head to separate if from the mid-section (**Figure 19**).

CAUTION
Lift the power head slowly and maintain support to ensure the power head lifts straight off the mid-section. The drive shaft and other components may be damaged if the power head is lifted or lowered at an angle.

18. Slowly lift the power head from the mid-section. Mount the power head securely to a suitable work stand (**Figure 1**).

19. Remove and discard the power head mounting gasket (2, **Figure 20**). Clean then inspect the power head mounting surfaces for deep scratches or extensive corrosion damage. Replace the housing if there are defects.

20. Inspect the locating pins (1, **Figure 20**) for damaged or missing pins. Inspect the pin holes in the mid-section and power head for elongation or cracked areas. Replace defective components as required. If the pins are found in the power head openings, remove them and place them in the corresponding opening in the mid-section.

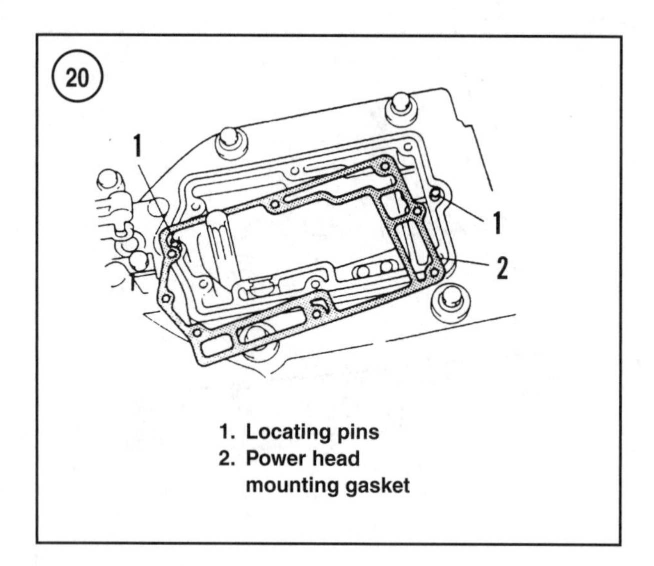

1. Locating pins
2. Power head mounting gasket

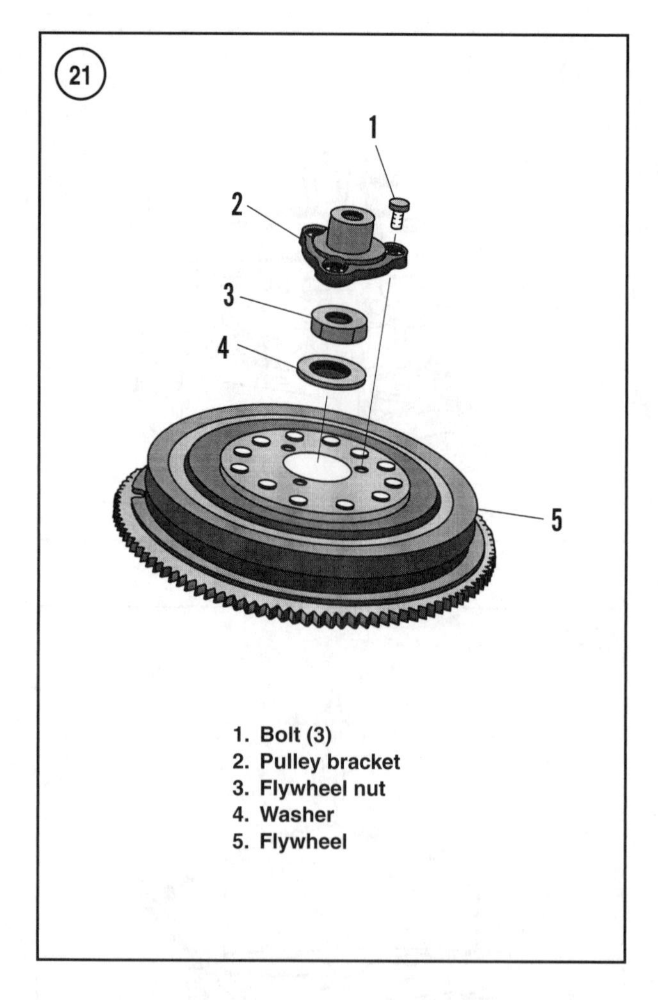

1. Bolt (3)
2. Pulley bracket
3. Flywheel nut
4. Washer
5. Flywheel

Power Head Installation

1. Relocate all hoses, wires, cables and linkages remaining on the engine away from the power head mating surfaces. Tie them back with plastic locking type clamps as needed.
2. *200-250 hp (3.1 liter) models*—Thread the lifting hook (Yamaha part No. YB-6202) into the crankshaft threads (**Figure 16**). The threaded end of the crankshaft must pass completely through the hook casting to ensure adequate thread engagement.
3. Attach an overhead hoist onto the lifting hook. Lift the power head enough to take all slack out of the lifting cable or chain. Remove the fasteners to free the power head from the work stand. Position the power head directly over the mid-section.
4. Apply a coating of Loctite Gasket Maker onto both surfaces of the mounting gasket.
5. Install the new mounting gasket (2, **Figure 20**) onto the mid-section and fit the corresponding openings over the locating pins (1).
6. Coat the lower drive shaft seal lips and the drive shaft splines with Yamaha All-Purpose Grease.
7. With assistance, slowly lower the power head onto the mid-section. Guide the drive shaft into the crankshaft seal opening. Keep the flywheel as level as possible to avoid damaging the drive shaft, gasket and mating surfaces. Rotate the flywheel to align the crankshaft and drive shaft splines. Then lower the power head onto the mating surface.
8. Apply Loctite 572 to the mounting fastener threads.
9. Thread the fasteners into the power head or mounting studs.
   a. *80 Jet, 105 Jet, 115-130 hp and 150-200 hp (2.6 liter) models*—Twelve fasteners attach the power head onto the adapter. Two nuts and washers are located at the front power head mounting surface. Two bolts are located at the rear power head mounting surface. Four bolts are located on each side of the power head mounting surface.
   b. *200-250 hp (3.1 liter) models*—Fourteen bolts attach the power head onto the adapter. Two bolts are located at the front power head mounting surface. Four bolts are located at the rear power head mounting surface. Four bolts are located on each side of the power head mating surface.
10. Tighten the mounting bolts in a crossing pattern to the specification in **Table 1**.
11. Connect the ground wire (A, **Figure 11**) to the power head. Connect the water stream hose (B, **Figure 11**) to the lower engine cover fitting.
12. Connect the electric trim motor wires to the trim relay. Connect the trim position sender harness to the engine

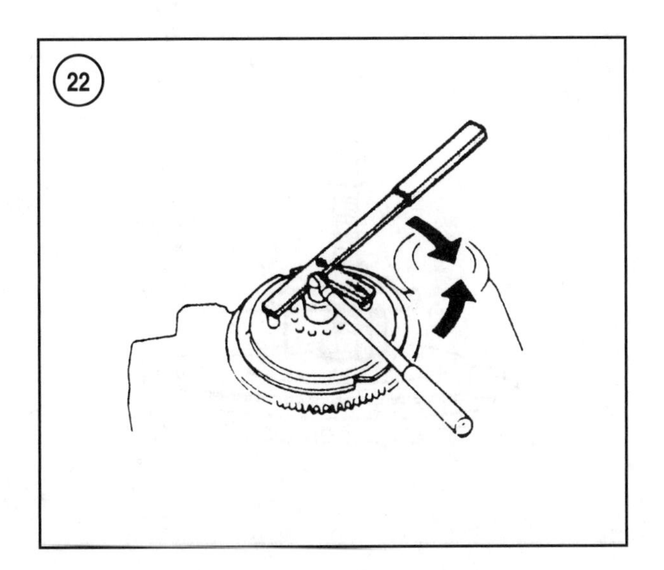

wire harness. Connect the engine cover mounted trim switch to the engine harness.

13. *Oil injected models*—Connect the oil supply hose to the engine mounted reservoir fitting. Refer to the diagrams in Chapter Twelve to identify the proper hose connection point.

14. *HPDI models*—Connect the shift position switch wires to the engine wire harness. See Chapter Six.

15. Push the water hose onto the water pressure relief valve fitting (**Figure 12**). Secure the hose with a clamp.

16. Connect the throttle and shift cables to the linkages (**Figure 10**, typical). Refer to Chapter Four to identify the linkages.

17. Connect the battery cable terminals to the starter relay (A, **Figure 8**) and the engine ground (B).

18. Plug the main engine harness into the remote control instrument harness (**Figure 7**).

19. Install the retainer (**Figure 9**) to secure the hoses and cables to the lower engine cover.

20. Install both aprons (**Figure 13**) onto the drive shaft housing. Securely tighten the apron screws.

21. Connect the fuel supply hose (**Figure 14**) to the fuel filter, fuel pump or other connection point. Refer to Chapter Five.

22. Connect both battery cables.

23. Perform all applicable adjustments as described in Chapter Four. Start the engine and immediately check for fuel, water or exhaust leakage. Correct any leakage before putting the engine into service.

FLYWHEEL

WARNING

Wear protective eyewear when removing or installing the flywheel or other components of the engine. Never use a hammer without using protective eyewear.

CAUTION

Use only the appropriate tools and procedures to remove the flywheel. Never strike the flywheel. The magnets may break and cause poor ignition system performance or potential damage to other engine components.

Removal

Make sure the engine or power head is securely mounted before removing the flywheel. Use a suitable flywheel holder and a flywheel removal tool to remove the flywheel.

1. Disconnect both cables from the battery. Ground the spark plug lead to prevent accidental starting or arcing.
2. Remove the flywheel cover (A, **Figure 13**).
3. *HPDI models*—Remove the drive belt for the high pressure fuel pump and the drive belt pulley from the flywheel as described in Chapter Five. Remove the three bolts (1, **Figure 21**), then carefully pry the pulley bracket (2) from the flywheel.
4. Fit the pins of the flywheel holder (Yamaha part No. YB-06139-90890-06522) into the openings in the flywheel (**Figure 22**). Engage an appropriate size socket and breaker bar onto the flywheel nut. Hold the flywheel stationary and rotate the breaker bar counterclockwise to loosen the flywheel nut (**Figure 22**).
5. Loosen the flywheel nut until the top edge is flush or slightly below the threads on the crankshaft. Install the flywheel removal tool (Yamaha part No. YB-06117/90890-06521) to the flywheel. Secure the puller to the flywheel with the included bolts. Make sure the puller bolts are threaded several turns into the flywheel and the puller is level with the flywheel surface (**Figure 23**). Check to make sure the large puller bolt is contacting the top end of the crankshaft and not the flywheel nut.
6. Use the flywheel holder to keep the flywheel from rotating. Turn the puller bolt in a clockwise direction until the bolt is difficult to turn (**Figure 24**).
7. Support the flywheel and lightly tap on the puller bolt (**Figure 25**). Tighten the puller bolt and again tap on the puller bolt. Continue until the flywheel pops free from the crankshaft.

8. Remove the flywheel puller. Remove and discard the flywheel nut and washer. Remove the woodruff key from the slot in the crankshaft (**Figure 26**). Retrieve the key from the flywheel magnets if not found in the key slot. Inspect the woodruff key for corroded, bent or marked surfaces. Replace the key if not in excellent condition.
9. Use a solvent soaked shop towel to clean debris and contaminants from the crankshaft threads, key slot and flywheel taper. Use compressed air to remove debris or contaminants from the flywheel. Do not use solvent to clean the flywheel. Solvent may soften the magnet adhesive. Wipe oily deposits from the flywheel with a shop towel.
10. Inspect the flywheel for cracked or damaged magnets, rough surfaces in the tapered opening and worn or damaged flywheel teeth. Replace the flywheel if these or other defects are present.
11. Inspect the crankshaft taper for corroded, worn or rough surfaces. Minor surface corrosion can be cleaned by light polishing with crocus cloth. Replace the crankshaft if the taper is discolored, deeply pitted or has rough surfaces.
12. Inspect the crankshaft threads for corroded or damaged threads. Repair minor thread damage with an appropriate die. Replace the crankshaft if the threads cannot be returned to like-new condition.

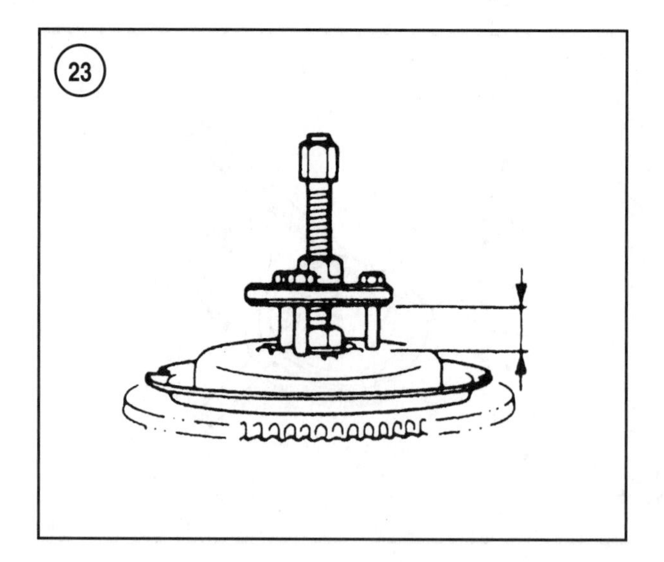

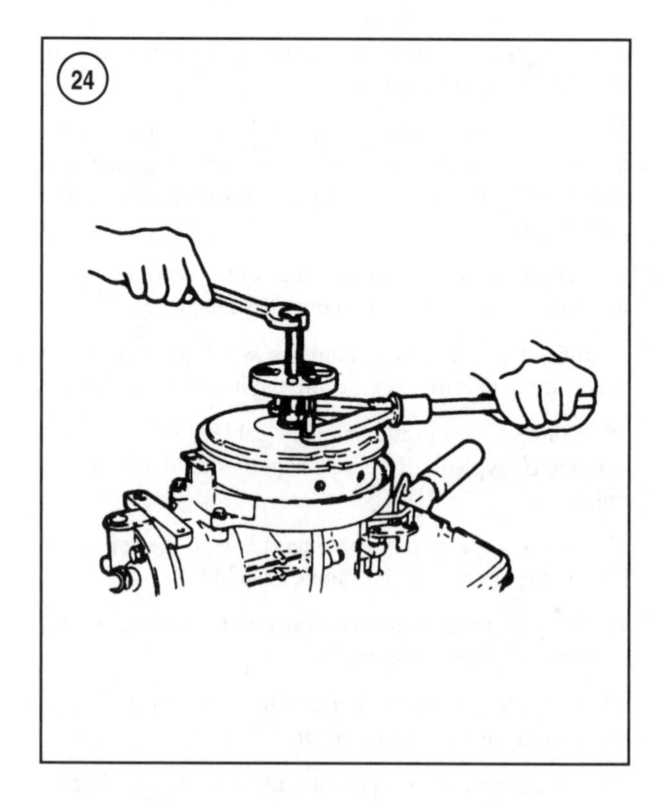

Installation

Make sure the engine or power head is securely mounted before installing the flywheel. Use a suitable flywheel holder (Yamaha part No. YB-01639-90890-06522) to install the flywheel.

1. Place the woodruff key into the key slot as shown in **Figure 26**.
2. Position the flywheel onto the crankshaft taper. Slowly rotate the flywheel to align the flywheel key slot with the woodruff key. The flywheel drops when the key slot aligns with the key. Rotate the flywheel clockwise while observing the threaded end of the crankshaft. If the crankshaft does not rotate with the flywheel, remove the flywheel, reposition the woodruff and repeat the process.
3. Apply a light coating of outboard engine oil to the crankshaft threads. Place a new washer on the crankshaft and seat it against the flywheel. Hand-thread the new flywheel nut onto the crankshaft.
4. Fit the pins of the flywheel holder (Yamaha part No. YB-06139-90890-06522) into the flywheel openings (**Figure 22**). Hold the flywheel stationary and tighten the flywheel nut to the specification in **Table 1** with an appropriate size socket and torque wrench on the flywheel nut.
5. *HPDI models*—Fit the pulley bracket (2, **Figure 21**) onto the flywheel. Align the openings. Then thread the three bolts (1, **Figure 21**) into the bracket and flywheel. Evenly tighten the bolts to the general torque specification in Chapter One. Install the drive belt pulley and drive belt for the high pressure fuel pump as described in Chapter Five.

6. Connect the spark plug leads and battery cables. Install the flywheel cover.

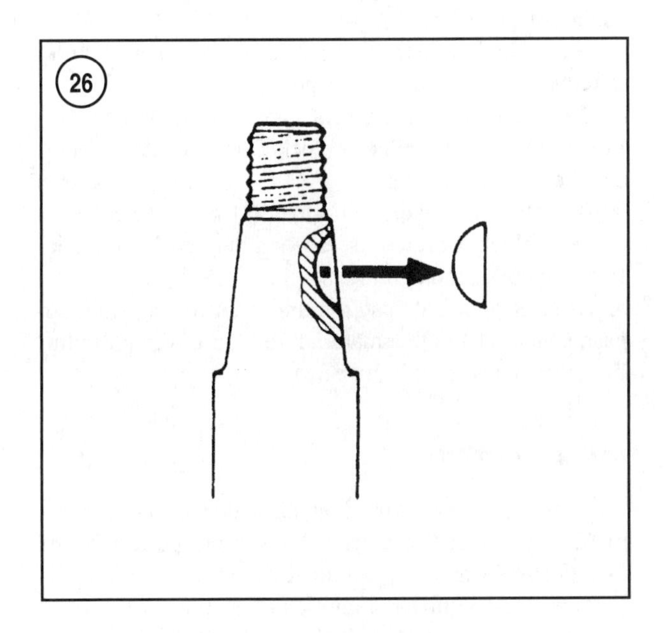

POWER HEAD DISASSEMBLY

Power head overhaul gasket sets are available for all models. It is often more economical and simpler to order the gasket set. Replace *every* gasket, seal and O-ring during a power head assembly.

Dowel pins locate the crankcase halves to each other and locate some crankshaft bearings. The dowel pins do not have to be removed if they are securely seated in a bore on either side of the crankcase halves or in the bearing. However, they must be accounted for during disassembly and assembly. If a dowel pin can be readily removed from the bore, it should be removed and stored with the other internal components until assembly begins.

Clean and inspect all power head components before any assembly occurs. If the power head has had a major failure, it may be more economical (in time and money) to replace the basic power head as an assembly.

Special tools and the part number are listed in the repair instructions.

A large number of fasteners of different lengths and sizes are used in a power head. Plastic sandwich bags and/or cupcake tins are excellent for keeping small parts organized. Tag all larger internal parts for location and orientation. Use a felt-tipped permanent marker to mark components after cleaning them. Avoid scribing or stamping internal components as the marking process may damage or weaken the component.

If only partial disassembly is required to access the faulty components (such as the thermostat, exhaust or water jacket gaskets), perform the disassembly procedures until the component is accessible. Refer to the section in the power head assembly section that describes the faulty component(s) to reassemble the gearcase.

Always make notes, drawings or photographs of all external power head components *before* beginning the disassembly process. Although illustrations are provided throughout this manual, drawings and photographs save a great deal of time during the assembly process. Correct wire and hose routing is important for proper engine operation. An incorrectly routed wire or hose may interfere with linkage movement and cause poor throttle or shift control. Hoses or wires may chafe and short circuit if allowed to contact sharp edges or moving parts. Other components such as fuel pumps can often be mounted in two or more positions. Mark or make note of the top and forward direction *before* removing such components.

If possible, remove a cluster of components that share common wires or hoses. This step will reduce the time required to disassemble and assemble the power head and reduce the chance of improper connections on assembly.

Bearings and some other internal components of the power head can be removed and reused if in good condition. The *Power Head Components (Cleaning and Inspection)* section of this chapter describes which components must be replaced. If needle bearings must be reused, make certain they are installed in the same location during assembly.

The piston rings must be replaced and the cylinder bore must be honed if the piston(s) are taken out of the cylinder bore. Always replace piston pin lockrings if removed. The cost of many of these components is small compared to the expense to repair the damage to other components should they fail.

External Component Removal

1. Remove the flywheel as described in this chapter.
2. Make note of the connections and mounting locations, then remove all electrical and ignition components. Refer to Chapter Six for removal procedures.
3. Make note of all hose, wire and linkage connections, then remove all fuel system components. Refer to Chapter Five for removal procedures.
4. *Oil injection models*—Drain oil from the engine mounted reservoir, then remove all oil injection system components as described in Chapter Twelve.
5. Loosen the pivot bolt (**Figure 27**), then remove the throttle arm, as an assembly, from the power head.
6. Remove the intake manifold/reed block housing from the power head as described in Chapter Five.

Model Variations

Refer to the appropriate illustrations (**Figures 28-30**) to assist with component orientation and mounting arrangement.

Water Jacket Cover, Exhaust Cover, Water Pressure Valve and Thermostat Removal

Outboard engines are normally exposed to a corrosive operating environment. Corrosion on the components, especially the fasteners, will be more prevalent if the engine is operated in saltwater or heavily polluted water. Corroded fasteners are difficult to remove. Refer to Chapter One in this manual for useful tips on removing stubborn fasteners.

Loosen water jacket or exhaust covers gradually and in a crossing pattern to help prevent plate warpage. Use caution when it is necessary to pry a component loose. Avoid using any sharp object that can damage the mating surfaces. Castings or pry points (**Figure 31**) are usually present and allow the components to be separated without damaging the to mating surfaces.

Upper water jacket cover removal

An upper water jacket is used on all models covered in this manual. The jacket cover is located on the top of the cylinder block and directly behind the flywheel. The flywheel must be removed to access the cover bolts. On HPDI models, the high pressure mechanical fuel pump and fuel rails must also be removed.

1. Remove the flywheel as described in this chapter.

2. Note the mounting location and orientation of any hose clamps (2, **Figure 32**, typical) that are secured with the cover bolts (1).
3. Gradually loosen the cover bolts, in a crossing pattern, to reduce the chance of warpage.
4. Using a blunt tip instrument, carefully pry the cover (5, **Figure 32**) from the cylinder block. Do not scratch or damage the mating surfaces.
5. Remove the gasket (6, **Figure 32**) from the cover or cylinder block. Carefully scrape residual gasket material from the cover and cylinder block.
6. Use a Scotch-brite pad to remove corrosion, stubborn gasket material or other material from the cover and cylinder block surface (7, **Figure 32**).

Exhaust cover removal

An exhaust cover is used on all models covered in this manual. The exhaust cover is located on the rear of the cylinder block and between the cylinder bank. The ECU (engine control unit) and ignition coils must be removed to access the cover. On HPDI models, the high pressure mechanical fuel pump and fuel rails must also be removed.

1. Remove the ECU and ignition coils as described in Chapter Six.
2. *HPDI models*—Remove the high pressure fuel pump and fuel rails as described in Chapter Five.
3. Gradually loosen the cover bolts, in a crossing pattern, to reduce the chance of warpage.
4. Locate pry points on the sides of the cover (6, **Figure 33**). Use a blunt tip pry bar to carefully pry the cover loose. Do not damage the mating surfaces. If removal is difficult, check for overlooked bolts. Apply moderate heat to the mating surface if the cover or plate is seized to the cylinder block.

POWER HEAD COMPONENTS (80 JET AND 115-130 HP MODELS)

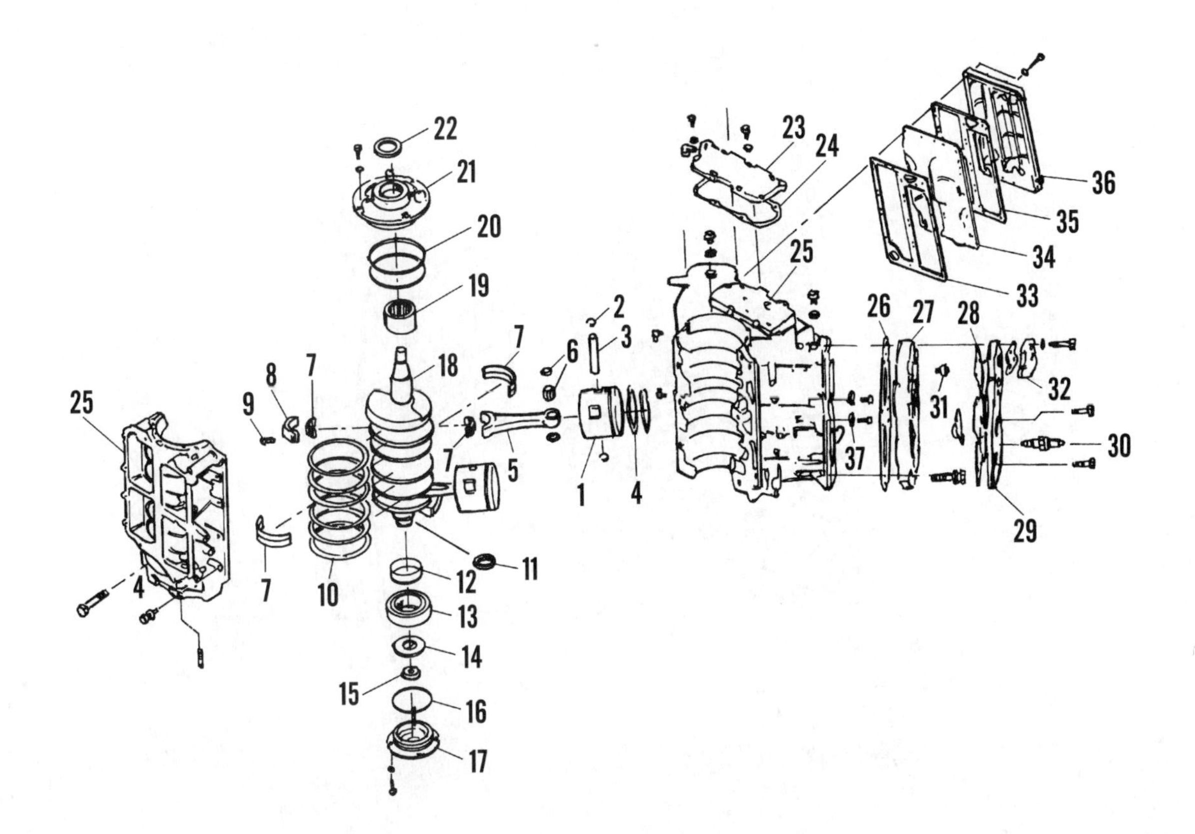

1. Piston
2. Lockring
3. Piston pin
4. Piston rings
5. Connecting rod
6. Bearing and thrust washers
7. Bearing set
8. Connecting rod cap
9. Rod bolt
10. Crankcase sealing ring
11. Snap ring
12. Oil injection drive gear
13. Lower main bearing (ball bearing)
14. Seal
15. Seal
16. O-ring
17. Bearing/cover/seal housing (lower)
18. Crankshaft
19. Upper main bearing (roller)
20. O-rings
21. Bearing/cover/seal housing (upper)
22. Seal
23. Water jacket cover
24. Gasket
25. Cylinder block/crankcase cover
26. Head gasket
27. Cylinder head
28. Gasket
29. Water jacket cover
30. Spark plug
31. Thermostat
32. Thermostat cover
33. Gasket
34. Exhaust plate
35. Gasket
36. Exhaust cover
37. Anode

(29)

POWER HEAD COMPONENTS
(105 JET AND 150-200 HP MODELS [2.6 LITER])

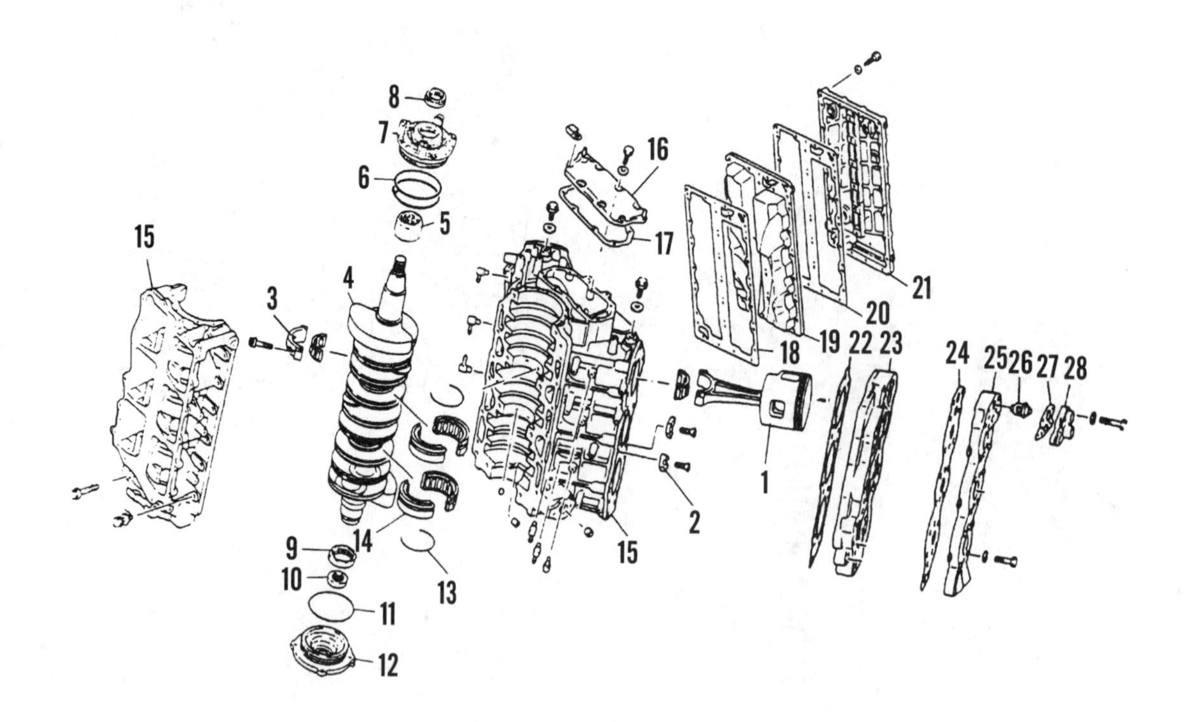

1. Piston/connecting rod assembly
2. Anode
3. Connecting rod cap
4. Crankshaft
5. Upper main bearing (roller)
6. O-rings
7. Bearing/cover/seal housing (upper)
8. Seal
9. Oil injection drive gear
10. Seal
11. O-ring
12. Bearing/cover/seal housing (lower)
13. Retaining ring
14. Center main bearing
15. Cylinder block/crankcase cover
16. Water jacket cover
17. Gasket
18. Gasket
19. Exhaust plate
20. Gasket
21. Exhaust cover
22. Cylinder head gasket
23. Cylinder head
24. Gasket
25. Water jacket cover
26. Thermostat
27. Gasket
28. Thermostat cover

30

POWER HEAD COMPONENTS (200-250 HP MODELS [3.1 LITER])

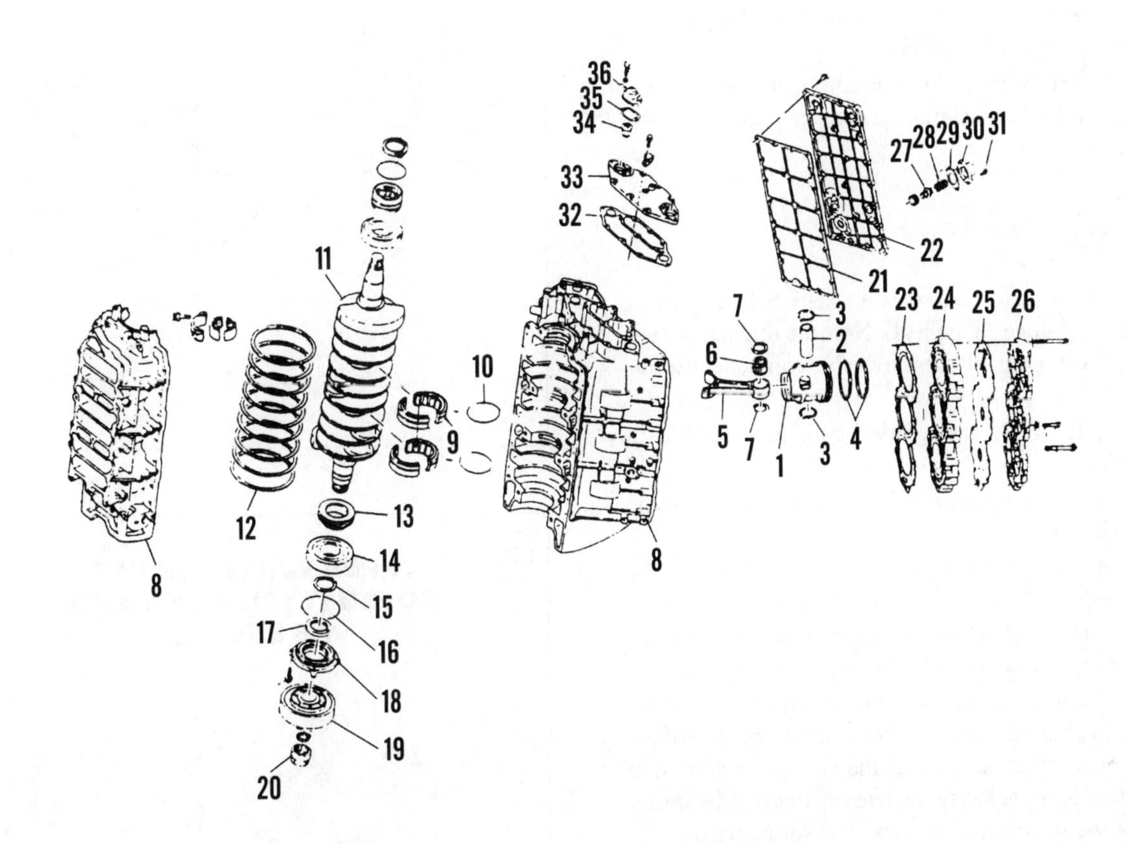

1. Piston
2. Piston pin
3. Lockring
4. Piston rings
5. Connecting rod
6. Piston pin bearing
7. Thrust washer
8. Cylinder block/crankcase cover
9. Center main bearing
10. Retaining ring
11. Crankshaft
12. Crankcase sealing rings
13. Oil injection drive gear
14. Lower main bearing (ball bearing)
15. Snap ring
16. O-ring
17. Seal
18. Bearing/cover/seal housing (lower)
19. Torsional dampener
20. Nut
21. Gasket
22. Exhaust cover
23. Cylinder head gasket
24. Cylinder head
25. Gasket
26. Water jacket cover
27. Water pressure relief valve
28. Spring
29. Gasket
30. Cover
31. Bolt
32. Gasket
33. Water jacket cover
34. Thermostat
35. Gasket
36. Thermostat cover

5. Locate a pry point near the top of the exhaust plate (8, **Figure 33**). Carefully pry the plate from the cylinder block. Do not damage the mating surfaces.

6. Carefully scrape the gaskets (7 and 9, **Figure 33**) from the cylinder block, cover and exhaust plate. Never reuse the cover gasket or plate gasket.

7. Use a Scotch-brite pad to clean carbon deposits, corrosion, residual gasket material or other material from the exhaust plate and cover.

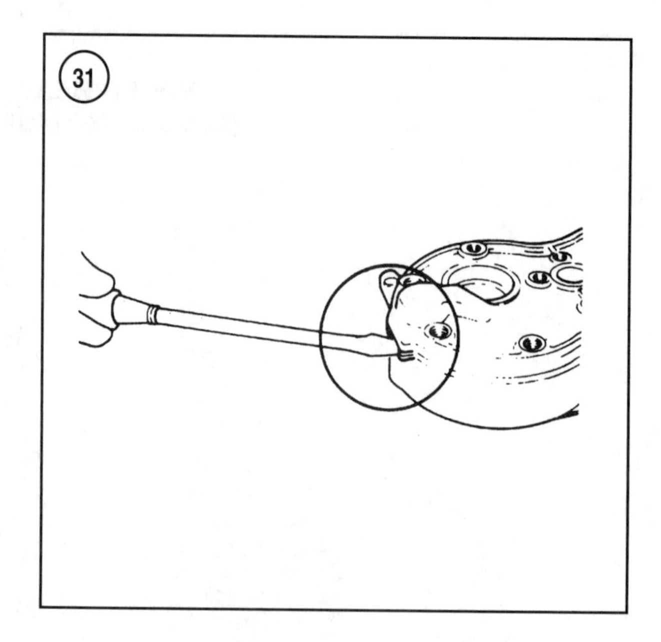

Water pressure valve removal

The water pressure valve is located below exhaust cover (see 4, **Figure 33**, typical). Note the orientation and mounting location for all pressure valve components prior to removal.

1. Remove the ECU and ignition coils as described in Chapter Six.

2. Remove the clamp, then pull the water hose from the pressure valve cover.

3. Gradually loosen the two bolts. Then carefully pry the pressure valve cover (1, **Figure 33**) from the exhaust cover. Maintain pressure on the cover to prevent the components from inadvertently falling out of the opening.

4. Remove the cover, then pull the spring (3, **Figure 33**) and valve (4) from the opening. Note the orientation of the valve prior to removal. Inspect the spring for corrosion damage, bent loops or lost spring tension. Replace the spring if it is damaged, corroded or it has lost spring tension.

5. Use needlenose pliers to remove the valve seat (5, **Figure 33**) from the opening.

6. Remove the gasket (2, **Figure 33**) from the pressure valve or exhaust cover. Work carefully to avoid damaging the mating surfaces.

7. Clean all corrosion and salt or mineral deposits from the opening, spring, valve and cover. If necessary, use a Scotch-brite pad to clean the surfaces.

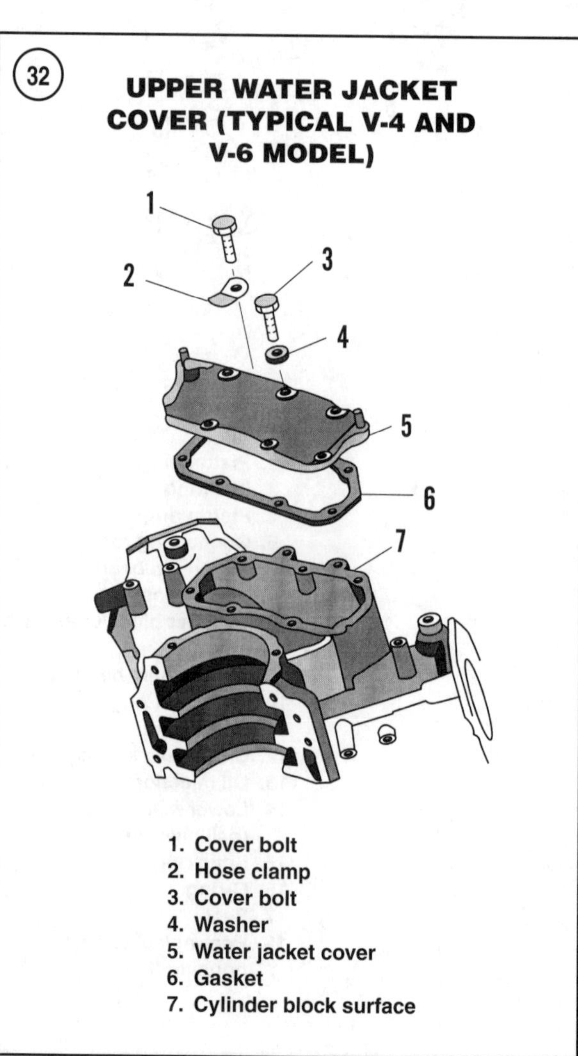

1. Cover bolt
2. Hose clamp
3. Cover bolt
4. Washer
5. Water jacket cover
6. Gasket
7. Cylinder block surface

Thermostat removal

The thermostats are located beneath a cover (**Figure 34**) on each cylinder head water jacket.

1. Remove the two bolts from the thermostat cover.

2. Carefully pry the thermostat cover from the cylinder head, cylinder block or exhaust cover. Use a blunt tip pry bar and work carefully to avoid damaging the mating surfaces.

3. Use needlenose pliers to pull the thermostat from the cylinder head water jacket.

33

EXHAUST COVER/PLATE AND WATER PRESSURE VALVE (TYPICAL V-4 AND V-6 MODEL)

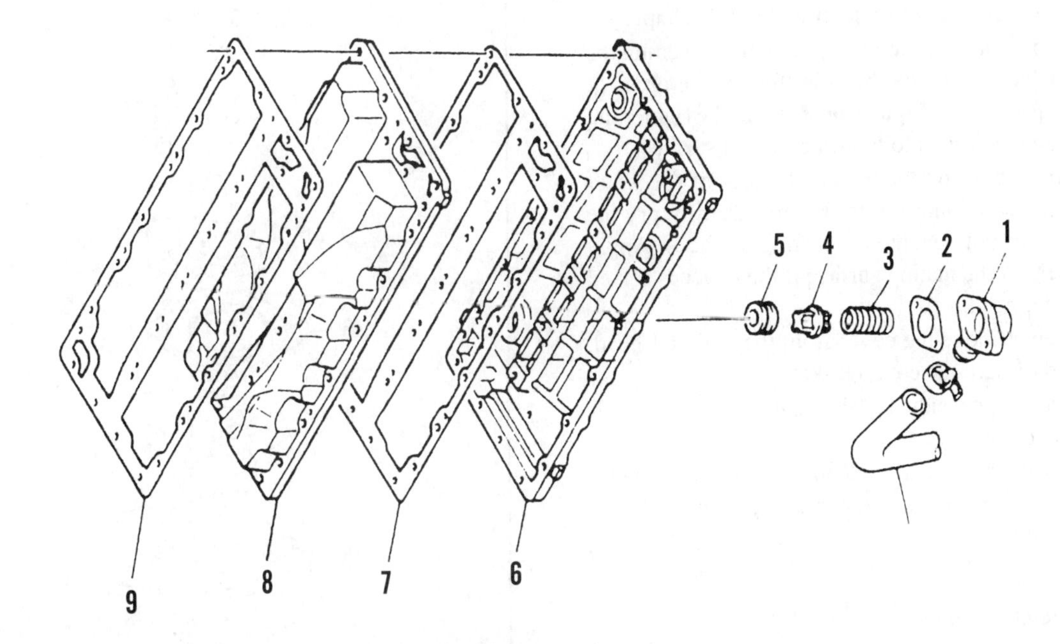

1. Cover
2. Gasket
3. Spring
4. Pressure valve
5. Seat
6. Exhaust cover
7. Gasket
8. Exhaust plate
9. Gasket

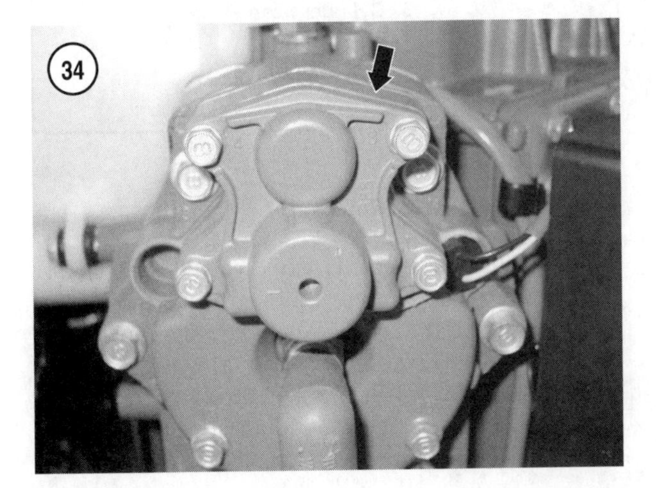

34

4. Remove the gasket (3, **Figure 35**) from the cover or cylinder head water jacket. Work carefully to avoid damaging the mating surfaces.
5. Clean all corrosion and salt or mineral deposits from the opening, spring, valve and cover. If necessary, use a Scotch-brite pad to clean the surfaces.
6. Test the thermostat as described in Chapter Two. Replace the thermostat if faulty, corroded, or coated with salt or mineral deposits.

Water jacket cover removal (cylinder head mounted)

Refer to the appropriate illustration (**Figures 28-30**) to assist with component orientation and mounting location.

Locate the torque sequence numbers (**Figure 36**) on the water jacket surface prior to removing the jacket. To help prevent warpage of the surfaces, gradually loosen each bolt in the reverse order of the tightening sequence. Always replace the water jacket gasket anytime the mounting bolts are loosened.

1. Remove the thermostat as described in this chapter.
2. Support the water jacket while gradually loosening the mounting bolts. Loosen the bolts in the reverse order of the tightening sequence **Figure 36**. Refer to the tightening sequence numbers cast into the water jacket surface.
3. Locate pry points on the top and bottom of the water jacket cover. Use a blunt tip pry bar to carefully pry the cover loose. Do not damage the mating surfaces. Apply moderate heat to the mating surface if the cover is seized to the cylinder head.
4. When free, remove the cover from the cylinder head. Remove and discard the cover gasket.
5. Carefully scrape residual gasket material from the cylinder head and cover.
6. Use a Scotch-brite pad to clean salt crystals, mineral deposits, corrosion or other material from the cylinder head and cover.

Cylinder Head Removal

Refer to the appropriate illustration (**Figures 28-30**) to assist with component orientation and mounting location. Locate the torque sequence numbers (**Figure 36**) on the cylinder head surface prior to removing the cylinder head. To help prevent the surfaces from warping, gradually loosen each cylinder head bolt in the reverse of the tightening sequence. Always replace the cylinder head gasket anytime the mounting bolts are loosened.

1. Mark the cylinder head to identify the mounting location (port or starboard side) on the cylinder block.
2. Support the cylinder head while gradually loosening the mounting bolts. Loosen the bolts in the reverse of the tightening sequence. Refer to the tightening sequence numbers cast into the cylinder head.
3. Tap lightly on the end of the cylinder head with a plastic or rubber mallet to free it from the cylinder block. Use a blunt tip instrument and take all necessary precaution to avoid damaging the mating surfaces.
4. When free, remove the cylinder head and discard the cylinder head gasket.
5. Repeat this procedure to remove the remaining cylinder head.
6. Clean and inspect the cylinder head as described in this chapter. See *Power Head Component Cleaning and Inspection*.

(35) THERMOSTAT AND COVER (TYPICAL)

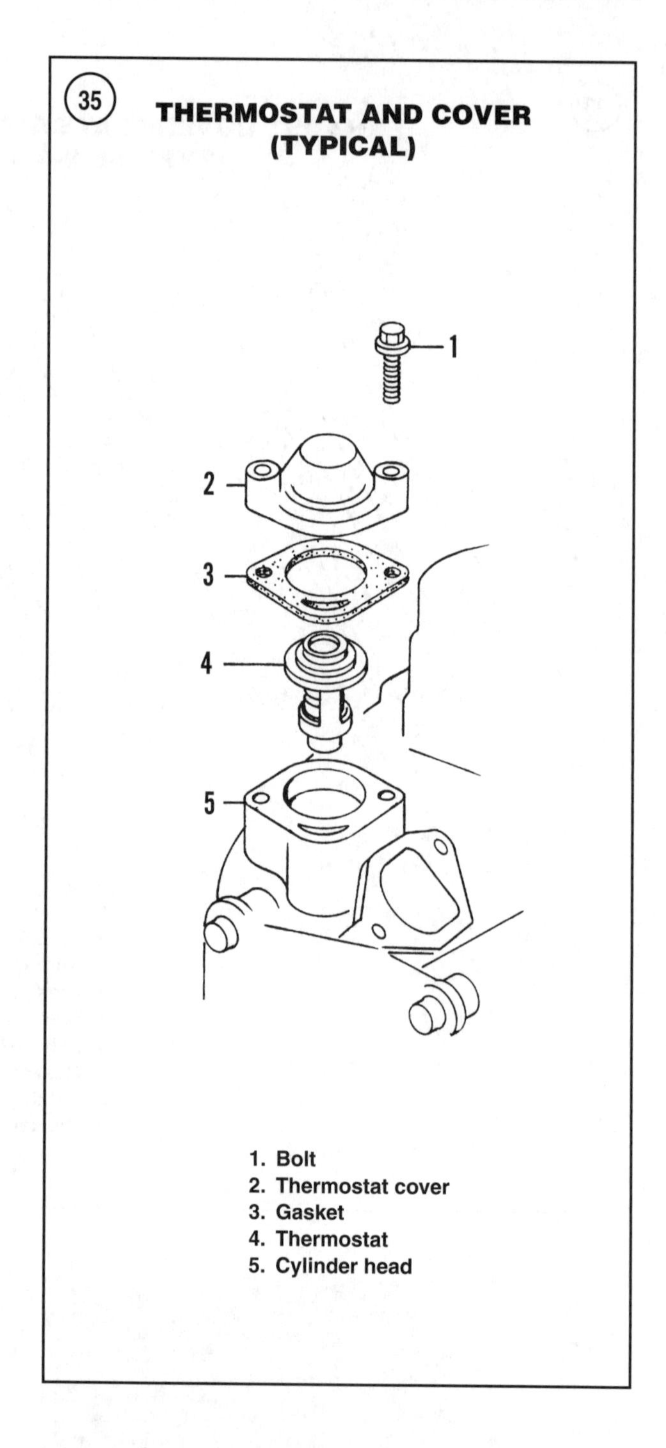

1. Bolt
2. Thermostat cover
3. Gasket
4. Thermostat
5. Cylinder head

Torsional Dampener Removal

The torsional dampener is used on all 200-250 hp (3.1 liter) models. The dampener is located on the lower end of the crankshaft. Use a flywheel holding tool (Yamaha part No. YB-01639/90890-06522) and a universal puller (Yamaha part No. YB-6117/90890-06521) to remove the dampener. Refer to **Figure 30** to assist with component orientation and mounting location.

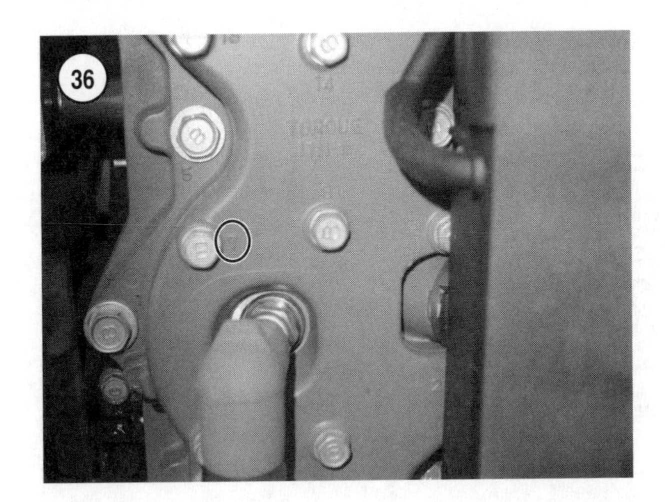

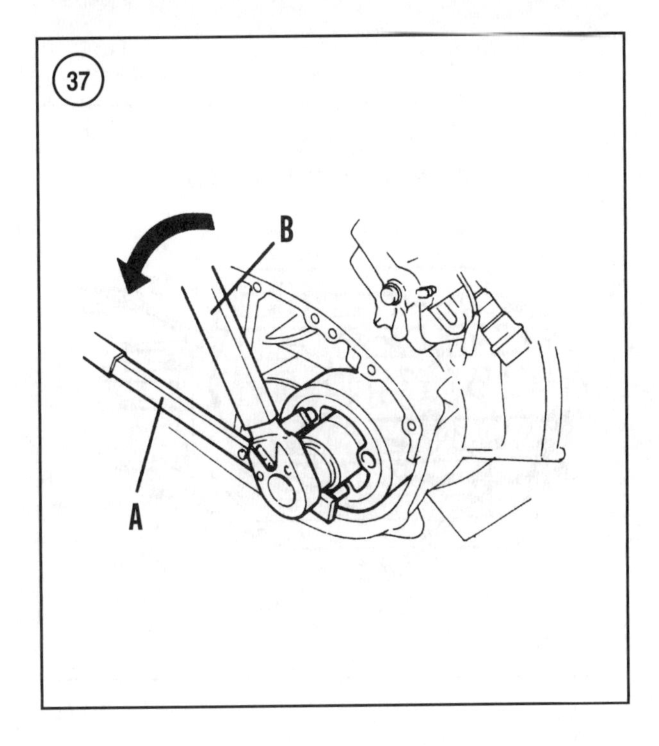

1. Fit the pins of the flywheel holding tool (A, **Figure 37**) into the openings in the torsional dampener.

2. Hold the flywheel holding tool securely, then turn the nut (20, **Figure 30**) counterclockwise. Continue until the nut is free from the crankshaft.

3. Place padding on the block (**Figure 38**) to protect the block surfaces from the puller. Align the puller with the dampener as shown in **Figure 38**. Then thread the puller bolts through the puller and into the dampener. Tighten the bolts until the puller just seats against the padding. Make sure the puller bolt contacts the tapered opening in the crankshaft.

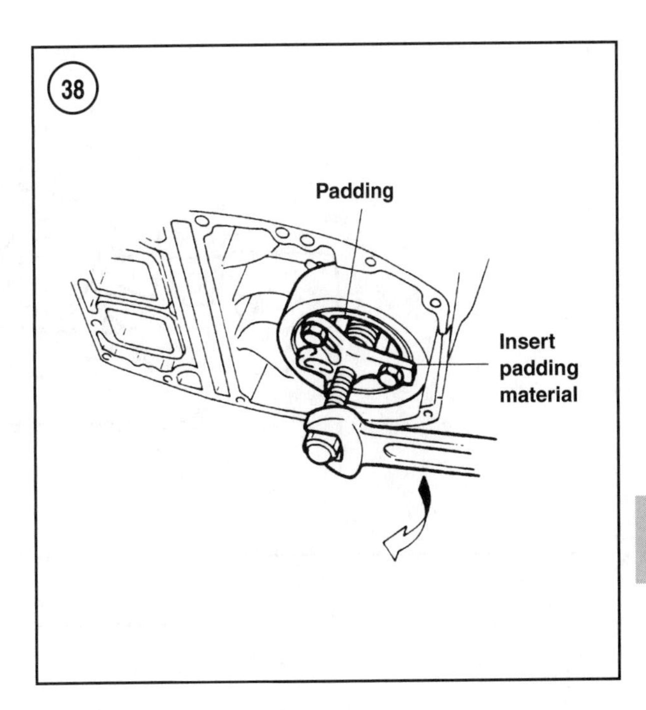

4. Tighten the puller bolt to remove the dampener from the crankshaft. Remove the puller from the dampener.

Crankcase Cover Removal

Refer to the appropriate illustration (**Figures 28-30**) to assist with component orientation and mounting location. To help prevent warpage of the cover, gradually loosen each bolt in a crossing pattern sequence starting from the ends and working toward the middle.

1. Remove the bolts from the upper and lower bearing/cover/seal housing (**Figure 39**, typical). To help prevent damage to the housing, do not remove the bearing/cover/seal housing at this time. The component is easily removed after removing the crankcase cover.
2. Locate all of the fasteners on the crankcase cover (**Figure 40**, typical).
   a. *80 Jet and 115-130 hp models*—14 bolts are used.
   b. *105 Jet and 150-200 hp (2.6 liter) models*—20 bolts are used.
   c. *200-250 hp models (3.1 liter)*—22 bolts are used.
3. Gradually loosen the crankcase cover bolts in a crossing pattern.
4. Locate suitable pry points (**Figure 41**) and carefully pry the crankcase cover from the cylinder block. If removal is difficult, check for overlooked bolts. Apply moderate heat to the mating surface if the cover is seized onto the cylinder block. Take all necessary precautions to prevent damage to the mating surfaces.

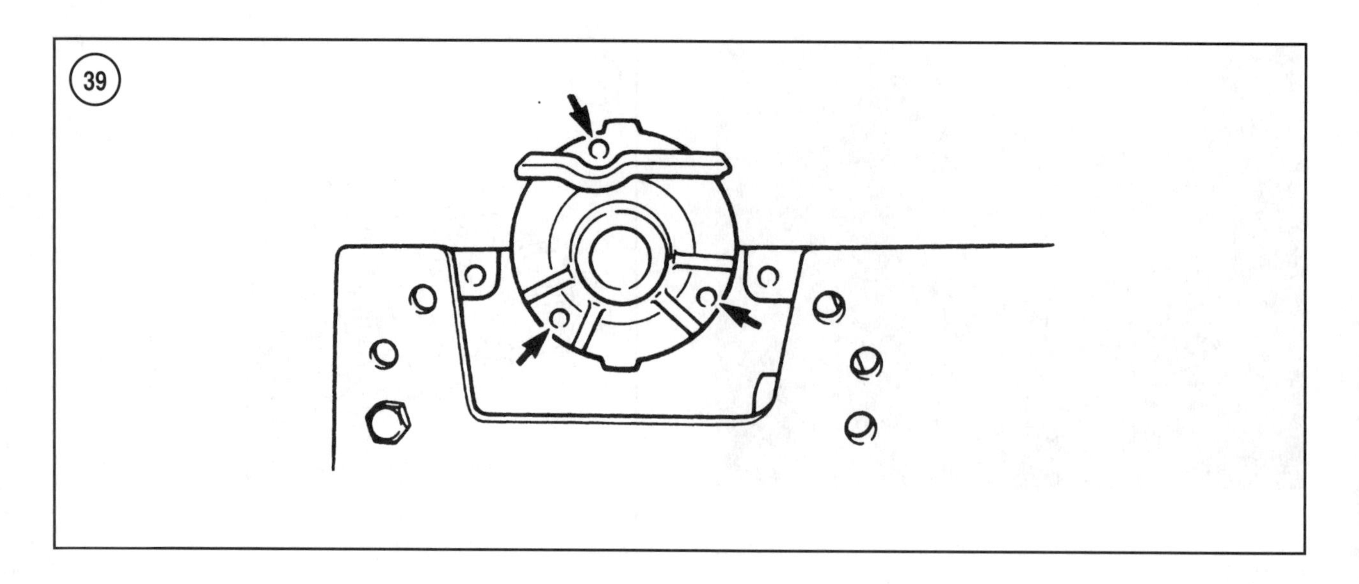
39

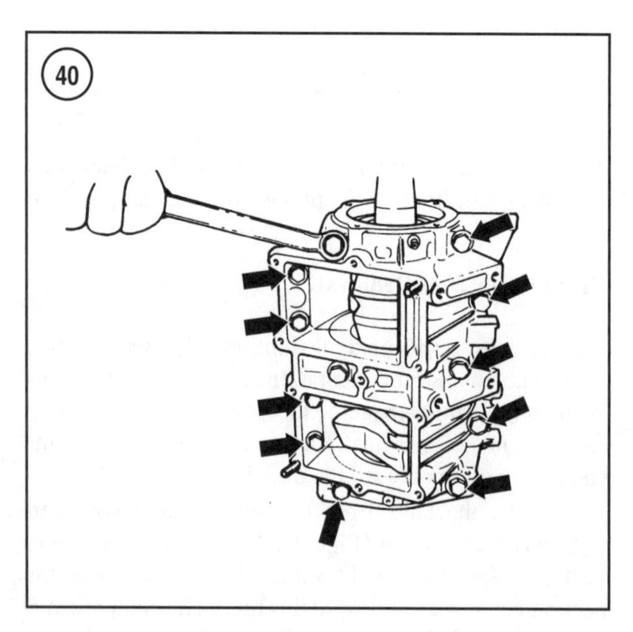
40

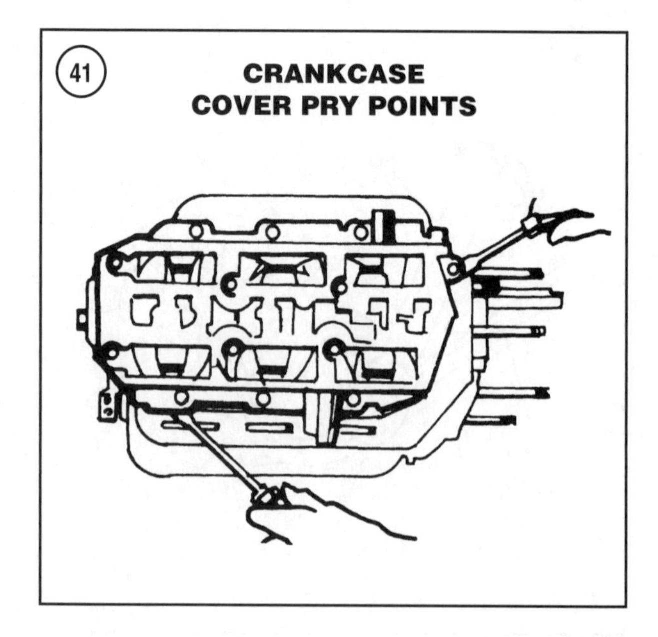
41

Piston and Connecting Rod Removal

Always keep components organized. The piston/connecting rod and other components must be installed into the same cylinder unless the component is replaced. Mark the cylinder number on the piston and connecting rods. Use an ink marker that can withstand solvent washing or a scratch awl. The number one cylinder is at the top starboard side as viewed from the cylinder head side of the block. Even numbered cylinders are located on the port side cylinder bank. The connecting rod cap is matched to the connecting rod. Install the cap only onto the corresponding rod and in the original position. Never allow the connecting rod cap to become switched with a different rod.

1. Thread two of the mounting bolts through the bearing/cover/seal housing and into the cylinder block (**Figure 42**) to prevent the crankshaft from becoming dislodged from the cylinder block during piston removal.
2. Rotate the crankshaft until the No. 1 cylinder is at the bottom of the stroke. Loosen the connecting rod bolts (**Figure 43**) in three even steps. Gently tap on the rod cap to loosen the cap.
3. Mark the flywheel side of the connecting rod to ensure proper orientation of the rod on assembly.
4. Remove the rod bolts (1, **Figure 44**). Then carefully pull the rod cap (2, **Figure 44**) from the connecting rod.

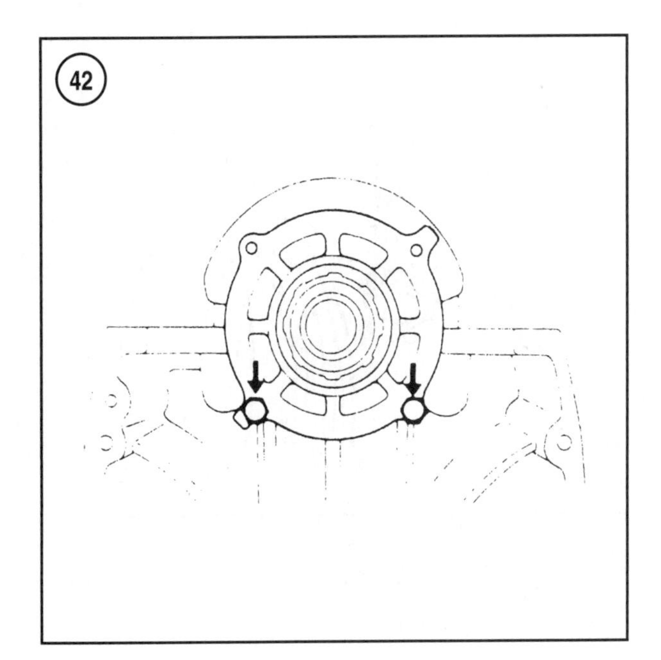

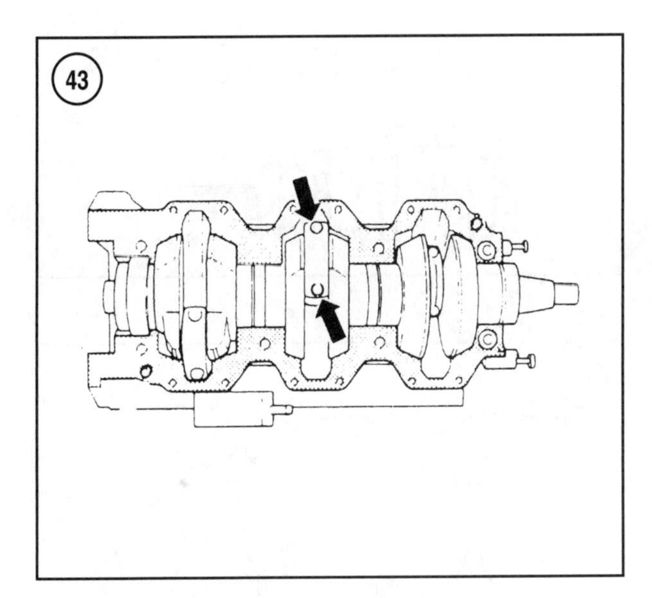

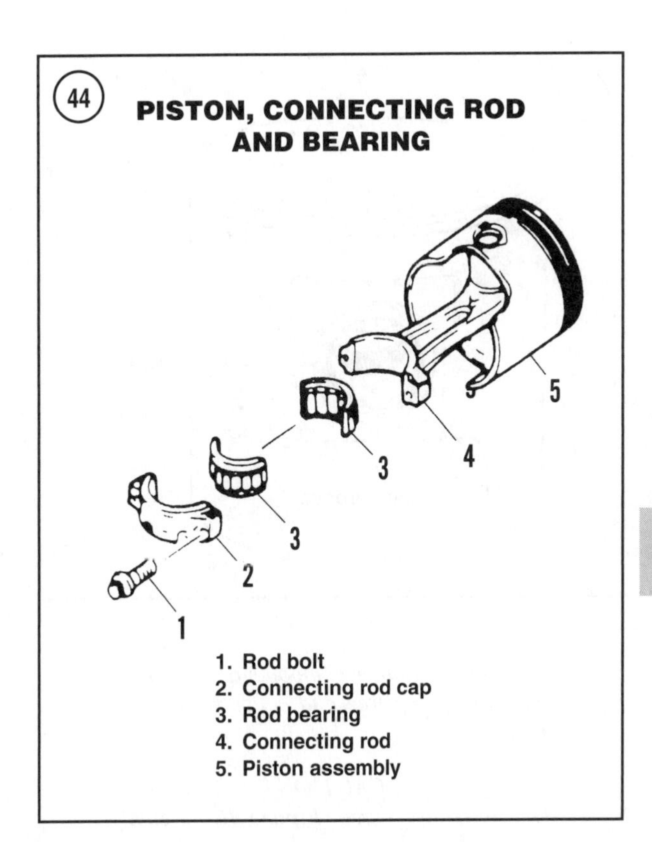

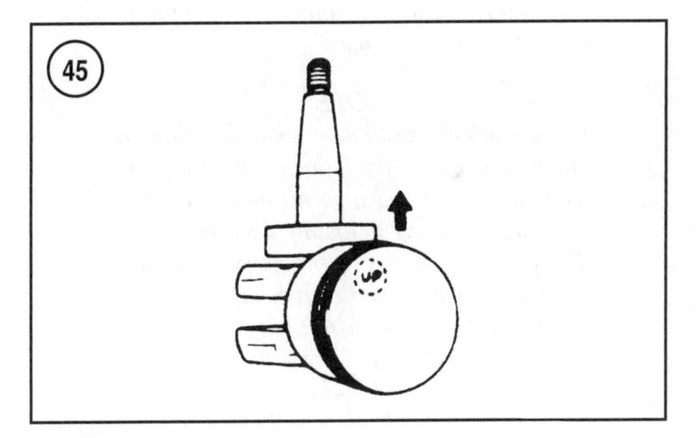

5. Remove the rod bearings (3, **Figure 44**). If the bearings must be reused, mark the flywheel side to ensure they turn in the same direction after assembly.

6. Use a large wooden dowel or hammer handle to push the connecting rod and piston assembly out of the cylinder head side of the cylinder block. Provide continuous support to the piston and connecting rod to prevent it from falling from the cylinder block.

7. Note the *UP* marking on the piston dome (**Figure 45**), then position the rod cap onto the connecting rod with the up side correctly oriented. Install the rod bolts and hand-tighten. Mark the cylinder number on the inside of the piston skirt and the I-beam portion of the connecting rod.

8. Repeat Steps 2-7 for the remaining pistons and connecting rods.

Piston Disassembly

Refer to the appropriate illustration (**Figures 28-30**) to assist with component orientation and mounting location.

WARNING
Use protective eyewear and gloves when working with the power head. Piston pin

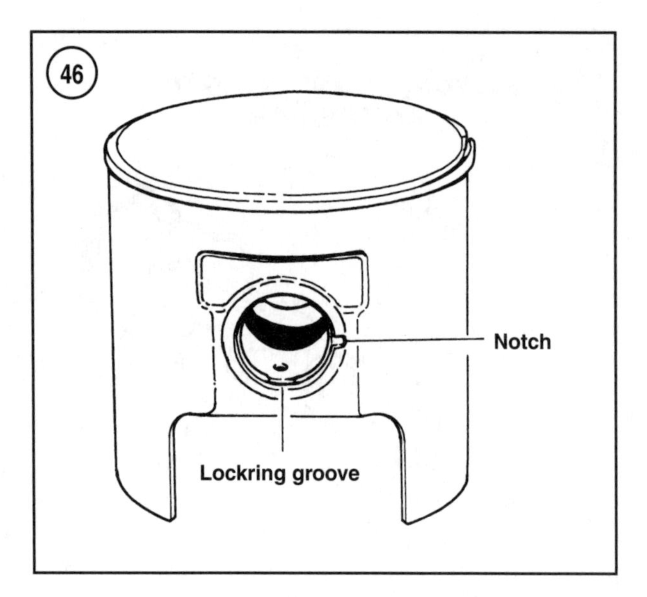

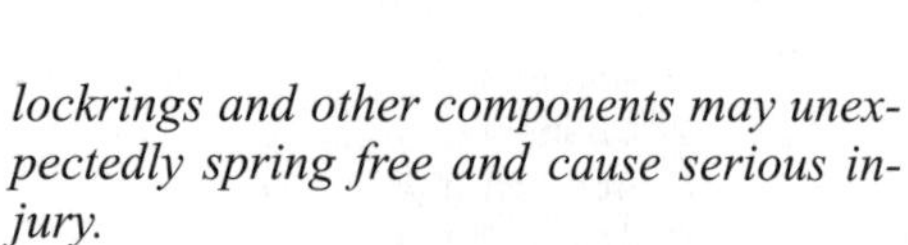
lockrings and other components may unexpectedly spring free and cause serious injury.

CAUTION
*The lockring groove (**Figure 46**) is easily damaged by the removal tool. Work carefully and avoid any unnecessary contact with the piston surfaces.*

NOTE
Some models use loose needle bearings at the piston pin end of the connecting rod. Work carefully and take all necessary precautions to prevent losing needles during the removal process. If the needle bearings must be reused, install them into the same connecting rod and piston pin.

NOTE
If piston pin removal is difficult, heat the piston to approximately 60° C (140° F) with a portable hair dryer. The piston pin bore expands when heated, allowing easier pin removal.

1. Use a scribe to carefully pry the lockrings from the piston (**Figure 47**). Hold a protected thumb over the lockring to prevent it from springing free upon removal. Use the notch opposite the lockring gap on models with two notches in the lockring groove. Discard the lockring upon removal.
2. Select a suitable steel rod, socket or section of tubing to push the piston pin from the piston. The removal tool must be slightly smaller in diameter than the piston pin. Piston pin diameters are listed in **Table 3**.

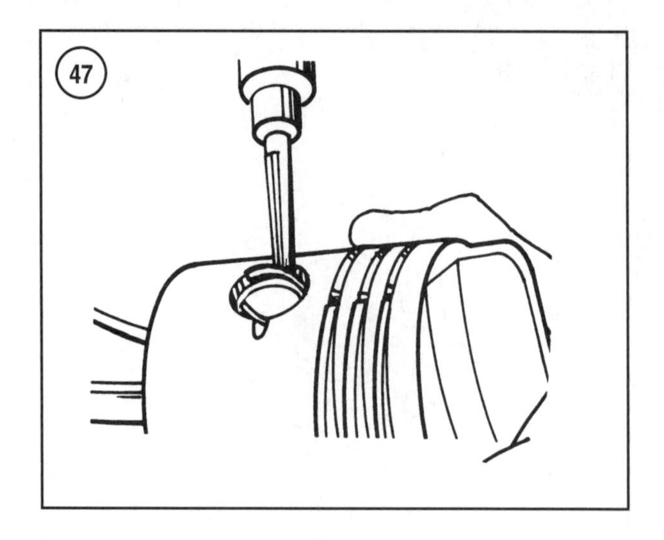

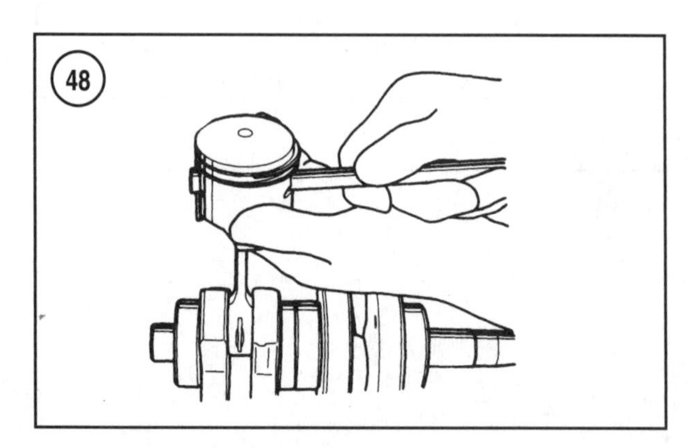

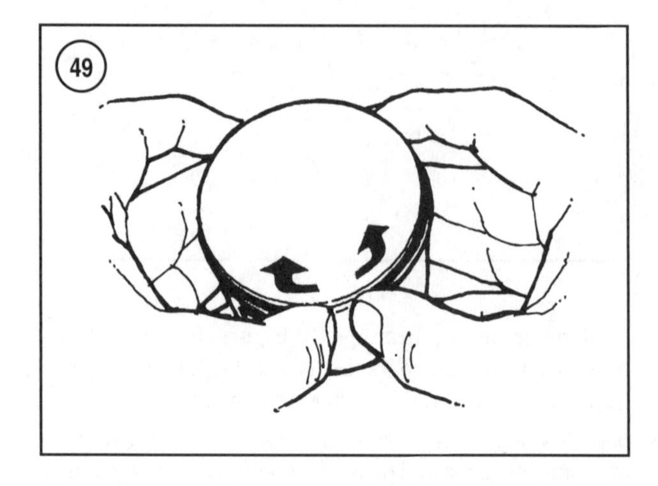

3. Insert the removal tool into the pin bore (**Figure 48**). Push on the tool until the pin extends from the opposite side and is clear from the connecting rod.
4. Hold a hand under the piston and carefully remove the removal tool, then pull the piston from the connecting rod. Retrieve the piston pin bearing and two thrust washers.

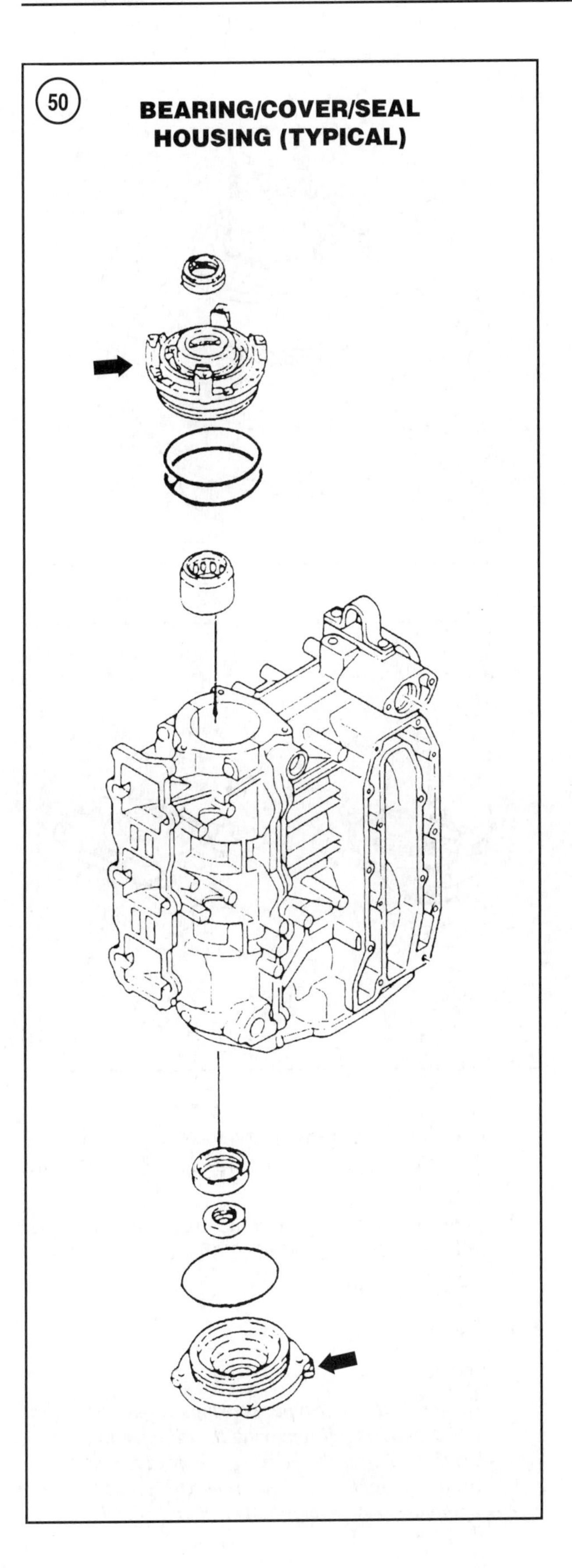

5. Repeat this step for each piston. Mark the cylinder number on each piston prior to removal. If the piston pin bearing or thrust washers must be reused, store them in containers that identify the cylinder number. They must be installed into the original piston and cylinder bore.
6. Mark the top side and note the groove in which the ring is installed prior to removal. Spread the piston rings (**Figure 49**) just enough to clear the ring grooves. Then slide them from the piston. Avoid scratching or damaging the piston. Tag the rings to indicate the original cylinder number. Do not discard the rings at this time.

Bearing/Cover/Seal Housing Removal and Disassembly

Refer to the appropriate illustration (**Figures 28-30**) to assist with component orientation and mounting location. An upper bearing/cover/seal housing is not used on 200-250 hp (3.1 liter) models. The upper crankshaft seal is mounted into the upper crankshaft roller bearing.
1. Remove the mounting bolts used to retain the crankshaft into the cylinder block during piston removal.
2. Carefully pull the housing(s) from the crankcase. If necessary, carefully pry the housing(s) from the crankcase (**Figure 50**). Work carefully and use a blunt tip instrument to prevent damage to the mating surfaces.
3. Remove the O-rings from the housing. Discard the O-rings.
4. Remove the seals from the housing only if they must be replaced. The removal process damages the seals. Removal instructions follow:
   a. Clamp the housing into a vise with protective jaws.
   b. Note the seal lip directions prior to removal.
   c. Use a blunt tip instrument to carefully pry the seal from the housing. Work carefully and avoid any tool contact with the seal bore in the housing.
   d. Discard the seals.
5. *80 Jet, 105 Jet, 115-130 hp and 150-200 hp models (2.6 liter)*—Remove the roller bearing(s) from the upper bearing/cover/seal housing only if they must be replaced. The removal process damages the bearing(s). Instructions follow:
   a. Select an appropriately sized socket, section of tubing or rod to use for a bearing removal tool. The tool must be of sufficient diameter to contact on the bearing rollers and be small enough to be inserted through the crankshaft opening at the top of the housing.
   b. Place the housing under a press with the flywheel side facing upward. The area below the bearings must be of sufficient diameter and depth to accom-

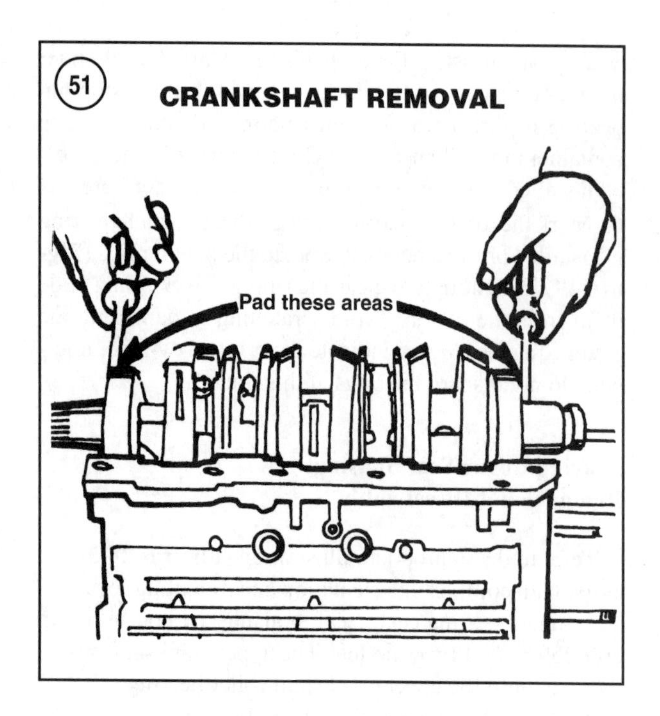

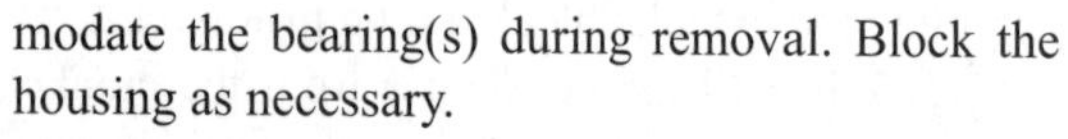

modate the bearing(s) during removal. Block the housing as necessary.

c. Insert the removal tool through the upper opening and seat it against the ends of the bearing rollers.
d. Press against the removal tool to remove the bearing(s). Discard the bearing(s).

Crankshaft

Refer to the appropriate illustration (**Figures 28-30**) to assist with component orientation and mounting location. An upper bearing/cover/seal housing is not used on 200-250 hp (3.1 liter) models. The upper crankshaft seal is mounted into the upper crankshaft roller bearing.

Removal

1. Remove the crankcase cover and bearing/cover/seal housing(s) as described in this chapter.
2. Use cloth or wooden pads (**Figure 51**) to protect the cylinder block and crankshaft. Then carefully pry the crankshaft from the cylinder block. Avoid using excessive force.
3. Position the crankshaft on a suitable stand or clamp into a vise with protective jaws.
4. *200-250 hp models (3.1 liter)*—Pull the roller bearing from the flywheel end of the crankshaft.

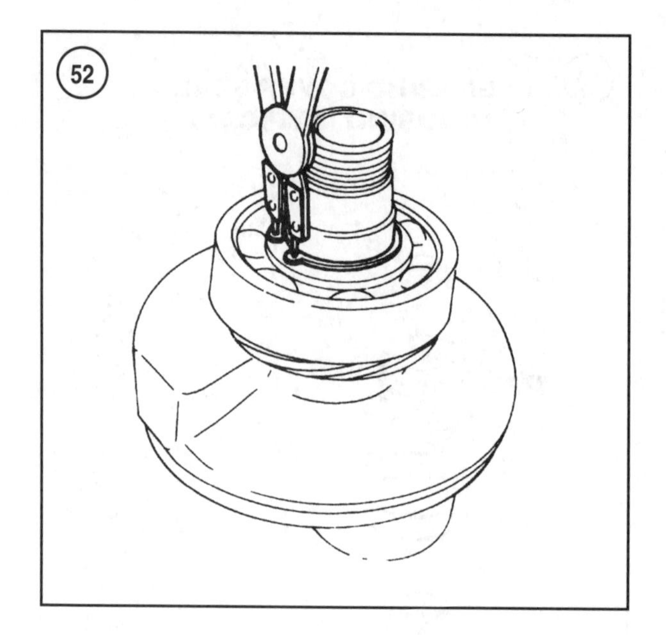

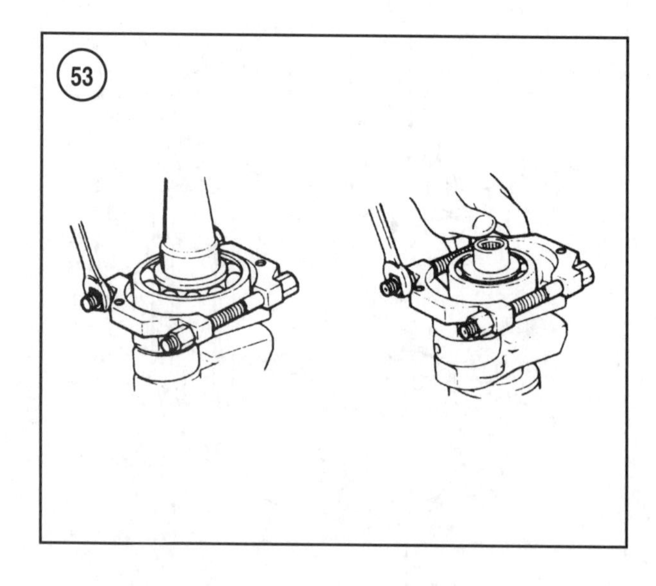

a. Use a blunt tip instrument to carefully pry the seal from the upper crankshaft roller bearing. Discard the seal.
b. Remove the O-ring from the groove on the upper roller bearing. Discard the O-ring.

Disassembly

NOTE
Remove the ball-type bearings from the crankshaft only if removing the oil injection drive gear or if the bearing is defective and must be replaced. The removal process damages the bearing. Always replace the

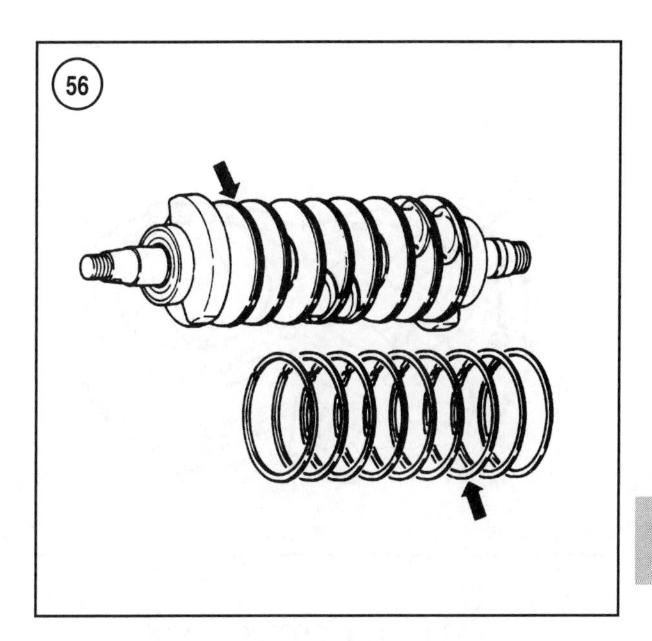

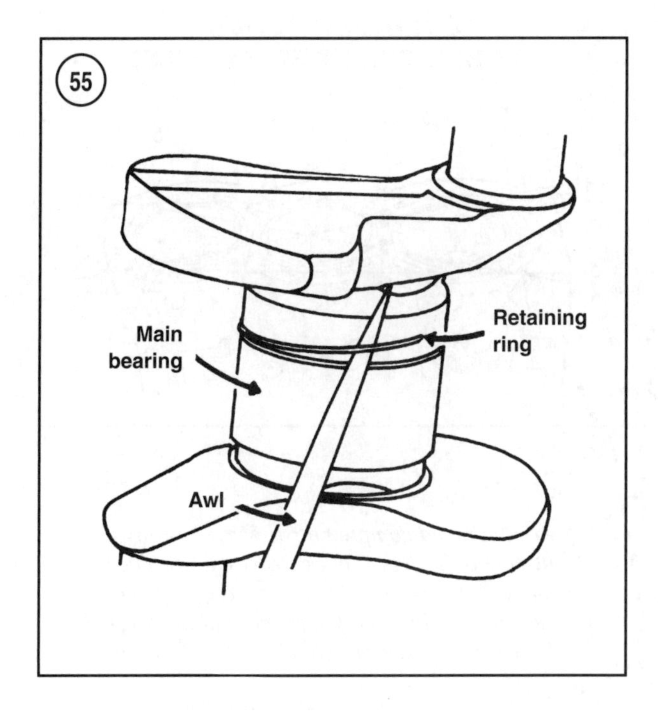

bearing if removed. Refer to ***Power Head Components Cleaning and Inspection*** *in this chapter to determine the need for replacement.*

1. Inspect the lower ball-type crankshaft bearing for excessive wear or defects as described in this chapter. See *Power Head Components Cleaning and Inspection*. Do not remove the bearing unless it or the oil injection drive gear must be replaced. The removal process damages the bearing. Removal procedures follow:

   a. Remove the snap ring (**Figure 52**) from the crankshaft.
   b. Install a bearing separator onto the roller bearing (**Figure 53**). Tighten the bolts on the separator until just snug. Position the separator so that the sharp edges contact between the bearing and the crankshaft (**Figure 54**).
   c. Use a piece of hard plastic or wood to protect the crankshaft from the ram of the press. Support the bearing separator in a manner that allows adequate travel for the crankshaft as the bearing is removed.
   d. Provide continual support for the crankshaft while pressing the bearing from the crankshaft.
   e. Discard the bearing upon removal.

2. Mark the center main bearing(s) mounting location and flywheel end reference prior to removal. Use an awl to remove the retaining ring from the bearing cases (**Figure 55**). Be prepared to capture the bearing rollers. Then separate the halves of the bearing cases. Repeat this process for the remaining center main bearings. Keep each set of bearings in a container that identifies the mounting location on the crankshaft.

3. Carefully spread and remove the crankcase sealing rings from the grooves on the crankshaft (**Figure 56**). Replace the sealing rings if any wear is noted, steps have formed on the surfaces or if any are cracked or broken.

4. Slide the oil injection drive gear from the bottom of the crankshaft. If the gear cannot be removed by hand, remove the gear using the same method used to remove the ball bearing from the crankshaft. Replace the gear if it must be pressed from the crankshaft.

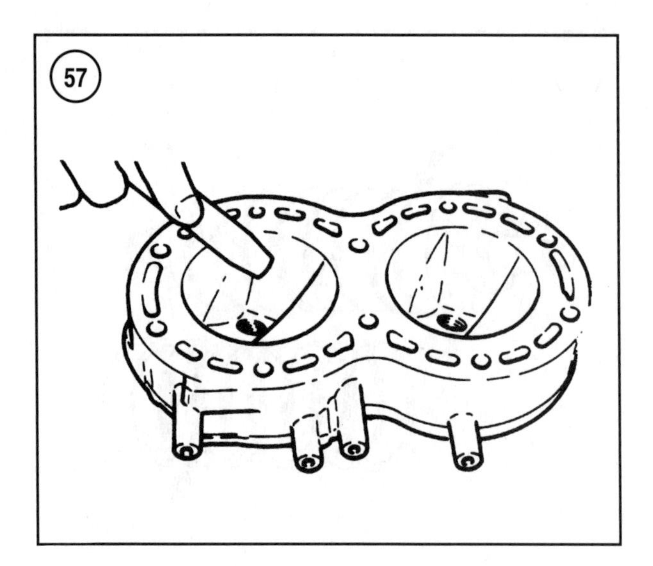

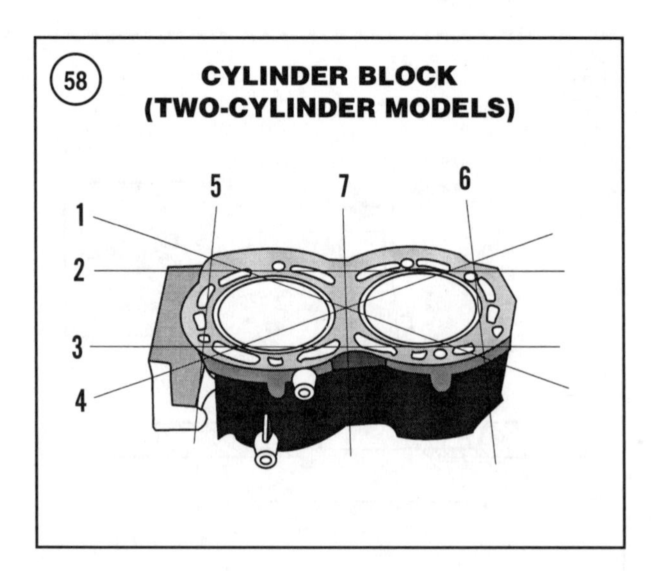

POWER HEAD COMPONENTS CLEANING AND INSPECTION

This section includes component cleaning procedures, inspecting for visual defects and measuring the components.

Take all precautions necessary to maintain the marks indicating the original position of each component. If reused, all components must be installed into the same cylinder or position on the power head. Wear patterns form on any contacting surfaces. Maintaining the wear patterns helps ensure a durable and reliable repair.

Review the following sections: *Lubricants, Sealants and Adhesives; Sealing Surfaces; Fasteners and Torque.* All are located at the beginning of this chapter.

The manufacturer recommends replacing all seals, O-rings, gaskets, piston pin lock rings, piston rings and all piston pin bearings any time a power head is disassembled.

Perform the cleaning and inspection procedure in each of the following sections that applies to the engine *before* beginning assembly procedures.

Cylinder Block and Crankcase

Yamaha outboard cylinder blocks and crankcase covers are matched, align-bored assemblies. For this reason, do not attempt to assemble an engine with parts salvaged from other blocks. If the following inspection procedure indicates that the block or cover requires replacement, replace the cylinder block and crankcase cover as an assembly.

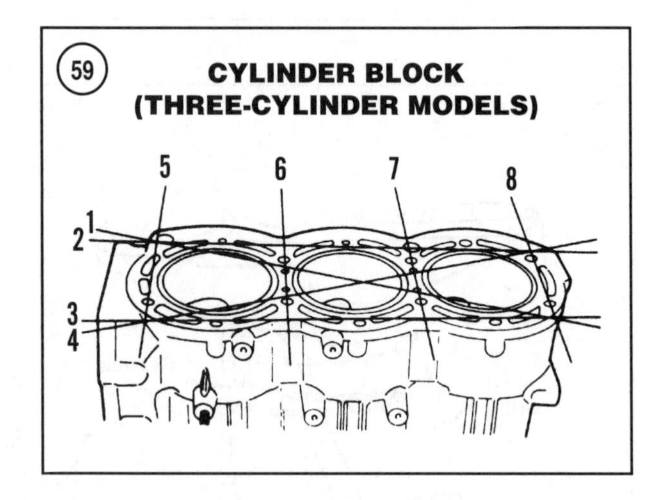

NOTE

All fuel bleed components (hoses, T-fittings, threaded fittings, check valves and check valve carriers) must be removed if it is necessary to submerge the block and/or cover in a strong cleaning solution. See Chapter Five.

1. Clean the cylinder block and crankcase cover thoroughly with clean solvent using a parts washing brush. Carefully remove all gasket and sealant material from the mating surfaces.
2. Inspect the sacrificial anodes as described in this chapter. See *Anode Replacement.* Replace deteriorated anodes before beginning the assembly procedures.
3. Remove all carbon and varnish deposits from the combustion chambers, exhaust ports and exhaust cavities with a carbon removing solvent, such as Yamaha Combustion Chamber Cleaner. A hardwood dowel or plastic scraper

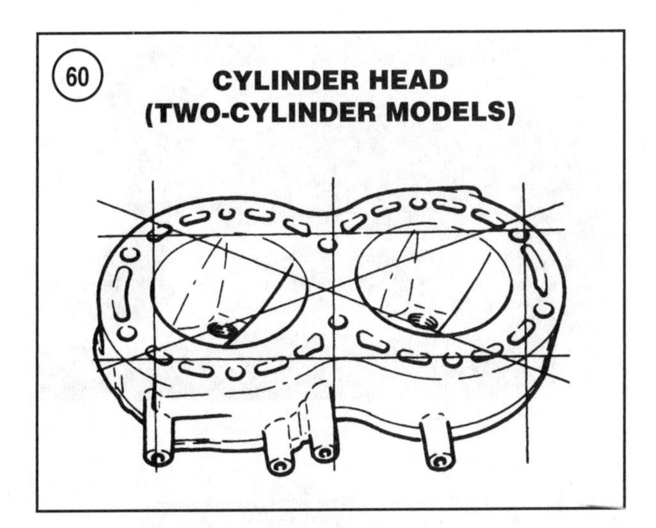

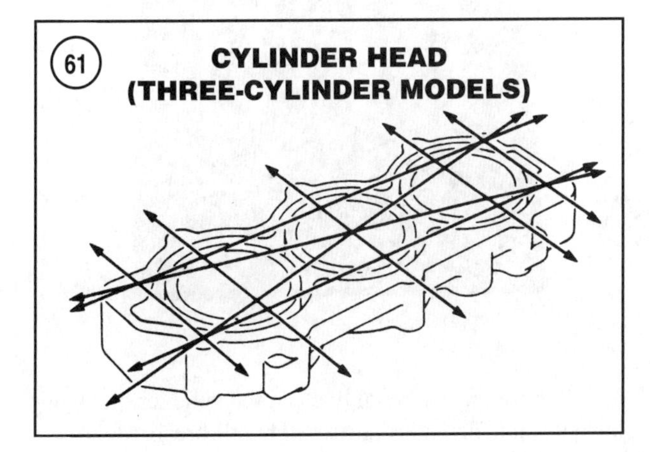

can be used to remove stubborn deposits. See **Figure 57**. Do not scratch, nick or gouge the combustion chambers or exhaust ports.

WARNING
Use suitable hand and eye protection when using muriatic acid products. Avoid breathing the vapors. Use only in a well-ventilated area.

CAUTION
Do not allow muriatic acid to contact the aluminum surfaces of the cylinder block.

4. If the cylinder bore(s) has aluminum transfer from the piston(s), clean loose deposits using a stiff bristle brush. Apply a *small* quantity of diluted muriatic acid to the aluminum deposits. A bubbling action indicates the aluminum is dissolving. Wait 1-2 minutes, and then thoroughly wash the cylinder with hot water and detergent. Repeat this procedure until the aluminum deposits are removed. Lightly oil the cylinder wall to prevent rusting.
5. Check the cylinder block and crankcase cover for cracks, fractures, stripped threads or other damage.
6. Inspect gasket mating surfaces for nicks, grooves, cracks or distortion. Any defects may cause leakage. Check the surfaces for distortion (warpage) as described in *Sealing Surfaces* in this chapter. Warpage that exceeds 0.1 mm (0.004 in.) is cause for component replacement, unless otherwise specified. Smaller imperfections can be removed by lapping the component as described under *Sealing Surfaces*. **Figures 58-61** show typical directions in which to check for warpage on the cylinder head and cylinder block surfaces.
7. Check all water, oil and fuel bleed passages in the block and cover for obstructions. Make sure all pipe plugs are installed tightly. Seal pipe plugs with Loctite 567 PST pipe sealant.

Cylinder bore inspection

Inspect the cylinder bores for scoring, scuffing, grooving, cracks or bulging and any other mechanical damage. Inspect the cylinder block (casting) and cast-iron liner for separation from the aluminum cylinder block. There must be no gaps or voids between the aluminum casting and the liner. Remove any aluminum deposits (aluminum transfer from the pistons) as described in this section. If the cylinders are in a acceptable condition, hone the cylinders as described in *Cylinder bore honing* in this section. If the cylinders are in an unacceptable condition, rebore the defective cylinder bore(s) or replace the cylinder block and crankcase cover as an assembly.

NOTE
Rebore only the cylinders that are defective. A mixture of standard and oversize cylinders on a given power head as long as the correct piston (standard or oversize) is used to match each bore is acceptable. Always check the manufacturer's parts catalog for oversize piston availability and bore sizes, before over-boring the cylinder(s).

Cylinder bore honing

The manufacturer recommends using only a rigid cylinder hone to deglaze the bore to aid in the seating of new piston rings. If the cylinder has already been bored, the rigid hone is used in two steps, a rough (deburring) hone to remove the machining marks and a finish hone to establish the correct cross-hatch pattern in the cylinder bore.

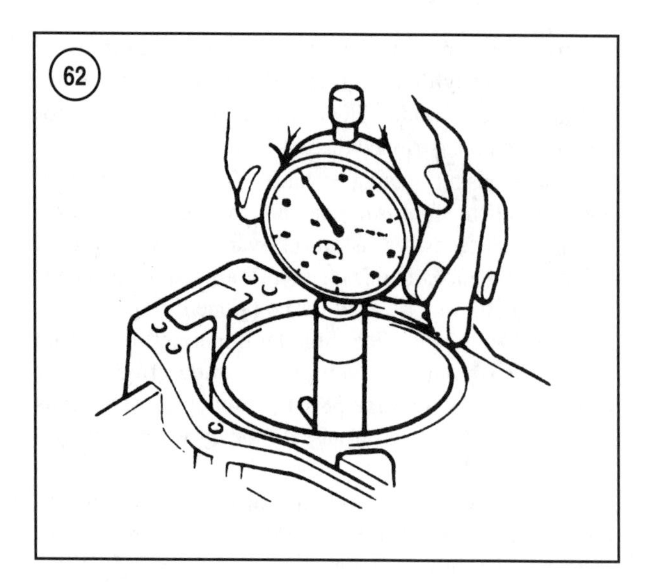

Flex (ball type) hones and spring-loaded hones are not acceptable, as they do not produce a straight and perfectly round bore.

NOTE
If there is uncertainty with the correct use of a rigid cylinder hone, cylinder bore honing should be performed at a qualified machine shop or dealership.

If the cylinders are in a acceptable condition, prepare the cylinder bore for new piston rings and remove any glazing, light scoring and/or scuffing by lightly honing the cylinders as follows:

1. Follow the rigid hone manufacturer's instructions when using the hone. Make sure the correct stones for the bore (cast-iron) are installed onto the hone.
2. Pump a continuous flow of honing oil into the bore during the honing operation. If an oil pumping system is not available, enlist the aid of an assistant to keep the cylinder walls flushed with honing oil.
3. If the hone loads (slows down) at one location in the bore, this indicates the narrowest portion of the bore. Localize the stroking in this location to remove stock until the hone maintains the same speed throughout the entire bore.
4. Frequently remove the hone from the cylinder bore and inspect the bore. Do not remove any more material than necessary.
5. Attempt to achieve a stroke rate of approximately 30 cycles per minute, adjusting the speed of the hone to achieve a cross-hatch pattern with an intersecting angle of approximately 30°. Do not exceed a cross-hatch of more than 45°.

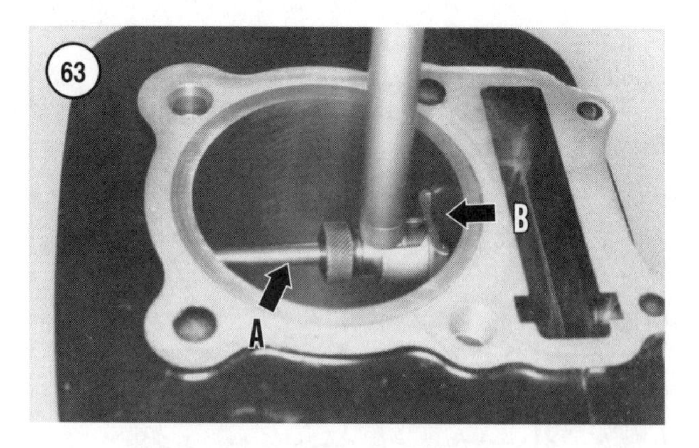

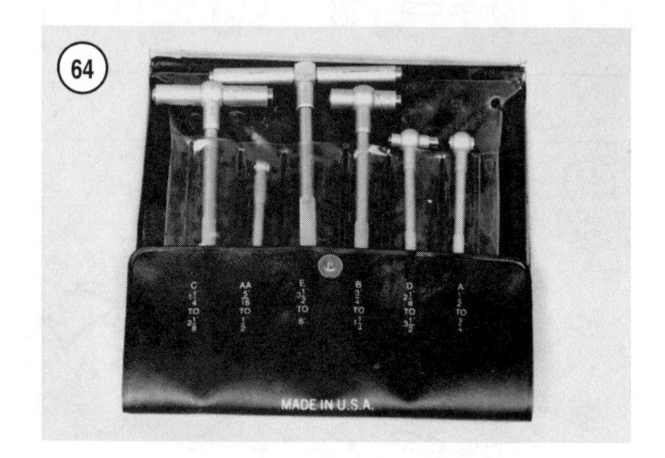

6. After honing, thoroughly clean the cylinder block using hot water, detergent and a stiff bristle brush. Make certain to remove all abrasive material from the honing process. After washing and flushing, coat the cylinder walls with a film of outboard motor oil to prevent rusting.
7. Proceed to *Cylinder bore measurements* to determine if the cylinder bores are within the manufacturer's specifications for wear, taper and out-of-round.

Cylinder bore measurements

Measure each cylinder bore as follows. Oversize bore specifications are simply the standard bore specification *plus* the oversize dimension (check parts catalog for available oversize dimensions). All standard bore specifications, maximum taper and out-of-round specifications are listed in **Table 2**.

Use a cylinder bore gauge (**Figure 62**), inside micrometer (**Figure 63**), or a telescoping gauge (**Figure 64**) and a regular micrometer (**Figure 65**) to measure the entire area of ring travel in the cylinder bore. Three sets of readings must be taken at the top, middle and bottom of the ring travel area (**Figure 66**).

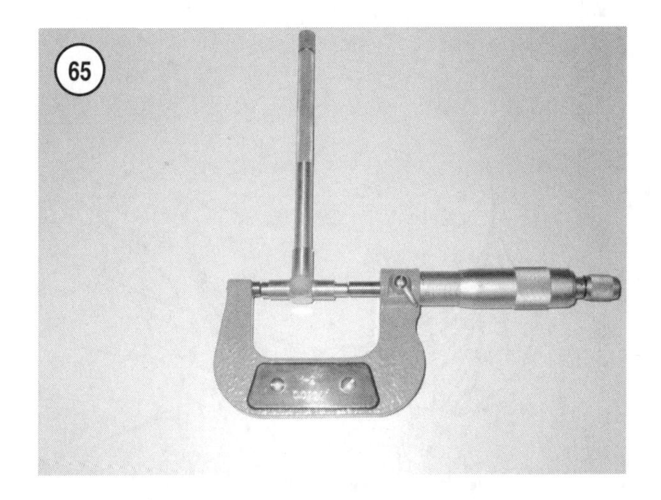
65

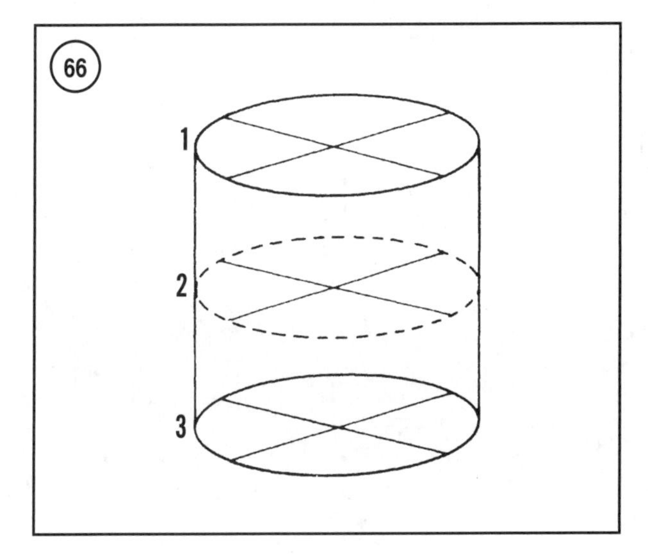

66

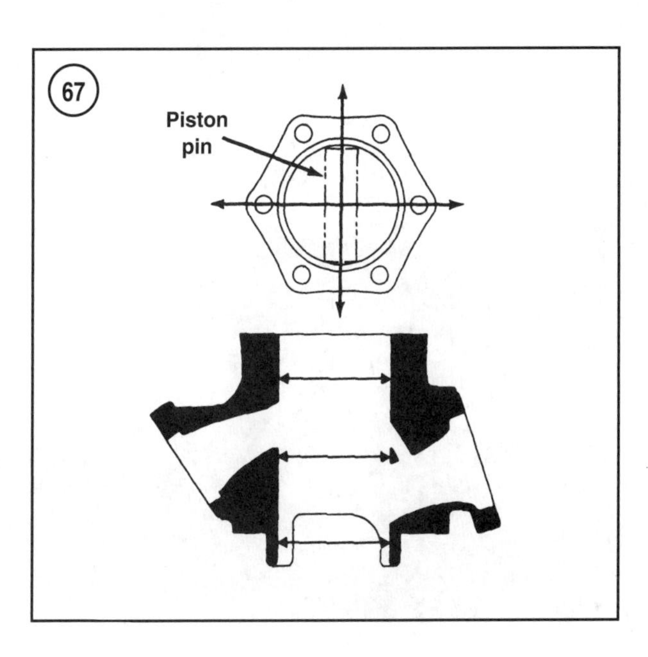

67

1. Take the first reading at the top of the ring travel area at a depth in the bore at approximately 10 mm (0.4 in.) from the top of the cylinder bore (**Figure 67**). The gauge must be aligned with the crankshaft centerline. Record the reading. Then, turn the gauge 90° to the crankshaft centerline and record another reading.
2. The difference between the two (or highest and lowest) readings is the cylinder out-of-round. The reading cannot exceed the specification in **Table 2**.
3. Take a second set of readings at a depth in the bore at approximately 5 mm (0.2 in.) above the exhaust port (**Figure 67**) using the same alignment points described in Step 1. Record the readings. Calculate the cylinder out-of-round by determining the difference between the two (or highest and lowest) readings. The reading cannot exceed the specification in **Table 2**.
4. Take a third set of readings at a depth in the bore at approximately 5 mm (0.2 in.) below the lowest transfer port (**Figure 67**) using the same alignment points described in Step 1. Record the readings. Calculate the cylinder out-of-round by determining the difference between the two (or highest and lowest) readings. The reading cannot exceed the specification in **Table 2**.
5. To determine the cylinder taper subtract the readings taken at the top of the cylinder bore (Step 1) from the readings taken at the bottom of the cylinder bore (Step 4). The difference in these readings is the cylinder taper. The reading cannot exceed the specification in **Table 2**.
6. Repeat Steps 1-5 for each remaining cylinder.
7. If any cylinder exhibits excessive out-of-round, taper or any of the measurements exceed the maximum bore diameter specification, the cylinder(s) must be bored oversize or the cylinder block and crankcase cover replaced as an assembly.

Piston

The piston and piston pin are serviced as an assembly. If either is damaged, replace them together. Piston pins must be reinstalled in the pistons in which they were removed.

CAUTION
Do not use an automotive ring groove cleaning tool as it can damage the ring grooves and loosen the ring locating pins.

Cleaning

1. Clean the piston(s), piston pin(s), thrust (locating) washers and the piston pin needle bearing assemblies thoroughly with cleaning solvent using a parts washing brush. Do not wire brush the piston as metal from the wire

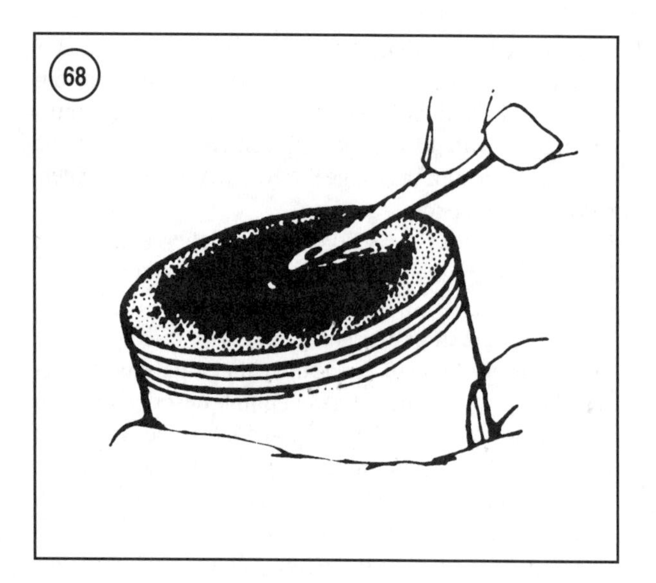

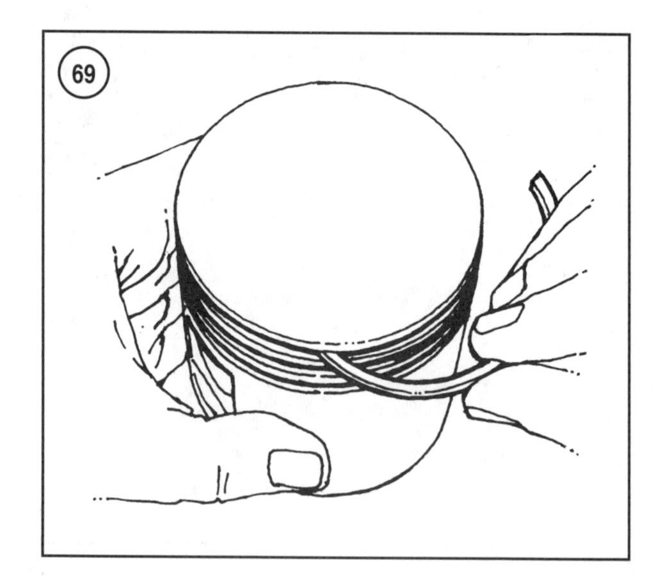

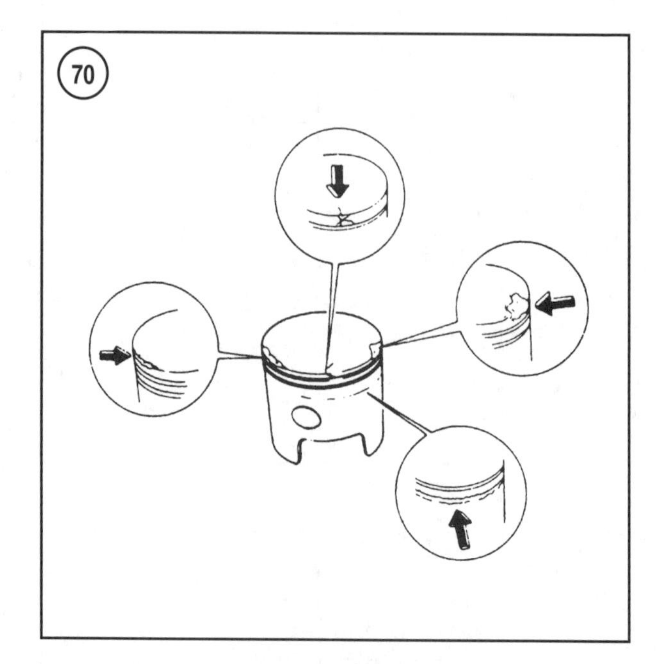

wheel may become imbedded in the piston. This can lead to preignition and detonation damage.

2. Remove all carbon and varnish deposits from the top of the piston, piston ring groove(s) and under the piston crown with a carbon removing solvent, such as Yamaha combustion chamber cleaner. Use a piece of hardwood or a plastic scraper (**Figure 68**) to remove stubborn deposits. Do not scratch, nick or gouge any part of the piston. Do not remove any stamped or cast identification marks.

CAUTION

The locating pin in the ring grooves may be loosened within the bore if care is not taken when cleaning the ring grooves. If loosened, the pin may dislodge during operation and cause serious power head damage. Use solvent and a small plastic brush to remove carbon deposits from the locating pin.

NOTE

On keystone and semi-keystone rings, it is necessary to grind off enough of the ring taper to allow the inside edge of the broken ring to reach the inside diameter of the ring groove.

3. Clean stubborn deposits from the ring groove(s) as follows:

a. Fashion a ring cleaning tool from the original piston ring(s). Rings are shaped differently for each ring groove. Make sure to use the correct original ring for each ring groove.

b. Break off approximately 1/3 of the original ring. Grind a beveled edge onto the broken end of the ring.

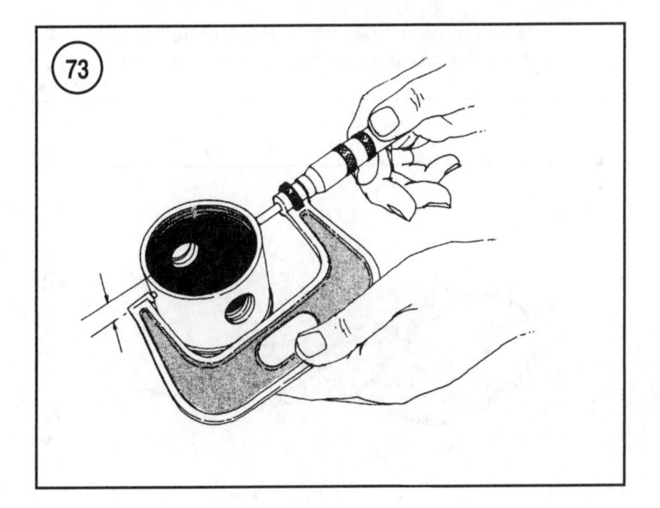

c. Use the ground end of the ring to gently scrape the ring groove clean (**Figure 69**). Be careful to only remove the carbon. Do not gouge the metal and do not damage or loosen the piston ring locating pin(s).

4. Polish any nicks, burrs or sharp edges on and around the piston skirt with crocus cloth or 320 grit carborundum cloth. Do not remove any cast or stamped identification markings. Wash the piston thoroughly to remove all abrasive grit.

5. Inspect the piston(s) overall condition for scoring, cracks, worn or cracked piston pin bosses and any other mechanical damage (**Figure 70**). Carefully inspect the crown and the top outer diameter for burning (**Figure 71**), erosion, evidence of ring migration and mechanical damage. Inspect the piston skirt for discoloration, highly polished surfaces and scoring (**Figure 72**). Replace the piston and pin if these or other defects are noted.

6. Check piston ring grooves for wear, erosion, distortion and loose ring locating pins.

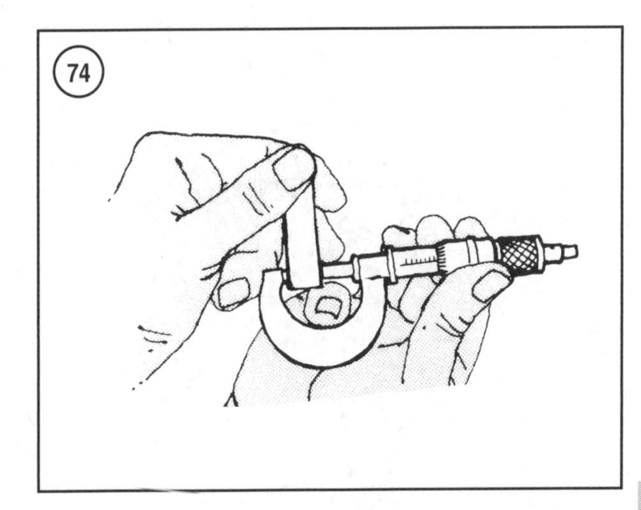

7. Inspect the piston pin for water etching, pitting, scoring, heat discoloration, excessive wear, distortion and mechanical damage. Roll the pin across a machinist's surface plate to check the pin for distortion. Replace the piston and piston pin as necessary.

8. Inspect the thrust (locating) washers and needle bearings for water damage, pitting, scoring, overheating, wear and mechanical damage. Replace damaged washers and needle bearings.

Piston measurements

The pistons used in these engines are cam shaped. The piston is built out-of-round. The piston is engineered to fit the bore perfectly when at operating temperature and fully expanded, which makes the engine run quietly and efficiently. The piston must be measured at the specified point(s) or the readings will be inaccurate.

Measure each piston skirt with a micrometer as described in the following procedure and compare the readings to the specifications listed in **Table 3**.

To calculate the specified skirt dimension on oversize pistons, simply add the oversize dimension to the standard skirt diameter listed in **Table 3**.

1. Using a micrometer, measure each piston skirt diameter at a 90° angle to the piston pin bore (**Figure 73**) and at the distance of 10 mm (0.39 in.) from the bottom of the piston skirt. Measure all pistons and record each measurement by cylinder number. If the piston(s) is not within specification, the piston(s) must be replaced.

2. Measure the piston pin diameter (**Figure 74**) at the pin bore and needle bearing contact surfaces. If the pin diameter at all locations is not within the specification in **Table 3**, replace the piston and piston pin assembly.

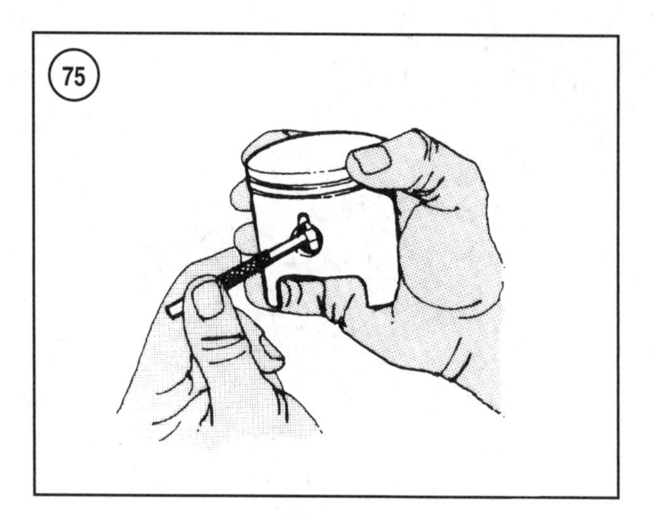

3. Measure the piston pin bore diameter (**Figure 75**) on both sides of the piston and in several locations around the diameter. If the pin bore diameter is not within the specification in **Table 3** at all locations, replace the piston and piston pin assembly. Perform these measurements for each piston and pin.
4. Measure the piston ring side clearance as follows:
   a. Note the shape of the rings and ring grooves in the piston (**Figure 76**, typical). Some models use keystone shaped piston rings in both locations.
   b. Carefully install new rings onto the pistons (**Figure 77**). Use an appropriate ring expander to install the rings. To help prevent ring breakage, spread the ring just enough to clear the piston dome. Install the bottom ring first. Then install the top ring. Do not scratch the piston surfaces. Make sure the piston ring shape matches the ring groove and the ring marking (**Figure 78**) faces toward the piston dome.
   c. Rotate the ring to align the ring gap with the locating pin as indicated in **Figure 79**.
   d. Insert a feeler gauge between the piston ring and the side of the piston ring groove (**Figure 80**). The feeler gauge must contact the square side of the piston ring (**Figure 81**) for accurate measurement.
   e. The ringside clearance equals the thickness of the feeler gauge that can be passed between the ring and the side of the groove with a slight drag. Measure and record the side clearance for each piston ring groove.
   f. Replace the piston if either of the side clearance measurements is greater or less than the specification in **Table 4**.
   g. Repeat this step for each piston.
   h. Carefully remove the rings from the piston. Do not scratch the piston surfaces.

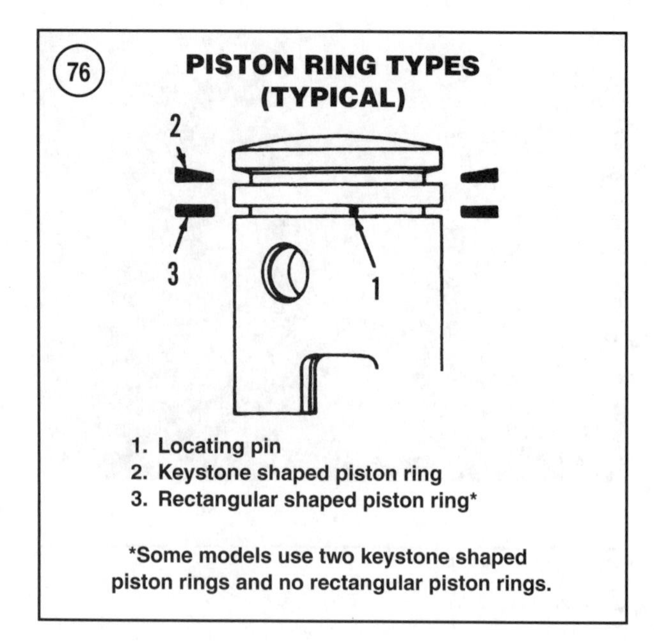

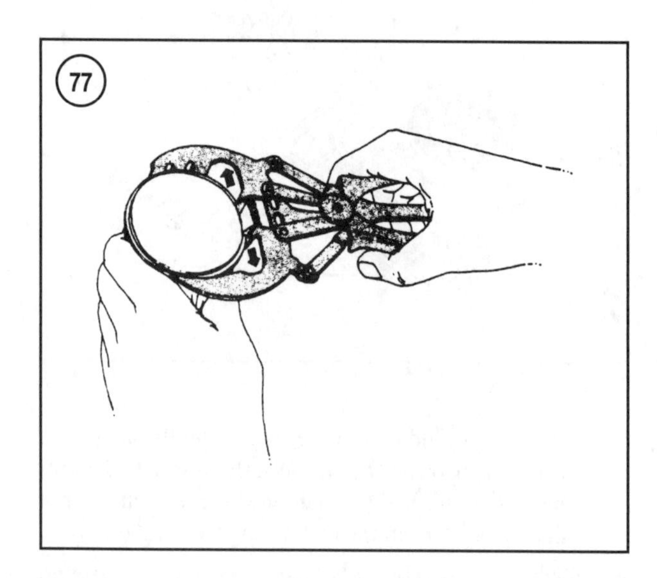

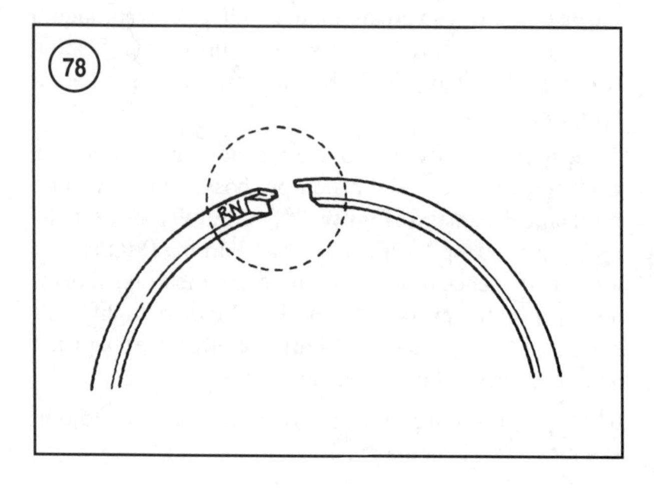

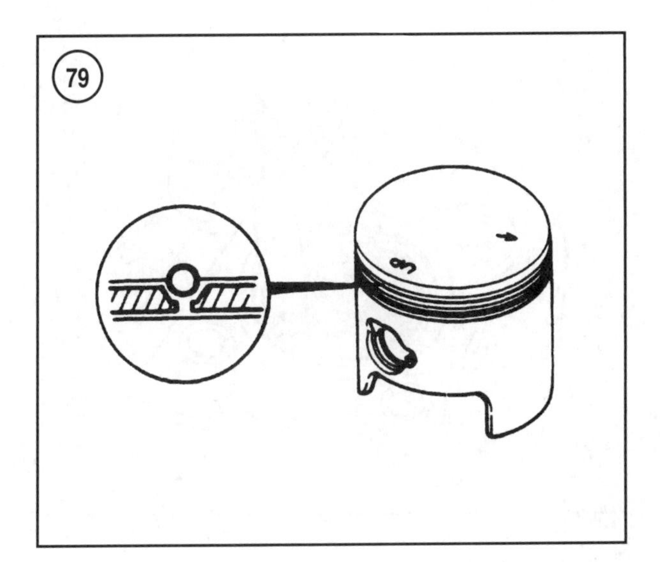

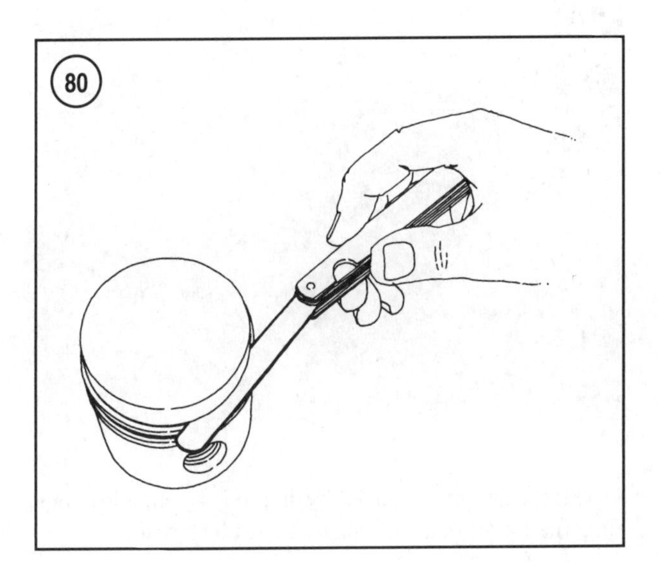

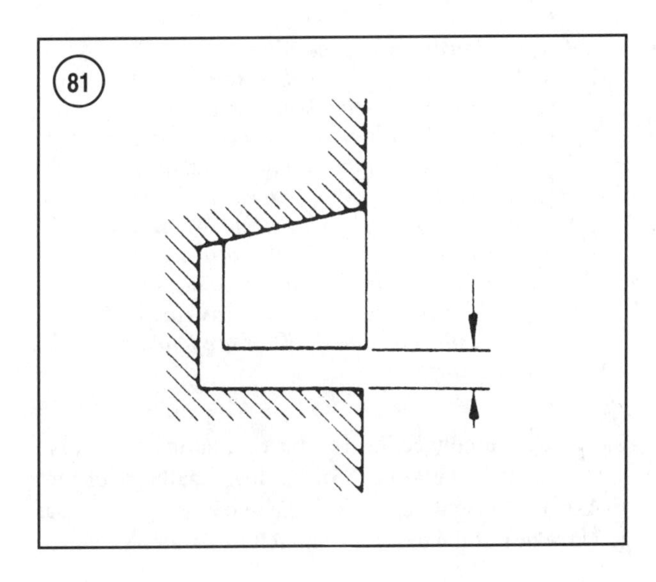

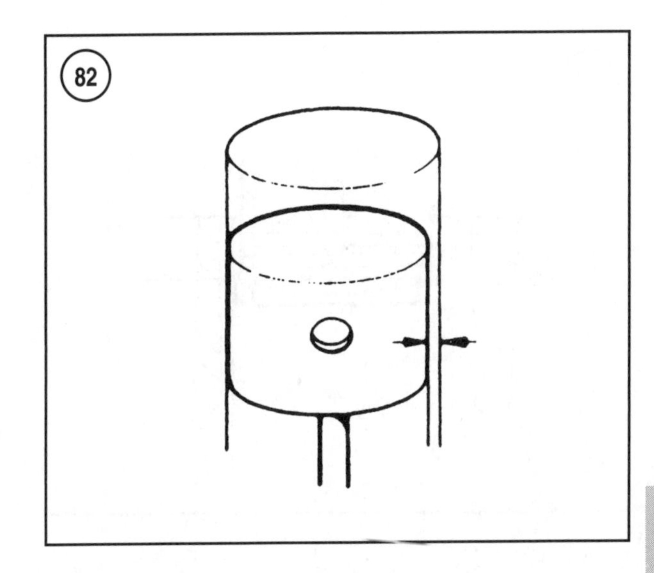

5. Determine the piston skirt clearance (**Figure 82**) by subtracting the measured piston diameter from the smallest bore diameter measurement for the corresponding cylinder. The skirt clearance must be within the specification in **Table 3**.

   a. *Skirt clearance is less than the specification*—Repeat the piston and cylinder bore measurements. If the piston measurement is correct, have a machinist hone the cylinder to attain the specified skirt clearance.

   b. *Skirt clearance is greater than the specification*—Repeat the piston and cylinder bore measurements. If both measurements are correct, repeat the measurements using a new piston. New pistons usually measure at the upper limit of the specification. If the skirt clearance remains greater than the specification, bore the cylinder and install the next larger oversize piston.

6. Measure the piston ring end gap after the cylinder bore and piston diameter are confirmed to be correct, as follows:

   a. Carefully compress and install the ring into the cylinder bore. Use a piston without rings to push the new ring to the depth specified in **Table 4**. See **Figure 83**.

   b. Use a feeler gauge to measure the ring end gap (**Figure 84**).

   c. Compare the feeler gauge measurement to the specification in **Table 4**.

   d. *The gap is within the specification*—Tag the ring to identify the cylinder number.

   e. *The gap is not within the limit*—Try a different (new) ring in the cylinder and remeasure.

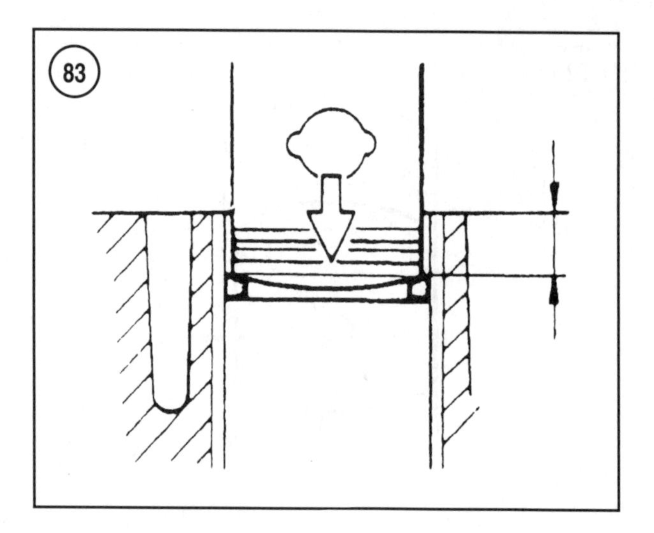

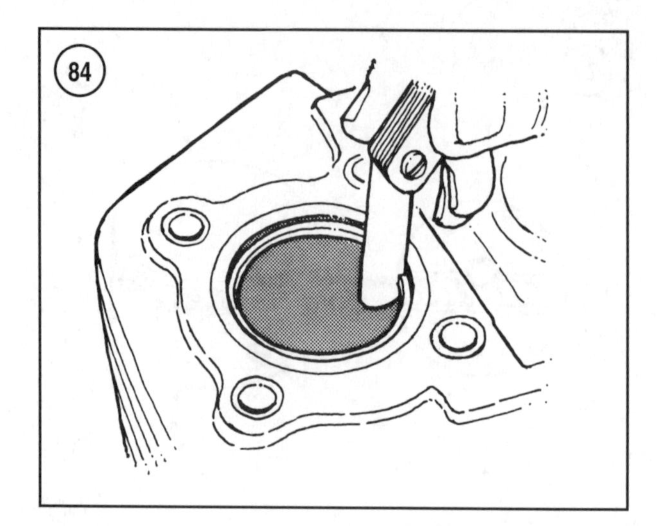

f. Continue until new rings with the correct gap are selected for each piston. Tag all rings to identify the cylinder number.

Crankshaft and Connecting Rod

WARNING
Never allow bearings or other components to spin while drying with compressed air. The bearings or other components may rotate at extremely high speeds and fly apart. Serious injury or death can occur by flying debris.

CAUTION
Never spin the bearings with the compressed air. Spinning the bearings at high speed without lubrication causes overheating and serious damage to bearings and the surfaces they contact.

Thoroughly wash the crankshaft and the main and connecting rod bearing assemblies with clean solvent and a parts washing brush. Thoroughly dry the crankshaft with compressed air.

Perform the following inspections and measurements.

Crankshaft visual inspection

1. Inspect the drive shaft splines, flywheel taper, flywheel key groove and flywheel nut threads for corrosion, cracks, excessive wear and mechanical damage. Replace the crankshaft if any defects are present.
2. Inspect the upper and lower seal and all bearing surfaces for excessive grooving, pitting, nicks or burrs. The seal surfaces may be polished with crocus cloth as necessary. If the defect can be felt by dragging a pencil lead or a fingernail over it, the crankshaft must be replaced.

NOTE
Normal bearing surface coloration is silver with very fine lines in the surface. Discolored surfaces occur when excessive heat is applied. Highly polished surfaces occur from inadequate lubrication, excessive engine operating speed or excessive bearing wear. Chatter marks may occur where the needle bearings contact the bearing surfaces. Chatter marks resemble the surface of a washboard. Replace the crankshaft if chatter marks or other defects are found on bearing surfaces.

3. Check the connecting rod bearing surfaces for rust, water damage (**Figure 85**), pitting, spalling, chatter marks, heat discoloration and excessive or uneven wear (**Figure 86**). If the defect can be felt by dragging a pencil

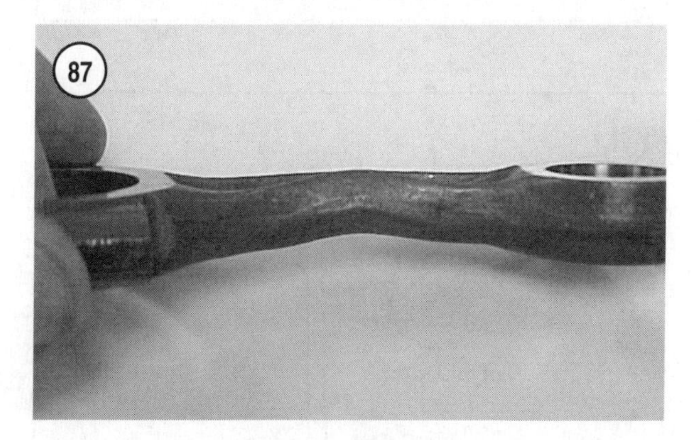

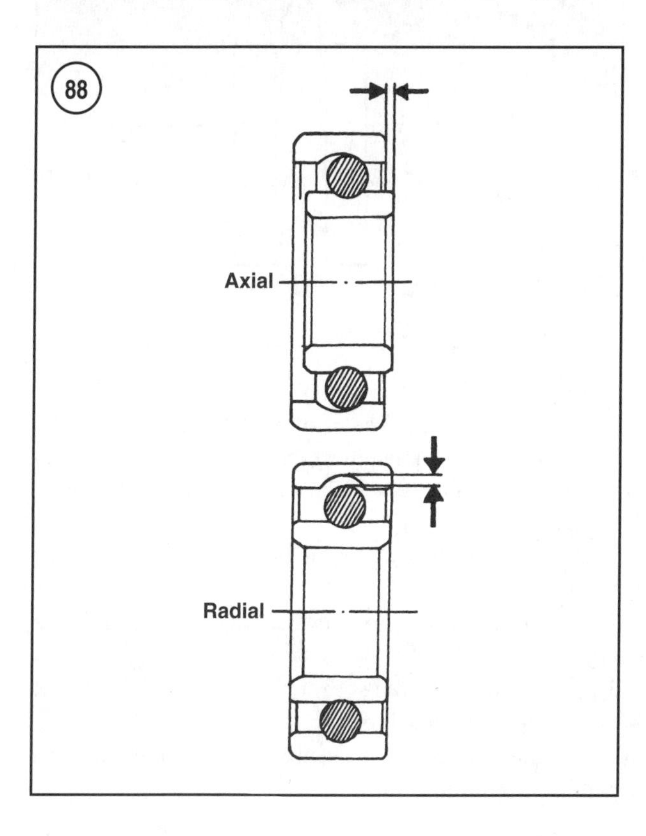

lead or a fingernail over it, the crankshaft must be replaced. Stains or marks that cannot be felt can be polished with crocus cloth. Do not remove any more material than absolutely necessary.

4. Visually check for bending or twisting of the I-beam section of the connecting rod(s). See **Figure 87**. Replace the connecting rod or crankshaft and connecting rod assembly if defective.

5. Inspect the oil pump drive gear for worn or chipped teeth, heat damage or any other damage. Replace the oil pump drive if it is damaged. Replace the oil pump driven gear (see Chapter Twelve) when replacing the drive gear.

6. If the crankshaft is visually acceptable, lightly oil the crankshaft to prevent rusting.

Crankshaft bearing inspection

Inspect the crankshaft main bearings as follows:

1. *Ball bearing*—Rotate the bearing. The bearing must rotate smoothly with no rough spots, catches or noise. There must be no discernible end or axial play (**Figure 88**) between the inner and outer races of the bearing. If the bearing shows any visible signs of wear, corrosion or deterioration, replace the bearing.

2. *Roller/needle bearings*—Inspect the rollers and/or needles for water etching, pitting, chatter marks, heat discoloration and excessive or uneven wear. Inspect the cages for wear and mechanical damage. Replace bearings as an assembly—do not attempt to replace individual rollers or needles.

Crankshaft and connecting rod measurement

NOTE

Crankshaft bearing surface diameters are not provided by the manufacturer. Replace the crankshaft if these surfaces are damaged, have transferred bearing material or have discolored. The hardening process used does not allow for machining of the crankshaft or the use of undersized bearings.

1. Measure the crankshaft runout as follows:
   a. Support the crankshaft on V- blocks or other means as shown in **Figure 89**.
   b. Position a dial indicator to place the indicator stem in contact with the crankshaft main bearings as shown in **Figure 89**.
   c. Slowly rotate the crankshaft while observing the needle on the dial indicator. The needle indicates the amount of crankshaft runout.

d. Measure the runout at all main bearing surfaces. Replace the crankshaft if any of the runout measurements exceed the specification in **Table 5**.

2. Measure the connecting rod axial play as follows:
   a. Install the connecting rods without the piston onto the crankshaft. Use new needle bearings between the crankshaft and the connecting rod. Make sure the flywheel side of the connecting rod faces the flywheel end of the crankshaft. Hand-tighten the connecting rod bolts. Do not torque the bolts for this measurement.
   b. Secure the crankshaft and rod assembly in a horizontal position.
   c. Use a sturdy mount to position a dial indicator perpendicular to the connecting rod as shown in **Figure 90**. Place the pointer in contact with the connecting rod on the flat surface that aligns with the center of the piston pin bore.
   d. Observe the dial indicator while moving the connecting rod in the direction shown in **Figure 90**. Note the dial indictor reading when the rod reaches each limit of travel. The amount of needle movement indicates the axial play measurement.
   e. Record the measurement and repeat the measurement for each connecting rod.
   f. Compare the measurements with the specification in **Table 5**. Replace the corresponding connecting rod if any of the measurements exceed the maximum specification in **Table 5**. Repeat the measurement after replacing the connecting rod. Replace the crankshaft if the axial play measurement exceeds the specification with a new connecting rod.

3. Use feeler gauges to measure the connecting rod side clearance (**Figure 91**) for each connecting rod. Select the feeler gauge that can pass between the flyweight and the connecting rod with a slight drag. All side clearance measurements must be within the specification in **Table 5**. Check for improper connecting rod cap alignment if the side clearance is below the specification. Replace the corresponding connecting rod if any of the measurements exceed or are below the specification in **Table 5**. Repeat the measurement after replacing the connecting rod. Replace the crankshaft if the side play measurement exceeds or is below the specification with a new connecting rod.

4. Remove the connecting rods, end caps and bearings. Make sure each rod cap is reinstalled onto the original connecting rod after removal.

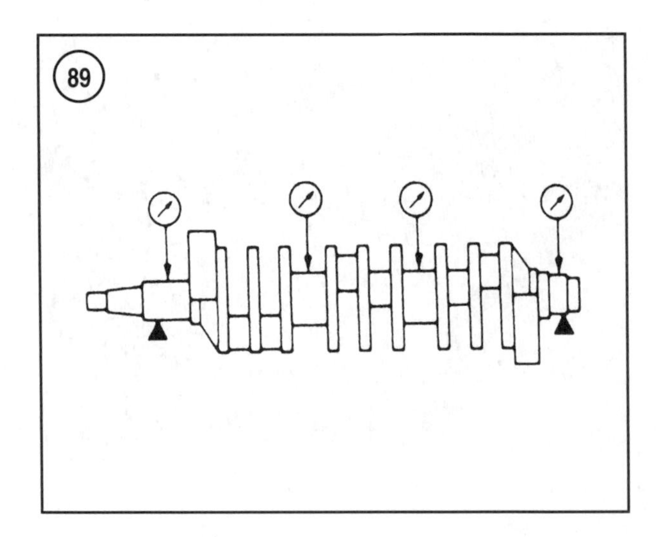

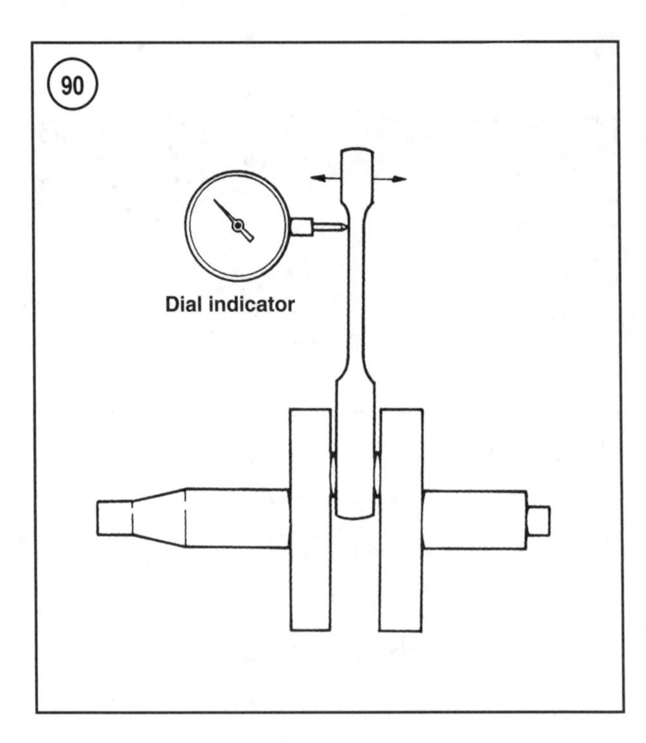

Oil pump driven gear bushing removal/installation

Inspect the oil pump driven gear bushing (**Figure 92**) for worn, discolored damaged or corroded surfaces. Replace the bushing if these or other defects are noted. *Do not* remove the bushing unless it must be replaced. The removal process damages the bushing. Remove the bushing as follows:

1. Mark the depth at which the bushing is installed in the block.
2. Carefully thread a properly sized lag bolt into the bushing and pull to remove the bushing.
3. Use a properly sized socket and extension to push the new bushing into the block.
4. Measure to ensure the bushing is installed to the same depth as the original bushing.

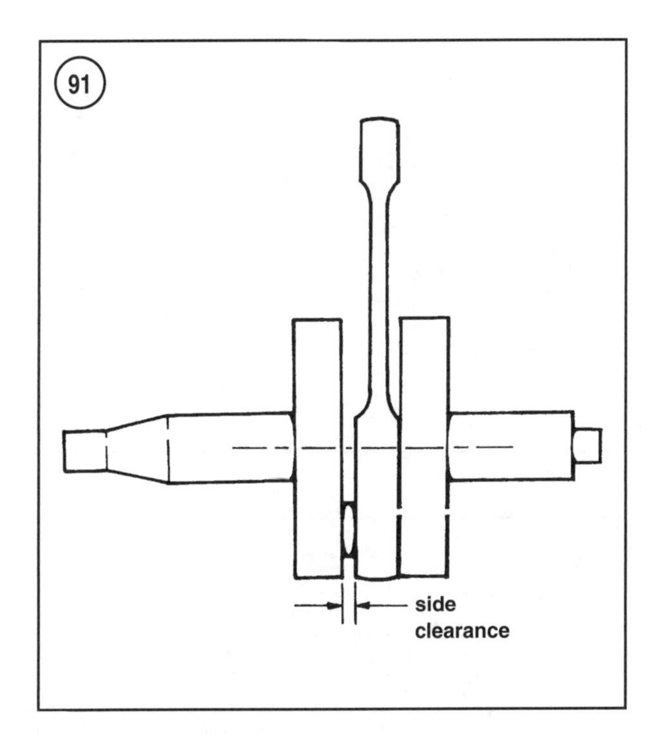

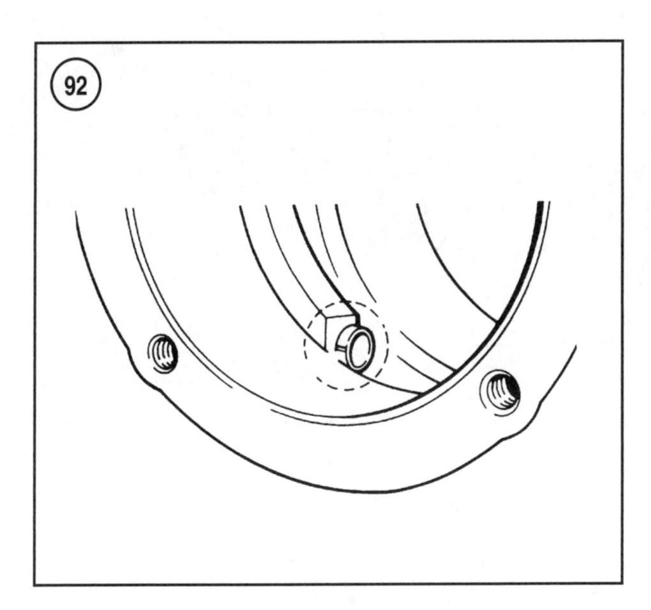

5. Apply a coating of Yamaha outboard oil to the bushing bore prior to assembling the power head.

Seal housing/covers

1. Clean the seal housing/covers(s) thoroughly with clean solvent and a parts washing brush. Carefully remove all sealant material from the mating surfaces.

2. Inspect the seal bore(s) for nicks, gouges or corrosion around the outer diameter. Replace the housing or cover if the seal bore is damaged.

3. Inspect the end cap mating surface and O-ring groove for nicks, grooves, cracks, corrosion or distortion. Replace the end cap(s) if a defect is found.

Cylinder head/block cover(s)

1. Clean the cylinder head(s) or block cover thoroughly with clean solvent and a parts washing brush. Carefully remove all gasket and sealant material from the mating surfaces.

2. Remove all carbon and varnish deposits from the combustion chambers with a carbon removing solvent, such as Yamaha Combustion Chamber Cleaner. A sharpened hardwood dowel or plastic scraper can be used to remove stubborn deposits. Do not scratch, nick or gouge the combustion chambers.

3. Check the cylinder head(s) and block cover for cracks, fractures, distortion or other damage. Check the cylinder head(s) for stripped or damaged threads. Refer to *Sealing Surfaces* in this chapter and check the cylinder head(s) for warpage. Minor imperfections can be removed as described in *Sealing Surfaces* in this chapter.

4. Inspect all gasket surfaces or O-ring and water seal grooves for nicks, grooves, cracks, corrosion or distortion. Replace the cylinder head or block cover if the defect is severe enough to cause leakage.

5. Check all water, oil and fuel bleed passages in the head(s) for obstructions. Make sure all pipe plugs are installed tightly. Seal all pipe plugs with Loctite 567 PST pipe sealant.

Exhaust plate/cover

1. Clean the exhaust cover, manifold (and plate) thoroughly with clean solvent and a parts washing brush. Carefully remove all gasket and sealant material from the mating surfaces.

2. Remove all carbon and varnish deposits with a carbon removing solvent, such as Yamaha Combustion Chamber Cleaner. A hardwood dowel or plastic scraper can be used to remove stubborn deposits. Do not scratch, nick or gouge the mating surfaces.

3. Inspect the component and all gasket surfaces for nicks, grooves, cracks, corrosion or distortion. Replace the cover/plate if the defect is severe enough to cause leakage.

Thermostat cover

1. Clean the thermostat cover and mating surface with a suitable solvent and a stiff plastic bristle brush. Carefully remove all mineral deposits, gasket material and sealant from the cover.
2. Inspect the cover for corrosion, cracking or other damage. Replace the cover as needed.

POWER HEAD ASSEMBLY

Before beginning assembly, make sure to complete all applicable sections of the *Power Head Components Cleaning and Inspection* in this chapter.

Review the following sections: *Sealing Surfaces; Fasteners and Torque* and *Lubricants, Sealants, and Adhesives* in this chapter.

The manufacturer recommends replacing all seals, O-rings, gaskets, piston pin lock rings, piston rings and all needle bearings any time a power head is disassembled. If reusing the original needle bearings, make sure to reinstall them in the original positions.

Any identification (dot or letter) mark on a piston ring (**Figure 93**) must face up when installed. Some pistons use a combination of ring styles. Rings may be rectangular, semi-keystone, or full-keystone.

Rectangular and full key-stone rings fit the grooves in either direction, but must be installed with the identification mark facing up.

Semi-keystone rings are beveled 7-10° on the upper surface only. These rings do not fit the groove correctly if installed upside down. Carefully examine the construction of the rings and look for identification marks before installation. The beveled side must face up (matching the ring groove) and is identified by the mark (dot or letter) on the upper surface.

Lubricate the needle and roller bearings, pistons, rings and cylinder bores with two-stroke outboard engine oil during assembly.

Clean the cylinder block thoroughly after boring or honing to remove any contaminants. Use hot, soapy water under pressure. *Do not* use solvent to remove the contaminants. After cleaning, use compressed air to dry the cylinder block. Wipe the cylinder bores with a white shop towel. Stop cleaning when the towel remains clean and white after wiping the cylinder bores. To prevent corrosion, thoroughly coat the cylinder bores with Yamalube two-cycle outboard oil after cleaning.

CAUTION
A selection of torque wrenches is essential for correct assembly and to ensure maximum longevity of the power head assembly. Failing to torque items as specified causes premature power head failure.

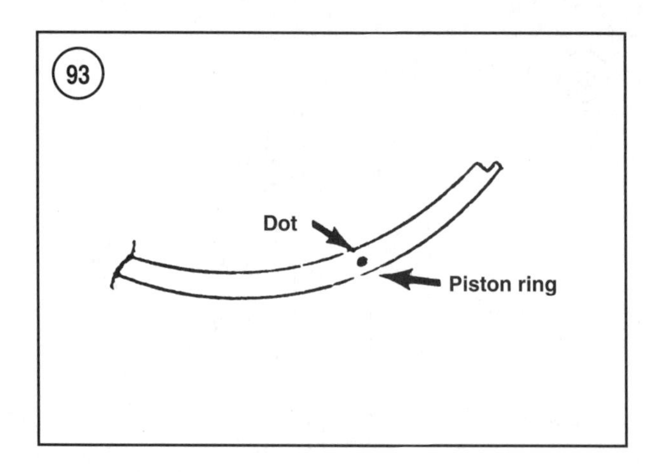

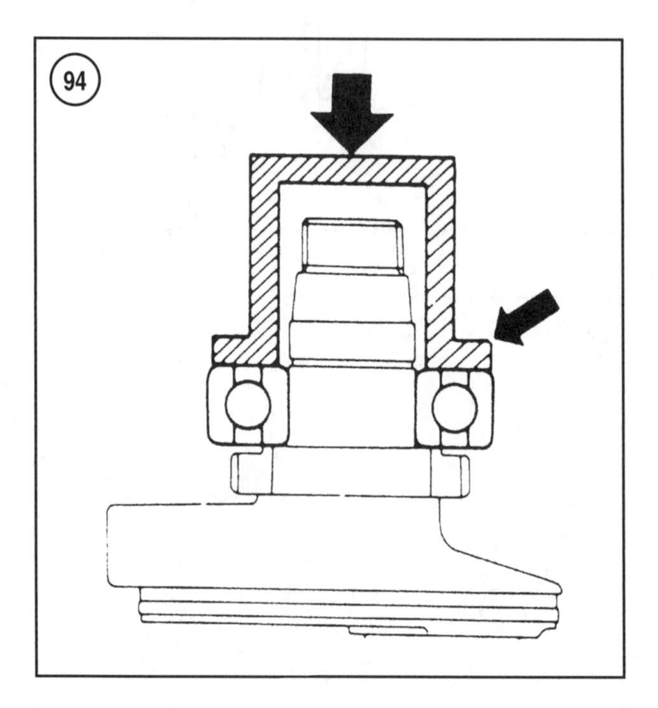

Power head torque specifications are listed in **Table 1**. General torque specifications are listed in Chapter One. Use the general torque specification for fasteners not listed in **Table 1**.

Mating surfaces must be absolutely free of gasket material, sealant residue, dirt, oil, grease or any other contaminants. Lacquer thinner, acetone, isopropyl alcohol and similar solvents are excellent oil, petroleum and wax-free solvents to use for the final preparation of mating surfaces.

Refer to **Figures 28-30**.

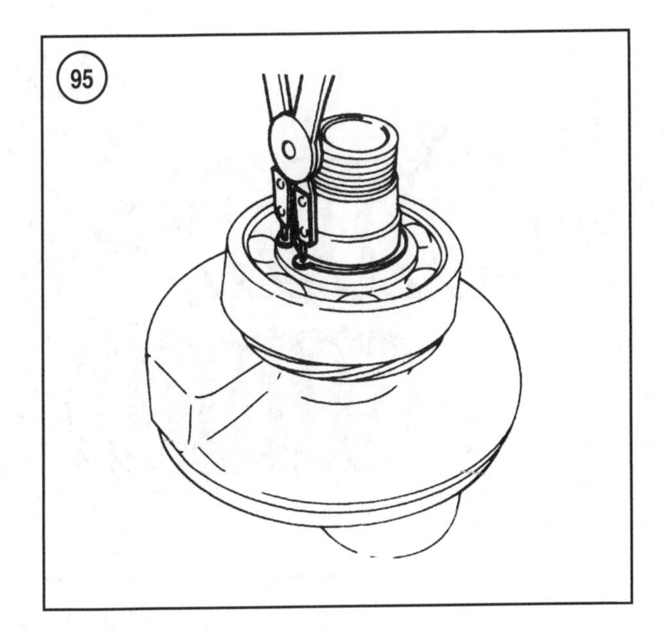

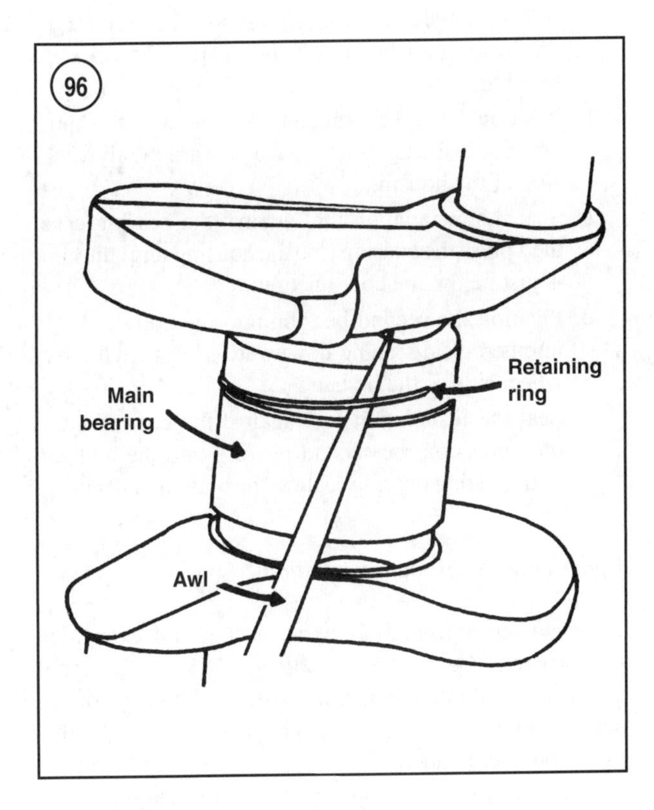

Crankshaft Assembly

An upper bearing/cover/seal housing is not used on 200-250 hp (3.1 liter) models. The upper crankshaft seal is mounted into the upper crankshaft roller bearing.

1. If removed, install the oil injection drive gear onto the crankshaft as follows:
   a. Select a section of tubing or other suitable tool to press the drive gear onto the crankshaft. The tool must be of sufficient diameter to slide over the crankshaft and only contact the flat surfaces of the gear. The tool must be of sufficient length to prevent the press from contacting the end of the crankshaft.
   b. Apply two-stroke outboard oil to the surfaces, then slide the oil injection drive gear onto the crankshaft. The stepped side of the gear hub must face away from the crankshaft flyweight with the gear teeth closest to the flyweight. Rotate the gear to align the flat drive surface in the gear bore with the corresponding surface on the crankshaft.
   c. Support the lower flyweight on the table of a press with the flywheel end facing downward.
   d. Press on the gear until it just contacts the step on the crankshaft.
2. If removed, install the lower bearing onto the crankshaft as follows:
   a. Select a section of tubing or other suitable tool to press the bearing onto the crankshaft. The tool must be of sufficient diameter to slide over the crankshaft and only, yet contact the inner race of the bearing. The tool must be of sufficient length to prevent the press from contacting the end of the crankshaft.
   b. Support the lower flyweight on the table of the press with the flywheel end facing downward.
   c. Apply two-stroke outboard engine oil to the surfaces, then slide the bearing over the crankshaft. The numbered side of the bearing must face upward or away from the flywheel end of the crankshaft. Rest the bearing on the step.
   d. Slide the installation tool over the crankshaft and seat it against the inner bearing race (**Figure 94**).
   e. Press on the tool until the bearing seats against the step on the crankshaft.
   f. Install the snap ring into the groove in the crankshaft (**Figure 95**).
3. If removed, install the center main bearings as follows:
   a. Coat the bearing and crankshaft surfaces with Yamalube two-cycle outboard oil, then install the roller bearings and bearing cases onto the crankshaft center main surfaces (**Figure 96**). If reusing the bearings, install them into the original location on the crankshaft with the components oriented to the flywheel side.
   b. Carefully wind the retaining ring into the groove on the bearing cases. Do not spread the ring more than necessary for installation.
   c. Check for proper alignment of the bearing cases. The mating surface of the bearing must be undetectable when passing a sharp pencil or scribe across

the parting line. Remove the bearing and check for debris in the bearing or improper installation if misaligned.

4. Carefully spread, then install the crankcase sealing rings into the grooves on the crankshaft (**Figure 97**). Spread the rings just enough to pass the ring over the crankshaft from the upper or lower ends. Release the rings into the groove. Do not spread the rings more than necessary for installation.

Bearing/Cover/Seal Housing Assembly

Refer to the appropriate illustration (**Figures 28-30**) to assist with component orientation and mounting location. An upper bearing/cover/seal housing is not used on 200-250 hp (3.1 liter) models. The upper crankshaft seal is mounted into the upper crankshaft roller bearing.

Roller bearing installation (upper housing)

A single upper drive shaft bearing is used on 80 Jet and 115-200 hp models (except HPDI). Two upper drive shaft bearings are used on 150-250 hp HPDI models.

The bearing(s) must be installed with the numbered side facing toward the flywheel side of the housing.

1. Select an appropriately sized socket and extension, section of tubing or other suitable tool to press the upper crankshaft bearings into the bearing/cover/seal housing. The tool must be of sufficient diameter to adequately contact the bearing case, yet not contact the bore in the housing during installation.

2. Support the upper bearing/cover/seal housing on the table of a press with the flywheel end facing downward. Place a suitable block under the housing and in contact with the flat housing surfaces to prevent the housing from contacting the table.

3A. *Models using a single bearing*—Install the bearing as follows:

   a. Apply Yamalube two-cycle outboard engine oil onto the bearing surfaces and the bearing bore in the housing.
   b. Position the bearing into the bore with the numbered side facing downward or toward the flywheel side of the housing.
   c. Seat the installation tool against the bearing case, then press the bearing into the housing until fully seated.

3B. *Models using two bearings*—Install the bearings as follows:

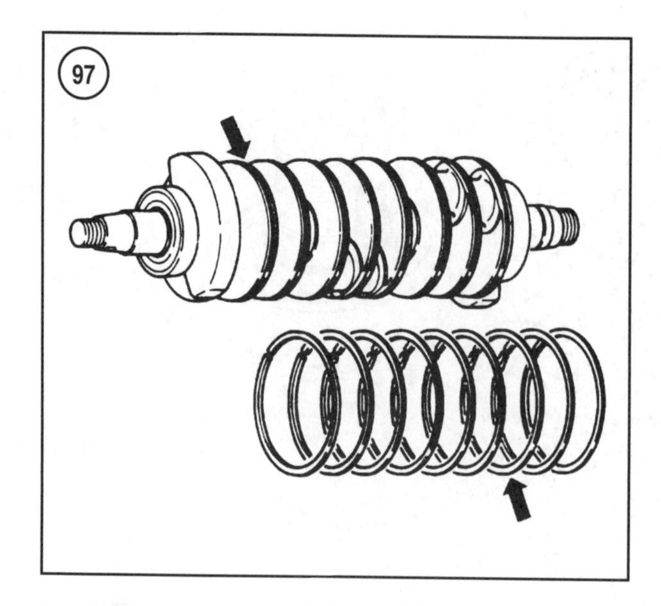

   a. Apply Yamalube two-cycle outboard engine oil to the bearing surfaces and the bearing bore in the housing.
   b. Position the first bearing into the bore with the numbered side facing downward or toward the flywheel side of the housing.
   c. Seat the installation tool against the bearing case, then press the bearing into the housing until the case is just below the bore opening.
   d. Position the second bearing into the bore with the numbered side facing downward or toward the flywheel side of the housing.
   e. Seat the installation tool against the bearing case, then press on the second bearing into the housing until the first bearing seats in the bottom of the bore.

Upper crankshaft seal installation

1. Select an appropriately sized socket and extension, section of tubing or other suitable tool to press the upper crankshaft seal into the bearing/cover/seal housing or upper bearing (3.1 liter models). The tool must be of sufficient diameter to adequately contact the seal case, yet not contact the bore in the housing during installation.

2A. *80 Jet, 105 Jet, 115-130 and 150-200 hp (2.6 liter) models*—Install the upper crankshaft seal into the bearing/cover/seal housing as follows:

   a. Support the upper bearing/cover/seal housing on a press table with the flywheel end facing upward (**Figure 98**). Place a suitable block under the housing and in contact with the flat housing surfaces to prevent the housing from contacting the table.

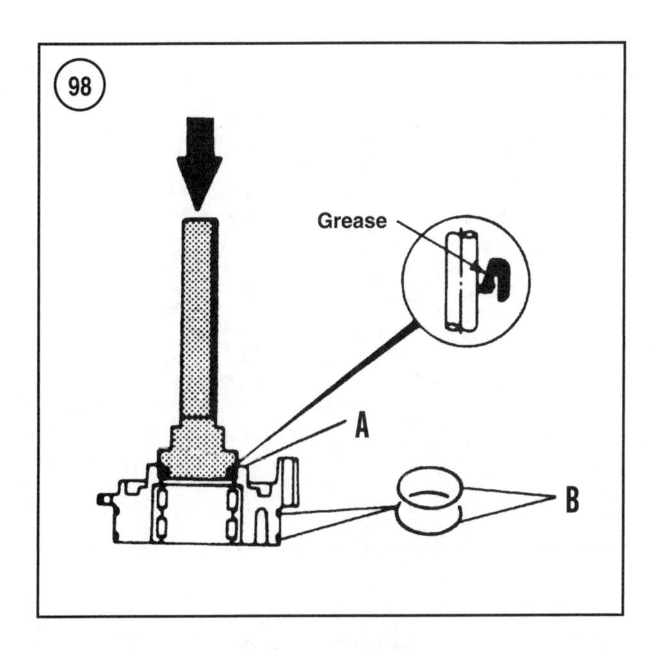

b. Position the seal (A, **Figure 98**) into the bore with the lip (open) side facing downward or away from the flywheel when installed.
c. Seat the installation tool against the seal casing, then press the seal into the housing until fully seated.

2B. *200-250 hp (3.1 liter) models*—Install the upper crankshaft seal into the upper crankshaft bearing as follows:

a. Place the upper bearing onto a press table with the seal bore opening facing upward.
b. Position the seal into the bore with the lip (open) side facing downward or away from the flywheel when installed.
c. Seat the installation tool against the seal casing, then press the seal into the bearing until fully seated in the recess.

3. Apply a coating of Yamaha All-Purpose Grease to the seal lip.
4. Apply a coating of Yamalube two-cycle outboard oil to the surfaces, then install new O-rings (B, **Figure 98**) into the respective grooves on the bearing/cover/seal housing or upper crankshaft bearing (3.1 liter models).

Lower crankshaft seal installation

1. Select two appropriately sized sockets and extension, sections of tubing or other suitable tools to press the lower crankshaft seals into the lower bearing/cover/seal housing. The tool must be of sufficient diameter to adequately contact the seal cases, yet not contact the bores in the housing during installation.
2. Place the lower bearing/cover/seal housing onto a press table with the seal bore (open) side facing upward.
3. Position the smaller diameter seal into the housing bore with the lip (open) side facing downward.
4. Seat the installation tool against the seal casing, then press the seal into the housing until fully seated in the recess.
5. Position the larger diameter seal into the housing bore with the lip (open) side facing downward.
6. Seat the installation tool against the seal casing, then press the seal into the housing until fully seated in the recess.
7. Apply a coating of Yamaha All-Purpose Grease to the seal lips.
8. Apply a coating of Yamalube two-cycle outboard oil to the surfaces, then install a new O-ring into the groove in the housing.

Crankshaft Installation

The crankshaft and supporting bearings are installed into the cylinder block prior to installing the pistons and connecting rods. The individual piston and connecting rod assemblies are installed on the crankshaft through the cylinder openings.

1. Apply a generous coating of Yamalube two-cycle outboard to the crankshaft bearings and surfaces where the crankshaft bearings contact the cylinder block. Position the cylinder block with the crankcase side facing upward.

   200-250 hp (3.1 liter) models—Apply a generous coating or Yamalube two-cycle outboard oil onto the rollers of the upper crankshaft bearing. Apply a coating of Yamaha All-Purpose Grease to the upper crankshaft seal lip. Guide the tapered end of the crankshaft through the bearing and seal. Align the bearing with the bearing contact surfaces on the crankshaft.
2. Position the crankshaft assembly over the crankcase with the flywheel taper end toward the top of the cylinder block.
3. Have an assistant rotate the sealing rings so that the gaps face away from the cylinder bores and are aligned with the crankshaft centerline (**Figure 99**).
4. Carefully guide the locating pins into the recesses in the cylinder block or main bearings while lowering the crankshaft into position.
5. Use hand pressure to seat the crankshaft main bearings into the cylinder block. Do not strike the crankshaft. Carefully move the main bearings to align the locating pins. Continue to work with the bearings until the crankshaft fully seats in the crankcase.

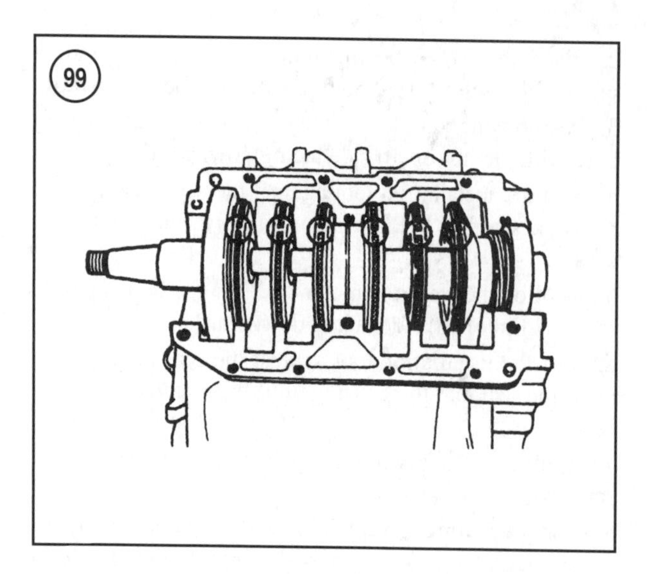

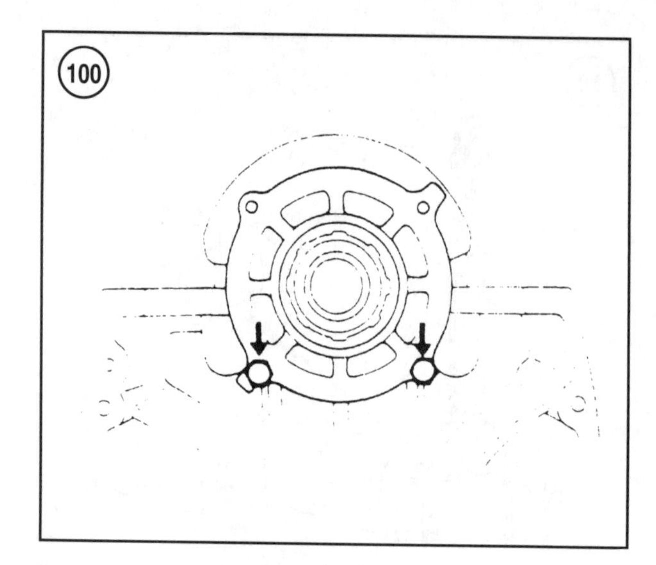

6. Slowly rotate the crankshaft within the cylinder block. Remove the crankshaft and inspect all bearing and alignment pins for damage if there is any binding or roughness.
7. Maintain the cylinder block with the crankcase side facing upward until the bearing/cover/seal housing(s) are installed.

Bearing/cover/seal housing installation

1. Apply a light coating of Loctite Gasket Maker to the mating surfaces of the upper and lower seal housings. An upper housing is not used on 200-250 hp (3.1 liter) models.
2. Apply a generous coating or Yamalube two-cycle outboard oil to the rollers of the upper crankshaft bearing.
3. Apply a coating of Yamaha All-Purpose Grease to the upper and lower crankshaft seal lips.
4. *80 Jet, 105 Jet, 115-130 hp and 150-200 hp (2.6 liter) models*—Install the upper bearing/cover/seal housing as follows:
   a. Guide the tapered end of the crankshaft through the upper bearing/cover/seal housing while installing the lower cover/seal housing onto the cylinder block. Work carefully to avoid damaging the O-ring. Lift the crankshaft slightly if necessary to fit the housing into the crankshaft bore. Do not disturb the main bearing alignment with the locating pins while lifting the crankshaft.
   b. Rotate the seal housing to align the bolt openings. To help secure the crankshaft into the crankcase, install the two bolts that retain the lower bearing/cover/seal housing onto the cylinder block (**Figure 100**). Hand-tighten the bolts. The bolts are fully tightened after installing the crankcase cover.

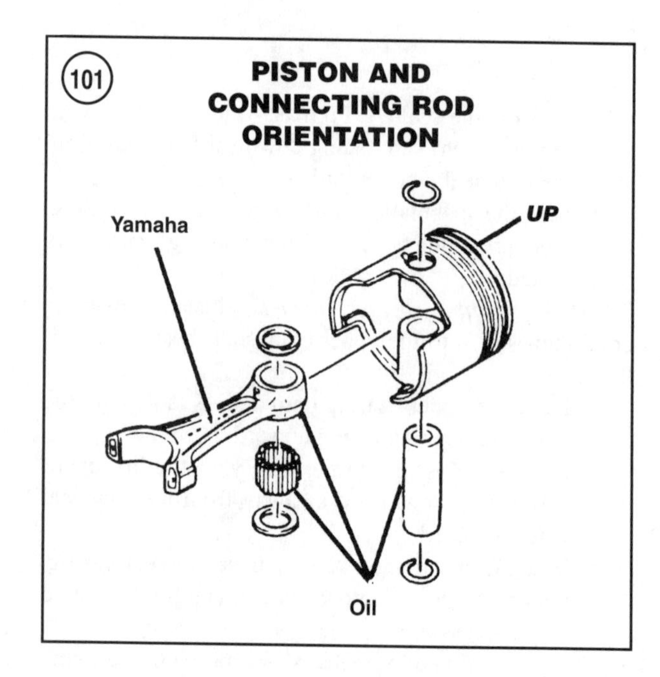

5. Install the lower bearing/cover/seal housing as follows:
   a. Guide the lower end of the crankshaft through the seal openings while installing the lower bearing/cover/seal housing onto the cylinder block. Work carefully to avoid damaging the O-ring. Lift the crankshaft slightly if necessary to fit the housing into the crankshaft bore. Do not disturb the main bearing alignment with the locating pins while lifting the crankshaft.
   b. Rotate the seal housing to align the bolt openings. To help secure the crankshaft into the crankcase, install the two bolts that retain the lower bear-

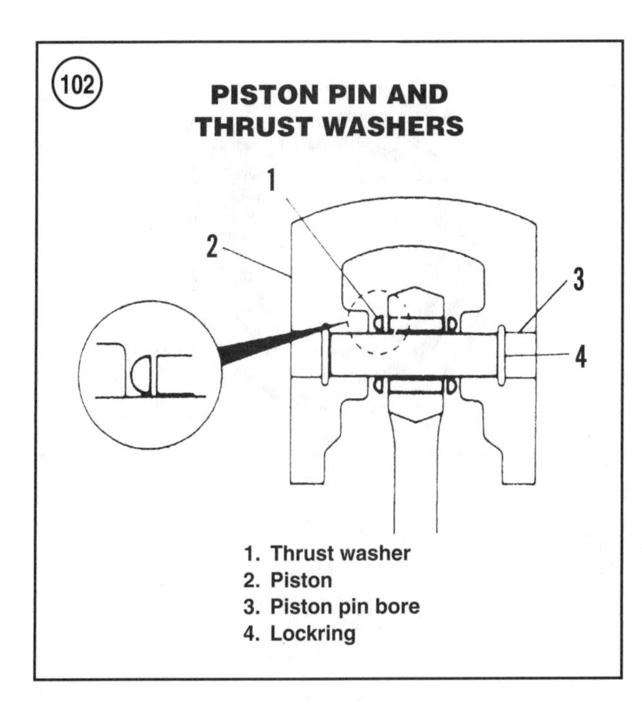

1. Thrust washer
2. Piston
3. Piston pin bore
4. Lockring

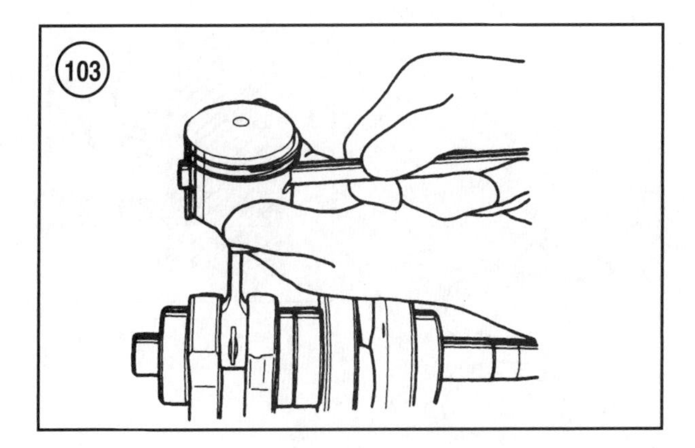

ing/cover/seal housing onto the cylinder block (**Figure 100**). Hand-tighten the bolts. The bolts will be fully tightened after installing the crankcase cover.

Piston and Connecting Rod Assembly

WARNING

Use protective eyewear when working with the power head. Piston pin lockrings and other components may unexpectedly spring free and cause serious injury.

NOTE

If piston pin installation is difficult, heat the piston to approximately 60° C (140° F) with a portable hair dryer. The piston pin bore expands when heated, allowing easier pin installation.

1. Use a section of tubing or rod as a piston pin installation tool. The tool must be slightly smaller in diameter than the piston pin. Piston pin diameters are listed in **Table 3**.
2. Clamp the No. 1 connecting rod into a vise with protective jaws. The Yamaha marking on the connecting rod (**Figure 101**) must face upward.
3. Lubricate the piston pin, needle bearing assembly, small end bore in the connecting rod and thrust washers with Yamalube two-cycle outboard oil.
4. Insert the caged needle bearing assembly into the small end bore in the connecting rod. Hold the bearing into position until the piston is installed to the connecting rod.
5. Position the thrust washers (1, **Figure 102**) to each side of the connecting rod bore. The flat side of the thrust washer must contact the connecting rod.
6. Fit the No. 1 piston over the connecting rod and align the piston pin bores in the piston and connecting rod. The up marking on the piston (**Figure 101**) must be orientated toward the flywheel end of the crankshaft. Verify that the needle bearing and thrust washers are centered with the connecting rod bore, then slide the installation tool through the pin bores, washers, bearing and connecting rod. Hold the installation tool in the piston until the piston pin is fully installed.
7. Lubricate the piston pin with two-stroke outboard oil. Maintain pin bore alignment and carefully push the piston pin into the bores while guiding the installation tool out of the piston (**Figure 103**). Work carefully to avoid damaging the bearings. Push the piston pin in until the installation tools exits the bore on the other side of the piston and the lockring grooves are exposed on each side.
8. Use needle nose pliers to insert the lockring into the groove in the piston pin bore (**Figure 104**). Rotate the lockring to position the tang into the tang recess as shown in **Figure 105**. Repeat this step on the other side of the piston.
9. Inspect the piston to verify the lockrings are fully seated into the grooves.
10. Repeat Steps 2-9 for the remaining pistons. Each piston must be installed onto the corresponding connecting rod.

Piston ring installation

CAUTION

Use caution when spreading the rings for installation onto the piston. Spread the gap ends only enough to allow the ring to slide

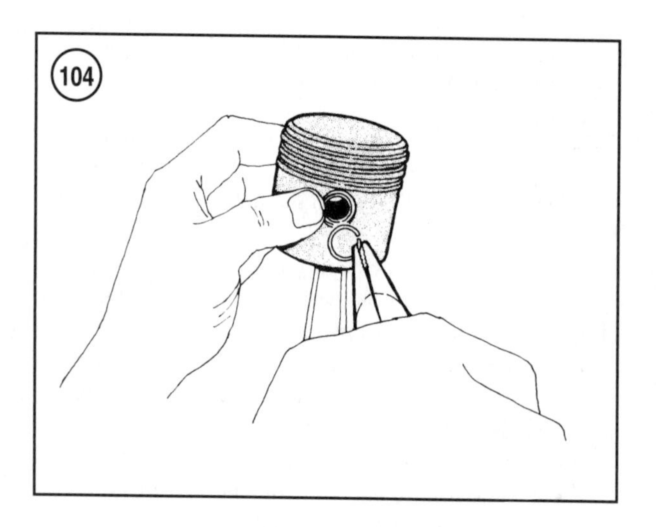

104

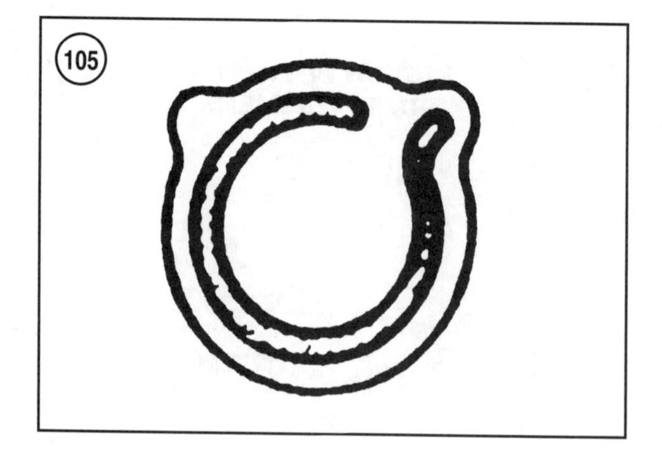

105

over the piston. The ring will break or crack if spread too much when installing.

NOTE

*Various types of piston rings are used. A rectangular ring (2, **Figure 106**), full keystone or half keystone (3) may be used. The piston may use one or a combination of types. Inspect the piston ring and ring groove prior to installing any of the rings to ensure the proper ring installation. The marking on the rings (4, **Figure 106**) must face toward the piston dome.*

1. Select the rings for the No. 1 cylinder and the No. 1 piston.
2. Use a ring expander to spread the bottom ring open enough to clear the top of the piston (**Figure 107**). The marking on the ring (4, **Figure 106**) must face upward. Avoid scratching the piston. Slide the ring over the piston and release the ring into the bottom ring groove. Repeat this procedure for the top ring. The marking on the top ring must face upward. Apply a coating of two-stroke outboard engine oil onto all surfaces of the piston. Repeat Step 2 and Step 3 for the remaining piston(s). The rings must be installed onto the corresponding piston(s).

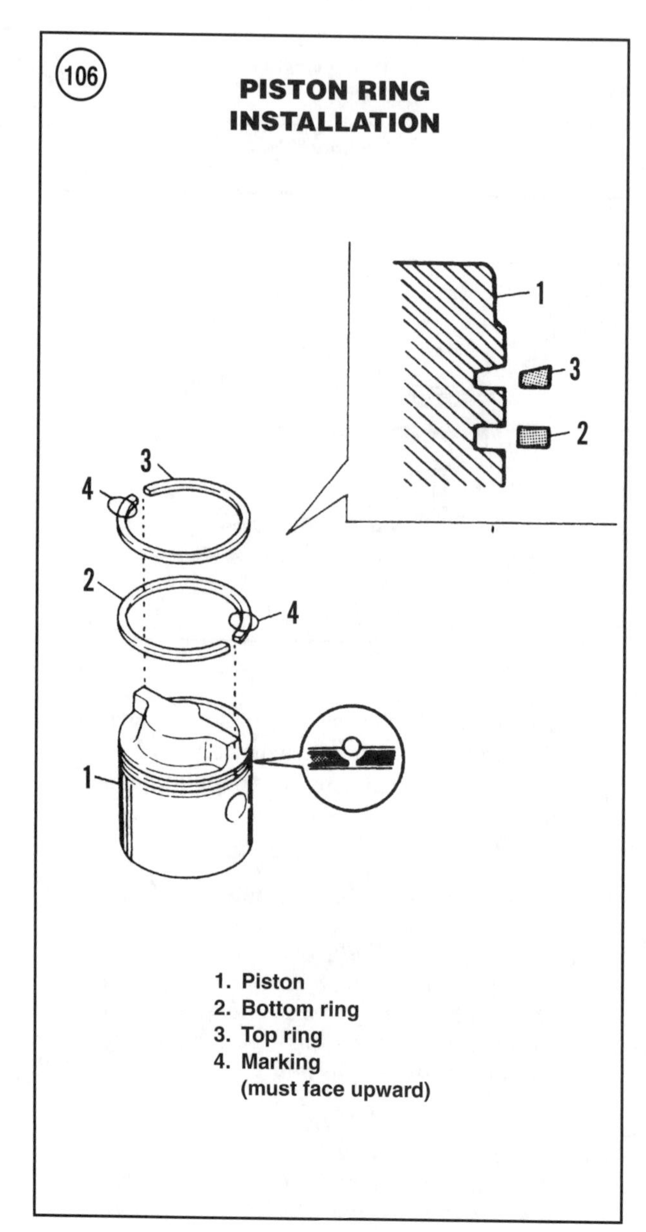

106

PISTON RING INSTALLATION

1. Piston
2. Bottom ring
3. Top ring
4. Marking (must face upward)

Piston and Connecting Rod Installation

CAUTION

Work carefully and patiently when installing the pistons into the cylinder bores. Never use excessive force to insert the piston. Excessive force will damage the piston, rings or cylinder bore. Remove the piston and check for improper ring gap alignment and excessive or inadequate ring compressor

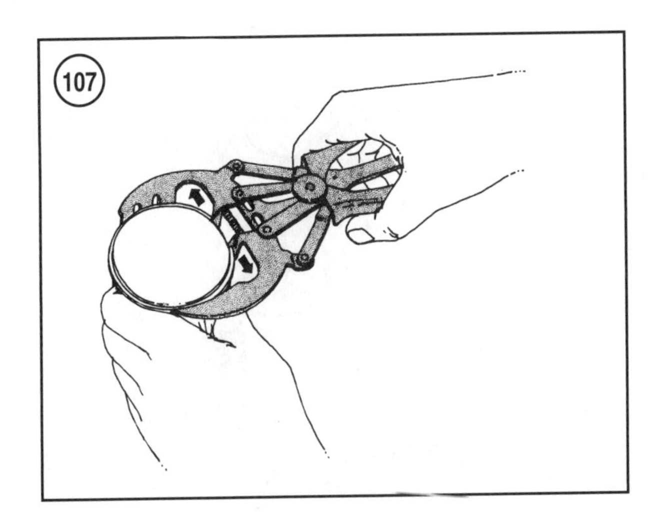

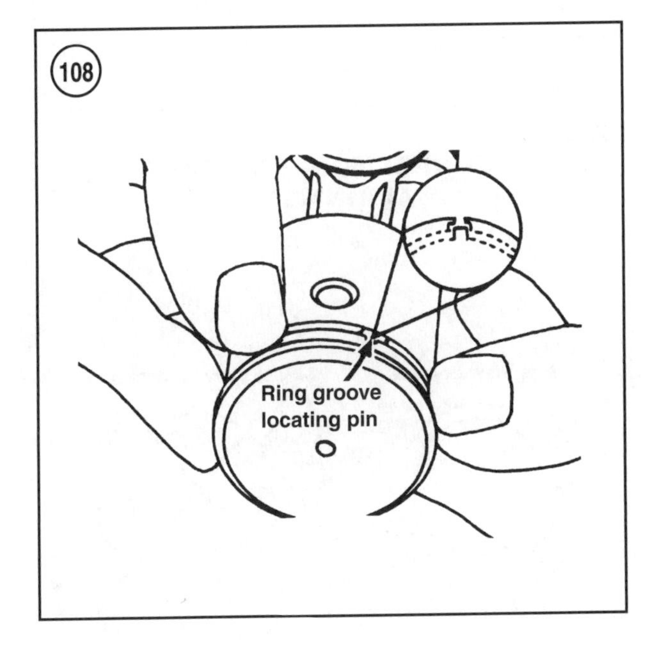

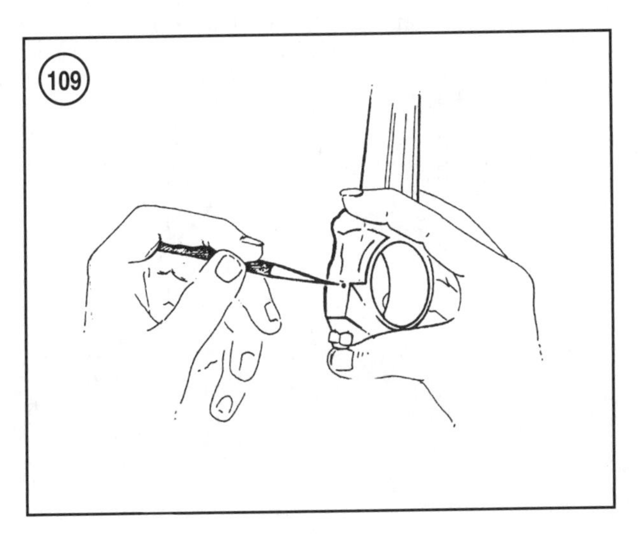

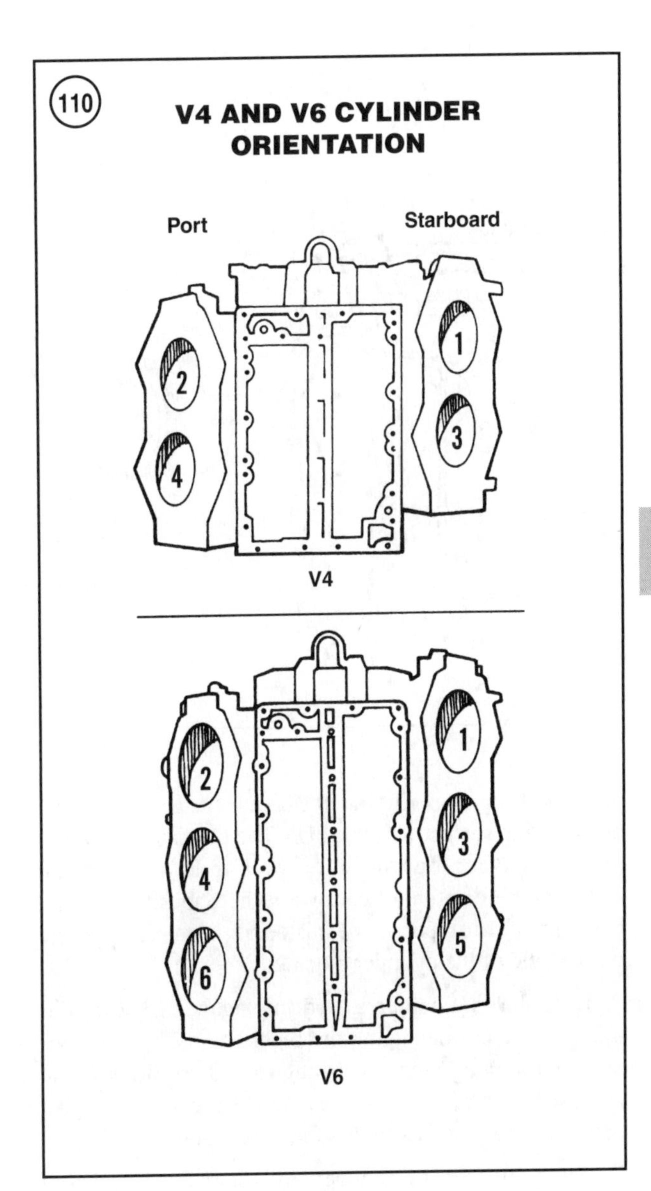

pressure if the piston(s) do not slide easily into the bore.

1. Apply a generous coating of two-stroke outboard oil to the cylinder bores, crankshaft bearings and connecting rod bearings.

2. Rotate the crankshaft until the No. 1 crankshaft journal is at the bottom of the stroke.

3. Position the ring gaps over the locating pins as indicated in **Figure 108**. The gaps must remain over the locating pins during piston installation.

4. Mark the rod cap for orientation (**Figure 109**) and remove it from the rod. Refer to **Figure 110** and locate the No. 1 cylinder bore in the block.

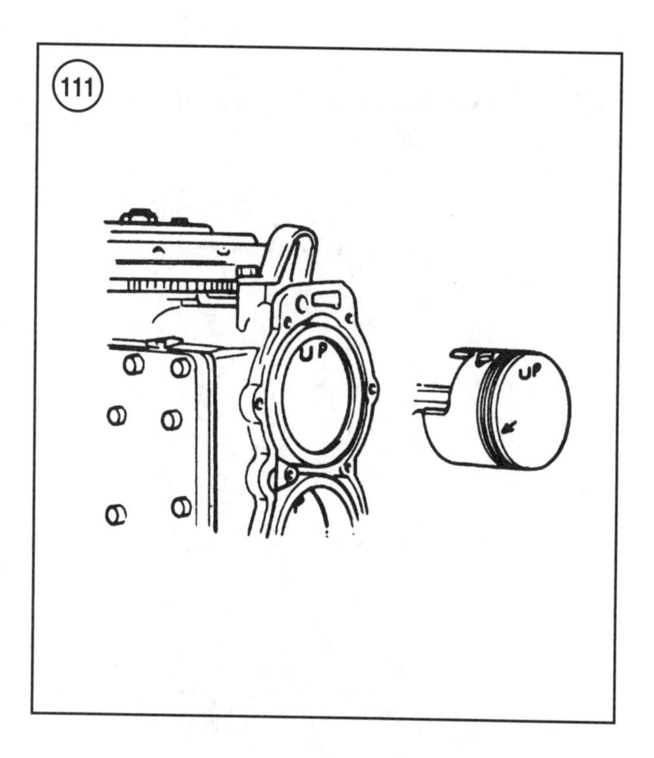

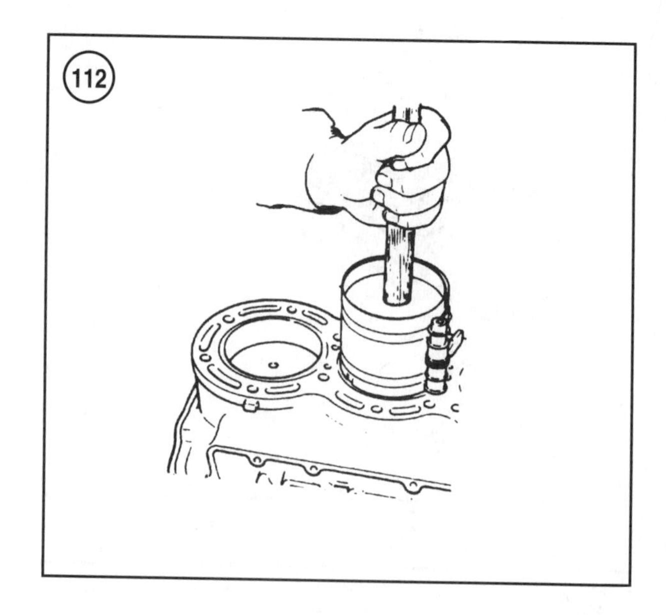

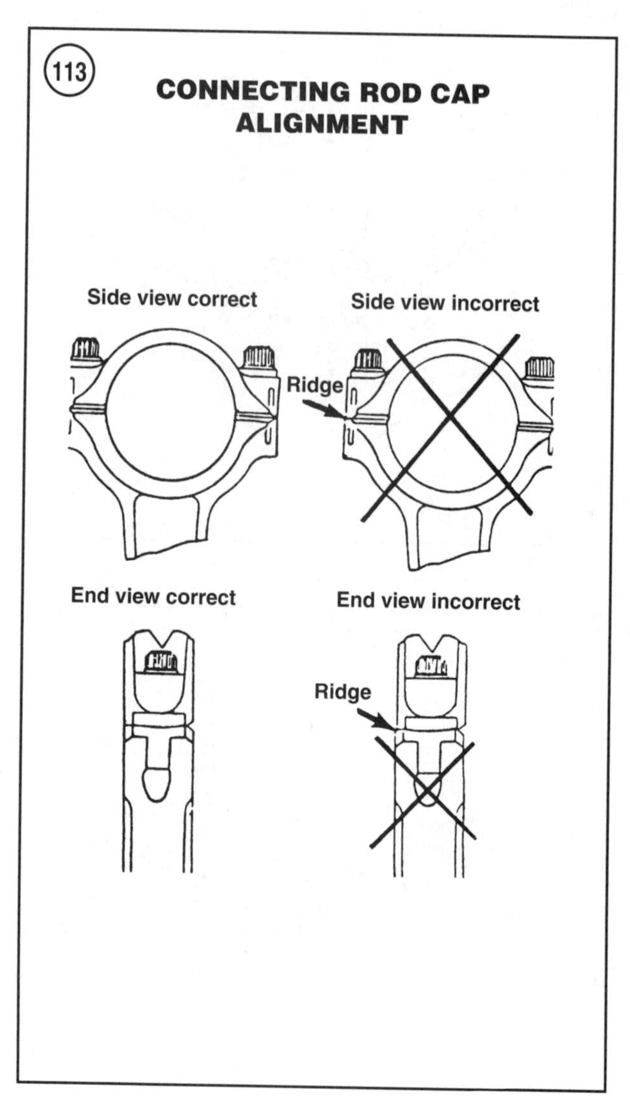

5. Insert the piston skirt into the bore at the No. 1 cylinder head mating surface (**Figure 111**). The *UP* marking on the piston must face toward the flywheel. If the arrow on the piston dome does not face toward the opposite cylinder bank or exhaust ports. the piston is installed into the wrong side of the cylinder block.

6. Hold the ring compressor firmly against the cylinder block. To avoid damaging the block, crankshaft or connecting rod, guide the connecting rod into position in the crankshaft while installing the piston. Use a large wooded dowel to push the piston into the bore (**Figure 112**).

7. Apply two-stroke outboard oil onto the crankshaft journal and install both halves of the bearing onto the crankshaft. Pull the connecting rod into the crankcase until it contacts the bearing.

8. Install the rod cap onto the connecting rod. Make sure the connecting rod is oriented correctly and install the used rod bolts. Torque the rod bolts in two steps to the specification in **Table 1**.

9. Check the alignment as indicated in **Figure 113**. Pass a sharp pencil point or other sharp pointed object across the rod and cap mating surfaces (**Figure 109**). If the point catches an edge, the rod is not aligned properly. If the visual inspection or sharp point indicates misalignment, remove the cap and check for debris between the mating surfaces. Carefully align the rod while reinstalling the cap and bolts. If misalignment persists, replace the connecting rod.

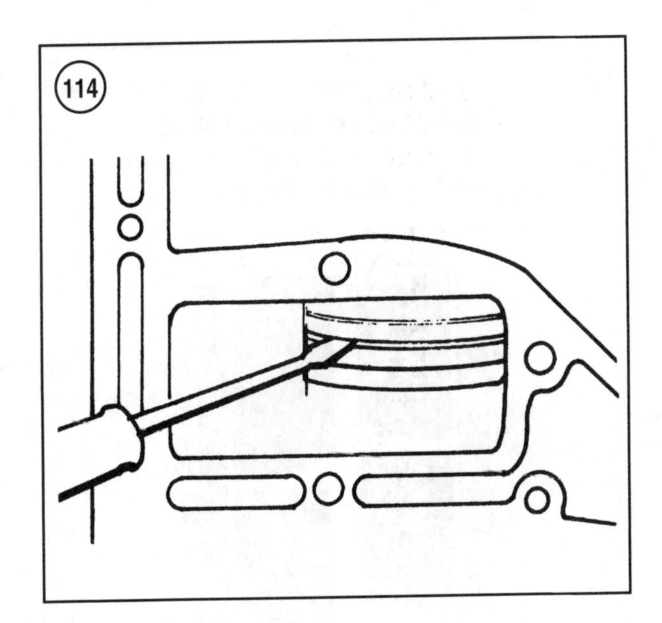

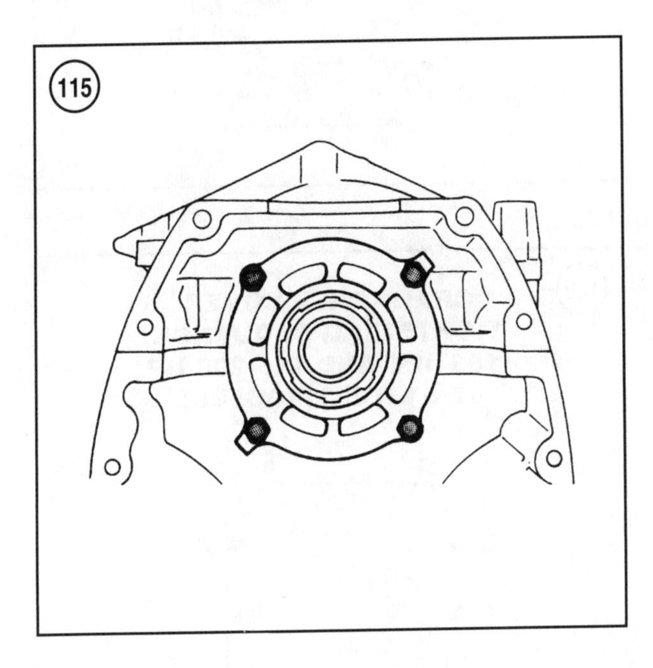

10. Loosen and remove one of the rod bolts. Apply a light coating or Yamalube two-cycle engine oil to the threads. Then install a new rod bolt. Tighten the new bolt in two steps to the specification in **Table 1**. Remove, replace and tighten the other rod bolt as well. Rotate the crankshaft until the No. 2 cylinder is at the bottom of the stroke. Repeat this procedure for the remaining cylinders. Check for proper rod cap alignment for all cylinders as described in Step 9. Correct misalignment before continuing.

11. Maintain downward pressure on the crankshaft to prevent the crankshaft from moving out of the cylinder block during this procedure. Look into the exhaust opening and slowly rotate the crankshaft until the piston rings for the No. 1 cylinder span the exhaust port (**Figure 114**). Use a blunt tip screwdriver to carefully press *in* on each ring. The ring must spring back when the screwdriver is pulled back. If either of the rings fail to spring back, remove the crankshaft, connecting rod and piston assembly and check for broken piston rings. Repeat this test for the remaining cylinder(s).

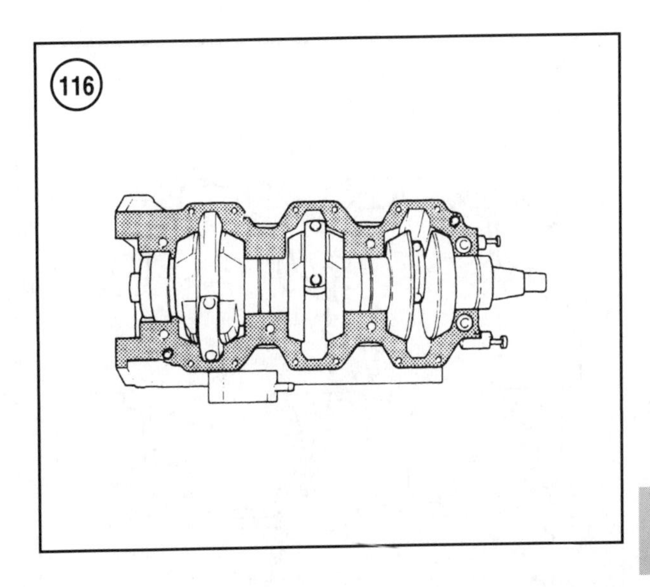

Crankcase Cover Installation

CAUTION

Make sure the crankcase cover to cylinder block mating surface is completely cleaned of oil or other contaminants prior to applying the Gasket Maker sealant. Oil, debris or contaminants left on the surfaces may prevent proper sealing and cause crankcase leakage.

1. Loosen the bolts (**Figure 115**) that retain the upper (except 3.1 liter models) and lower bearing/cover/seal housings onto the cylinder block. This allows for easier installation of the crankcase cover and easier alignment of the cover bolt openings.
2. Make sure the crankcase cover alignment pins (**Figure 116**) are in position in the cylinder block.
3. Apply a continuous bead of liquid gasket maker to the crankcase cover mating surfaces of the cylinder block. Apply the sealant bead on the inside of the crankcase bolt opening as shown in **Figure 117**. Use enough sealant to form a film on all mating surfaces when the bead is compressed by the crankcase cover. Avoid using too much of sealant.

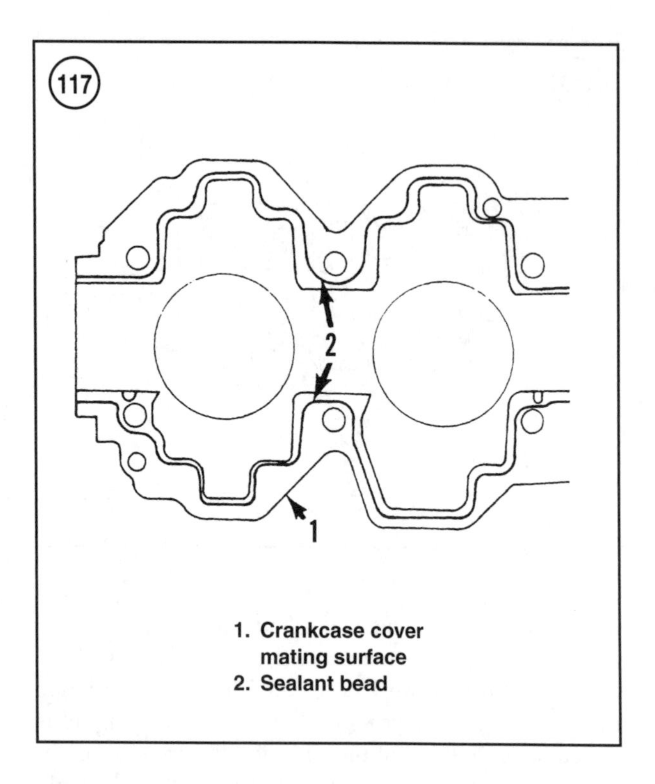

1. Crankcase cover mating surface
2. Sealant bead

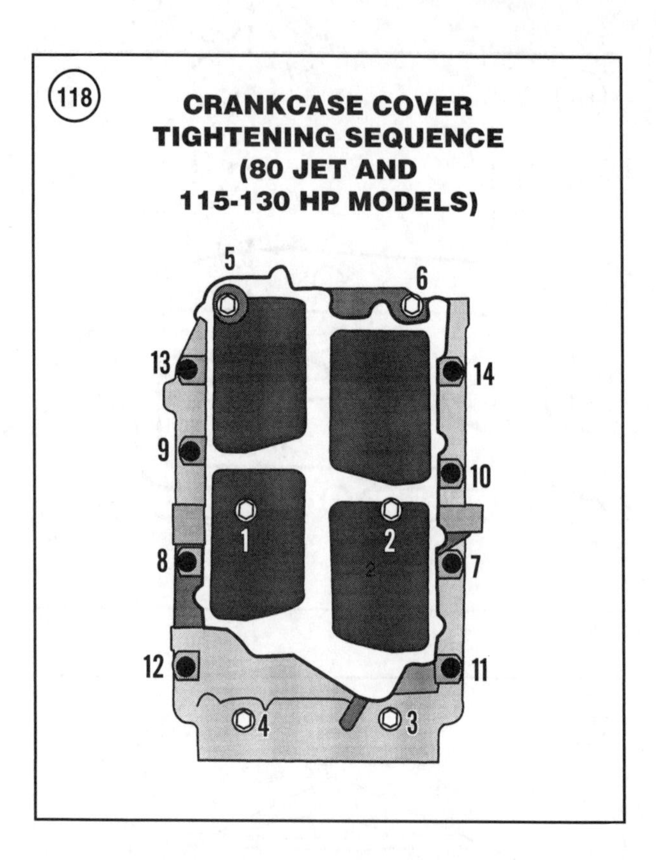

4. Install the crankcase cover and seat it onto the cylinder block. Push on the upper and lower bearing/cover/seal housings until it seats against the cylinder block and crankcase cover.

5. Apply Loctite 242 to the crankcase cover bolts. Hand-thread the bolts into the cover. Thread the remaining bolts into the upper and lower bearing/cover/seal housings.

6. Slowly rotate the crankshaft. If binding or roughness is noted, remove the cover and check for improper bearing installation. Check for broken piston rings if there is a clicking noise. The mating surface must be cleaned and a new bead of sealant applied each time the crankcase is removed.

NOTE

Tightening sequence numbers are cast into the exhaust cover, water jacket covers and cylinder heads. Tightening sequence numbers are not present on the crankcase cover. Refer to the appropriate illustrations for the tightening sequence.

7. Refer to the appropriate illustration (**Figures 118-120**) and tighten the cover bolts in two steps to the specification in **Table 1**.

8. Tighten the upper and lower bearing/cover/seal housing bolts to the general torque specification Chapter One.

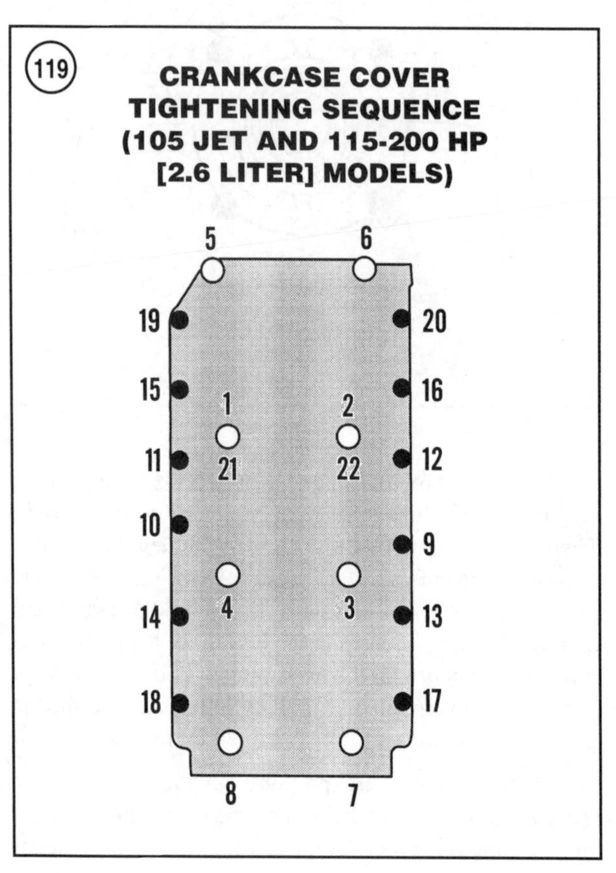

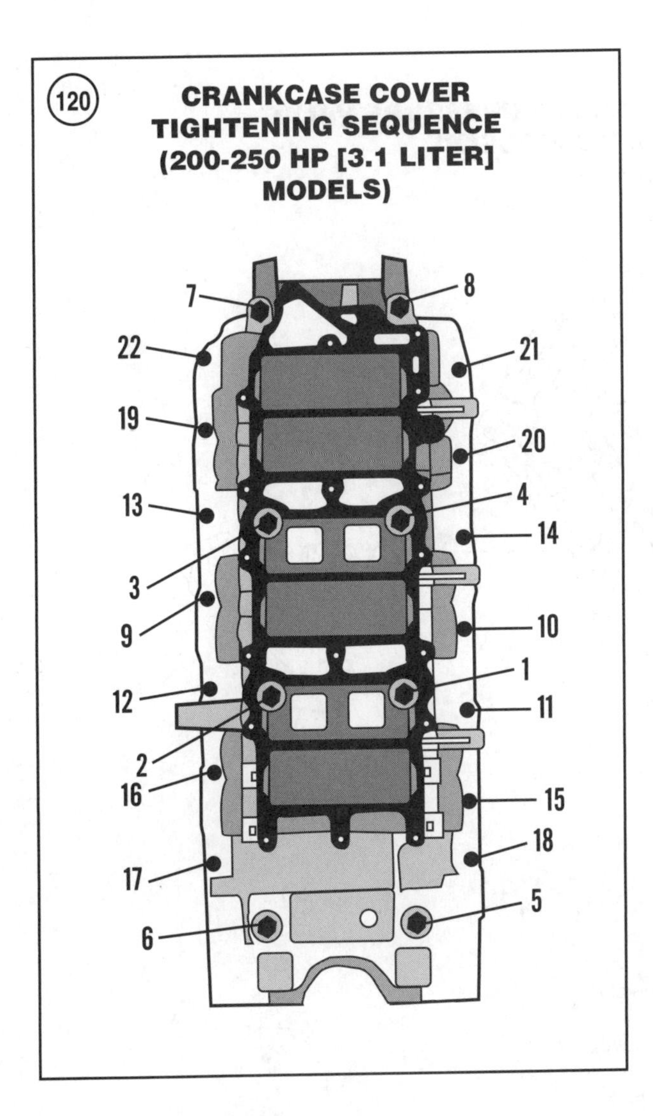

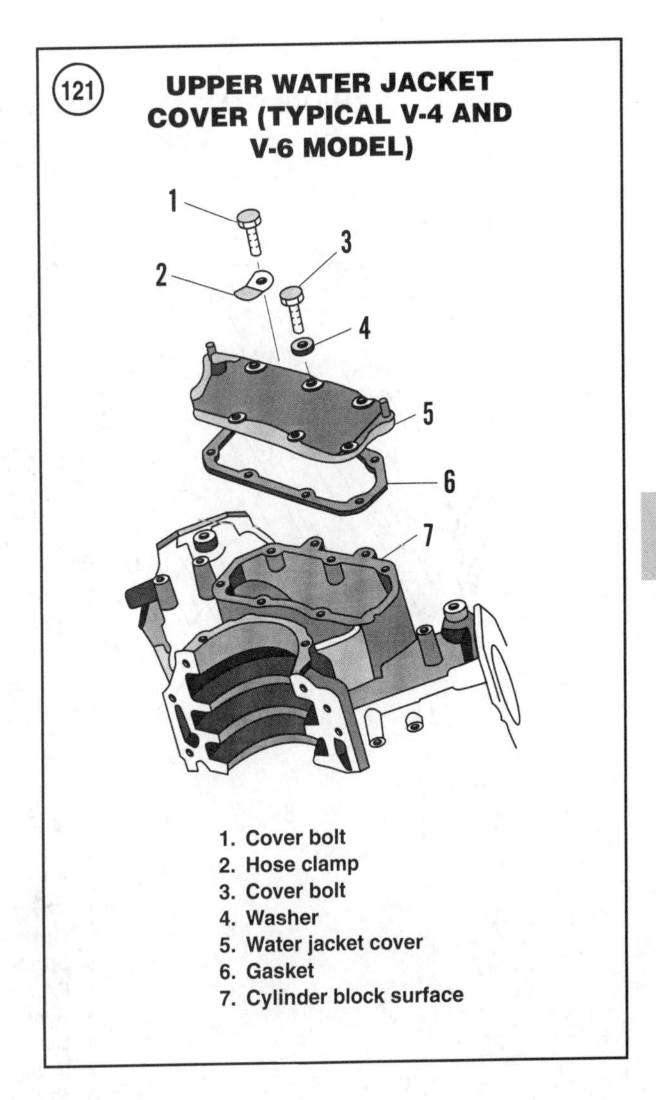

Water Jacket, Exhaust Cover, Water Pressure Valve and Thermostat Installation

Refer to the appropriate illustration (**Figures 28-30**) to assist with component orientation and mounting location. Apply Loctite 572 to the threads of all cover bolts prior to assembly. Hand thread the bolts, then tighten the bolts in sequence and to the specification in **Table 1** or general torque specifications in Chapter One.

Upper water jacket cover installation

An upper water jacket is used on all models covered in this manual.

1. Install the new upper water jacket cover gasket (6, **Figure 121**) onto the cylinder block. Align the gasket and cylinder block bolt openings. Fit the upper water jacket cover (5, **Figure 121**) onto the cylinder block. Align the bolt openings.
2. Apply Loctite 572 to the threads, then hand-thread the cover bolts into the cover and cylinder block. If equipped, install the hose clamp(s) (2, **Figure 121**) onto the correct bolt during installation.
3. Tighten the cover bolts in a crossing pattern, working from the center out, to the general torque specification in Chapter One.

Exhaust cover installation

An exhaust cover is used on all models covered in this manual.

1. Install the new cylinder block to the exhaust plate gasket (9, **Figure 122**) onto the exhaust plate. Do not apply sealant onto the gasket or sealing surfaces.

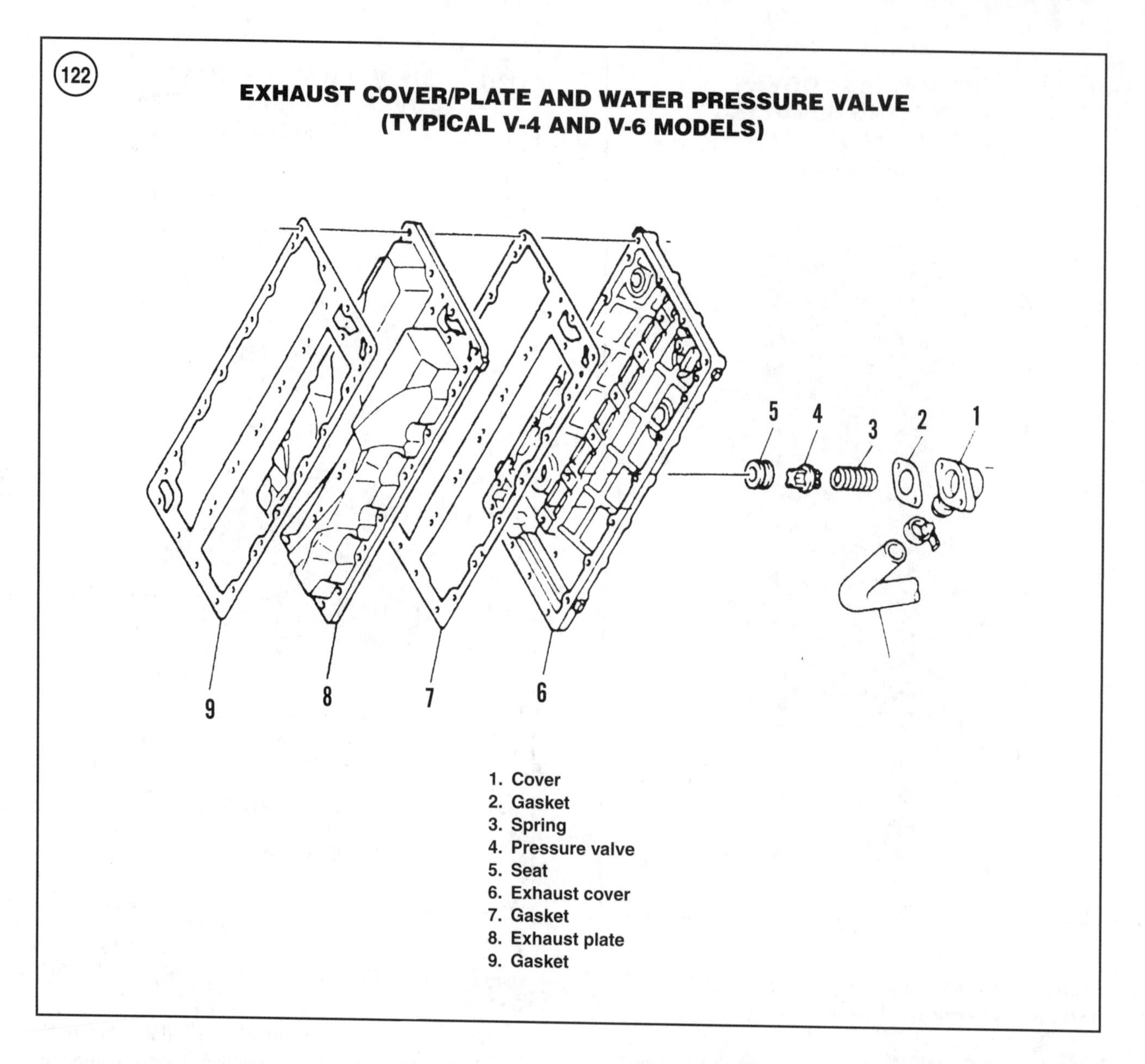

2. Install a new exhaust cover to exhaust plate gasket (7, **Figure 122**) onto the exhaust plate. Do not apply sealant to the gasket or sealing surfaces.

3. Fit the exhaust cover (6, **Figure 122**) onto the exhaust plate. Insert two of the mounting bolts into the plate, cover and gasket openings to help maintain alignment of the components during installation.

4. Install the cover, plate and gaskets as an assembly onto the cylinder block. Thread the two bolts into the block to secure the components.

5. Apply Loctite 572 to the remaining mounting bolts, then hand-thread them into the cover, plate and cylinder block.

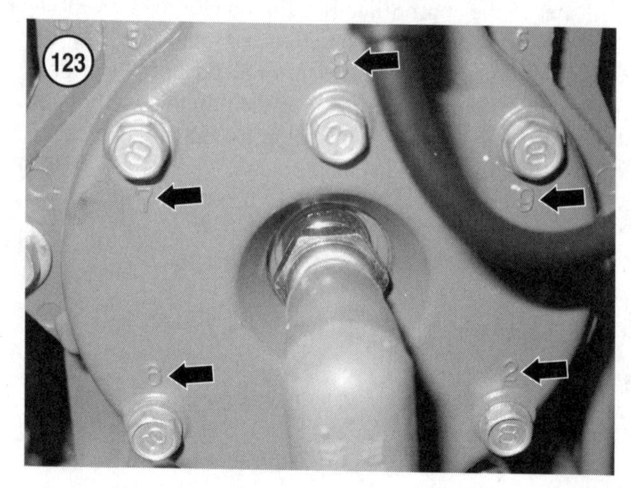

124 THERMOSTAT AND COVER (TYPICAL)

1. Bolt
2. Thermostat cover
3. Gasket
4. Thermostat
5. Cylinder head

6. Remove the two bolts used to hold the components in position. Then apply Loctite 572 onto the threads. Thread the two bolts into the block.
7. Tighten the exhaust cover bolts in two steps to the specification in **Table 1**. Tighten the bolts in the sequence corresponding to the tightening sequence numbers cast into the cover (**Figure 123**).

Water pressure valve installation

1. Fit the valve seat (5, **Figure 122**) into the recess in the exhaust cover.
2. Insert the pressure valve (4, **Figure 122**) into the valve seat. The side with the longer X shaped end must fit into the seat.
3. Install the spring into the bore in the pressure valve cover (1, **Figure 122**).
4. Install a new gasket onto the cover. Do not apply sealant to the gasket.
5. Guide the spring onto the pressure valve while seating the cover and gasket onto the exhaust cover.
6. Apply Loctite 572 to the threads, then thread the bolts through the cover and into the exhaust cover.
7. Tighten the cover bolts in two steps to the specification in **Table 1**.

Thermostat installation

1. Insert the thermostat (4, **Figure 124**) into the cylinder head water jacket opening with the smaller diameter side facing inward.
2. Install a new gasket (3, **Figure 124**) onto the cylinder head. Do not apply sealant to the gasket or sealing surfaces.
3. Fit the cover (2, **Figure 124**) onto the gasket and seat against the cylinder head water jacket.
4. Apply Loctite 572 to the threads, then hand-thread the cover bolts into the cover and cylinder head water jacket.
5. Tighten the cover bolts in two steps to the specification in **Table 1**.

Cylinder head water jacket installation

1. Fit the new water jacket gasket onto the cylinder head. Do not apply sealant onto the gasket or sealing surfaces.
2. Fit the cylinder head water jacket to the cylinder head. Check for water jacket bolts that pass through the water jacket and cylinder head.
   a. *Bolts pass through the cylinder head*—Tighten the bolts after the cylinder head and new gasket are installed onto the cylinder block.
   b. *Bolts do not pass through the cylinder head*—Apply Loctite 572 to the threads, then hand-thread the bolts into the jacket and cylinder head. Tighten the bolts in the sequence corresponding to the tightening sequence numbers cast into the water jacket (**Figure 123**). Tighten the bolts in two steps to the specification in **Table 1**.

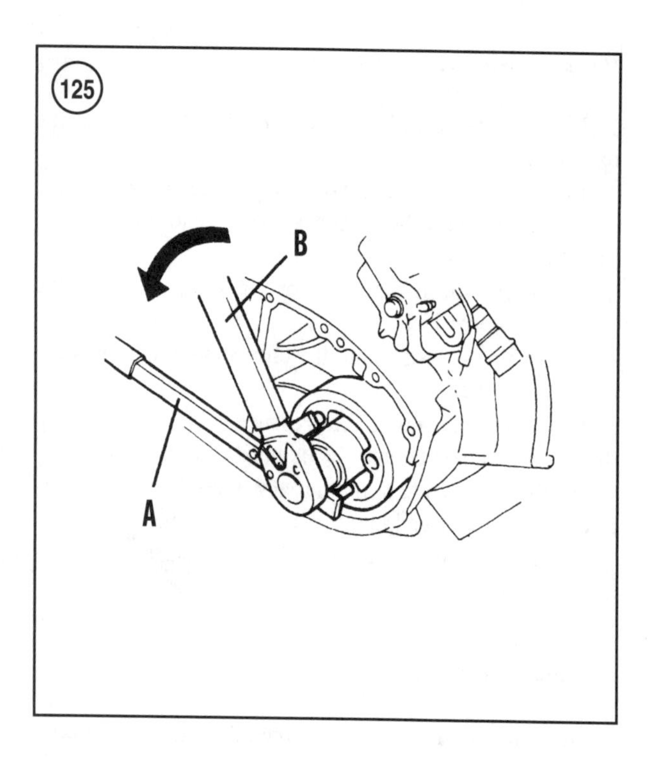

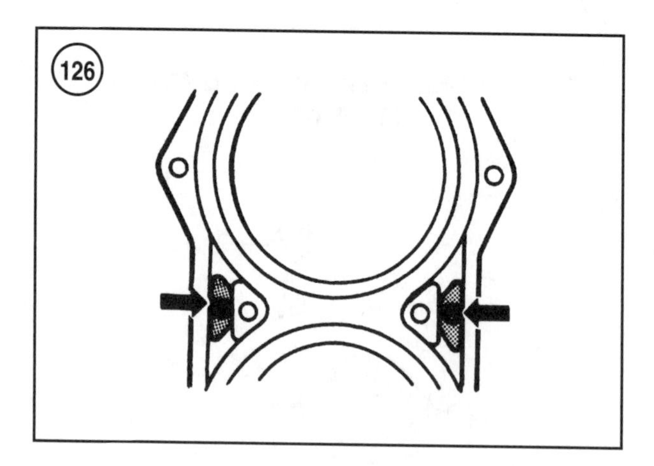

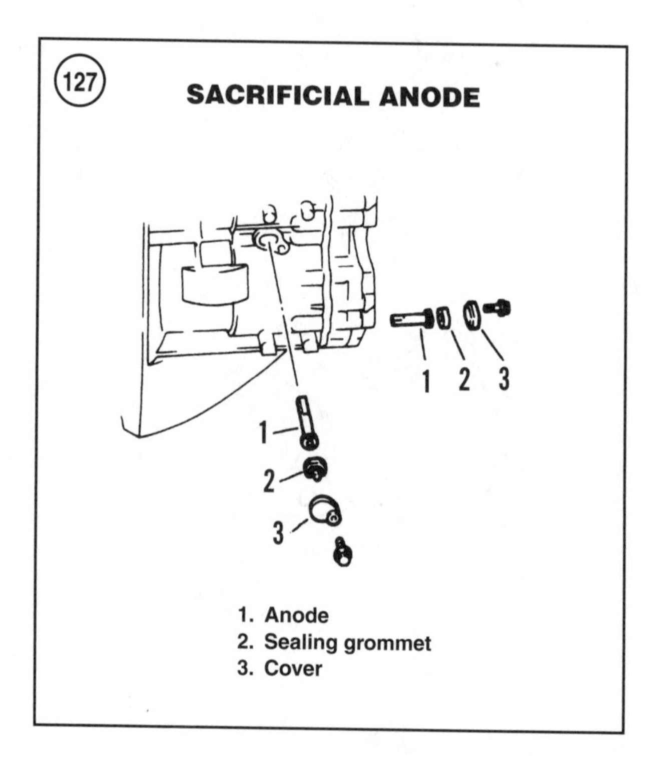

Cylinder Head Installation

Refer to the appropriate illustration (**Figures 28-30**) to assist with component orientation and mounting locations. On some models, the bolts that secure the water jacket also pass through and secure the cylinder head.

1. Fit a new gasket onto the cylinder head. Do not apply sealant onto the gasket or sealing surfaces.
2. Install the cylinder head and water jacket assembly onto the cylinder block.
3. Apply a light coating of Yamalube two-cycle outboard oil to the threads, then hand-thread the bolts into the water jacket (on some models), cylinder head and cylinder block.
4. Tighten the bolts in the sequence corresponding to the tightening sequence numbers cast into the water jacket and/or cylinder head (**Figure 123**). Tighten the bolts in two steps to the specification in **Table 1**.
5. Re-tighten the cylinder head/water jacket bolts after running the engine long enough to reach operating temperature. Stop the engine and allow the engine to cool for approximately 30 minutes. Loosen the bolts a quarter turn, one at a time, and in the tightening sequence, then retighten to the specification in **Table 1**.

Torsional Dampener Installation

This component is used on 200-250 hp (3.1 liter) models. The dampener mounts onto the lower end of the crankshaft. Use a flywheel holding tool (Yamaha part No. YB-01639/90890-06522) to install the dampener. Refer to **Figure 30** to assist with component orientation and mounting location.

1. Apply Yamalube two-cycle outboard oil to the lower end of the crankshaft and the opening in the torsional dampener.
2. Align the flat drive surfaces while fitting the dampener onto the lower end of the crankshaft. Carefully tap on the dampener until the nut can be threaded onto the crankshaft.
3. Fit the pins of the flywheel holding tool (A, **Figure 125**) into the openings in the torsional dampener.

4. Hold the flywheel holding tool securely, then turn the nut (20, **Figure 30**) clockwise with a suitable tool (B, **Figure 125**) to the torque specification in **Table 1**.

External Component Installation

1. Install the intake manifold/reed block housing from the power head as described in Chapter Five.
2. Install the throttle arm, as an assembly, onto the power head. Securely tighten the pivot bolt.
3. Install the wire harness and all electrical components as described in Chapter Six. Make sure all wiring is routed to avoid interference with moving components.
4. Install all fuel system components as described in Chapter Five.
5. *Oil injected models*—Install all oil injection system components as described in Chapter Twelve.
6. Install the flywheel as described in this chapter.
7. Install the power head as described in this chapter.
8. Perform all applicable adjustments as described in Chapter Four.
9. Perform the break-in procedures as described in Chapter Three.

Anode Replacement

The anodes are located in the water jacket between the cylinder bores (**Figure 126**) or under a cylinder block cover (**Figure 127**). Refer to **Figure 28** and **Figure 29** to assist with locating the anodes. On 200-250 (3.1 liter) models, the anodes are located between the waterjacket cover and cylinder head. On some models, the cylinder heads must be removed to access the anodes.

1A. *Anode mounted behind the cylinder head*—Remove the cylinder head as described in this chapter. Remove the retaining screws. Then pull the anodes from the water jacket.
1B. *Anode mounted beneath the cylinder block cover*—Remove the screw and cover (3, **Figure 127**), then pull the sealing grommet (2) and anode (1) from the cylinder block.
2. Inspect the anode for deep pitting, significant loss of material or a dark gray coating. Replace the anode if not in excellent condition.
3. Thoroughly clean the anode mounting bolts, mounting surface in the cylinder block or water jacket, covers and sealing grommets. If the anode must be reused, carefully sand any coating or foreign material from the surfaces.
4A. *Anode mounted behind the cylinder head*—Fit the anodes into the water jacket and secure them with the mounting screws. Install the cylinder head as described in this chapter.
4B. *Anode mounted beneath the cylinder block cover*—Carefully insert the anode into the cylinder block opening. Fit the sealing grommet into the opening, then install the cover over the grommet. Secure the cover with the screw.

Table 1 POWER HEAD TORQUE SPECIFICATIONS

Fastener	N•m	in.-lb.	ft.-lb.
Connecting rod bolt			
80 Jet and 115-130 hp			
First step	12	106	–
Final step	37	–	27
105 Jet and 150-200 hp (2.6 liter)			
First step	19	168	14
Final step	37	–	27
200-250 hp (3.1 liter)			
First step	28	–	21
Final step	45	–	33
Crankcase cover			
80 Jet and 115-130 hp			
Short bolts (30 mm)			
First step	10	89	–
Final step	18	156	13
Long bolts (60 mm)			
First step	20	177	14.5
Final step	40	–	29

(continued)

7

Table 1 TIGHTENING TORQUE (continued)

Fastener	N·m	in.-lb.	ft.-lb.
Crankcase cover (continued)			
105 Jet and 150-200 hp (2.6 liter)			
Short bolts (30 mm)			
First step	10	89	–
Final step	18	156	13
Long bolts (60 mm)			
First step	20	177	14.5
Final step	39	–	28
200-250 hp (3.1 liter)			
Short bolts (30 mm)			
First step	4	35	–
Final step	8	70	–
Medium length bolts (70 mm)			
First step	20	177	14.5
Final step	40	–	29
Long bolts (90 mm)			
First step	20	177	14.5
Final step	40	–	29
Cylinder head and water jacket			
80 Jet and 115-130 hp			
Short bolts (30 mm)			
First step	4	35	–
Final step	8	70	–
Medium length bolts (40 mm)			
First step	4	35	–
Final step	8	70	–
Long bolts (60 mm)			
First step	15	133	11
Final step	30	–	22
105 Jet and 150-200 hp (2.6 liter [except HPDI])			
Short bolts (30 mm)			
First step	4	35	–
Final step	8	70	–
Medium length bolts (40 mm)			
First step	4	35	–
Final step	8	70	–
Long bolts (60 mm)			
First step	15	133	11
Final step	30	–	22
150-250 hp (HPDI models)			
First step	15	133	11
Final step	30	–	22
200-250 hp (3.1 liter)			
Short bolts (35 mm)			
First step	4	35	–
Final step	8	70	–
Medium length bolts (70 mm)			
First step	15	133	11
Final step	28	–	21
Long bolts (80 mm)			
First step	15	133	11
Final step	28	–	21
Exhaust cover			
First step	4	35	–
Final step	8	70	–
Flywheel nut	190	–	140
Power head mounting			
80 Jet, 105 Jet and 115-130 hp	21	189	15

(continued)

Table 1 TIGHTENING TORQUE (continued)

Fastener	N•m	in.-lb.	ft.-lb.
Power head mounting (continued)			
150-200 hp (2.6 liter)	21	189	15
200-250 hp (3.1 liter)			
Long bolts	21	189	15
Short bolts	18	156	13
Thermostat cover			
First step	4	35	–
Final step	8	70	–
Torsional dampener	100	–	74
Water pressure valve cover			
First step	4	35	–
Final step	8	70	–

Table 2 CYLINDER BORE SERVICE SPECIFICATIONS

Model	Specification
Standard bore diameter	90.00-90.10 mm (3.543-3.547 in.)
Maximum cylinder taper	0.08 mm (0.003 in.)
Maximum cylinder out-of-round	0.05 mm (0.002 in.)

Table 3 PISTON SERVICE SPECIFICATIONS

Model	Specification
Standard piston diameter	
80 Jet and 115-130 hp	89.920-89.935 mm (3.5402-3.5407 in.)*
105 Jet and 150-200 hp (2.6 liter [except HPDI])	
Cylinder No. 1	89.845-89.869 mm (3.5372-3.5381 in.)*
Cylinder No. 2-6	89.895-89.915 mm (3.5392-3.5400 in.)*
150-200 hp (2.6 liter [HPDI])	89.845-89.869 mm (3.5372-3.5381 in.)*
200-250 hp (3.1 liter)	89.840-89.860 mm (3.5370-3.5378 in.)*
Piston skirt to cylinder clearance	
80 Jet and 115-130 hp	0.080-0.085 mm (0.0031-0.0033 in.)
105 Jet and 150-200 hp (2.6 liter [except HPDI])	
Cylinder No. 1	0.150-0.156 mm (0.0059-0.0061 in.)
Cylinder No. 2-6	0.100-0.106 mm (0.0039-0.0042 in.)
150-200 hp (2.6 liter [HPDI])	0.150-0.156 mm (0.0059-0.0061 in.)
200-250 hp (3.1 liter)	0.155-0.161 mm (0.0061-0.0063 in.)
Piston pin bore	
80 Jet and 115-130 hp	23.074-23.085 mm (0.9084-0.9089 in.)
105 Jet and 150-200 hp (2.6 liter)	23.074-23.085 mm (0.9084-0.9089 in.)
200-250 hp (3.1 liter)	26.004-26.015 mm (1.0238-1.0242 in.)
Piston pin diameter	
80 Jet and 115-130 hp	23.065-23.070 mm (0.9081-0.9083 in.)
105 Jet and 150-200 hp (2.6 liter)	23.065-23.070 mm (0.9081-0.9083 in.)
200-250 (3.1 liter)	25.995-26.000 mm (1.023-1.024 in.)

*Measured 10 mm (0.39 in.) from the bottom of the piston skirt.

Table 4 PISTON RING SERVICE SPECIFICATIONS

Model	Specification
Ring end gap	0.30-0.60 mm (0.012-0.024 in.)
Ring end gap measuring depth	
80 Jet and 115-130 hp	20 mm (0.787 in.)*

(continued)

Table 4 PISTON RING SERVICE SPECIFICATIONS (continued)

Model	Specification
Ring end gap measuring depth (continued)	
105 Jet and 150-200 hp (2.6 liter)	20 mm (0.787 in.)*
200-250 hp (3.1 liter)	5 mm (0.196 in.)
Side clearance	0.02-0.06 mm (0.001-0.0024 in.)

*Measuring point from top (head side) of the cylinder bore.

Table 5 CRANKSHAFT SERVICE SPECIFICATIONS

Measurement	Specification
Connecting rod side clearance	0.12-0.26 mm (0.005-0.010 in.)
Maximum axial play	2.0 mm (0.078 in.)
Maximum runout	0.05 mm (0.002 in.)

Chapter Eight

Gearcase

This chapter provides lower gearcase removal/installation, rebuilding resealing procedures for all 115-250 hp Yamaha outboards.

The gearcases covered in this chapter require different service procedures. The chapter is arranged in a normal disassembly/assembly sequence. When only partial repair is required, follow the procedure(s) to the point where the faulty parts can be replaced, then jump ahead to reassemble the assembly procedures.

This chapter covers a large range of models, the gearcases shown in the accompanying illustrations are the most common models. While it is possible that the components shown in the pictures may not be identical with those being serviced, the step-by-step procedures cover each model. Gearcase service specifications are located in **Table 1** and **Table 2** at the end of the chapter.

GEARCASE OPERATION

The gearcase transfers the rotation of a vertical drive shaft to a horizontal propeller shaft. See **Figure 1**. A pinion (drive) gear on the drive shaft is in constant mesh with forward and reverse (driven) gears in the lower gearcase housing. These gears are spiral bevel cut to change the vertical power flow into the horizontal flow required by the propeller shaft. The spiral bevel design also provides quiet operation.

All gearcases use precision shimmed gears. This means that the gears are precisely located in the gear housing by the use of thin metal spacers, called shims (**Figure 2**). After assembly, correct shimming of the gears is verified by measuring the *gear lash*, also referred to as *backlash*. Gear lash is the measurement of the clearance (air gap) between a tooth on the pinion gear and two teeth on the forward or reverse gear.

Excessive gear lash indicates that the gear teeth are too far apart. This causes excessive gear noise (whine) and a reduction in gear strength and durability due to gear teeth insufficiently overlapping.

Insufficient gear lash indicates that the gear teeth are too close together. Operation with insufficient gear lash leads to gear failure since there will not be enough clearance to maintain a film of lubricant. Heat expansion only compounds the problem.

All lower gearcases incorporate a water pump to supply cooling water to the power head. All models require gearcase removal to service the water pump. Water pump removal and installation procedures are covered in this chapter.

Standard Rotation (RH) Gearcase

A sliding clutch, splined to the propeller shaft, engages the spinning forward or reverse gear. See **Figure 3**. This creates a direct coupling of the drive shaft to the propeller shaft. Since this is a straight mechanical engagement, shifting should only be done at idle speed. Shifting at higher speeds results in gearcase failure.

When neutral is selected (**Figure 3**), the shift mechanism positions the clutch midway between the driven gears. This allows the propeller shaft to freewheel or remain stationary. No propeller thrust is delivered.

When forward is selected (**Figure 3**), the shift mechanism moves the clutch to engage the front mounted gear. This mechanical engagement causes clockwise rotation of the propeller. These models use only a right-hand propeller and forward propeller thrust is delivered.

When reverse is selected (**Figure 3**), the shift mechanism moves the clutch to engage the rear mounted gear. This mechanical engagement causes counterclockwise rotation of the propeller. These models use only a right-hand propeller and reverse propeller thrust is delivered.

NOTE
Models with an L in the model designation have a counter rotation gearcase. The gearcase provides forward thrust when the propeller shaft rotates in the left-hand, or counterclockwise, direction (as viewed from the rear). A left-hand propeller must be used on these models.

CAUTION
Never use a left-hand propeller on a gearcase designed to use only a right-hand propeller. Gearcase component failure may occur from continued operation in the wrong direction for forward thrust.

Counter Rotation (LH) Gearcase

Left-hand, or counterclockwise, propeller shaft rotation is used for forward thrust on models with an *L* in the model designation code. Left-hand units are fused along with a right-hand unit on dual engine applications. The use of the left-hand unit allows for balanced propeller torque from the two engines. A control box is used to provide the opposite direction of shift cable movement versus right-hand units.

A sliding clutch, splined to the propeller shaft, engages the spinning forward or reverse gear. See **Figure 3**. This creates a direct coupling of the drive shaft to the propeller

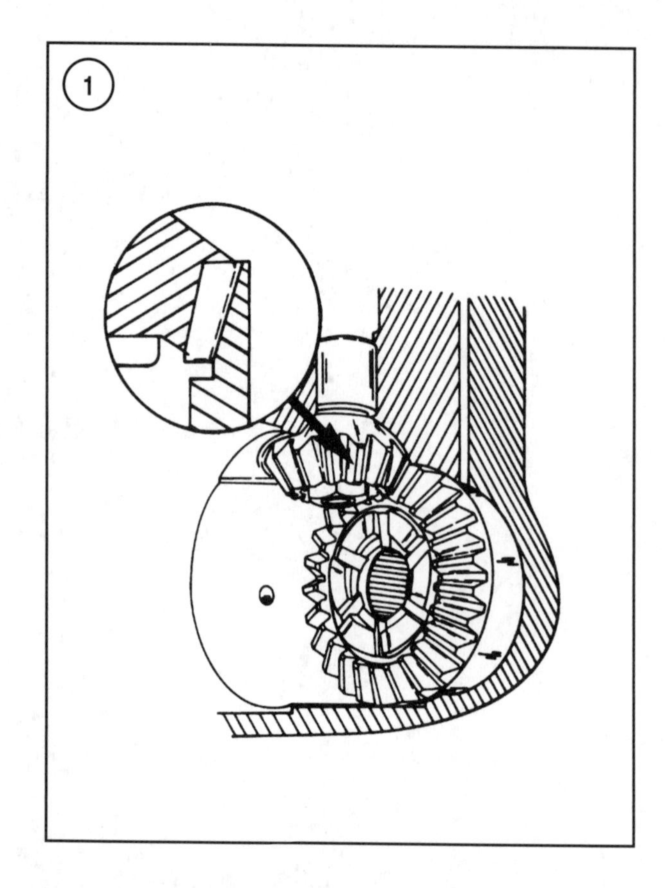

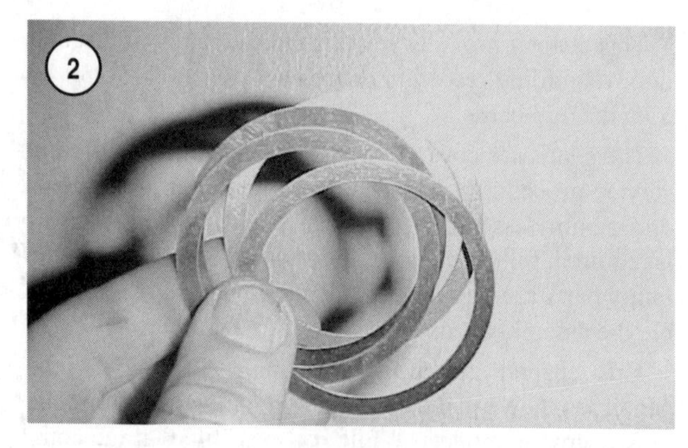

shaft. Since this is a straight mechanical engagement, shifting should only be done at idle speed. Shifting at higher speeds results in gearcase failure.

When neutral is selected (**Figure 3**), the shift mechanism positions the clutch midway between the driven gears. This allows the propeller shaft to freewheel or remain stationary. No propeller thrust is delivered.

When forward is selected (**Figure 3**), the shift mechanism moves the clutch to engage the rear mounted gear. This mechanical engagement causes counterclockwise rotation of the propeller. These models use only a

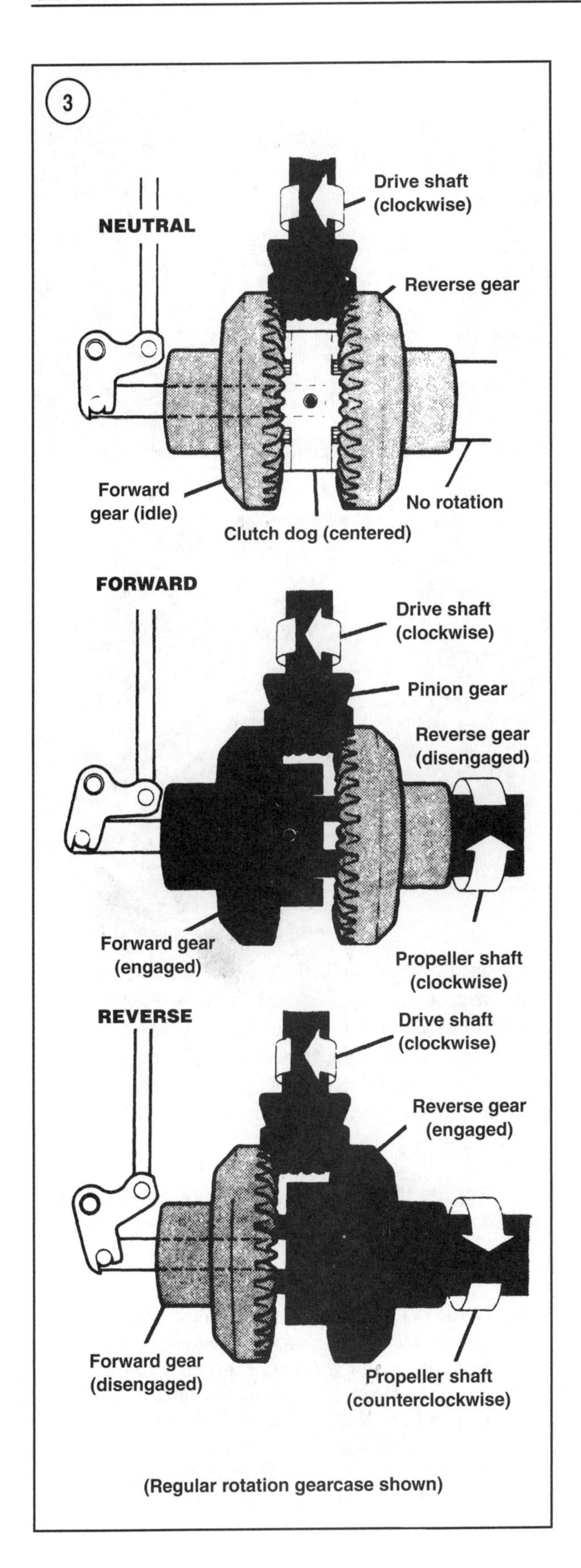

left-hand propeller and forward propeller thrust is delivered.

When reverse is selected (**Figure 3**), the shift mechanism moves the clutch to engage the front mounted gear. This mechanical engagement casues clockwise rotation of the propeller. These models use only a left-hand propeller and reverse propeller thrust is delivered.

Twin Propeller Gearcase

Twin counter-rotating propellers (**Figure 4**) are used on 150 hp models with a *D* in the model designation code. The use of two propellers and one engine provides balanced propeller torque and improved performance versus a single propeller gearcase. Both propellers are used for forward thrust and one propeller is used for reverse thrust.

PROPELLER REMOVAL/INSTALLATION

All models covered in this manual, except models a twin propeller gearcase, use a thrust hub design propeller (**Figure 5**). The propeller is driven via a splined connection of the propeller shaft to the rubber thrust hub. The rubber thrust hub is pressed into a bore in the propeller (**Figure 6**) and provides a cushion effect when shifting. It also provides some protection for the gearcase components in the event of underwater impact. The front mounted thrust washer (1, **Figure 5**) directs the propeller thrust to a tapered area on the propeller shaft.

CAUTION

Use light force if necessary to remove the propeller from the propeller shaft. Using excessive force causes damage to the propeller, propeller shaft and internal components of the gearcase. If it is too hard to remove the propeller by normal means, have a reputable marine repair shop or propeller repair shop remove the propeller.

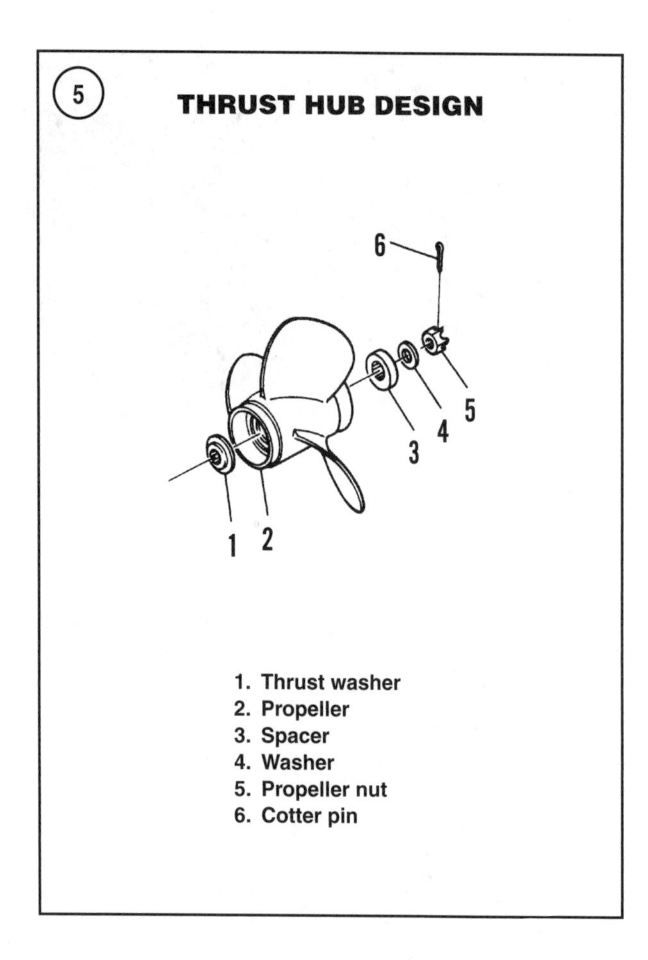

Removal (Except Twin Propeller Gearcase)

Upon removal, inspect the propeller for the black rubber material in the drive hub area (**Figure 6**). Have the hub inspected or replaced at a propeller repair facility if this material is found. It normally indicates the hub has spun in the propeller bore. Satisfactory performance is not possible with a spun propeller hub.

1. Disconnect the battery cables and ground the spark plug leads to prevent accidental starting.
2. Use pliers to straighten and remove the cotter pin (**Figure 7**). To prevent propeller rotation, place a block of wood between the propeller and the gearcase as shown in **Figure 8**.
3. Turn the propeller nut counterclockwise. Remove the nut and washer (4, **Figure 5**) and spacer (3).
4. Carefully pull the propeller from the propeller shaft. Use a block of wood as a cushion and carefully drive the propeller from the shaft if necessary. Use light force to avoid damaging the propeller or gearcase components.
5. Mark the propeller side of the thrust washer. Remove the thrust washer (**Figure 9**). Tap the washer lightly if seized onto the propeller shaft.
6. Clean all debris and old grease from the propeller shaft. Inspect the propeller shaft for twisted splines (**Figure 10**) or excessively worn surfaces. Pay particular attention to the tapered surface that contacts the thrust washer. Rotate the propeller shaft look for shaft deflection. Replace the propeller shaft if there are worn surfaces, twisted splines or shaft deflection.

7. Inspect the thrust washer (1, **Figure 5**) for worn or cracked surfaces and replace if necessary.

CAUTION

Never use a damaged, worn or cracked propeller thrust hub. An excessively worn hub may allow the propeller to contact the gearcase during operation. A cracked hub may unexpectedly fail, allowing the propeller to contact the gearcase. The gearcase will suffer irreparable damage if the engine operates with the propeller contacting the gearcase.

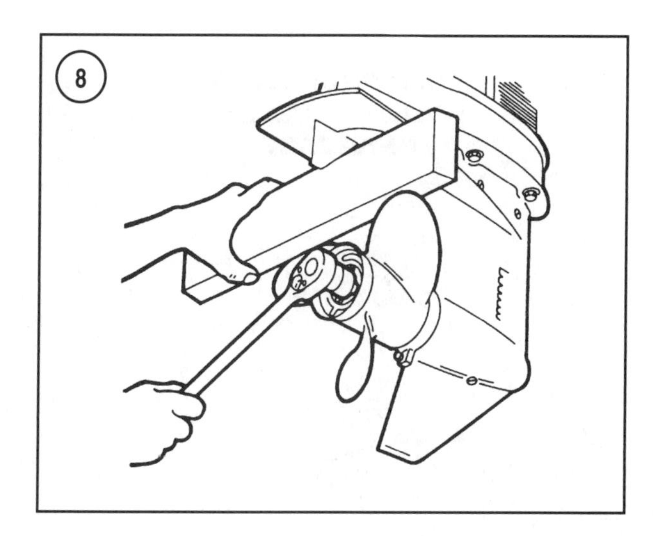

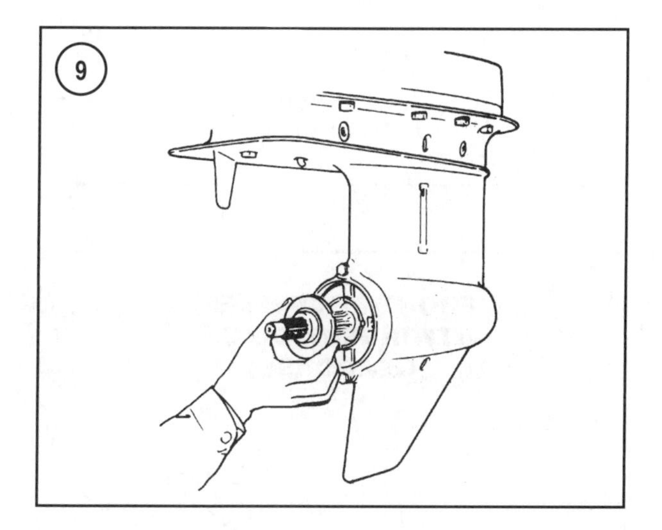

Installation (Except Twin Propeller Gearcase)

1. Apply a light coating of Yamaha All-Purpose Grease or equivalent to the propeller shaft tapered surface and threaded end. Apply grease to the propeller shaft and the shaft bore in the propeller.
2. Slide the thrust washer onto the propeller shaft. The tapered surface of the thrust washer must contact the tapered surface of the propeller shaft.
3. Slide the propeller onto the propeller shaft. Rotate the propeller to align the propeller splines with the propeller shaft splines. Seat the propeller against the thrust washer.
4. Refer to **Figure 5** and install the washer and spacer. Hand tight the propeller nut onto the propeller shaft.
5. To prevent propeller rotation, place a block of wood between the propeller and the gearcase as shown in **Figure 8**. Tighten the propeller nut to the specification in **Table 1**.

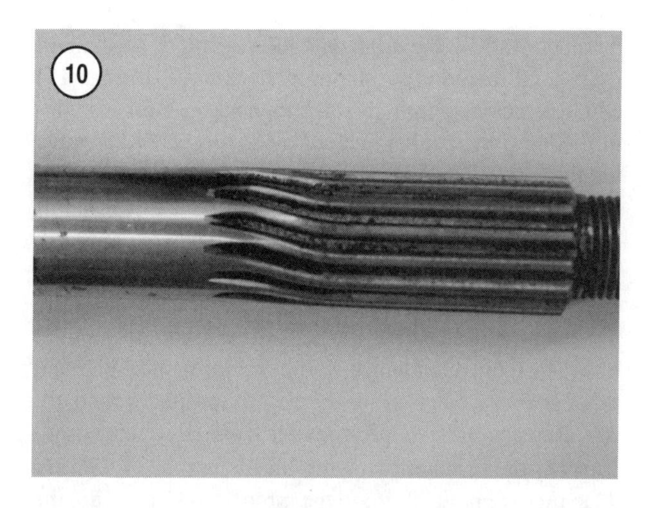

6. Align the cotter pin slot in the propeller nut with the hole in the propeller shaft. Tighten the propeller nut an additional amount as necessary to align the holes. Do not over-tighten the nut. Never loosen the nut to align the cotter pin holes.
7. Insert a new cotter pin into the holes and bend the ends over.
8. Connect the battery cables and spark plug leads.

Removal (Twin Propeller Gearcase)

1. Disconnect the battery cables and ground the spark plug leads to prevent accidental starting.
2. To prevent propeller rotation, place a block of wood between the rear propeller and the gearcase as shown in **Figure 8**.
3. Use pliers to straighten and remove the cotter pin (**Figure 7**). Discard the cotter pin. The threaded end of the rear propeller shaft is a standard right-hand thread. Turn the propeller nut (2, **Figure 11**) counterclockwise for removal and clockwise for installation. Use a breaker bar and socket to loosen and remove the rear propeller nut.
4. Remove the washer (3, **Figure 11**) and spacer (4) from the propeller shaft. Pull the rear propeller from the shaft. A block of wood can be used as a cushion to carefully drive the propeller from the shaft if necessary. Use light blows to avoid damaging the propeller, propeller shaft and other gearcase components. Inspect the propeller for damage. Replace or have the propeller repaired if necessary.
5. Note the orientation of the thrust washer (6, **Figure 11**), then remove it from the propeller shaft. Tap the washer lightly if seized onto the propeller shaft.
6. Carefully bend all of the tabs of the lockwasher (2, **Figure 12**) away from the front propeller nut (1). To prevent propeller rotation, place a block of wood between the front propeller and gearcase as shown in **Figure 8**. The

threaded end of the front propeller shaft is a standard right-hand thread. Turn the propeller nut (1, **Figure 12**) counterclockwise for removal and clockwise for installation. Use a breaker bar and socket to loosen and remove the rear propeller nut.

7. Remove the lockwasher. Replace the lockwasher if corroded or if the tabs are cracked or otherwise damaged. Use heavy gloves to pull the front propeller from the shaft. A block of wood can be used as a cushion to carefully drive the propeller from the shaft if necessary. Use light blows to avoid damaging the propeller, propeller shaft and other gearcase components. Inspect the propeller for damage. Replace or have the propeller repaired if necessary.

8. Note the orientation of the thrust washer (4, **Figure 12**), then remove it from the propeller shaft. Tap the washer lightly if seized onto the propeller shaft.

9. Inspect the thrust washers (6, **Figure 11** and 4, **Figure 12**) for worn, cracked or damaged surfaces and replace as necessary.

10. Clean all debris and old grease from the propeller shafts. Inspect the propeller shafts for twisted splines (**Figure 10**) or excessively worn surfaces. Pay particular attention to the tapered surface that contacts the thrust washer. Look for shaft deflection while rotating the propeller shafts. Replace the propeller shaft(s) if there are excessively worn surfaces, twisted splines or shaft deflection.

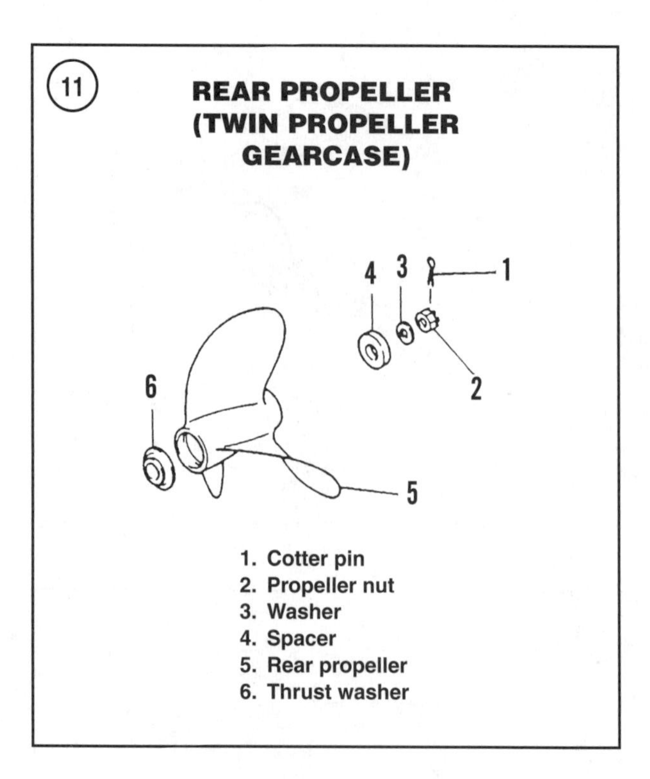

1. Cotter pin
2. Propeller nut
3. Washer
4. Spacer
5. Rear propeller
6. Thrust washer

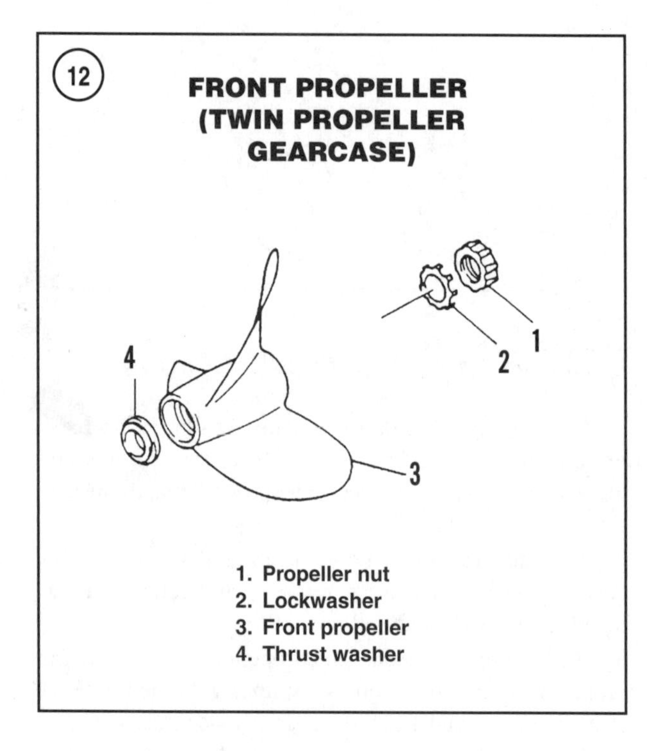

1. Propeller nut
2. Lockwasher
3. Front propeller
4. Thrust washer

Installation (Twin Propeller Gearcase)

Clean all corrosion or other contaminants from the propellers, thrust washers, spacers, propeller nuts and exposed portions of the propeller shafts. Thoroughly clean the splined bores of the propellers. Apply Yamaha All-Purpose Grease to the splined and tapered sections of the propeller shafts prior to propeller installation.

1. Install the front propeller thrust washer (4, **Figure 12**) onto the propeller shaft. The tapered bore must mate with the tapered surface on the front propeller shaft.

2. Slide the front propeller (3, **Figure 12**) over the propeller shaft. Slightly rotate the propeller to align the splines. Then push the propeller onto the shaft until firmly seated against the thrust washer. Place a block of wood between the front propeller and the gearcase to prevent propeller rotation during installation.

3. Install the lockwasher (2, **Figure 12**) onto the propeller with the bent ends of the tabs facing away from the propeller. Thread the front propeller nut (1, **Figure 12**) onto the propeller shaft with the flat surface facing the propeller. Tighten the propeller nut to the specification in **Table 1**. Bend two or more of the tabs into recesses in the propeller nut and two or more of the tabs down into recesses in the propeller.

4. Install the rear propeller thrust washer (6, **Figure 11**) onto the propeller shaft. The tapered bore of the washer must mate with the tapered area of the rear propeller shaft.

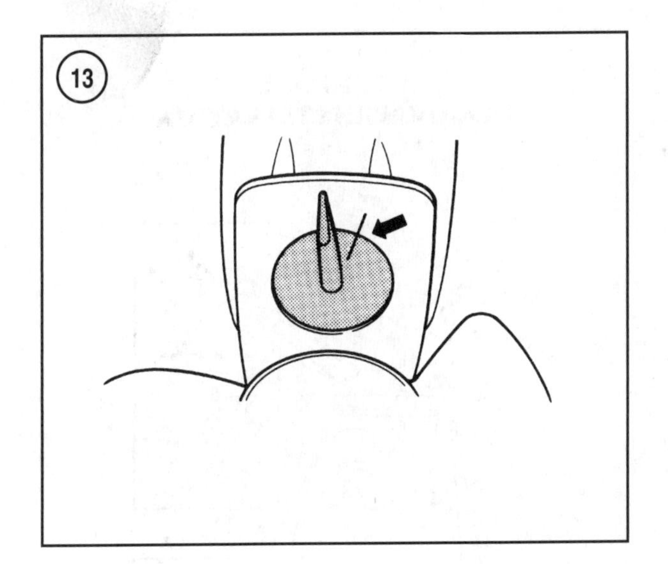

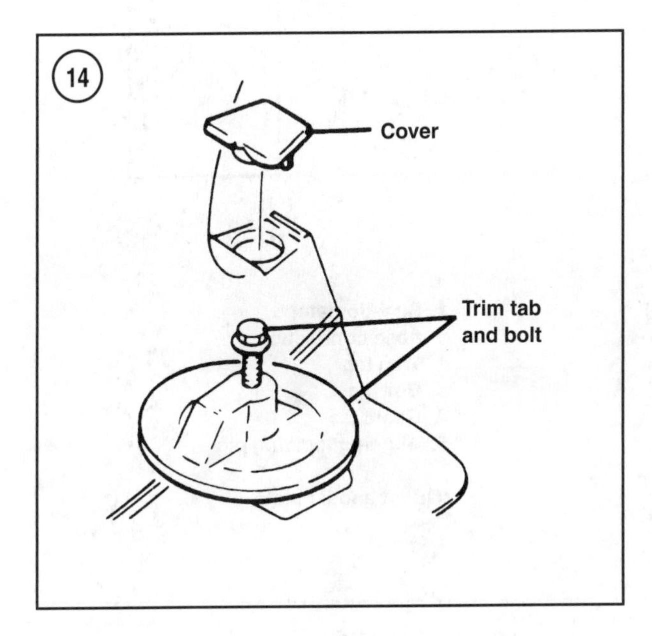

5. Slide the rear propeller onto the propeller shaft. Slightly rotate the propeller to align the splines. Then push the propeller onto the shaft until firmly seated against the thrust washer. Place a block of wood between the front propeller and the gearcase to prevent propeller rotation during installation.

6. Install the spacer and washer (3 and 4, **Figure 11**) onto the propeller shaft. Thread the propeller nut onto the propeller shaft with the cotter pin slots facing outward. Tighten the propeller nut to the specification in **Table 1**.

7. Install a new cotter pin through the slot in the propeller nut and opening in the propeller shaft. Tighten the propeller nut slightly more if necessary to align the opening with the slot. Never loosen the nut to align the opening. Bend over the ends of the cotter pin.

GEARCASE REMOVAL/INSTALLATION

Always remove the propeller prior to removing or installing the gearcase. Refer to the procedures described in this chapter. To prevent accidental starting or starter engagement, disconnect all spark plug leads and the battery cables prior to gearcase removal.

Drain the gearcase lubricant prior to removal if the gearcase requires disassembly. Refer to Chapter Three for procedures.

CAUTION
Work carefully if using a blunt tip pry bar to separate the gearcase from the drive shaft housing. Always ensure that all fasteners are removed before prying the housings apart. Use the pry points near the front and rear mating surfaces.

NOTE
Apply moderate heat to the gearcase and drive shaft housing mating surfaces if corrosion prevents easy removal of the gearcase.

Removal

1. Disconnect the battery cables and ground the spark plug leads to prevent accidental starting.
2. Use a marker to reference the trim tab setting (**Figure 13**). Do not use a scribe, as the gearcase will corrode where the paint is scratched.
3. Remove the bolt access cover (**Figure 14**). Use a socket and extension to remove the trim tab bolt. Carefully tap the trim tab loose. Then remove it from the gearcase.
4. Cut the plastic locking clamp and disconnect the speedometer hose from the connector (**Figure 15**). Heat the speedometer hose with a portable hair dryer for easier removal from the connector. If the hose cannot be pulled from the fitting, cut the hose as close as possible to the fitting. Slit the hose to remove it from the fitting. Do not inadvertently cut the fitting.
5. Shift the engine into NEUTRAL gear.
6. Remove the bolt (3, **Figure 16**) from the trim tab mounting cavity. Support the gearcase while removing the three bolts and washers (**Figure 17**) from each side of the gearcase.
7. Verify that all bolts are removed. Then lower the gearcase from the drive shaft housing. If necessary, care-

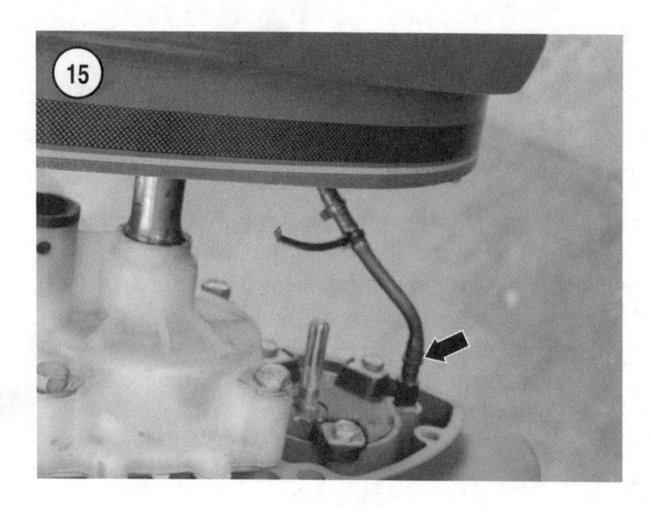

fully tug or pry the gearcase from the housing. Pry at points near the front and rear of the gearcase-to-drive shaft housing mating surface.

8. Reposition the water tube and upper sealing grommet if they were dislodged during gearcase removal.
9. Clean corrosion and other contaminants from the gearcase mating surfaces, shift shafts and drive shaft.
10. Inspect the grommet that connects the water tube to the water pump for damage or deterioration. Replace the grommet are in questionable condition.

Installation

CAUTION
Never apply grease to the top of the drive shaft or connection at the crankshaft. The grease may cause a hydraulic lock on the shaft that can cause failure of the gearcase, power head or both.

1. Apply a light coating of Yamaha All-Purpose Grease or an equivalent to the splined section at the upper end of the drive shaft and the grommet that connects the water tube into the water pump (**Figure 18**).
2. Install the shift handle (Yamaha part No. YB-6052) to the lower shift shaft (**Figure 19**). If this tool is not available, grip the lower shift shaft with pliers and a shop towel. (**Figure 20**).
3. Rotate the drive shaft in the clockwise direction (**Figure 19**) while observing the propeller shaft. Rotate the lower shift shaft until the propeller shaft rotates clockwise then counterclockwise. Note the shift shaft position for each gear. Position the shift shaft at the point midway between the location where clockwise and counterclockwise rotation occurs. Rotate the drive shaft to verify the gearcase is in neutral gear. Remove the shift handle.

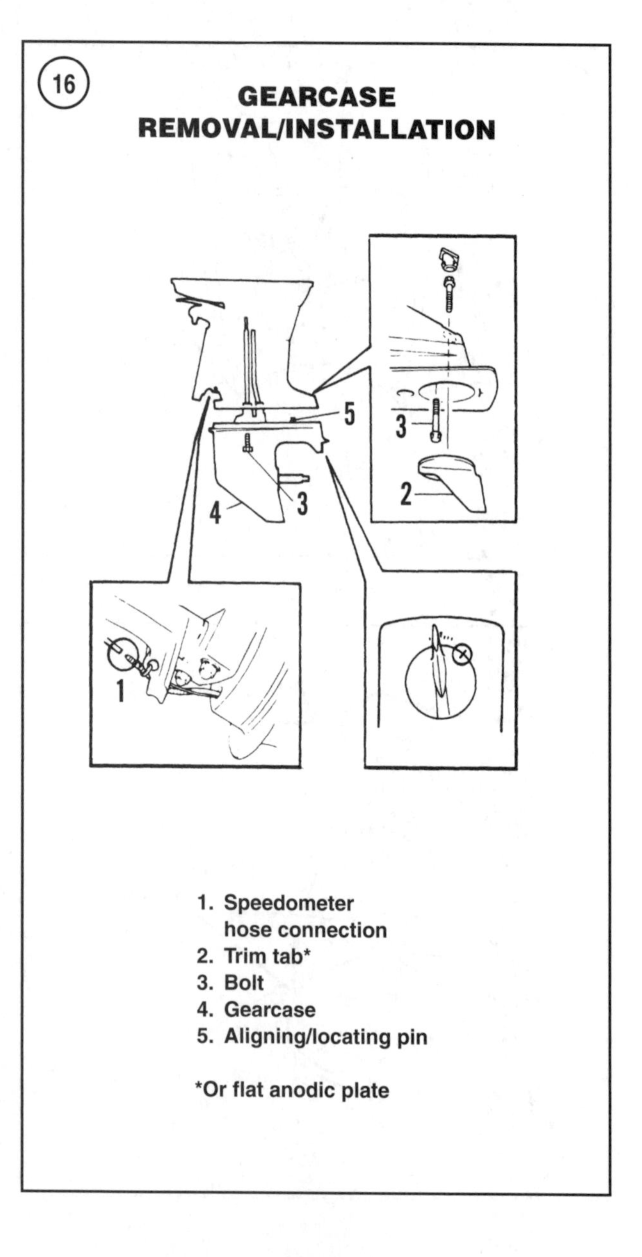

GEARCASE REMOVAL/INSTALLATION

1. Speedometer hose connection
2. Trim tab*
3. Bolt
4. Gearcase
5. Aligning/locating pin

*Or flat anodic plate

CAUTION
Never rotate the propeller shaft to align the drive shaft splined connection with the crankshaft splines. The water pump impeller can suffer damage that leads to overheating of the engine.

CAUTION
Work carefully when installing the upper end of the drive shaft into the crankshaft. The lower seal on the crankshaft can be dislodged or damaged by the drive shaft. Make sure the shafts are properly aligned before inserting the drive shaft into the crankshaft. Never force the gearcase in position. Rotate

17

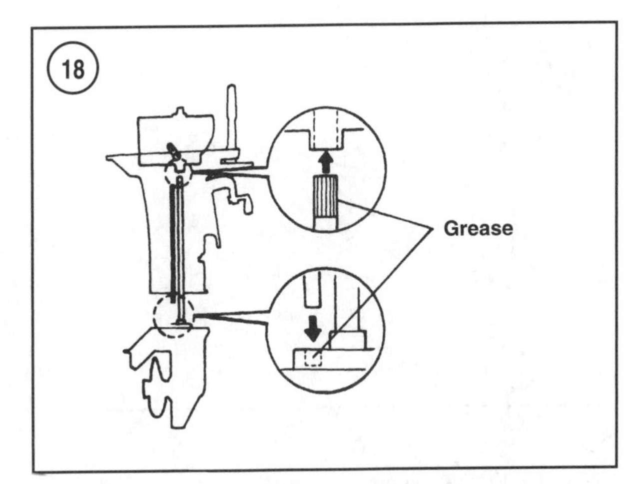
18
Grease

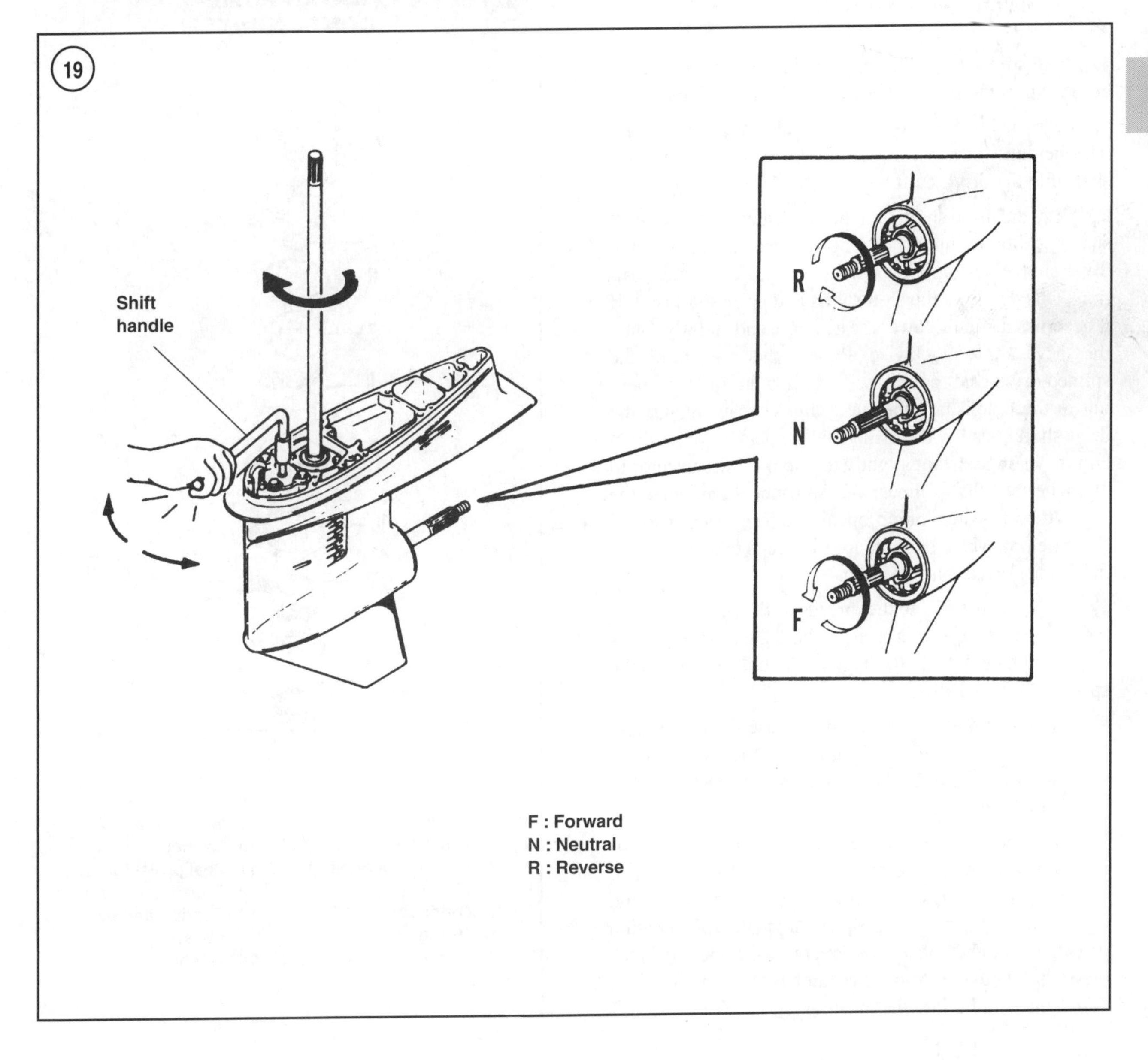
19
Shift
handle
R
N
F
F : Forward
N : Neutral
R : Reverse

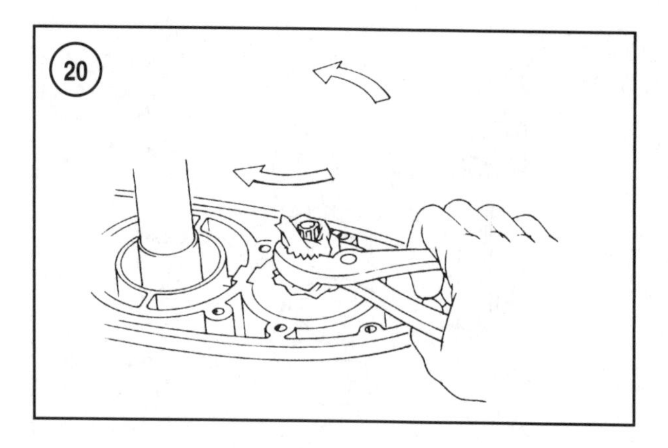

the drive shaft slightly to install the gearcase if necessary.

4. Make sure all locating pins (5, **Figure 16**) are in the bores within the gearcase and drive shaft housings.

5. Align the lower end of the water tube (**Figure 18**) with the opening in the water pump while guiding the drive shaft into the drive shaft housing.

6. Align the drive shaft with the crankshaft and both shift shafts while attempting to seat the gearcase against the drive shaft housing. Never force the gearcase into position. This damages the shift shafts and drive shaft seal. If it becomes difficult, lower the gearcase and slightly rotate the drive shaft clockwise. Repeat this step until the splined drive shaft connection engages the splined opening in the crankshaft and the gearcase seats against the drive shaft housing. If shift shaft alignment becomes difficult, have an assistant slightly toggle the shift selector to align the shift shaft splines. Verify proper alignment of the water tube-to-water pump opening during each installation attempt. Make sure the locating pins enter the openings in the drive shaft housing.

7. Apply Loctite 572 to the threads of the seven gearcase mounting bolts. Have an assistant hold the gearcase in position and install the bolts. Tighten the bolts evenly to the specification in **Table 1**.

8. Heat the end of the speedometer hose with a portable hair dryer and quickly slip it fully over the connector (1, **Figure 16**). Secure the hose onto the connector with a plastic locking clamp.

9. Install the trim tab onto the gearcase and hold it in position. Align the reference marking (**Figure 13**) made prior to removal. Use a socket and extension to install and tighten the trim tab bolt (3, **Figure 16**) to the specification in **Table 1**. Fit the bolt access cover into the opening in the drive shaft housing. Apply soap and water to the cover for easier installation into the opening.

WATER PUMP COMPONENTS

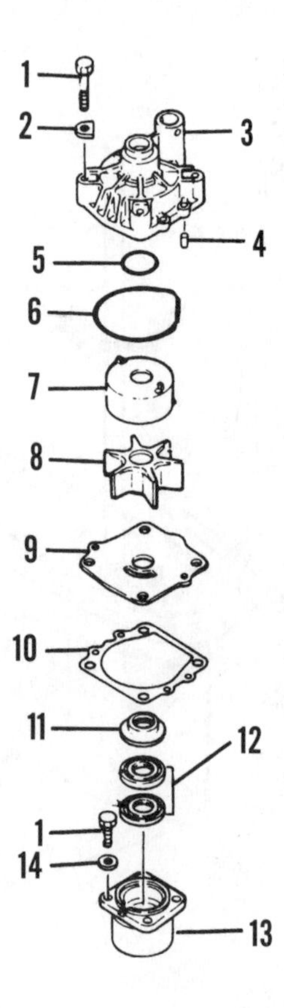

1. Bolt
2. Washer
3. Water pump body
4. Locating/alignment pin
5. O-ring
6. O-ring
7. Insert
8. Impeller
9. Wear plate
10. Gasket
11. Seal protector
12. Drive shaft seals
13. Bearing and seal housing
14. Washer

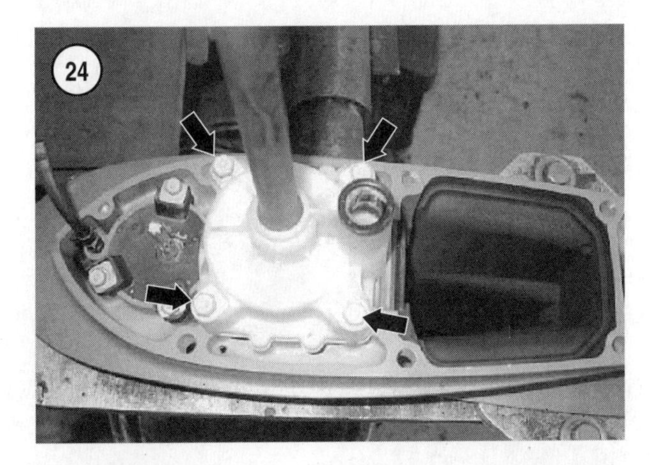

10. Adjust the shift linkages and cables as described in Chapter Four.

11. Connect the battery cables and spark plug leads.

12. Run the engine to check for proper shifting and cooling system operation. Water must exit the telltale opening (**Figure 21**). Otherwise, remove the gearcase and check

for a dislodged water tube or grommet. Check and correct the linkage adjustment if there is improper shift operation.

WATER PUMP

Service the water pump if the engine is running warmer than normal and at the intervals listed in Chapter Three.

Always replace the impeller, seals, O-rings and all gaskets when servicing the water pump. Never compromise the operation of the water pump. Overheating and extensive power head damage can result from operation with faulty water pump components.

Disassembly

Refer to **Figure 22** for this procedure. The appearance of the components may differ slightly from the illustration.

1. Remove the gearcase as described in this chapter. To ease impeller removal and reduce the chance of contaminating the gearcase, clean all corrosion and debris from the exposed surfaces of the drive shaft.
2. Carefully pry the water tube grommet (**Figure 23**) and spacer from the water pump body.
3. Remove the four screws and washers from the pump body (**Figure 24**). Carefully pry the water pump body from the wear plate (**Figure 25**). Work carefully to avoid damaging mating surfaces. Slide the body up and over the drive shaft.
4. Remove the O-ring (6, **Figure 22**) from the pump body or wear plate (9). Discard the O-ring.
5. Carefully pull the bushing, collar and washers from the water pump impeller (**Figure 26**). Then slide them off the drive shaft. If removal is difficult, carefully split the collar with a chisel and remove it from the shaft. Do not damage the drive shaft.

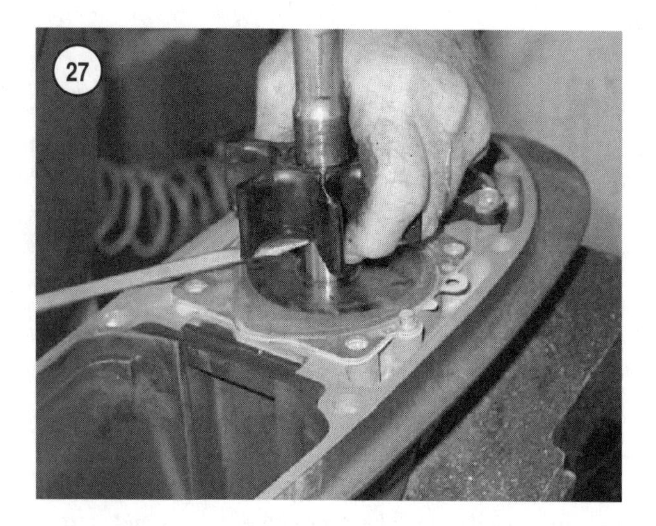

6. Mark the *up* side of the impeller (8, **Figure 22**) if it must be reused. Carefully pry the impeller away from the wear plate using a blunt tip pry bar (**Figure 27**). Work carefully to avoid damaging the wear plate. Remove the impeller. Carefully split the inner hub of the impeller with a sharp chisel if the impeller is seized to the drive shaft. Work carefully to avoid damaging the drive shaft surfaces. To help prevent drive shaft damage, cut the impeller hub at the drive key slot.

7. Remove the drive key (**Figure 28**) from the slot in the drive shaft.

8. Remove the water pump insert (7, **Figure 22**) only if it must be replaced. Refer to *Component Inspection* in this section to determine the need for replacement. The removal process damages the insert. Use a punch to drive the insert from the water pump body through the drive shaft bore. Discard the insert after removal.

9. If the wear plate must be reused, mark the impeller side of the wear plate with a permanent marker. Do not scratch the plate. Carefully pry the wear plate (9, **Figure 22**) loose from the pump base. Work carefully to avoid damaging the pump base or dislodging the locating pins (4, **Figure 22**). Slide the wear plate. Remove the gasket (10, **Figure 22**) from the wear plate or gearcase housing. Discard the gasket. Insert the locating pins into the respective opening(s) in the gearcase housing, if dislodged.

10. Inspect the seal protector (11, **Figure 22**) for melted or damaged surfaces. If melted or damaged, replace the seal protector as described in this chapter (See *Drive Shaft Bearing and Seal Housing Removal* and *Assembly).*

11. Clean all corrosion and contaminants from the water pump body, wear plate and mating surface on the gearcase.

12. Inspect all water pump components as described in this section. See *Component Inspection*.

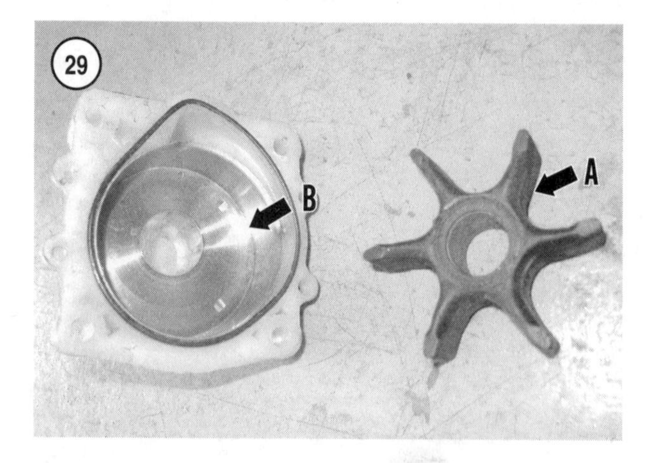

Component Inspection

Inspect the impeller (A, **Figure 29**) for brittle, missing, burnt or tightly curled vanes. Replace the impeller if these or other defects are evident. Squeeze and release the vanes toward the hub. If the vanes do not spring back to the extended position when released, the impeller material has lost flexibility and must be replaced.

Inspect the water pump insert (B, **Figure 29**) for worn or damaged surfaces. Remove and replace the insert if not in excellent condition. Operating the engine with rough insert surfaces will quickly wear the impeller.

Inspect the water pump body for melted plastic or impeller material transferred onto the insert. Replace the water pump body and wear plate if either defect is evident.

Inspect the water tube grommet for cracked, brittle or distorted surfaces and replace as required.

Inspect the wear plate for grooved, worn or rough surfaces. Replace the wear plate if these or other defects are evident. Operating the engine with a worn, grooved or damaged wear plate will quickly wear the impeller.

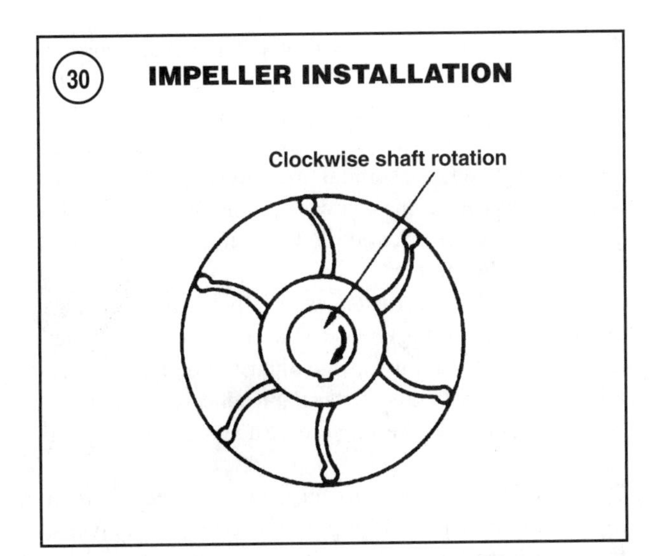

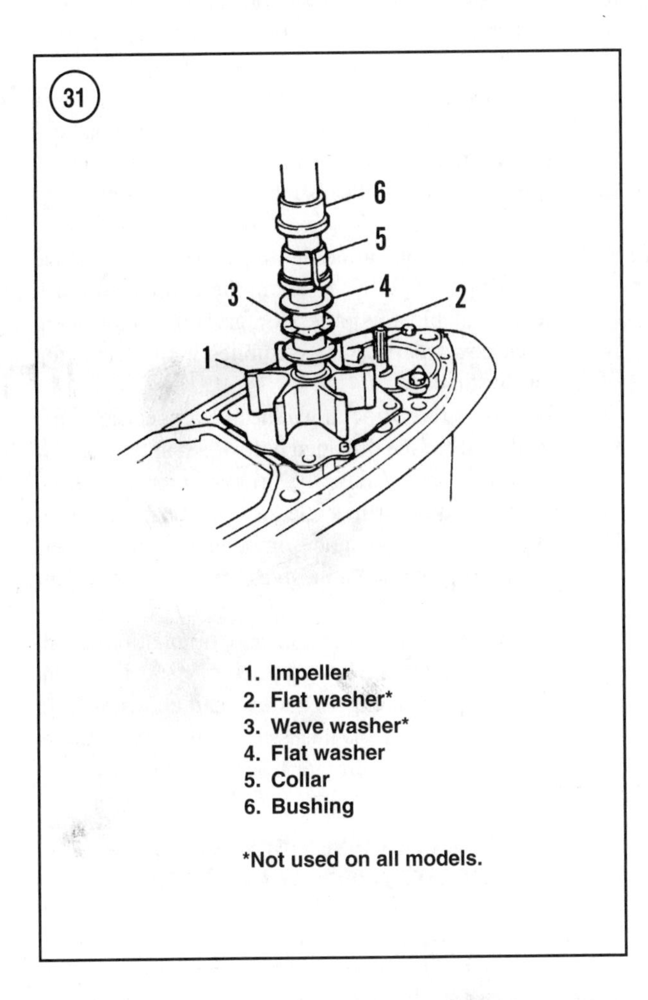

1. Impeller
2. Flat washer*
3. Wave washer*
4. Flat washer
5. Collar
6. Bushing

*Not used on all models.

Use a suitable solvent along with a Scotch-Brite pad to remove all carbon and salt or mineral deposits from the exposed portions of the drive shaft. The impeller must slide freely into position on the drive shaft.

Assembly

Refer to **Figure 22** for this procedure.

1. Install a new gasket (10, **Figure 22**) and the wear plate (9) onto the gearcase housing. The openings in the gasket and wear plate must fit over the locating/alignment pins (4, **Figure 22**). Seat the gasket and wear plate against the housing.
2. Apply a light coating of Yamaha All-Purpose Grease to the slot. Then fit the drive key (**Figure 28**) into the drive shaft. Apply a light coating of the same grease to the exposed surfaces of the wear plate and the impeller contacting area of the drive shaft.
3. Slide the impeller over the drive shaft. Align the impeller drive key slot with the drive key. Then seat the impeller against the wear plate. If reusing the impeller, it must be installed with the set (curled vanes) facing in the original direction or as shown in **Figure 30**.
4. Install the washers, collar and bushing onto the drive shaft as shown in **Figure 31**. Seat these components against the impeller. Some models do not use the washers.
5. Install a new O-ring (6, **Figure 22**) into the groove in the water pump body. Use a light coating of 3M Weather-strip Adhesive to hold the O-ring in position. Apply a light coating of Yamaha All-Purpose Grease to the surfaces of the insert. Slide the water pump body over the drive shaft. Align the bolt and locating/alignment pin openings, then rest the body on the impeller vanes.
6. Rotate the drive shaft in the clockwise direction as viewed from the top while lightly pressing down on the water pump body. Continue until the impeller vanes enter the insert in the water pump body. Rotate the water pump body counterclockwise as needed to align the bolt and locating/alignment pin openings, then seat the body against the wear plate.
7. Apply Loctite 572 onto the threads of the four screws and washers, then thread them into the water pump body and gearcase housing. Tighten the four bolts in a crossing pattern, and in two steps, to the general torque specification in Chapter One.
8. Apply Yamaha all purpose grease, then insert the spacer and grommet into the water tube opening in the water pump body.
9. Install the gearcase as described in this chapter. Run the engine to check for proper cooling system operation. If water does not exit the telltale opening, remove the gearcase and check for a dislodged water tube or grommet. If there is no fault with the tube or grommet, disassemble the water pump and check for improperly assembled or damaged components.

GEARCASE REPAIR

If the engine is equipped with a LH or twin propeller gearcase, refer to *Counterotation Gearcase* or *Twin Propeller Gearcase* in this chapter for special disassembly and re-assembly instructions.

Once the gearcase is disassembled, refer to *Component Inspection* to determine the need to replace gearcase components.

The gears must be shimmed and the gear lash (clearance) between forward gear and the pinion gear must be verified before continuing with assembly. The assembly procedures describe the shimming operation.

Special gauging fixtures and precision measuring instruments are required to shim the gearcase. Have the shimming operation performed by a qualified technician if the special tools are not available or you are unfamiliar with them. The cost of the special tools is far greater than the labor charge to have a professional perform the procedure.

Many of the disassembly and assembly operations require special service tools. The Yamaha part No. of required tools is listed in the model specific disassembly/assembly procedures. Purchase or rent any special tool from a Yamaha outboard dealership.

Note the mounting location and orientation of all components prior to removing them from the gearcase. Mark components that can be installed in different directions accordingly.

Upon disassembly, clean and inspect all gearcase components as described in *Component Inspection* in this chapter.

Service Precautions

When working on a gearcase, there are several good procedures to keep in mind that make work easier, faster and more accurate.

1. Never use elastic locknuts more than twice. Replace such nuts each time they are removed. Never use an elastic locknut that can be turned by hand.
2. Use special tools where noted. Makeshift tools can damage components and cause serious injury. The expense to replace damaged components can easily exceed the cost of the special tool.
3. Use the appropriate fixtures to hold the gearcase housing whenever possible. Use a vise with protective jaws to hold smaller housings or individual components. If protective jaws are not available, insert blocks of wood or similar padding on each side of the housing or component before clamping.
4. Remove and install pressed-on parts with an appropriate mandrel, support and press (arbor or hydraulic). Do not pry or hammer press-fit components on or off.
5. Refer to tables at the end of the chapter for torque values. Proper torque is essential to ensure long life and satisfactory service from gearcase components. Use the general torque specification in Chapter One for fasteners not listed in **Table 1**.
6. To help reduce corrosion, especially in saltwater areas, apply Yamaha Marine Grease or an equivalent to all external surfaces of bearing carriers, housing mating surfaces and fasteners when no other sealant, adhesive or lubricant is recommended. Make sure grease does not contact gears or bearings.
7. Discard all O-rings, seals and gaskets during disassembly. Apply Yamaha All-Purpose Grease or an equivalent to new O-rings and seal lips to provide initial lubrication.
8. Tag all shims with the location and thickness of each shim removing them from the gearcase. Shims are reusable as long as they are not damaged or corroded. Follow shimming instructions closely and carefully. Shims control gear location and/or bearing preload. Incorrectly shimming a gearcase will cause the gears and/or bearings to fail.
9. Work in an area with good lighting and sufficient space for component storage. Keep an ample number of clean containers available for parts storage. When not being worked on, cover parts and assemblies with clean shop towels or plastic bags.
10. Whenever a threadlocking adhesive is specified, first spray the threads of the threaded hole or nut and the screw with Locquic Primer. Allow the primer to air dry before proceeding. Locquic primer cleans the surfaces and allows better adhesion. Locquic primer also accelerates the cure rate of threadlocking adhesives from an hour or longer to 15-20 minutes.
11. Note the mounting location and orientation of all components prior to removing them. Mark the front and/or upper side of components that can be installed in different directions. This is necessary to maintain the existing wear pattern on sliding or bearing surfaces.

GEARCASE DISASSEMBLY (STANDARD RH ROTATION GEARCASE)

See *Gearcase Repair* in this chapter before proceeding Perform each procedure in this section for complete gearcase disassembly. If only partial disassembly is required to access the faulty components, perform the disassembly procedures until the component is accessible. Then refer to the section in the gearcase assembly section to reassemble the gearcase.

Refer to **Figure 32** or **Figure 33** during the gearcase disassembly procedures.

Perform the following prior to gearcase disassembly.

1. Drain the gearcase lubricant as described in Chapter Three.
2. Remove the propeller then gearcase as described in this chapter.
3. Remove all water pump components as described in this chapter. Do not remove the bearing and seal housing until instructed to do so in the disassembly procedures.

Bearing Carrier Removal

1. Rotate the shift shaft while rotating the drive shaft clockwise, as viewed from the top, until neutral gear is obtained. The gearcase must remain in neutral gear until the propeller shaft is removed from the gearcase.

2A. *115-130 hp and 150-200 hp (2.6 liter) models*—Carefully bend the tabs on the locking tab washer away from the cover nut (**Figure 34**). Use the correct cover nut tool and turn it counterclockwise to remove the cover nut.

   a. *For 115-130 hp models*—Use the cover nut tool part No. YB-34447/90890-06511.
   b. *For 150-200 hp (2.6 liter) models*—Use the cover nut tool part No. YB-34447/90890-06512.

NOTE

Apply heat if necessary to help remove the cover nut. If corrosion prevents removal of the cover nut, drill two holes approximately 25 cm (1 in.) apart into the surface of the cover nut. Drill the holes to the depth of the cover nut thickness. Use a chisel to break a small section from the cover nut at the drilled holes, then remove the cover nut with the cover nut tool as indicated.

2B. *200-250 hp (3.1 liter) models*—Remove the two bolts, then remove the retaining ring (**Figure 35**). Remove the two bolts (**Figure 36**) that retain the bearing carrier into the gearcase housing.

3. Use the puller claws and plate (Yamaha part No. YB-6207 and YB-6117) to pull the bearing carrier from the gearcase. Fit the puller and plate onto the carrier and propeller shaft as shown in **Figure 37**. Turn the puller bolt clockwise to move the carrier. When free, remove the puller and manually pull the bearing carrier assembly from the gearcase. Remove the thrust washer (22, **Figure 32**) from the reverse gear or propeller shaft.

4. Remove the large O-ring from the bearing carrier and discard it. Retrieve the locating key from the slot in the bearing carrier or gearcase housing (**Figure 38**). Refer to **Figure 32** or **Figure 33** to determine if a locating key is used on the gearcase.

CAUTION

Locate the bearing carrier locating key, if so equipped, before cleaning the gearcase. The key can get easily dislodged when removing the bearing carrier. In many instances it falls into the gearcase and lost during the cleaning process. In may become lodged in a recess within the gearcase and lead to gearcase failure if it becomes dislodged at a later time.

Propeller Shaft and Drive Shaft Removal

NOTE

Drive shaft removal can be difficult. When necessary, clamp the drive shaft into a vise with protective jaws. Place a large sponge into the gearcase opening to prevent components from falling from the gearcase. Support the gearcase and use a block of wood for a cushion to carefully drive the gearcase from the drive shaft.

1. Remove the two bolts (**Figure 39**), then carefully pry the shift shaft seal housing from the gearcase. Carefully pull the shift shaft and seal housing from the gearcase. Do not rotate the propeller shaft after removing the shift shaft.

2. Use care to avoid rotating the propeller shaft during removal. Pull the propeller shaft (C, **Figure 40**) from the gearcase. Place the propeller shaft assembly on a stable work surface. Make sure the assembly cannot roll from the work surface.

3. Use a suitable socket wrench to engage the pinion nut. Use shop towels to protect the inner surfaces of the gearcase from damage. Attach a socket and breaker bar to the drive shaft adapter (Yamaha part No. YB-06201/90890-06520). Place the adapter onto the splined section of the drive shaft (**Figure 41**). Make sure the socket is properly engaged to the pinion nut. Then rotate the drive shaft in the counterclockwise direction to loosen and remove the pinion nut.

4. Remove the four bolts that retain the drive shaft bearing and seal housing into the gearcase (**Figure 42**).

5. Support the pinion gear, then carefully pull the drive shaft along with the bearing and seal housing from the gearcase. Remove the pinion gear from the housing. Remove the forward gear from the gearcase (**Figure 43**).

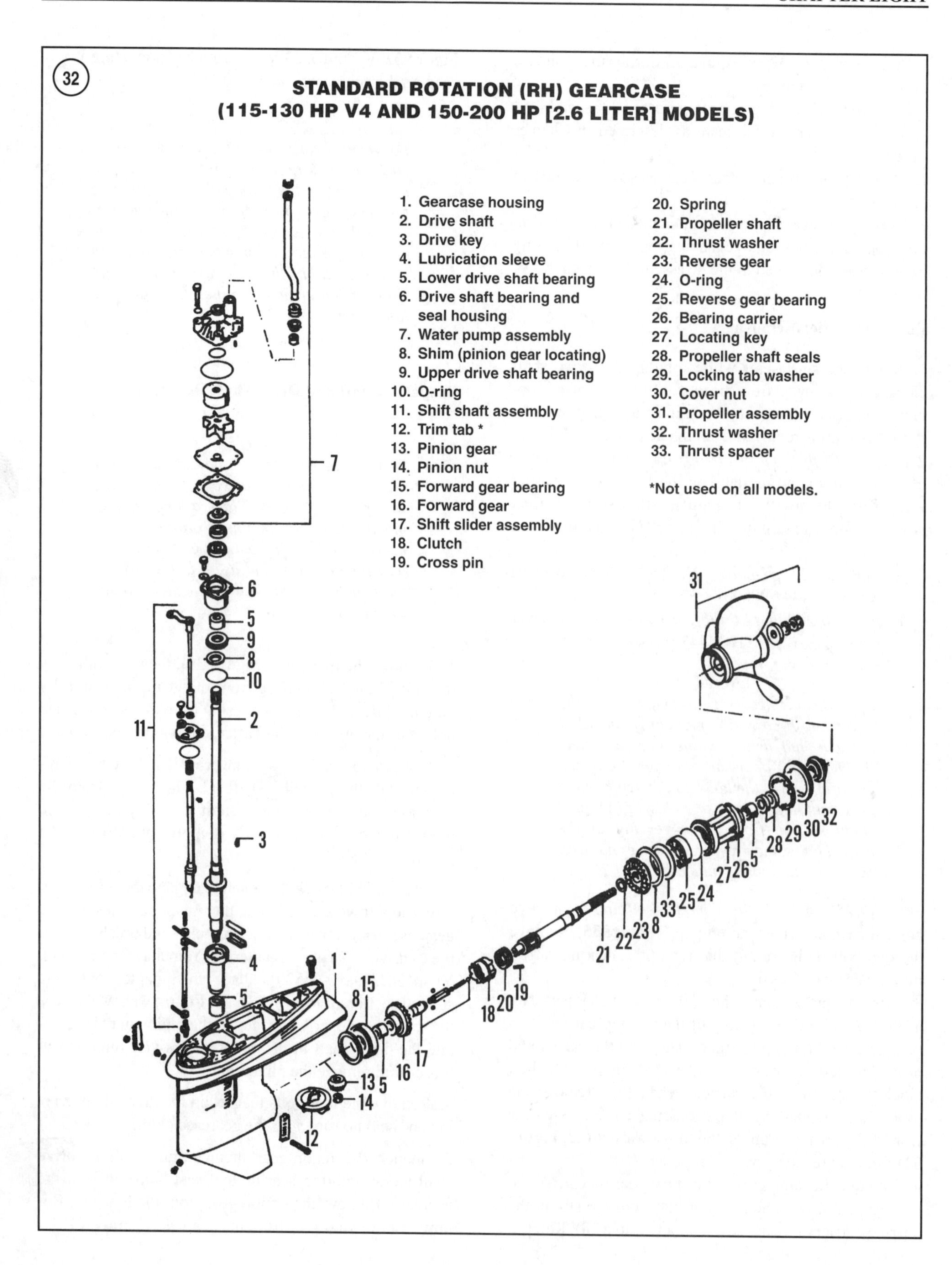
32
STANDARD ROTATION (RH) GEARCASE
(115-130 HP V4 AND 150-200 HP [2.6 LITER] MODELS)
1. Gearcase housing
2. Drive shaft
3. Drive key
4. Lubrication sleeve
5. Lower drive shaft bearing
6. Drive shaft bearing and seal housing
7. Water pump assembly
8. Shim (pinion gear locating)
9. Upper drive shaft bearing
10. O-ring
11. Shift shaft assembly
12. Trim tab *
13. Pinion gear
14. Pinion nut
15. Forward gear bearing
16. Forward gear
17. Shift slider assembly
18. Clutch
19. Cross pin
20. Spring
21. Propeller shaft
22. Thrust washer
23. Reverse gear
24. O-ring
25. Reverse gear bearing
26. Bearing carrier
27. Locating key
28. Propeller shaft seals
29. Locking tab washer
30. Cover nut
31. Propeller assembly
32. Thrust washer
33. Thrust spacer
*Not used on all models.

33

STANDARD ROTATION (RH) GEARCASE (200-250 HP MODELS [3.1 LITER])

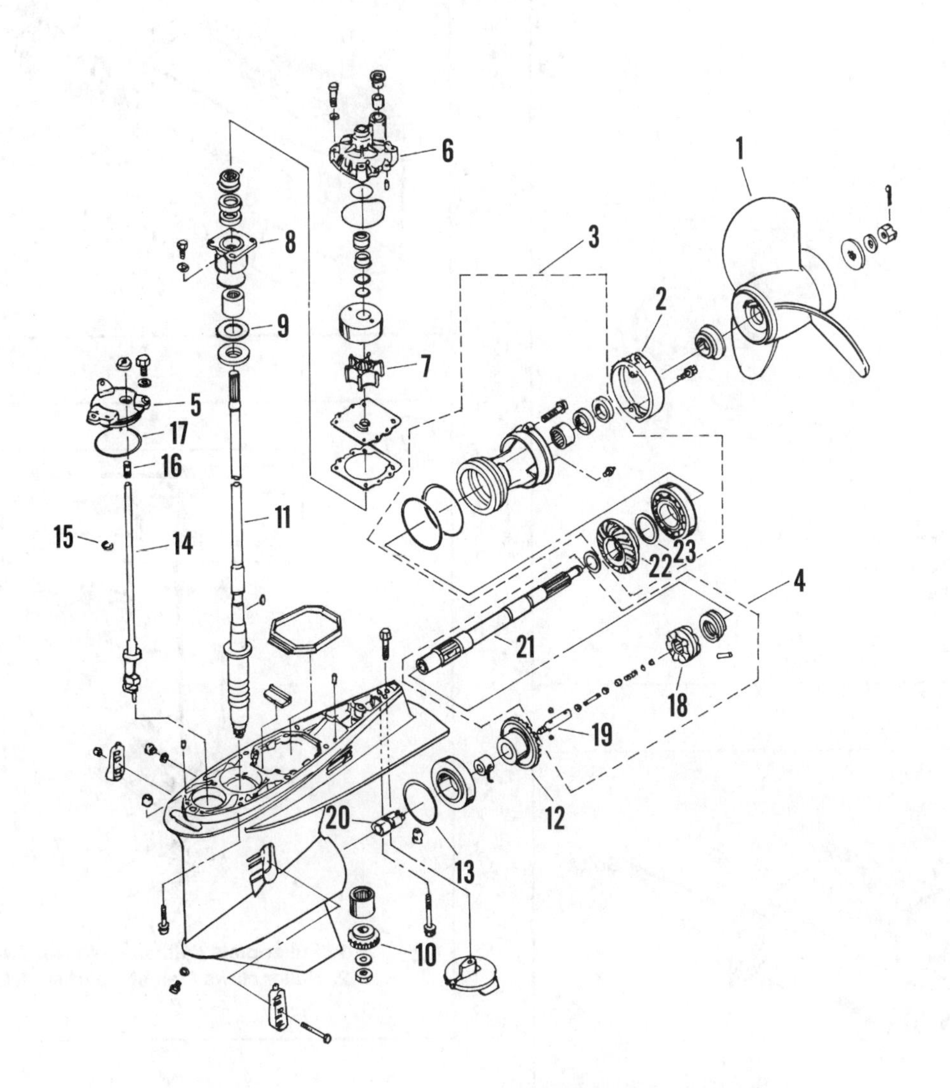

1. Propeller assembly
2. Retaining ring
3. Bearing carrier assembly
4. Shift slider and clutch assembly
5. Shift shaft seal housing
6. Water pump body
7. Impeller
8. Drive shaft bearing and seal housing
9. Shim (pinion gear locating)
10. Pinion gear
11. Drive shaft
12. Forward gear
13. Shim (forward gear locating)
14. Shift shaft
15. E-clip
16. Spring
17. O-ring
18. Clutch
19. Shift slider
20. Shift cam
21. Propeller shaft
22. Reverse gear
23. Shim (reverse gear locating)

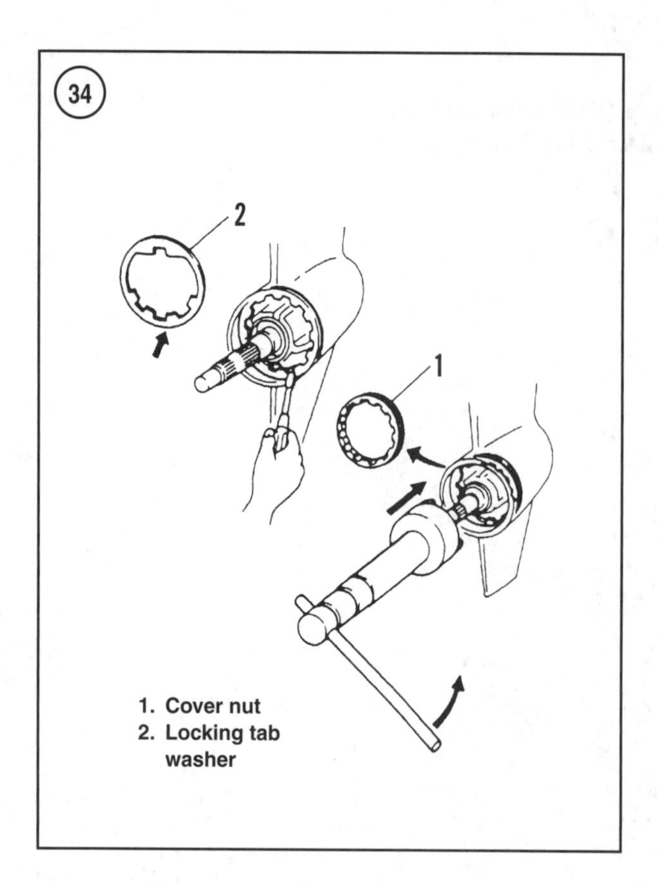

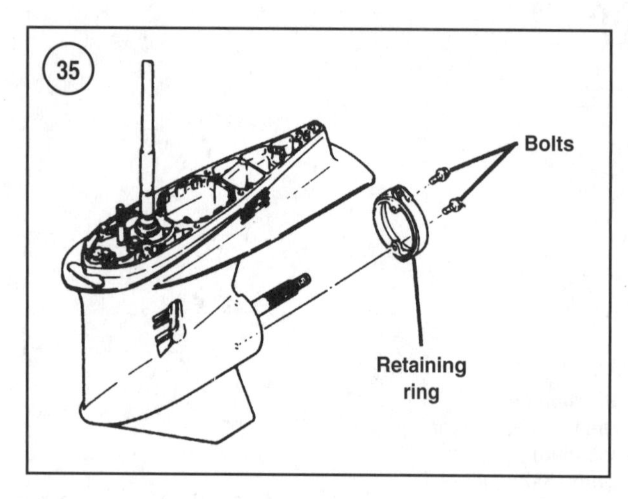

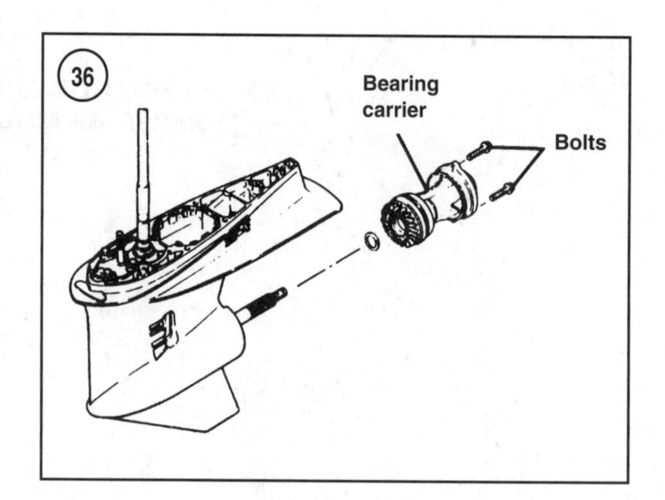

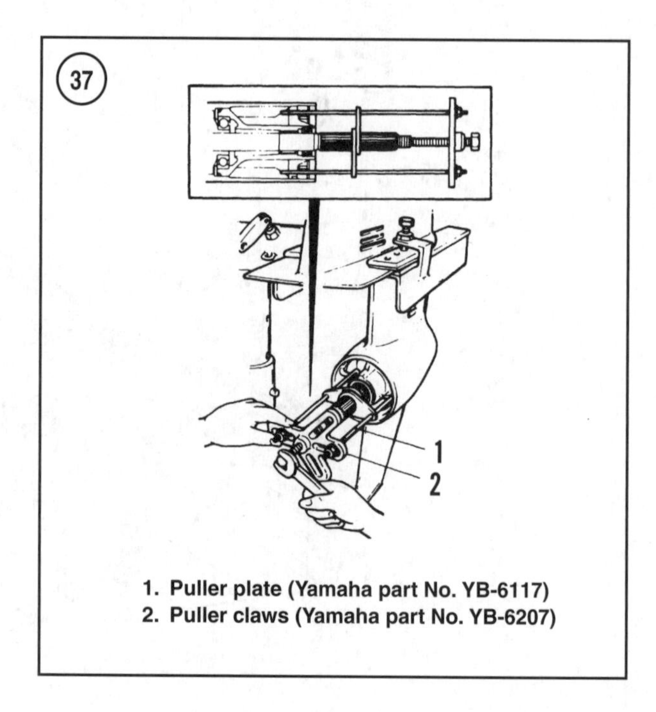

1. Puller plate (Yamaha part No. YB-6117)
2. Puller claws (Yamaha part No. YB-6207)

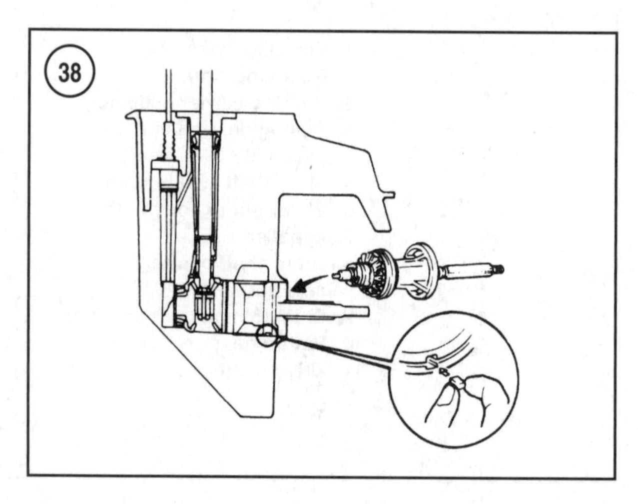

Drive Shaft Bearing and Seal Housing Removal

Remove the bearing and seal housing (2, **Figure 42**) from the drive shaft. Note the location and orientation of the thrust bearing (4, **Figure 42**) and shim (3) before removing them from the drive shaft. If reused, these components must be installed in the same location and orientation as removed.

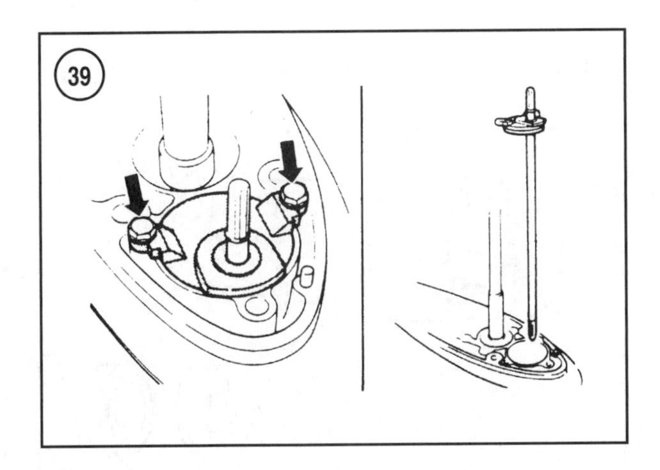
39

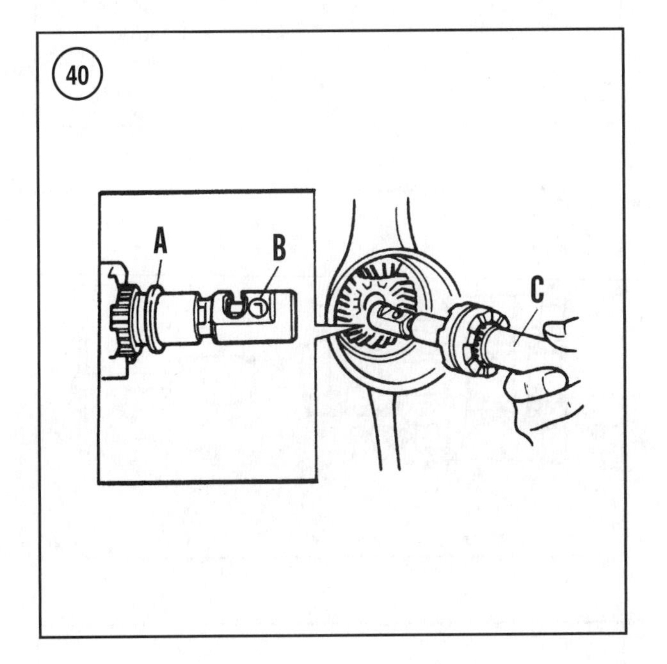
40
A
B
C

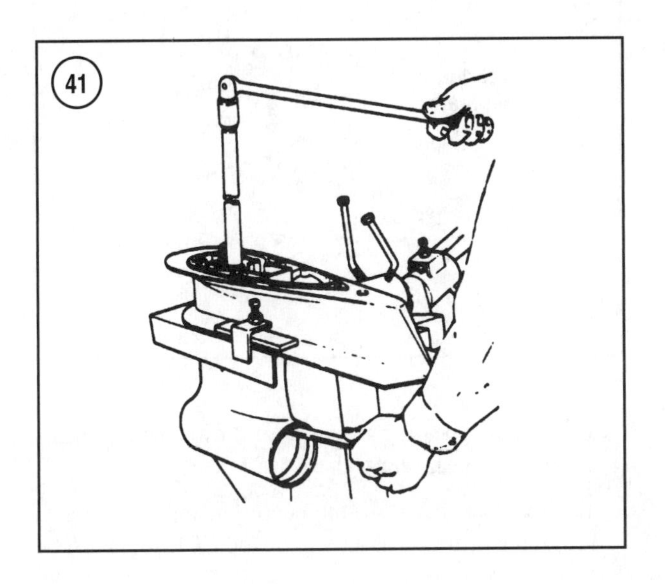
41

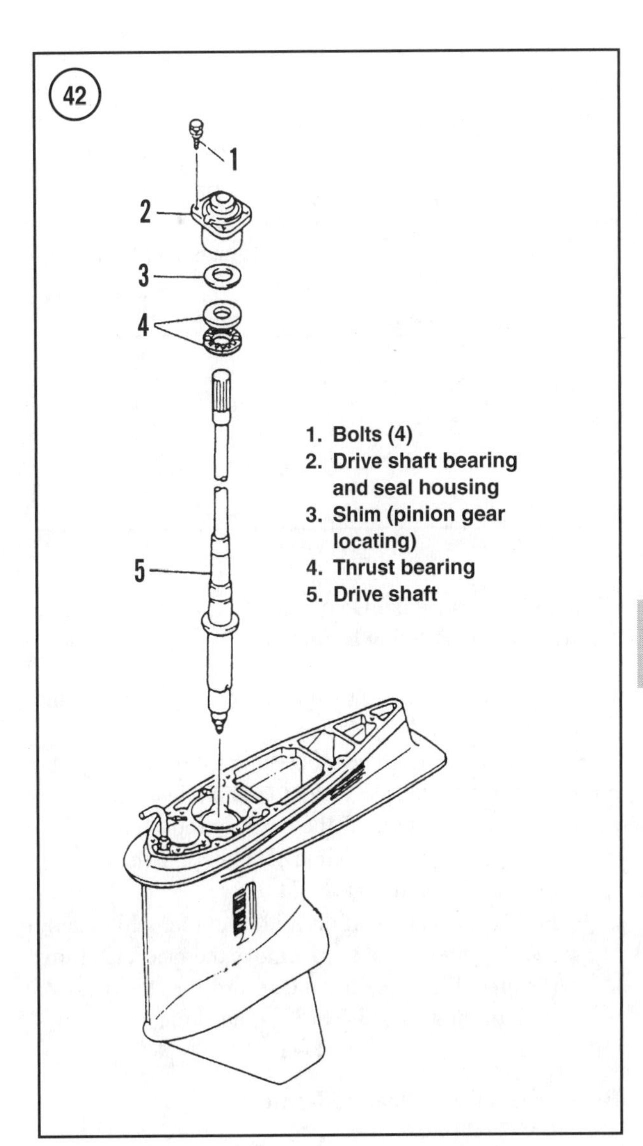
42
1
2
3
4
5
1. Bolts (4)
2. Drive shaft bearing and seal housing
3. Shim (pinion gear locating)
4. Thrust bearing
5. Drive shaft

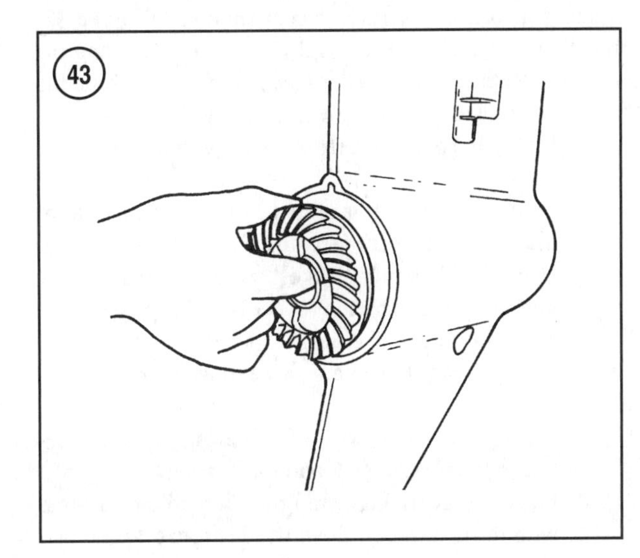
43

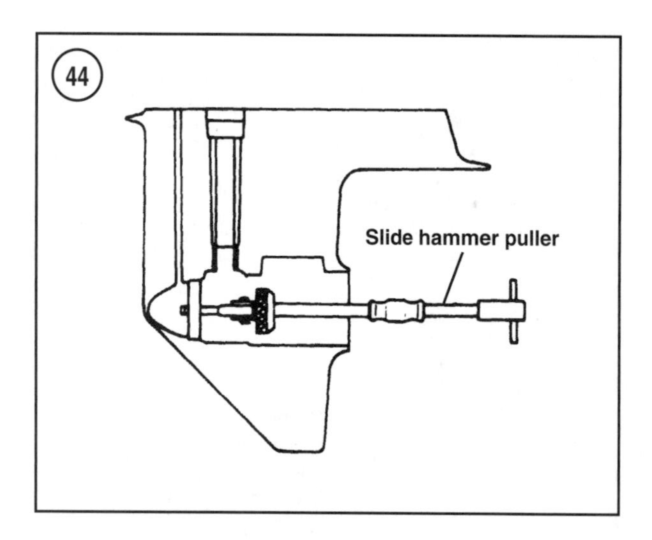

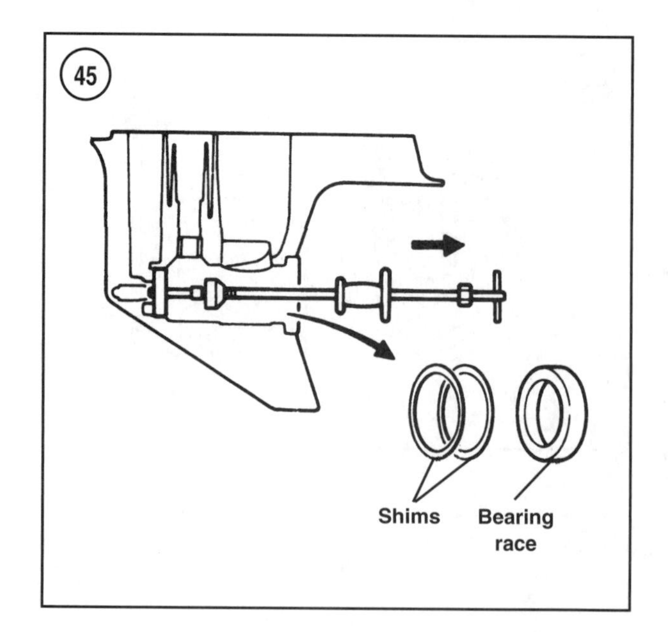

Pinion Gear, Forward Gear and Forward Gear Bearing Removal

1. Remove the pinion gear from the housing. Remove the forward gear from the gearcase.
2. Remove the forward gear bearing race only if it or the housing must be replaced. Refer to *Component Inspection* in this chapter to determine the need for replacement.
   a. Engage the jaws of a slide hammer onto the bearing race as shown in **Figure 44**.
   b. Pull on the handle of the slide hammer while using short hammer strokes to remove the race and shims (**Figure 45**). Measure and record the thickness of each of the forward gear locating shims.

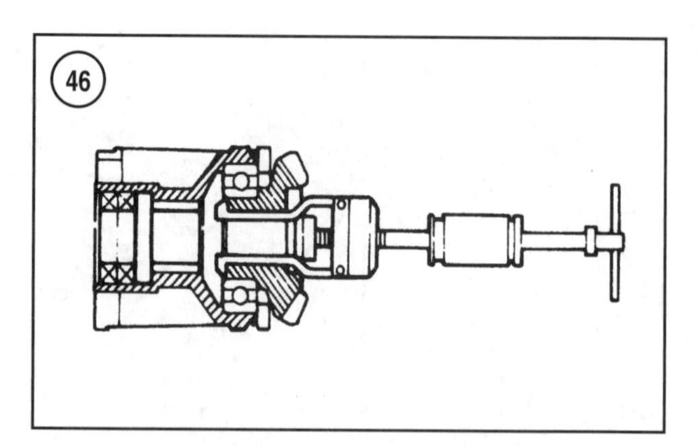

Reverse Gear and Bearing Removal

1. Remove the reverse gear only if the reverse gear, reverse gear bearing or reverse gear shim (23, **Figure 33**) must be replaced. The removal process may damage the bearing. Refer to *Component Inspection* to determine the need for replacement. Removal instructions follow:
   a. Clamp the bearing carrier into a vise with protective jaws.
   b. Engage the jaws of a slide hammer into the hub of the reverse gear (**Figure 46**).
   c. Use short hammer strokes to remove the reverse gear from the carrier.
2. In some instances, the gear is pulled along with the bearing. In this event, separate the bearing and gear as follows:
   a. Engage the sharp edge of a bearing separator between the reverse gear and the bearing.
   b. Place the gear and bearing on the table of a press with the gear facing downward (**Figure 47**).

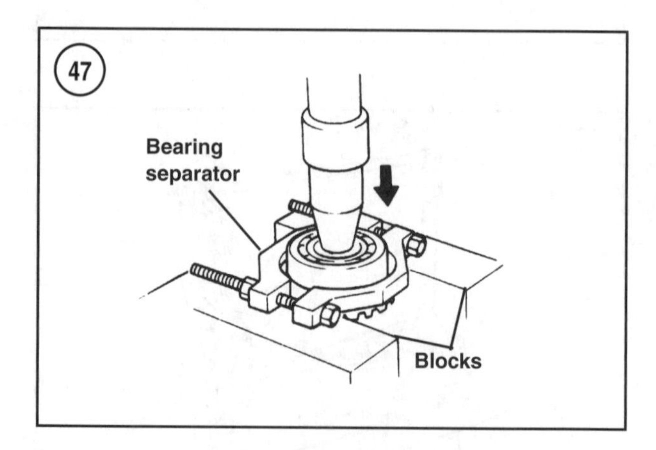

   c. Block the bearing separator to allow adequate clearance for gear removal.
   d. Press the gear from the bearing using a section of tubing or socket as a removal tool. The tool must be

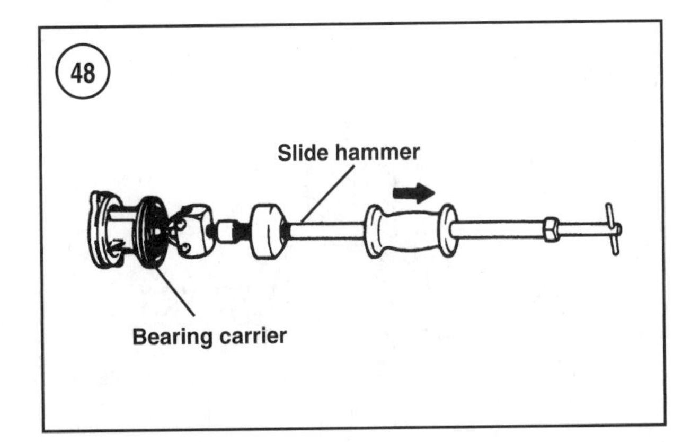

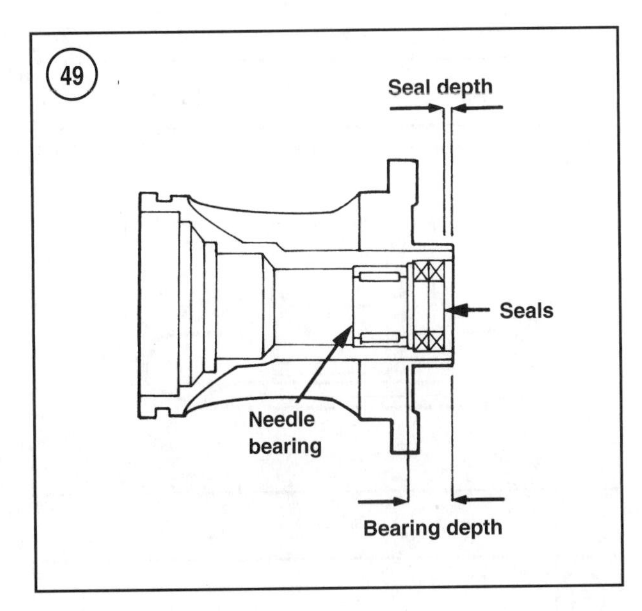

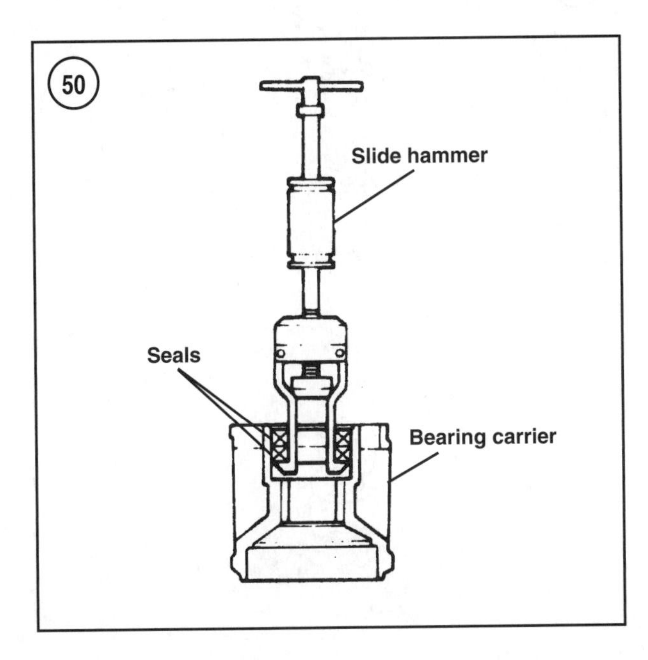

of sufficient diameter to contact the hub of the gear, yet not contact the bearing bore.

3. Remove the reverse gear bearing only if the bearing must be replaced. The removal process may damage the bearing. Refer to *Component Inspection* to determine the need for replacement. Removal instructions follow:

   a. Clamp the bearing carrier into a vise with protective jaws.
   b. Engage the jaws of a slide hammer beneath the inner bearing race (**Figure 48**).
   c. Use short hammer strokes to remove the bearing.
   d. *200-250 hp (3.1 liter) models*—Remove the reverse gear shim (23, **Figure 33**) from the reverse gear or bearing.

Propeller Shaft Seals and Needle Bearing Removal

1. Remove the propeller shaft seals only if they must be replaced. The removal process damages the seals. Refer to *Component Inspection* to determine the need for replacement. Remove the seals as follows:

   a. Measure and record the seal depth as indicated in **Figure 49**.
   b. Clamp the bearing carrier into a vise with protective jaws. The gear side of the carrier must face downward.
   c. Engage the jaws of a slide hammer beneath the casing of the inner seal (**Figure 50**).
   d. Use short hammer strokes to remove the seals.
   e. Discard the seals.

2. Remove the needle bearing from the bearing carrier only if it must be replaced. The removal process damages the bearing. Refer to *Component Inspection* to determine the need for replacement. Removal instructions follow:

   a. Measure and record the needle bearing depth as indicated in **Figure 49**.
   b. Place the carrier on the table of a press with the gear side facing upward. Block the carrier or place it over an opening to ensure adequate clearance for the bearing during removal.
   c. Select a section of tubing or socket and extension to use as a removal tool. The tool must be of sufficient diameter to adequately contact the casing of the bearing; yet not contact the bearing carrier bore during removal.
   d. Seat the tool against the bearing, then slowly tap the bearing from the bore (**Figure 51**).
   e. Discard the bearing.

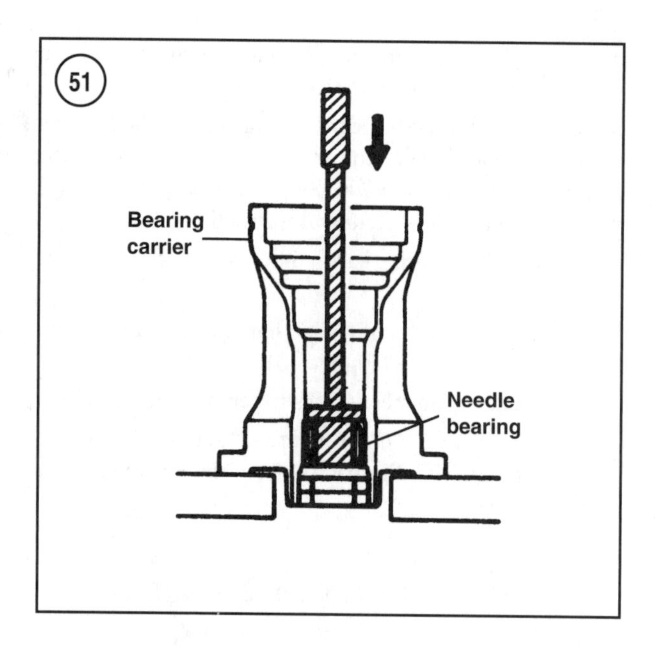

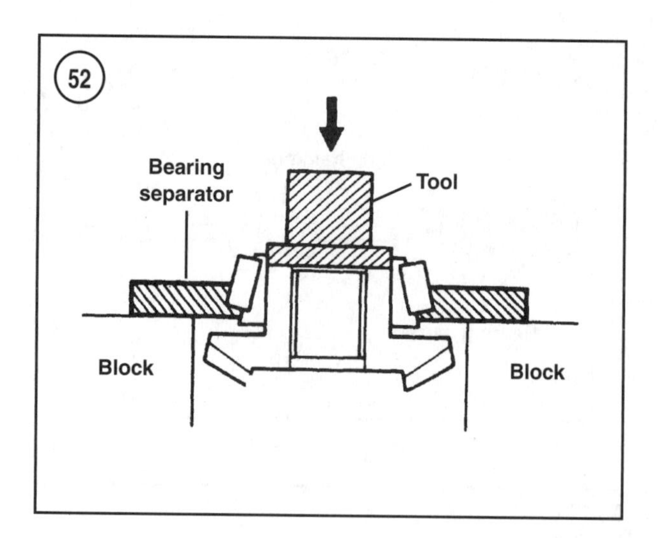

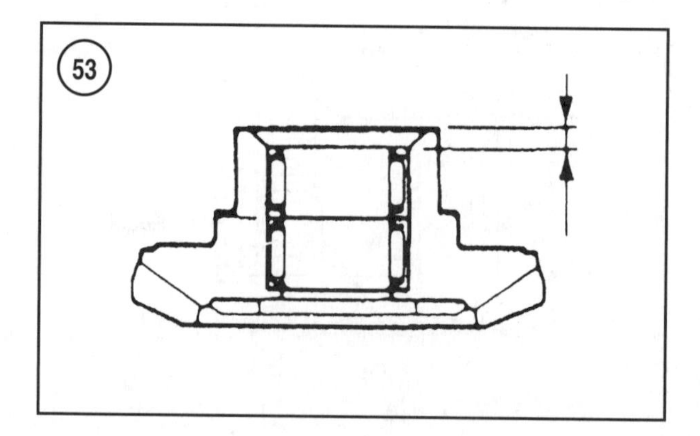

Forward Gear Bearing Removal

1. Remove the tapered roller bearing from the forward gear only if the bearing or forward gear must be replaced. The removal process damages the bearing. Refer to *Component Inspection* to determine the need for replacement. Removal instructions follow:
   a. Engage the sharp edge of a bearing separator between the bearing and forward gear (**Figure 52**).
   b. Block the sides of the separator to ensure adequate travel for the gear.
   c. Select a section of tubing or socket to use as a removal tool. The tool must be of sufficient diameter to contact the gear hub, yet not contact the inner race of the bearing.
   d. Press on the tool to remove the gear from the bearing. Discard the bearing.

NOTE
The propeller shaft needle bearing arrangement in the forward gear varies by model. Either a single or double needle bearing is used.

2. Remove the needle bearing(s) from the forward gear only if replacement is required. The removal process damages the bearing. Refer to *Component Inspection* to determine the need for replacement.
   a. Prior to removal, use a depth micrometer to measure the bearing(s) depth in the bore at the points indicated in **Figure 53**. This is required to ensure the bearing will contact the propeller shaft at the proper location on assembly. Record the measurement.

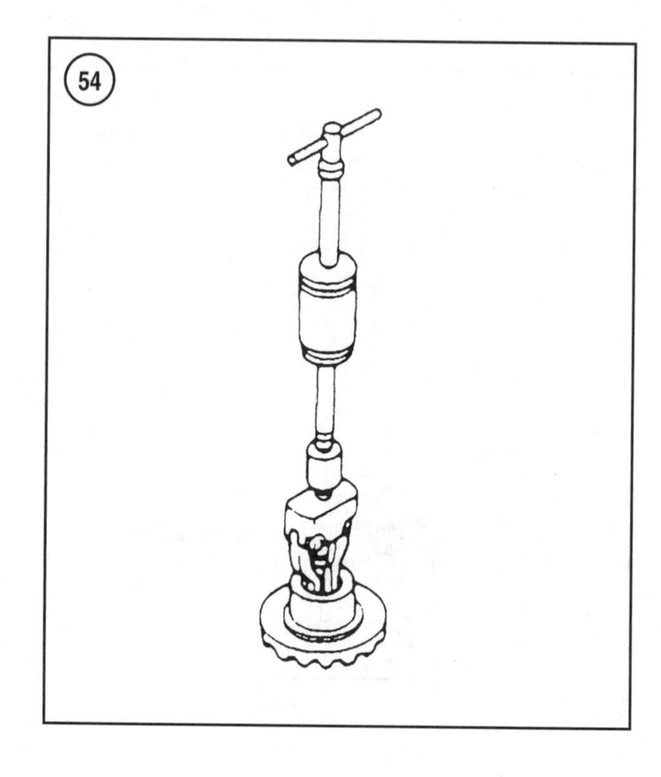

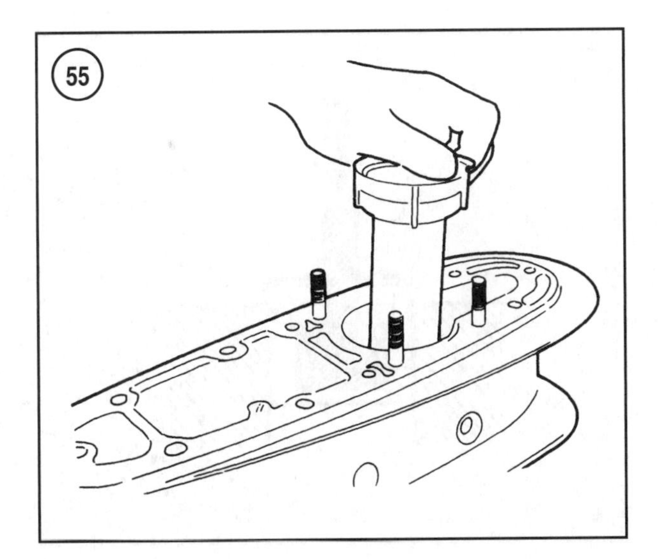

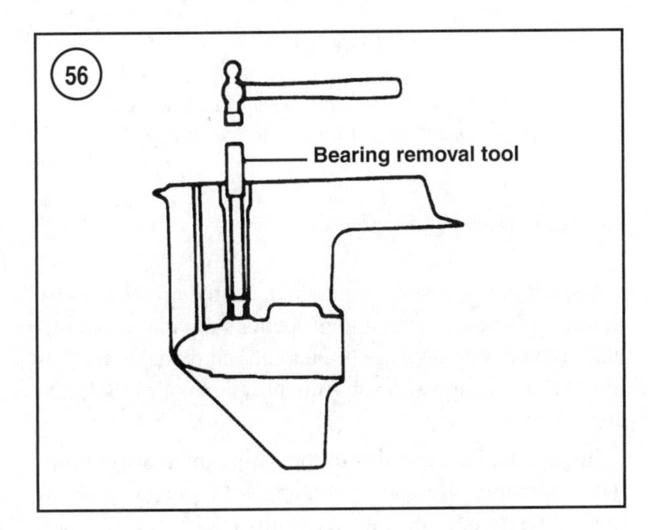

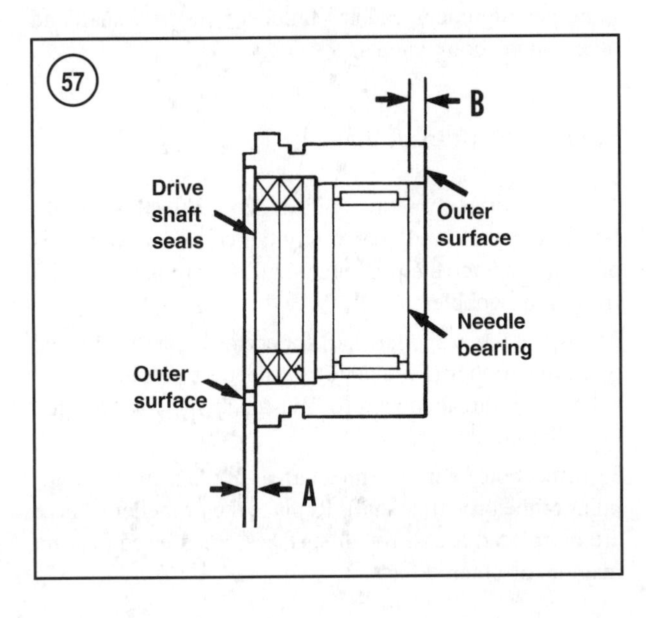

b. Carefully secure the forward gear in a vise with protective jaws. Do not over-tighten the vise.

c. Engage the jaws of a slide hammer onto the needle bearing case (**Figure 54**). Use short hammer strokes to remove the bearing(s). Note the orientation (forward or rearward facing) of the numbered side of the bearing(s). Then discard the bearings.

Drive Shaft Bearing and Seal Removal

1. Remove the lower drive shaft bearing only if it must be replaced. The removal process damages the bearing. Refer to *Component Inspection* to determine the need for replacement. Removal instructions follow:

a. Measure the distance from the gearcase to the drive shaft housing mating surface to the topside of the lower drive shaft bearing case. Record the measurement. This step is necessary to ensure the bearing contacts the drive shaft at the proper location when assembled.

b. *115-130 hp and 150-200 hp (2.6 liter) models*—Remove the lubrication sleeve from the drive shaft bore (**Figure 55**).

c. Select a section of tubing or socket and extension to use as a removal tool. The tool must be of sufficient diameter to contact the case of the bearing, yet not contact the drive shaft bore in the gearcase housing.

d. Insert the removal tool into the drive shaft bore and seat it against the bearing (**Figure 56**).

e. Carefully tap the bearing from the gearcase housing. Discard the bearing upon removal.

2. Replace the seals in the drive shaft bearing and seal housing anytime the housing requires service.

a. Carefully pry the seal protector from the topside of the housing.

b. Prior to removal, use a depth micrometer to measure the seal depth at the points indicated (A, **Figure 57**). This is required to ensure the seals contact the drive shaft at the proper location when assembled. Record the measurement.

c. Note the seal lip direction, then use a suitable blunt tip pry bar to pry the seals from the housing. Work carefully to avoid damaging the seal bore.

d. Discard the seals.

3. Remove the bearing from the drive shaft bearing and seal housing only if it must be replaced. The removal process damages the bearing. Refer to *Component Inspection*in this chapter to determine the need for replacement.

a. Place the housing on a sturdy surface with an opening below the housing that is of sufficient size for the bearing.

b. Use an appropriately sized section of tubing or socket and extension to remove the bearing. The tubing or socket must be of sufficient diameter to contact the bearing case while not contacting the housing during bearing removal.

c. Keep the removal tool in firm contact with the bearing (**Figure 58**) while carefully driving the bearing from the housing.

Propeller Shaft and Shifter Disassembly

A dual action cam-type shifter (9, **Figure 59**) or (1, **Figure 60**) is used to move the clutch. Detent balls located on the shift slider (8, **Figure 59** or 3, **Figure 60**) engage slots with the shifter bore to provide smooth shifting and positive engagement. Disassemble the propeller shaft over a box or small parts pan as small parts will likely fall from the propeller shaft. Disassemble the propeller shaft as follows:

1. Use a small screwdriver to unwind the spring from the clutch (**Figure 61**).
2. With the spring removed, push the cross pin from the clutch (**Figure 62**).
3. Slide the clutch from the propeller shaft (**Figure 63**).
4. Working over a container, carefully pry the shifter from the propeller shaft (**Figure 64**).
5. Remove the balls, springs, washers and the shift slider from the propeller shaft.
6. Carefully slide the shift shaft seal housing from the shift shaft. Remove the seal from the shift shaft seal housing anytime the housing requires service. Use a small tip pry bar to carefully pry the seal from the housing. Work carefully to avoid damaging the housing.

COMPONENT INSPECTION

Never compromise a proper repair by using damaged or questionable components.

Prior to inspection, thoroughly clean all components using clean solvent. Note component orientation before cleaning when necessary. Use compressed air to dry all components then arrange them in an orderly fashion on a clean work surface. Never allow bearings to spin while using compressed air to dry them.

Make sure all components are removed. Then use pressurized water to clean the gearcase housing. Inspect all passages and crevices for debris or contaminants. Use compressed air to thoroughly dry the gearcase.

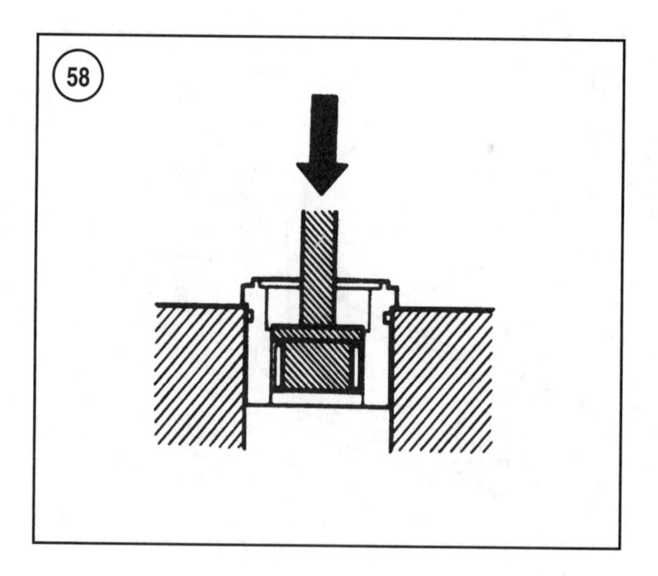

WARNING
Never allow bearings to spin when using compressed air to dry them. The bearing may fly apart or explode causing injury.

Gearcase Housing Inspection

Inspect the gearcase for cracked, dented or excessively pitted surfaces. A reputable propeller shop can economically repair skeg damage. Other surface damage may allow water leakage and subsequent failure of the internal components.

Inspect the locating pins for bent pins and worn or elongated openings. Replace a housing with elongated openings. Otherwise, the drive shaft may operate with improper alignment, causing failure of the drive shaft and other engine components.

Propeller Shaft Inspection

1. Position the propeller shaft on V-blocks (**Figure 65**). Rotate the shaft and note if any deflection or wobble is present. Replace the propeller shaft if there is visible deflection or wobble.
2. Inspect the propeller shaft for corrosion, and damaged or worn surfaces (**Figure 66**). Inspect the propeller shaft splines and threaded area for twisted splines or damaged propeller nut threads.
3. Inspect the bearing contact areas at the front and midpoint of the propeller shaft. Replace the propeller if there are discolored areas, rough surfaces, transferred bearing material or other defects.

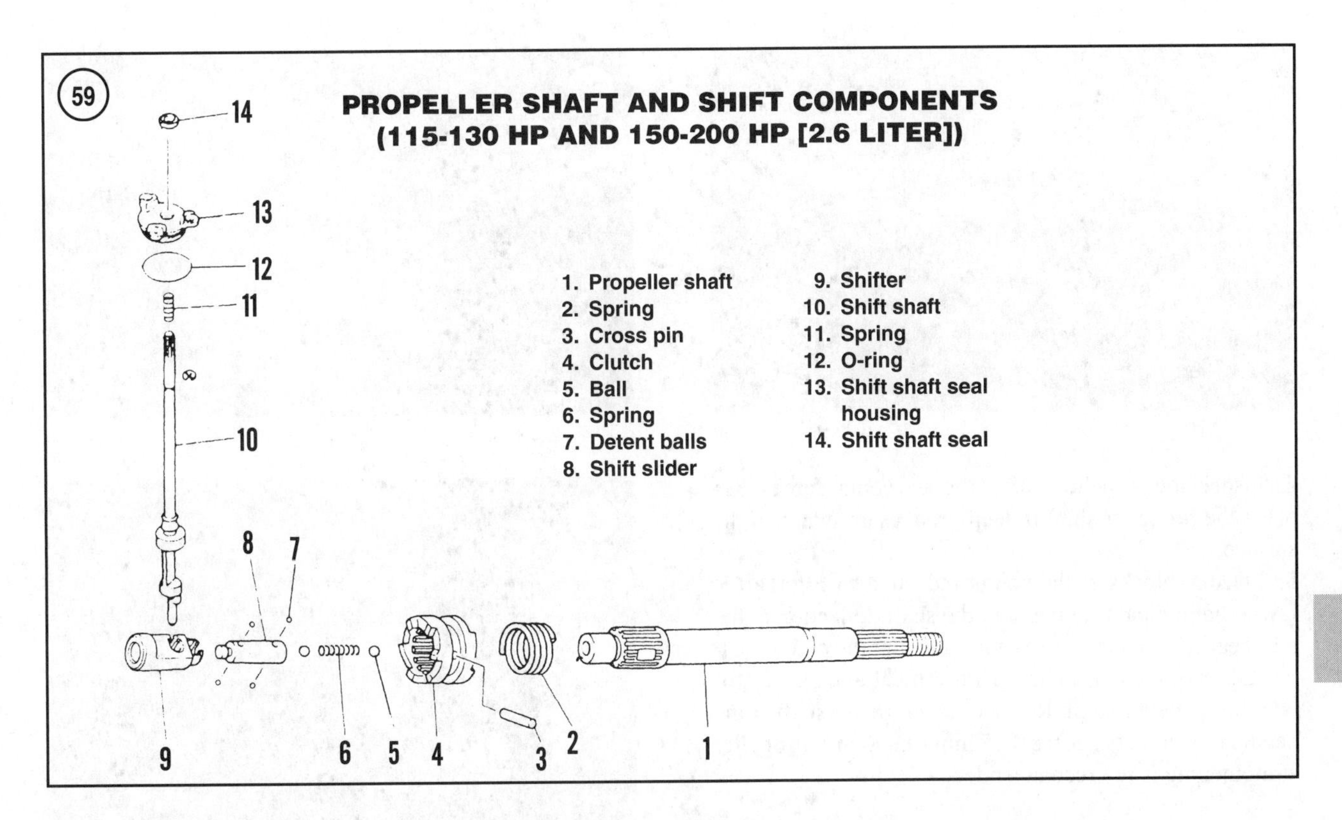
59
PROPELLER SHAFT AND SHIFT COMPONENTS
(115-130 HP AND 150-200 HP [2.6 LITER])
1. Propeller shaft
2. Spring
3. Cross pin
4. Clutch
5. Ball
6. Spring
7. Detent balls
8. Shift slider
9. Shifter
10. Shift shaft
11. Spring
12. O-ring
13. Shift shaft seal housing
14. Shift shaft seal
14
13
12
11
10
8
7
9
6
5
4
3
2
1

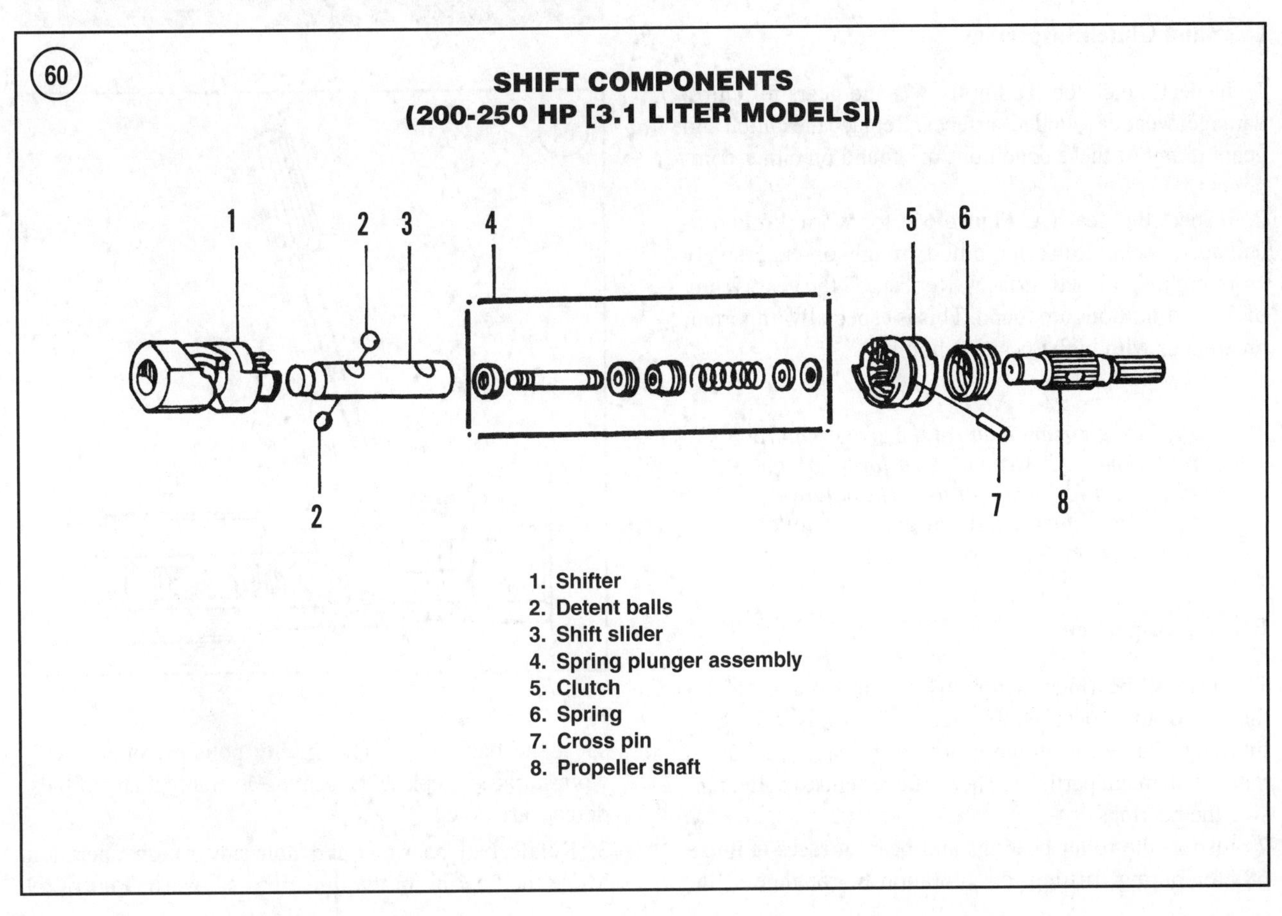
60
SHIFT COMPONENTS
(200-250 HP [3.1 LITER MODELS])
1
2
3
4
5
6
2
7
8
1. Shifter
2. Detent balls
3. Shift slider
4. Spring plunger assembly
5. Clutch
6. Spring
7. Cross pin
8. Propeller shaft

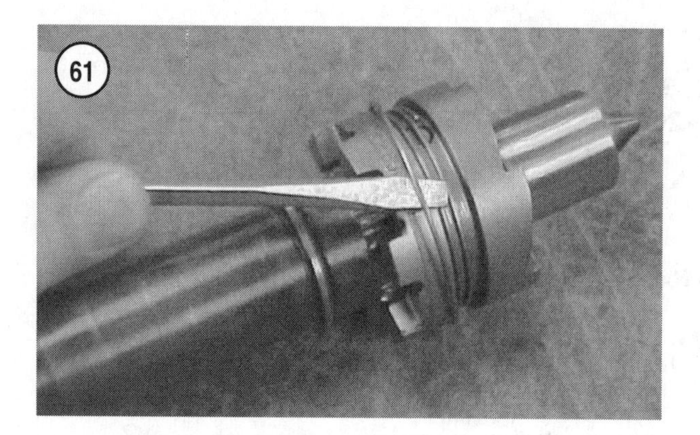

4. Inspect the propeller shaft at the seal contact areas. Replace the propeller shaft if deep grooves are worn in the surface.
5. Place V-blocks at the points indicated in **Figure 65**. Use a dial indicator to measure the shaft deflection at the rear bearing support area. Securely mount the dial indicator. Observe the dial indicator movement and slowly rotate the propeller shaft. Replace the propeller shaft if the needle movement exceeds 0.15 mm (0.006 in.). Propeller straightening is not recommended.

Gear and Clutch Inspection

1. Inspect the clutch (B, **Figure 67**) and gears for chips, damage, wear or rounded surfaces. Replace the clutch and gears if any of these conditions are found on either component.
2. Inspect the gear (A, **Figure 67**) for worn, broken or damaged teeth. Look for pitted, rough or excessively worn highly polished surfaces. Replace all the gears if any of these conditions are found. This is especially important on engines with high operating hours.

NOTE

Replace all gears if any of the gears require replacement. A wear pattern forms on the gears in a few hours of use. The wear patterns are disturbed if a new gear is installed with used gears.

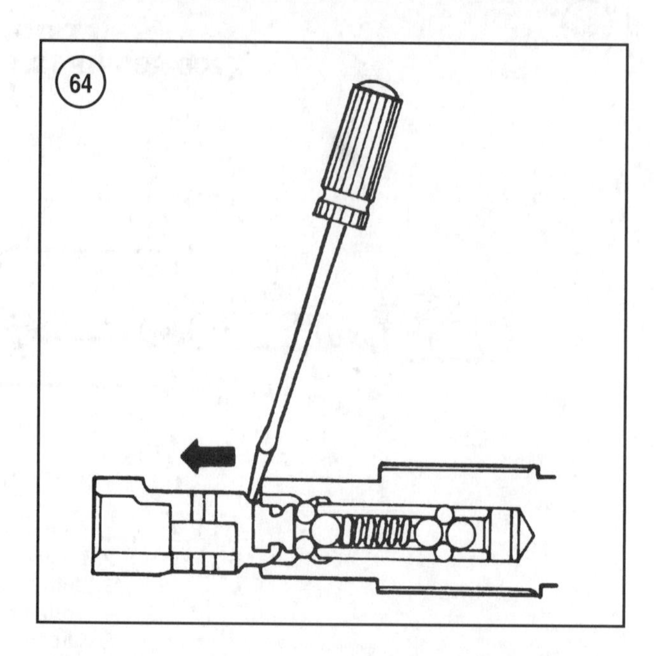

Bearing Inspection

1. Clean all bearings thoroughly in solvent and air dry them prior to inspection. Replace the bearings if the gear lubricant drained from the gearcase is heavily contaminated with metal particles. The particles tend to collect inside the bearings.
2. Inspect the roller bearings and bearing races (**Figure 68**) for pitting, rusting, discoloration or roughness. Inspect the bearing race for highly polished or unevenly worn surfaces. Replace the bearing assembly if any of these defects are noted.
3. Rotate ball bearings and note any rough operation. Move the bearing in the directions shown in **Figure 69**.

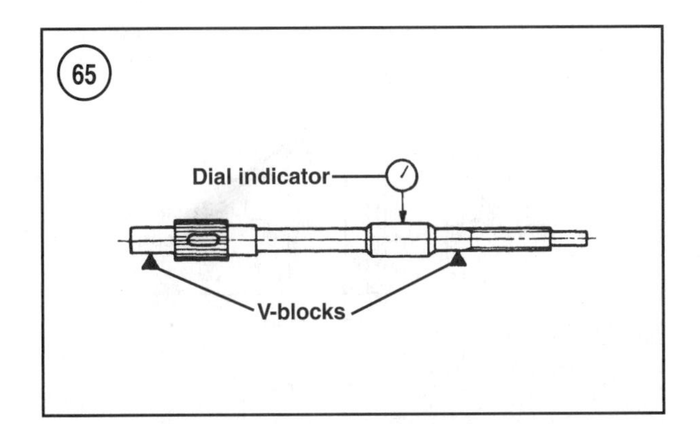

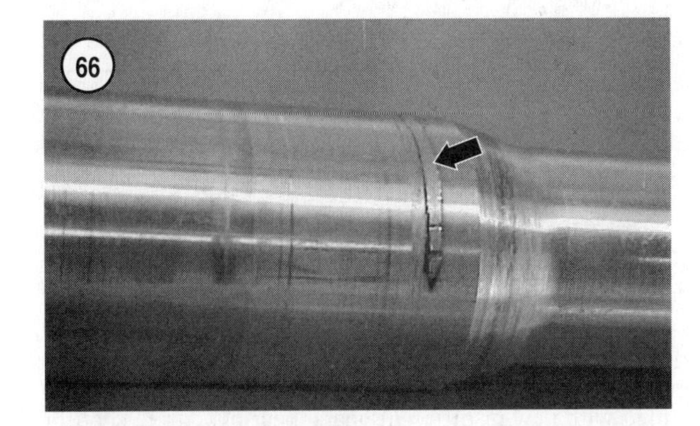

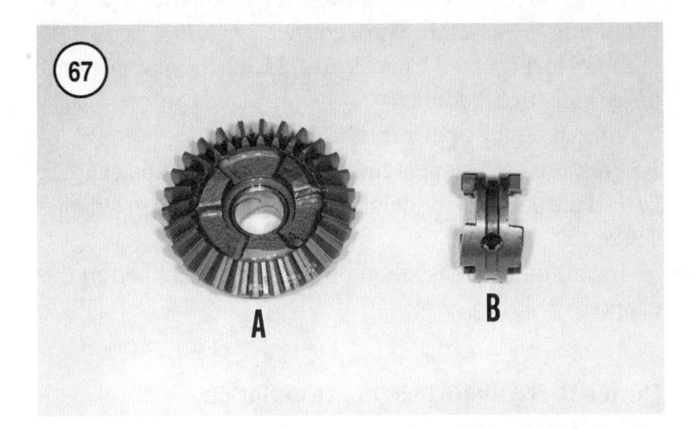

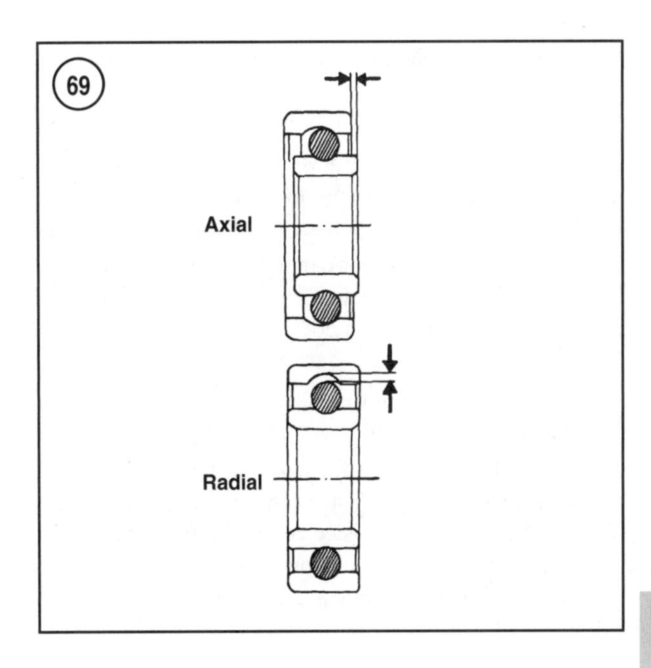

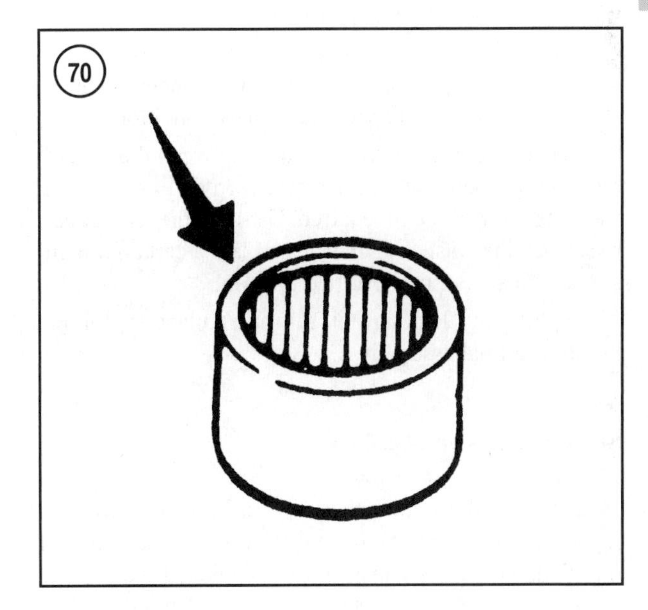

Note the presence of *axial* or *radial* looseness. Replace the bearing if there is rough operation or looseness.

4. Inspect the needle bearings (**Figure 70**) located in the bearing carrier, forward gear, drive shaft bore in the gear housing and drive shaft seal and bearing housing. Replace the bearing if there are flattened rollers, discoloration, rusting, roughness or pitting is noted.

5. Inspect the propeller shaft and drive shaft at the bearing contact area. Replace the drive shaft or propeller shaft along with the needle bearing if there is discoloration, pitting, transferred bearing material or roughness.

8

Shift Cam and Related Components Inspection

1. Inspect the bore in the propeller shaft for debris, damage or excessive wear. Clean debris from the bore.
2. Inspect the clutch spring for damage, corrosion or weak spring tension and replace if there are any defects.
3. Inspect the cross pin for damage, roughness or excessive wear. Replace as required. Inspect the detent balls and spring for damage or corrosion and replace as required.
4. Inspect the shifter for cracks, broken or worn areas. Replace any worn or defective components.
5. Inspect the shift slider for worn or damaged surfaces and replace as required. Inspect the shift shaft for excessive wear and a bent or twisted condition. Inspect the shift shaft seal housing for cracks or a worn shift shaft bore. Replace the housing if there are any defects.

Shims, Spacers, Fasteners and Washers Inspection

1. Inspect all shims for bent, rusted or damaged surfaces. Replace any shim not appearing in new condition.
2. Inspect the thrust washer located between the reverse gear and propeller shaft for worn, corroded or damaged surfaces and replace as required. Use only the correct part to replace the washer. In most cases it is a certain dimension and made with a specified material.
3. Replace any locking nut. Always replace the pinion nut during final assembly.

Seals, O-rings and Gaskets

Replace seals anytime they are removed and if replacing a shaft that contacts the seal. Never use a damaged or questionable seal. A damaged seal will likely leak and lead to extensive damage to internal gearcase components.

Replace O-rings anytime they are removed. Never use a damaged or questionable O-ring.

Never reuse gaskets. A gasket is made of a material that forms into small imperfections in a mating surface. With a few exceptions, gaskets are simply not designed to maintain good sealing after the initial installation.

Grommets

Replace grommets that are split, deformed or hard and brittle. Check for improper installation if the removed grommet is deformed.

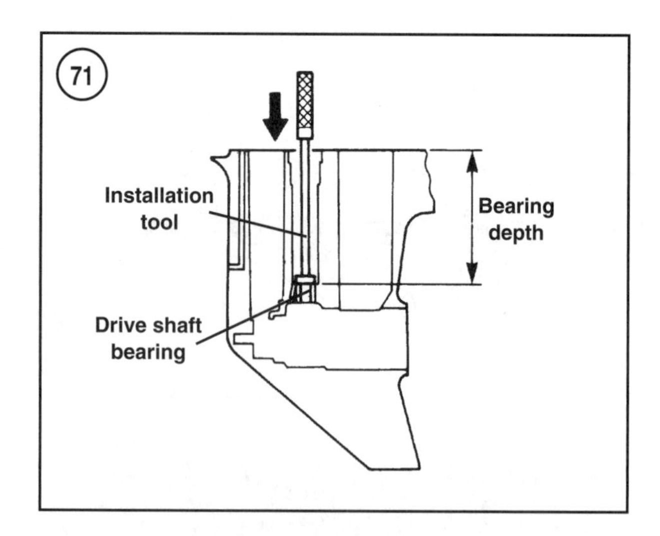

GEARCASE ASSEMBLY (STANDARD RH ROTATION GEARCASE)

Perform each procedure in this section for complete gearcase assembly. If only partial disassembly was required to access the faulty components, refer to the section that describes the faulty component(s) and perform the remaining procedures. If the gearcase will be assembled with new or different gears, bearings, drive shaft or housing, refer to *Shim Selection* for instructions prior to beginning the assembly procedure.

Refer to **Figure 32** or **Figure 33** during the gearcase disassembly procedures.

After assembly, perform the following:

1. Pressure test the gearcase as described in this chapter.
2. Fill the gearcase with lubricant as described in Chapter Three.
3. Install the gearcase then propeller as described in this chapter.

Lower Drive Shaft Bearing Installation (115-130 hp Models)

CAUTION

The lower drive shaft needle bearing must be installed to the correct depth in the drive shaft bore. Otherwise, the bearing may contact the drive shaft on a non-bearing surface during operation. The resulting bearing failure damages the drive shaft and gearcase housing.

1. If removed, install a new lower drive shaft needle bearing as follows:
   a. Select a section of tubing or socket and extension to use as a bearing installation tool. The tool must be

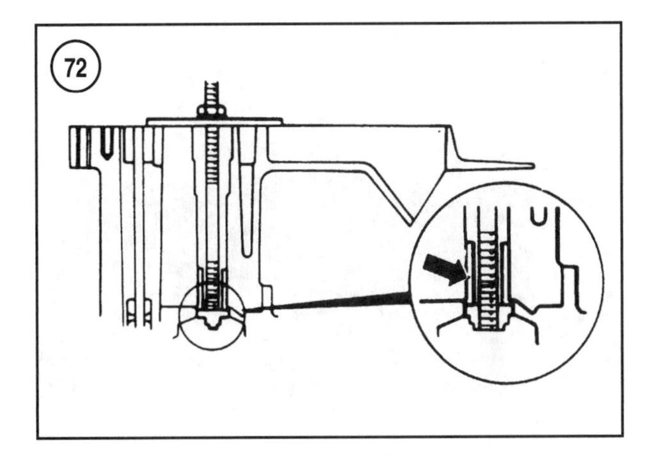

of sufficient diameter to provide adequate contact with the bearing case, yet not contact the drive shaft bore during installation.

b. Working through the drive shaft bore opening, insert the needle bearing into the drive shaft bore. The numbered side of the bearing case must face upward or toward the water pump when installed.

c. Insert the installation tool into the drive shaft bore (**Figure 71**) and seat it against the numbered side of the bearing.

d. Stop frequently for measurement while slowly tapping the bearing into the bore. Stop when the bearing reaches the depth (**Figure 71**) measured prior to removal.

Lower Drive Shaft Bearing Installation (150-250 hp Models)

Use the recommended special tools to install the lower drive shaft bearing.

150-200 hp (2.6 liter) models—Use Yamaha part No. YB-06246/90890-06636, YB-06029/90890-06523 and YB-06247.

200-250 hp (3.1 liter) models—Use Yamaha part No. YB-06432/90890-06655, YB06029/90890-06523 and YB-06213.

1. Apply gearcase lubricant to the outer surfaces of the needle bearing and the lower drive shaft bore.
2. Position the lower drive shaft bearing and installation tools into the drive shaft bore as shown in **Figure 72**. Fit the needle bearing over the installation tool with the numbered side of the case facing upward.
3. Install the threaded rod and nut through the plate and into the drive shaft bore. Thread the rod into the installation tool.
4. Tighten the nut until the bearing installation tool bottoms on the housing and the nut becomes snug. Remove the installation tools.

Forward Gear Bearing Race Installation

NOTE

When driving a bearing race into the gearcase housing, use light taps and listen when the driver is struck. Strike the driver enough to slightly move the bearing or race. Stop driving when the pitch changes or a sharp ring is heard when striking the driver. This indicates the bearing or race is seated. Continued driving may bounce the bearing or race out of the bore.

Use the manufacturer recommended tool for this procedure.

115-130 hp models—Use Yamaha part No. YB-06199/90890-06620 and YB-6071-90890-06605.

150-200 hp (2.6 liter) models—Use Yamaha part No. YB-6258/90890-06619 and YB-6071/90890-06605.

200-250 hp (3.1 liter) models—Use Yamaha part No. YB-06432/90890-06658 and YB-6071-90890-06605.

If this tool is not available, use a section of tubing or suitable mandrel and driver. The tool must be of sufficient diameter to contact near the outer diameter of the bearing race, yet not contact the gear housing during installation.

1. Apply a light coating of gearcase lubricant onto the race surfaces and the corresponding bore in the gearcase housing.
2. Place the shim(s) into the housing. The shim(s) must rest in the bearing race bore.
3. Install the bearing race into the bore with the tapered opening facing outward.
4. Maintain firm contact with the bearing race while gently driving it into the bore (**Figure 73**). Use light force and do not allow the race to tilt in the bore. Continue driving until the race fully seats in the housing.

Forward Gear Bearing(s) Installation

1. If removed, install a new tapered roller bearing onto the forward gear as follows:

   a. Place the forward gear on the table of a press with the gear teeth side facing downward.

   b. Select a section of tubing or large diameter socket to use as a bearing installation tool. The tool opening must contact the inner race of the bearing.

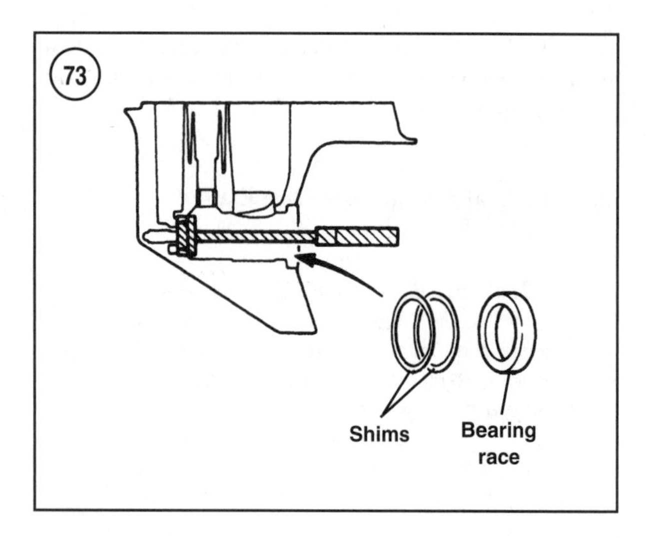

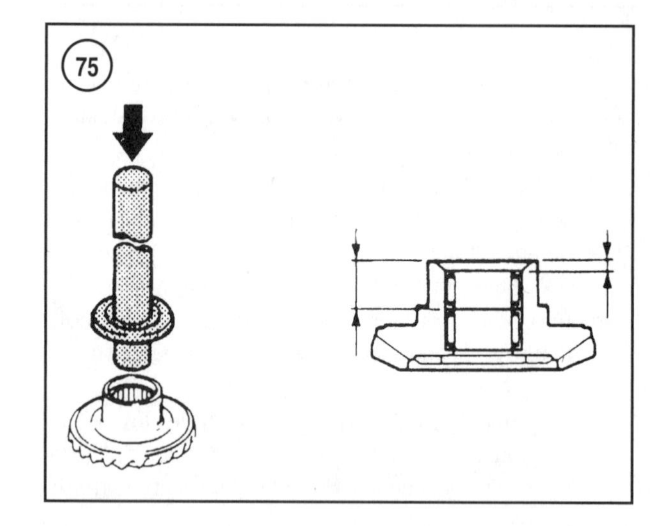

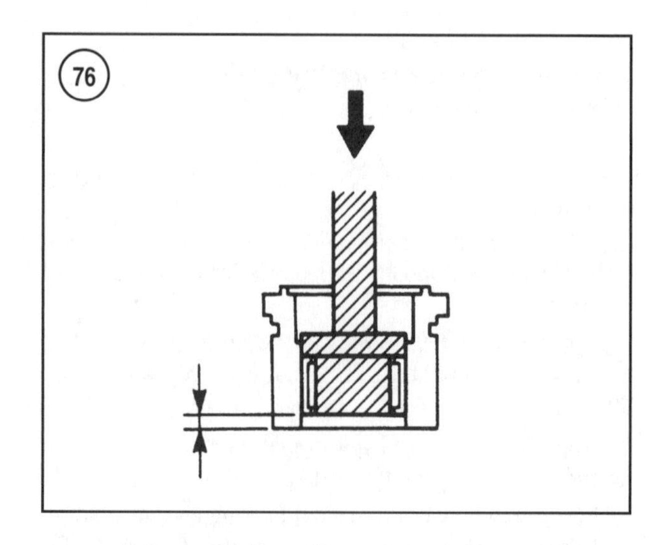

c. Fit the bearing onto the hub of the gear with the tapered side facing upward. Place the installation tool onto the bearing (**Figure 74**).

d. Press the bearing onto the gear hub until fully seated.

2. If removed, install the new needle bearing(s) into the forward gear as follows:

a. Select an appropriately sized socket and extension or section or pipe, tubing or rod to install the bearing. The tool must contact the bearing case; yet not contact the bearing bore in the forward gear.

b. Apply gearcase lubricant to the bearing surfaces and the bearing bore in the forward gear.

c. *Single needle bearing units*—Fit the bearing into the bore with the numbered side facing outward. Carefully press the bearing into the bore until it just reaches the exact depth (**Figure 75**) measured prior to removal.

d. *Two needle bearing units*—Press the first bearing into the bore until the surface is just below the bore opening. Fit the second bearing into the bore. Press on the second bearing until it just reaches the exact depth (**Figure 75**) measured prior to removal.

Drive Shaft Bearing and Seal Housing Assembly

NOTE

The upper drive shaft bearing must be installed to the proper depth within the bearing and seal housing to ensure that the bearing contacts the drive shaft at the proper location. The housing does not provide a step or shoulder for the bearing. Slowly tap the bearing into the housing. Stop and check the depth frequently during installation.

1. Select an appropriately sized socket or section of pipe or tubing to use as a bearing installation tool. The tool

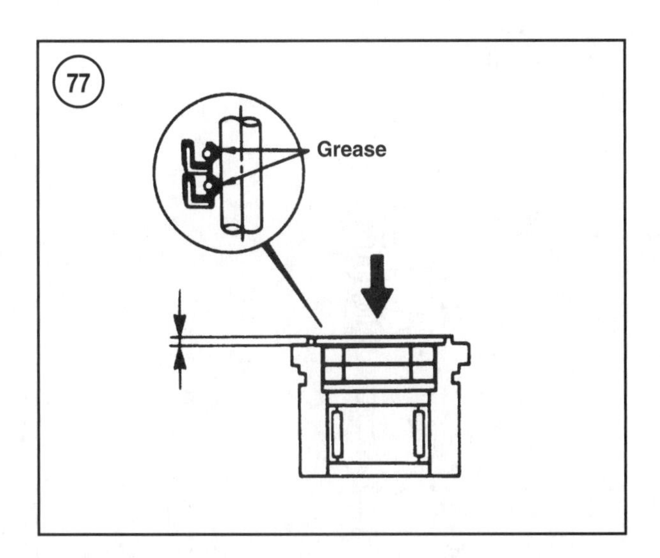

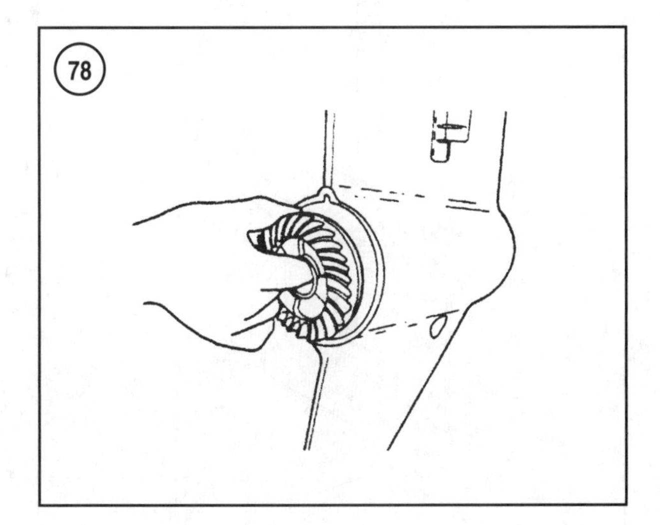

must contact the bearing case, yet not contact the bearing and seal housing during installation.

2. Apply gearcase lubricant onto the bearing surface and the bearing bore in the housing.

3. Position the bearing at the opening to the housing with the numbered or lettered side facing *up* or toward the oil seal bore.

4. Using the installation tool, slowly tap the bearing into the bore. Stop when the bearing reaches the exact depth (**Figure 76**) measured prior to removal.

5. Select an appropriately sized socket or section of pipe or tubing to use as a seal installation tool. The tool must contact the seal case yet not contact the bearing and seal housing during installation.

6. Clean the seal bore in the housing and the outer diameter of the seals with Loctite Primer T to remove all contaminants.

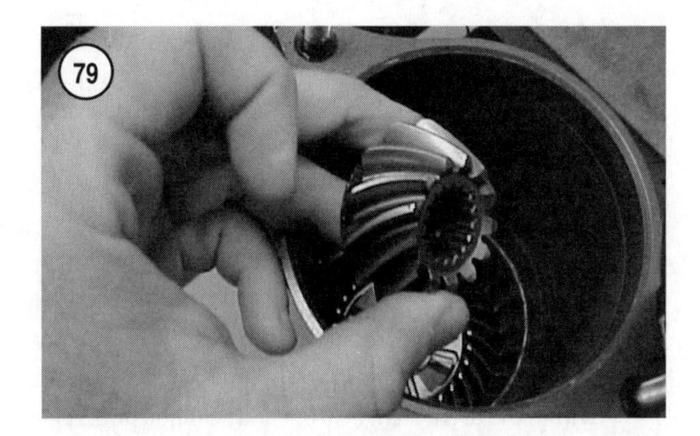

7. Apply a light coating or Loctite 271 to the seal bore and the outer diameter of the seal cases.

8. Place the first seal into the bore with the lip side (open side) facing upward. Use the installation tool to drive the seal into the bore until the seal case is just below the bore opening.

9. Place the second seal into the bore with the lip side (open side) facing upward. Use the installation tool to drive both seals into the bore until the second seal case just reaches the exact depth (**Figure 77**) measured prior to removal.

10. Wipe excess Loctite from the seals. Then apply Yamaha All-Purpose Grease to the seal lips.

Pinion Gear, Forward Gear and Drive Shaft Installation

1. Align the protrusion(s) or tab on the sleeve with the notches in the drive shaft bore, then insert the sleeve into the drive shaft bore (**Figure 55**). Seat the sleeve in the bore.

2. Install the forward gear and bearing into the housing (**Figure 78**). Seat the gear and bearing against the bearing race.

3. Install the pinion gear into the housing (**Figure 79**). Mesh the pinion gear teeth with the driven gear teeth. Align the splined opening in the pinion gear with the drive shaft bore.

4. Apply a coating of gearcase lubricant onto the lower surfaces of the drive shaft. Carefully guide the drive shaft into the bore. Make sure the drive shaft passes through the lower bearing. Slowly rotate the drive shaft to align the drive shaft splines with the pinion gear splines. The shaft drops into the pinion gear when the splines align.

5. Install the drive shaft bearing and seal housing as follows:

   a. Apply Yamaha All-Purpose Grease to the surfaces, then install the new O-ring onto the bearing and seal housing. Fit the O-ring onto the surface just below the mounting flange.

b. Apply gearcase lubricant to the thrust bearing, thrust bearing race and the flange on the drive shaft.

c. Slide the thrust bearing and race (4, **Figure 80**) over the drive shaft. Seat the bearing and race against the flange on the drive shaft (5).

d. Slide the pinion gear locating shim (3, **Figure 80**) over the drive shaft. Seat the shim against the thrust bearing race.

e. Apply gearcase lubricant to the needle bearing within the drive shaft bearing and seal housing. Apply Yamaha All-Purpose grease to the lips of the grease seals.

f. Carefully guide the drive shaft through the bearing and seals while installing the housing. Rotate the housing to align the bolt openings, then seat the housing onto the gearcase.

g. Apply Loctite 572 to the threads, then install the four bolts (1, **Figure 80**). Evenly tighten the bolts in a crossing pattern to the general torque specifications in Chapter One.

6. Install the pinion nut as follows:

a. Apply Loctite 271 onto the threads of the new pinion nut. Hand-thread the nut onto the drive shaft.

b. Install the splined drive shaft adapter (Yamaha part No. YB-06201/90890-06520) onto the upper end of the drive shaft.

c. Use a suitable socket wrench to properly engage the pinion nut (**Figure 81**). Use shop towels to protect the housing surface from the socket wrench.

d. Rotate the drive shaft clockwise, as viewed from the top while tightening the pinion nut to the specification in **Table 1**.

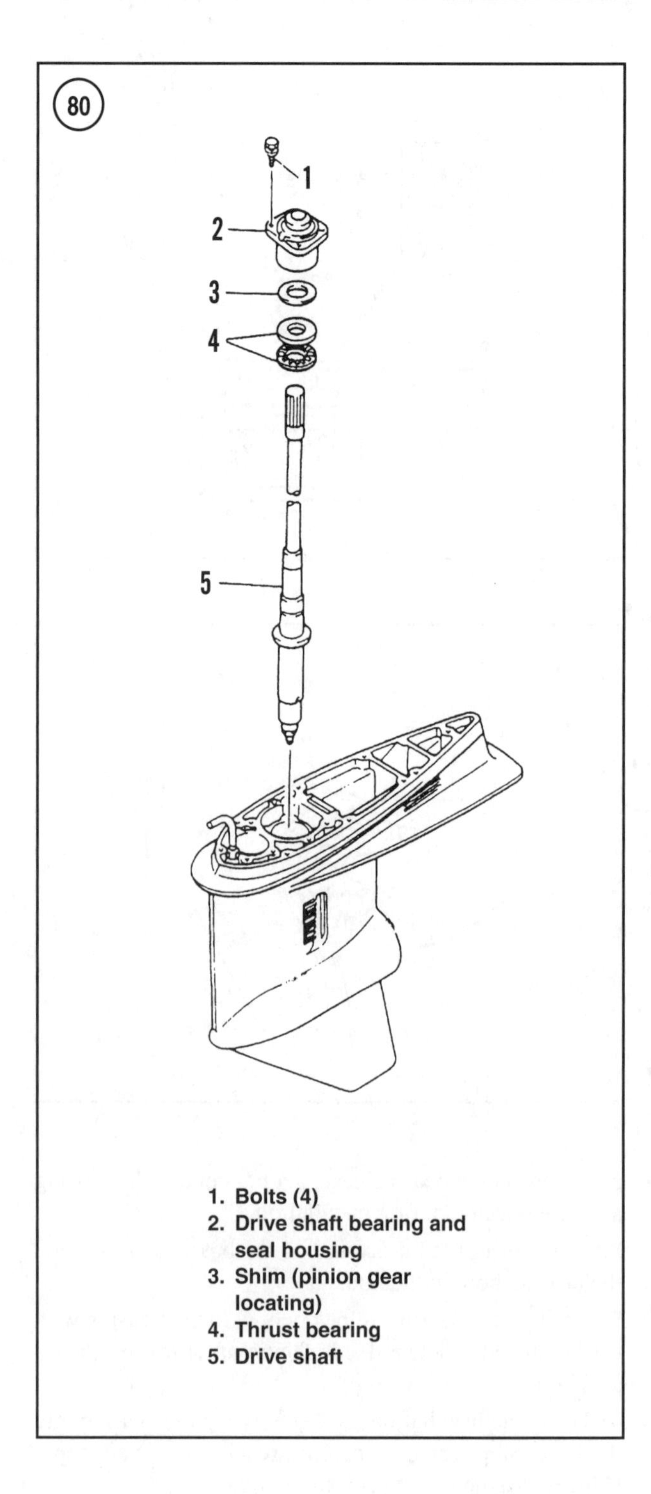

1. Bolts (4)
2. Drive shaft bearing and seal housing
3. Shim (pinion gear locating)
4. Thrust bearing
5. Drive shaft

Propeller Shaft and Shifter Assembly

Refer to **Figure 59** or **Figure 60** for this procedure.

1A. *115-130 hp and 150-200 hp (2.6 liter) models*—Install the shift slider and shifter into the propeller shaft as follows:

a. Install the larger diameter ball bearing (5, **Figure 59**), spring (6) then second ball bearing into the propeller shaft bore.

b. Apply grease to the surfaces to help retain the ball bearings. Then fit the four detent balls (7, **Figure 59**) into the respective recesses in the shift slider (8).

c. Fit the slot in the shifter (9, **Figure 59**) over the protrusion on the end of the slider.

d. Carefully insert the slider into the propeller shaft bore. Use a screwdriver or other suitable object to press the detent balls into the recesses during installation.

1B. *200-250 hp (3.1 liter) models*—Install the shift slider into the propeller shaft as follows:

a. Slide the spring plunger components (4, **Figure 60**) into the aft end of the shift slider (3). The components must be arranged as shown in **Figure 60**.

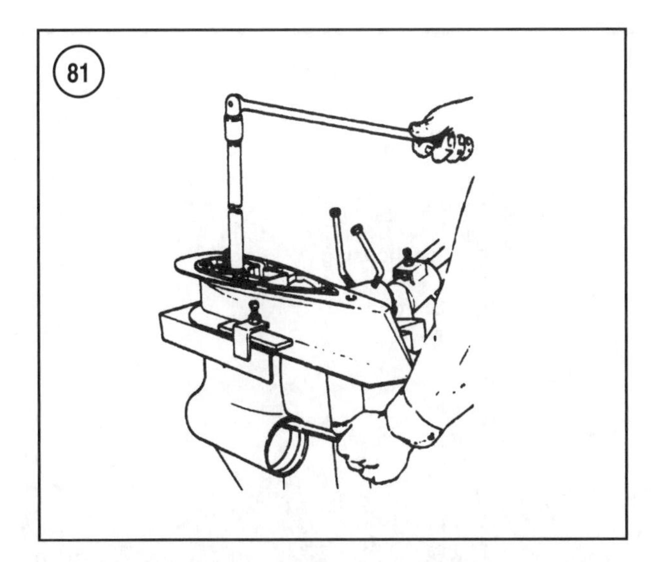

b. Apply grease to the surfaces to help retain the ball bearings. Then fit the detent balls (2, **Figure 60**) into the respective recesses in the shift slider.

c. Fit the slot in the shifter (1, **Figure 60**) over the protrusion on the end of the slider.

d. Carefully insert the slider into the propeller shaft bore. Use a screwdriver or other suitable object to

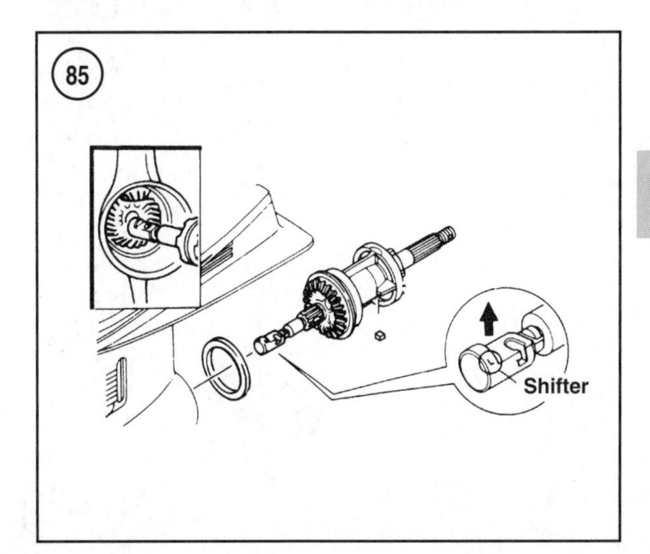

press the detent balls into the recesses during installation.

2. Rotate the shift slider to align the cross pin opening in the slider with the slot in the propeller shaft (**Figure 82**).

3. Align the cross pin opening in the clutch with the slot in the propeller shaft (**Figure 83**), then slide the clutch over the forward side of the propeller shaft. The *F* marking on the clutch face, if so equipped, must face forward.

4. Insert the cross pin through the clutch and slider opening. Carefully wind the clutch spring onto the clutch (**Figure 84**). The spring must fit tightly on the clutch and loops must span both ends of the cross pin. Each loop must lay flat against the clutch.

Propeller Shaft Installation

1. Guide the propeller shaft through the forward gear needle bearing with the flat surface of the shifter facing upward (**Figure 85**). *Do not* rotate the propeller shaft until the shift shaft engages the shifter.

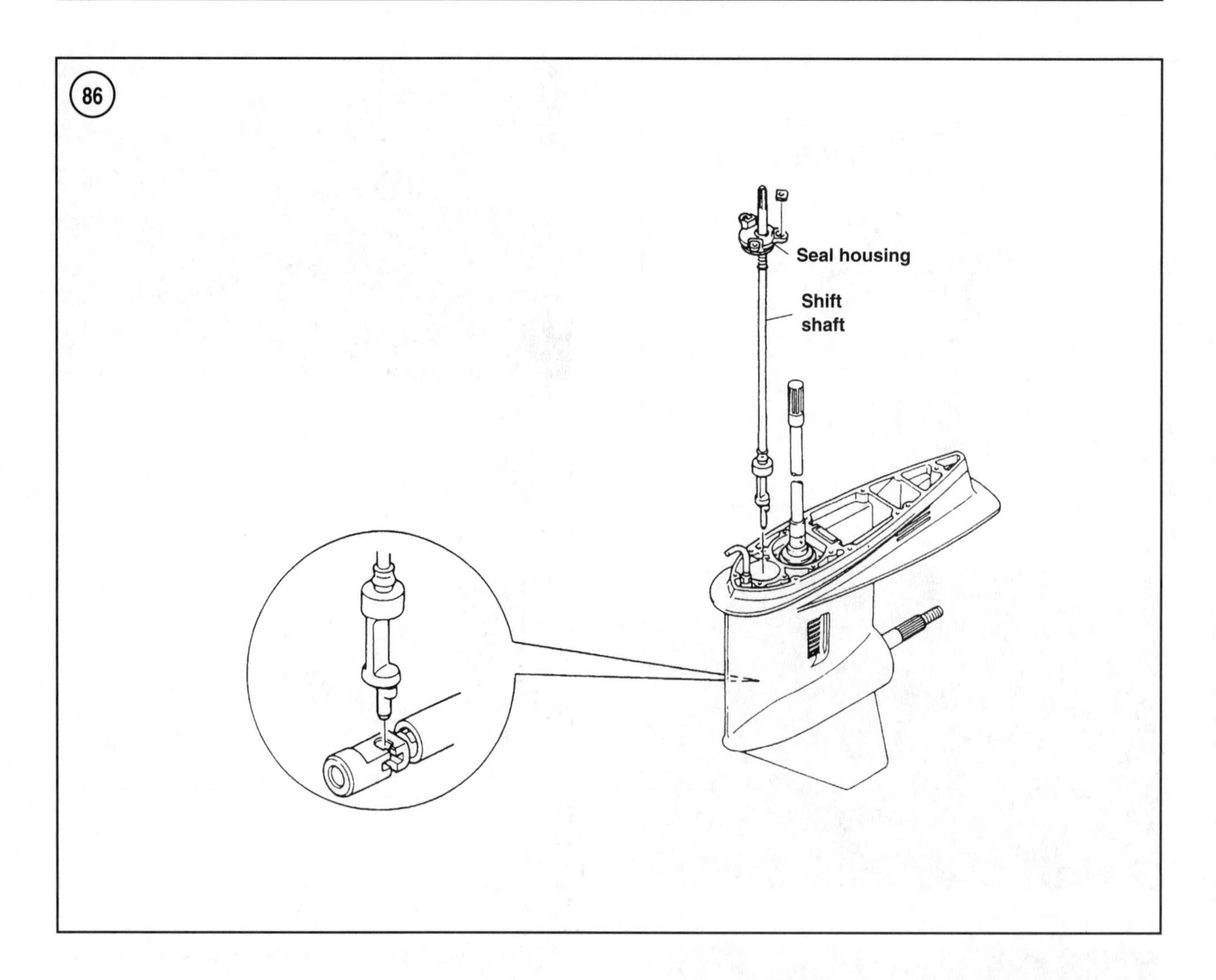

2. If removed, install a new seal into the shift shaft seal housing as follows:

a. Use Loctite primer T to clean contaminants from the seal bore in the housing and the outer diameter of the new seal.
b. Apply a coating of Loctite 271 into the seal bore and to the outer diameter of the seal casing.
c. Fit the seal into the bore with the lip (open) side facing downward.
d. Use an appropriately sized socket to press the seal into the housing until seated in the bore.
e. Apply a coating of Yamaha All-Purpose Grease to the seal lip.

3. Slide the spring over the shift shaft, then insert the shift shaft through the shift shaft seal and housing.

4. Install a new O-ring onto the shift shaft seal housing. Seat the O-ring onto the surface just below the mounting flange.

5. Guide the shift shaft through the shift shaft bore. Carefully rotate the shift shaft until the tip at the lower end of the shift shaft enters the opening in the shifter (**Figure 86**).

6. Seat the seal housing onto the gearcase housing. Rotate the seal housing to align the bolt openings. Apply Loctite 572 to the bolts, then thread the bolt into the seal housing and gearcase. Evenly tighten the bolts to the general torque specifications in Chapter One.

Bearing Carrier and Reverse Gear Assembly

1. If removed, install a new needle bearing into the bearing carrier as follows:

a. Apply a light coating of gearcase lubricant to the outer surfaces of the bushing or bearing bore in the bearing carrier.
b. Place the bearing carrier onto a sturdy work surface with the propeller side facing upward.

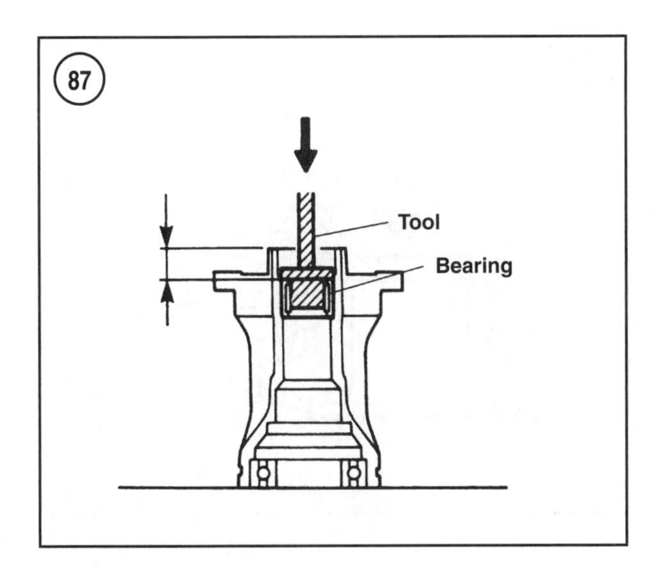

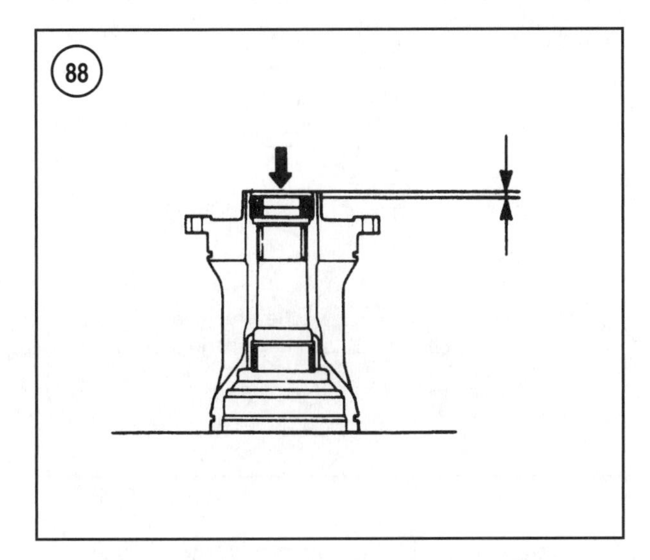

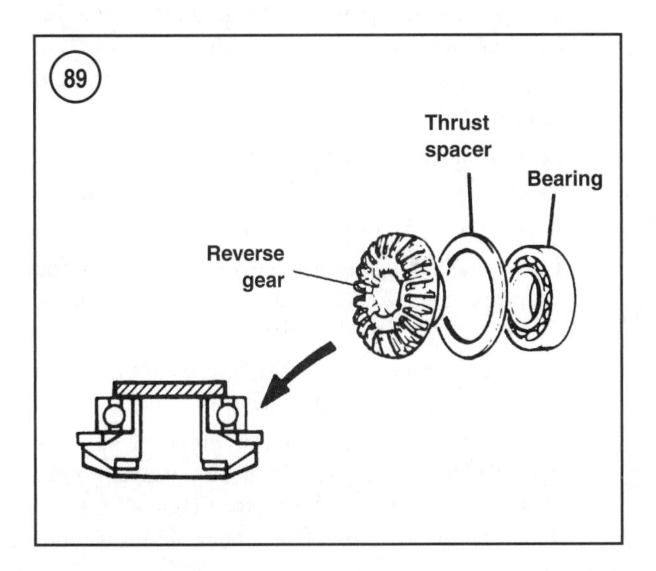

c. Insert the needle bearing into the bore with the numbered side facing upward.

d. Select a section of tubing, socket and extension or other tool to use as an installation tool. The tool must be 0.25-0.50 mm (0.010-0.020 in.) smaller in diameter than the bearing case.

e. Stop frequently for measurement and slowly tap the bearing into the bore. Stop when the bearing reaches the exact depth as measured from the bearing carrier surface to the bearing case (**Figure 87**) recorded prior to removal. The depth should be within the specified range:

115-130 hp models—24.75-25.25 mm (0.974-0.994 in.);

150-200 hp (2.6 liter) models—24.75-25.25 mm (0.974-0.994 in.);

200-250 hp (3.1 liter) models—25.05-25.55 mm (0.986-1.006 in.).

2. If removed, install new propeller shaft seals into the bearing carrier as follows:

a. Apply a coating of Loctite 271 to the outer diameter of the seal casings and the seal bore in the bearing carrier.

b. Select a section of tubing, socket and extension or other suitable tool to use as a seal installation tool. The tool must contact the seal casing, yet not contact the seal bore in the bearing carrier.

c. Place the bearing carrier on a sturdy work surface with the propeller side facing upward.

d. Place the first seal into the bore with the lip side facing upward. Carefully drive the seal into the bore until the seal case is just below the bore opening.

e. Place the second seal into the bore with the lip side facing upward. Carefully drive the seal into the bore until seated against the first seal.

f. Stopping frequently for measurement, slowly tap both seals into the bore until the outer seal reaches the exact depth in the bore opening (**Figure 88**) recorded prior to removal. The seal depth should 4.75-5.25-mm (0.187-0.207 in.) for all models covered in this manual.

g. Wipe excess Loctite from the seal bore. Apply a coating of Yamaha All-Purpose Grease to the seal lips.

3. If removed, install the reverse gear and bearing into the bearing carrier as follows:

a. Place the reverse gear onto the table of a press with the gear teeth facing downward. Apply a light coating of gearcase lubricant to the hub of the reverse gear.

b. *115-130 hp and 150-200 hp (2.6 liter) models*—Fit the thrust spacer (**Figure 89**) over the hub of the re-

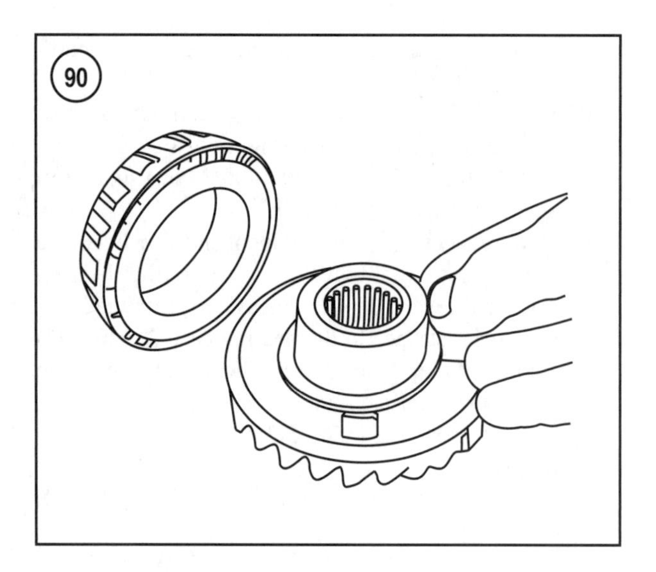

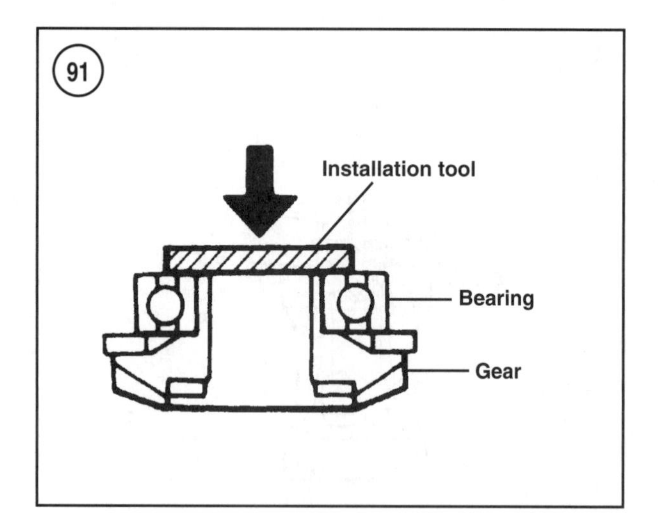

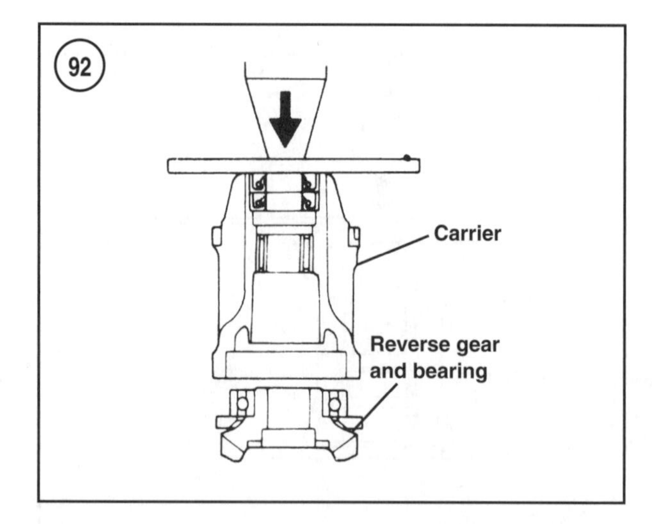

verse gear. The beveled side must face toward the reverse gear.

c. *200-250 hp (3.1 liter) models*—Install the reverse gear shim onto the hub of the reverse gear (**Figure 90**).
d. Select a flat piece of steel or other suitable driver to use as a bearing installation tool. The tool must contact only the inner race of the bearing (**Figure 91**).
e. Fit the bearing onto the hub of the gear with the numbered side facing away from the gear. Press the bearing onto the gear until fully seated.
f. Apply a light coating of gearcase lubricant into the bearing bore in the bearing carrier. Position the bearing carrier onto the bearing as shown in **Figure 92**.
g. Place a block onto the propeller side of the bearing carrier for protection, then press the carrier onto the reverse gear bearing until the bearing fully seats in the bore. Do not allow the carrier to tilt on the bearing.

4. Apply a coating of Yamaha All Purpose Grease to the surfaces, then fit the new O-ring(s) into the groove(s) in the bearing carrier.

Bearing Carrier Installation

1. Slide the thrust washer over the propeller shaft and seat it against the step on the propeller shaft.
2. *115-130 and 150-200 hp (2.6 liter) models*—Insert the reverse gear shim into the gearcase opening and seat it against the step in the housing.
3. Apply a coating of Yamaha All-Purpose Grease to the O-rings, propeller shaft seals and all surfaces of the carrier that contact the gearcase housing.
4. Carefully slide the bearing carrier and reverse gear over the propeller shaft. Rotate the drive shaft while guiding the carrier into the gearcase opening.

5A. *115-130 hp and 150-200 hp (2.6 liter) models*—Secure the bearing carrier as follows:

a. Seat the carrier in the gearcase.
b. Rotate the carrier to align the slots for the locating key (**Figure 93**). Insert the key into the slot.
c. Install the locking tab washer (**Figure 94**) onto the bearing carrier. The raised areas on the carrier must fit into the slots in the washer.
d. Apply grease onto the threads, then hand-thread the cover nut (**Figure 94**) into the gearcase.
e. Slip the cover nut tool over the propeller nut. Engage the lugs of the cover nut tool onto the cover nut lugs. Use cover nut tool part No. YB-34447/ 90890-06511 for 115-130 hp models. Use cover nut

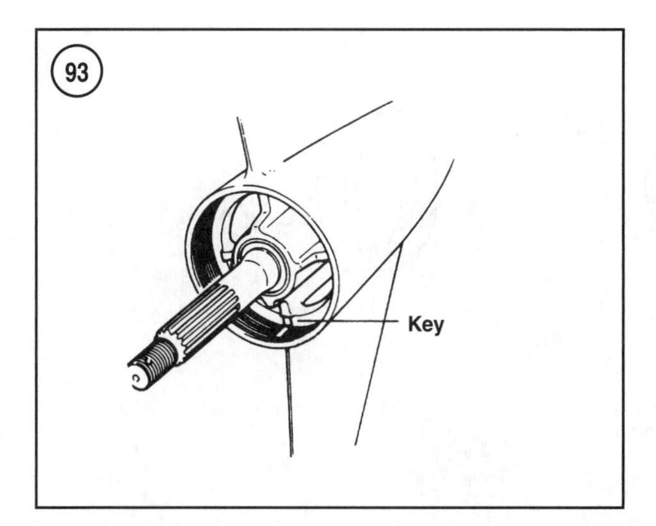

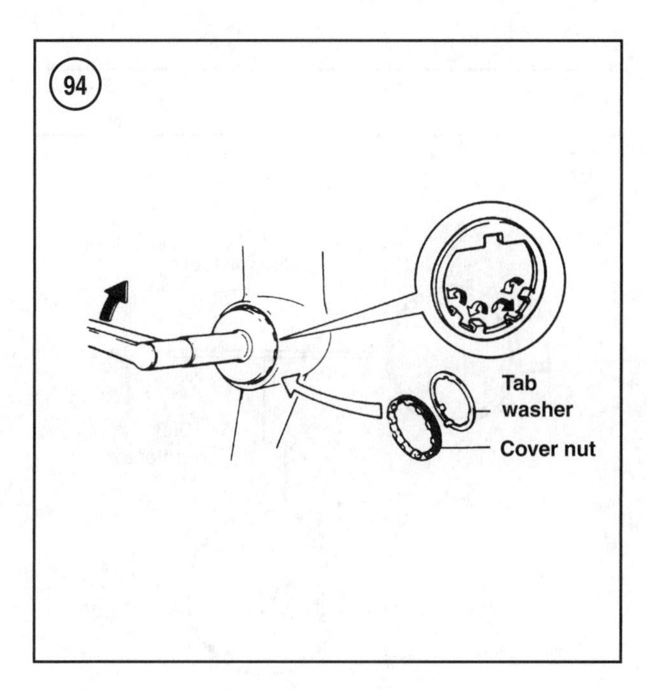

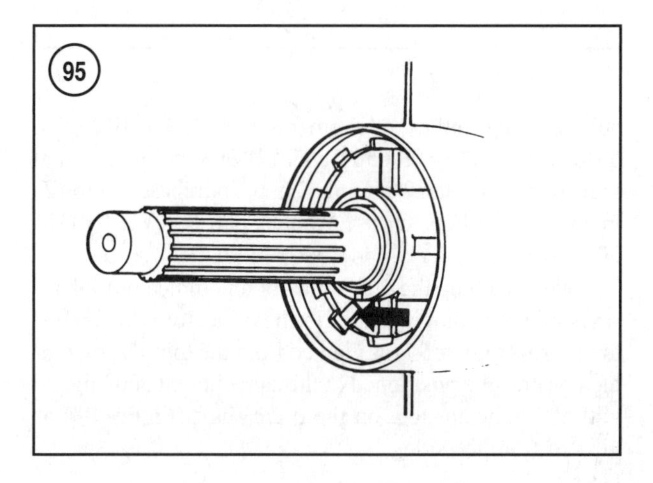

tool part No. YB-34447/90890-06512 for 150-200 hp (2.6 liter) models.

f. Turn the cover nut clockwise while tightening the cover nut to the specification in **Table 1**.

g. Use of slip joint pliers to bend at least one of the locking tabs (**Figure 95**) into a slot between the cover nut lugs. Bend tabs down that do not align with the slots. Tighten the cover nut slightly if necessary to align the tab and slot. Never loosen the cover nut to align the tab. Bend the remaining tabs down.

5B. *200-250 hp (3.1 liter) models*—Secure the bearing carrier as follows:

a. Align the two bolts holes, then seat the carrier in the housing.

b. Apply a coating of Loctite 572 onto the threads, then thread the two bolts (3, **Figure 96**) into the carrier and gearcase housing. Tighten the two bolts to the specification in **Table 1**.

c. Fit the retaining ring (5, **Figure 96**) onto the carrier.

d. Align the screw openings, then seat the ring onto the carrier.

e. Apply Loctite 572 onto the threads, then thread the two screws (6, **Figure 96**) into the ring and bearing carrier. Tighten the screws to the general tightening torque specification in Chapter One.

f. Lubricate the mating surfaces of the carrier to the gearcase by pumping in Yamaha All-Purpose Grease into the grease fitting (4, **Figure 96**) until it flows from the housings.

SHIM SELECTION (ALL MODELS)

Proper pinion gear engagement and forward and reverse gear backlash are necessary for smooth operation and long service life. A shim adjustment is required to achieve the proper tolerances when replacing major internal components such as gears, bearings, drive shaft or the gearcase housing. The marine dealership customarily uses specialized tools and precise measuring equipment to select the proper shims for the bearing, gear, drive shaft and housing combination. If at all possible, have a Yamaha dealership perform the measurements and select shims if replacing any major gearcase components.

If this measurement cannot be performed, there is a less preferable but usually effective method of shim selection. Use the same shim thickness as removed from all shim locations. Assemble the gearcase. Then measure the gear backlash. Gear backlash measurements give a fairly accurate indication of what shim changes are required. Refer to *Measuring Gear Backlash* in this section for procedures.

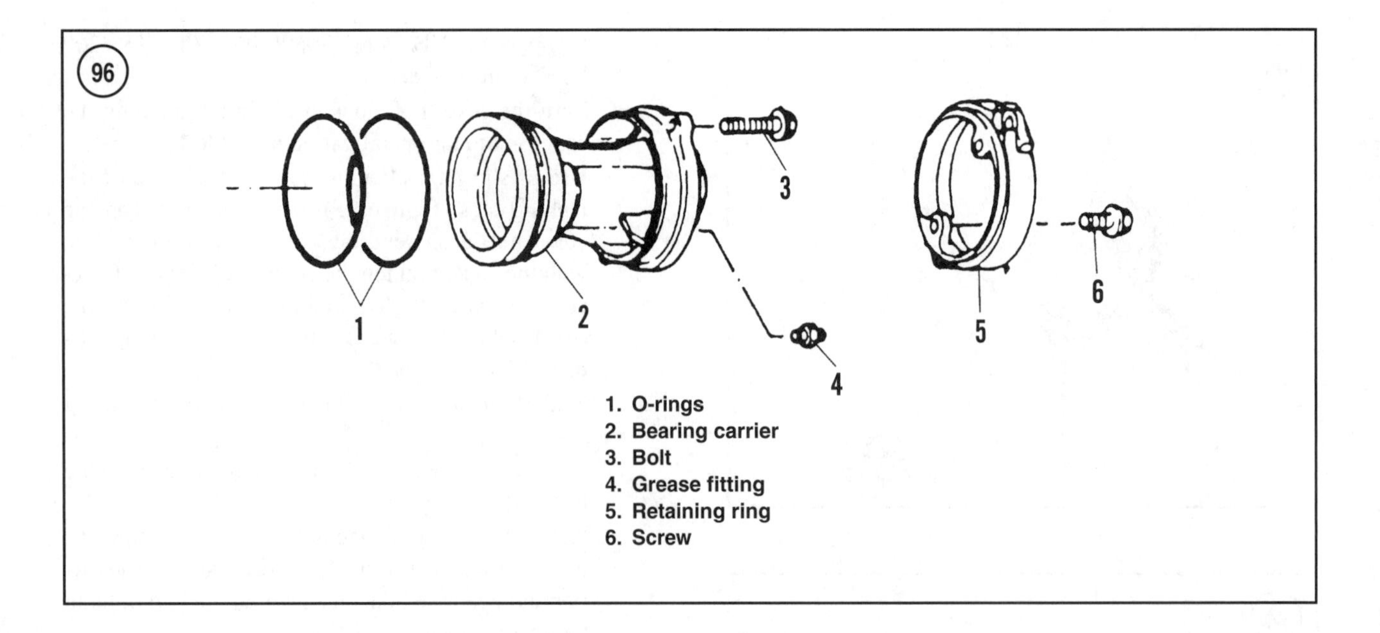

If replacing the gearcase housing, do not rely solely on the gear backlash measurement to determine shim change requirements. Have a Yamaha dealership select shims for the combination of components used if a new or different housing will be used.

Gear Backlash Measurement

CAUTION

Use very light force when moving the drive shaft during backlash measurements. Inaccurate measurement is certain if excessive force is used.

NOTE

Remove all water pump components prior to measuring gear backlash. The drag created by the water pump components will hinder accurate measurements.

NOTE

Make sure all gearcase lubricant is drained from the gearcase prior to measuring gear backlash. Inaccurate readings will result from the cushion effect of the lubricant on the gear teeth.

Gear backlash measurements and measuring procedures vary by model and type of gearcase used.

Use a dial indicator, sturdy mount for the dial indicator, bearing puller assembly and backlash indicator tool to accurately measure backlash.

Use only the recommended backlash indicator tool (Yamaha part No. YB-06265/90890-06706). Use a suitable bearing puller or Yamaha part No. YB-06234, YB06117, YB-90890/6501 and YB-90890-6504. A dial indicator and sturdy mount can be purchased through most tool suppliers. Or use Yamaha part No. YU-03097/90890-01252 and YU-34481/90890-06705.

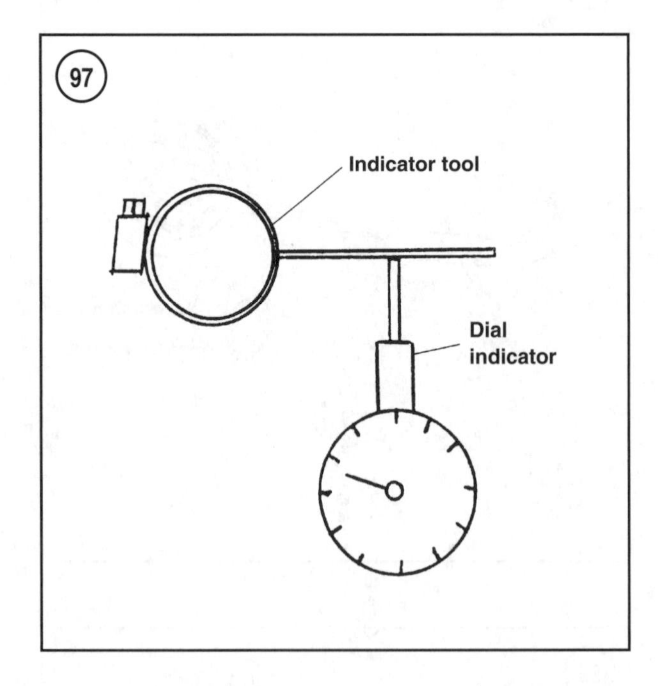

When attaching the dial indicator and indicator tool, always align the dial indicator to the indicator arm at a 90° to the arm (**Figure 97**) as viewed from the top. The dial indicator must be positioned at the same height and aligned with the indicator tool on the drive shaft (**Figure 98**) as viewed from the side.

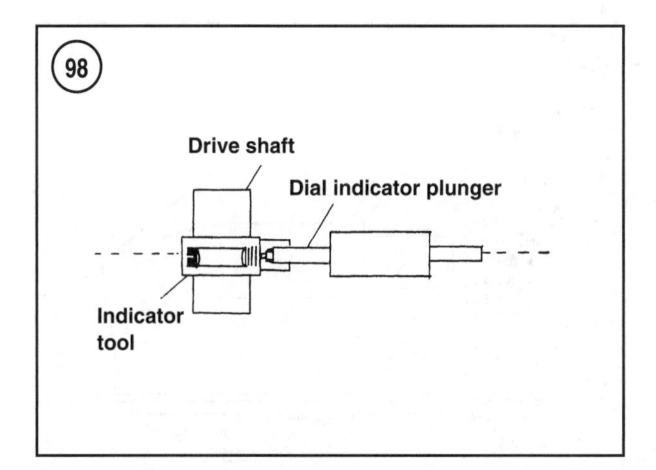

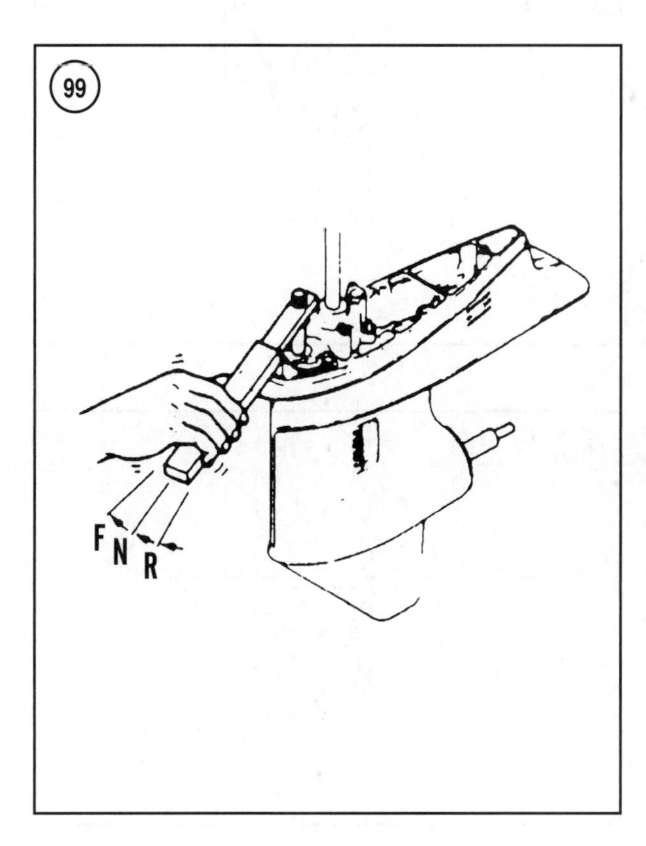

To measure gear backlash, load the propeller shaft in the fore or aft position using hand pressure or a puller to prevent shaft movement as described in the procedures. Position the gearcase with the drive shaft facing upward or downward as described in the procedures. Lightly rotate the drive shaft clockwise and counterclockwise while observing the dial indicator. The free movement of the drive shaft indicates the amount of gear lash or tooth clearance that is present. Read the amount of gear lash on the dial indicator. Gear backlash specifications are listed in **Table 2**.

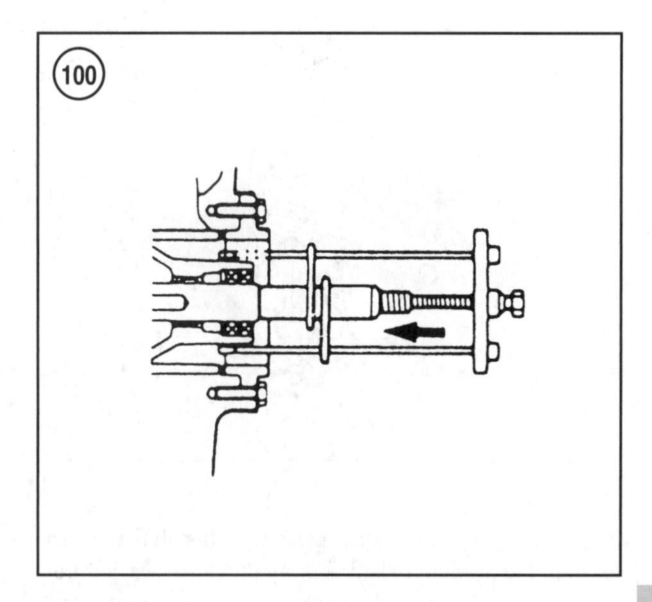

Standard (RH) and counterclockwise (LH) rotation gearcase

1. Rotate the drive shaft clockwise, as viewed from the top, and move the shift shaft (**Figure 99**) until the propeller shaft rotates clockwise. Use the shift shaft wrench (Yamaha part No. YB-06052) to rotate the shift shaft.
2. Install the puller assembly onto the bearing carrier and propeller shaft as indicated in **Figure 100**. Tighten the center bolt to 5.0 N•m (44.0 in. lb.). This will prevent free propeller shaft movement and load the propeller shaft forward.
3. Slide the backlash indicator onto the drive shaft. Secure the dial indicator and indicator mount onto the gearcase. Align the dial indicator stem with the marking on the indicator. Position the gearcase with the drive shaft facing downward.
4. Pull down lightly on the drive shaft, or away from the gearcase. Then gently rotate the drive shaft in the clockwise and counterclockwise directions (**Figure 101**). Note the dial indicator readings when the drive shaft freely moves in each direction.
   a. *Standard rotation (RH) gearcase*—Record the amount of dial indicator movement as *forward gear backlash*.
   b. *Counter rotation (LH) gearcase*—Record the amount of dial indicator movement as *reverse gear backlash*.
5. Loosen the backlash indicator from the drive shaft. Remove the puller assembly from the propeller shaft.
6. Rotate the drive shaft clockwise, as viewed from the top, and move the shift shaft (**Figure 99**) until the propeller shaft rotates counterclockwise.

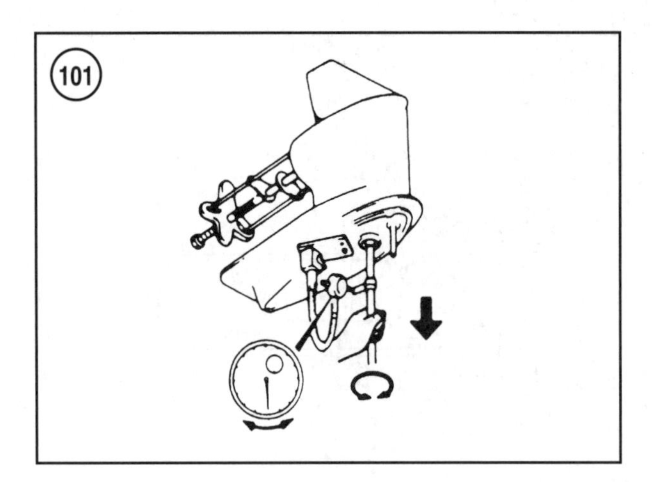

7A. *Standard rotation (RH) gearcase*—Install the propeller onto the propeller shaft *without* the thrust washer normally located forward of the propeller. Tighten the propeller nut to 5.0 N•m (44.0 in. lb.).

7B. *Counter rotation (LH) gearcase*—Preload the rear mounted gear as follows:

a. Temporarily remove the backlash indicator and position the gearcase with the drive shaft facing upward.
b. Rotate the drive shaft clockwise while moving the shift shaft until the propeller shaft rotates counterclockwise.
c. Stop rotating the drive shaft and move the shift shaft to the neutral gear position.
d. Rotate the drive shaft counterclockwise approximately 30°.
e. Using the shift shaft wrench, move the shift shaft toward the direction noted in substep b. This pushes the clutch dogs against the dogs on the gear to preload the rear mounted gear. Maintain constant pressure on the shift shaft wrench during the backlash measurement. This step prevents gear rotation and seats the gear against the support bearings.

8. Pull down lightly on the drive shaft or away from the gearcase. Then gently rotate the drive shaft in the clockwise and counterclockwise directions (**Figure 101**). Note the dial indicator readings when the drive shaft freely moves in each direction.

a. *Standard rotation (RH) gearcase*—Record the amount of dial indicator movement as *reverse gear backlash.*
b. *Counter rotation (LH) gearcase*—Record the amount of dial indicator movement as *forward gear backlash.*

9. Remove the propeller, dial indicator, mount and the backlash indicator.

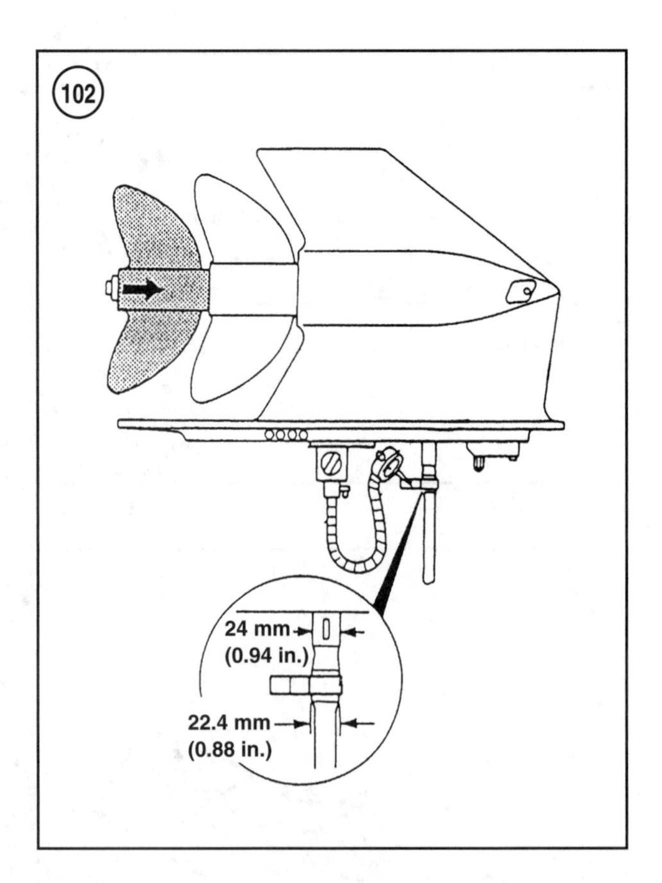

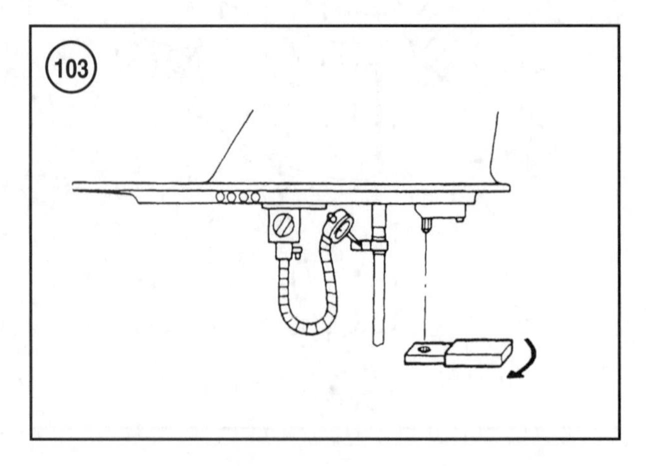

Twin propeller gearcase

1. Install both propellers onto the gearcase as described in this chapter.
2. Position the gearcase with the drive shaft facing downward. Install the backlash indicator onto the large diameter portion of the drive shaft (**Figure 102**).
3. Place the shift shaft wrench (Yamaha part No. YB-06052) on the shift shaft. Rotate the shift shaft tool in the reverse gear direction as indicated in **Figure 103**.

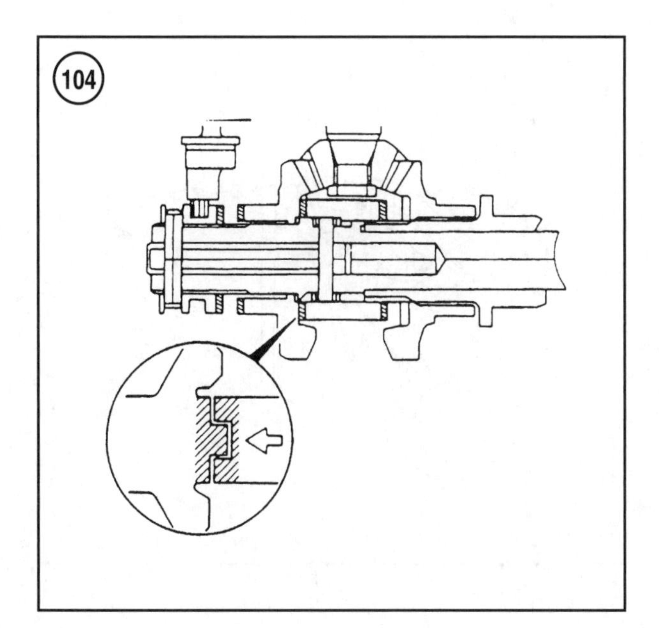

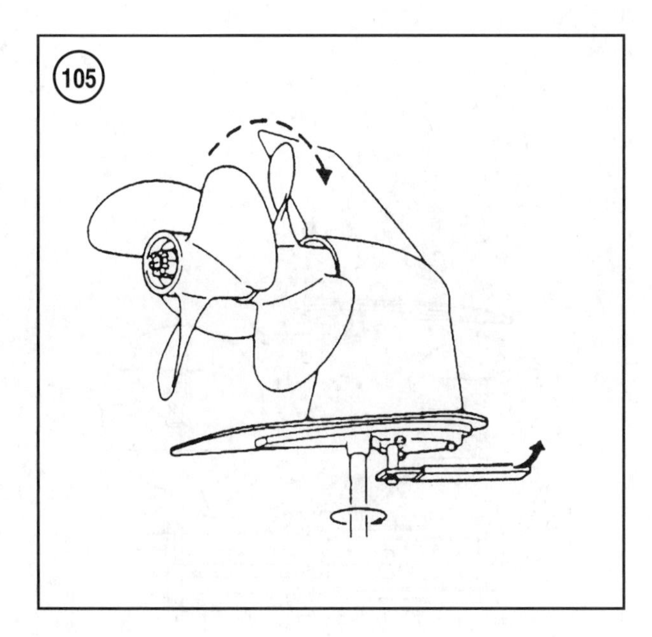

4. Use a block of wood, a suitable scale and an assistant to apply 400 N (90 lb.) of forward pressure to the rear propeller (**Figure 102**). Do not apply any pressure to the front propeller.

5. With the pressure applied to the rear propeller, slowly turn the drive shaft clockwise until the clutch engages the gear (**Figure 104**).

6. Shift the gearcase into neutral gear. Rotate the front propeller approximately 30° clockwise. Move the shift shaft toward the reverse gear direction (**Figure 105**). The movement toward reverse gear should be difficult. If not,

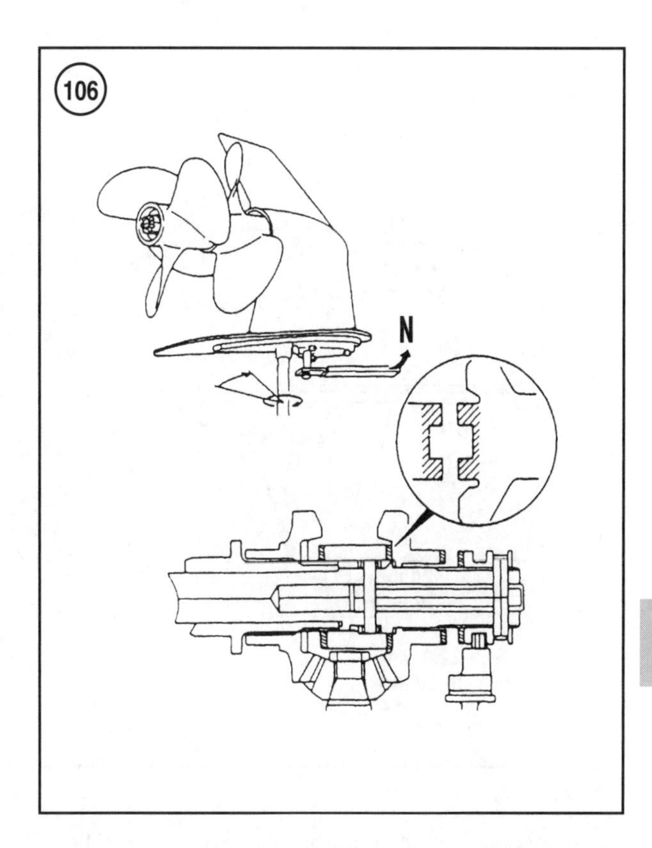

repeat Steps 3-6 until the shift shaft movement is difficult when shifted toward reverse gear.

7. Rotate the drive shaft approximately 20° clockwise (**Figure 106**). Install the backlash indicator onto the drive shaft and align the dial indicator stem with the marking on the indicator. Move the shift shaft toward reverse gear.

8. While maintaining the force applied to the rear propeller and the shift shaft held toward the reverse gear position, carefully rotate the drive shaft clockwise and counterclockwise. Not the dial indictor readings when the drive shaft reaches the free movement limit in each direction. Record the amount of dial indicator movement as *front gear backlash*.

9. Remove the backlash indicator from the drive shaft. Remove the force from the rear propeller and apply 400 N (90 lb.) to the front propeller (**Figure 107**).

10. Move the shift shaft toward the forward gear position (**Figure 108**). Watch for propeller movement and rotate the drive shaft clockwise until the clutch engages and both propellers begin to move (**Figure 109**). Make sure the front propeller rotates in the direction indicated in **Figure 109**.

11. Move the shift shaft to the neutral gear position (**Figure 110**). Slowly rotate the drive shaft clockwise until the front propeller rotates approximately 20° clockwise (**Figure 110**).

8

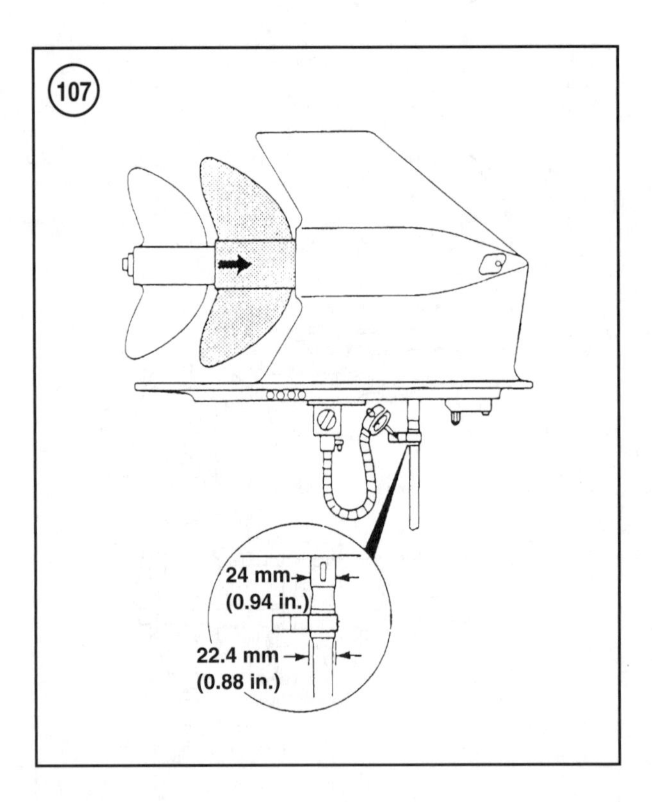

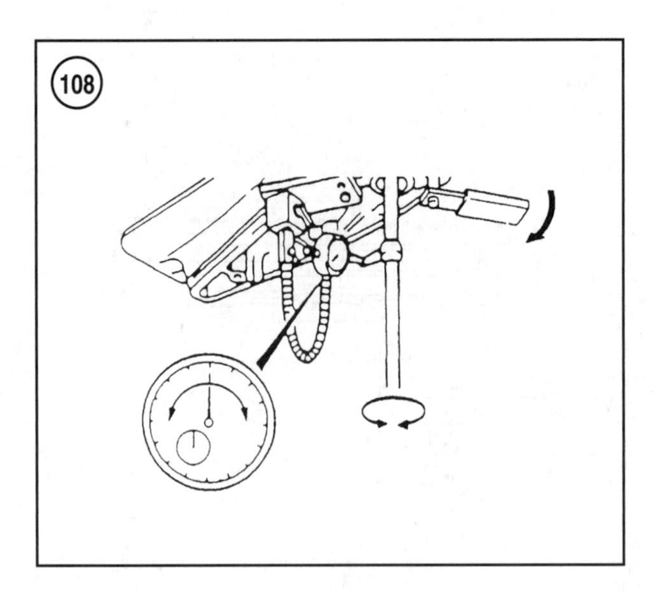

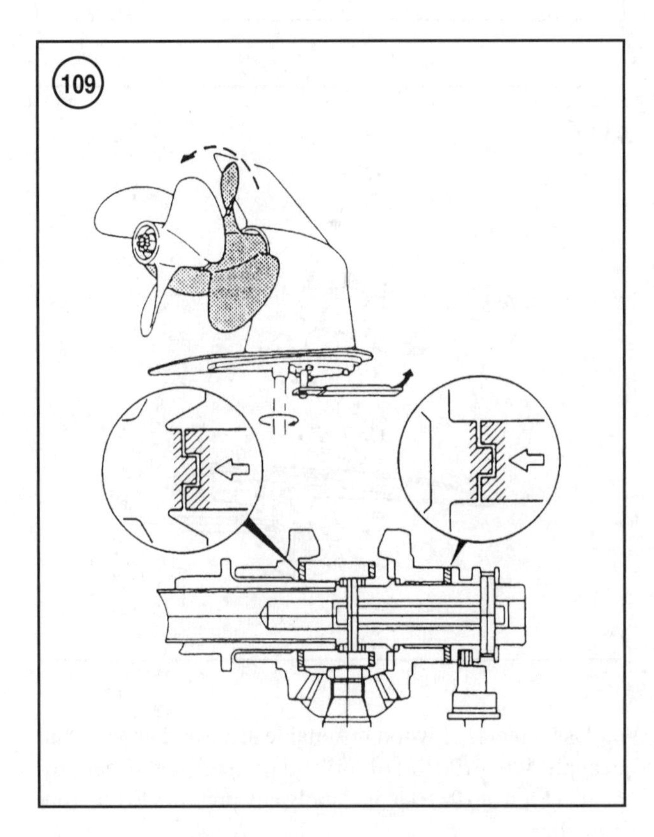

12. Move the shift shaft toward the forward gear direction and maintain pressure on the shift shaft. Align the dial indicator stem with the marking on the backlash indicator. Tighten the dial indicator onto the large diameter section of the drive shaft (**Figure 107**). Gently rotate the drive shaft in the clockwise and counterclockwise directions (**Figure 111**). Note the dial indicator readings when the drive shaft reaches the free movement limit in each direction. Record the amount of needle movement on the dial indicator as *rear gear backlash*.

Gear Backlash Adjustment

Compare the backlash measurements with the specifications in **Table 2**. Shim changes are required when one or both backlash measurements are beyond the specification. Refer to the description of the backlash readings for recommended shim changes.

NOTE
Shim locations in the gearcase vary by model. Always note the location of all shims during disassembly to ensure that all shims are installed in the same location.

Make all shim changes in small increments. Aim for the middle of the backlash specification anytime shim changes are required.

Partial or complete gearcase disassembly is required to access the shims. Refer to the disassembly procedures in this chapter.

After any shim change, assemble the gearcase and repeat the backlash measurements. Make required changes until both measurements are within the specifications. Several shim changes may be required.

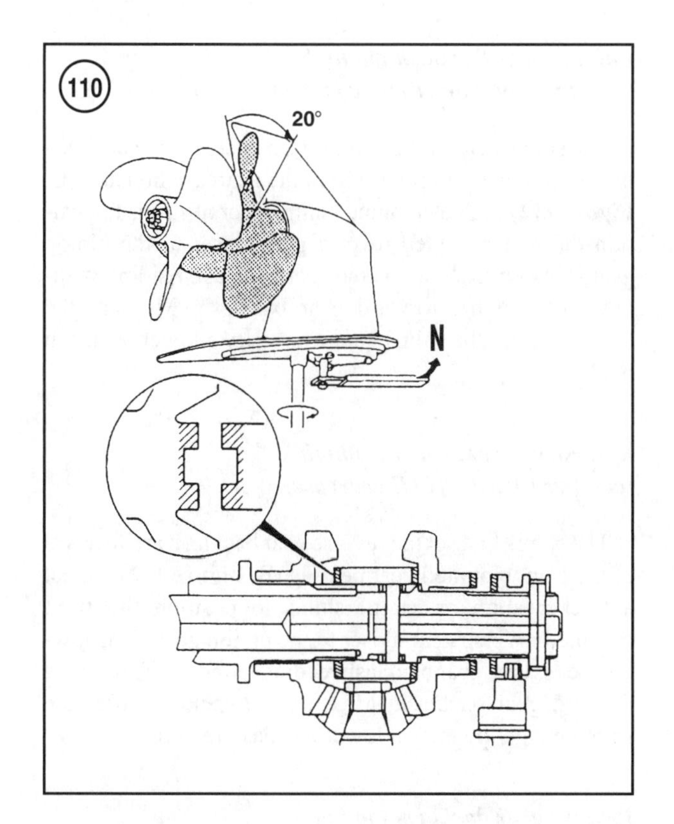

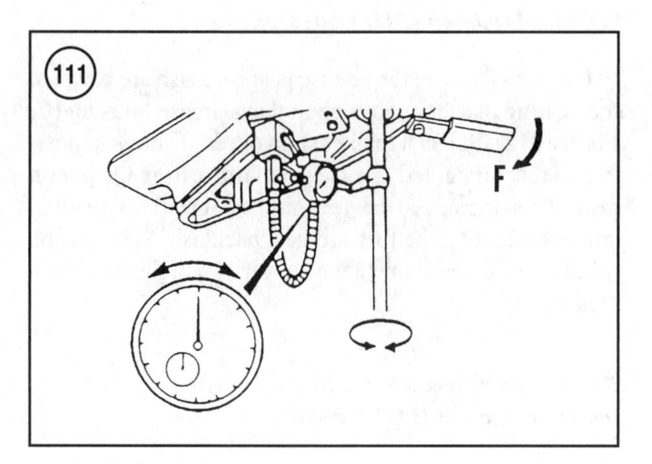

Both measurements too high

The pinion gear is probably positioned too high in the gearcase when both backlash measurements are above the maximum specification. Shim changes are required to lower the pinion gear closer to both driven gears.

The shims that set the pinion gear height are above the upper drive shaft bearing and within the drive shaft bearing and seal housing (A, **Figure 112**). Select a thicker shim or set of shims to position the pinion gear closer to the driven gears. Measure the backlash to verify lower readings.

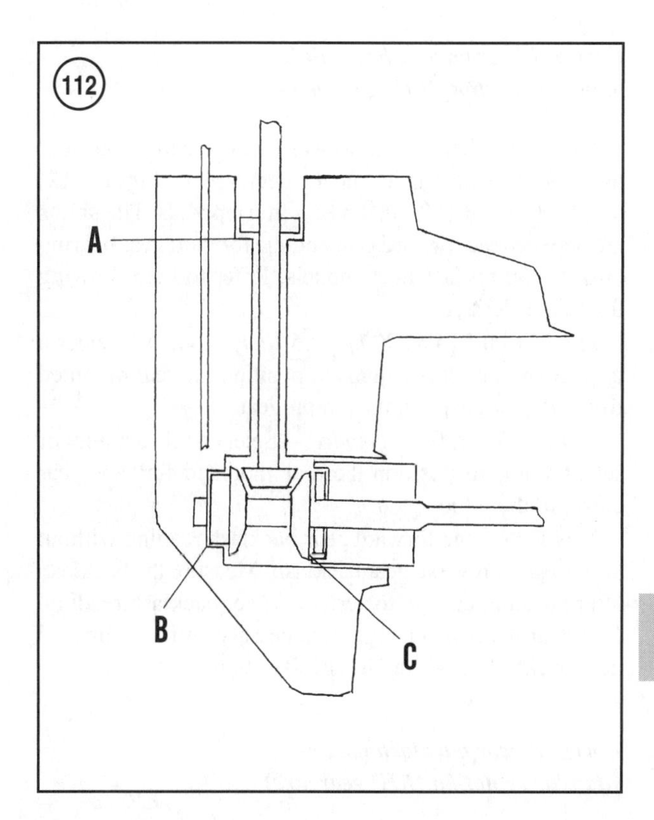

Both measurements too low

The pinion gear is probably positioned too low in the gearcase when both backlash measurements are below the minimum specification. Shim changes are required to raise the pinion gear away from both driven gears.

The shims that set the pinion gear height are above the upper drive shaft bearing and within the drive shaft bearing and seal housing (A, **Figure 112**). Select thinner shim or set of shims to position the pinion gear farther from the driven gears. Measure the backlash to verify higher readings.

Forward gear backlash too high (standard rotation [RH] gearcase)

The shims that set the forward gear backlash are located just forward of the front mounted gear bearing (B, **Figure 112**). Select a thicker shim or set of shims to position the forward gear closer to the pinion gear. This should lower the forward gear backlash reading without affecting the reverse gear backlash. Measure the backlash after the shim change to verify a lower backlash reading.

Forward gear backlash too high (counter rotation [LH] gearcase)

The shims that set the forward gear backlash are between the bearing carrier and the gearcase (C, **Figure 112**) on 115-130 and 150-200 hp (2.6 liter) models. The shims are between the forward gear and the forward gear bearing on 200-250 hp (3.1 liter) models. Refer to the following for shim selection.

115-130 and 150-200 hp (2.6 liter) models—Select a thinner shim or set of shims to position the rear mounted forward gear closer to the pinion gear.

200-250 hp (3.1 liter) models—Select a thicker shim or set of shims to position the rear mounted forward gear closer to the pinion gear.

This lowers the forward gear backlash reading without affecting the reverse gear backlash. Measure the backlash after the shim change to verify a lower backlash reading. Make further shim changes if necessary until the forward gear backlash is within the specification.

Forward gear backlash too low (standard rotation [RH] gearcase)

The shims that set the forward gear backlash are just forward of the front mounted gear bearing (B, **Figure 112**). Select a thinner shim or set of shims to position the forward gear farther from the pinion gear. This increases the forward gear backlash reading without affecting the reverse gear backlash. Measure the backlash after the shim change to verify a higher backlash reading.

Forward gear backlash too low (counter rotation [LH] gearcase)

The shims that set the forward gear backlash are between the bearing carrier and a step in the gearcase (C, **Figure 112**) on 115-130 and 150-200 hp (2.6 liter) models. The shims are between the forward gear and the forward gear bearing on 200-250 hp (3.1 liter) models. Refer to the following for shim selection.

115-130 and 150-200 hp (2.6 liter) models—Select a thicker shim or set of shims to position the rear mounted forward gear farther from the pinion gear.

200-250 hp (3.1 liter) models—Select a thinner shim or set of shims to position the rear mounted forward gear farther from the pinion gear.

This increases the forward gear backlash reading without affecting the reverse gear backlash. Measure the backlash after the shim change to verify a higher backlash reading.

Reverse gear backlash too high (standard rotation [RH] gearcase)

The shims that set the reverse gear backlash are between the bearing carrier and a step in the gearcase housing (C, **Figure 112**). Select a thinner shim or set of shims to position the rear mounted reverse gear closer to the pinion gear. This lowers the reverse gear backlash reading without affecting the forward gear backlash. Measure the backlash after the shim change to verify a lower backlash reading.

Reverse gear backlash too high (counter rotation [LH] gearcase)

The shims that set the reverse gear backlash are forward of the front mounted gear bearing (B, **Figure 112**). Select a thicker shim or set of shims to position the front mounted reverse gear closer to the pinion gear. This lowers the reverse gear backlash reading without affecting the forward gear backlash. Measure the backlash after the shim change to verify a lower backlash reading.

Reverse gear backlash too low (standard rotation [RH] gearcase)

The shims that set the reverse gear backlash are between the bearing carrier and a step in the gearcase housing (C, **Figure 112**). Select a thicker shim or set of shims to position the rear mounted reverse gear farther from the pinion gear. This increases the reverse gear backlash reading without affecting the forward gear backlash. Measure the backlash after the shim change to verify a higher backlash reading.

Reverse gear backlash too low (counter rotation [LH] gearcase)

The shims that set the reverse gear backlash are forward of the front mounted gear bearing (B, **Figure 112**). Select a thinner shim or set of shims to position the front mounted reverse gear farther from the pinion gear. This increases the reverse gear backlash reading without affecting the forward gear backlash. Measure the backlash after the shim change to verify a higher backlash reading.

Front gear backlash too high (twin propeller gearcase)

The shims that set the front gear backlash are between the front gear bearing and the step in the gearcase housing

(B, **Figure 112**). Select a thicker shim or set of shims to position the front mounted gear closer to the pinion gear. This lowers the front gear backlash readings without affecting the rear gear backlash. Measure the backlash after the shim change to verify a lower backlash reading.

Front gear backlash too low (twin propeller gearcase)

The shims that set the front gear backlash are between the front gear bearing and the step in the gearcase housing (B, **Figure 112**). Select a thinner shim or set of shims to position the front mounted gear farther from the pinion gear. This increases the front gear backlash readings without affecting the rear gear backlash. Measure the backlash after the shim change to verify a higher backlash reading.

Rear gear backlash too high (twin propeller gearcase)

The shims that set the rear gear backlash are between the bearing carrier and the step in the gearcase housing (B, **Figure 112**). Select a thinner shim or set of shims to position the rear mounted gear closer to the pinion gear. This lowers the rear gear backlash without affecting the front gear backlash. Measure the backlash after the shim change to verify a lower backlash reading.

Rear gear backlash too low (twin propeller gearcase)

The shims that set the rear gear backlash are between the bearing carrier and the step in the gearcase housing (B, **Figure 112**). Select a thicker shim or set of shims to position the rear mounted gear farther from the pinion gear. This increases the rear gear backlash without affecting the front gear backlash. Measure the backlash after the shim change to verify a higher backlash reading.

COUNTER ROTATION (LH) GEARCASES

Disassembly

Many of the components used on counter rotation (LH) gearcases are identical or similar to the components used on standard rotation (RH) gearcases. When applicable, you will be directed to the information and procedures for the standard rotation gearcases. Refer to the appropriate illustration (**Figure 113-115**) during the repair procedures.

Water pump

Water pump component removal, inspection and assembly are identical to the standard rotation (RH) models of the same horsepower. Refer to the *Water Pump* section in this chapter.

Bearing carrier removal

CAUTION
Avoid damaging the shifter and other shift components during the propeller shaft removal process. Always remove the shift shaft from the gearcase prior to pulling the propeller shaft from the gearcase.

1. Remove the propeller from the propeller shaft following the procedures for the standard rotation (RH) model of the same horsepower.
2. Rotate the shift shaft while rotating the drive shaft clockwise, as viewed from the top, until neutral gear is obtained. The gearcase must remain in neutral gear until the propeller shaft is removed from the gearcase.
3. Remove the shift shaft following the procedures for the standard (RH) models of the same horsepower. See *Propeller shaft and drive shaft removal*.
4A. *115-130 hp and 150-200 hp (2.6 liter) models*—Remove the cover nut following the procedures described for the standard rotation (RH) gearcases. See *Bearing carrier removal*.
4B. *200-250 hp (3.1 liter) models*—Remove the retainer and carrier retaining bolts following the procedures described for the standard rotation (RH) gearcases. See *Bearing carrier removal*.
5A. *115-130 hp models*—Use a suitable puller or Yamaha part No. YB-06117, YB-06501 and YB-06504 to pull the forward gear and bearing carrier from the gearcase (**Figure 116**).
5B. *150-250 hp models*—Use a slide hammer (Yamaha part No. YB-06096) and puller head (Yamaha part No. YB-6335) to remove the carrier, propeller shaft and forward gear assembly. Thread the puller head onto the propeller shaft. Thread the slide hammer onto the puller. Use quick strokes of the slide hammer (**Figure 117**) to remove the assembly. Make certain to retain the locating key from the carrier (2.6 liter model).
6. *200-250 hp (3.1 liter) models*—Pull the front propeller shaft from the gearcase. Remove the thrust washer from the propeller shaft or front gear.
7. *115-130 hp and 150-200 hp (2.6 liter) models*—Use a suitable puller (or Yamaha part No. YB-06207) to push the rear propeller shaft and forward gear from the bearing carrier as shown in **Figure 118**.

8

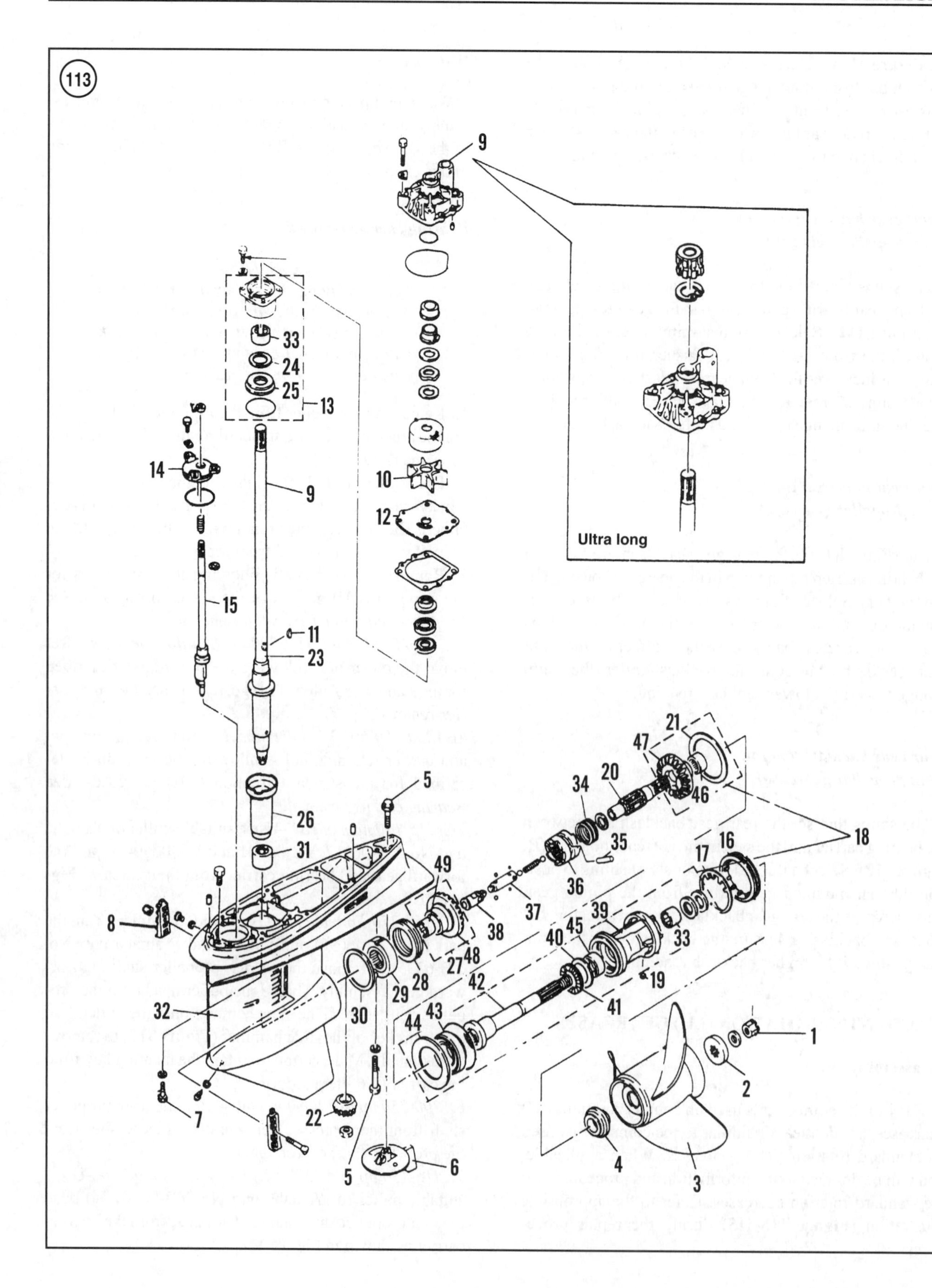
113
9
Ultra long
33
24
25
13
14
9
10
12
15
11
23
21
47
20
34
46
18
5
26
35
16
31
17
49
36
37
39
8
38
45
33
40
27
48
28
42
19
41
29
30
32
43
44
1
2
7
22
6
4
5
3

COUNTER ROTATION (LH) GEARCASE (115-130 HP MODELS)

1. Propeller nut
2. Spacer
3. Propeller
4. Thrust washer
5. Bolt
6. Trim tab *
7. Bolt and washer
8. Water inlet screen
9. Drive shaft
10. Impeller
11. Drive key
12. Wear plate
13. Drive shaft bearing and seal housing
14. Shift shaft seal housing
15. Shift shaft
16. Cover nut
17. Locking tab washer
18. Bearing carrier and forward gear assembly
19. Locating key
20. Rear propeller shaft
21. Thrust spacer
22. Pinion gear
23. Drive shaft impeller contact surface
24. Shim
25. Bearing
26. Lubrication sleeve
27. Reverse gear assembly
28. Bearing carrier
29. Reverse gear bearing
30. Shim
31. Lower drive shaft bearing
32. Gearcase housing
33. Upper drive shaft bearing
34. Spring
35. Cross pin
36. Clutch
37. Shift slider
38. Shifter
39. Bearing carrier
40. Shim
41. Bearing
42. Front propeller shaft
43. Forward gear bearing carrier
44. Thrust spacer
45. Needle bearing
46. Forward gear assembly
47. Forward gear
48. Reverse gear
49. Needle bearing

*Not used on all models.

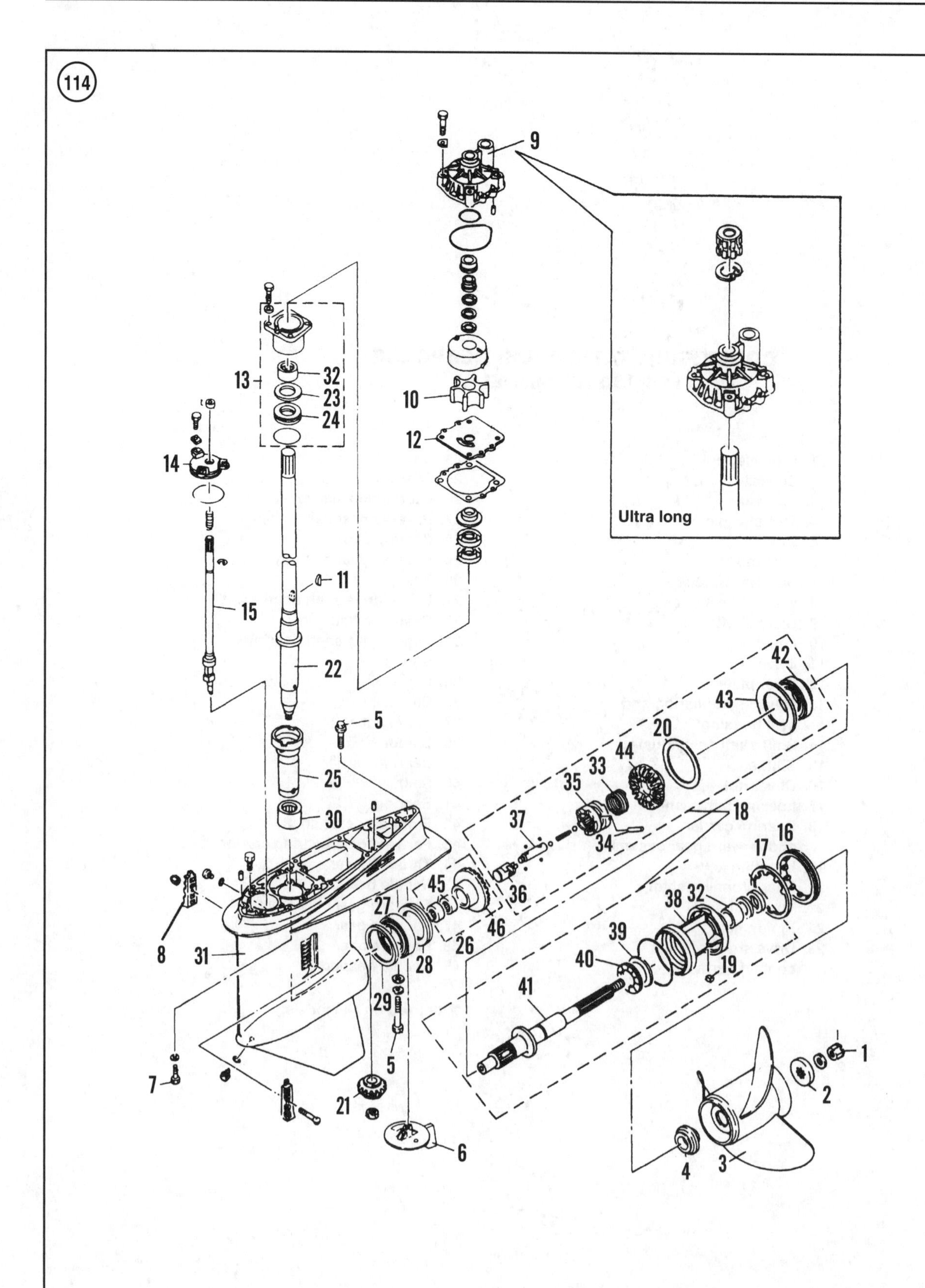
114
Ultra long
1
2
3
4
5
6
7
8
9
10
11
12
13
14
15
16
17
18
19
20
21
22
23
24
25
26
27
28
29
30
31
32
33
34
35
36
37
38
39
40
41
42
43
44
45
46

COUNTER ROTATION (LH) GEARCASE
(150-200 HP MODELS [2.6 LITER])

1. Propeller nut
2. Spacer
3. Propeller
4. Thrust washer
5. Bolt
6. Trim tab *
7. Bolt
8. Water inlet screen
9. Water pump body
10. Impeller
11. Drive key
12. Wear plate
13. Drive shaft bearing and seal housing
14. Shift shaft seal housing
15. Shift shaft
16. Cover nut
17. Locking tab washer
18. Bearing carrier and forward gear assembly
19. Locating key
20. Shim
21. Pinion gear
22. Drive shaft
23. Shim
24. Bearing
25. Lubrication sleeve
26. Reverse gear assembly
27. Reverse gear bearing
28. Reverse gear bearing carrier
29. Shim
30. Lower drive shaft bearing
31. Gearcase housing
32. Needle bearing
33. Spring
34. Cross pin
35. Clutch
36. Shifter
37. Shift slider
38. Forward gear and propeller shaft bearing carrier
39. Shim
40. Bearing
41. Propeller shaft
42. Forward gear bearing
43. Thrust spacer
44. Forward gear
45. Needle bearing
46. Reverse gear

*Not used on all models.

115

COUNTER ROTATION (LH) GEARCASE (200-250 HP MODELS [3.1 LITER])

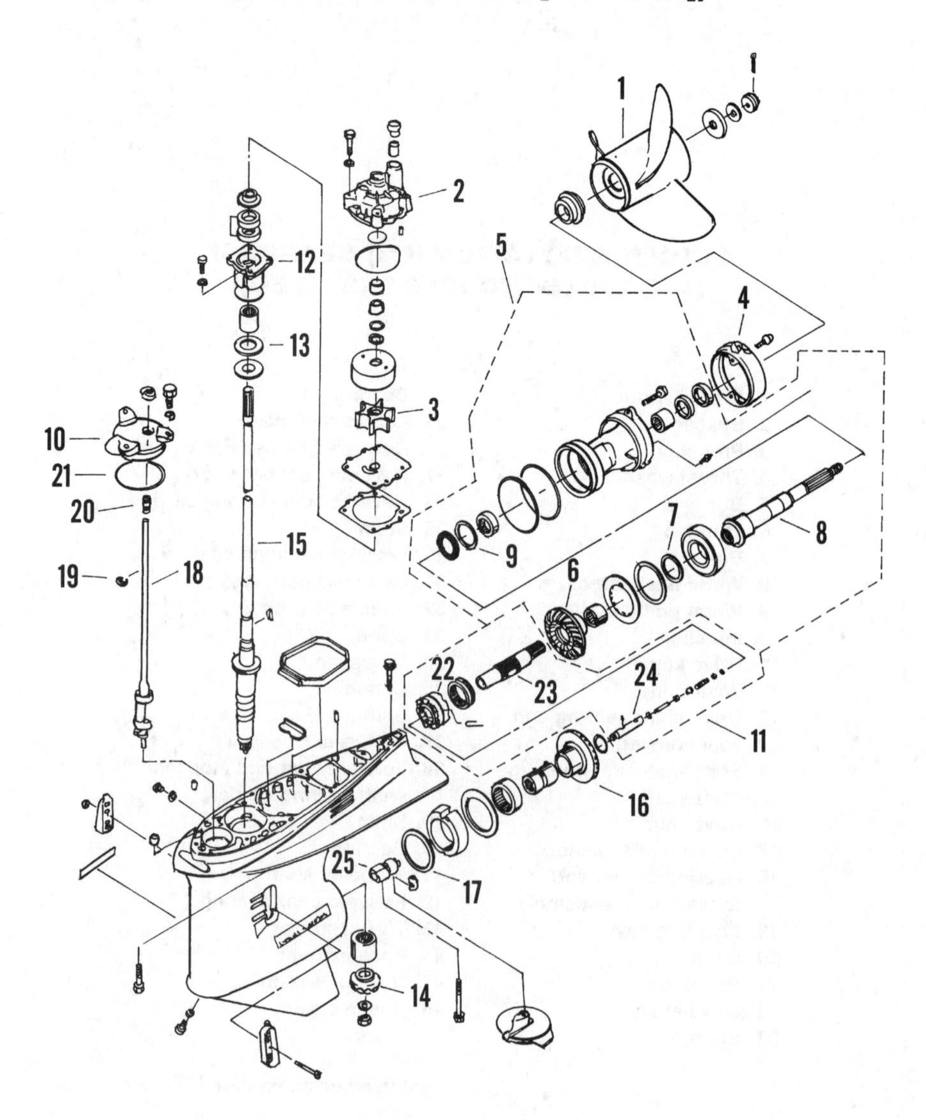

1. Propeller
2. Water pump body
3. Impeller
4. Retainer
5. Bearing carrier assembly
6. Forward gear
7. Shim
8. Rear propeller shaft
9. Shim
10. Shift shaft seal housing
11. Shift slider assembly
12. Drive shaft bearing and seal housing
13. Shim
14. Pinion gear
15. Drive shaft
16. Reverse gear
17. Shim
18. Shift shaft
19. Clip
20. Spring
21. O-ring
22. Clutch
23. Front propeller shaft
24. Shift slider
25. Shifter

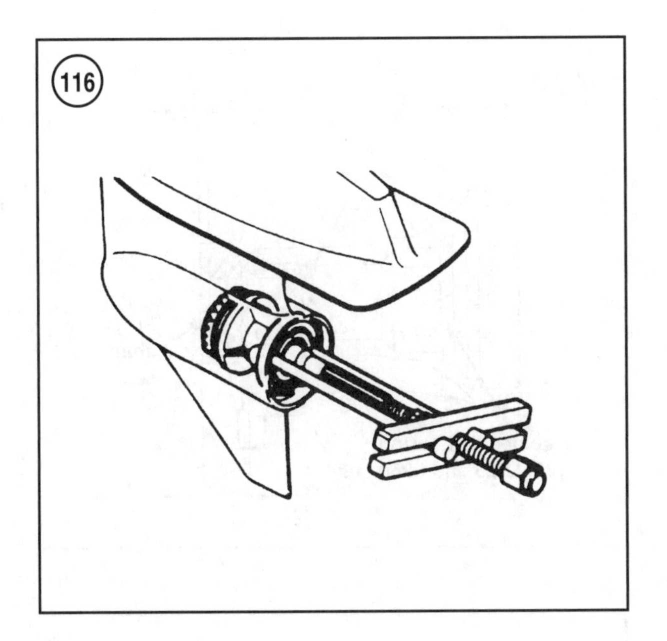

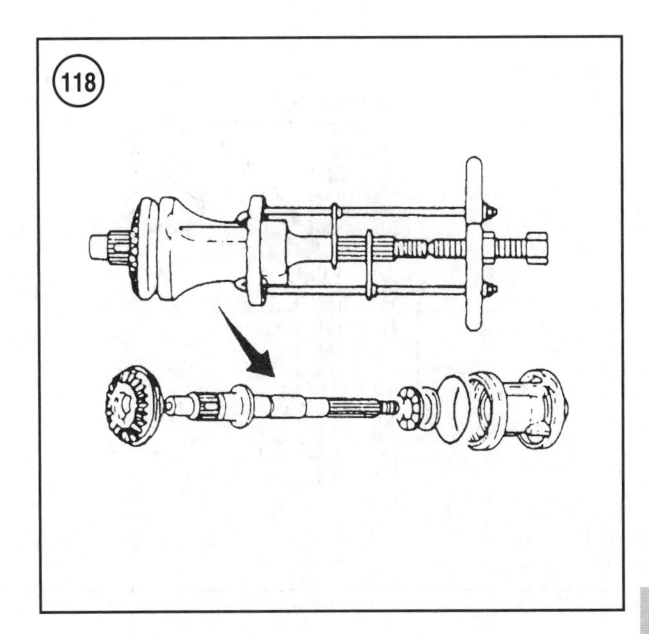

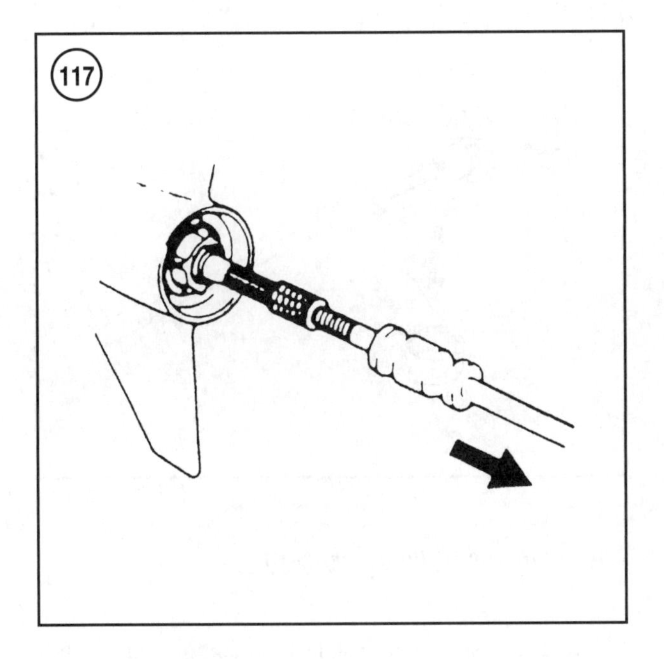

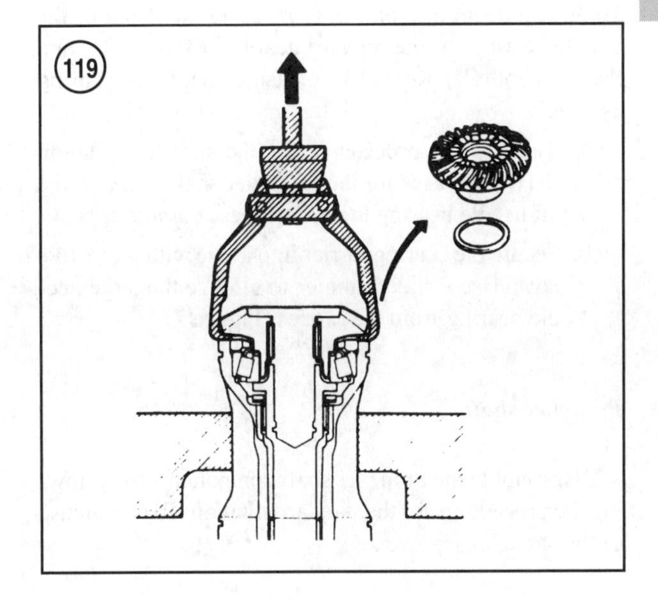

8. *200-250 hp (3.1 liter) models*—Secure the bearing carrier into a vise with protective jaws. Use a slide hammer to remove the forward gear from the carrier as shown in **Figure 119**.

9. *200-250 hp (3.1 liter) models*—Carefully bend the locking tab away from the ring nut. With the bearing carrier secured in a vise, use the removal tool (Yamaha part No. YB-06048/90890-06510) to loosen and remove the ring nut (**Figure 120**). Use a suitable puller (or Yamaha part No. YB-06207) to push the rear propeller shaft and bearing from the bearing carrier as shown in **Figure 118**. Note the location and orientation of all components before removing.

NOTE

Do not remove the needle bearings from the bearing carrier unless they must be replaced. The removal process damages the bearings. Refer to ***Component Inspection*** *to determine the need for replacement. As with the standard rotation (RH) gearcases, always measure the depth of the seals and bearings within the bearing carrier before removing. Record the measurements to use during assembly.*

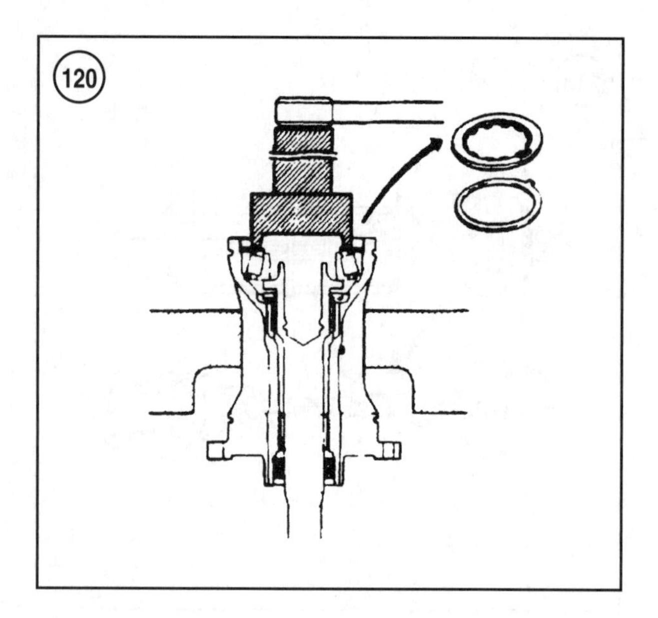

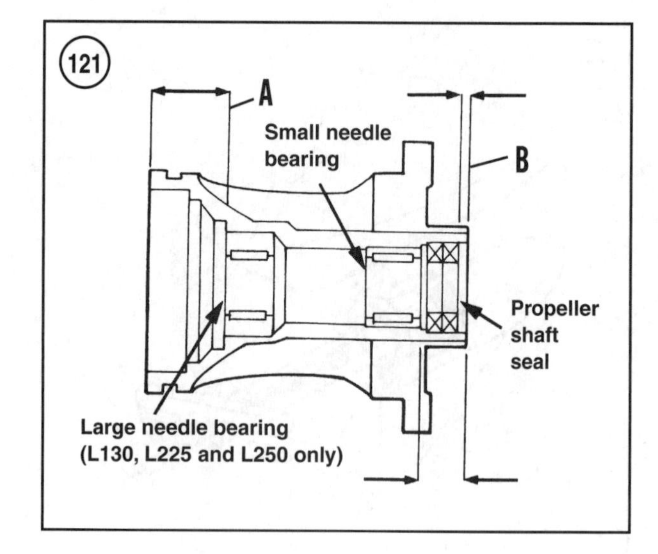

10. Use a depth micrometer to measure the depth of the needle bearing for the forward gear (A, **Figure 121**) and the seal depth (B). Record the measurement for use during assembly.
   a. Refer to the procedures for the standard rotation (RH) gearcases for the propeller shaft seal and the aft needle bearing to remove these components.
   b. Secure the bearing carrier into a vise with protective jaws. Use a slide hammer to remove the large needle bearing from the carrier (**Figure 122**).

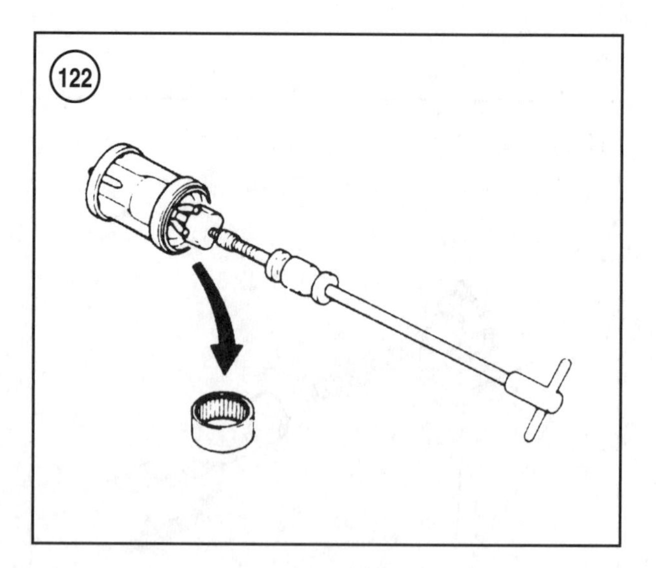

Propeller shaft

Disassemble the propeller shaft components by following the procedures for the standard rotation (RH) gearcase of the same horsepower.

Gears and bearings

Refer to the procedures for the standard rotation (RH) gearcases for removal and inspection of the gears and bearings. As with the standard rotation gearcases, measure the installed depth of all needle bearings before removing.

Remove the drive shaft and pinion gear following the procedures for the standard rotation (RH) gearcase of the same horsepower. Do not remove any bearings from the gears unless replacing them.

Use a slide hammer (**Figure 123**) to remove the reverse gear bearing carrier and shims as required.

Shift shaft and shifting components

Remove, inspect and install these components by following the procedures for the standard rotation (RH) gearcase of the same horsepower.

Drive shaft and pinion gear

Remove and inspect the drive shaft and pinion gear following the procedures for the standard rotation (RH) gearcase of the same horsepower. Disassemble the drive shaft bearing and seal housing following the same procedures as with the standard rotation (RH) gearcases.

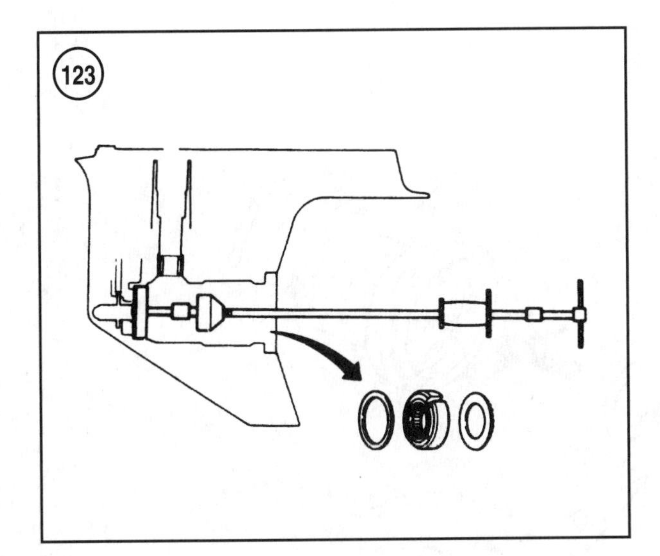

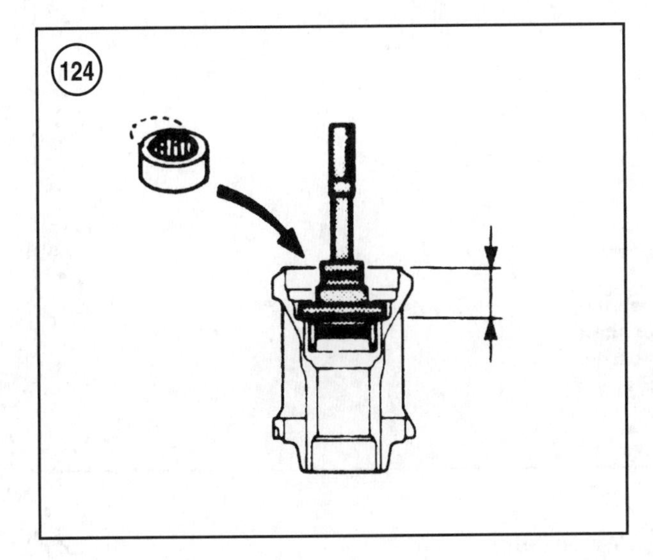

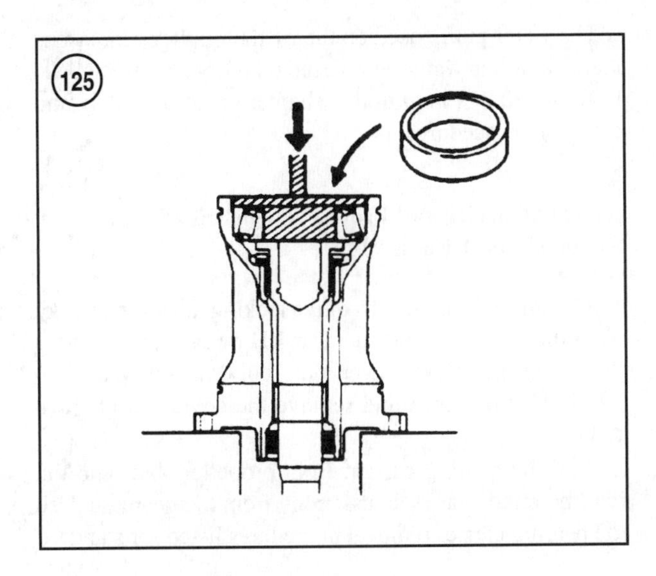

Assembly

Assembly is the reverse of disassembly. Refer to the procedures for the standard rotation (RH) gearcases when the components are identical or similar. Note the following:

1A. *115-130 hp and 150-200 hp (2.6 liter) models*—Assemble the bearing carrier as follows:

a. Use an appropriately sized socket, section of pipe or tubing to drive the large roller/needle bearing into the bearing carrier (**Figure 124**). The numbered or lettered side of the bearing case must face outward. Install the bearing to the exact depth recorded prior to removal.
b. Install all bearings, washers and shims to the same location and orientation as noted prior to removal.
c. Slide or press the propeller shaft and forward gear into the bearing carrier.

1B. *200-250 hp (3.1 liter) models*—Assemble the bearing carrier as follows:

a. Install the forward gear shims, then use Yamaha part No. YB-06071 and YB-06430 to drive the bearing into position (**Figure 125**).
b. Install all shims washers and bearings into the carrier (**Figure 126**).
c. Press the propeller shaft into the tapered roller bearing. Install the ring nut with the same tools used to remove it (**Figure 127**). Tighten the ring nut to 108 N·m (80 ft.-lb.), then bend the locking tab into the slot on the ring nut. Additional tightening of the ring nut may be required to align the tab with an opening.
d. Install the shim onto the hub of the gear, then press the forward gear into the tapered bearing.
e. Install the shims into the step at the front of the gearcase. Use a suitable driver to carefully drive the reverse gear roller bearing into the gearcase (**Figure 128**). The tool used must contact the bearing case; yet not contact the gearcase during assembly.
f. Lubricate the thrust bearing with gearcase lubricant, then install the bearing on the reverse gear bearing surface.

2. Complete the assembly using the procedures for the standard rotation (RH) gearcase when applicable.

3. Upon assembly, use a dial indicator and suitable mount to measure the amount of propeller shaft end play (**Figure 129**). Move the propeller shaft in the fore and aft direction while observing the dial indicator. If the end play is not 0.25-0.40 mm (0.01-0.016 in.), first check for proper orientation of the shims and washers. If the components are installed correctly, increase or decrease the thickness of the shim next to the thrust bearing to correct the end play.

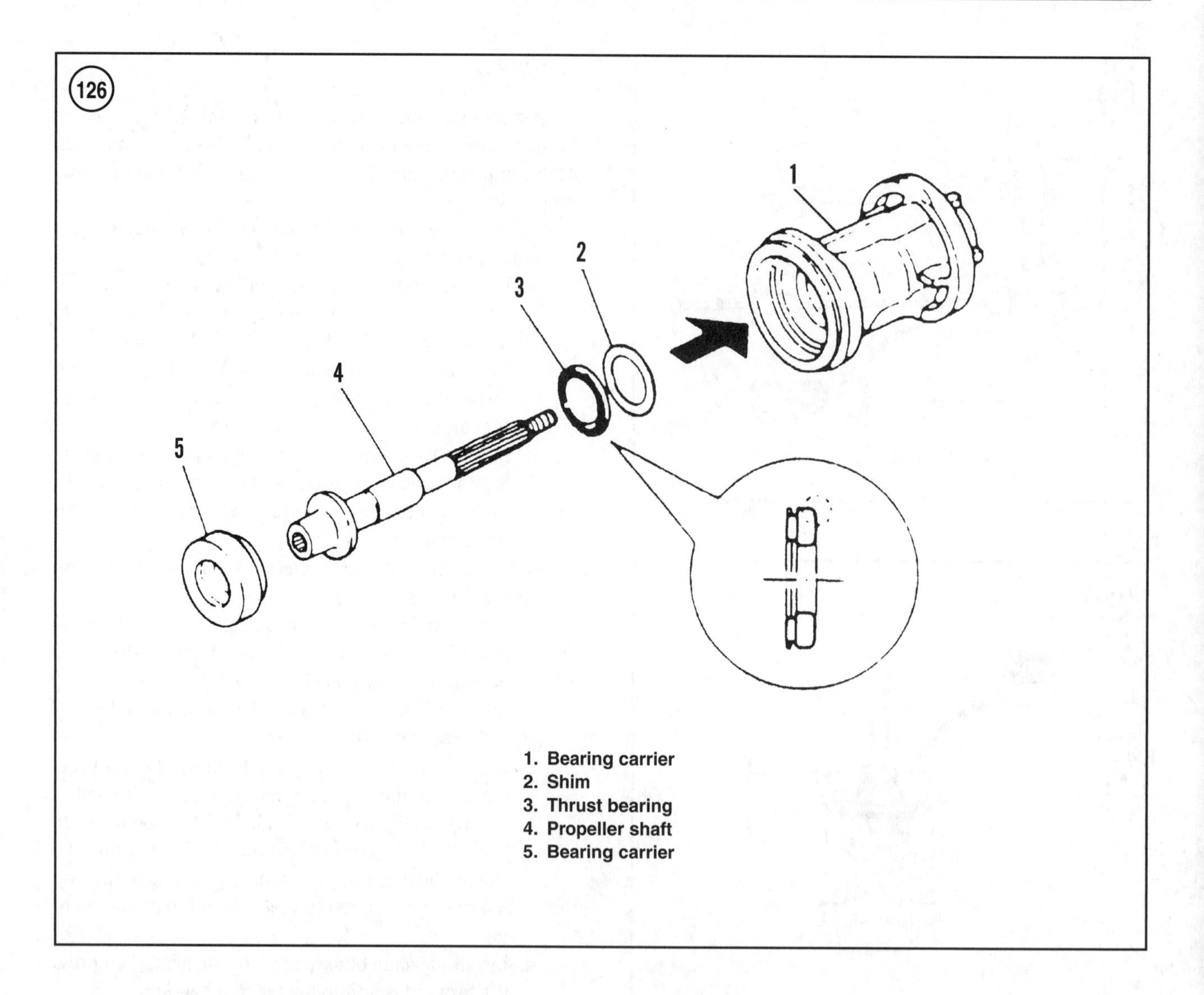

1. Bearing carrier
2. Shim
3. Thrust bearing
4. Propeller shaft
5. Bearing carrier

4. Measure the gear backlash as described in this chapter. Correct the backlash as necessary.
5. Install the water pump components as described in this chapter. Install the gearcase as described in this chapter. Fill the gearcase with lubricant as described in Chapter Three.

D150 (Twin Propeller) Model Gearcase

Many of the components used on twin propeller gearcases are identical or similar to the components used on standard rotation (RH) gearcases. This section described procedures specifically for twin propeller gearcase. When applicable, you will be directed to the procedures for the standard rotation gearcases. To assist with component identification and orientation, refer to **Figure 130**.

Water Pump

The water pump used on the twin propeller gearcase is identical to the water pump used on other 150 hp models. Refer to *Water Pump* in this chapter for disassembly and assembly procedures.

Bearing Carrier and Front Propeller Shaft Removal and Disassembly

1. Carefully bend the tabs of the locking tab washer away from the cover nut. Standard or RH threads are used for the cover nut. Use cover nut tool Yamaha part No. YB-42223 to loosen and remove the cover nut (**Figure 131**).
2. Pull the bearing carrier, front propeller shaft and the rear mounted gear as an assembly from the gearcase. Use the bearing carrier removal procedures listed for standard

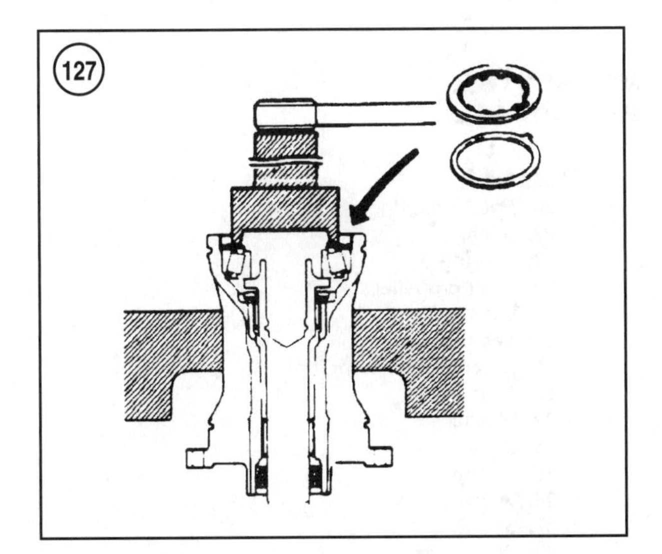
(127)

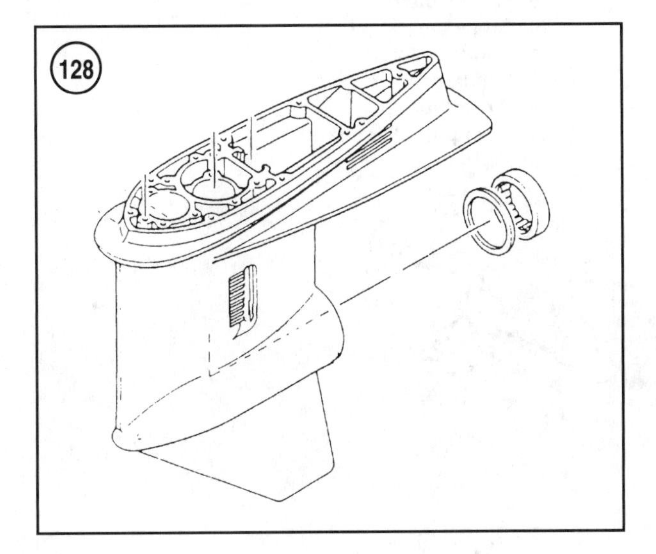
(128)

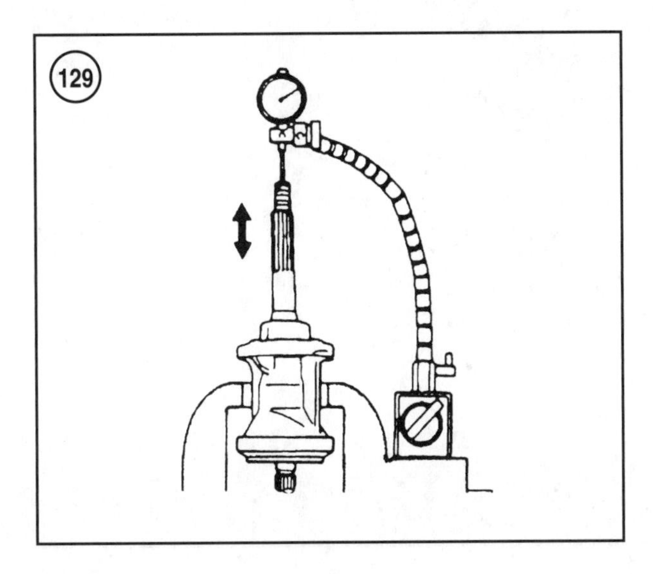
(129)

rotation (RH) gearcases if the assembly cannot be easily pulled from the gearcase.

3. Clamp a bearing separator onto the rear mounted gear as shown in **Figure 132**. Place the bearing carrier assembly onto the press table as shown in **Figure 133**. Support the propeller shaft while carefully pressing the gear and shaft from the carrier.

4. Note the orientation of the components prior to removal. Remove the rear mounted gear and bearing (1, **Figure 134**) from the propeller shaft (2). Slide the thrust bearing (3, **Figure 134**) and spacer (4) off the propeller shaft.

5. Use a depth micrometer to measure the seal depth at the points indicated in **Figure 135**. Record the measurement. Then carefully pry the seal from the carrier. Do not damage the seal bore in the carrier.

6. Use a depth micrometer to measure the depth of the rear needle bearing at the points indicated in **Figure 136**. Measure the front needle bearing at the points indicted in **Figure 137**. Record all measurements.

7. Note the orientation of the numbered side of the bearing cases prior to removal. Securely clamp the bearing carrier into a vise with protective jaws. Then use a slide hammer to remove the needle bearings (**Figure 138**).

Drive Shaft and Pinion Gear Removal

Refer to the procedures for the standard rotation (RH) gearcase. Then remove the drive shaft seal and bearing housing (**Figure 139**). Use the special pinion nut tool (Yamaha part No. YB-42224) and drive shaft holder (Yamaha part No. YB-06201) to loosen and remove the pinion nut (**Figure 140**). If necessary, use a thin wrench to grip the pinion nut. Pull the drive shaft from the gearcase. Then remove the pinion nut and pinion gear.

Rear Propeller Shaft Removal and Disassembly

The pinion nut, pinion gear and drive shaft must be removed before removing of the rear propeller shaft. Refer to the instructions for the standard rotation 150-200 hp (2.6 liter) models, then remove the drive shaft from the gearcase.

1. Carefully pull the shift shaft from the gearcase.

2. Grasp the propeller shaft (8, **Figure 139**). Carefully pull the shaft and front gear assembly from the gearcase without rotating.

3. Use a small blade screwdriver to unwind the spring (1, **Figure 141**) from the front clutch (2). Push the cross pin from the clutch, then slide the front clutch from the pro-

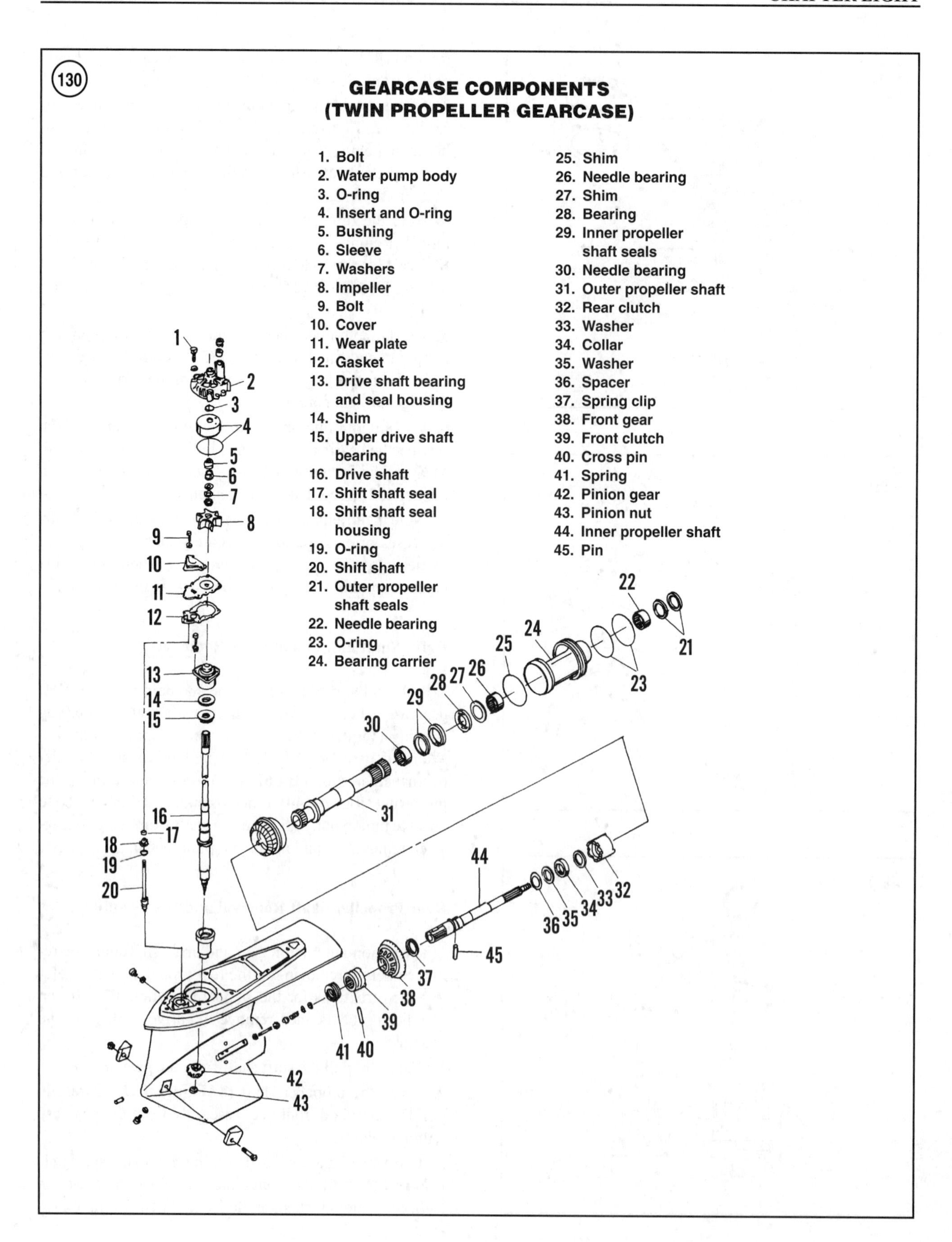
130
GEARCASE COMPONENTS
(TWIN PROPELLER GEARCASE)
1. Bolt
2. Water pump body
3. O-ring
4. Insert and O-ring
5. Bushing
6. Sleeve
7. Washers
8. Impeller
9. Bolt
10. Cover
11. Wear plate
12. Gasket
13. Drive shaft bearing and seal housing
14. Shim
15. Upper drive shaft bearing
16. Drive shaft
17. Shift shaft seal
18. Shift shaft seal housing
19. O-ring
20. Shift shaft
21. Outer propeller shaft seals
22. Needle bearing
23. O-ring
24. Bearing carrier
25. Shim
26. Needle bearing
27. Shim
28. Bearing
29. Inner propeller shaft seals
30. Needle bearing
31. Outer propeller shaft
32. Rear clutch
33. Washer
34. Collar
35. Washer
36. Spacer
37. Spring clip
38. Front gear
39. Front clutch
40. Cross pin
41. Spring
42. Pinion gear
43. Pinion nut
44. Inner propeller shaft
45. Pin

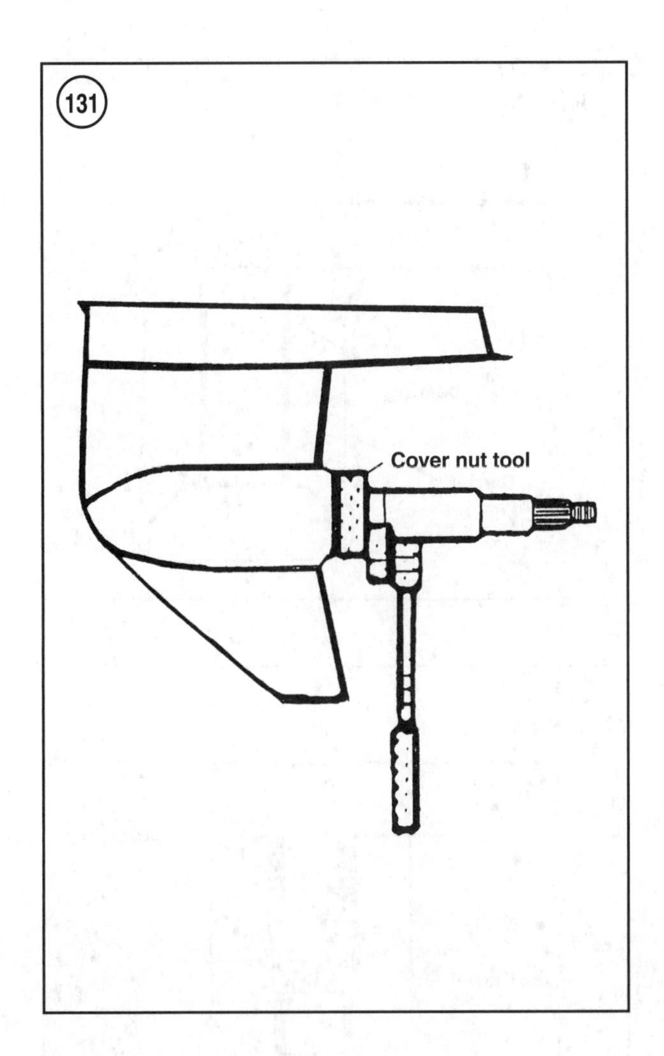

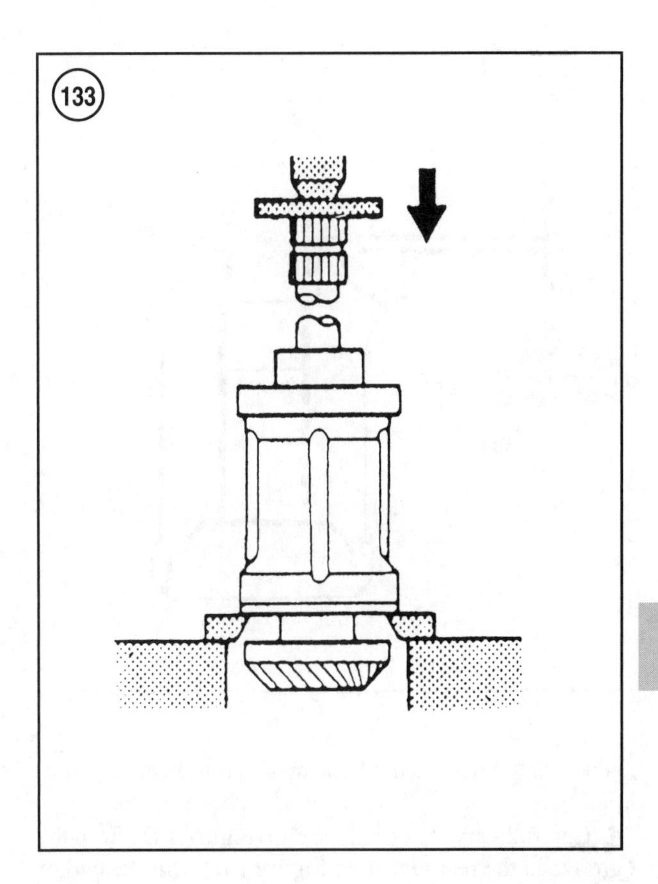

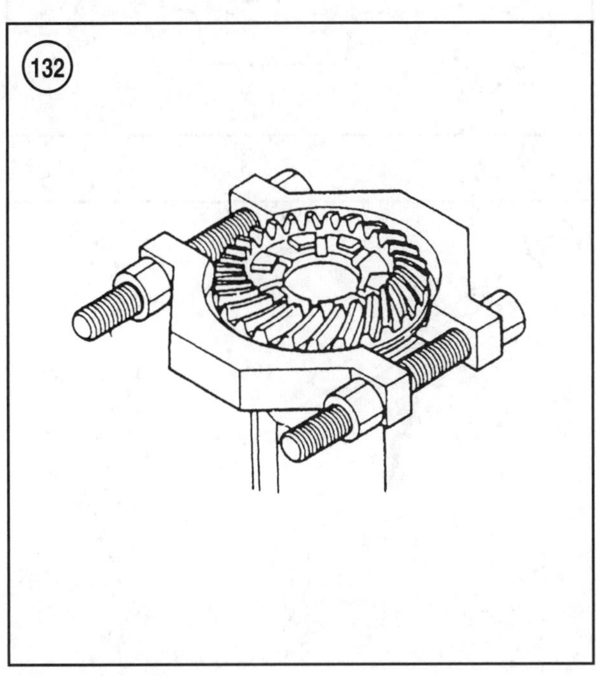

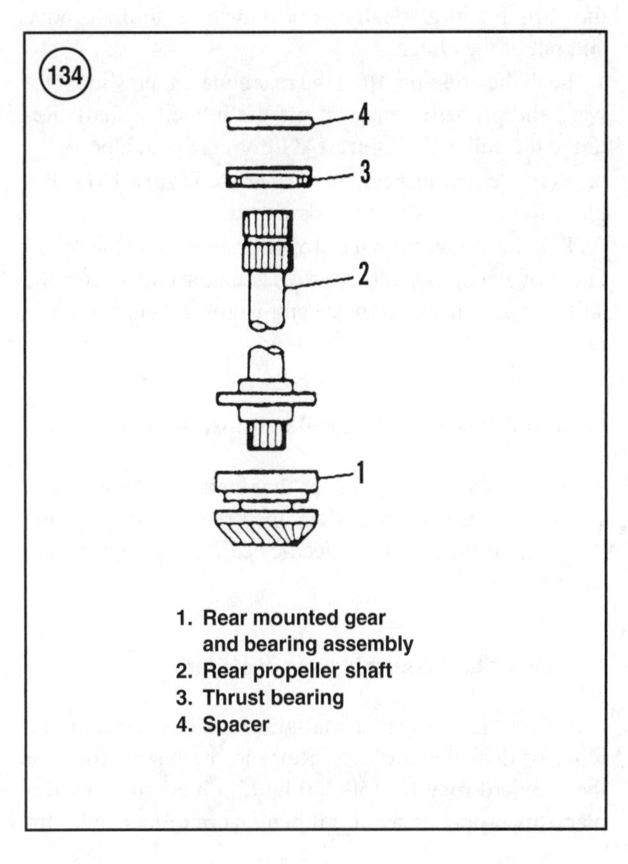

1. Rear mounted gear and bearing assembly
2. Rear propeller shaft
3. Thrust bearing
4. Spacer

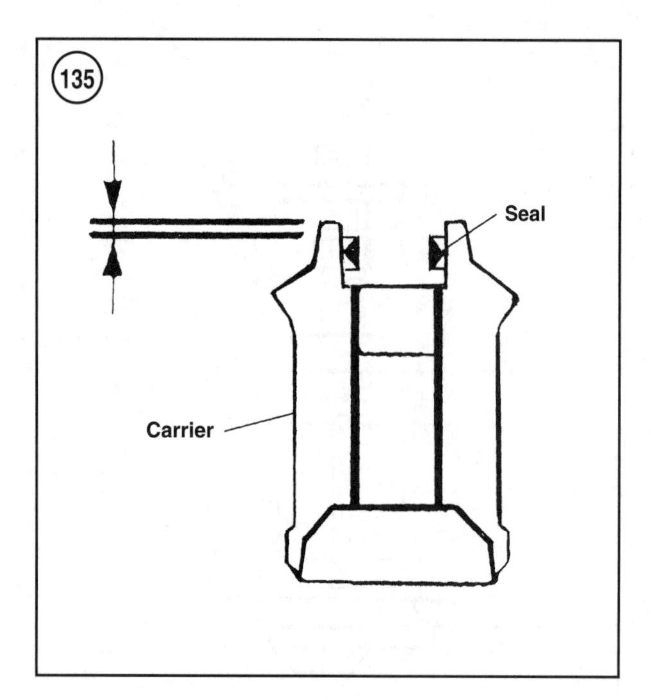

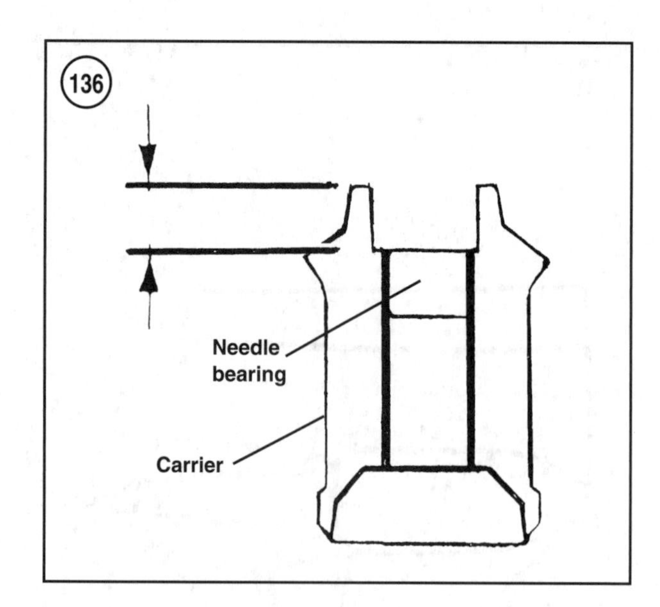

peller shaft. Slide the front gear and bearing assembly from the shaft.

4. Carefully pry the end of the clip (**Figure 142**) from the groove in the rear clutch (6, **Figure 141**). Lift the end of the clip up and gradually work it away from the groove and out of the clutch.
5. Push the cross pin from the rear clutch, then slide it toward the propeller end and off the propeller shaft. Remove the collar (8, **Figure 141**) from the clutch bore.
6. Remove and inspect the washer (4, **Figure 141**). Replace the washer if worn or damaged.
7. Pull the shifter from the propeller shaft. Note the orientation of all components, then disassemble and inspect the shifter components. Replace any worn or damaged components.

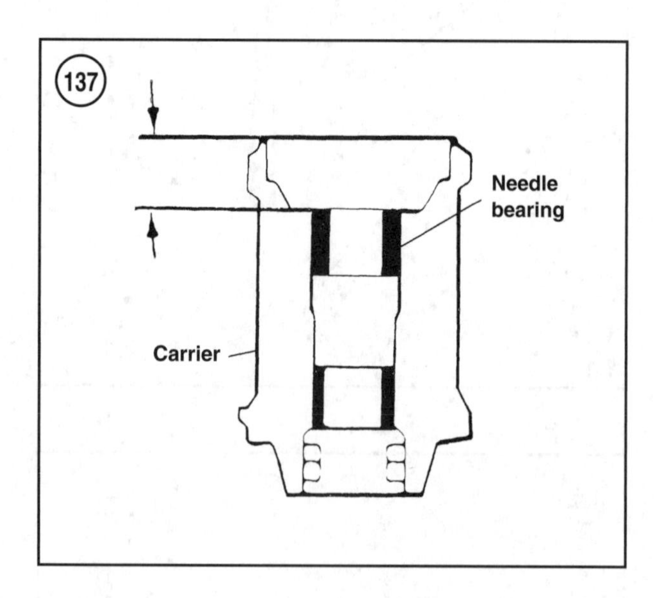

Gear and Bearings Removal and Installation

Refer to the instructions for the standard rotation (RH) 150-200 hp (2.6 liter) models, then remove, inspect and install the front gear, race, needle bearing and lower drive shaft bearing.

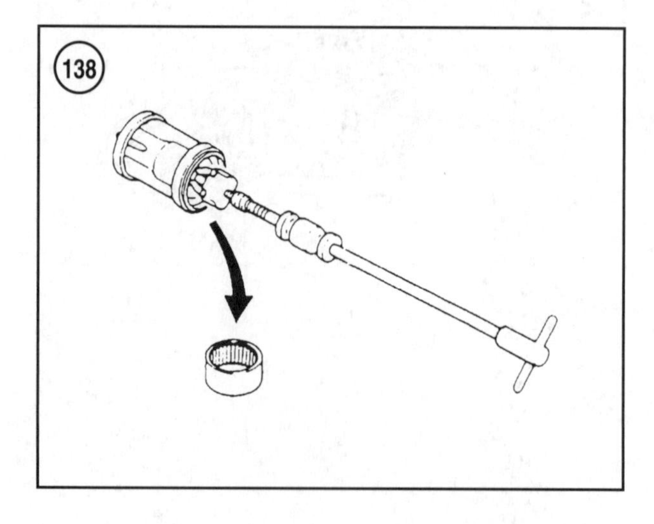

Propeller Shaft Assembly and Installation

Inspect both propeller shafts for a bent condition and worn or damaged surfaces. Refer to the instructions for the standard rotation 150-200 hp (2.6 liter) models. Replace the propeller shaft(s) if bent, worn or damaged. In-

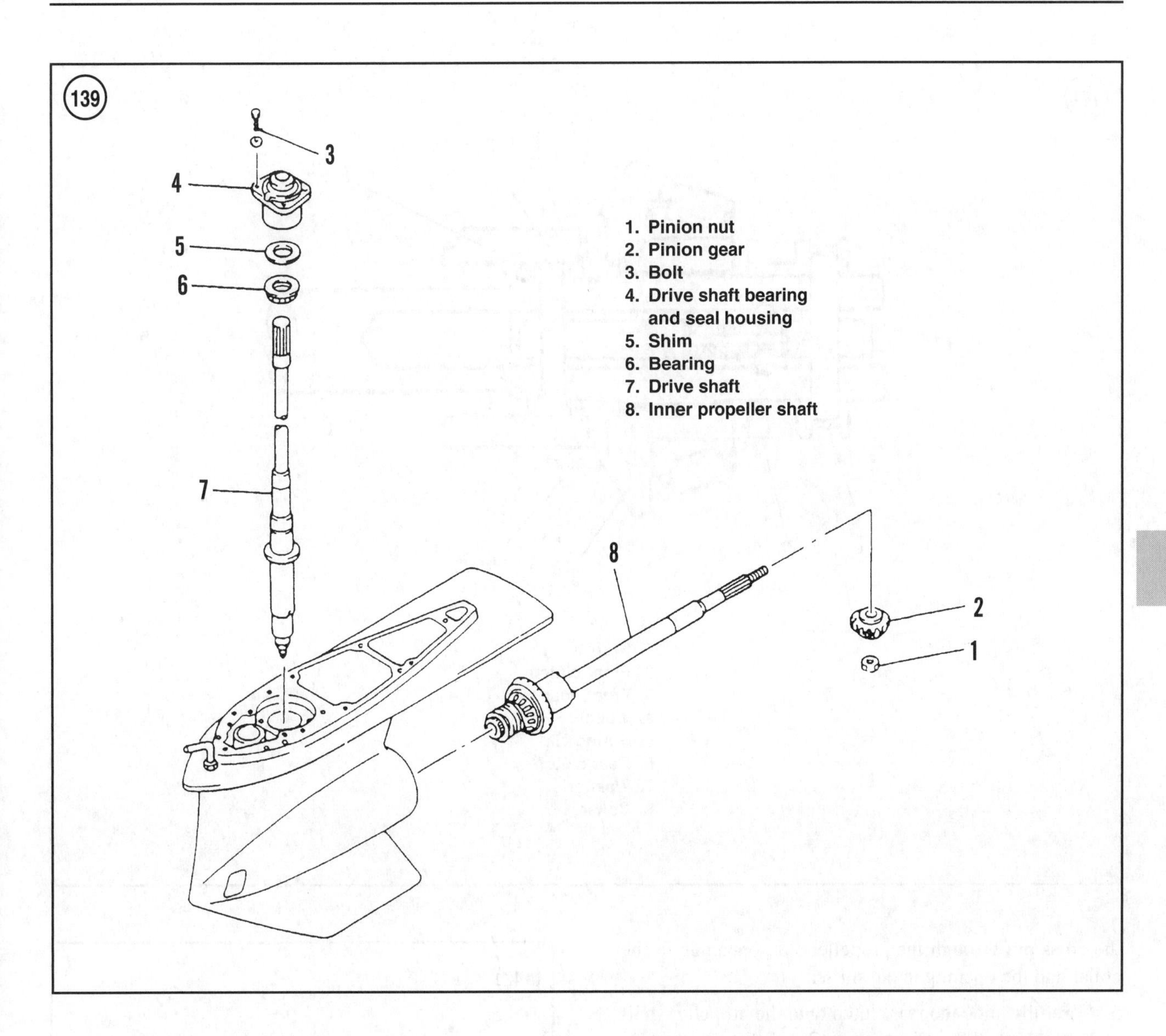

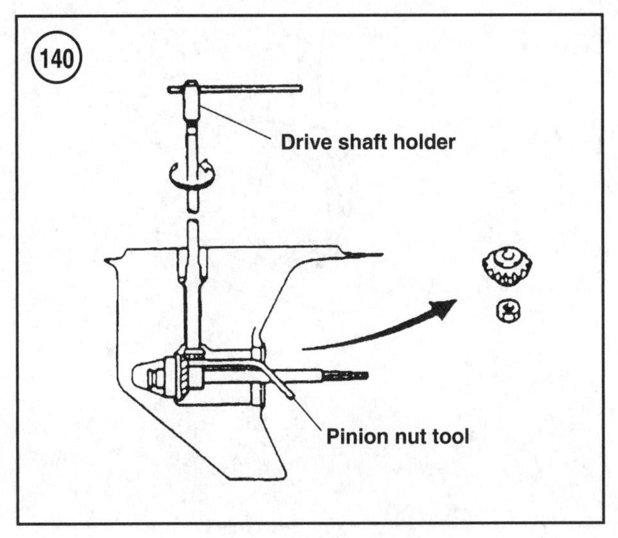

spect the cross pin retaining springs for damaged or corroded surfaces and lost spring tension. Replace the spring(s) if the condition is questionable.

1. Slide the spring clip, followed by the washers (5 and 7, **Figure 143**) over the aft end of the propeller shaft.

2. Guide the collar (8, **Figure 143**) over the propeller shaft with the cross pin openings aligned with the cross pin slot. Slide the second washer (7, **Figure 143**) over the propeller shaft and seat it against the collar.

3. Lubricate with Yamaha All-Purpose Grease, then insert the shifter components into the bore of the propeller shaft. Make sure each detent ball is positioned into respective recess in the propeller shaft bore.

4. Align the cross pin bore of the shifter shaft with the slot in the propeller shaft and the openings in the collar. Insert

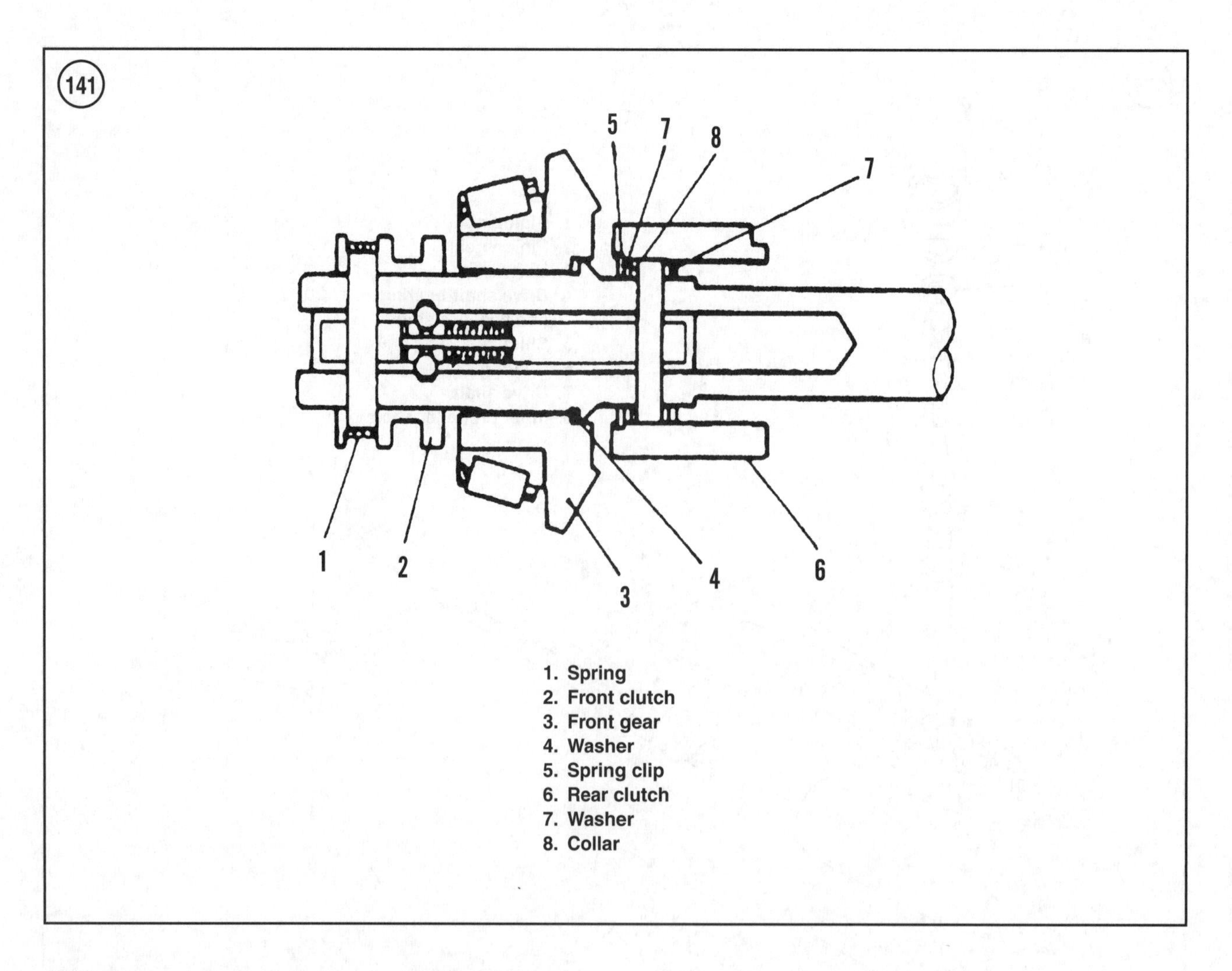

the cross pin through the propeller slot, openings in the collar and the opening in the shifter.

5. Carefully slide the rear clutch onto the propeller shaft and over the collar and cross pin. Carefully push the clip into the groove in the rear clutch as shown in **Figure 144**. Make sure the clip fully seats in the groove.

6. Place the washer (5, **Figure 143**) onto the front end of the propeller shaft. Seat the washer against the step on the propeller shaft. Slide the front gear and bearing (3, **Figure 143**) over the propeller shaft and seat against the washer.

7. Align the front clutch (2, **Figure 143**) onto the splined section of the propeller shaft so that the opening for the cross pin aligns with the slot in the propeller shaft. The dog side of the clutch must face toward the front gear as shown in **Figure 143**.

8. Push the cross pin through the opening in the clutch and slot in the propeller shaft. Carefully manipulate the shifter to align the cross pin opening in the shifter with the cross pin, then push the cross pin through the shifter.

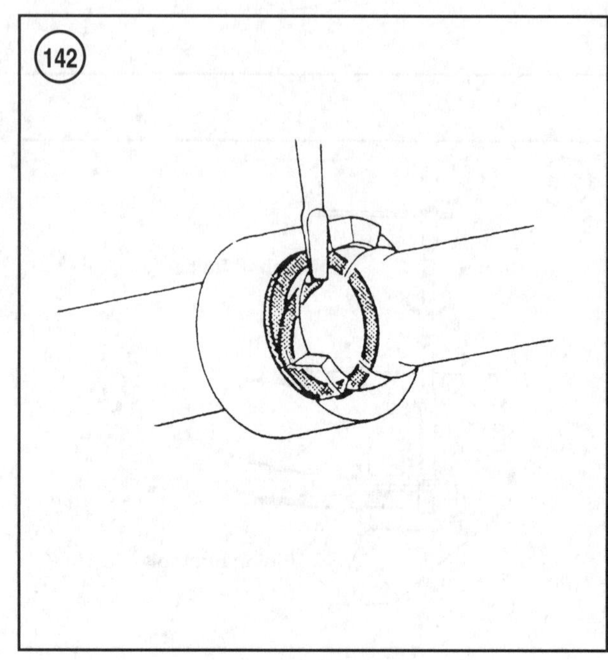

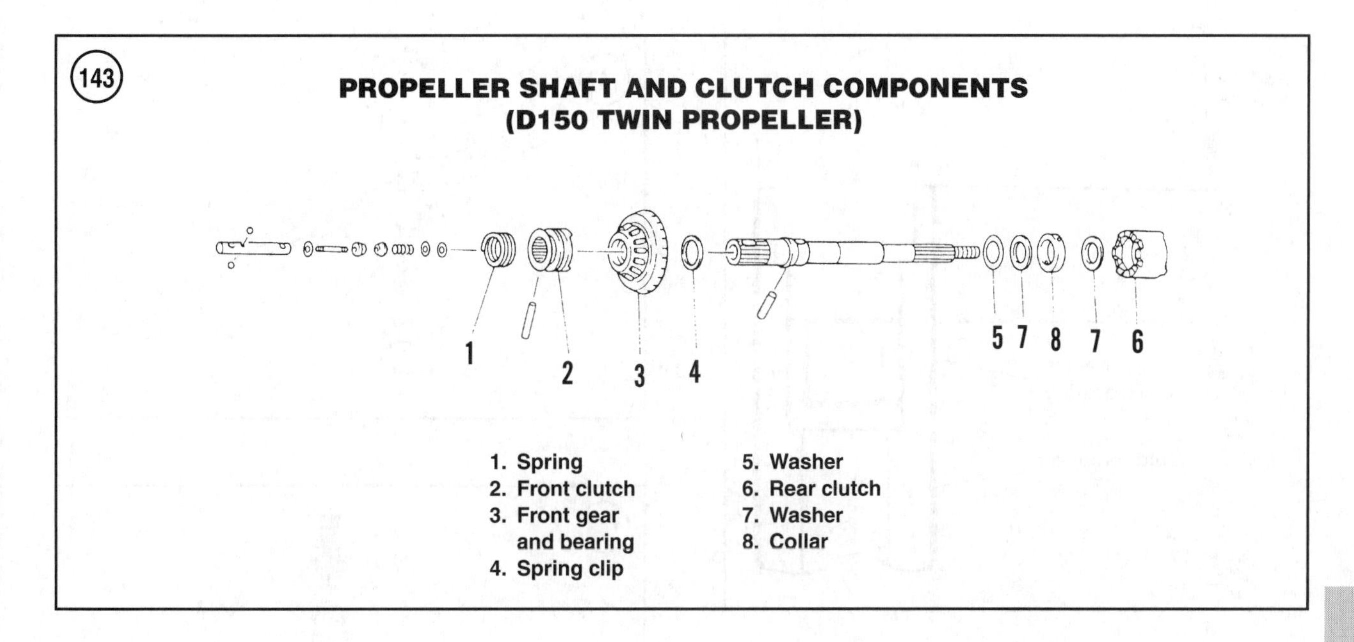

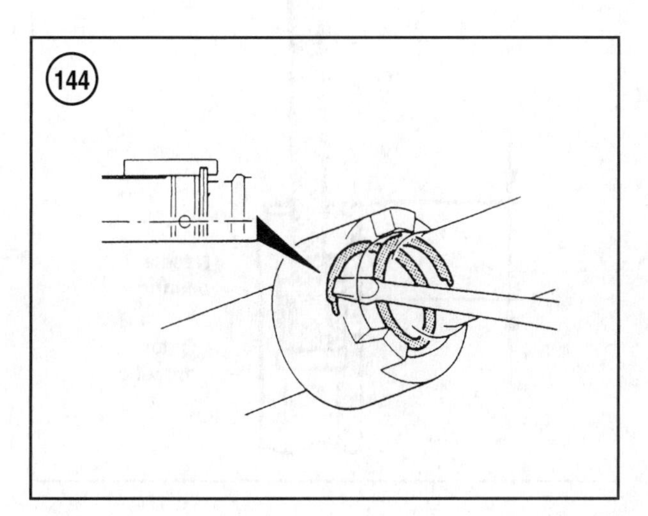

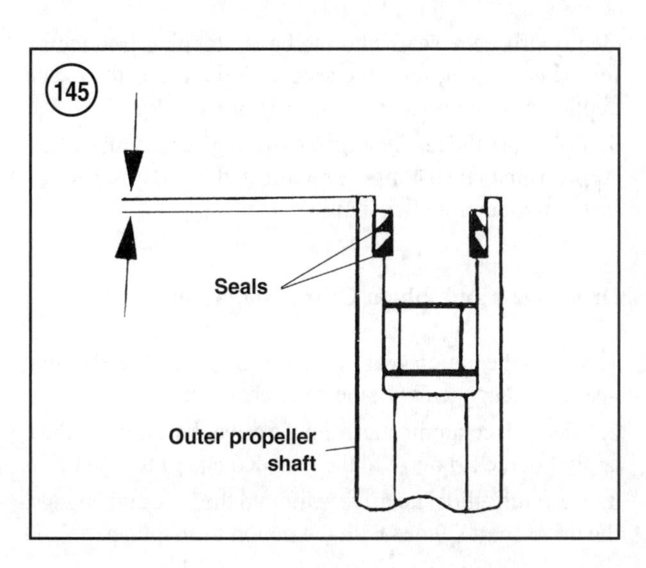

9. Carefully wind the spring (1, **Figure 143**) onto the clutch and into the groove covering the cross pin. If the spring loops cross over one another and the spring does not fit snug against the clutch, replace the spring.

10. Carefully guide the propeller shaft and front gear assembly into the gearcase. Refer to instructions for the standard rotation (RH) 150-200 hp (2.6 liter) models gearcases. Then install the shift shaft into the gearcase. Rotate the shift shaft until it drops into position and the tip on the lower end of the shaft enters the opening in the shifter.

Rear (Outer) Propeller Shaft Disassembly and Assembly

NOTE
Remove the needle bearing from the outer propeller shaft only if it must be replaced. The removal process damages the bearing.

1. Use a depth micrometer to measure the installation depth of the seals at the points indicated in **Figure 145**. Record the seal depth, then use a blunt tip instrument to carefully pry the seals from the shaft. Do not damage the seal bore. Discard the seals.

2. Use a depth micrometer to measure the installation depth of the propeller shaft needle bearing at the points indicated in **Figure 146**. Record the bearing depth. Note if the numbered side of the bearing case is facing the propeller end of the shaft prior to removal.

3. Clamp the propeller shaft into a vise with protective jaws, then use a slide hammer to remove the bearing

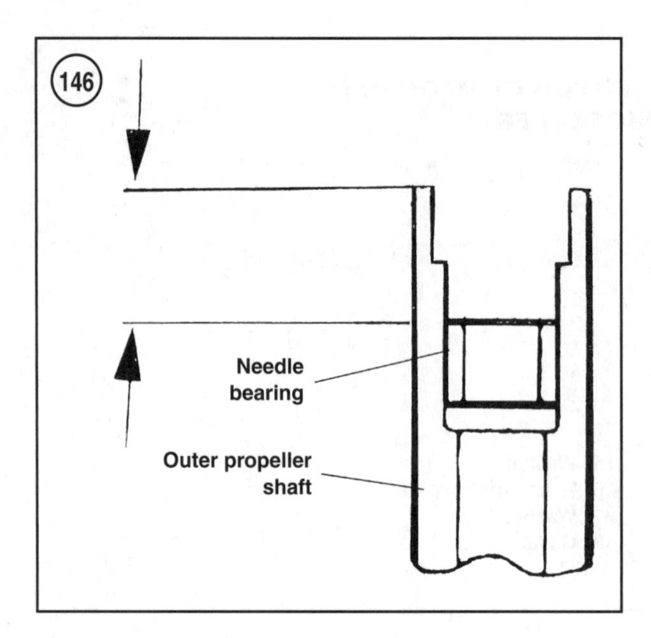

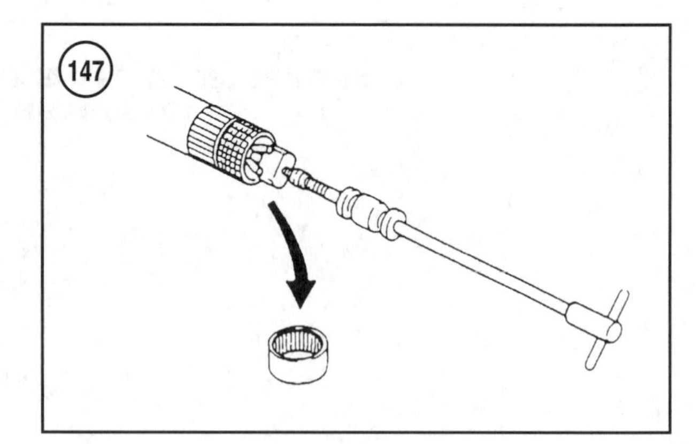

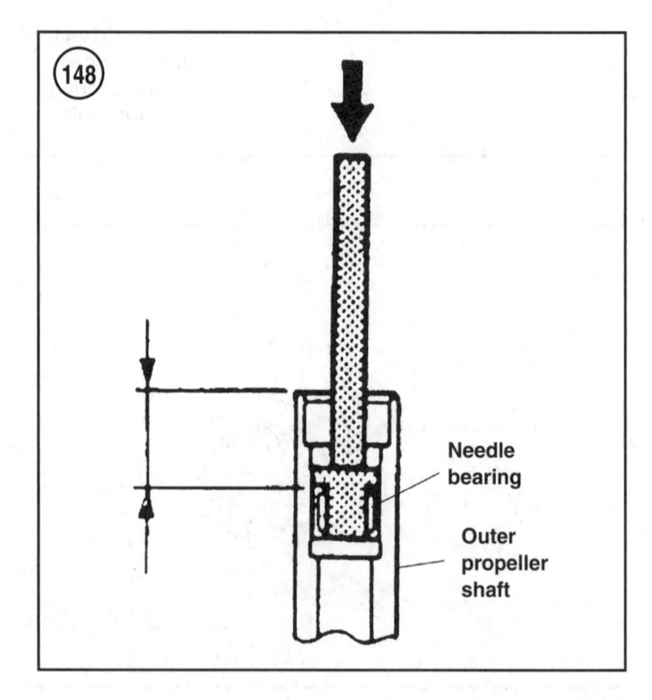

(**Figure 147**). Discard the bearing. Thoroughly clean the propeller shaft.

4. Apply gearcase lubricant onto the needle bearing and the bore in the propeller shaft. Place the needle bearing into the bore from the propeller side. The numbered side must be facing outward as noted prior to removal. Use an appropriately sized socket, section of tubing or rod to drive the bearing into the bore. The tool must be of sufficient diameter to adequately contact the bearing case, yet not contact the bearing bore in the propeller shaft during installation. Gently drive the bearing into the bore, stopping frequently for measurement, until the bearing reaches the exact depth measured prior to removal (**Figure 148**). The bearing must be installed to the propeller depth to ensure the inner propeller shaft contacts the bearing at the proper location.
5. Apply Loctite Primer T to the seal bore in the end of the propeller shaft and the outer diameter of the new seals. When the primer dries, apply Loctite 271 into the seal bore in the propeller shaft and the outer diameter of the seals.
6. Use an appropriately sized socket, section of tubing or rod to drive the seals into the bore. The tool must be of sufficient diameter to adequately contact the seal cases; yet not contact the seal bore in the propeller shaft during installation.
7. Place the first seal into the bore with the open end facing downward or away from the propeller end of the shaft. Carefully drive the seal into the bore until just below the seal bore opening. Wipe excess Loctite from the seals.
8. Place the second seal into the bore with the open end facing upward or toward the propeller end of the shaft. Gently drive the seals into the bore, stopping frequently for measurement, until the second seal reaches the exact depth measured prior to removal (**Figure 149**).
9. Lubricate the needle bearings with gearcase lubricant. Apply Yamaha All-Purpose Grease to the seal lips prior to installing the propeller shaft.

Drive Shaft and Pinion Gear Installation

1. Install the pinion gear into the gearcase with the pinion gear teeth engaged with the front gear teeth.
2. Clean all contaminants from the drive shaft splines, then apply Loctite Primer T to the threaded end of the shaft.
3. Carefully slide the drive shaft into the bore and engage the drive shaft splines with the pinion gear splines.

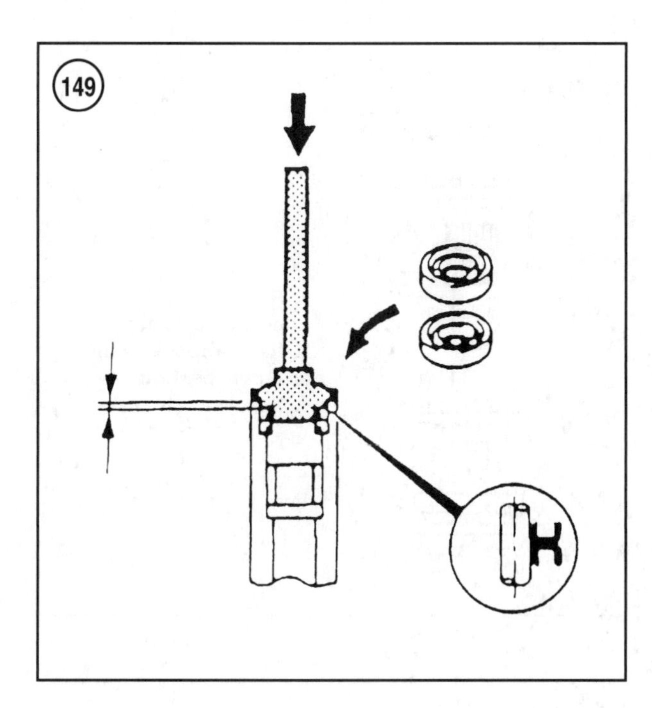

149

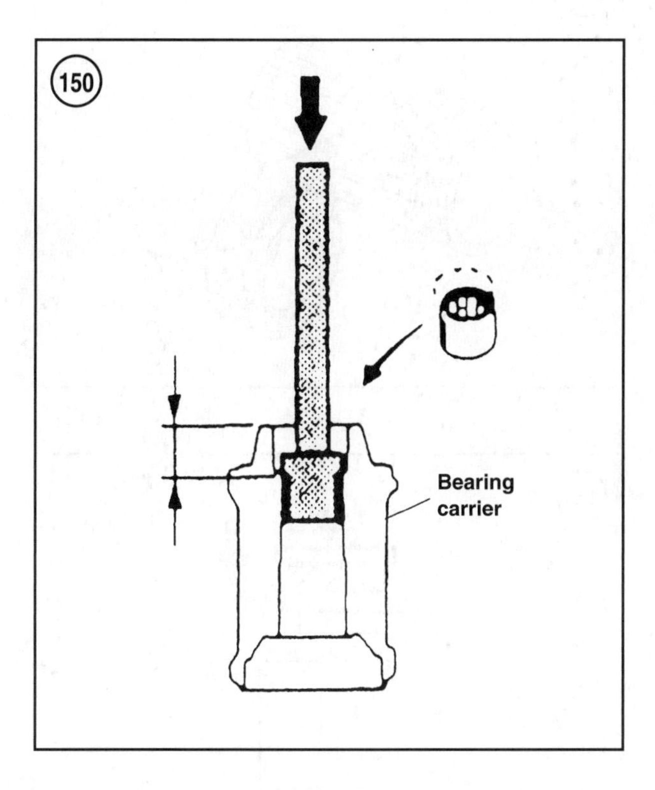

150

4. Apply Loctite 271 to the threaded end of the drive shaft and the threads of the new pinion nut. Thread the pinion nut onto the drive shaft.

5. Refer to the instructions for the standard rotation (RH) 150-200 hp (2.6 liter) models, then tighten the pinion nut to the specification in **Table 1**.

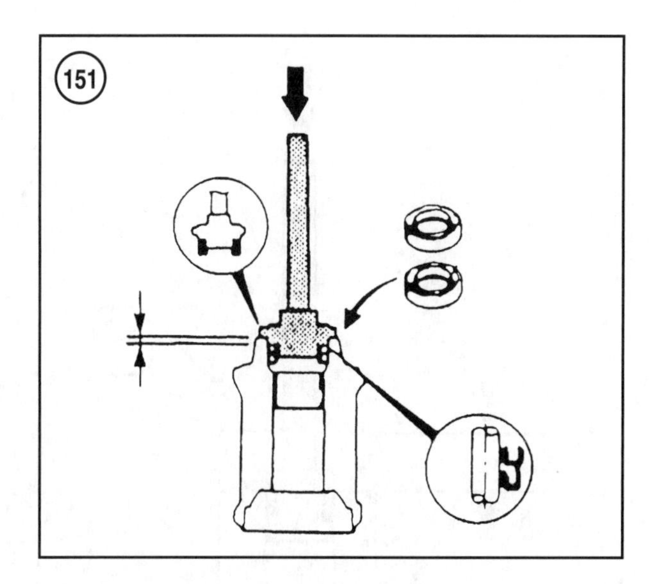

151

Bearing Carrier Assembly and Installation

1. Apply gearcase lubricant to the aft needle bearing and the bore in the bearing carrier. Place the needle bearing into the bore from the propeller side. The numbered side must be facing outward or toward the propeller. Use an appropriately sized socket, section of tubing or rod to drive the bearing into the bore. The tool must contact the bearing case; yet not contact the bearing bore in the bearing carrier during installation. Gently drive the bearing into the bore, stopping frequently for measurement, until the bearing reaches the exact depth measured prior to removal (**Figure 150**). The bearing must be installed to the propeller depth to ensure the outer propeller shaft contacts the bearing at the proper location.

2. Apply Loctite Primer T to the seal bore in the bearing carrier and the outer diameter of the new seals. When the primer dries, apply Loctite 271 into the seal bore in the bearing carrier and the outer diameter of the seals.

3. Use an appropriately sized socket, section of tubing or rod to drive the seals into the bore. The tool must be of sufficient diameter to adequately contact the seal cases, yet not contact the seal bore in the carrier during installation.

4. Place the first seal into the bore with the open end facing upward or toward the propeller end of the carrier. Carefully drive the seal into the bore stop just below the seal bore opening.

5. Place the second seal into the bore with the open end facing upward or toward the propeller end of the carrier. Gently drive the seals into the bore, stopping frequently for measurement, until the second seal reaches the exact depth measured prior to removal (**Figure 151**). Wipe excess Loctite from the seals.

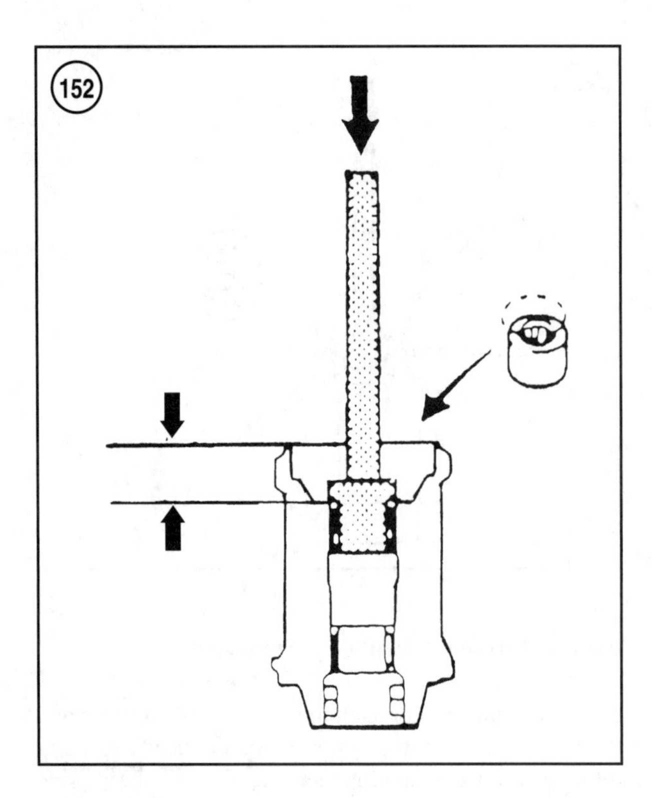

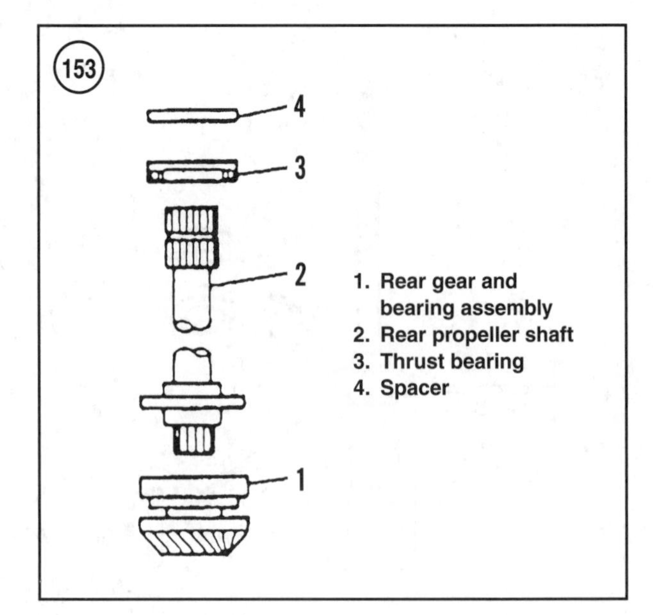

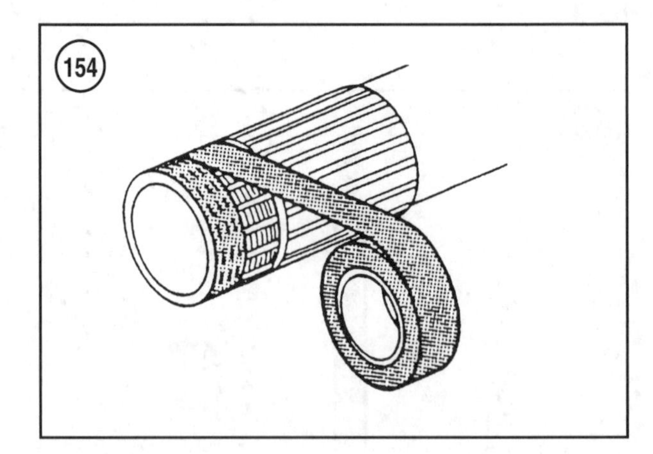

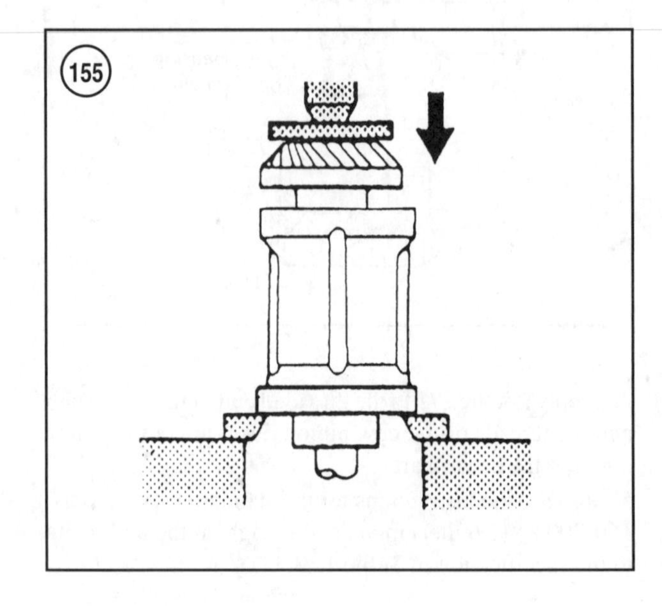

6. Lubricate the needle bearings with gearcase lubricant. Apply Yamaha All-Purpose Grease to the seal lips prior to installing the propeller shaft.

7. Apply gearcase lubricant to the needle bearing that fit into the forward bore in the bearing carrier. Place the needle bearing into the bore from the gear side. The numbered side must face outward, or toward the gears when installed. Use an appropriately sized socket, section of tubing or rod to drive the bearing into the bore. The tool must be of sufficient diameter to adequately contact the bearing case, yet not contact the bearing bore in the bearing carrier during installation. Gently drive the bearing into the bore, stop frequently to measure, until the bearing reaches the exact depth measured prior to removal (**Figure 152**). The bearing must be installed to the propeller depth to ensure the outer propeller shaft contacts the bearing at the proper location.

8. Install the spacer (4, **Figure 153**) into the large opening in the carrier. Place the thrust bearing (3, **Figure 153**) into the opening and position it next to the spacer.

9. Wrap the propeller nut threads and splines with tape (**Figure 154**) to protect the seals from damage while installing the shaft into the carrier.

10. Insert the propeller shaft into the bearing carrier until the flange seats against the thrust bearing. Position the rear gear assembly into the carrier and rotate it until the splines in the gear and propeller shaft align.

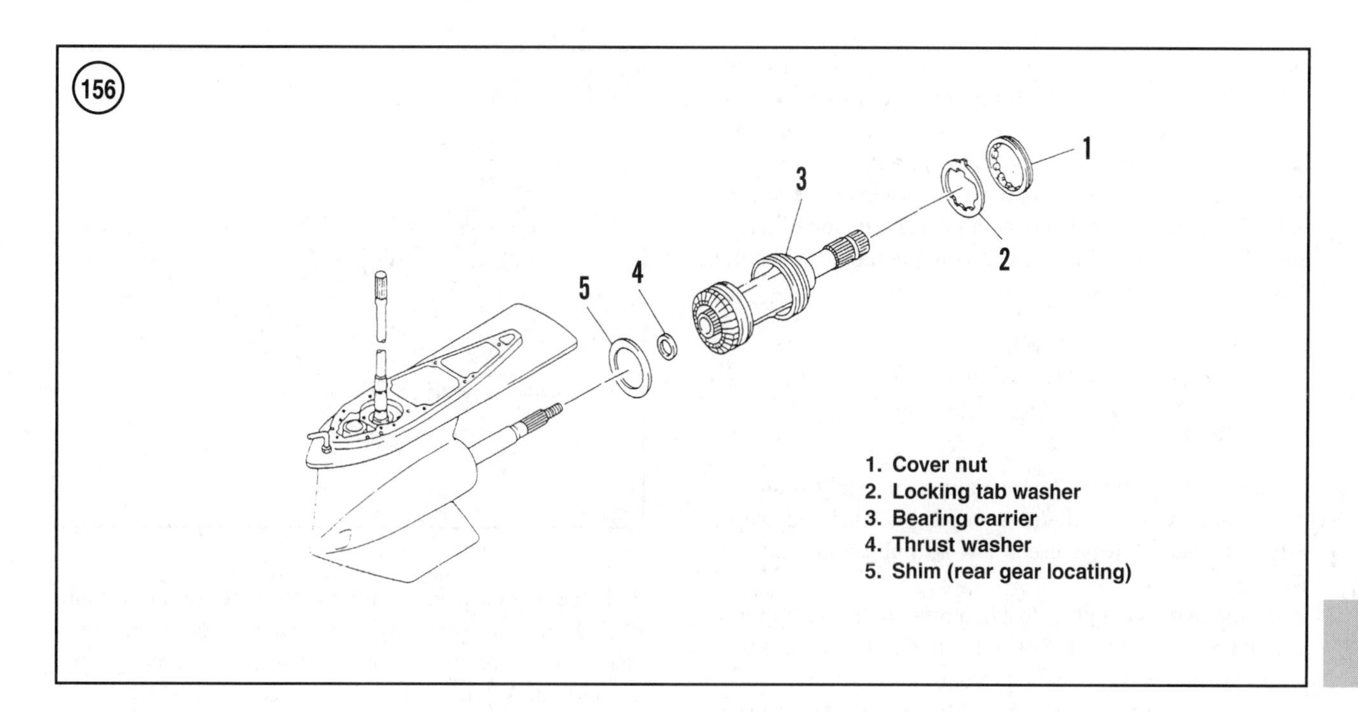

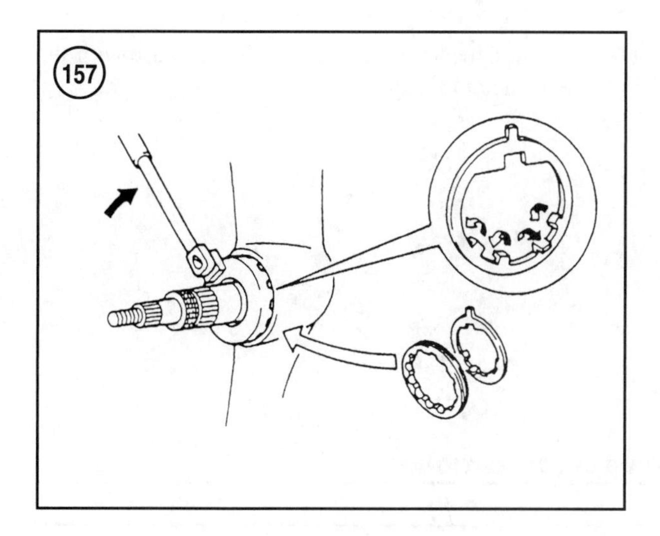

11. Place the carrier onto a press table with the gear side facing upward (**Figure 155**). The seal side of the carrier must be adequately supported prior to pressing against the gear. Using a suitable cushion, press on the gear until it fully seats in the carrier.

12. Place the rear gear shim (5, **Figure 156**) into the gearcase. Make sure the shim fits onto the step in the gearcase bore. Place the thrust washer (4, **Figure 156**) onto the rear (inner) propeller shaft.

13. Lubricate the bearing carrier and carrier contact surfaces with Yamaha All-Purpose Grease. Position the *UP* marking on the carrier facing upward, then carefully slide the carrier, propeller shaft and rear gear assembly into the gearcase. Rotate the propeller shaft to mesh the gear teeth, then hand press the carrier into the gearcase.

14. Place the locking tab washer (2, **Figure 156**) onto the bearing carrier with the tab inserted into the opening in the gearcase housing. Lubricate the cover nut (1, **Figure 156**) and the threads in the gearcase with Yamaha All-Purpose Grease, then hand-thread the cover nut into the gearcase. Use the cover nut tool (Yamaha part No. YB-42223) and a torque wrench to tighten the cover nut (**Figure 157**) to the specification in **Table 1**.

15. Bend one tab into a gap on the cover nut. Bend the other tabs down to prevent restricted exhaust flow. Tighten the cover nut a slight amount if necessary to align the tab with the cover nut slot. Never loosen the cover nut to align the tab.

16. Measure and correct the gearcase backlash as described in this chapter.

17. Install the water pump components as described in this chapter. Refer to standard rotation (RH) 150-200 hp (2.6 liter) model instructions.

GEARCASE PRESSURE TEST

Pressure test the gearcase anytime the gearcase is disassembled. If the gearcase fails the pressure test, find and correct the source of the leakage. Failure to correct any leakage will cause major gearcase damage from water entering the gearcase or lubricant leaking out.

Pressure testers are available from most tool suppliers. If necessary, a pressure tester can be fabricated using a common fuel primer bulb, air pressure gauge, fittings and hoses. If necessary, use the fitting from a gearcase lubricant pump. If using a fabricated pressure tester, clamp the hoses shut with locking pliers after applying pressure with the primer bulb. Otherwise, air may leak past the check valve in the primer bulb, giving a false indication of leakage. If at all possible, use a commercially available gearcase pressure tester.

NOTE
The gearcase lubricant must be drained before pressure testing. Refer to Chapter Three.

1. Verify the gearcase lubricant is completely drained. Then, make sure the fill/drain plug is installed and properly tightened. Always use a new sealing washer on the fill plug.
2. Remove the vent plug. Install a pressure tester into the vent hole. See **Figure 158**. Tighten the tester. Always use a new sealing washer on the pressure tester fitting.
3. Pressurize the gearcase to 69 kPa (10 psi) for at least five minutes. During this time, periodically rotate the propeller and drive shafts and move the shift linkage through the full range of travel.

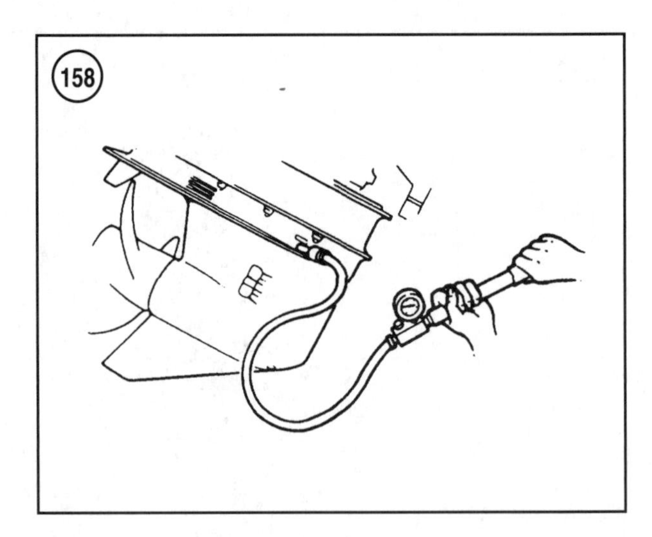

4. If the gearcase does not hold pressure for a minimum of five minutes, pressurize the gearcase again and spray soapy water on all sealing surfaces or submerge the gearcase in water to locate the source of the leak.
5. Correct any leakage before proceeding.
6. Refer to Chapter Three and fill the gearcase with the recommended lubricant.

Table 1 GEARCASE TORQUE SPECIFICATIONS

Fastener	N·m	in.-lb.	ft.-lb.
Bearing carrier bolts			
200-250 hp (3.1 liter)	24	–	17
Cover nut	145	–	107
Gearcase mounting bolts			
115-130 hp	40	–	29
150-200 hp (2.6 liter)	40	–	29
200-250 hp (3.1 liter)	48	–	35
Pinion nut			
115-130 hp	95	–	70
150-200 hp (2.6 liter)	95	–	70
200-250 hp (3.1 liter)	145	–	107
Propeller nut			
Right and left gearcase	55	–	40
Twin propeller gearcase			
Rear propeller	55	–	40
Front propeller	65	–	48

(continued)

Table 1 TIGHTENING TORQUE (continued)

Fastener	N•m	in.-lb.	ft.-lb.
Trim tab/anodic plate bolt			
115-130 hp	43	–	32
150-200 hp (2.6 liter)	40	–	29
200-250 hp (3.1 liter)	43	–	32

Table 2 GEARCASE BACKLASH SPECIFICATIONS

Model	Backlash specification
115 and 130 hp	
Standard (RH) rotation gearcase	
Forward gear backlash	0.32-0.50 mm (0.013-0.020 in.)
Reverse gear backlash	0.80-1.17 mm (0.031-0.046 in.)
Counter (LH rotation gearcase	
Forward gear backlash	0.32-0.45 mm (0.013-0.0177 in.)
Reverse gear backlash	0.80-1.12 mm (0.031-0.044 in.)
150-200 hp (2.6 liter [except P150 and D150)	
Standard (RH) rotation gearcase	
Forward gear backlash	0.25-0.46 mm (0.010-0.018 in.)
Reverse gear backlash	0.74-1.29 mm (0.029-0.050 in.)
Counter (LH rotation gearcase	
Forward gear backlash	0.21-0.43 mm (0.008-0.0169 in.)
Reverse gear backlash	0.97-1.29 mm (0.038-0.050 in.)
P150	
Forward gear backlash	0.71-1.01 mm (0.028-0.040 in.)
Reverse gear backlash	0.79-1.38 mm (0.032-0.054 in.)
D150 (twin propeller)	
Front gear backlash	0.19-0.59 mm (0.007-0.023 in.)
Rear gear backlash	0.39-0.70 mm (0.015-0.027 in.)
200-250 hp (3.1 liter)	
Standard (RH) rotation gearcase	
Forward gear backlash	0.19-0.40 mm (0.007-0.0157 in.)
Reverse gear backlash	0.64-0.93 mm (0.026-0.036 in.)
Counter (LH rotation gearcase	
Forward gear backlash	0.32-0.52 mm (0.013-0.020 in.)
Reverse gear backlash	0.64-0.93 mm (0.026-0.036 in.)

Chapter Nine

Jet Drive

Jet drives (**Figure 1**) offer significant advantages for shallow water operation. The absence of the propeller and gearcase allows the engine to operate in areas much too shallow for a standard propeller drive gearcase.

A drawback is the substantial reduction in efficiency. The performance is generally equivalent to 70 percent of the same basic engine of a standard propeller drive gearcase.

When a jet drive is installed:

1. A 115 hp model becomes an 80 Jet.
2. A 150 hp model becomes an 105 Jet.

Service to the power head, ignition, electrical, fuel and power trim and tilt system is the same as on propeller-driven outboard models. Refer to the appropriate chapter and service section for the engine model using a propeller gearcase. Refer to **Table 1** for specific torque specifications at the end of this chapter and general torque specifications in Chapter One. Use the general tightening specification for fasteners not listed in **Table 1**.

JET DRIVE OPERATION

Jet Drive Components

The major components of the jet drive include the impeller (1, **Figure 2**), intake opening (2), volute tube (3) and the outlet nozzle (4). The impeller connects to a shaft supported by bearings on one end and the crankshaft on the other end. Anytime the engine is running the impeller turns.

The rotating impeller pulls water through the intake opening and pushes it through the volute tube (3, **Figure 2**). The volute tube directs the water toward the outlet nozzle (4, **Figure 2**). The water flowing from the outlet nozzle provides thrust to move the boat. Thrust increases with increased impeller speed.

Thrust Control

Operational direction is controlled by a thrust gate. The thrust gate is controlled by an engine-mounted directional

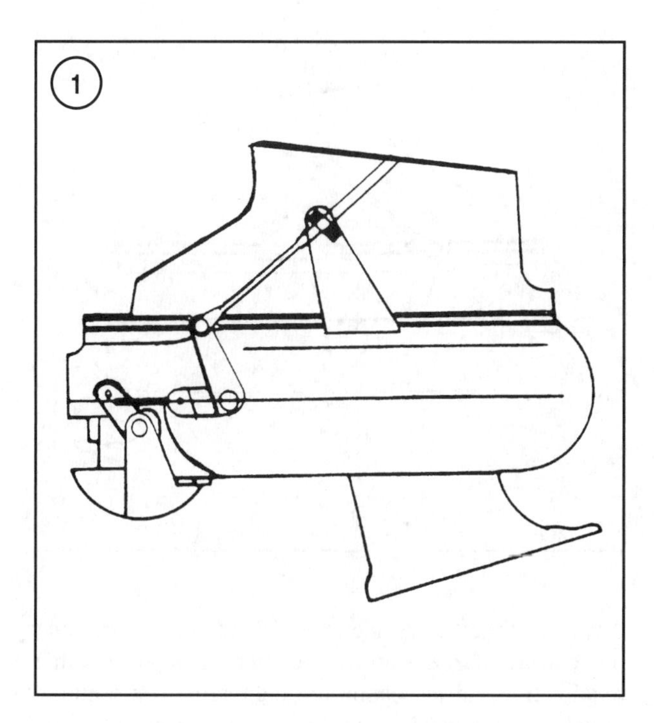

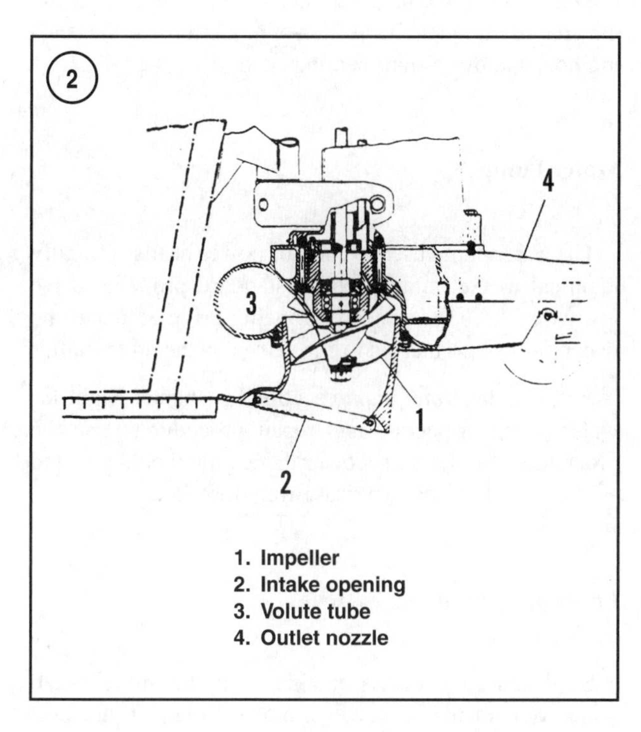

1. Impeller
2. Intake opening
3. Volute tube
4. Outlet nozzle

control cable. When the directional control lever is placed in the full forward position, the thrust gate should completely uncover the jet drive housing outlet nozzle (**Figure 3**). The thrust gate must seat securely against the rubber pad on the jet drive pump housing. This gate position allows water flow from the outlet nozzle for maximum forward thrust. When the directional control lever is

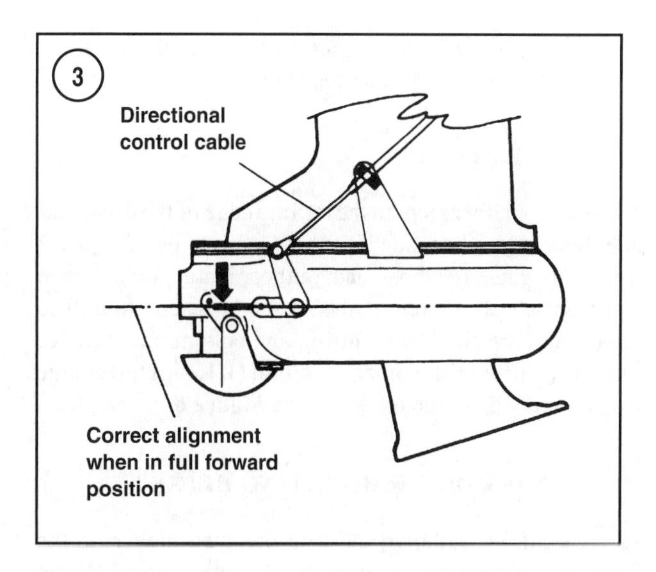

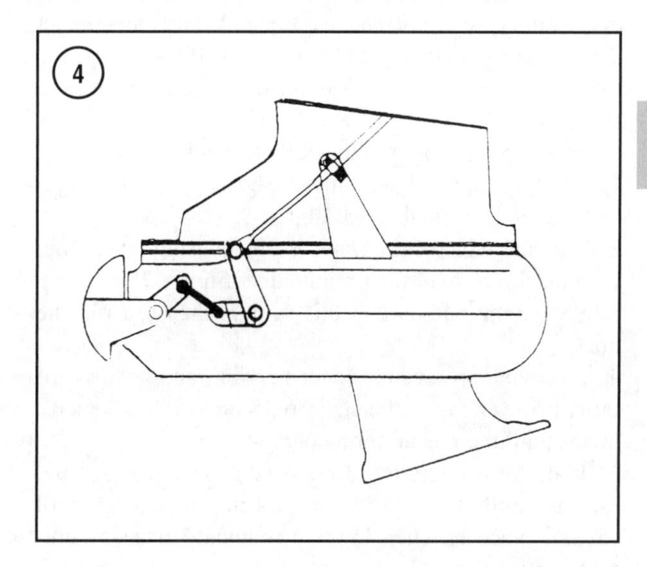

placed in full reverse position (**Figure 4**), the thrust gate should completely block the pump housing outlet nozzle. This directs water under the pump for reverse thrust. Reverse thrust is far less efficient than forward thrust. Neutral thrust gate position is midway between full forward and full reverse position (**Figure 5**). This position directs some of the water exiting the nozzle under the jet drive for balanced thrust and little if there is any boat movement.

Steering the boat is accomplished as the engine pivots port or starboard as with the standard propeller drive gearcases.

WARNING

Steering response is limited and unpredictable when reverse thrust is selected. Practice common backing maneuvers in open water until it feels comfortable using the jet drive backing characteristics.

Steering Torque

A minor adjustment to the trailing edge of the drive outlet nozzle may be made if the boat tends to pull in one direction when the boat and outboard are pointed in a straight-ahead direction. Should the boat tend to pull to the starboard side, bend the top and bottom trailing edge of the jet drive outlet nozzle 1.6 mm (1/16 in.) toward the starboard side of the jet drive. See **Figure 6**.

JET DRIVE MOUNTING HEIGHT

A jet drive outboard must be mounted higher on the transom plate than an propeller-driven outboard. However, if the jet drive is mounted too high, air enters the jet drive causing cavitation and power loss. If the jet drive is mounted too low, excessive drag, water spray and loss in speed occur.

Set the initial height of the outboard as follows:

1. Place a straightedge against the boat bottom (not keel) and the jet drive intake. See **Figure 2**.
2. The fore edge of the water intake housing must align with the top edge of the straightedge (**Figure 2**).
3. Secure the outboard at this setting, then test run the boat.
4. If cavitation occurs (over-revving and/or loss of thrust), lower the outboard in 6.35 mm (1/4 in.) increments until there is uniform operation.
5. If uniform operation occurs with the initial setting, raise the outboard in 6.35 mm (1/4 in.) increments until cavitation occurs. Then, lower the outboard to the last uniform setting.

CAUTION
A slight amount of cavitation in rough water and while turning is normal. However, excessive cavitation damages the impeller and can cause power head overheating.

CAUTION
The outboard should be in a vertical position when the boat is on plane. Adjust the trim setting as needed. If the outboard trim setting is altered, the outboard height must be checked and adjusted.

JET DRIVE REPAIR

Outboards with jet drives typically operate in shallow water. Certain components are susceptible to wear. A jet drive may ingest a considerable amount of sand, rock and other debris during normal operation. Components that require frequent inspection are the engine water pump, impeller and intake housing liner. Other components that may require frequent inspection or repair include the bearing housing, drive shaft bearings and drive shaft.

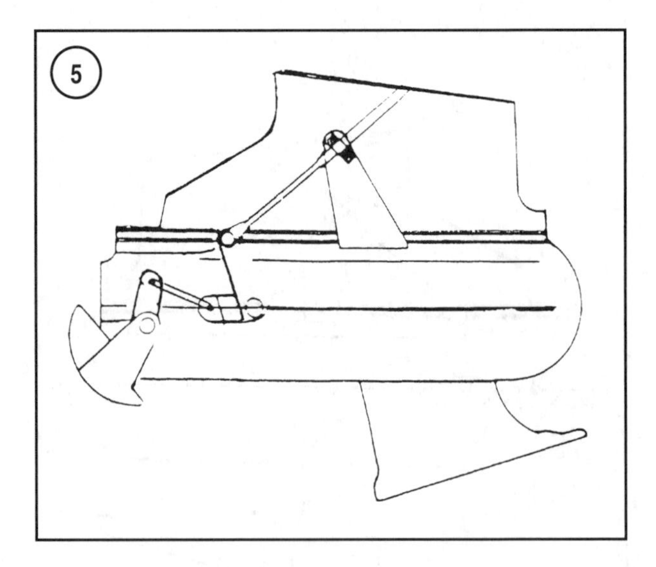

Water Pump

The water pump used to cool the power head is virtually identical to the pump used on standard propeller drive gearcases. Like the standard propeller drive gearcase, the water pump is connected to and driven by the drive shaft.

Refer to the *Water Pump* section in Chapter Eight for water pump inspection and repair procedures. Use the procedures for the horsepower the engine would be rated as if a propeller drive gearcase were installed.

Flushing the Cooling System

Sand and other debris quickly wear the impeller and other components of the water pump. Frequently inspect the water pump if operating the engine in sand or silt laden water. Sand and other debris may collect in the power head cooling water passages. These deposits eventually lead to overheating and increased corrosion. Flush the power head after each period of operation. This important maintenance step can add years to the life of the outboard. Refer to Chapter Three for power head flushing procedures and water pump maintenance intervals.

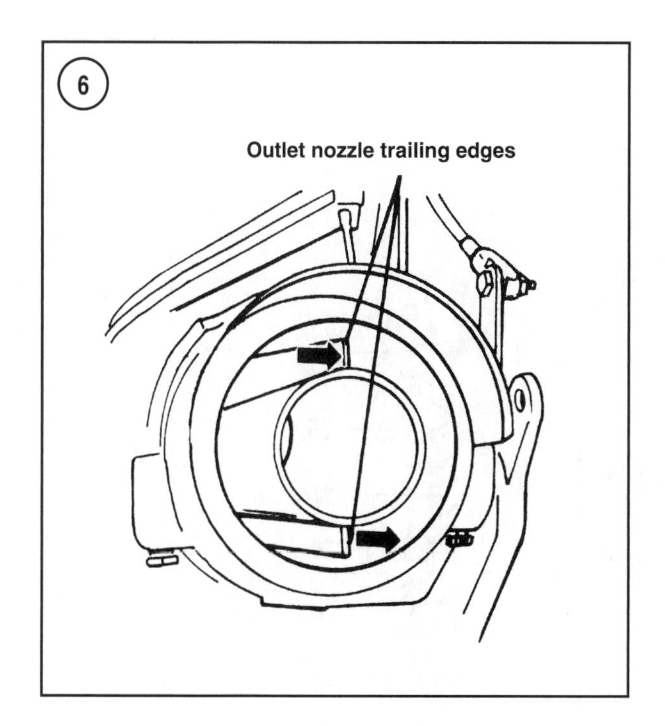

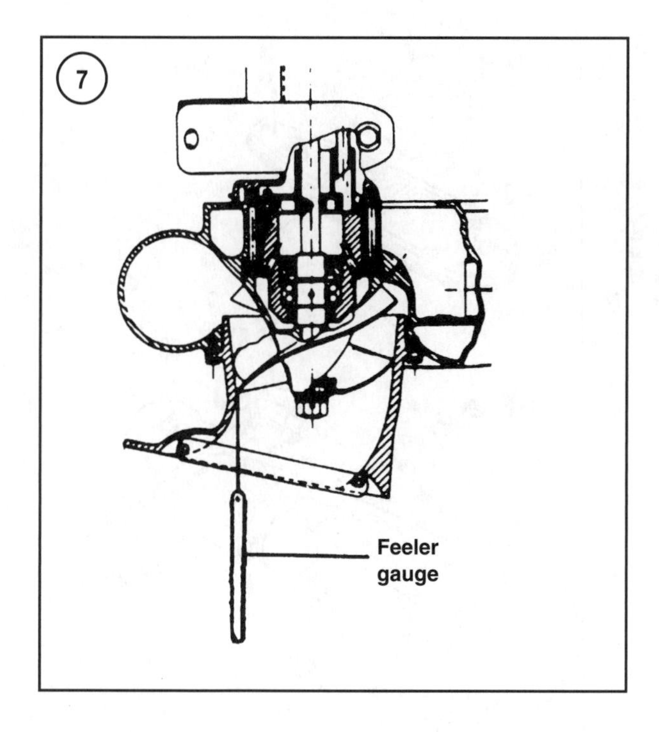

Bearing Lubrication

Lubricate the jet pump bearing(s) after *each* operating period, after every 10 hours of operation and prior to storage. In addition, after every 30 hours of operation, pump additional grease into the bearing(s) to purge any moisture. Refer to Chapter Three for bearing housing lubrication procedures.

Impeller to Housing Clearance Measurement

Worn or damaged surfaces on the impeller or the intake housing allow water to slip past the impeller and cause decreased efficiency. Measure the impeller clearance if there is increased engine rpm with a top speed or power loss.

1. Remove all spark plug leads and disconnect both battery cables from the battery.
2. Locate the intake grate on the bottom of the jet drive. Use long feeler gauges to measure the clearance between the impeller edges and the intake housing as indicated in **Figure 7**. Carefully rotate the flywheel and check the clearance in several locations. Determine the *average* clearance measured.
3. Correct average clearance is approximately 0.8 mm (0.030 in.). Small variations are acceptable as long as the impeller is not contacting the intake housing and the engine is not exceeding the maximum rated engine speed. Refer to *Impeller shimming* if there is excessive clearance or the impeller is contacting the intake housing.
4. Attach the spark plug leads. Connect the battery cables to the battery.

Jet Drive Removal and Installation

Disconnect the directional control cable (**Figure 3**) from the thrust gate lever. Remove the intake housing to gain access to the jet drive mounting bolts. Refer to *Intake housing removal and installation* in this chapter for procedures. Follow the procedures listed in Chapter Eight for removal and installation for the propeller drive gearcase and remove the jet drive from the engine. Disregard the references to the shift shaft when removing or installing a jet drive unit.

Install the intake housing as described in this chapter. Check and correct the directional control cable adjustment after installation. Adjustment procedures are described in Chapter Four. Always check for proper cooling system operation after installing the jet drive.

Intake Housing Removal and Installation

Remove the intake housing if shimming the impeller or if the impeller and housing require inspection for worn or damaged components. To assist with component identification and orientation, refer to **Figure 8** during the disassembly and assembly of the jet drive.

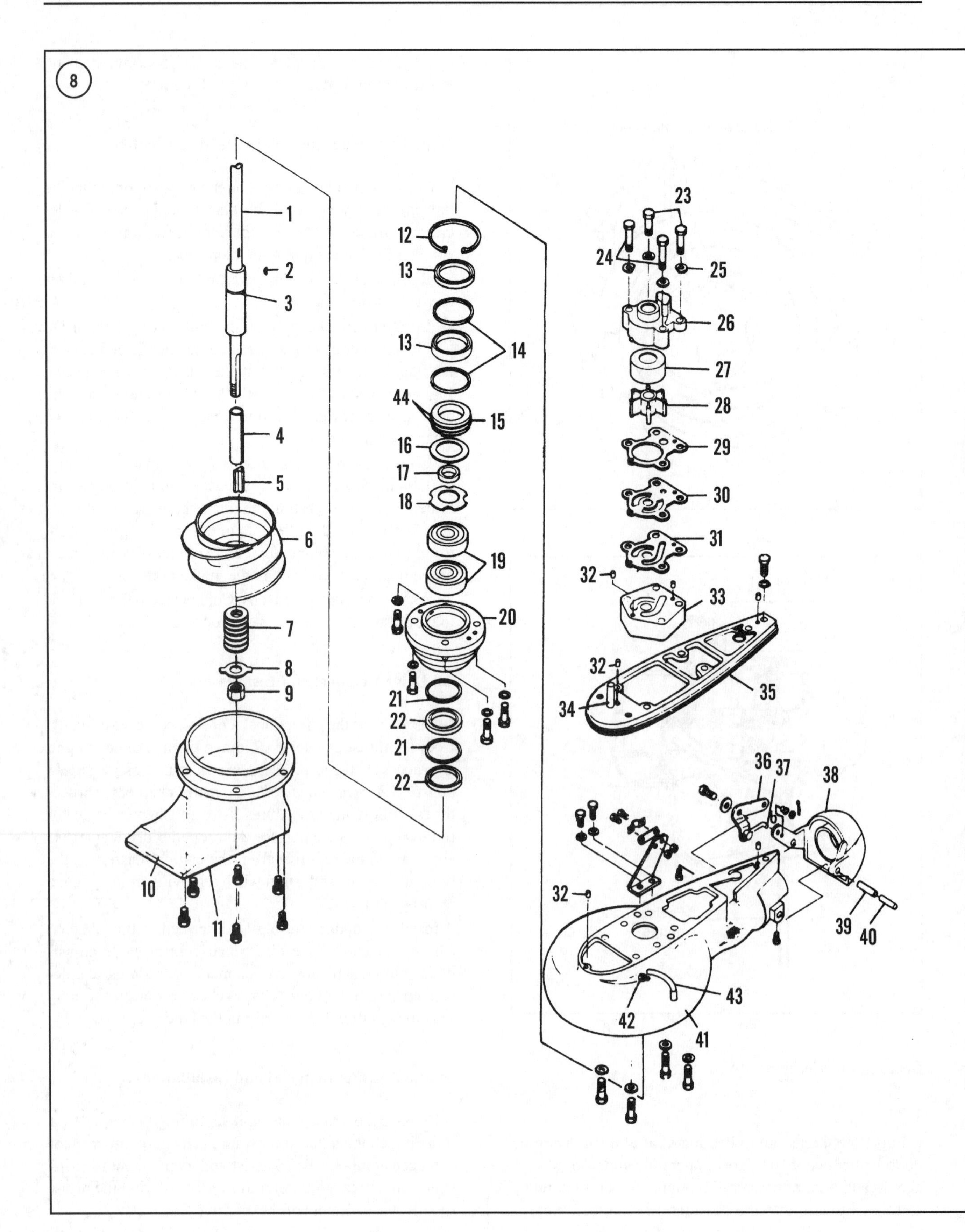
8
1
2
3
4
5
6
7
8
9
10
11
12
13
13
14
44
15
16
17
18
19
20
21
22
21
22
23
24
25
26
27
28
29
30
31
32
33
32
34
35
36
37
38
32
39
40
43
42
41

JET DRIVE COMPONENTS (80 JET AND 105 JET MODELS)

1. Drive shaft
2. Drive key (engine water pump)
3. Thrust ring
4. Sleeve
5. Impeller drive key
6. Impeller
7. Shims
8. Tab washer
9. Impeller nut
10. Intake housing
11. Intake grate opening
12. Snap ring
13. Seal
14. Seal retainer
15. Seal carrier
16. Washer
17. Collar
18. Thrust washer
19. Bearings
20. Bearing housing
21. Seal retainer
22. Seal
23. Bolts (rear)
24. Bolts (front)
25. Washer
26. Water pump cover
27. Insert
28. Impeller
29. Gasket
30. Wear plate
31. Gasket
32. Locating pin
33. Spacer
34. Locating pin
35. Adapter plate
36. Gate lever
37. Linkage
38. Gate
39. Sleeve
40. Pin
41. Jet drive housing
42. Grease fitting
43. Vent hose
44. O-rings

1. Disconnect and ground the spark plug leads. Disconnect the battery cables.

2. Remove the six intake housing mounting bolts. Pull the intake housing from the jet drive. Carefully tap the housing loose with a rubber mallet if necessary.

3. Clean and inspect the housing mounting bolts and bolt openings. Replace corroded or damaged bolts and renew threaded openings as needed.

4. Clean debris from the intake housing and jet drive mating surfaces.

5. Inspect the inner surfaces of the intake housing liner for deep scratches, eroded or damaged areas. Replace the liner if these conditions are present and if excessive impeller to housing clearance was measured. Refer to *Intake housing Liner Replacements.*

6. Install the intake housing onto the jet drive housing with the lower end facing the rear. Apply Yamaha All-Purpose Grease to the threads of the bolts. Tighten them to the specification in **Table 1**.

7. Connect all spark plug leads. Connect the battery leads to the battery. Run the engine and check for proper operation before putting the engine into service.

Intake Housing Liner Replacement

Removing the intake grate may be necessary for proper access to the liner.

1. Refer to *Intake housing removal and installation* for procedures, then remove the intake housing from the jet drive.

2. Locate the bolts on the side of the intake housing that retain the liner into the housing. Note the locations. Then remove the bolts.

3. Tap the liner loose with a long punch through the intake grate opening. Position the punch tip on the edge of the liner and carefully drive it from the housing.

4. Apply a light coating of Yamaha All-Purpose Grease to the liner bore in the intake housing. Carefully slide the new liner into position. Align the boltholes in the liner with the boltholes in the housing.

5. Apply Yamaha All-Purpose Grease to the threads of the bolts. Install the bolts and evenly tighten them to the specification in **Table 1**. Use a file to remove any burrs or bolt material protruding into the intake housing.

6. Install the intake housing onto the jet drive as described in this chapter.

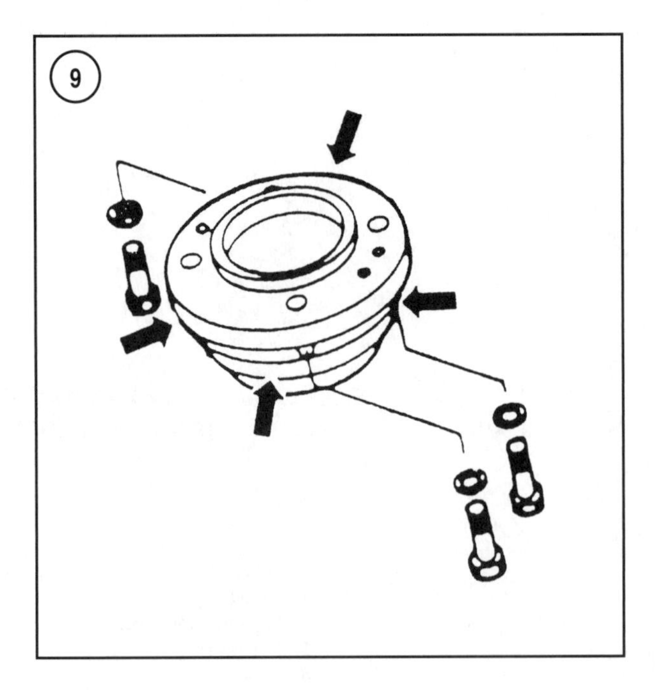

Impeller Removal and Installation

The jet drive must be removed from the engine for this procedure. Access to the upper end of the drive shaft is required to remove and install the impeller nut.

1. Remove the jet drive from the engine as described in this chapter.
2. Remove the intake housing as described in this chapter.
3. Use the drive shaft adapter (Yamaha part No. YB-6201) to grip the splined end of the drive shaft when removing or installing the impeller nut.
4. Use a screwdriver to bend the tabs of the tab washer (8, **Figure 8**) away from the impeller nut. Note the mounting location and orientation of all components prior to removal. Use a suitable socket and adapter to remove the impeller nut, tab washer and lower shims. Discard the tab washer.
5. Carefully pull the impeller, drive key, sleeve and upper shims from the drive shaft. Use a rubber mallet to carefully tap the impeller loose from the shaft if necessary.
6. Inspect the impeller for worn or damaged surfaces and replace the impeller if there are defects.
7. Place the sleeve and drive key (4 and 5 **Figure 8**) onto the drive shaft. Apply Yamaha All-Purpose Grease to the drive key and the impeller bore. Carefully slide the upper shims, then the impeller onto the drive shaft.
8. Place the lower shims, new tab washer (8, **Figure 8**) and impeller nut (9) onto the drive shaft. Make sure the shims and tab washer are in the proper position. Use the drive shaft adapter and suitable socket to tighten the im-

peller nut to the general torque specification in Chapter One.

9. Bend the tabs up to secure the impeller nut. Install the intake housing onto the jet drive as described in this chapter. Refer to *Measuring impeller to housing clearance* for procedures, then measure and correct the clearance. Install the jet drive onto the engine as described in this chapter.

Impeller Shimming

Shimming the impeller is necessary to ensure proper impeller to housing clearance. The intake housing and impeller must be removed for each shim change. Refer to the appropriate section in this chapter for removal and installation instructions. For component identification and orientation, refer to **Figure 8**.

Refer to *Impeller to Housing Clearance Measurement* after each shim change for measurement procedures. Repeat the process until the clearance is correct.

Nine shims are used on all models covered in this manual. All shims are 0.8 mm (0.030 in.) thick and are located both above and below the impeller.

1. Remove the jet drive as described in this chapter.
2. Remove the intake housing from the jet drive as described in this chapter.
3. Note the location and orientation of all components, then remove the impeller from the drive shaft. Refer to *Impeller Removal and Installation* for procedures.
4. If too much impeller clearance is present, remove a shim from below the impeller and place it above the impeller.
5. If too little clearance is present, remove a shim from above the impeller and place it below the impeller.
6. Install the impeller and intake housing onto the jet drive as described in this chapter. Repeat steps 1-5 until the clearance is correct.

Bearing Housing Removal

Drive shaft bearing inspection and/or replacement is required when contaminants or a significant amount of water is found while performing routine maintenance to the jet drive. Replace all seals, gaskets and O-rings anytime they are removed. The bearing housing must be removed to access bearings, shafts, seals and other components.

1. Remove the jet drive from the engine as described in this chapter.
2. Remove the intake housing and impeller following the procedures described in this chapter.
3. Remove all engine water pump components. Refer to *Water Pump* in Chapter Eight for procedures. Use the procedures for the horsepower the engine would be rated if a propeller drive gearcase was installed.
4. Remove the spacer plate (33, **Figure 8**) from the jet drive housing. Locate and remove the bolts and washers securing the bearing housing (**Figure 9**) onto the jet drive housing.
5. Carefully pull the bearing housing from the jet drive housing and place it on a clean work surface.

Drive Shaft Bearing Removal and Installation

WARNING
The components of the bearing housing become very hot during disassembly and assembly. Take all necessary precautions to avoid injury.

WARNING
Never use a flame or allow sparks in an area where any combustible material is present. Always have a suitable fire extinguisher nearby when using a flame or other heat producing device.

CAUTION
Pay strict attention to the direction the bearings are installed onto the drive shaft. The inner race surfaces are wider on one side. The inner race surfaces must be installed correctly to ensure a durable repair.

Refer to **Figure 11** during disassembly and assembly procedures. Note the location and orientation of all components before removing them from the bearing housing. Rotate the drive shaft to check the bearings for rough operation. Replace the bearings if they feel rough or loose. Move the drive shaft to check for excessive axial or radial play (**Figure 10**). Replace the bearings if there is excessive play.

Do not remove the bearing from the drive shaft unless replacement is required. The removal process may damage the bearings. Bearing removal requires applying heat to the bearing housing. Replace the seals as they are usually damaged from the heat or removal process.

1. Carefully pry the snap ring (1, **Figure 11**) from the groove in the bearing housing bore (12).
2. Secure the bearing housing into a vise with protective jaws. The impeller end of the housing must face upward. Thread the impeller nut onto the drive shaft to protect the threads. Apply moderate heat to the areas noted with arrows (**Figure 9**). Use a block of wood and hammer to tap

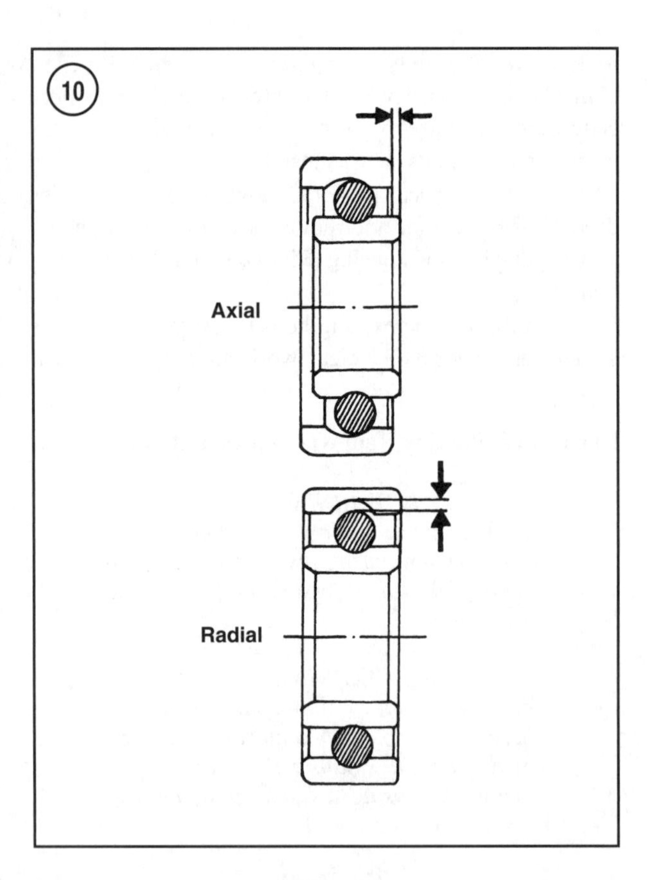

on the installed nut end while applying heat. Continue until the shaft and bearing slide from the housing.

3. Remove the seal carrier (6, **Figure 11**) from the drive shaft. Note the orientation of the seal lips. Then carefully pry the grease seals (2 and 4, **Figure 11**) and both retaining rings (3 and 5) from the carrier. Remove the O-rings from the seal carrier and discard them.

4. Note the location and orientation of the shim, collar and thrust washer (7-9, **Figure 11**), then remove them from the bearing housing bore.

5. Note the seal lip direction, then carefully pry the retaining rings and both seals (13-16, **Figure 11**) from the bearing housing bore. Discard the seals. Press the bearing(s) from the drive shaft. See **Figure 12**, typical.

6. Inspect the drive shaft for damaged surfaces or excessive wear on the seal contact areas. Replace the drive shaft if there are any defects. A deeply grooved shaft allows water into the bearings that causes eventual component failure.

7. Place the thrust washer (9, **Figure 11**) onto the drive shaft. Note the correct orientation of the bearing thrust surfaces (wide side of the inner race) as indicated in **Figure 13** or **Figure 14**, and slide the bearings(s) onto the shaft. Use a section of tubing or pipe to press the new

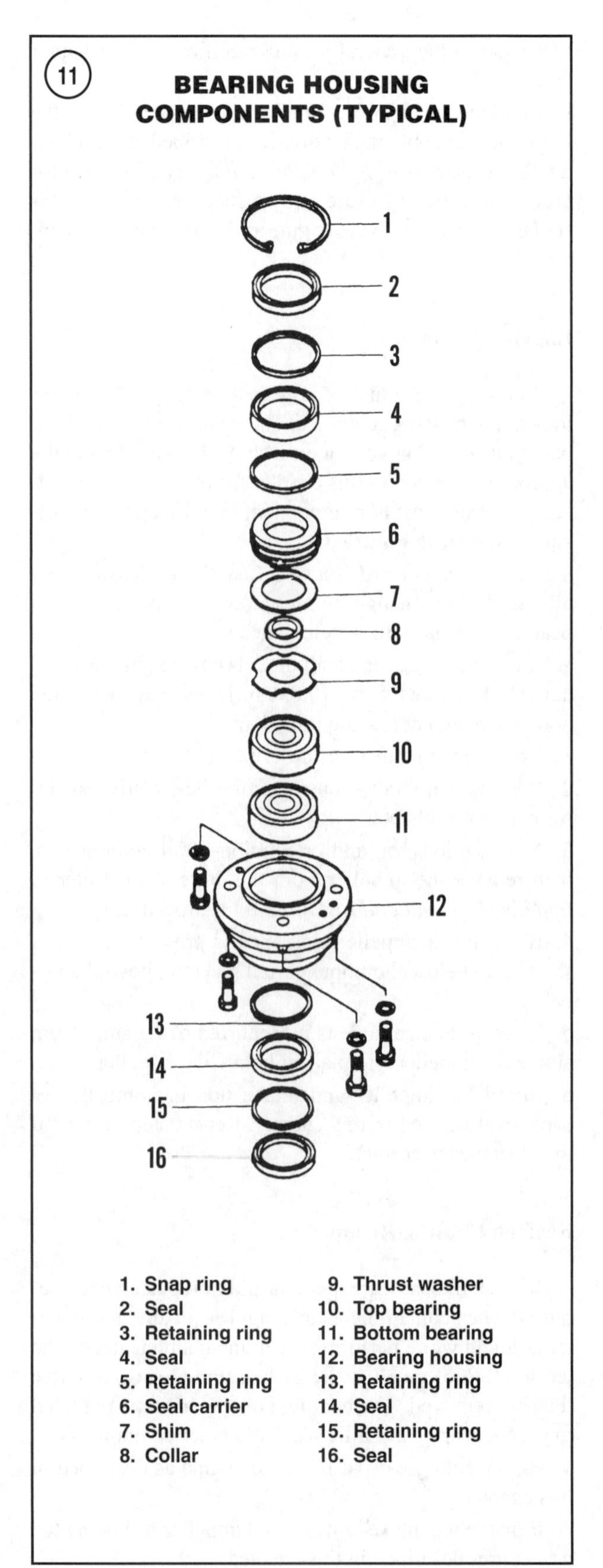

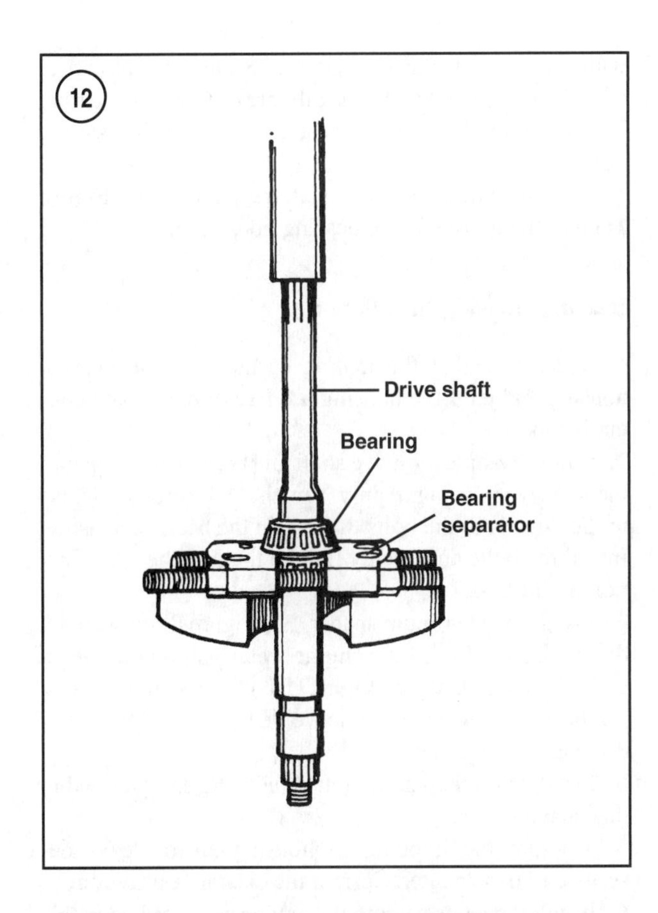

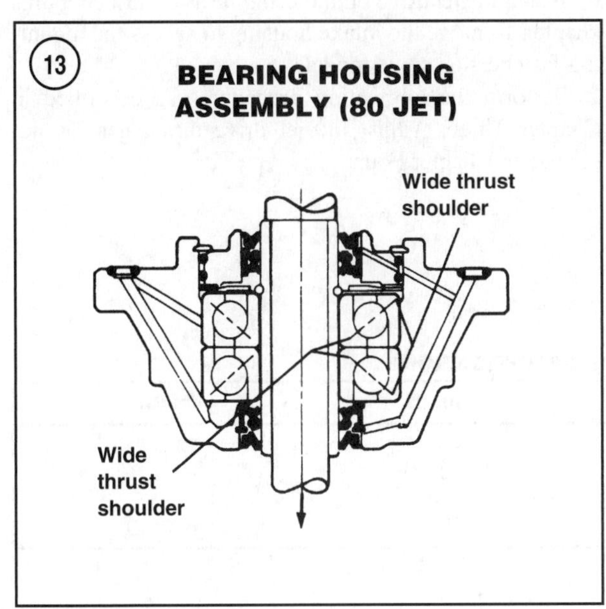

bearing(s) onto the drive shaft. The tool used must touch only on the inner bearing race.

8. Use Loctite Primer T to clean all contaminants from the seal bore in the bearing housing. Select a large socket

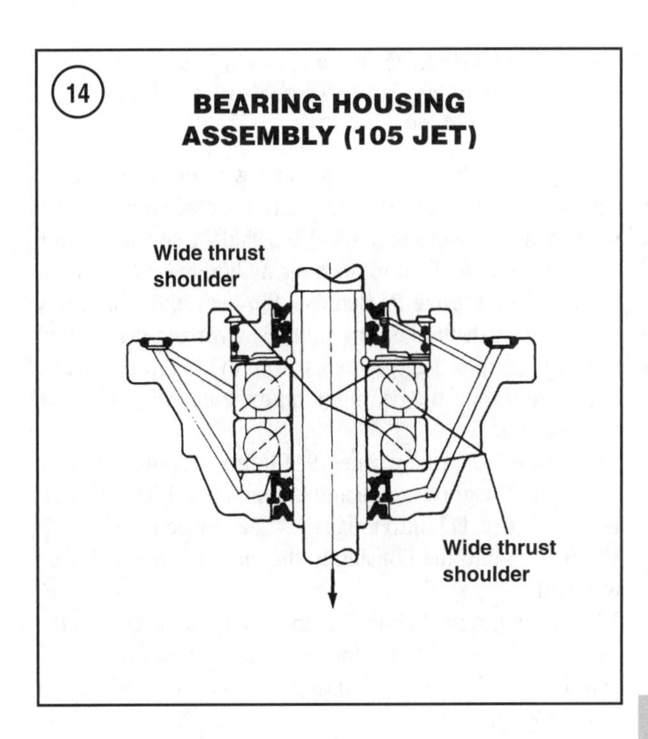

or section of tubing to install the seals into the housing. The tool selected must contact the seal; yet not contact the housing during seal installation.

9. Apply Loctite Primer T to the outer surface of the seal. When dry, apply Loctite 271 onto the seal bore in the housing and the outer diameter of the seal. Place the seal into the housing with the seal lip (open side) facing in the direction noted prior to removal. Use the selected tool to press the seal into the bearing housing until it fully seats.

10. Install the retaining ring into the groove in the housing bore. Apply Loctite 271 onto the outer diameter of the second seal. Place the seal into the housing with the seal lip (open side) facing in the direction noted prior to removal. Use the selected tool to press the seal into the bearing housing until it fully seats against the retainer. Place the second retaining ring into the groove in the housing bore. The notched area must align with the small hole in the retaining ring groove. Wipe excess Loctite from the seal area.

11. Apply Yamaha All-Purpose Grease to the seal lips and retaining ring before installing the drive shaft and bearings.

CAUTION

Apply heat to the bearing housing unit to allow it to expand and slide over the bearing(s). Excessive heat damages the seal and retainers. Use an appropriately sized section of tubing and a press, if necessary, to fully seat the bearing(s) into the housing.

Press only on the outer race of the bearing. Use light pressure. Never drive the bearing(s) into the housing.

12. Place the drive shaft into a vise with protective jaws. The impeller side of the drive shaft must face upward. Set the bearing housing onto the drive shaft. Use a heat lamp or torch to apply heat to the bearing housing in the areas indicated in **Figure 9**. Remove the heat and use heavy gloves to push the bearing housing over the drive shaft bearings. Apply heat gradually until the bearing housing slides fully onto the bearing(s) and the bearing(s) seat in the housing.
13. Install the thrust washer (9, **Figure 11**) onto the drive shaft with the gray side facing the impeller. Place the collar (8, **Figure 11**) onto the drive shaft. Place the shim (7, **Figure 11**) into the housing in the direction noted before removal.
14. Apply Loctite Primer T to the seal bore surfaces in the carrier (6, **Figure 11**) and the outer diameter of the seals (2 and 4). Install the first retaining ring into the seal carrier bore groove.
15. Place the seal into the seal bore opening with the seal lip (open side) facing in the direction noted prior to removal or as indicated in **Figure 13** or **Figure 14**. Use an appropriately sized socket or section of tubing to push the first seal into the carrier. The tool must contact the seal, yet not contact the seal bore during installation. Push until the seal seats fully against the retainer.
16. Place the second retainer into the groove in the seal carrier bore. Push the second seal into the seal carrier bore as described in Step 15. Place new O-rings onto the seal carrier. Lubricate the seal lips and all seal carrier surfaces with Yamaha All-Purpose Grease. Slide the seal carrier over the drive shaft and carefully press it into the bearing housing bore. The carrier must seat fully in the bearing housing.
17. Use snap ring pliers to install the snap ring (1, **Figure 11**) into the groove in the bearing housing bore.

Bearing Housing Installation

1. Make sure that the mating surfaces of the bearing housing and jet drive housing are free of debris and contaminants.
2. Carefully slide the drive shaft and bearing housing into the jet drive housing. Apply Yamaha All-Purpose Grease to the threads of the bolts that retain the bearing housing. Install the bolts and evenly tighten them to the specification in **Table 1**.
3. Install the aluminum spacer (33, Figure 8) onto the jet drive housing. Install the engine water pump onto the jet drive. Refer to Chapter Eight. Use the procedures listed for the horsepower the engine would be rated if a propeller drive gearcase was installed.
4. Install the impeller and intake housing as described in this chapter.
5. Measure the impeller to housing clearance as described in this chapter. Correct the clearance as required.
6. Install the jet drive onto the engine as described in this chapter. Remove the intake housing to access the mounting fasteners on some models.
7. Perform the lubrication procedures as described in Chapter Three. Adjust the jet drive thrust gate as described in Chapter Four.

Table 1 JET PUMP TORQUE SPECIFICATIONS

Fastener	N·m	in.-lb.	ft.-lb.
Jet drive mounting bolts	31.2	–	23
Intake housing mounting bolts	10.9	98	–
Intake housing liner bolts	11.3	100	–
Bearing housing mounting bolts	7.9	70	–

Chapter Ten

Power Tilt/Trim

This chapter provides removal, installation and minor repair procedures for the power tilt and trim systems.

Table 1 lists torque specifications for most trim/trim system components. Use the general torque specifications (Chapter One) for fasteners not listed in **Table 1**. **Table 2** lists electric trim motor specifications. **Table 1** and **Table 2** are at the end of the chapter.

WARNING
*Never work under any part of the engine without having suitable support. The engine mounted tilt lock or hydraulic system may unexpectedly collapse causing the engine to drop. Support the engine with suitable blocks or an overhead cable **before** working under the engine.*

WARNING
The trim system may contain fluid under high pressure. Always use protective eyewear and gloves when working with the trim system. Never remove components or plugs without first bleeding the pressure from the system. Follow instructions carefully.

HYDRAULIC TILT/TRIM SYSTEM REMOVAL AND INSTALLATION

Fastener location and component appearance vary by model. Mark the mounting location and orientation of all components prior to removal to ensure proper assembly. Make a sketch of the trim motor wire routing before removal. Improper routing may allow the wire to become pinched during engine operation. Apply Yamaha All-Purpose Grease to all bushings, pivot points and sliding surfaces during assembly. Apply Loctite 242 to the threads of the trim system fasteners. Refer to **Figure 1** for this procedure.

1. Disconnect the battery cables. Disconnect and ground *all* spark plug leads to prevent accidental starting.

2. Trace the wires (1, **Figure 1**) leading from the electric trim motor to the connections on the power head. Disconnect the wires from the relays or engine wire harness and route them out of the lower engine cover. Remove the plastic tie clamps (3, **Figure 1**) from the wire.

3. Locate the access opening for the manual release valve (**Figure 2**) in the starboard or port side clamp bracket. Rotate the valve counterclockwise 2-3 turns (**Figure 3**). With the aid of an assistant, carefully tilt the engine to the full up position. Use an overhead hoist or other suitable means

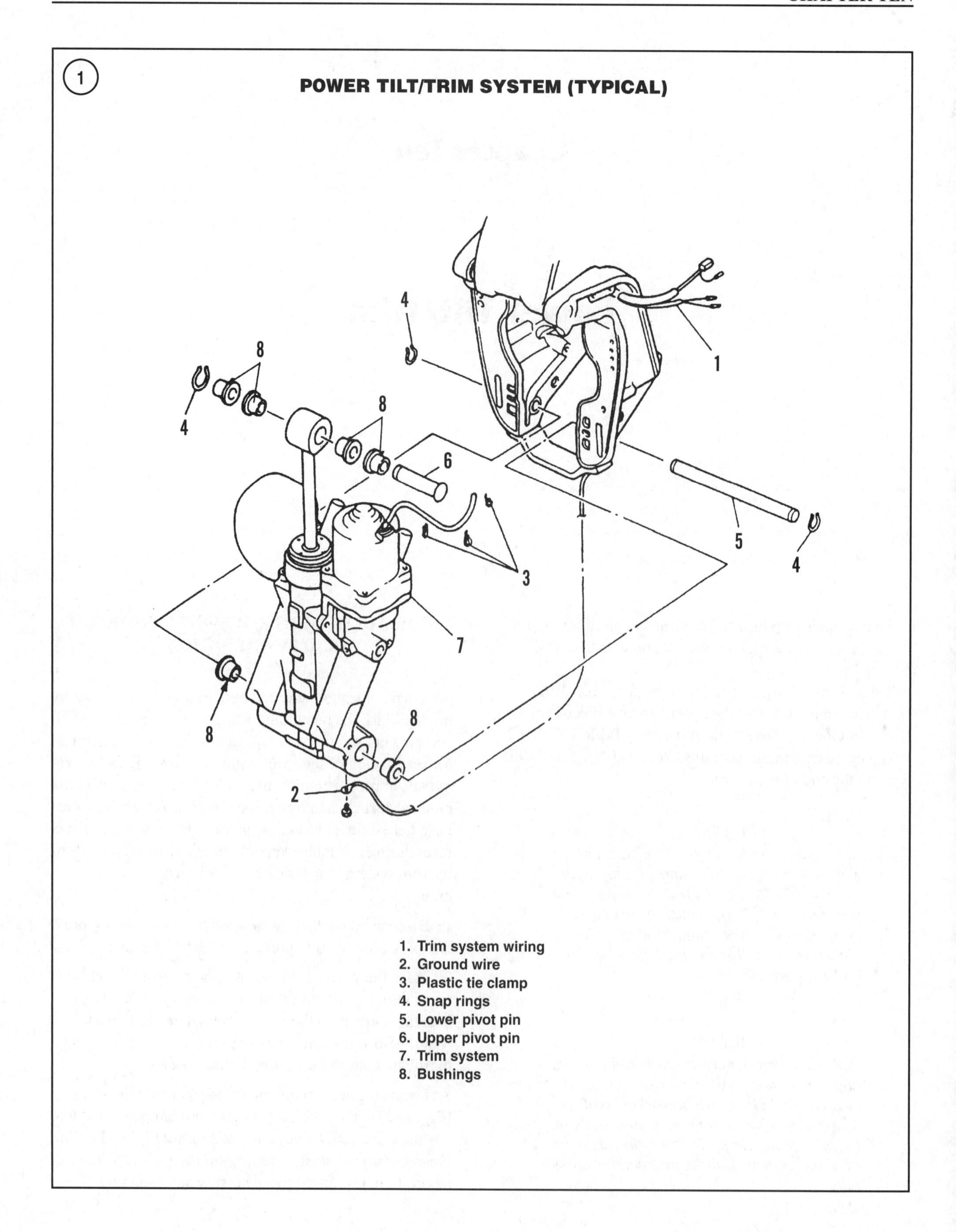
1
POWER TILT/TRIM SYSTEM (TYPICAL)
1
2
3
4
4
4
5
6
7
8
8
8
8
1. Trim system wiring
2. Ground wire
3. Plastic tie clamp
4. Snap rings
5. Lower pivot pin
6. Upper pivot pin
7. Trim system
8. Bushings

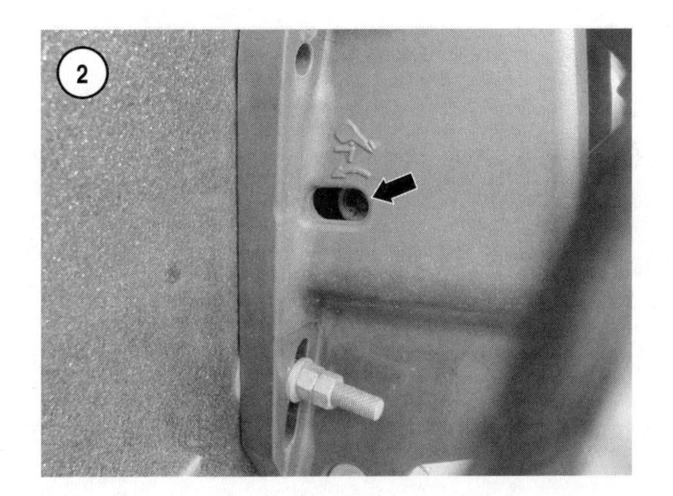

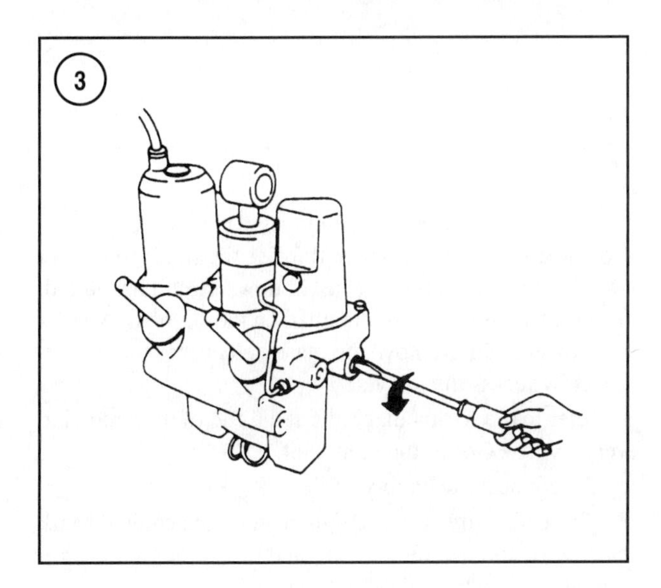

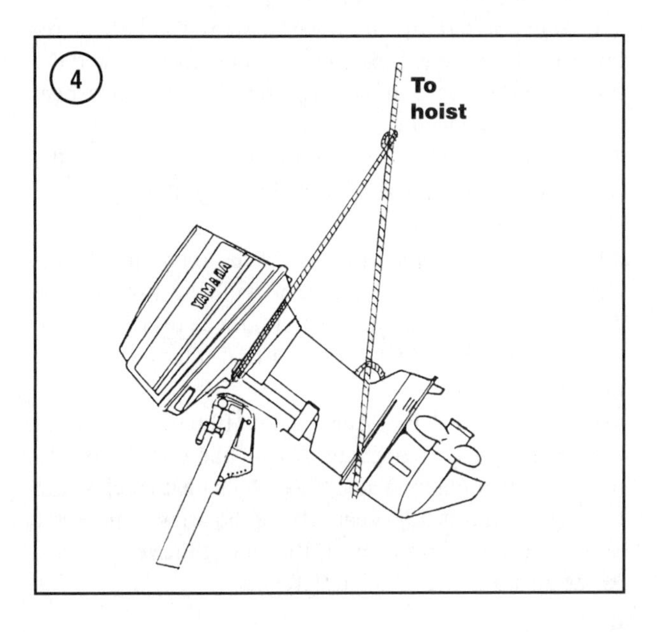

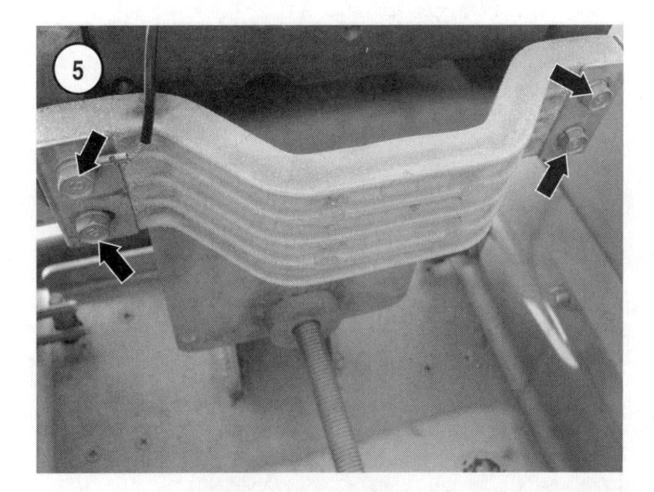

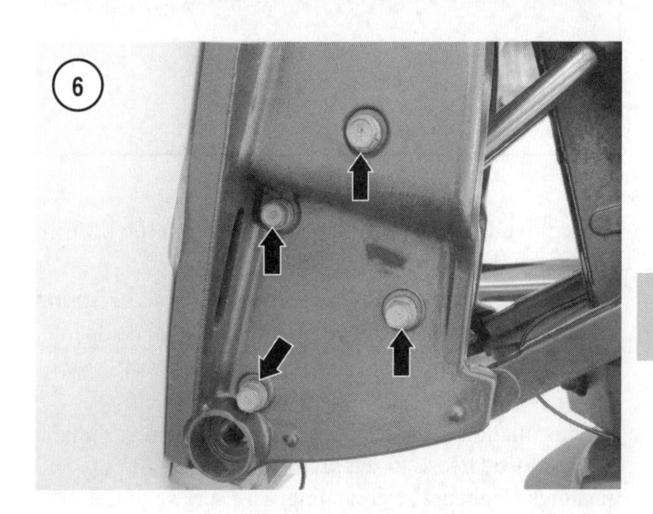

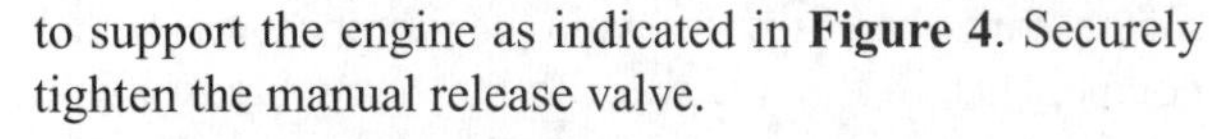

to support the engine as indicated in **Figure 4**. Securely tighten the manual release valve.

4. Disconnect the grounding wires (2, **Figure 1**). Clean and inspect the grounding wire mounting locations and screw threads. Replace damaged wires and renew corroded or damaged threaded openings.

5. *200-250 hp (Saltwater series [3.1 liter]) models*—Remove the four bolts, then pull the anode brackets and anode (**Figure 5**) from the clamp brackets.

6. Remove the snap rings (4, **Figure 1**) from the lower pivot pin. Carefully drive the pivot pin from the clamp brackets and tilt/trim system.

7. *200-250 hp (Saltwater series [3.1 liter]) models*—Remove the eight bolts (four on each side) that secure the trim system onto the clamp brackets (**Figure 6**).

8. Remove the snap ring (4, **Figure 1**) from the port side of the upper pivot pin (6). Support the tilt/trim system and carefully drive the upper pivot pin from the swivel housing. Pull the trim system from the clamp brackets.

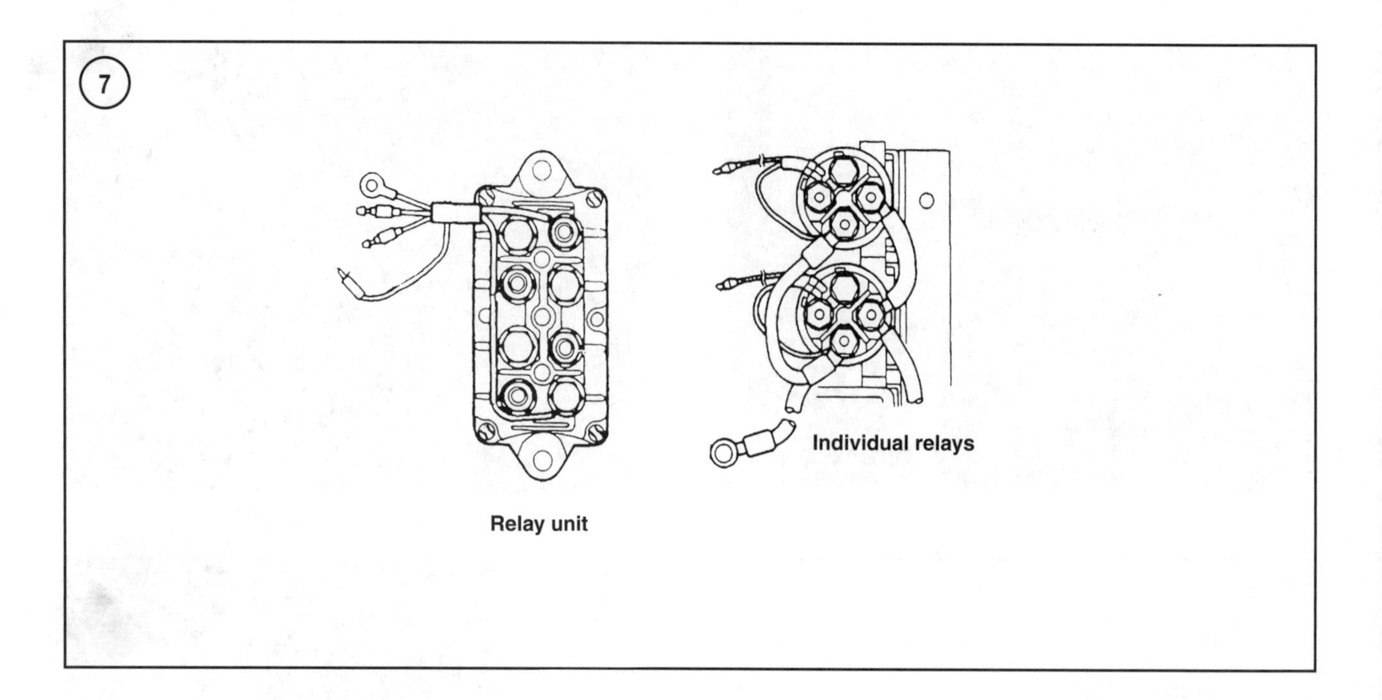

9. Remove the bushings (8, **Figure 1**) from the tilt/trim system, swivel housing and clamp brackets.

10. Inspect the snap rings for corrosion and weak spring tension. Replace the snap rings if they are questionable. Inspect all bushings and pivot pins for worn, corroded or damaged surfaces. Replace if required.

11. Installation is the reverse of removal. Apply Yamaha All-Purpose Grease to the pivot surfaces of all pins and bushings. Tighten the grounding wire screws to the general torque specification in Chapter One. Route the trim motor wires carefully to avoid interference with other components. Secure the wires to the clamp bracket with new plastic tie clamps (3, **Figure 1**). Make sure the snap rings (4, **Figure 1**) fit snugly into the pivot pin grooves.

12. *200-250 hp (Saltwater series [3.1 liter]) models*—Tighten the eight mounting screws (**Figure 6**) evenly to the torque specification in **Table 1**.

13. Check for proper tilt/trim system operation before putting the engine into service.

TILT/TRIM RELAY REPLACEMENT

The models covered in this manual are equipped with either a single relay unit or two separate tilt/trim relays (**Figure 7**).

80 Jet, 115, 105, 130 and C150 hp models—Two separate tilt/trim relays are used. One relay controls the up circuit. The remaining relay controls the down circuit. The relays may be replaced individually.

150-250 hp models—A single relay unit is used to control the up and down circuits. Replace the single relay unit as an assembly if either the up or down circuit has failed. The tilt/trim relay(s) are mounted on the starboard side of the power head. Remove the plastic electrical component cover to access the relay(s).

Refer to the wiring diagrams at the end of the manual to verify wire color of the relay unit.

1. Disconnect the battery cables.

2. Trace the wires to the component on the engine. Mark each wire connection location and orientation before removal. Disconnect all relay wires.

3A. *Single relay unit*—Remove the screws and lift the relay from the engine. Clean the mounting location and the threaded holes for the mounting screws. Clean and inspect all terminal connections.

3B. *Individual relays*—Carefully pull the relay from the rubber mounting sleeve. Twist the relay slightly in the sleeve to ease removal.

4A. *Single relay unit*—Reconnect the relay unit wires. Reinstall the relay and secure it with the two screws. Tighten the screws to the general torque specification in Chapter One. Route all wires to avoid interference with other components.

4B. *Individual relays*—Apply a light film of soapy water into the relay sleeve opening. Do not allow water to contaminate other components. Carefully slide the relay into the sleeve until it fully seats. Twist the relay to properly orient the wire terminal post. Reconnect the relay wires. Route all wires to avoid interference.

5. Connect the battery cables. Check for proper operation of the tilt/trim system before putting the engine into service.

TILT/TRIM SENDER REPLACEMENT

The tilt/trim sender (**Figure 8**) is used on all models incorporating a dash mounted tilt/trim gauge. It provides a varying voltage signal to the gauge indicating the engine tilt/trim position. Two screws secure the sender onto the swivel bracket. The tilt tube mounted arm (**Figure 9**) contacts the sender lever and functions to operate the sender. This arrangement allows easier adjustment compared to earlier designs with the sender mounted on the clamp bracket.

Make a sketch of the sender wire routing prior to removal to ensure proper routing.

NOTE
Always adjust the tilt/trim sender (Chapter Four) after replacement. An improperly adjusted tilt/trim sender causes inaccurate gauge readings.

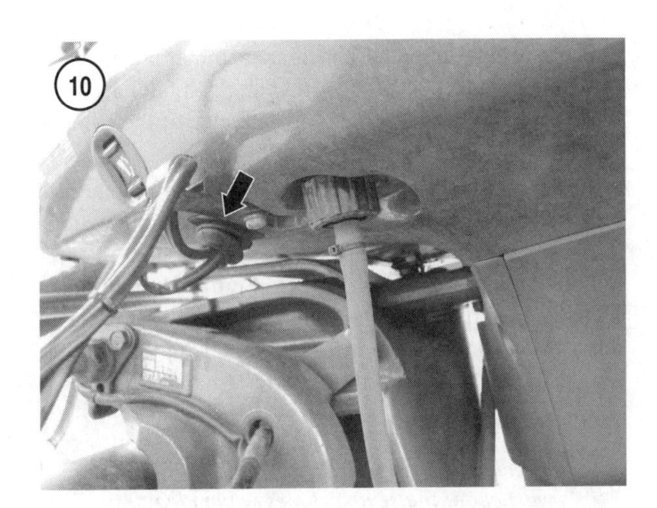

1. Position the engine to the full tilt and engage the tilt lock lever. Support the engine with blocks or a suitable overhead cable (**Figure 4**).
2. Trace the sender wires (**Figure 10**) to the engine wire harness connection. Disconnect the sender wires. Route the wires out of the lower engine cover. Remove the plastic tie clamps to allow removal of the sender.
3. Remove both mounting screws and the sender. Route the wires through the opening in the port clamp bracket and remove the assembly.
4. Clean corrosion and contamination from the sender mounting location and mounting screw openings.
5. Install the replacement sender onto the swivel housing. Install and securely tighten the mounting screws.
6. Route the sender wire through the clamp bracket and lower engine cover opening. Connect the sender wire to the engine wire harness.
7. Route the wires to prevent interference with other components. Secure the sender wires to the clamp bracket with new plastic tie clamps.
8. Adjust the sender as described in Chapter Four.

ELECTRIC TILT/TRIM MOTOR REMOVAL AND INSTALLATION

WARNING
The trim system may contain fluid under high pressure. Always use protective eyewear and gloves when working with the trim system. Never remove components or plugs without first bleeding the pressure from the system. Follow the instructions carefully.

The electric trim motor appearance, mounting arrangement and replacement procedures vary by model. Refer to **Figure 11** and **Figure 13**. The appearance of the components may vary slightly from the illustration.

Port clamp bracket removal may be required to access the electric motor mounting screws. On most applications it is easier to remove the entire trim system rather than removing the clamp bracket. Refer to **Figure 11** and **Figure 13** the appropriate section in this chapter for trim system removal and installation instructions. Clamp bracket removal and installation is described in Chapter Eleven.

CAUTION
Never direct pressurized water (pressure washer) at the seal surfaces when cleaning debris or contaminants from the trim system. The water may pass by the seal and contaminate the fluid.

The trim system must operate with clean fluid. Thoroughly clean the trim system external surfaces with soapy water prior to disassembly. Use compressed air to dry the trim system.

Work in a clean area and use lint free towels to wipe debris or fluid from the components. Cover any openings immediately after removal to prevent fluid contamination.

CAUTION
To avoid unnecessary disassembly and potential wire interference, always note the orientation of the electric trim motor and wire harness prior to disassembly. Use a paint dot or piece of tape to mark the location. Never scratch the housings, as it will cause corrosion.

80 Jet, 105 Jet and 115-200 hp Models (Except 3.1 Liter Saltwater Series)

1. Disconnect the battery cables and spark plug leads.
2. Trace the wires leading from the electric trim motor to the connections to the power head. Disconnect the wires from the relays or engine wire harness and route them out of the lower engine cover. Remove the plastic tie clamps.
3. Locate the manual release valve (**Figure 2**) in the port side clamp bracket. Rotate the valve counterclockwise 2-3 turns (**Figure 3**). Carefully tilt the engine to the full up position. Use an overhead hoist or other suitable means to support the engine as indicated in **Figure 4**. Securely tighten the manual release valve.
4. Make reference markings on the electric motor and corresponding mounting location on the trim system

(11) TILT/TRIM ELECTRIC MOTOR MOUNTING (80 JET AND 115-200 HP [2.6 LITER] MODELS [EXCEPT SALTWATER SERIES WITH XL SHAFT LENGTH])

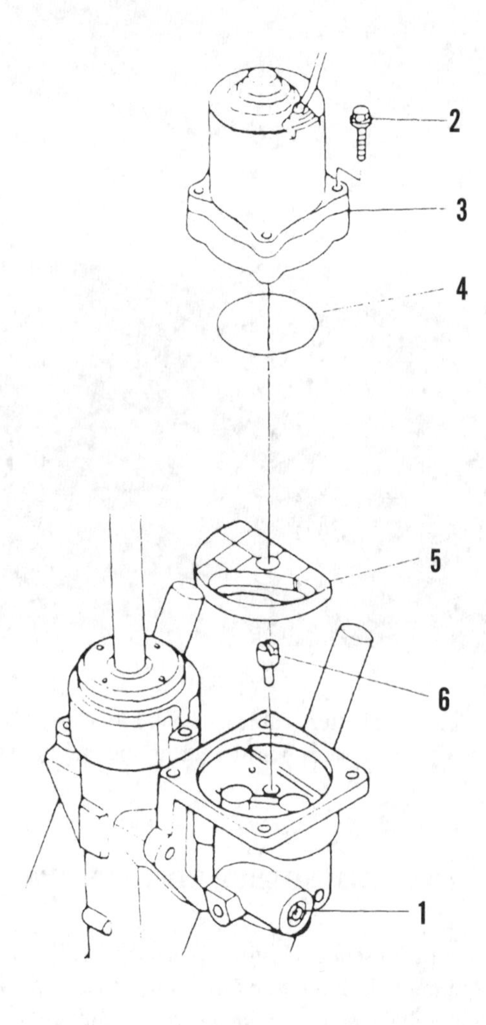

1. Manual release valve
2. Electric motor mounting screws
3. Electric motor
4. O-ring
5. Filter
6. Coupling

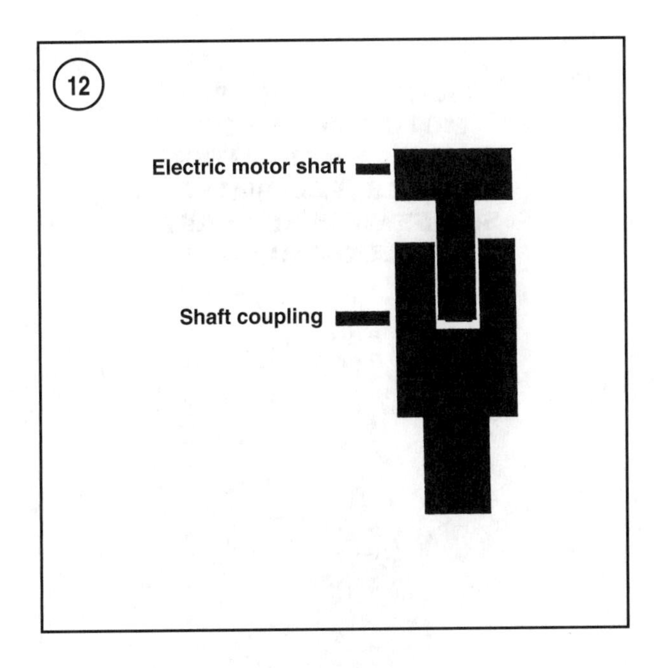

housing prior to removal. Use a paint dot or tape. Do not scratch the surfaces.

5. Remove the mounting screws (2, **Figure 11**) and carefully lift the motor from the housing. If necessary, use a flat scraper or dull putty knife to pry the motor from the pump. Work carefully to avoid damaging the electric motor mounting surfaces. A fluid or water leak can occur if the surfaces are damaged.

6. Remove the O-ring (4, **Figure 11**) from the electric motor or trim system housing. Discard the O-ring. Retrieve the coupling (6, **Figure 11**) from the electric motor or hydraulic pump shaft. Note the top and bottom orientation of the coupling. The slot opening must face the electric motor and the protrusion must face the hydraulic pump.

7. Remove the filter (5, **Figure 11**) from the hydraulic pump. Clean the filter in solvent and air dry. Replace the filter if damaged or if debris remains on the filter after cleaning.

8. Clean and inspect the trim motor mounting surfaces for corroded, pitted or scratches. Replace any components with scratches that are felt with a fingernail.

9. Apply a light coating of Yamaha All-Purpose Grease and install a new O-ring (4, **Figure 11**) onto the electric motor to hydraulic pump mating surface. Apply a light coating of grease to the coupling. Install the coupling onto the electric motor shaft as indicated in **Figure 12**.

10. Insert the filter into the trim system housing and rest it on the hydraulic pump. The flat side of the filter must align with the flat side in the filter bore.

11. Position the electric motor onto the pump and rotate it until the coupling protrusion aligns with the hydraulic pump shaft slot. When aligned the motor should drop into position.

12. Slowly rotate the electric motor to obtain correct wiring orientation.

13. Apply Loctite 572 onto the threads, then install the mounting screws (2, **Figure 11**). Tighten the screws to the specification in **Table 1**.

14. Fill and bleed the hydraulic system as described in this chapter. Check for proper operation before putting the engine into service.

200-250 hp Models (3.1 Liter)

1. Disconnect the battery cables and spark plug leads.
2. Trace the wires leading from the electric trim motor to the connections on the power head. Disconnect the wires from the relays or engine wire harness and route them out of the lower engine cover. Remove the plastic tie clamps.
3. Locate the manual release valve (**Figure 2**) in the starboard side clamp bracket. Rotate the valve counterclockwise 2-3 turns (**Figure 3**). Carefully tilt the engine to the full up position. Use an overhead hoist or other suitable means to support the engine as indicated in **Figure 4**. Securely tighten the manual release valve.
4. Make reference markings on the electric motor and corresponding mounting location on the trim system housing prior to removal. Use a paint dot or tape. Do not scratch the surfaces.
5. Remove the mounting screws (2, **Figure 13**) and carefully lift the motor from the housing. If necessary, use a flat scraper or dull putty knife to pry the motor from the pump. Work carefully to avoid damaging the electric motor mounting surfaces. A fluid or water leak can occur if the surfaces are damaged.
6. Remove the O-ring (3, **Figure 13**) from the electric motor or trim system housing. Discard the O-ring. Retrieve the coupling (5, **Figure 13**) from the electric motor or hydraulic pump shaft. Note the top and bottom orientation of the coupling. The slot opening must face the electric motor and the protrusion must face the hydraulic pump.
7. Remove the filter (4, **Figure 13**) from the hydraulic pump. Clean the filter in solvent and air dry. Replace the filter if damaged or debris remains on the filter after cleaning.
8. Clean and inspect the trim motor mounting surfaces for corroded, pitted or scratched surfaces. Replace any components with scratches that are felt with a fingernail.
9. Apply a light coating of Yamaha All-Purpose Grease and install a new O-ring (3, **Figure 13**) onto the electric

motor to the hydraulic pump mating surface. Apply a light coating of grease to the coupling. Install the coupling onto the electric motor shaft as indicated in **Figure 12**.

10. Insert the filter into the trim system housing. The filter must rest on top of the hydraulic pump (7, **Figure 13**) with the larger diameter side facing the pump. Center the filter over the pump.
11. Position the electric motor onto the pump and rotate it until the coupling protrusion aligns with the hydraulic pump shaft slot. When aligned the motor should drop into position.
12. Slowly rotate the electric motor to obtain correct orientation of the wiring.
13. Apply Loctite 572 to the threads then install the mounting screws (2, **Figure 13**). Tighten the screws to **Table 1** specification.
14. Fill and bleed the hydraulic system as described in this chapter. Check for proper operation before putting the engine into service.

TILT/TRIM SYSTEM ELECTRIC MOTOR REPAIR

This section describes disassembly, inspection and assembly of the tilt/trim electric motor. The procedures vary slightly by models and instructions are provided for the individual motor variations.

Work in a clean environment to avoid contaminants. Use electrical cleaner to remove contaminants and debris from the motor components. Electrical contact cleaner is available at most electrical supply sources. It evaporates rapidly and leaves no residue to contaminate components. Avoid touching the brushes and commutator after cleaning. Naturally occurring oils will contaminate these components.

NOTE
*Mark the upper cover, frame and lower cover of the electric motor (**Figure 14**) prior to repair. Use paint dots or removable tape. Never scratch the components, as it will cause corrosion of the metal.*

The electric trim motor component appearance, disassembly and assembly procedures vary by model. Refer to **Figure 15** and **Figure 16**. The appearance of the components may vary slightly from the illustration.

CAUTION
Use caution when working around the permanent magnets in the frame assembly. These magnets are quite powerful. Fingers are easily pinched between components by

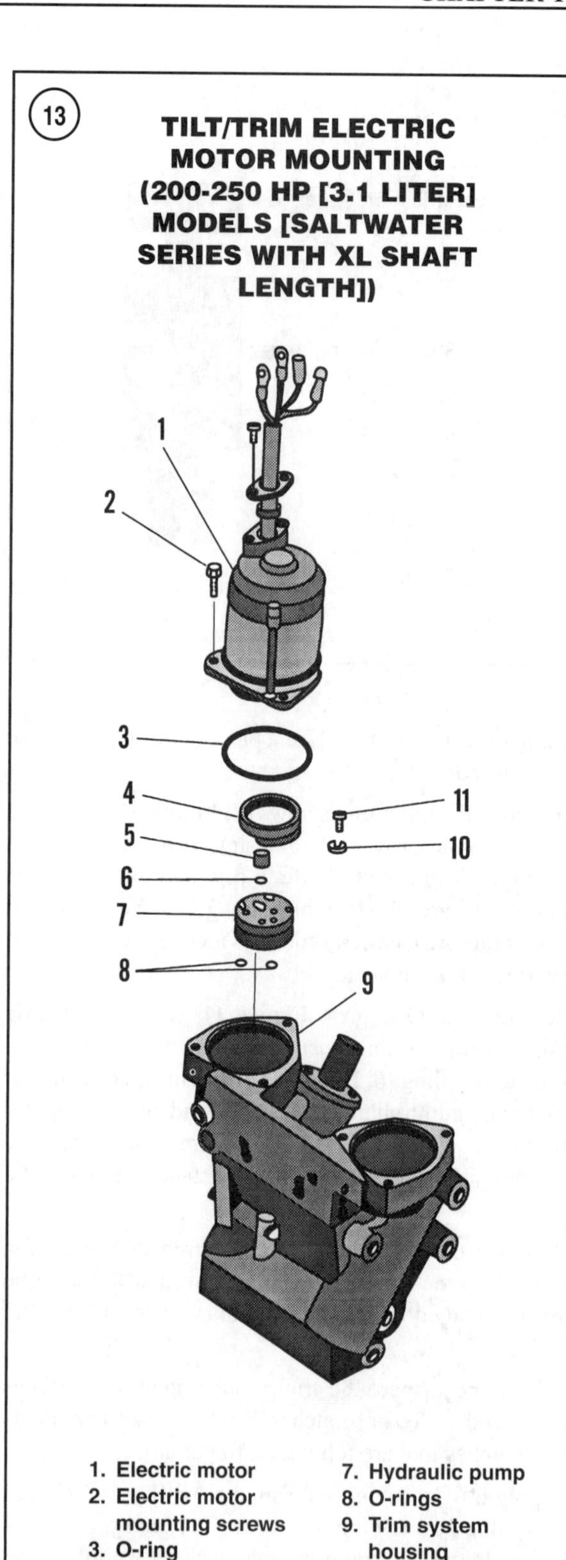

1. Electric motor
2. Electric motor mounting screws
3. O-ring
4. Filter
5. Coupling
6. O-ring
7. Hydraulic pump
8. O-rings
9. Trim system housing
10. Lockwasher
11. Screw

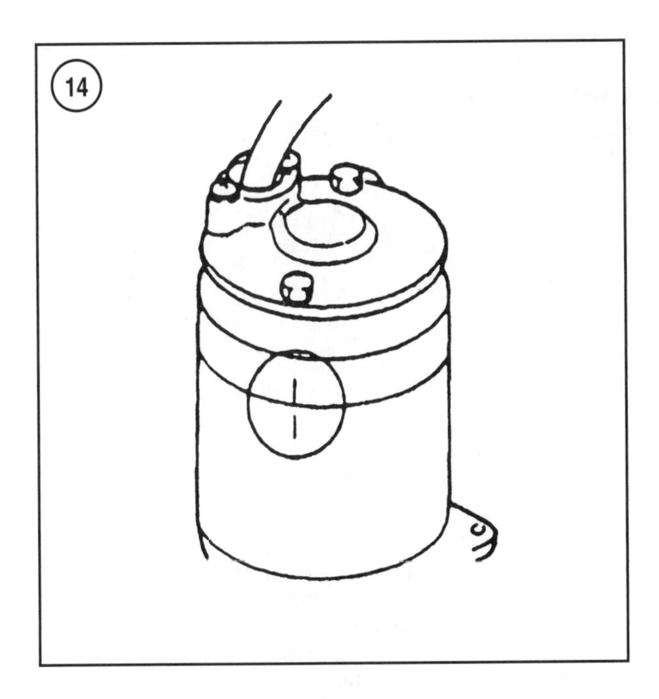

the magnetic force. Never drop or strike the frame assembly. The magnets may break and damage other components during operation.

Disassembly (80 Jet, 105 Jet and 150-200 hp Models [Except 3.1 Liter Saltwater Series])

Refer to **Figure 15**.
1. Remove the electric motor as described in this chapter.
2. Mark all components (**Figure 14**) prior to disassembly to ensure proper orientation during assembly.
3. Remove the screws that retain the frame onto the lower cover.

NOTE
The magnets in the frame assembly are quite powerful. Considerable effort may be required to remove the frame assembly from the armature. Make sure to remove all fasteners before removing the armature.

4. Grasp the armature shaft with pliers and a shop towel as shown in **Figure 17**. Pull the armature and lower cover from the frame. Remove and discard the O-ring from the frame or lower cover.
5. Unplug the frame wire connector from the bi-metal switch (8, **Figure 15**) and remaining brush lead. Using two screwdrivers, carefully push the brushes to collapse the brush springs (9, **Figure 15**). Then pull the armature (5, **Figure 15**) from the lower cover. Work carefully to avoid damaging the brushes or commutator. Remove the screws (6, **Figure 15**). Then lift the brush retainers (7, **Figure 15**), brushes and leads, springs (9) and bi-metal switch from the lower cover. Remove the thrust washer from the armature shaft or lower cover.

Disassembly (200-250 hp Models [3.1 Liter Saltwater Series])

Refer to **Figure 16**.
1. Remove the electric motor as described in this chapter.
2. Mark all components (**Figure 14**) prior to disassembly to ensure proper orientation during assembly.
3. Remove the throughbolts (1, **Figure 16**) that retain the top cover (14) to the frame (6) and bottom cover (2).
4. Pull the bottom cover from the frame. Remove and discard the O-ring (3, **Figure 16**). While holding the armature (8, **Figure 16**) firmly into the top cover, remove the frame assembly from the armature. Remove and discard the O-ring (7, **Figure 16**).
5. Use two small screwdrivers to collapse the brush springs (12, **Figure 16**) and move the brushes (11) away from the commutator. Work carefully to avoid damaging the brushes or commutator. Pull the armature from the cover while the brushes are away from the commutator.
6. Remove the washer (9, **Figure 16**) from the armature.
7. Use compressed air to blow debris from the components. Clean oily contaminants from the components with electrical contact cleaner.

Component Inspection and Testing

Prior to performing any test or measurement, clean debris and contaminants from all components. Inspect the magnets in the frame assembly for broken or loose magnets. Replace the frame assembly if there are any.
1. Calibrate a multimeter to the 1 ohm scale. Connect the positive test lead to the green wire connection on the cover assembly or frame assembly. Connect the negative test lead to the respective brush lead, bi-metal switch connector or brush screw terminal for that wire (**Figure 18**, typical). Repeat the test with the blue wire and respective connection. If the meter does not indicate *continuity* for each test, the lower frame or cover is faulty and must be replaced.
2. *80 Jet, 105 Jet and 150-200 hp models (except Saltwater series [3.1 liter])*—Connect the positive test lead to one of the terminals or brush lead connection at the bi-metal switch (**Figure 19**). Connect the negative test lead to the other terminal of the bi-metal switch. If the meter does not indicate *continuity*, the switch is faulty and must be replaced.

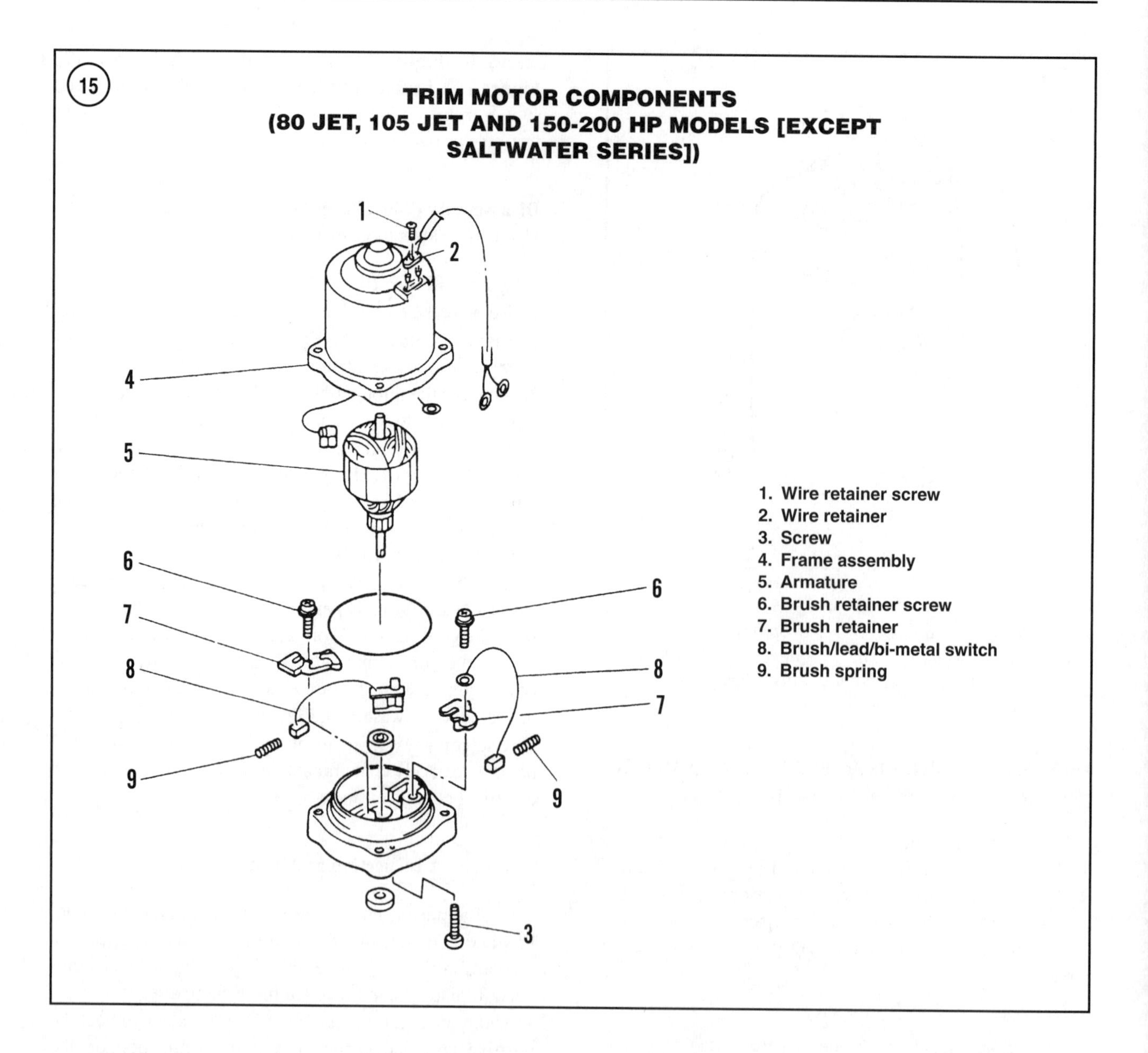

3. Carefully grip the armature in a vise with protective jaws (**Figure 20**). Excessive clamping force damages the armature. Use only enough force to lightly retain the component. Polish the commutator surfaces with 600 grit wet or dry carborundum. Periodically rotate the armature to polish evenly. Avoid removing too much material.

4. Use a disposable fingernail file to remove mica and brush material from the undercut surfaces (**Figure 21**).

5. Calibrate a multimeter to the 1 ohm scale. Touch the positive test lead to one of the segments on the commutator (**Figure 22**). Touch the negative test lead to a different segment. Repeat this test until all segments are tested. If the meter does not indicate *continuity* between each pair of commutator segments, the commutator is faulty and must be replaced.

6. Touch the positive test lead to one of the commutator segments. Touch the negative test lead to one of the laminated areas of the armature (**Figure 23**). Note the meter reading. Touch the negative test lead to the armature shaft (**Figure 23**) and note the meter reading. If the meter indicates *continuity* for each test, the armature is shorted and must be replaced.

7. Using a micrometer or vernier caliper, measure the commutator diameter (**Figure 24**). Compare the diameter with **Table 3**. Replace the armature if the diameter is less than the minimum specification.

(16)

TILT/TRIM MOTOR COMPONENTS (200-250 MODELS [SALTWATER SERIES WITH XL SHAFT LENGTH])

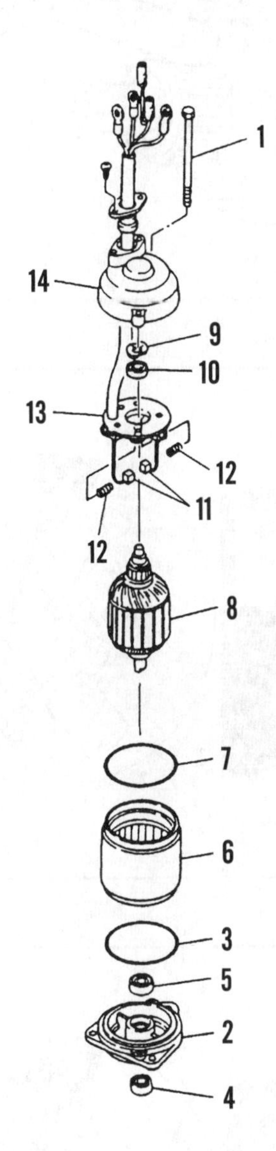

1. Throughbolts
2. Bottom cover
3. O-ring
4. Seal
5. Bearing
6. Frame assembly
7. O-ring
8. Armature
9. Wave washer
10. Bushing
11. Brushes
12. Brush spring
13. Brush plate
14. Top cover

(17)

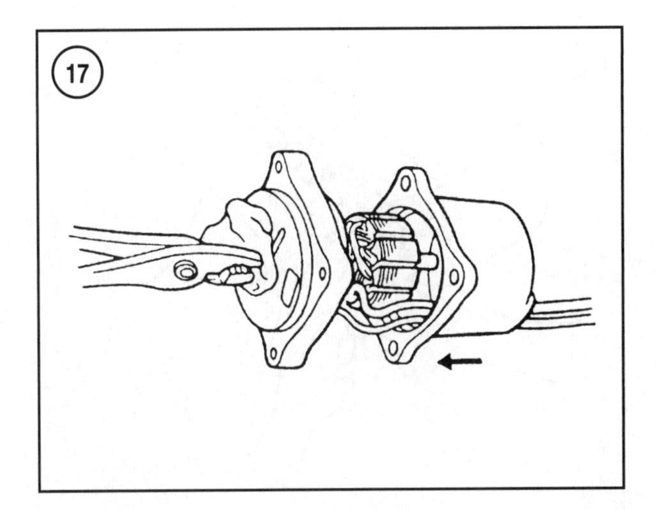

(18)

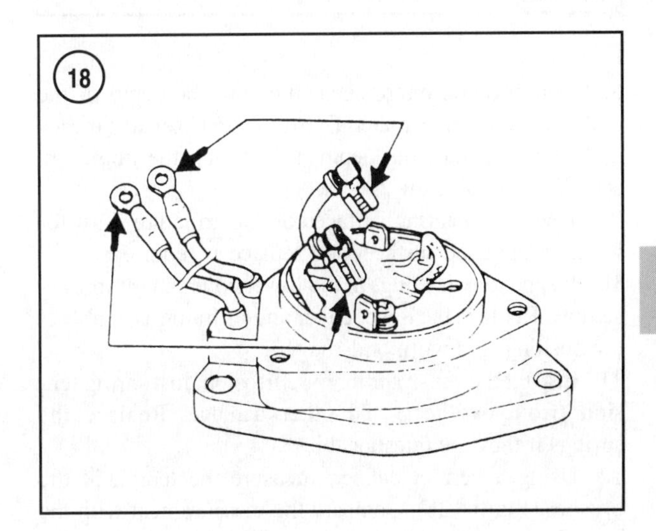

(19)

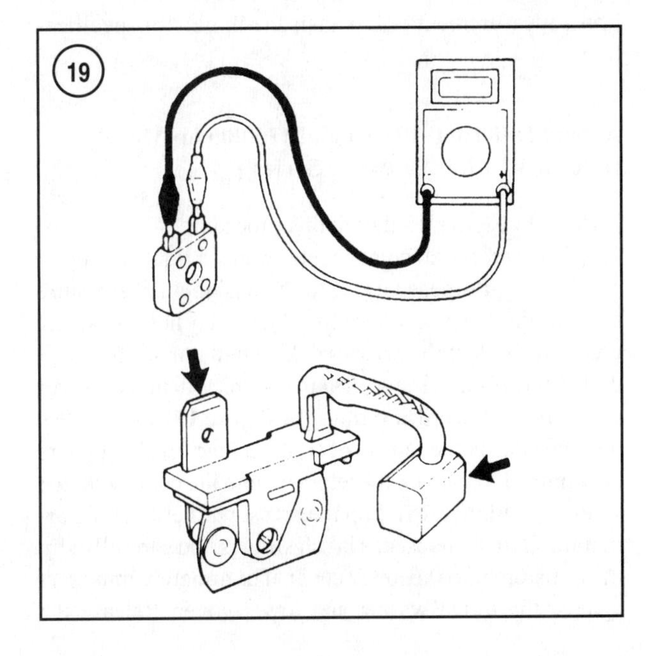

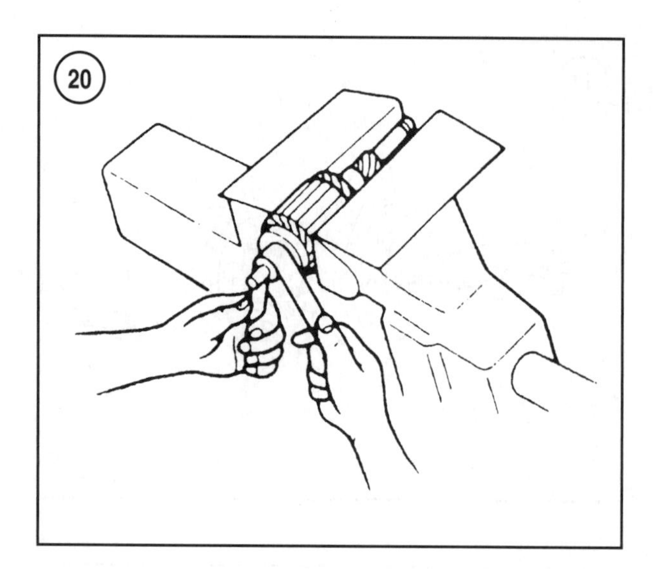

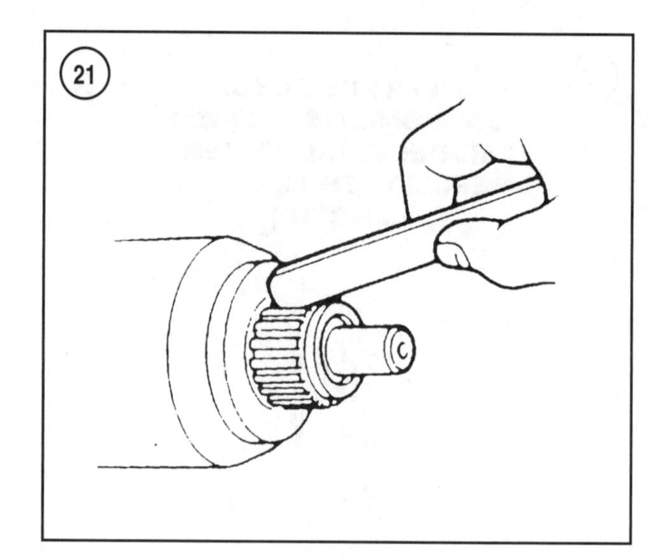

8. Using a depth micrometer, measure the depth of the undercut to the mica material (**Figure 25**). Replace the armature if the undercut depth is less than the minimum specification in **Table 2**.
9. Inspect the bearing surfaces on the armature shaft for worn or damaged surfaces and replace as required.
10. Inspect the bushings in the lower cover and frame assembly. Replace the lower cover and/or frame assembly if the bushings are worn or damaged.
11. Inspect the brush spring for corrosion, lost spring tension (from overheat) and other damage. Replace the springs if they are questionable.
12. Using a vernier caliper, measure the length of the brushes (**Figure 26**). Compare the measurement with the specification in **Table 2**. Replace both brushes if either brush measurement is less than the minimum specification.

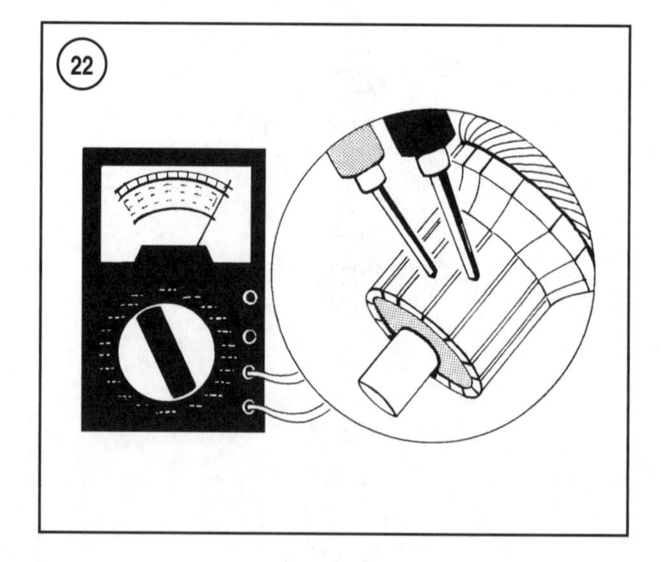

Assembly (80 Jet, 105 Jet and 115-200 hp Models [Except 3.1 Liter Saltwater Series])

Refer to **Figure 15** during this procedure.
1. Clean and dry all components. Apply a light coating of Yamaha All-Purpose Grease to the bushing and armature shaft at the bushing contact surfaces. Do not allow any grease to contact the brushes or commutator surfaces.
2. Fit the bi-metal switch into the recess in the lower cover. Insert the brush springs (9, **Figure 15**) and brushes into the slots in the lower cover. Install the brush retainers (7, **Figure 15**). Secure the retainers into the cover with the screws (6, **Figure 15**). Slip the thrust washer over the armature shaft. Collapse the brush springs and carefully slip the armature into the lower cover. The armature must seat against the thrust washer and lower cover. Release the

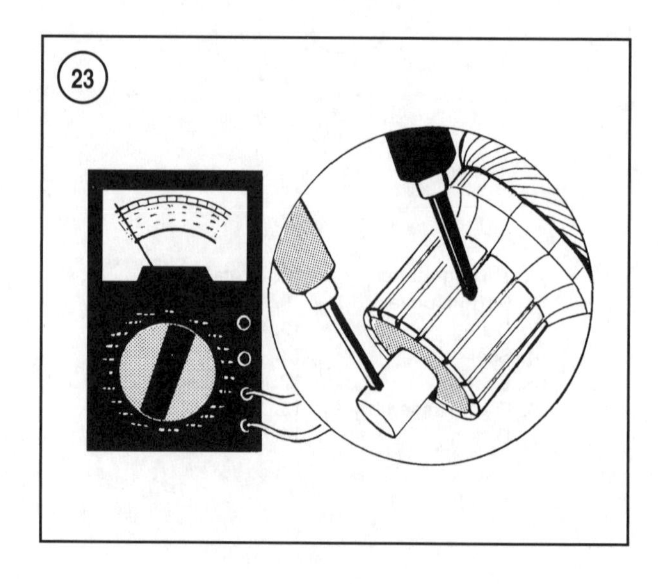

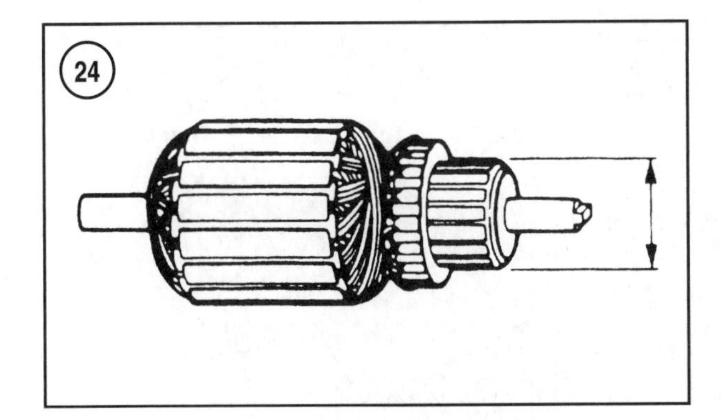

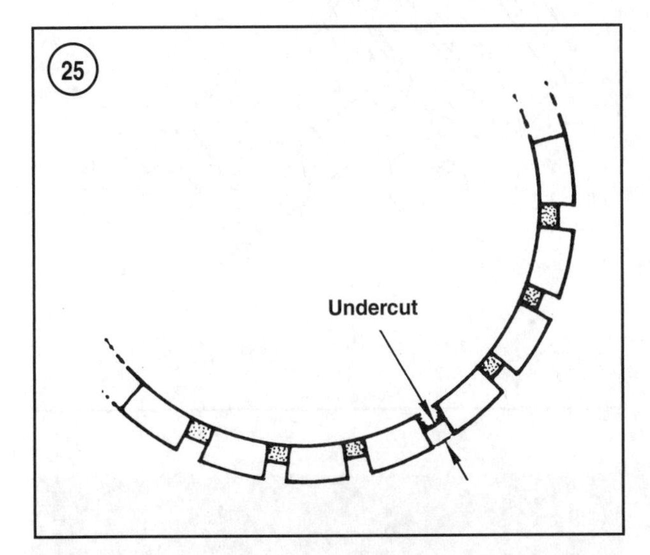

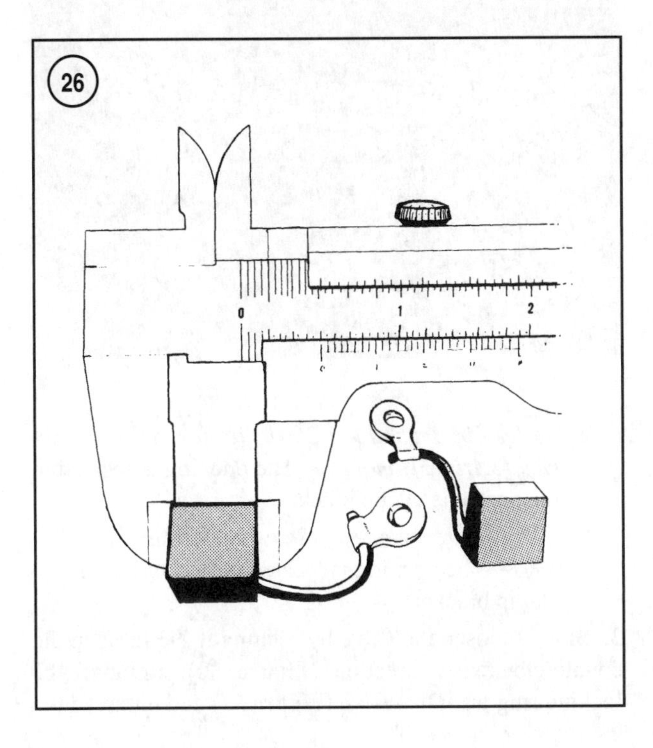

brush springs. Work carefully to avoid damaging the brushes and commutator.

3. Connect the push-on terminal extending from the frame onto the terminal of the bi-metal switch. Connect the remaining ring terminal frame lead onto the terminal provided on the remaining brush retainer.

4. Install a new O-ring onto the lower cover. Use pliers and a shop towel to maintain the position of the armature and the lower cover (**Figure 17**) while carefully guiding the armature into the frame. The upper end of the armature shaft must enter the bushing at the top of the frame.

5. Seat the frame against the lower cover. Make sure the O-ring remains in position and is not damaged. Water leakage will result if the O-ring is out of position or damaged.

6. Align the marks made prior to disassembly (**Figure 14**). Install the screws (3, **Figure 15**) that retain the lower cover to the frame assembly. Tighten the screws to the general toque specification in Chapter One. Rotate the armature shaft and check for binding. Disassemble the motor and check for improper assembly if binding occurs.

7. Install the electric trim motor as described in this chapter.

Assembly (200-250 hp Models [3.1 Liter Saltwater Series])

Refer to **Figure 16** during this procedure.

1. Clean and dry all components. Apply a light coating of Yamaha All-Purpose Grease to the bushing and armature shaft at the bushing contact surfaces. Do not allow any grease to contact the brushes or commutator surfaces.

2. Install the brush springs (12, **Figure 16**), brush plate (13) and brushes (11) into the top cover (14). Secure the brush plate into the top cover with the screws.

3. Use small screwdrivers to position the brushes fully into the brush holders (the springs are collapsed). Carefully position the commutator portion of the armature into the cover and release the brushes (**Figure 27**). Never force the armature into the cover as the brushes may become damaged.

4. Install a new O-ring (7, **Figure 16**) onto the upper cover. Install the frame assembly over the armature and seat it against the upper cover. Make sure the O-ring is in position between the cover and frame.

5. Install the washer, if so equipped, onto the armature shaft. Install a new O-ring (3, **Figure 16**) onto the bottom cover and carefully guide the bottom cover over the armature shaft. Seat the bottom cover against the frame.

6. Rotate the covers and frame to align the reference markings (**Figure 14**) made prior to disassembly. Install

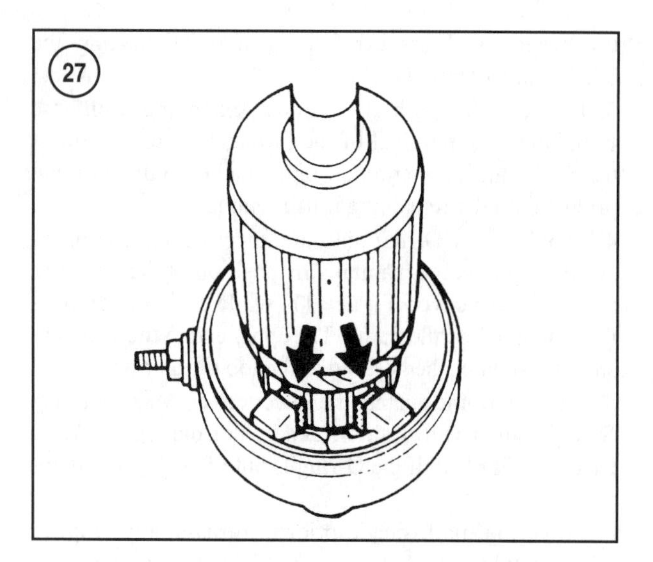

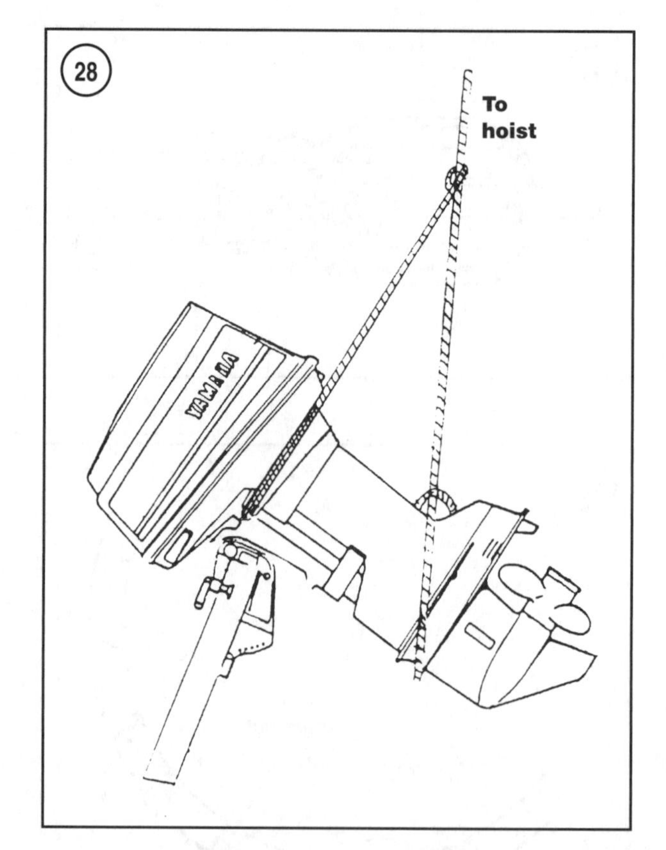

the throughbolts (1, **Figure 16**) and thread them into the lower cover. Do not tighten the bolts at this time.

7. Make sure the O-rings are properly positioned to prevent water leakage. Tighten the throughbolts to the general torque specification in Chapter One.
8. Rotate the armature shaft and check for binding. Disassemble the motor and check for improper assembly if binding occurs.
9. Install the electric trim motor as described in this chapter.

MANUAL RELEASE VALVE REMOVAL AND INSTALLATION

Replacement of the manual release valve is simple if the screwdriver slot is intact. If this is not the case, the valve can usually be removed by other means. Heat the tip of a screwdriver and then hold it against the remnants of the valve. The valve material will melt into the shape of the screwdriver tip. Allow the material to cool and use the same screwdriver to remove the valve. Never drill the valve out or the seating surfaces will suffer irreparable damage.

Inspect the O-rings on the valve even though they will be discarded. Problems may surface if large portions are missing or torn away from the O-rings. They usually migrate to a valve or other component with the tilt/trim system and cause the system to malfunction.

1. Position the engine in the full up tilt position. Engage the tilt lock lever and support the engine with blocks or an overhead cable (**Figure 28**).
2. Locate the access opening for the manual release valve (**Figure 29**).

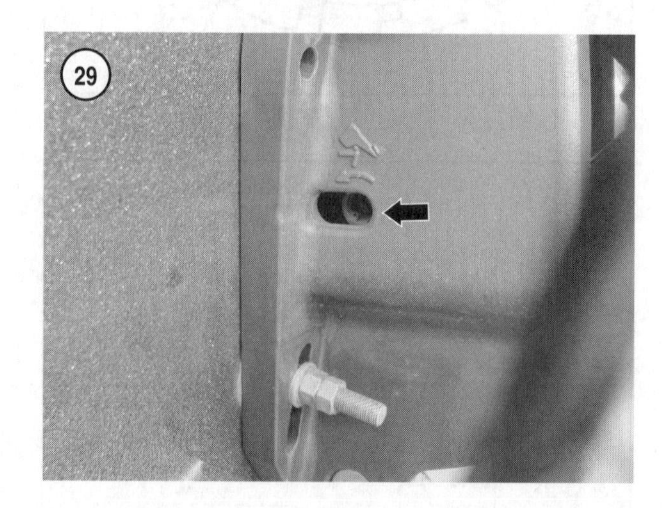

   a. *80 Jet, 105 Jet and 115-200 hp (except Saltwater series [3.1 liter]) models*—The opening is located in the port side clamp bracket.
   b. *200-250 hp models (Saltwater series [3.1 liter])*—The opening is located in the starboard side clamp bracket.
3. Slowly loosen the valve by rotating it 2-3 turns in the counterclockwise direction (**Figure 30**). Remove the locking ring from the valve (**Figure 31**) and unthread the

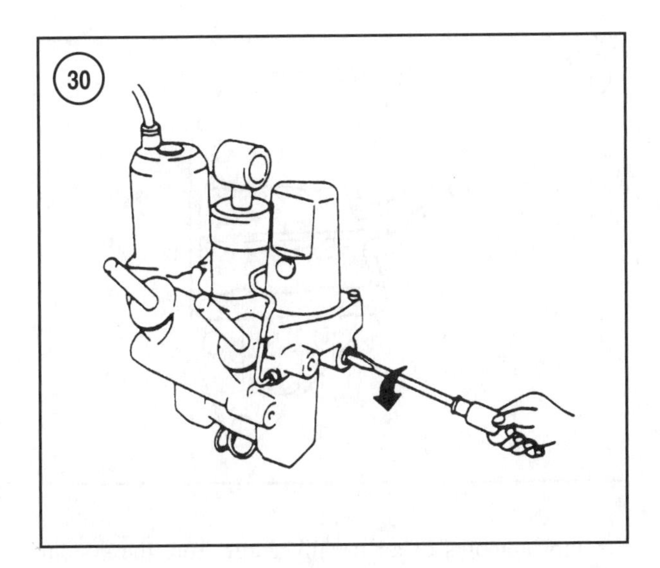

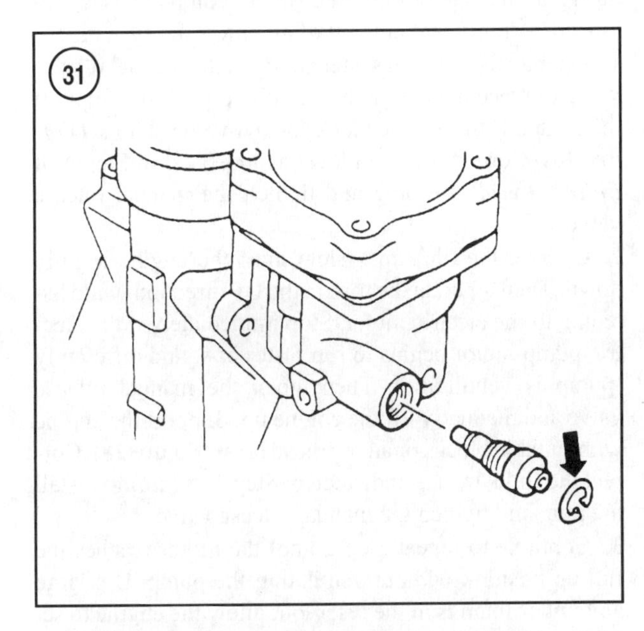

valve from the trim system. Remove the valve through the access opening.

4. Use a suitable light and small pick or screwdriver to remove any remnants of O-ring from the opening. Avoid damaging any of the machined surfaces in the opening.

5. Lubricate the new manual release valve with Dextron II automatic transmission fluid and install *new* O-rings onto the valve. Lubricate the new O-rings with Dextron II automatic transmission fluid and thread the valve into the opening. DO NOT tighten the valve at this time.

6. Rotate the valve clockwise until slight resistance is felt. Rotate the valve a quarter turn clockwise then one eighth turn counterclockwise. Repeat this process until the valve seats. Finally tighten the valve to **Table 1** specification.

7. Fill the system with fluid and bleed air from the system as described in this chapter.

HYDRAULIC SYSTEM FILLING AND BLEEDING

Refer to *Filling* if the unit has lost a large amount of fluid or after replacing a major component.

Refer to *Bleeding* if the tilt/trim system exhibits symptoms of air in the fluid.

WARNING

*Never work under any part of the engine without providing suitable support. The engine mounted tilt lock or hydraulic system may unexpectedly collapse and allow the engine to drop. Support the engine with suitable blocks or an overhead cable **before** working under the engine.*

Filling Procedure

Use Dextron II automatic transmission fluid in all tilt/trim systems covered in this manual.

1. Use the power tilt/trim system or open the manual release valve (**Figure 29**) to place the engine in the full up position. Engage the tilt/lock lever and support the engine with suitable blocks or overhead cable (**Figure 28**). Close the manual release valve.

2. Locate and clean the area around the fluid reservoir plug.

   a. *80 Jet, 105 Jet and 115-200 hp (except Saltwater series [2.6 liter]) models*—The cap is located on the rear of the fluid reservoir (**Figure 32**). The fluid reservoir mounts on the starboard side of the system.

   b. *200-250 hp (Saltwater series [3.1 liter]) models*—The plug is located on the rear of the fluid res-

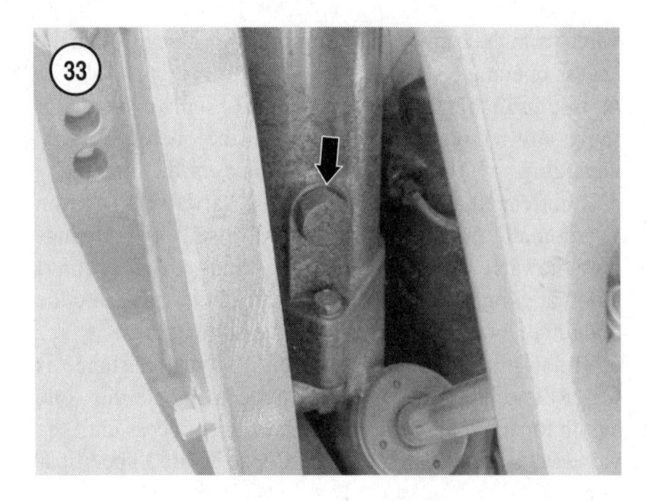

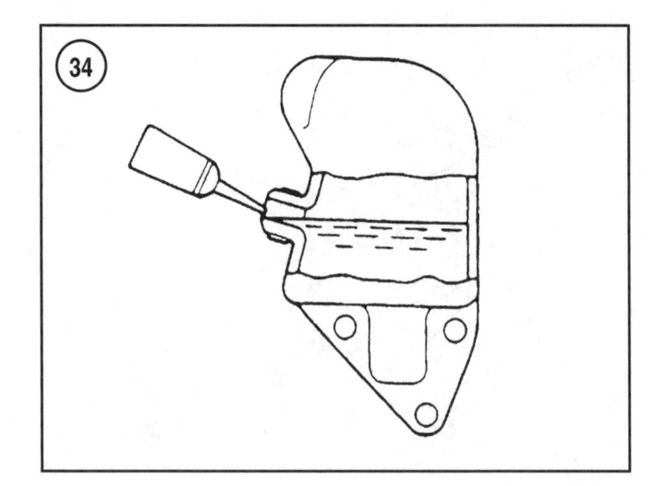

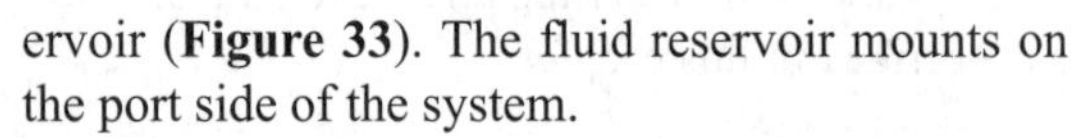

ervoir (**Figure 33**). The fluid reservoir mounts on the port side of the system.

3. Slowly loosen and remove the fluid fill plug. Inspect the plug gasket for damage and replace as necessary.
4. Fill the unit to the bottom of the plug opening (**Figure 34**) with the engine in the full up position. Reinstall and securely tighten to the general torque specifications in Chapter One.
5. Remove the blocks or overhead cable. Disengage the tilt lock lever.
6. Cycle the system to the full down then full up position several times. Check and correct the fluid level as necessary.
7. Tighten the reservoir cap to the specification in **Table 1**.

Bleeding

A spongy feel or inability to hold the trim position under load is a common symptom of air in the internal passages. The engine will usually tuck under when load is applied and tilt out when power is reduced.

Minor amounts of air usually purge from the system during normal operation. When major components are removed, a significant amount of air enters the system.

1. Operate the tilt/trim system or open the manual release valve to place the engine in the full up position. Support the engine with suitable blocks or an overhead hoist (**Figure 28**). Correct the fluid level as indicated in Step 4 of *Filling*. Install the plug and tighten the manual release valve.
2. Operate the tilt/trim system until the engine is fully down. Then, operate the trim in the UP direction while listening to the electric motor. Stop immediately if the electric pump motor begins to run faster or sound differently (pump is ventilating). Then, open the manual release valve and manually lift the engine up. Support the engine with suitable blocks or an overhead hoist (**Figure 28**). Correct the fluid level as indicated in Step 4 of *Filling*. Install the plug and tighten the manual release valve.
3. Continue to repeat Step 2 until the motor reaches the full up position without ventilating the pump. If a large amount of foam is in the reservoir, allow the engine to set for one hour and repeat the process.
4. Cycle the tilt/trim system up and down several times to purge the remaining air from the system.

Table 1 TILT/TRIM TORQUE SPECIFICATIONS

Fastener	N·m	in.-lb.	ft.-lb.
Electric motor mounting			
80 Jet, 105 Jet and 115-200 (2.6 liter) hp (Except Saltwater series [XL shaft length])	5	44	–
200-250 hp (3.1 liter) (Saltwater series [XL shaft length])	8	70	–

(continued)

Table 1 TILT/TRIM TORQUE SPECIFICATIONS (continued)

Fastener	N·m	in.-lb.	ft.-lb.
External hydraulic lines			
200-250 hp (Saltwater series [XL shaft length])	15	–	11
Fluid fill cap			
80 Jet, 105 Jet and 115-200 (2.6 liter) hp (except Saltwater series [XL shaft length])	8	70	–
Fluid reservoir to trim system housing			
80 Jet, 105 Jet and 115-200 (2.6 liter) hp (except Saltwater series [XL shaft length])	5	44	–
200-250 hp (3.1 liter) (Saltwater series [XL shaft length])	8	70	–
Manual release valve			
80 Jet, 105 Jet and 115-200 (2.6 liter) hp (except Saltwater series [XL shaft length])	4	35	–
200-250 hp (3.1 liter) (Saltwater series [XL shaft length])	2	18	–
Hydraulic pump mounting			
80 Jet, 105 Jet and 115-200 (2.6 liter) hp (except Saltwater series [XL shaft length])	6	53	–
200-250 hp (3.1 liter) (Saltwater series [XL shaft length])	4	35	–
Pump housing to trim system housing			
80 Jet, 105 Jet and 115-200 (2.6 liter) hp (except Saltwater series [XL shaft length])	9	80	–
Tilt cylinder end cap			
80 Jet, 105 Jet and 115-200 (2.6 liter) hp (except Saltwater series [XL shaft length])	130	–	96
200-250 hp (3.1 liter) (Saltwater series [XL shaft length])	90	–	66
Trim cylinder end cap			
80 Jet, 105 Jet and 115-200 (2.6 liter) hp (except Saltwater series [XL shaft length])	80	–	59
200-250 hp (3.1 liter) (Saltwater series [XL shaft length])	160	–	118

Table 2 TILT/TRIM MOTOR SERVICE SPECIFICATIONS

Measurement	Specification (minimum)
Brush length	
80 Jet, 105 Jet and 115-200 (2.6 liter) hp (except Saltwater series)	4.8 mm (0.189 in.)
200-250 (3.1 liter) hp (Saltwater series [XL shaft length])	4.0 mm (0.157 in.)
Commutator diameter	
80 Jet, 105 Jet and 115-200 (2.6 liter) hp (except Saltwater series)	21.0 mm (0.827 in.)
200-250 hp (3.1 liter) (Saltwater series [XL shaft length])	24.0 mm (0.945 in.)
Minimum commutator undercut	0.85 mm (0.033 in.)

Chapter Eleven

Midsection

Minor repair to the midsection involves replacing the lower motor mounts (**Figure 1**) or tilt lock mechanism (**Figure 2**).

Major repair may require removing the power head and gearcase followed by a complete disassembly of the midsection. Major repair is required when repairing or replacing the clamp bracket (**Figure 3**), swivel housing (**Figure 4**) or drive shaft housing (**Figure 5**).

Refer to Chapter Ten if it is necessary to remove the trim system components to access midsection component(s).

Apply Yamaha All Purpose Grease to all bushings, pins and pivot points during assembly. Apply Loctite 572 to the threads of all fasteners during assembly. **Table 1** is at the end of this chapter.

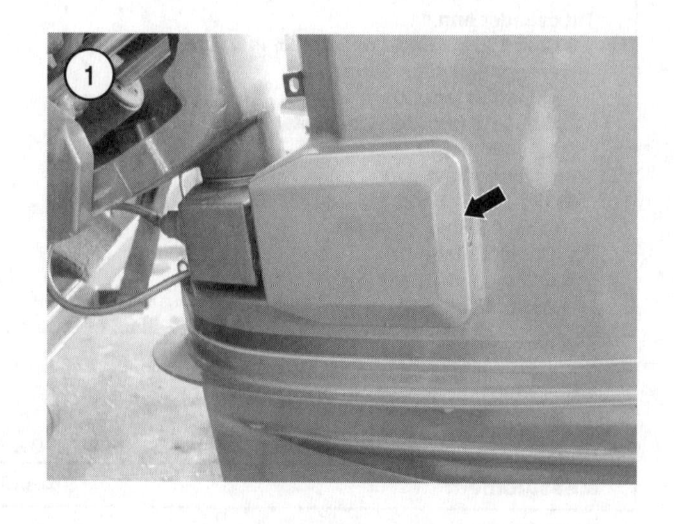

MIDSECTION

WARNING

Never work under any part of the engine without first providing suitable support. The engine-mounted tilt lock or hydraulic system may unexpectedly collapse and cause the engine to drop. Support the engine with blocks or an overhead cable before working under the engine.

Tilt Lock Lever Replacement

Refer to **Figure 6** during the tilt lock lever replacement.

1. Place the engine in the full up position.

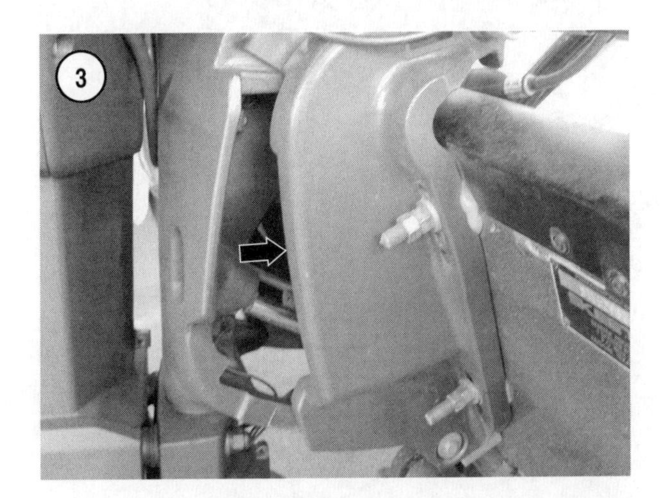

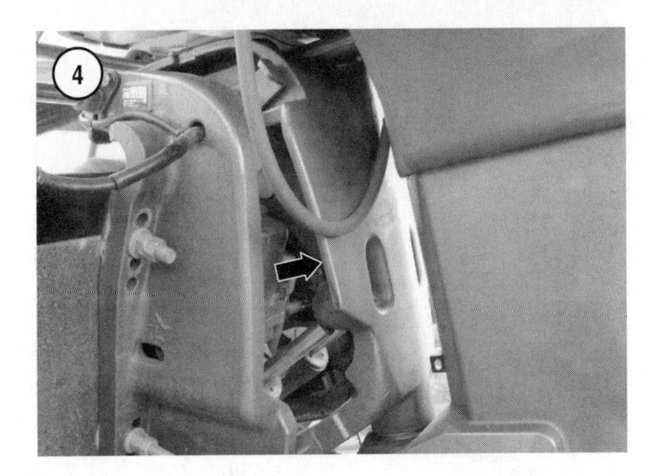

2. Use an overhead hoist to support the engine as indicated in **Figure 7**. Disconnect the battery cables and ground the spark plug leads to prevent accidental starting.

3. Disconnect the spring from the tilt lever coupler. The coupler connects to the inner side of each tilt lock lever.

4. Use a pin punch to drive the roll pin from the coupler and tilt lock lever. Pull the lever from the coupler and clamp bracket.

5. Inspect the lever and the respective bushing in the clamp bracket for worn or damaged surfaces and replace as required.

6. Lubricate the bushing with Yamaha All Purpose Grease. Then carefully slide the tilt lock lever through the bushing and into the coupler.

7. Rotate each tilt lock lever to the UP position. Align the roll pin opening in the tilt lock lever with the respective opening in the coupler. Carefully drive the roll pin into the lever and coupler until flush with the coupler surface. If the roll pin does not fit tightly in the opening, replace the roll pin. If a loose fit persists, replace the coupler and/or lock lever.

8. Reattach the spring to the tilt lever coupler. Make sure the tilt lock lever operates smoothly. Check for improper assembly if there is rough operation or binding.

9. Carefully remove the overhead support from the engine. Connect the battery cables and spark plug leads.

Anode Replacement

CAUTION

Applying grease or sealant to the anode bolt threads will insulate the threads from the threaded openings. The anode cannot protect the engine without maintaining electrical continuity to the engine ground.

1. Place the engine in the full up position. Use an overhead hoist to support the engine as shown in **Figure 7**.

2. Disconnect the battery cables and ground the spark plug leads to prevent accidental starting.

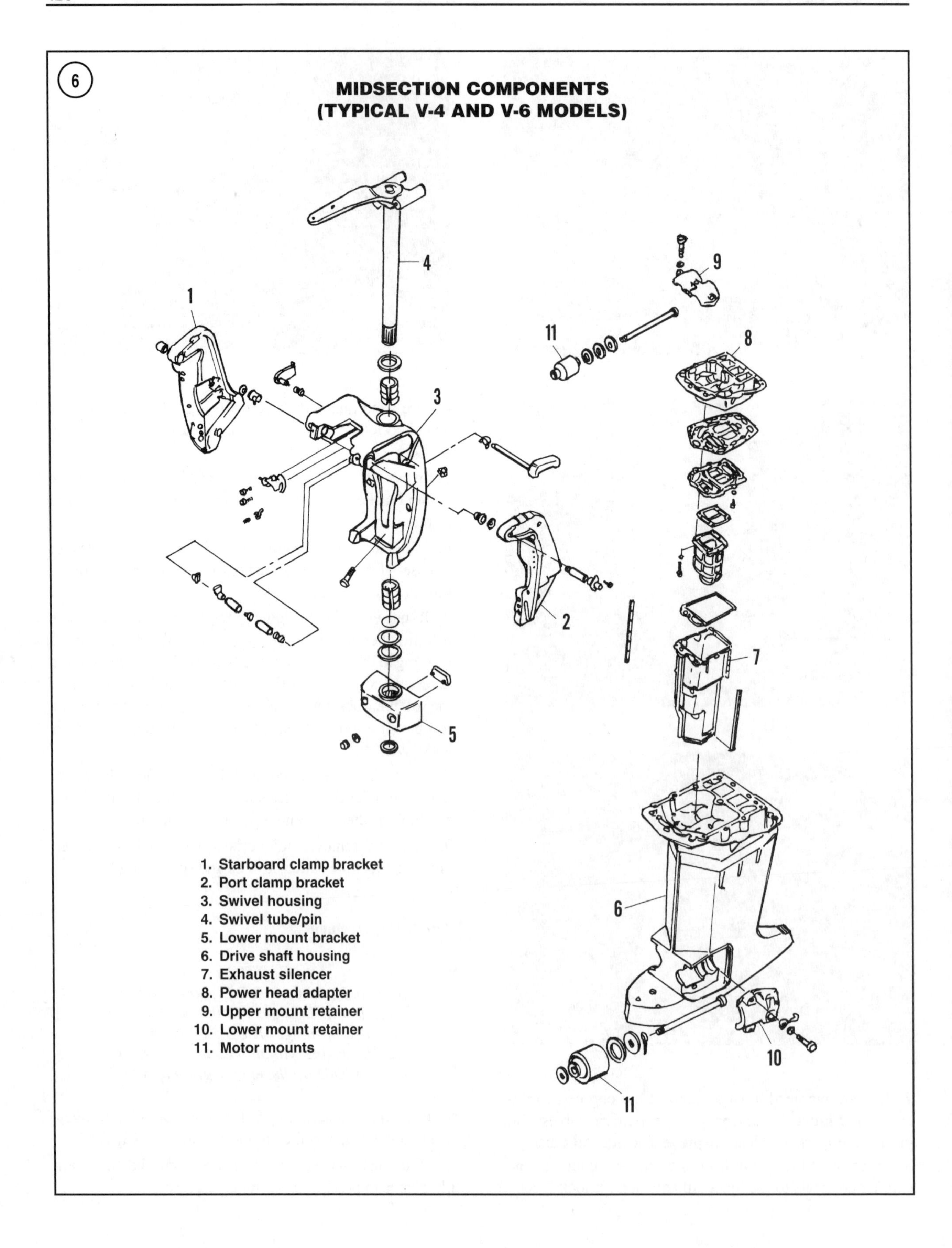
6
MIDSECTION COMPONENTS
(TYPICAL V-4 AND V-6 MODELS)
1
2
3
4
5
6
7
8
9
10
11
11
1. Starboard clamp bracket
2. Port clamp bracket
3. Swivel housing
4. Swivel tube/pin
5. Lower mount bracket
6. Drive shaft housing
7. Exhaust silencer
8. Power head adapter
9. Upper mount retainer
10. Lower mount retainer
11. Motor mounts

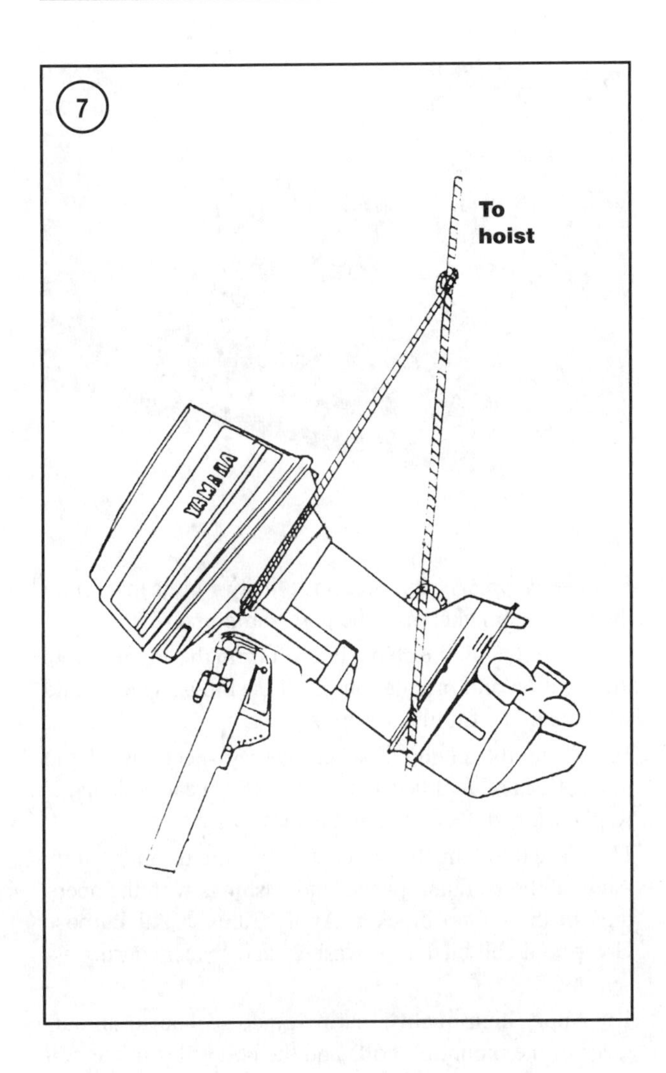

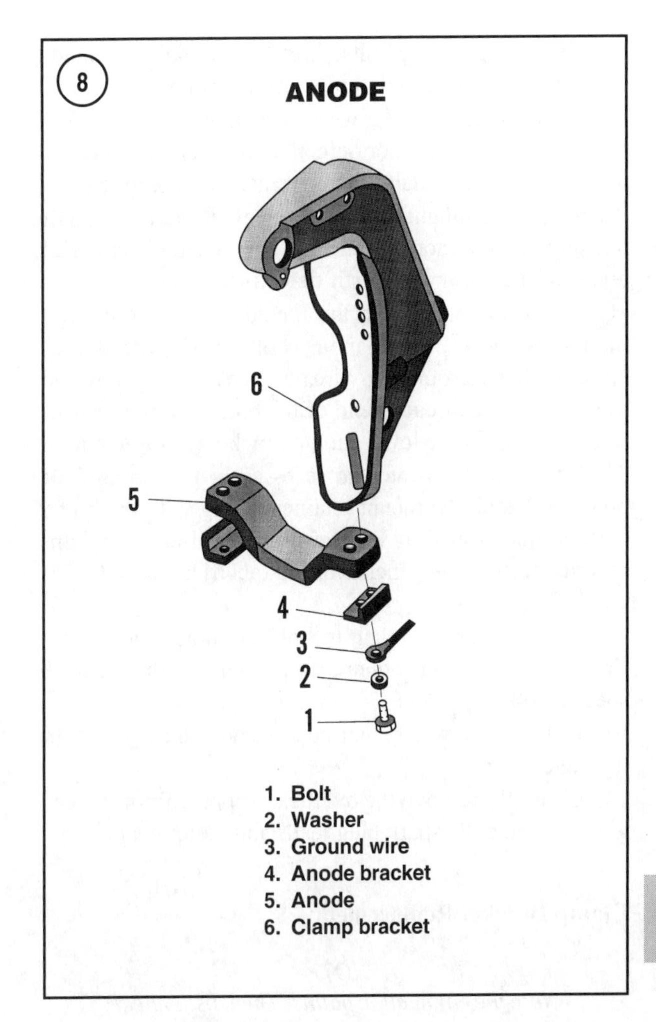

3. Remove the four bolts (1, **Figure 8**) and washers (2), then pull the grounding wire (3) from the anode bracket (4).
4. Carefully pry the two anode brackets (4, **Figure 8**) from the anode (5). Remove the anode from the clamp brackets. Tap the anode loose with a rubber mallet if necessary.
5. Clean all corrosion from the clamp brackets on the anode mating surface. Clean all corrosion from the threads of the anode bolt holes.
6. Clean all deposits from the anode surfaces. Sand the surfaces, if necessary, until they are clean.
7. Install the anode, brackets, bolts, washers and the grounding wire as shown in **Figure 8**. Evenly tighten the four mounting bolts to the general torque specification in Chapter One.
8. Use a multimeter to check for electrical continuity between the anode and a known ground wire on the power head. Remove the anode and clean the bolts, threads and grounding wire terminals if high resistance or no continuity is indicated.
9. Carefully remove the overhead support from the engine. Connect the spark plug leads and battery cables.

Lower Engine Mount Replacement

1. Place the engine in the full up position.
2. Use an overhead hoist to support the engine as indicated in **Figure 7**. Disconnect the battery cables and ground the spark plug leads to prevent accidental starting.
3. Remove the screws and carefully pull the mount covers (**Figure 1**) from the drive shaft housing.
4. Hold the head of the mount bolts with a suitable wrench while removing the locknut that secures the bolts to the lower mount bracket (5, **Figure 6**).
5. Remove the bolts and washer that secure the mount retainer (10, **Figure 6**) onto the drive shaft housing.
6. Carefully pull the mounts (11, **Figure 6**) from both sides of the drive shaft housing.

7. Clean the mounting bolts, threaded bolt holes, grounding wire terminals and mount contact surfaces.
8. Inspect all fasteners for wear, corrosion or damage. Replace any questionable or defective fasteners. Inspect the mounts for wear or damage and replace if defective.
9. Inspect the mount contact surfaces in the drive shaft housing and the mount retainers for cracked or damaged areas. Replace components if defective.
10. Apply Loctite 572 to the threads of all mount and mount fasteners. Slip the mount bolts along with the related washers, grounding wires and spacers through the mounts. Fit the threaded end of the bolts into the respective openings in the lower mount bracket (5, **Figure 6**).
11. Fit the mounts into the recesses in the drive shaft housing. Install the mount retainers along with the bolts, washers and grounding wire terminals. Tighten the mount retainer bolts to the general torque specification in Chapter One.
12. Use a suitable wrench to hold the bolt heads while tightening the lower mount bolt locknuts to the **Table 1** specification.
13. Install the lower mount covers and securely tighten the cover screws.
14. Carefully remove the overhead support from the engine. Connect the spark plug leads and battery cables.

Clamp Bracket Replacement

NOTE
Note the connection points. Then disconnect the grounding wires when necessary to remove other components. Clean all corrosion or contaminants from the wire contact surfaces. Make sure all grounding wires are securely tightened during assembly.

1. Place the engine in the full up position.
2. Use an overhead hoist to support the engine as indicated in **Figure 7**. Disconnect the battery cables and ground the spark plug leads to prevent accidental starting.
3. Refer to Chapter Ten and remove the tilt/trim system from the engine. If so equipped, loosen the screw that secures the trim position sender lever (**Figure 9**) onto the tilt tube.
4. Remove the anode as described in this chapter.
5. Remove the tilt lock lever as described in this chapter.
6. Disconnect and remove the steering linkage and cable from the swivel tube (4, **Figure 6**) and tilt tube. Remove the engine mounting bolts from the clamp bracket and boat transom.
7. *Starboard clamp bracket*—Remove the large nut that retains the tilt tube into the clamp bracket.

8. *Port clamp bracket*—Remove the two bolts that retain the tilt tube bracket onto the port clamp bracket.
9. Carefully pull the clamp bracket from the engine. Note the location and orientation of all bushings, spacers and washers in the tilt tube opening.
10. Clean the clamp bracket, then inspect the clamp bracket and related fasteners for cracks, wear or damage. Replace any defective components.
11. Place the clamp bracket in position on the midsection. Align all bolts, nuts, spacers and bushings with the openings in the clamp bracket. Apply Yamaha All Purpose Grease onto all bushings, washers and spacers during assembly.
12. Apply high quality marine-grade sealant to all surfaces of the mounting bolts and the bolt holes in the bolt transom. Install the engine mounting bolts through the clamp bracket(s) and boat transom. Securely tighten the mounting bolts.
13. *Starboard clamp bracket*—Install the large nut that retains the tilt tube into the clamp bracket. Tighten the tilt tube nut to **Table 1** specification.
14. *Port clamp bracket*—Install the two bolts that retain the tilt tube bracket onto the port clamp bracket. Tighten the bolts to the general torque specification in Chapter One.
15. Install the tilt/trim system as described in Chapter Ten.
16. Apply Yamaha All Purpose Grease to the steering cable and tilt tube bore. Attach the steering cable and linkage onto the tilt tube and swivel tube. Securely tighten all fasteners. Then engage any fastener locking devices.
17. Carefully remove the overhead support from the engine. Connect the spark plug leads and battery cables.
18. Adjust the trim position sender, if so equipped, as described in Chapter Four.

19. Operate the trim and steering system through the entire operating range. Inspect the midsection for improperly installed components if there are any unusual noises or binding. Correct any unusual noises, looseness or binding before operating the engine.

Upper Engine Mount Replacement

1. Remove the power head as described in Chapter Seven.
2. Provide continual support for the drive shaft housing during this step. Remove bolts that secure the upper engine mount retainers (9, **Figure 6**) onto the power head adapter. Fully loosen the mount bolts to free the drive shaft housing from the swivel tube (4, **Figure 6**).
3. Carefully pry the upper engine mounts, bolts, washers and spacers from the power head adapter.
4. Clean the mounting bolts, threaded bolt holes and mount contact surfaces.
5. Inspect all fasteners for wear, corrosion or damage. Replace any questionable or defective fasteners. Inspect the mounts for wear or damage and replace as needed.
6. Inspect the mount contact surfaces in the power head adapter and the mount retainers for cracked or damaged areas. Replace components with cracks or excessively worn areas.
7. Slip the mount bolts along with the related washers, grounding wires and spacers through the mounts. Fit the mounts into the openings in the power head adapter. Align the mount bolts. Then thread the bolts into the openings in the swivel tube (4, **Figure 6**).
8. Install the mount retainer onto the upper mounts. Apply Loctite 572 to the threads. Then install the mount retainer bolts and washers. Tighten the mount retainer bolts to the general torque specification in Chapter One.
9. Use a suitable wrench to tighten the upper mount bolts to the **Table 1** specification.
10. Install the power head as described in Chapter Seven.

Drive Shaft Housing Replacement

1. Remove the gearcase as described in Chapter Eight.
2. Remove the power head as described in Chapter Seven.
3. Remove the lower engine mounts as described in this chapter.
4. Provide continual support to the drive shaft housing while removing the upper engine mounts as described in this chapter.
5. Remove all bolts, washers and nuts that secure the power head adapter (8, **Figure 6**) onto the drive shaft housing, then carefully pry the adapter from the drive shaft housing. Remove and discard the adapter plate gaskets.
6. Remove the bolts and washers. Then carefully pry the exhaust plate from the power head adapter. Remove and discard the exhaust plate gasket.
7. Remove the bolts and washers that retain the exhaust silencer (7, **Figure 6**) onto the exhaust plate. Carefully pry the silencer from the plate. Remove and discard the silencer gasket.
8. Remove the exhaust tuner only if it must be replaced. Remove the mounting bolts. Then carefully pry the tuner from the exhaust plate. Remove and discard the tuner gasket.
9. Thoroughly clean all surfaces. Work carefully to avoid damaging the gasket sealing surfaces.
10. Inspect all removed components for cracked, worn or corrosion damaged surfaces. Replace any defective components.
11. Inspect all alignment pins and bores for worn or damaged pins or elongated holes. Replace any defective components.
12. Assembly is the reverse of disassembly. Note the following:
    a. Install new gaskets and seals at all locations during assembly. Gasket sealing compound is not required.
    b. Apply Loctite 572 onto the threads of all mounting bolts.
    c. Tighten all fasteners to the specification in **Table 1** or Chapter One.

Swivel Housing Replacement

1. Remove the tilt/trim system as described in Chapter Ten.
2. Remove the power head as described in Chapter Seven.
3. Remove the gearcase as described in Chapter Eight.
4. Remove the upper and lower engine mounts as described in this chapter.
5. Remove the port and starboard clamp brackets as described in this chapter.
6. Remove the trim position sender as described in Chapter Ten.
7. Clean the swivel housing. Inspect the housing or wear for damage. Replace the housing if there are defects.
8. Move the swivel tube through the full range of motion. Remove the swivel tube if it binds or feels loose.
9. Installation is the reverse of removal. Note the following:

Operate the tilt/trim and steering through the entire operating range. Inspect the midsection for improperly installed components. Correct any unusual noises, looseness or binding before operating the engine.

Swivel Tube Replacement

Refer to **Figure 6** for this procedure.

1. Remove the swivel housing as described in this chapter.
2. Remove the snap ring from the groove on the lower end of the swivel tube (4, **Figure 6**). Inspect the snap ring for corrosion, cracks or lost spring tension. Replace the snap ring if not in excellent condition.
3. Support the lower mount bracket (5, **Figure 6**) while pulling the swivel tube up and out of the swivel housing. Remove the washer from the upper side of the swivel bracket or swivel tube.
4. Remove the mount bracket, spacers and washers from the swivel housing.
5. Inspect the bushings in the swivel tube bore in the swivel housing. Remove and replace both bushings if either is worn or damaged.
6. Inspect the swivel tube for worn, cracked or corrosion damaged surfaces. Replace the swivel tube if these or other defects are noted.
7. Assembly is the reverse of disassembly. Note the following:
   a. Apply Yamaha All Purpose Grease to the swivel tube bushing and swivel tube during assembly.
   b. Install all washers and spacers as shown in **Figure 6**.
   c. Make sure the snap ring fits fully into the groove at the lower end of the swivel tube.

Tilt Tube Replacement

1. Place the engine in the full up position.
2. Use an overhead hoist to support the engine as indicated in **Figure 7**. Disconnect the battery cables and ground the spark plug leads to prevent accidental starting.
3. Disconnect the steering linkages and cables from the swivel tube and tilt tube.
4. Fully loosen the screw that secures the trim position sender lever onto the tilt tube.
5. Remove the large nut that secures the tilt tube into the starboard clamp bracket.
6. Remove the two bolts that secure the tilt tube onto the port clamp bracket.
7. Use a one-foot section of pipe or tubing to drive the tilt tube through the clamp brackets and swivel housing. The tool must be slightly smaller in diameter than the tilt tube.
8. Provide just enough overhead support to keep the tube from binding during removal. Do not use excessive force when driving the tube from the brackets. Excessive force can cause the end of the tube to flare out preventing removal. Use a wooden block as padding until the tube is flush with the clamp bracket surface.
9. Carefully drive the tube through the starboard clamp bracket, swivel housing and port clamp bracket. Support the engine. Then slowly remove the driver tool. Retain the washers as they drop from the clamp brackets and swivel housing.
10. Thoroughly clean the tilt tube. Inspect the tilt tube for excessive wear, corrosion, cracking or other damage. Replace the tilt tube if not in excellent condition.
11. Thoroughly clean all surfaces of the swivel housing and clamp brackets. Inspect the tilt tube bushings and washers for excessive wear, cracking or other defects. Replace any defective components.
12. Apply a coating of Yamaha All-Purpose Grease to the tilt tube bore in the swivel housing and clamp bracket bushings.
13. Place any washer between the clamp brackets and swivel housing and align them with the tilt tube bore.
14. Apply a coating of Yamaha All-Purpose Grease to the external surfaces of the tilt tube. Place the tilt tube into the opening in the port clamp bracket. Align the tilt tube bores in the clamp brackets, washers and swivel housing.
15. Using a block of wood for a cushion, carefully tap the tilt tube through the port clamp bracket, swivel housing and starboard clamp bracket.
16. Install the two bolts and bracket that retain the tilt tube onto the port clamp bracket. Tighten the bolts to the general torque specification in Chapter One.
17. Install the large nut that retains the tilt tube into the starboard clamp bracket. Tighten the nut to the **Table 1** specification.
18. Apply a coating of Yamaha All-Purpose Grease to the steering cable and tilt tube bore. Attach the steering linkage and cable onto the swivel tube and tilt tube. Securely tighten all fasteners and engage any fastener locking device.
19. Carefully remove the overhead support from the engine. Connect the spark plug leads and battery cables.
20. Operate the tilt/trim and steering through the entire operating range. Inspect the midsection for improperly installed components. Correct any unusual noises, looseness or binding before operating the engine.

Table 11 MIDSECTION TORQUE SPECIFICATIONS

Fastener	N·m	in.-lb.	ft.-lb.
Exhaust silencer to exhaust plate			
80 Jet and 115-130 hp	18	156	13
105 Jet and 150-200 hp (2.6 liter)	18	156	13
200-250 hp (3.1 liter)			
L drive shaft length	18	156	13
XL shaft length	21	185	15
Exhaust plate to power head adapter			
80 Jet and 115-130 hp	18	156	13
105 Jet and 150-200 hp (2.6 liter)	18	156	13
200-250 hp (3.1 liter)			
L drive shaft length	18	156	13
XL drive shaft length	21	185	15
Exhaust tuner pipe to exhaust plate			
80 Jet and 115-150 hp	18	156	13
105 Jet and 150-200 hp (2.6 liter)	18	156	13
200-150 hp (3.1 liter)			
L drive shaft length	18	156	13
XL drive shaft length	21	185	15
Tilt tube locknut	15	132	11
Trim ram striker plate	37	–	27
Upper engine mount to swivel housing			
80 Jet and 115-130 hp	53	–	39
105 Jet and 150-200 hp (2.6 liter)	53	–	39
200-250 hp (3.1 liter)			
L drive shaft length	53	–	39
XL drive shaft length	73	–	54
Lower engine mount to swivel housing	73	–	54

Chapter Twelve

Oil Injection System

This chapter describes removal, inspection and installation instructions for all oil injection components. Air bleeding instructions are described as well. Oil injection specifications are located in **Tables 1-3** at the end of this chapter.

All Yamaha outboards covered in this manual, except models C115 and C150, are factory equipped with Precision Blend oil injection. Refer to Chapter Two for testing and troubleshooting instructions for the oil injection and warning system components.

OIL INJECTION SYSTEM OPERATION

Oil injection eliminates the need to mix fuel and oil (after break-in). A gear-driven variable rate oil pump (**Figure 1**) delivers the correct amount of oil to the engine at all speeds. Changes to the fuel/oil ratio occur as the pump lever and shaft (1, **Figure 2**) are rotated by a linkage connected to the carburetors or throttle body linkage. This changes the pump stroke. Changing the pump stroke increases or decreases the volume of oil supplied to the engine.

Figure 3 shows a diagram of the typical oil injection system. On carburetor equipped models, oil is injected directly into the power head at the intake manifold. Air and fuel flowing into the engine disperses the oil throughout the power head. Oil is never introduced into the carburetors to reduce the formation of deposits in the carburetors during extended storage.

On OX66 EFI models, oil is injected into the vapor separating tank (**Figure 4**) where it mixes with the fuel sup-

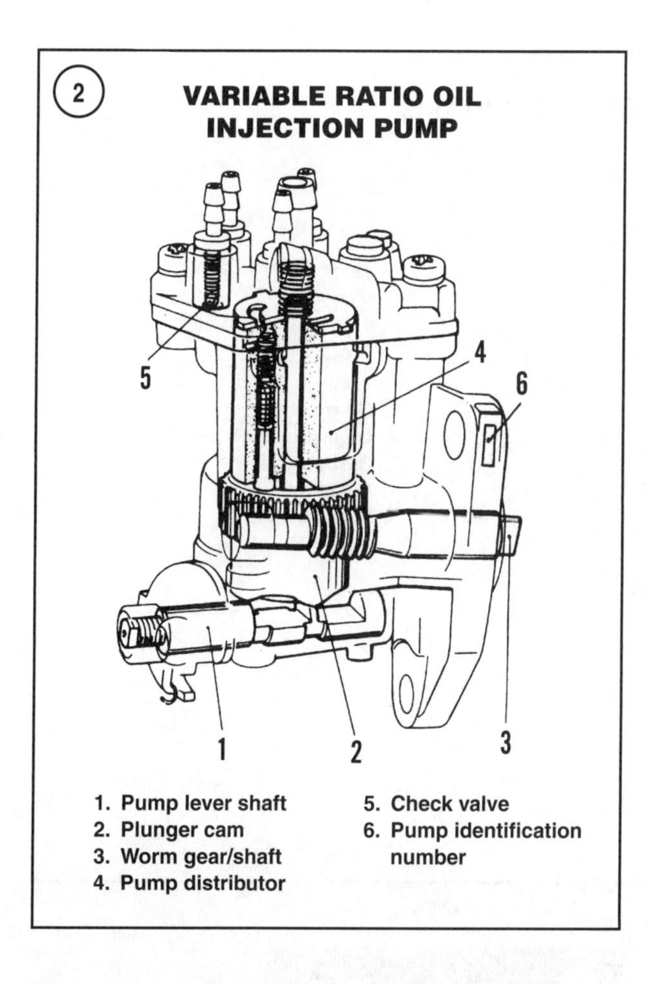

plied to the fuel injectors. The injectors spray the fuel/oil mixture into the power head through the throttle valve openings. Air and fuel flowing into the engine disperses the oil throughout the power head.

On HPDI models, oil is injected directly into the engine at the intake manifold. Air flowing into the engine disperses the oil throughout the power head. These models also incorporate an electric oil pump (A, **Figure 5**) to introduce oil via a hose and fitting (B) into the vapor separator tank. The oil mixes with the fuel where it lubricates the mechanical high-pressure fuel pump and fuel injectors. The engine control unit controls the electric pump operation to ensure these components receive the correct amount of oil for various operating conditions.

A warning system continuously monitors the oil level in the engine mounted oil reservoir (10, **Figure 6**) using the oil level sensor (8). As oil is consumed, the sensor mounted float drops until it activates the oil reservoir refill switch in the sensor assembly. The switch signals the CDI unit or engine control unit or oil injection module (80 Jet, 115 hp and 130 hp models) to switch on the electric oil pump in the onboard mounted oil reservoir (11, **Figure 6**).

Oil is then pumped from the on-board reservoir until the engine mounted reservoir is refilled, at which point the oil pump is switched off. This process repeats each time the oil drops below the threshold necessary to maintain an adequate oil level. If the system fails to operate, the oil level will drop below the normal level, causing the sensor to switch on the low oil level warning. The CDI unit or engine control unit sounds the warning horn and/or illuminates the dash mounted light to notify the operator. If the oil level drops below a critical level, the CDI unit or engine control unit reduces the power to help reduce power head damage.

If the oil level sensor, oil injection module, CDI unit or engine control unit fails to switch on the pump, use the emergency switch (**Figure 7**) to switch on the electric oil pump to refill the reservoir. Hold the switch in the RUN position until the reservoir is refilled. Do not overfill the reservoir. An internal spring toggles the switch to the off position when released. Continue to monitor the oil level during the return to shore and refill the reservoir as needed. Never run the engine with a low oil level in the engine mounted reservoir. If the reservoir cannot be refilled using the emergency switch, pour oil directly into the engine mounted reservoir using the oil level sensor opening. Refer to Chapter Two for specific warning system descriptions and troubleshooting procedures. Premixing the oil and fuel is required on oil injected models (except HPDI models) during the initial break-in period and after a major power head repair. Operating an HPDI model engine with premix during the break-in period provides no benefit as the fuel is injected directly into the cylinder; the fuel never contacts the crankcase components. If premix is used, it usually fouls the spark plugs and may cause power head damage. Refer to Chapter Three for oil and fuel premixing procedures.

OIL INJECTION SYSTEM

Oil Level Sensor Removal/Installation and Oil Strainer Service

Refer to Chapter Two for troubleshooting and testing of this component. Work only in a clean environment. Any debris in the system eventually migrates to the oil pump or check valves where it blocks oil flow or seizes moving components in the oil pump, causing in sufficient lubrication and eventual power head failure.

1. Locate the oil level sensor (**Figure 8**) at the top of the oil supply reservoir. Disconnect the sensor wires at the engine harness connector.
2. Grasp the oil level sensor between the forefinger and thumb (**Figure 9**). Squeeze the sensor and carefully pull

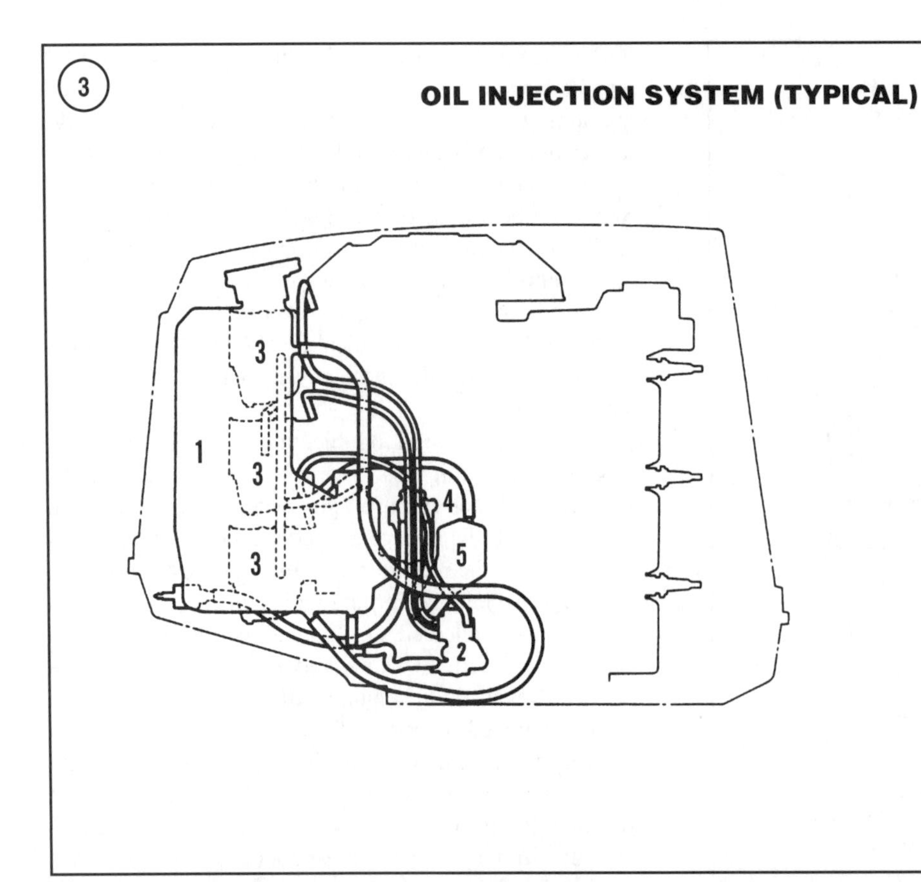

1. Oil reservoir
2. Gear driven oil pump
3. Carburetors
4. Fuel filter
5. Fuel pump

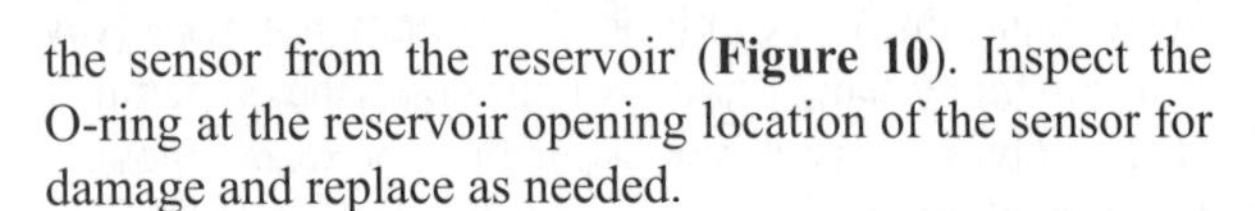

the sensor from the reservoir (**Figure 10**). Inspect the O-ring at the reservoir opening location of the sensor for damage and replace as needed.

3. Note the alignment of the oil strainer to the oil level sensor. Alignment markings are provided on most models. If necessary, mark the sensor and strainer to ensure proper alignment on assembly.

4. Carefully lower the strainer (**Figure 11**) from the sensor. Clean the strainer with a suitable solvent and air dry. Inspect the strainer and replace it if there are torn or damaged surfaces.

5. Inspect the gasket (3, **Figure 12**) and replace if torn, deteriorated or deformed.

CAUTION
Improper installation of the strainer onto the sensor prevents air from venting from the strainer. Without proper venting, the

6

OIL INJECTION SYSTEM/HOSE ROUTING (80 JET, 115 HP AND 130 HP)

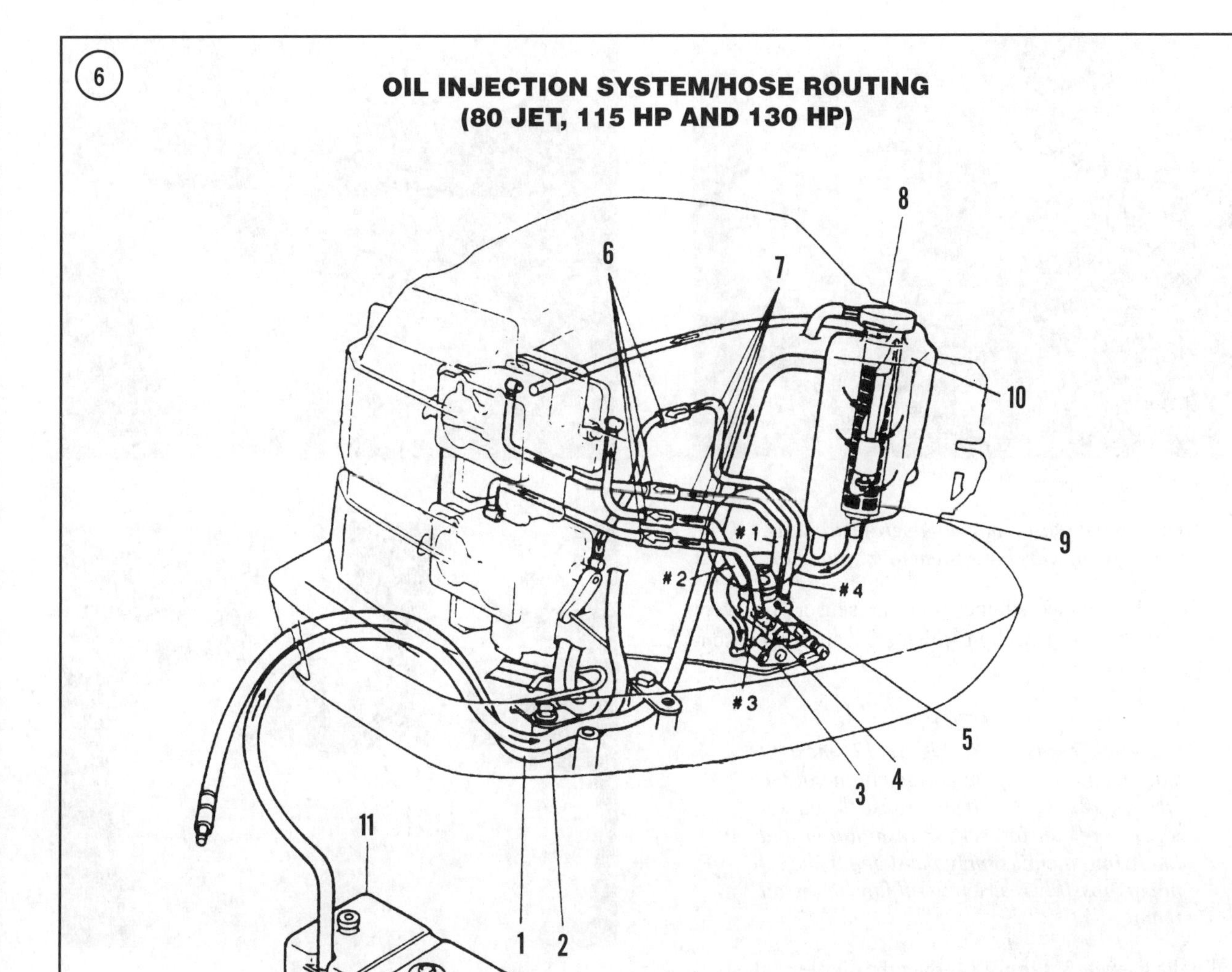

1. Fuel supply hose
2. Oil hose
3. Pump control lever (gear driven pump)
4. Gear driven oil pump
5. Oil outlet fitting
6. Oil line check valves
7. Oil outlet hoses
8. Oil level sensor
9. Oil strainer
10. Oil reservoir (engine mounted)
11. Oil reservoir (on-board mounted)

float in the sensor may not operate properly, causing the warning system to malfunction.

6. Install the cleaned strainer onto the sensor. The vent hole in the strainer must align with the flat area on the sensor (**Figure 13**).

CAUTION

*The rubber gasket (3, **Figure 12**) must fit snugly onto the nipple on the bottom of the strainer (2). A loose fit may cause the gasket to drop off during sensor installation and lodge into the oil supply passage for the oil pump, possibly restricting oil flow to the oil pump.*

7. Fit the gasket (3, **Figure 12**) over the nipple on the bottom of the strainer. Seat the gasket against the flat surface at the bottom of the strainer. The gasket must fit snugly on the nipple. If otherwise, replace the gasket.

CAUTION

Improper installation of the sensor may prevent oil flow toward the oil pump and subsequent power head failure due to a lack of sufficient lubrication. Always align the sensor cap and reservoir marking during sensor installation.

8. Note the alignment marking on the cap and the oil reservoir (**Figure 14**). Lubricate the O-ring on the sensor with Yamalube or equivalent. Avoid pinching the O-ring during sensor installation. Align the markings and carefully slip the sensor into the reservoir opening. The nipple on the bottom of the strainer must fit into the oil passage at the bottom of the reservoir. Carefully press the sensor into the reservoir until it fully seats.
9. Connect the oil lever sensor wires onto the engine wire harness connector. Fill the reservoir with oil as described in Chapter Three. Bleed air from the system as described in this chapter.

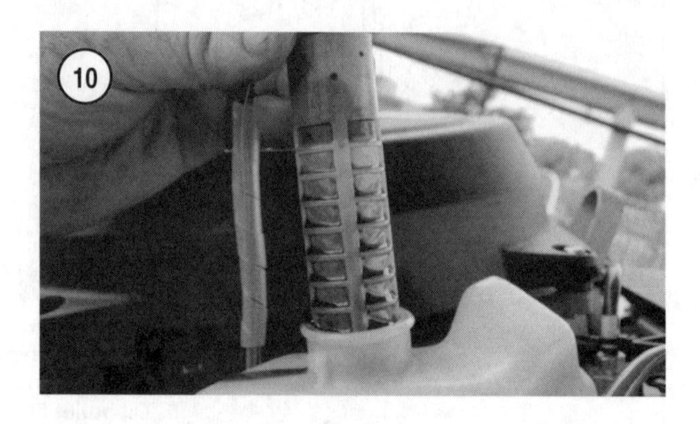

Oil Reservoir (Engine Mounted) Removal/Installation

Refer to **Figure 6** and **Figures 15-17**.

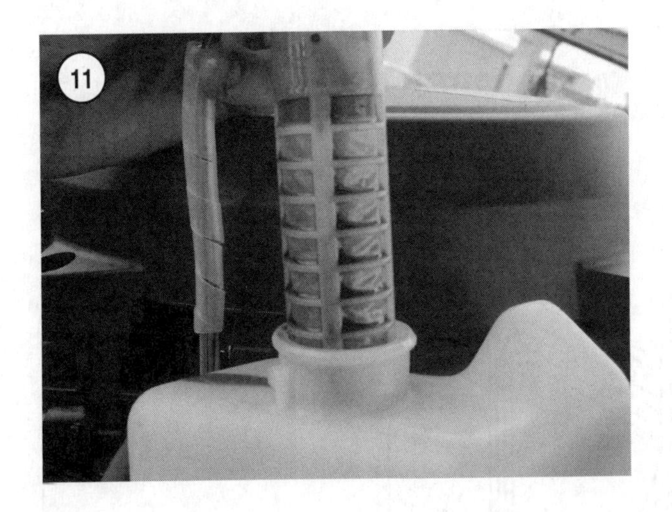

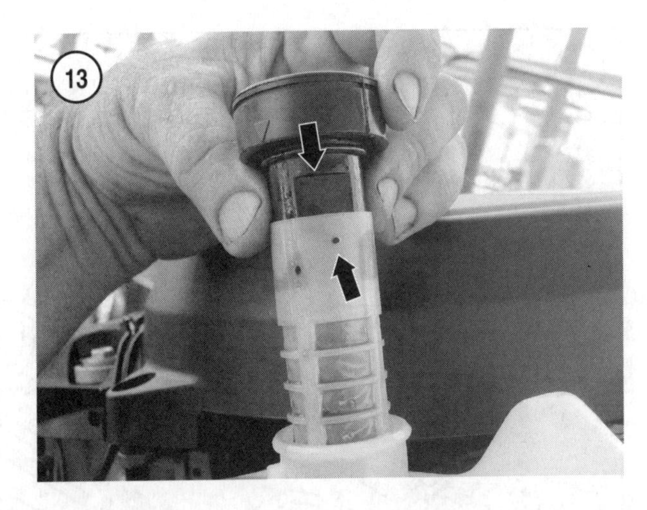

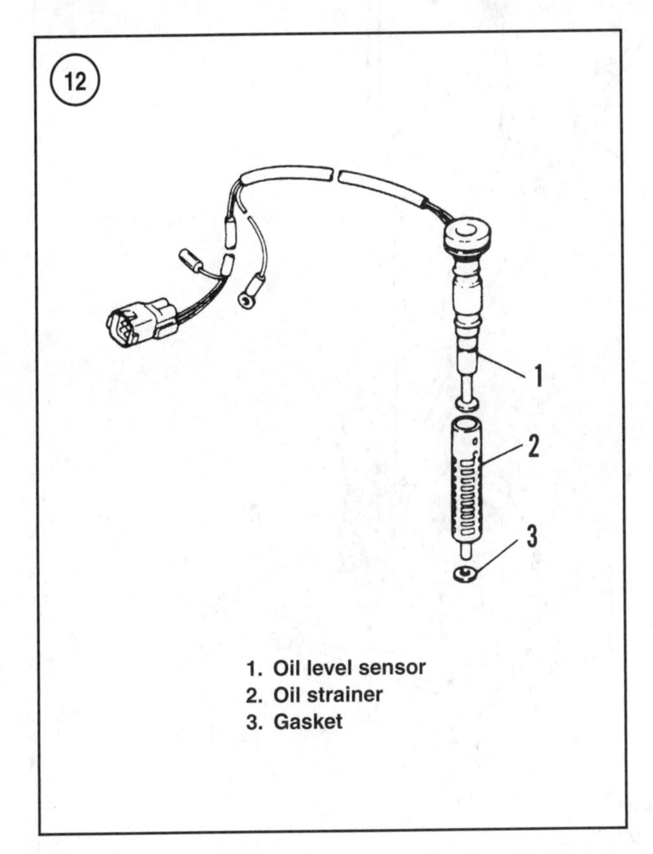

1. Oil level sensor
2. Oil strainer
3. Gasket

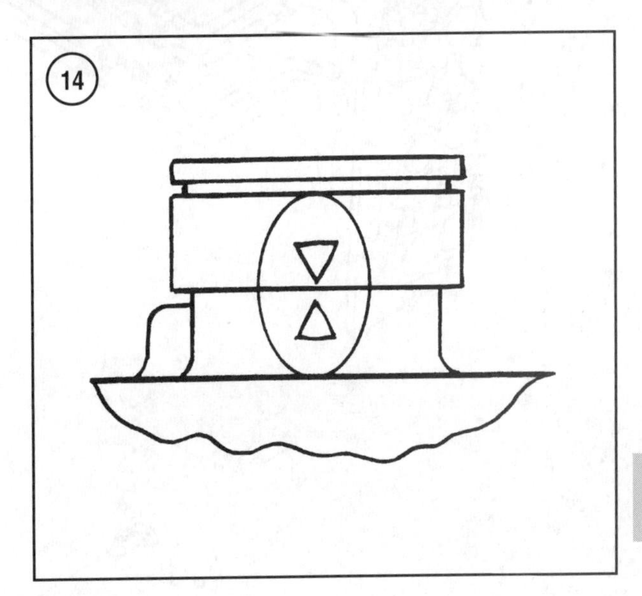

1. Disconnect the battery cables and ground all spark plug leads to prevent accidental starting.

2. Remove the oil level sensor as described in this chapter.

3. Remove the oil hose leading to the oil pump or the water drain hose (if so equipped) and quickly direct the hose into a container of suitable material and capacity to capture the oil. **Table 2** lists oil reservoir capacities.

4. Note the hose routing and connection points for all hoses connected onto the reservoir. Mark or make a sketch of the hose routing and connection points as needed. Replace hose clamps that are corroded, distorted or have weak spring tension.

5. Note the orientation of the grommets, washers and sleeves. Then, support the reservoir while removing the mounting bolts. Verify that all hoses are disconnected from the reservoir. Then remove the reservoir.

6. Clean debris and oil film from the reservoir and power head mating surfaces.

7. Clean the reservoir in a suitable solvent or warm, soapy water and air dry. Thoroughly rinse with clean water to remove soap residue or solvent. Inspect the reservoir for cracked, leaking or abraded surfaces. Replace the reservoir if any of these are found.

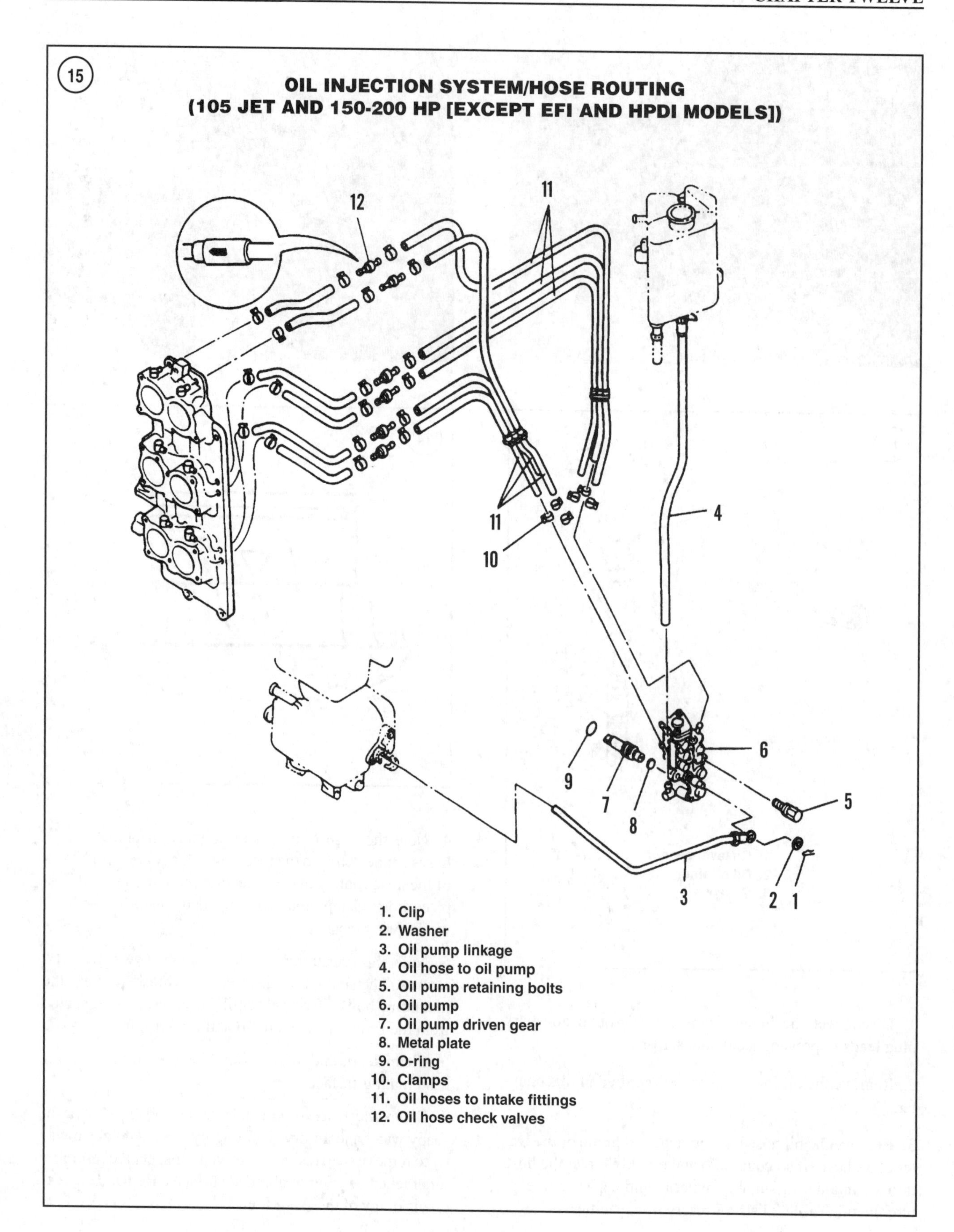

1. Clip
2. Washer
3. Oil pump linkage
4. Oil hose to oil pump
5. Oil pump retaining bolts
6. Oil pump
7. Oil pump driven gear
8. Metal plate
9. O-ring
10. Clamps
11. Oil hoses to intake fittings
12. Oil hose check valves

OIL INJECTION SYSTEM/HOSE ROUTING (150-250 HP [EFI MODELS])

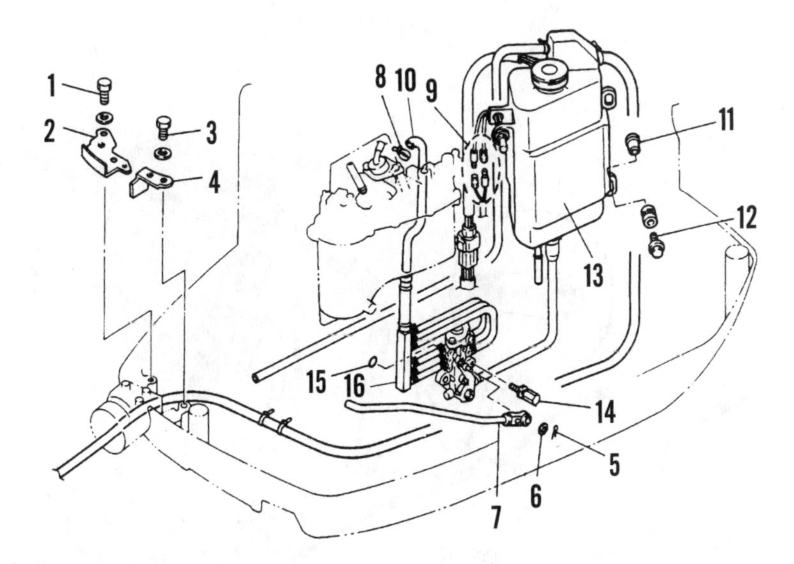

1. Bolt
2. Hose retainer
3. Bolt
4. Hose retainer
5. Clip
6. Washer
7. Oil pump linkage
8. Clamp
9. Emergency switch wire connectors
10. Oil hose to vapor separator tank
11. Grommet
12. Bolt
13. Oil reservoir
14. Oil pump retaining bolt
15. O-ring at oil pump mount
16. Oil hose manifold

8. Install the grommets. Carefully mount the reservoir onto the power head. Verify that no hose, wiring or other components are pinched between the reservoir and the power head. Install the sleeves, washers and mounting bolts. Again verify that no hoses, wiring or other components are pinched, then securely tighten the mounting bolts to the general torque specifications in Chapter One.

9. Refer to the appropriate illustration, sketch or markings and connect all hoses onto the reservoir. Secure the hoses with appropriate clamps.

10. Fill the reservoir with Yamalube or equivalent as described in Chapter Three. Reconnect the battery cables and spark plug leads.

11. Bleed air from the system as described in this chapter.

Oil Hoses and Check Valves Repair

Refer to **Figure 6** and **Figures 15-17**.

Clean all loose material and contaminants from the hose fittings prior to disconnecting the hoses to prevent contamination. The oil injection system must operate with clean oil.

Hose removal and inspection

CAUTION

Work carefully when removing plastic tie clamps from the oil hoses. The hose may be inadvertently cut as the clamp is cut and not

(17)

OIL INJECTION SYSTEM/HOSE ROUTING (HPDI MODELS)

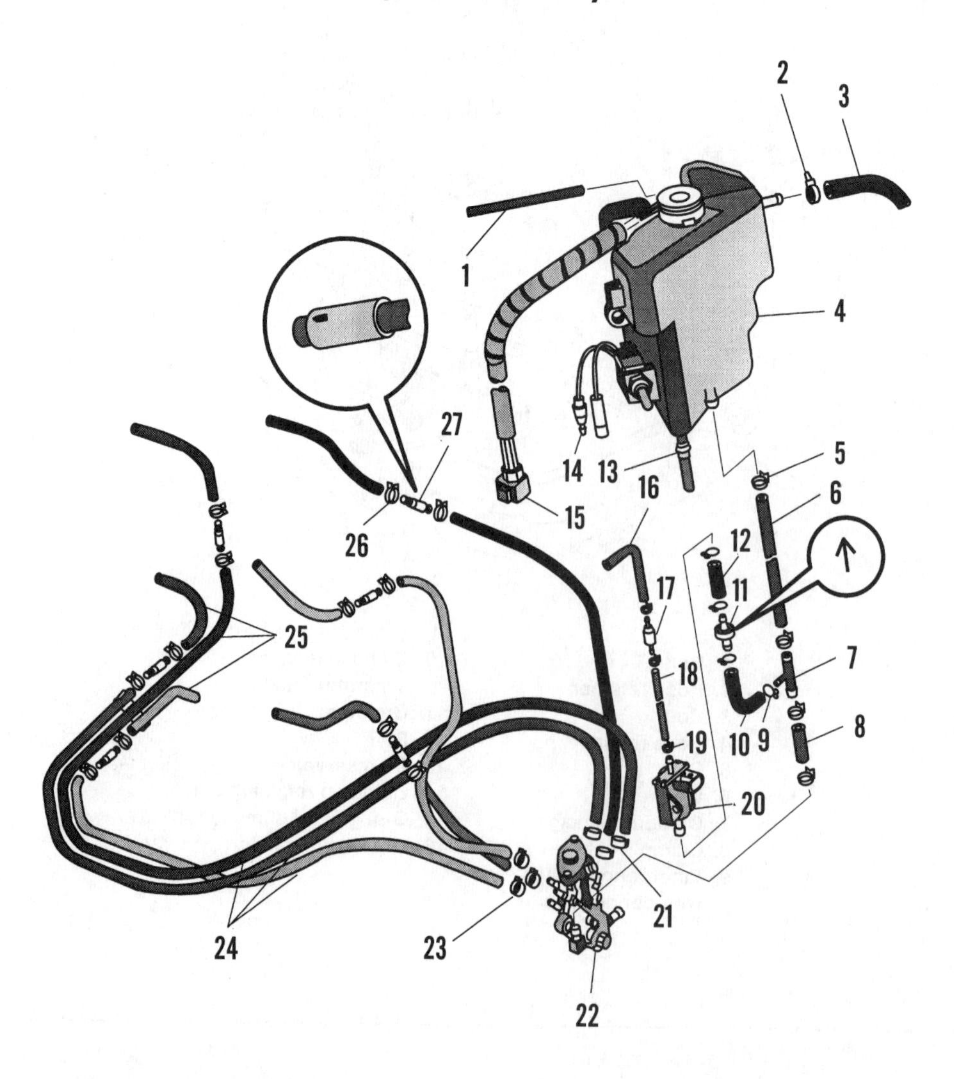

1. Vent hose (to silencer cover fitting)
2. Plastic locking type clamp
3. Oil hose (from on-board oil reservoir)
4. Oil reservoir (engine mounted)
5. Spring type clamp
6. Oil hose (from oil reservoir)
7. T-fitting
8. Oil hose (to gear driven oil pump)
9. Spring type clamp
10. Oil hose (to check valve)
11. Check valve (arrow faces upward)
12. Oil hose (to electric oil pump)
13. Water drain hose
14. Emergency switch wire connections
15. Oil level sensor wire connector
16. Oil hose (to vapor separator tank fitting)
17. Check valve (arrow faces upward)
18. Hose (electric oil pump to check valve)
19. Spring type clamp
20. Electric oil pump
21. Spring type clamp
22. Gear driven oil pump
23. Spring type clamp
24. Oil hoses (to check valves)
25. Oil hoses (to intake manifold fittings)
26. Spring type hose clamps
27. Oil hose check valve (arrow toward the intake manifold fitting)

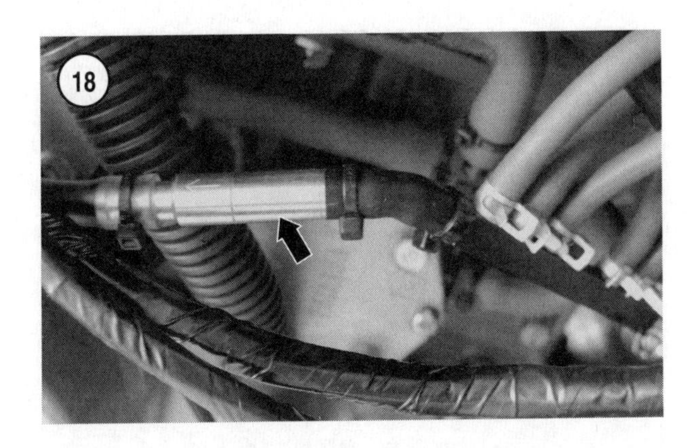

be apparent. The hose may eventually leak or fail at the cut area, causing oil leakage and/or insufficient lubrication of the power head. The power head will fail if the engine is operated with insufficient lubrication.

Carefully cut, remove and replace all plastic locking tie clamps when it is necessary to remove them. Work carefully to avoid damaging hoses or hose fittings. Inspect spring hose clamps for distortion, corrosion or weak spring tension. Replace damaged or questionable clamps.

Remove multiple hose connections in a cluster along with the oil injection system components whenever possible. This reduces oil leakage, incorrect hose connections during assembly and the time needed to bleed air from the system.

Inspect hoses for cut, cracked or abraded surfaces. Replace hoses with these defects. Replace hard or brittle hoses. Hard or brittle hoses eventually break and leak oil. Replace hoses that feel excessively soft or spongy. Check the type of oil that is being used (Chapter Three). Using the wrong type of oil may soften the oil hoses.

Clear hoses are often used at the oil pump. Inspect these hoses as described. These hoses must be transparent. Always replace clear hoses that have become cloudy. This condition is more prevalent if the engine is operated or stored in a warmer climate.

After hose installation, bleed air from the oil injection system as described in this chapter.

Check valves

CAUTION

Use caution when installing oil hose check valves. The engine will suffer serious power head damage if a check valve is installed in the wrong direction.

The oil line check valves (**Figure 18**) prevent oil from flowing out of the oil hoses when the engine is not running. Before removing them from the oil lines, note the direction of the arrow on the check valve body. The arrow indicates the direction the oil flows through the valve. The arrow must always lead toward the oil discharge fitting on the intake manifold, vapor separator tank (EFI models) or intake manifold (HPDI models).

Refer to **Figure 6** and **Figures 15-17**.

1. Disconnect the battery cables and ground the spark plug leads to prevent accidental starting.
2. Remove the clamps from the check valve fittings. Work carefully to avoid damaging the hoses. Inspect all oil hoses as described in this chapter.
3. Inspect spring hose clamps used on the check valve fittings for corrosion, distortion and weak spring tension. Always replace damaged or questionable clamps.
4. Test the check valve as described in Chapter Two. See *Oil Injection*.
5. Fit the hoses over the check valve fittings. The arrow on the check valve body must always point toward the oil hose leading to the discharge fitting on the intake manifold, vapor separator tank (EFI models) or intake mainifold (HPDI models).
6. Secure the hoses onto the fitting with the clamps.
7. Reconnect the battery cables and spark plug leads. Bleed air from the system as described in this chapter.

Gear Driven Oil Pump Removal/Installation

The oil pump is precisely matched to deliver the required amount of oil to the engine. Never use an oil pump that is not designated for use on the engine. Insufficient or excessive lubrication may occur. Most pumps have an identification number stamped on them. The typical location is near the pump lever or mounting boss. **Table 3** lists the identification for the models covered in this manual.

Replace the oil pump if faulty. Repair information or parts or are not provided for the oil pump. The drive and driven gears, seal, O-ring and mounting bolts are available.

Removal

Mark all hoses and make a sketch of hose routing prior to removal. Note the size and mounting location of all plastic locking tie clamps and replace them if removed.

1. Disconnect the battery cables and ground all spark plug leads to prevent accidental starting.

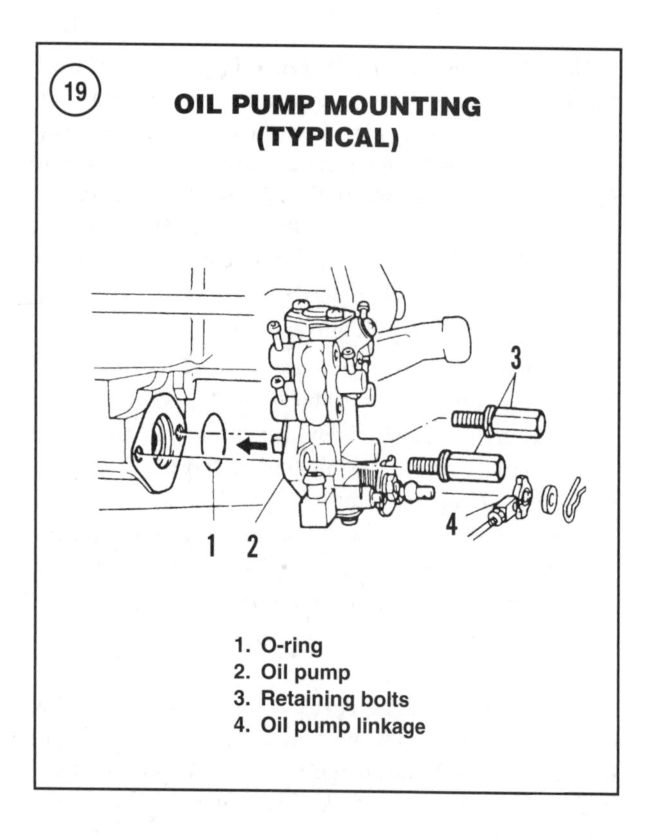

OIL PUMP MOUNTING (TYPICAL)

1. O-ring
2. Oil pump
3. Retaining bolts
4. Oil pump linkage

20

OIL PUMP DRIVE COMPONENTS

1. Oil pump
2. O-ring
3. Collar
4. Seal
5. Washer
6. Driven gear
7. Oil pump drive shaft

2. Carefully pry the linkage connector (4, **Figure 19**) from the pump lever. Do not disturb the link rod adjustment.

3. Remove the clamp and carefully pull the inlet hose from the oil pump. Quickly plug the hose with a suitable object such as a dowel rod or golf tee. Inspect spring hose clamps for corrosion, distortion and weak spring tension. Replace faulty or suspect clamps.

4. Note the cylinder number or outlet fitting on the oil hose manifold (EFI models) on all outlet hoses and pump fittings. Remove the clamps and carefully pull each outlet hose from the oil pump. Plug the hoses as described in Step 3.

5. Remove the two retaining bolts (3, **Figure 19**) and carefully pull the pump (2) from the cylinder block. Remove and discard the O-ring (1, **Figure 19**). Wipe oil residue and other contaminants from the oil pump mounting surfaces.

Installation

1. Apply Yamalube or equivalent onto the surfaces. Then, install a new O-ring (1, **Figure 19**) onto the oil pump. The O-ring must seat against the step on the oil pump mounting surface.

2. Position the oil pump onto the driven gear collar. Insert the oil pump drive shaft (7, **Figure 20**) into the driven gear (6). Rotate the pump on the collar until the oil pump shaft aligns with the slot in the driven gear. The oil pump seats against the collar when aligned. Never force the pump against the collar.

3. Hold the pump in contact with the collar while rotating the pump to align the retaining screw holes. Install the mounting bolts (3, **Figure 19**) and tighten to the **Table 1** specification.

4. Remove the plugs and fill the disconnected hoses with Yamalube. Quickly connect the oil hoses to the respective fittings on the oil pump. Secure the hoses with clamps.

5. Carefully snap the linkage connector (4, **Figure 19**) onto the oil pump lever. Operate the throttle and shift linkages to ensure the oil hoses do not contact moving components.

6. Adjust the oil pump linkage as described in Chapter Four.

7. Reconnect the battery cables and spark plug leads. Bleed air from the system as described in this chapter.

Driven Gear Removal/Installation

Upon removal, inspect the driven gear for damaged or worn gear teeth. Damaged or excessively worn gear teeth indicate the crankshaft mounted drive gear is suspect and

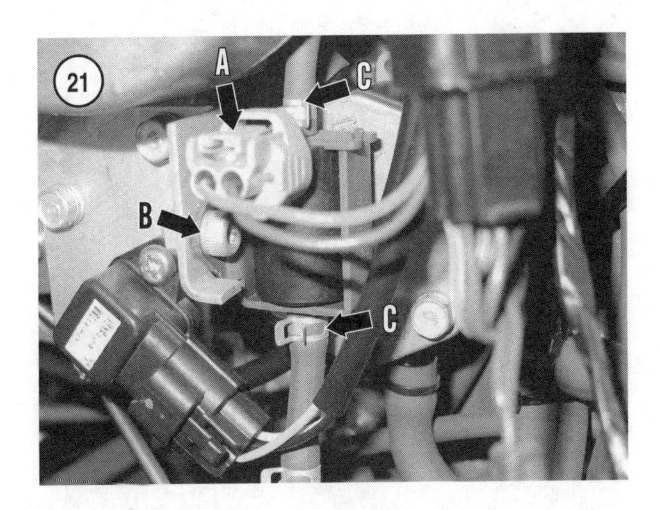

must be inspected. Refer to Chapter Seven for drive gear removal and installation instructions.

Removal

1. Remove the oil pump as described in this chapter.
2. Note the location and mounting location and orientation of the driven gear components before removing them.
3. Carefully remove the collar (3, **Figure 20**) and seal (4) from the power head. If necessary, pry the collar loose. Work carefully to avoid damaging the mating surfaces. Remove and discard the seal. Mark the pump side. Then remove the washer (5, **Figure 20**).
4. Use needle nose pliers to pull the driven gear from the power head. Pull the gear from the coupling end. Wipe all oil residue and other contaminants from the oil pump and collar mounting surfaces.
5. Clean the collar and gear with a suitable solvent and air dry. Replace any worn or damaged components.

Installation

1. Lubricate the driven gear, O-rings, seal and collar with Yamalube or equivalent. Use needle nose pliers to carefully insert the driven gear into the power head. Slowly rotate the gear during installation to mesh the driven gear teeth with the drive gear teeth.
2. Install the washer, seal and collar as shown in **Figure 20**. Seat the collar against the mating surface. Check for improper driven gear installation if the collar does not seat. Never force the collar against the mating surface.
3. Install the oil pump as described in this chapter.

Electric Oil Pump Removal/Installation

The electric oil pump is used only on 150-250 hp HPDI models. A single screw secures the electric oil pump onto the lower aft end of the vapor separator tank. Two screws secure the oil pump and air pressure sensor mounting bracket onto the vapor separator tank.
1. Disconnect the battery cables and ground all spark plug leads to prevent accidental starting.
2. Pinch the tab. Then carefully pull the connector (A, **Figure 21**) from the electric oil pump.
3. Move the spring hose clamps away from the oil pump hose fittings (C, **Figure 21**). Then carefully pull the hoses from the fittings. *Quickly* plug the disconnected hoses to prevent oil leaks. Fill the hoses with the recommended oil (see Chapter Three) to replenish any lost oil.
4. Remove the single mounting screw (B, **Figure 21**), then pull the oil pump from the mounting bracket.
5. Fit the replacement pump onto the mounting bracket. The wire harness connection must face outward. Tighten the single mounting screw to the **Table 1** specification.
6. Remove the plug, then *quickly* push each oil hose onto the electric oil pump fittings. The hose connected to the upper fitting must lead to the vapor separator tank fitting. The hose connected to the lower fittings must lead to the T-fitting in the oil hose below the oil reservoir.
7. Move the spring hose clamps over the oil pump fittings. Tug on the hoses to verify a secure connection. Replace the clamps if the hoses fit loosely.
8. Pinch the tab. Then carefully push the wire harness connection onto the oil pump connector. Release the tab. Then tug on the connector to verify a secure connection.
9. Reconnect the battery cables and spark plug leads. Bleed air from the system as described in this chapter.

Emergency Switch

Mounting location for the emergency switch varies by model:

80 Jet, 115 hp and 130 hp models—The emergency switch (**Figure 22**) is integrated into the oil injection module. Replace the module if the switch fails. Refer to *Oil injection module* in this chapter.

105 Jet and 150-200 hp (2.6 liter) models—The emergency switch (**Figure 23**) is located on the rear of the engine and on the port side of the engine control unit.

200-250 hp (3.1 liter) models—The emergency switch (**Figure 24**) mounts onto the engine mounted reservoir.

150-250 hp (HPDI) models—The emergency switch (**Figure 24**) mounts onto the engine mounted reservoir
1. Disconnect the battery cables and ground all spark plug leads to prevent accidental starting.

2. Disconnect the two emergency switch bullet connectors from the engine wire harness connectors.
3. Turn the hex shaped section of the gray rubber protector/retainer to free the switch from the mount opening.
4. Fit the replacement switch into the mount opening. Orient the switch so the toggle facing downward.
5. Fit the gray rubber protector over the toggle, then thread the hex shaped section onto the switch. Tighten the switch protector/retainer to the **Table 1** specification. Make sure the toggle faces downward.
6. Connect the switch bullet connectors onto the engine wire harness connectors. Route the wiring to prevent interference with other components.
7. Reconnect the battery cables and spark plug leads.

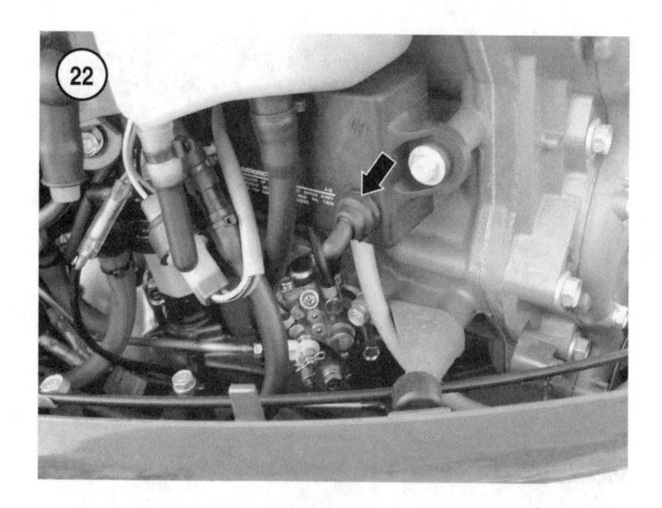

Oil Injection Module

This component is used on 80 Jet, 115 hp and 130 hp models. Two bolts secure the module (**Figure 25**) onto the port side of the power head.

1. Disconnect the battery cables and ground all spark plug leads to prevent accidental starting.
2. Disconnect both module wire harness connectors from the engine wire harness connectors.
3. Remove the two mounting bolts. Then carefully pull the module from the power head.
4. Remove the sleeves and grommets from the module.
5. Clean any corrosion, debris or other contaminants from the mounting bosses.
6. Fit the grommets into the replacement module. Insert the sleeves into the grommets. The wider diameter side of the sleeve must contact the mounting bosses on the power head.
7. Install the module onto the power head mounting bosses. Orient the module so the emergency switch toggle faces downward (**Figure 25**). The toggle returns to the off position when released.
8. Install the two mounting bolts and tighten to the general torque specification in Chapter One.
9. Connect the two module wire harness connectors to the engine wire harness connectors. Route the wiring to prevent interference with other components.
10. Reconnect the battery cables and spark plug leads.

On-Board Mounted Oil Reservoir Service

Service to the on-board mounted oil reservoir includes replacing the oil level sensor, replacing the oil supply pump or cleaning and inspecting the filter on the oil pickup. Clean and inspect the reservoir for leaks or physical damage. Work in a clean environment and clean all components thoroughly. Even a small particle can disable some oil injection components. Replace plastic locking hose clamps anytime they are removed. Replace spring hoses if corroded, distorted or they have weak spring tension.

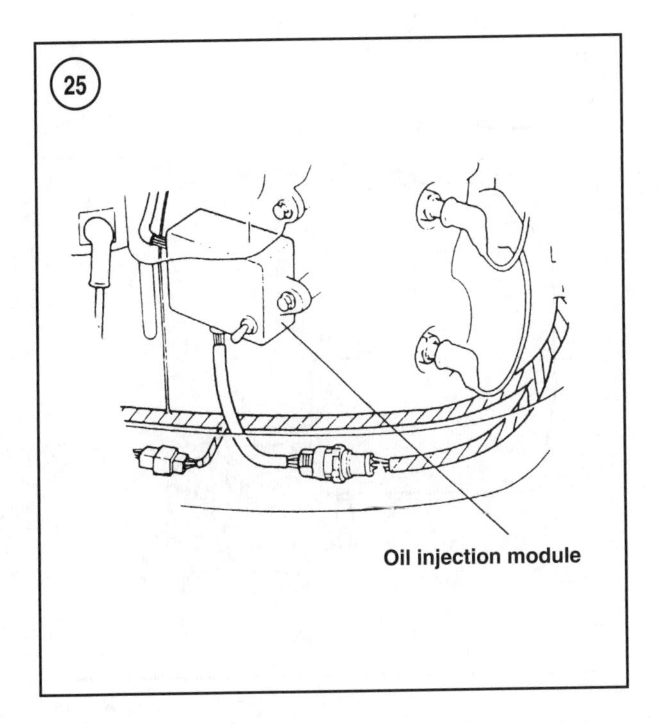

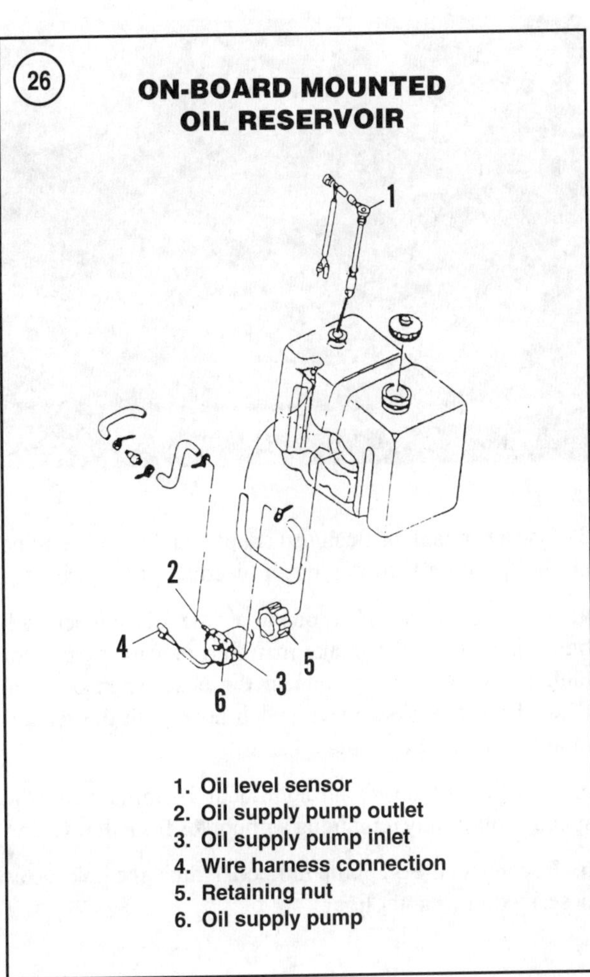

Oil level sensor removal/installation

Refer to **Figure 26** for this procedure.

1. Disconnect the sensor wire harness from the instrument wire harness extension.
2. Clean debris or contaminants from the oil reservoir before removing the sensor.
3. Grasp the top of the sensor (1, **Figure 26**) and carefully pull it from the reservoir. Gently rock the sensor from side to side to assist with removal.
4. Inspect the O-rings on the sensor for tears or other damage. Replace defective or questionable O-rings to prevent oil contamination.
5. Lubricate the O-rings on the sensor with Yamalube. Then carefully insert the sensor into the reservoir. Push down on the sensor until it seats in the opening.
6. Connect the sensor wire harness onto the instrument wire harness extension.
7. Check for proper operation of the dash mounted oil level gauge.

Oil supply pump removal/installation

1. Clean debris from the oil reservoir before removing the oil supply pump.
2. Lift the oil reservoir from the mount. Remove the fill cap and pour the oil into a suitable clean container.
3. Disconnect the oil hoses (2 and 3, **Figure 26**) from the pump. Disconnect the oil pump harness connector (4, **Figure 26**) from the engine wire harness extension.
4. Twist the retaining nut (5, **Figure 26**) counterclockwise to remove the oil supply pump from the reservoir.
5. Inspect the oil hose filter for debris or contaminants. Replace the filter if there are contaminants.
6. Install the replacement oil pump and securely tighten the retaining nut.
7. Refer to **Figure 26** while attaching the oil inlet and outlet hoses. Secure the hoses with plastic locking clamps. Tug on the hoses to verify a secure fit.
8. Connect the oil pump wire harness connector onto the engine wire harness extension. Install the reservoir onto the mount and secure with the mounting strap or bracket. Fill the tank with Yamalube or equivalent.
9. With the engine running in a test tank or with a flush test adapter, operate the emergency switch to purge air from the oil hoses. Continue holding the switch in the run position until there is an oil level increase in the engine mounted oil reservoir.

Bleeding Air From the System

Perform this procedure anytime an oil injection system component is removed, is disconnected or if the air has entered the system due to the oil running low. Air bleeding procedures varies by the type of fuel system used on the engine.

80 Jet, 105 Jet and 115-250 hp models (except 150-250 hp HPDI models)

1. Fill the oil reservoir as described in Chapter Three. Manually fill the oil hoses with Yamalube or equivalent.
2. Prepare a 50:1 fuel/oil mixture in a portable fuel tank as described in Chapter Three. Connect the portable fuel tank to the fuel inlet hose on the engine.
3. Use a test tank or flush/test adapter and run the engine on the fuel/oil mixture during the air bleeding procedure.
4. Locate the air bleed screw on the oil pump (**Figure 27**). The air bleed screw has a gasket under the screw head.
5. Position a shop towel under the screw to capture spilled oil. With the engine running at idle speed, loosen the bleed screw three to four turns counterclockwise. Fully tighten the screw clockwise when oil begins flowing from the screw opening.

6A. *Transparent oil outlet hoses*—Visually inspect the oil outlet hoses for air bubbles. Repeat Steps 1-5 until there are no bubbles in the outlet hoses.

6B. *Non-transparent oil outlet hoses*—One at a time, disconnect each hose from the intake manifold or oil hose manifold fitting (EFI models). Reconnect the hose when all air is purged and oil flows from the hose.

7. *EFI models*—Disconnect the oil outlet hose from the vapor separator tank fitting (10, **Figure 16**). *Quickly* reconnect the hose when oil flows from the hose.
8. Run the engine for an additional 10 minutes on the premix to purge any remaining air bubbles from the system. If transparent oil hoses are used, visually inspect them for air bubbles. Continue to run the engine on the premix until all bubbles are purged from the hoses.

150-250 hp (HPDI) models

1. Fill the oil reservoir as described in Chapter Three. Disconnect and manually fill the oil hoses with Yamalube or equivalent. Refer to **Figure 17** to assist with hose routing and connection points. Reconnect all hoses.
2. Disconnect the oil pump link rod (**Figure 28**) from the oil pump lever. Rotate the oil pump lever clockwise until the lever contacts the stop on the oil pump body.

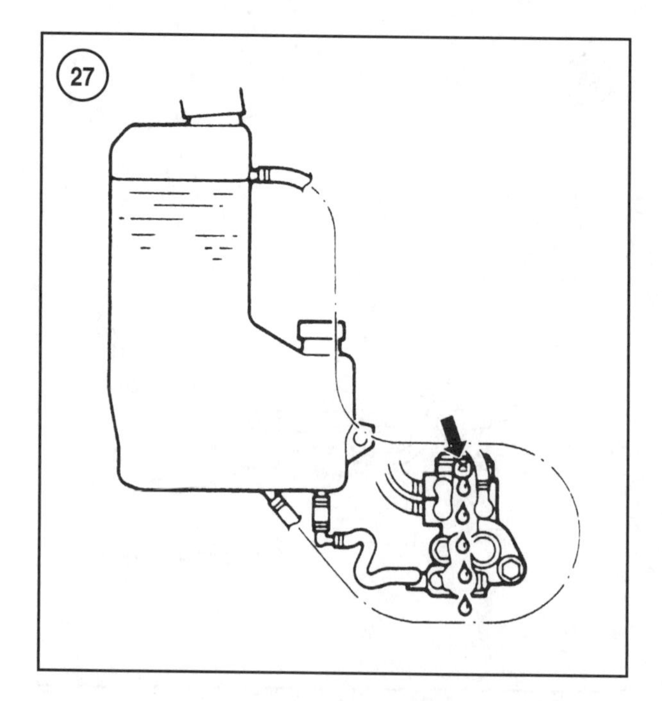

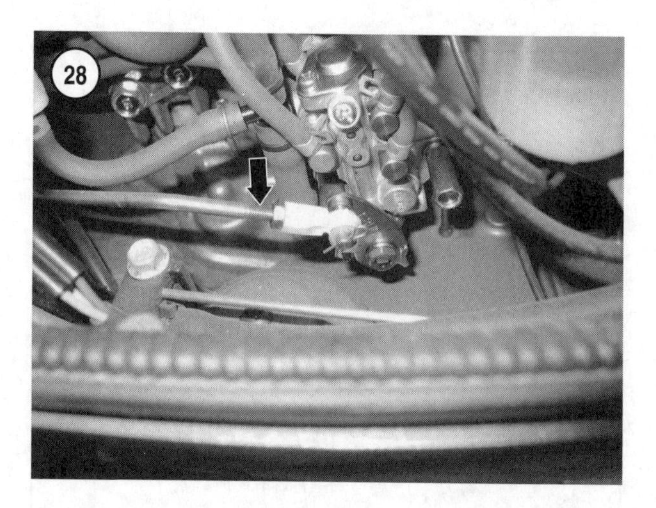

3. Use a test tank or flush/test adapter and run the engine at idle speed only during the air bleeding procedure.
4. Start the engine. Then one at a time, disconnect each outlet hose from the intake manifold or vapor separator tank fitting. *Quickly* reconnect the hose when only oil flows from the hose. Secure each hose with the recommended clamp. See **Figure 17**.
5. Run the engine for an additional 10 minutes at idle speed to purge any remaining air bubbles from the system.
6. Reconnect the oil pump link rod. Adjust the link rod as described in Chapter Four.

Table 1 OIL INJECTION TORQUE SPECIFICATIONS

Fastener	N•m	in.-lb.
Electric oil pump bracket		
150-250 hp HPDI models	8	70
Emergency switch	4	35
Oil pump (gear driven)	7	62
Oil pump (electric)		
150-250 hp HPDI	8	70

Table 2 OIL RESERVOIR CAPACITY

Model	Capacity
115 and 130 hp	
On-board reservoir	10.5 L (11.1 qt.)
Engine mounted reservoir	0.9 L (0.95 qt.)
150-200 hp (2.6 liter models)	
On-board reservoir	10.5 L (11.1 qt.)
Engine mounted reservoir	0.9 L (0.95 qt.)
200-250 hp (3.1 liter models)	
On-board reservoir	10.5 L (11.1 qt.)
Engine mounted reservoir	1.2 L (1.27 qt.)

Table 3 OIL PUMP IDENTIFICATION

Model	Identification marking
80 Jet and 115 hp	6N600
130 hp	6N700
105 Jet, 150 hp and 175 hp (except HPDI models)	6R400
150 hp HPDI	68H00
175 hp HPDI	68L00
200 hp (2.6 liter)	67H00
200 hp HPDI	68F00
200-250 hp (3.1 liter)	65L00

Chapter Thirteen

Remote Control

The remote control provides a means to control throttle, shifting and other engine operations from a location away from the engine (**Figure 1**). This chapter provides neutral throttle operating instructions, shift cable replacement procedures and remote control disassembly/assembly instructions. Torque specifications are located in **Table 1** at the end of this chapter.

REMOTE CONTROL

CAUTION
Always refer to the owner's manual for specific operating instructions for the remote control. Become familiar with all control functions before operating the engine.

Neutral Throttle Operation

Two common types of controls are used with Yamaha outboards. The most commonly used type is the surface mount 703 control or *pull for neutral throttle control* (**Figure 2**). To activate neutral throttle operation, position the handle in neutral. Pull the handle straight out or away from the control (**Figure 3**). The throttle can then be advanced without shifting the engine into gear. This feature allows easier starting and quicker engine warm-up. To return to normal operation, position the handle into neutral and push the handle toward the control.

The less commonly used type is the 705 panel mounted (**Figure 4**) or binacle mount (**Figure 5**) *push for neutral throttle control*. This control is used primarily with higher horsepower engines mounted onto larger boats. To activate neutral throttle operation, position the handle in neutral. Push in and hold the free accelerator button (**Figure 6**). Advance the throttle slightly and release the button. The button should remain depressed. For normal operation, move the handle to the neutral position. The free accelerator button returns to the normal position and operation with shifting can resume.

WARNING
A malfunction in the remote control can lead to lack of shift and throttle control. Never operate an outboard when there is any malfunction with the remote control. Damage to property, serious injury or death can result if the engine is operated without proper control. Check for proper remote control opera-

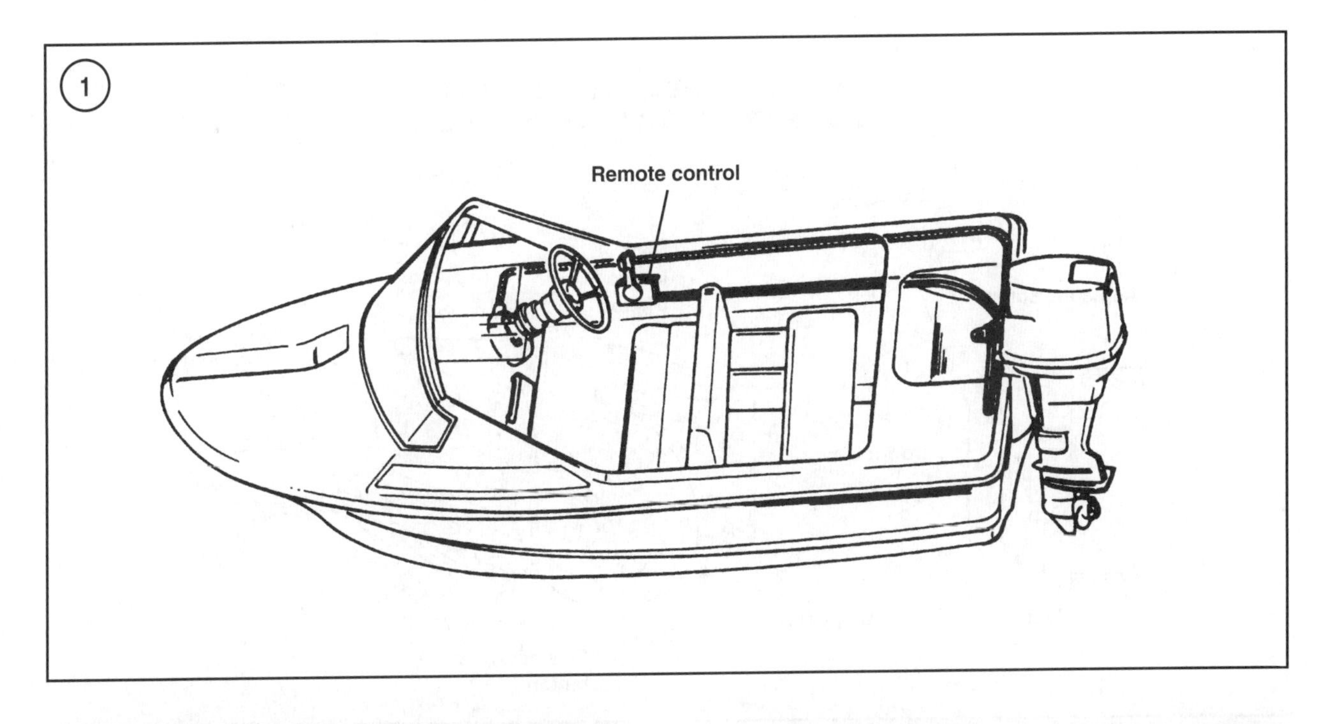
1
Remote control

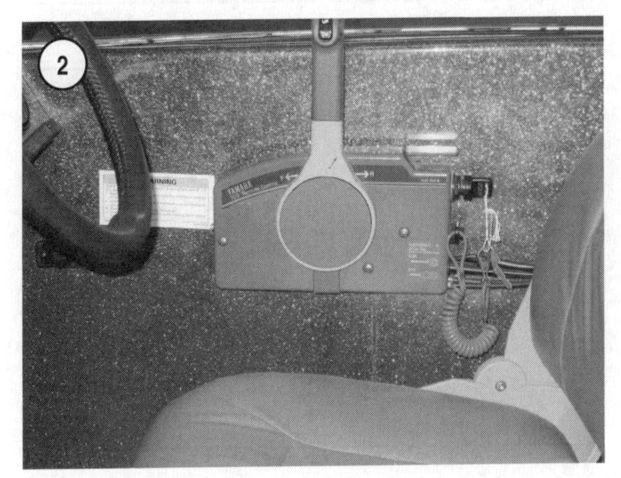
2

4

13

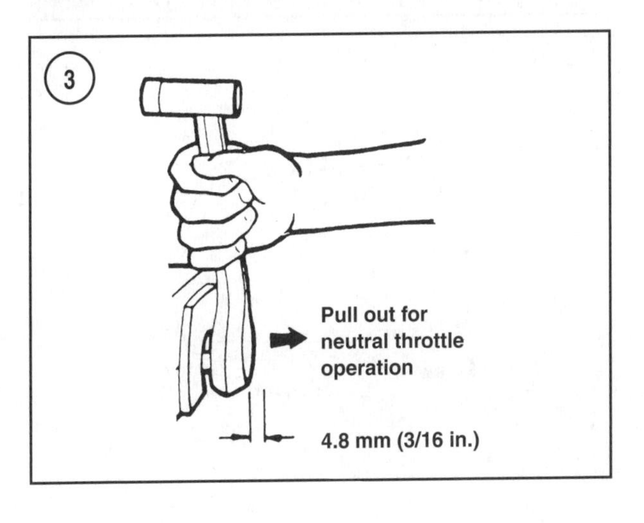
3
Pull out for
neutral throttle
operation
4.8 mm (3/16 in.)

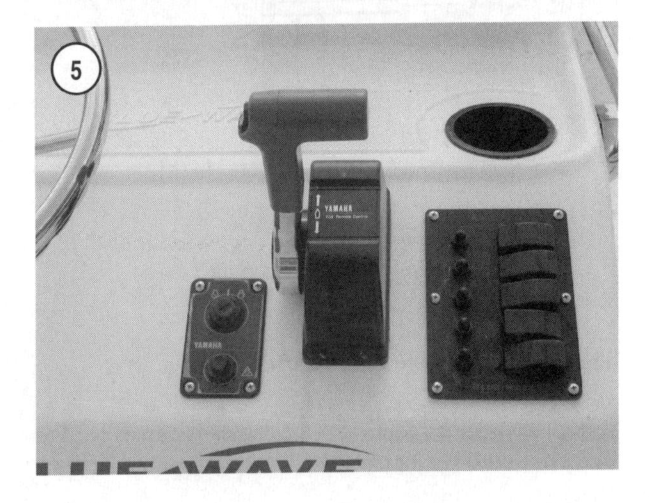
5

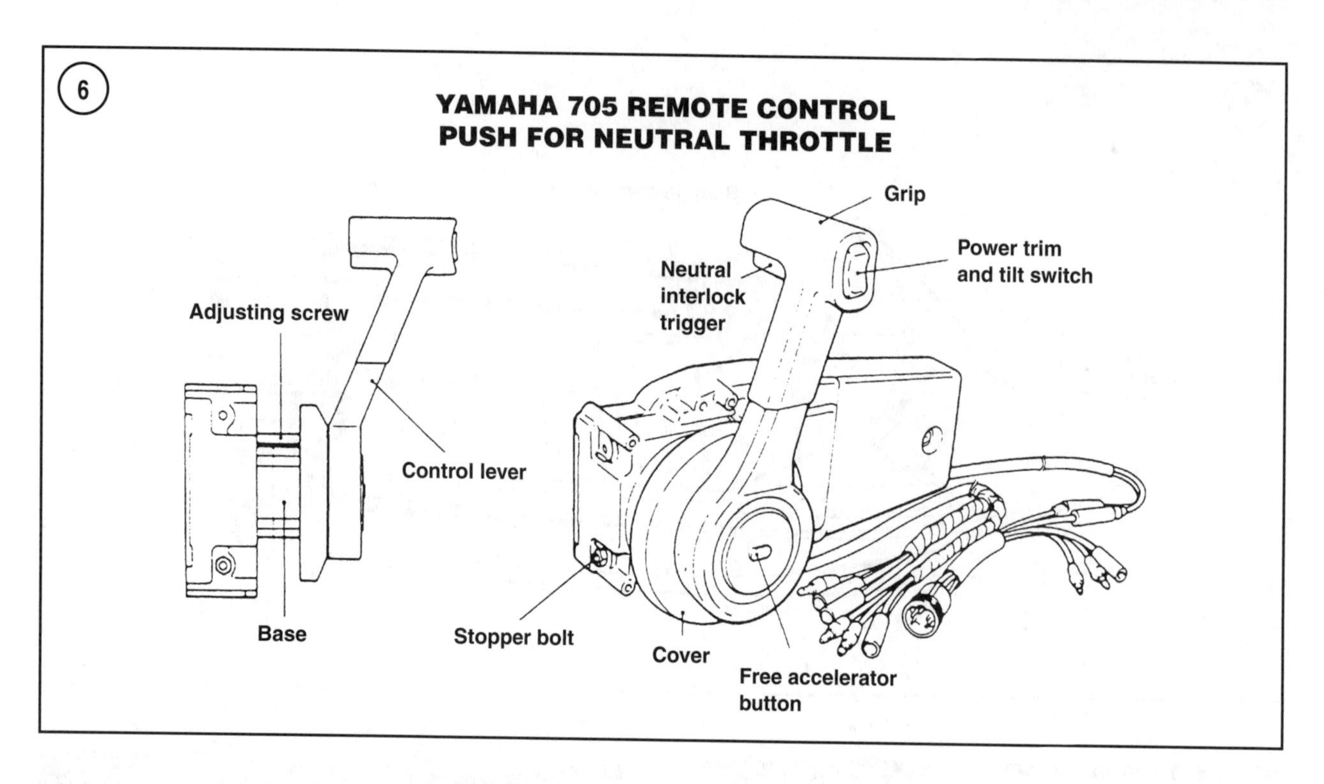

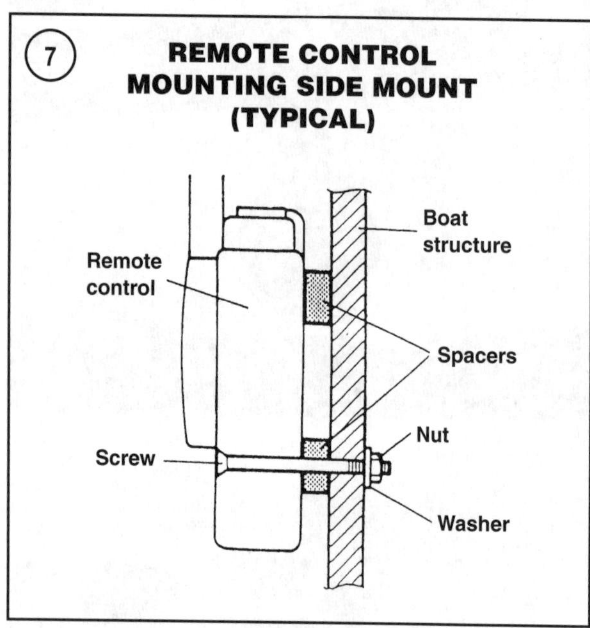

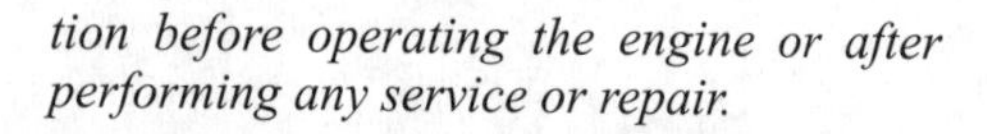
tion before operating the engine or after performing any service or repair.

Throttle/Shift Cable Removal/Installation

Replacement is required if the cables becomes hard to move or excessive play occurs due to cable wear. Replace

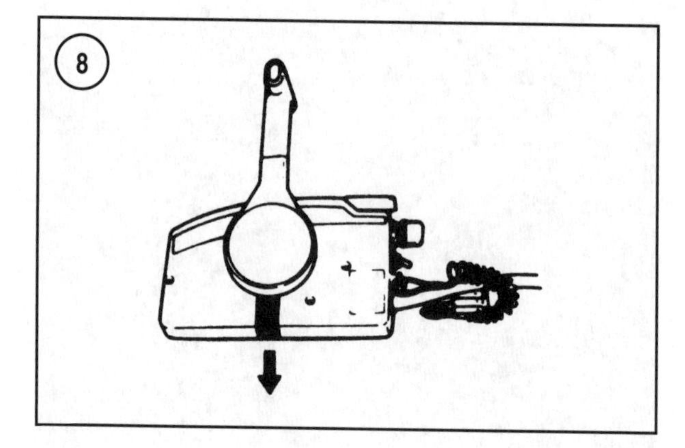

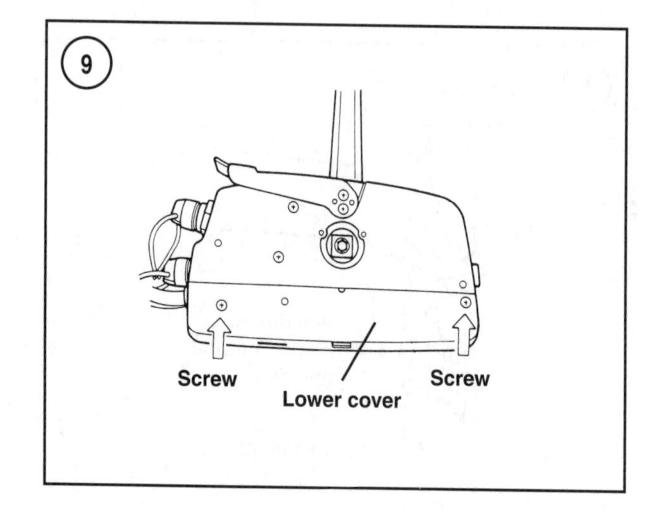

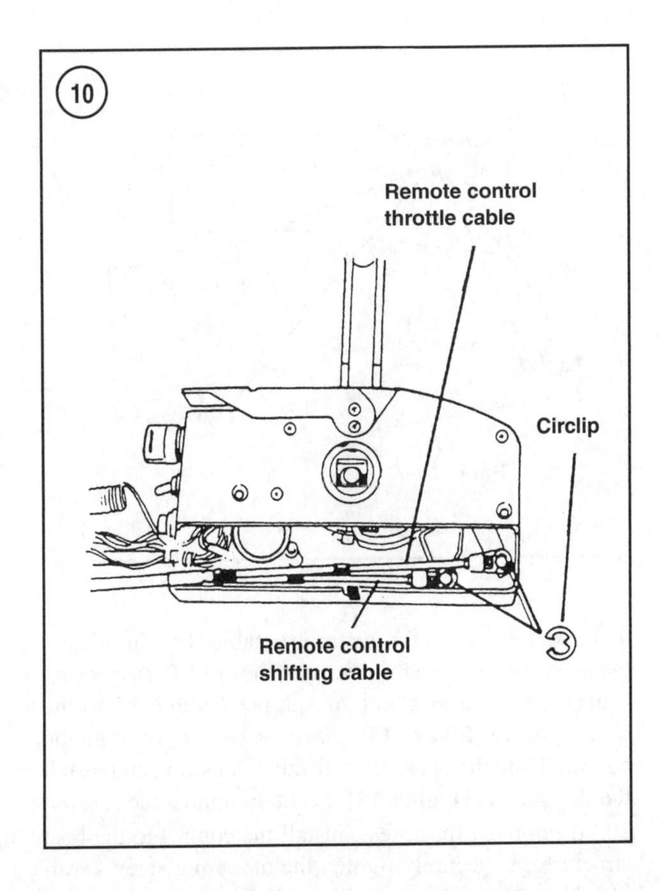

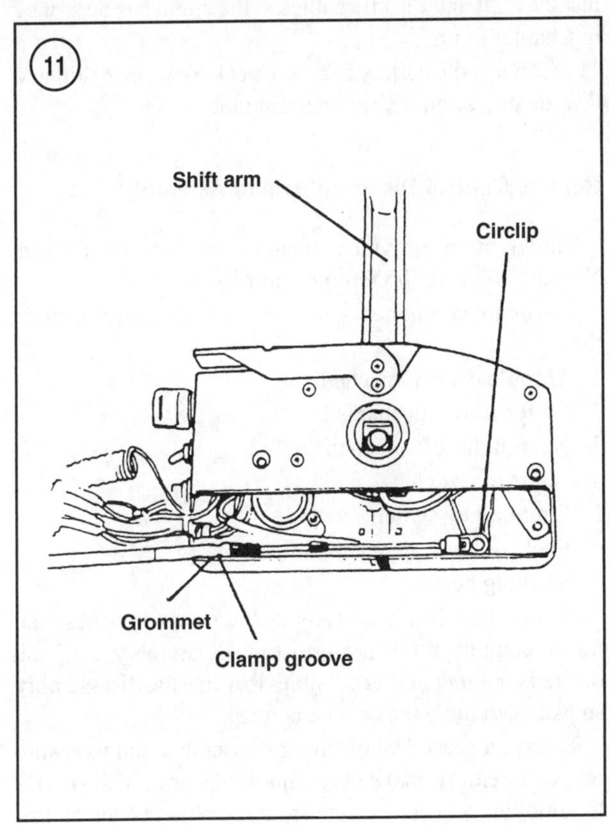

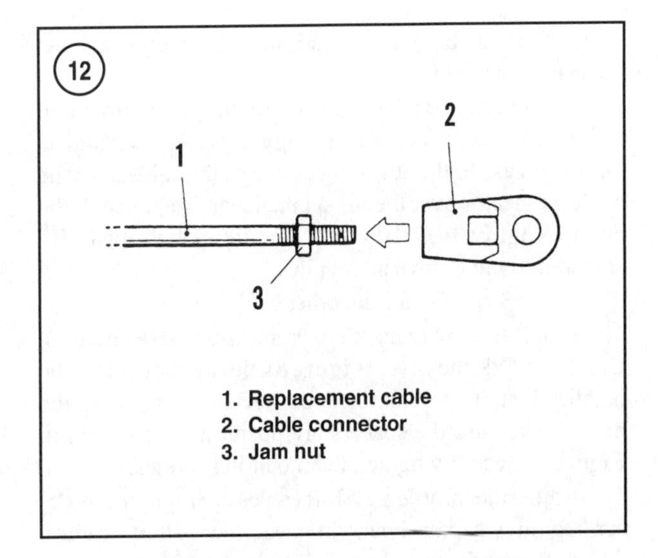

both the throttle and shift cables at the same time. The conditions that caused one cable to require replacement are likely in the other cable. Mark the cables prior to removal to ensure the cables are installed to the proper attaching points. Remove and attach one cable at a time to avoid confusion.

Procedures are provided for both the 703 and 705 remote controls. Refer to *Neutral Throttle Operation* to identify the control box.

Remote control (703 model)

1. Disconnect the battery cables. Disconnect and ground the spark plug leads to prevent accidental starting.
2. Remove the screws or bolts attaching the remote control to the boat structure. See **Figure 7**. Capture the spacers while removing the mounting screws.
3. Carefully pull the cover from the area below the handle (**Figure 8**).
4. Remove the two screws and the lower back cover from the control (**Figure 9**).
5. Identify the throttle and shift cables (**Figure 10**). Move the control handle to provide access to the circlip on one of the cables.
6. Note and/or mark the location of the cable and cable grommet to the clamp groove (**Figure 11**). Remove the circlip (**Figure 10**), then lift the cable from the clamp groove and pin on the lever. Inspect the circlip for corrosion, damage or lost spring tension. Replace if there are any defects.
7. Apply Yamaha marine grease or equivalent to the threaded end of the replacement cable (1, **Figure 12**). Thread the connector (2, **Figure 12**) onto the threaded end until 11.0 mm (0.4 in.) of the threaded end is in the con-

nector. Tighten the jam nut (3, **Figure 12**) securely against the cable connector.

8. Place the cable and grommet into the clamp groove in the control box as indicated in **Figure 11**. Apply Yamaha marine grease to the attaching points of the cables. Fit the cable connector over the pin on the lever. Then install the circlip (**Figure 10**). Make sure the circlip is properly installed into the groove in the pin.

9. Repeat Steps 5-8 for the other cable.

10. Install the lower back cover and screws (**Figure 9**). Carefully slide the cover (**Figure 8**) into the slot below the handle. Install the control and attaching screws to the boat. Make sure the spacers are positioned as shown in **Figure 7**. Securely tighten the mounting screws.

11. Adjust the throttle and shift cables at the engine as described in Chapter Four. Connect the battery cables. Check for proper throttle and shift operation before operating the engine.

705 Remote control (705 model)

1. Disconnect the battery cables. Disconnect and ground the spark plug leads to prevent accidental starting.
2. Remove the screws or bolts attaching the control to the boat structure and the control as necessary to access the back cover or the control.
3. Remove the five screws (**Figure 13**) from the back cover. Support the cables. Then carefully pry the back cover from the control.
4. Refer to **Figure 14** to identify the shift cable. Mark it accordingly.
5. Make note of the position of the shift cable and grommet (**Figure 14**) in the clamp groove.
6. Move the handle to provide access to the circlip on one of the shift cable connectors. Use needle nose pliers to remove the circlip from the shift arm attaching pin. Carefully lift the cable from the shift arm and clamp groove. Inspect the circlip for corrosion, damage or lost spring tension. Replace if there are any defects.
7. Apply Yamaha marine grease or equivalent to the threaded end of the replacement cable (1, **Figure 12**). Thread the connector (2, **Figure 12**) onto the threaded end until 11.0 mm (0.4 in.) of the threaded end is in the connector. Tighten the jam nut (3, **Figure 12**) securely against the cable connector.
8. Apply Yamaha marine grease to the attaching points for the cable. Place the cable and grommet into the clamp groove as indicated in **Figure 14**. Fit the cable connector over the pin on the shift arm. Install the circlip onto the pin. Make sure the circlip is properly installed in the groove in the pin.

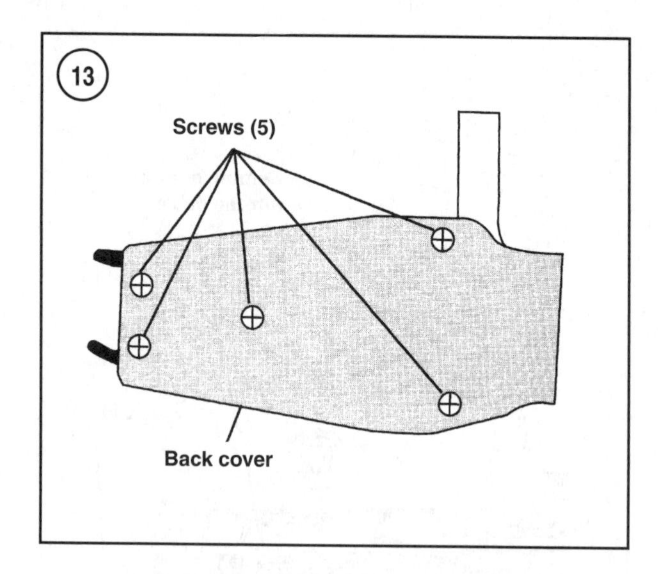

9. Repeat Steps 5-8 for the throttle cable. The throttle cable must attach to the throttle arm (**Figure 14**). Inspect the cables to ensure that both are properly aligned into the clamp groove (**Figure 14**). Place the back cover onto the control. Hold the back cover firmly in position and install the five screws (**Figure 13**). Securely tighten the screws.
10. If removed for access, install the control to the boat structure and securely tighten the mounting screws. Adjust the shift and throttle cables at the engine as described in Chapter Four.
11. Connect the battery cables. Check for proper shift and throttle operation before operating the engine.

Remote Control Disassembly and Assembly

This section provides separate procedures for the Yamaha 703 and 705 remote controls.

The major components or system of the control include the:

1. Throttle control mechanism.
2. Shift control mechanism.
3. Neutral throttle mechanism.
4. Tilt/trim switch.
5. Ignition/key switch.
6. Fuel enrichment switch.
7. Warning horn.

If complete disassembly is not required to access the faulty component(s), perform the disassembly until the desired component is accessible. Reverse the disassembly steps to assemble the remote control.

To save a great deal of time on assembly and to ensure proper assembly, make notes, markings or drawings prior to removing any component from the remote control. Im-

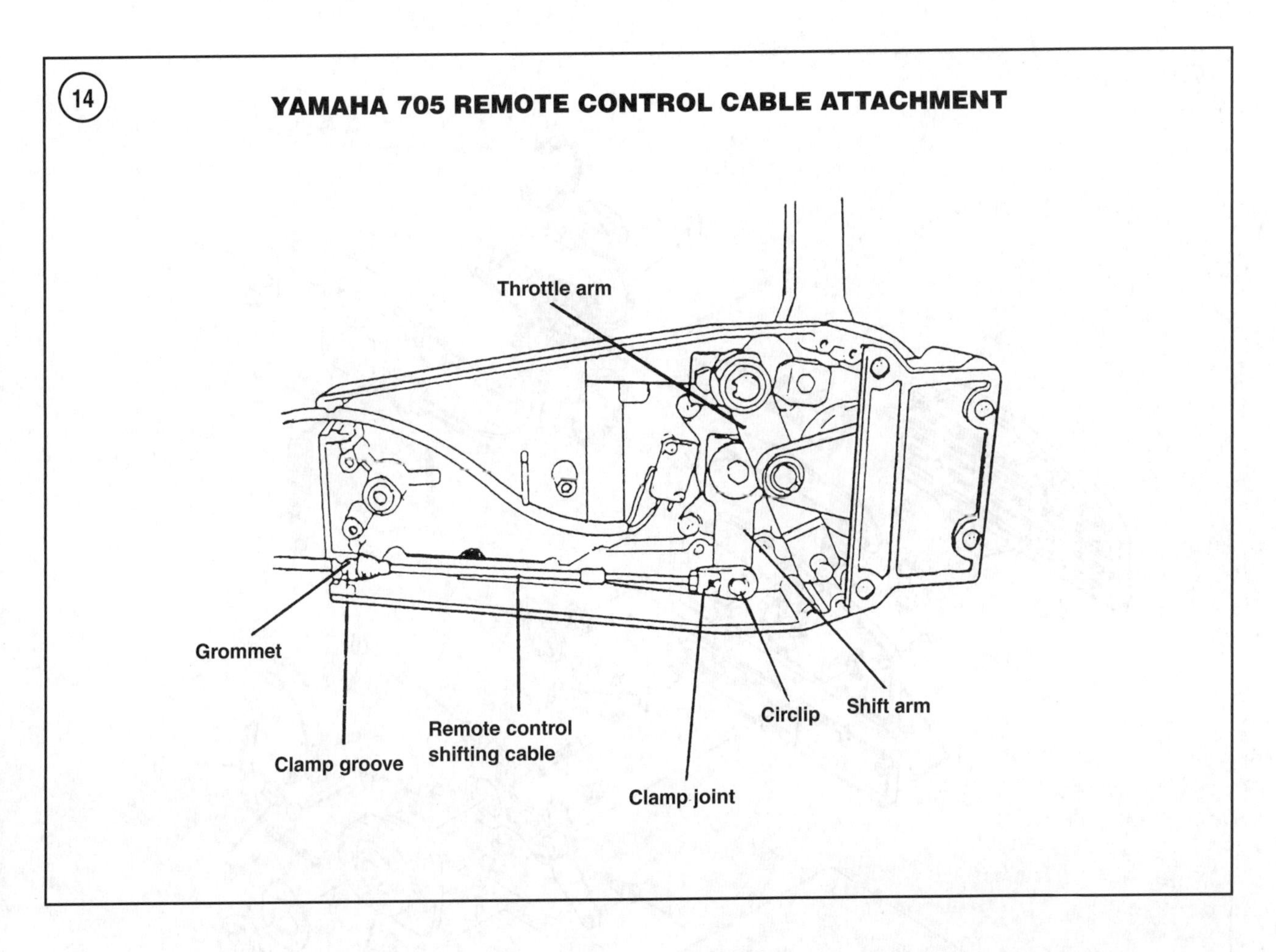

proper assembly can cause internal binding, reversed cable movement or improper control operation.

Clean all components, except electric switches and the warning horn, in a suitable solvent. Use compressed air to blow debris from the components. Inspect all components for damaged or worn surfaces. Replace any defective or suspect components. Apply Yamaha All-Purpose Grease or equivalent to all pivot points or sliding surfaces upon assembly. Test all electrical components if removed to ensure proper operation upon assembly. Test procedures for all electrical components are described in Chapter Two.

703 Model

Refer to **Figure 15**.

1. Disconnect the battery cables. Disconnect and ground the spark plug leads to prevent accidental starting.
2. Remove the screws or bolts that attach the control to the boat structure. Mark all wires leading to the control to ensure proper connection on installation. Disconnect all leads. Place the control handle in the neutral position.
3. Mark the cable location for reference, and then remove the throttle and shift cables as described in this chapter. Place the control on a clean work surface.
4. Remove the screws, then back cover plates (1, **Figure 15**). Remove the throttle-only lever and shaft (13 and 4, **Figure 15**). Remove the retaining bolt. Then lift the throttle arm and shift arm from the control housing. Remove any accessible bushings, grommets and retainers at this point.
5. Retain the retainer (5, **Figure 15**), then lift the retainer and neutral only key switch from the control housing.
6. Note the position of the gear (7, **Figure 15**). Then carefully lift it from the control housing. Remove any accessible bushings, grommets and retainers. Note the tilt/trim switch wire routing. Then remove the control handle.
7. Remove the neutral position lever from the handle. Note the wire routing, then pull the tilt/trim switch from the handle.
8. Note all wire routing and connection points prior to removal. Disconnect the wires. Then remove the key switch, lanyard switch, warning horn and fuel enrichment

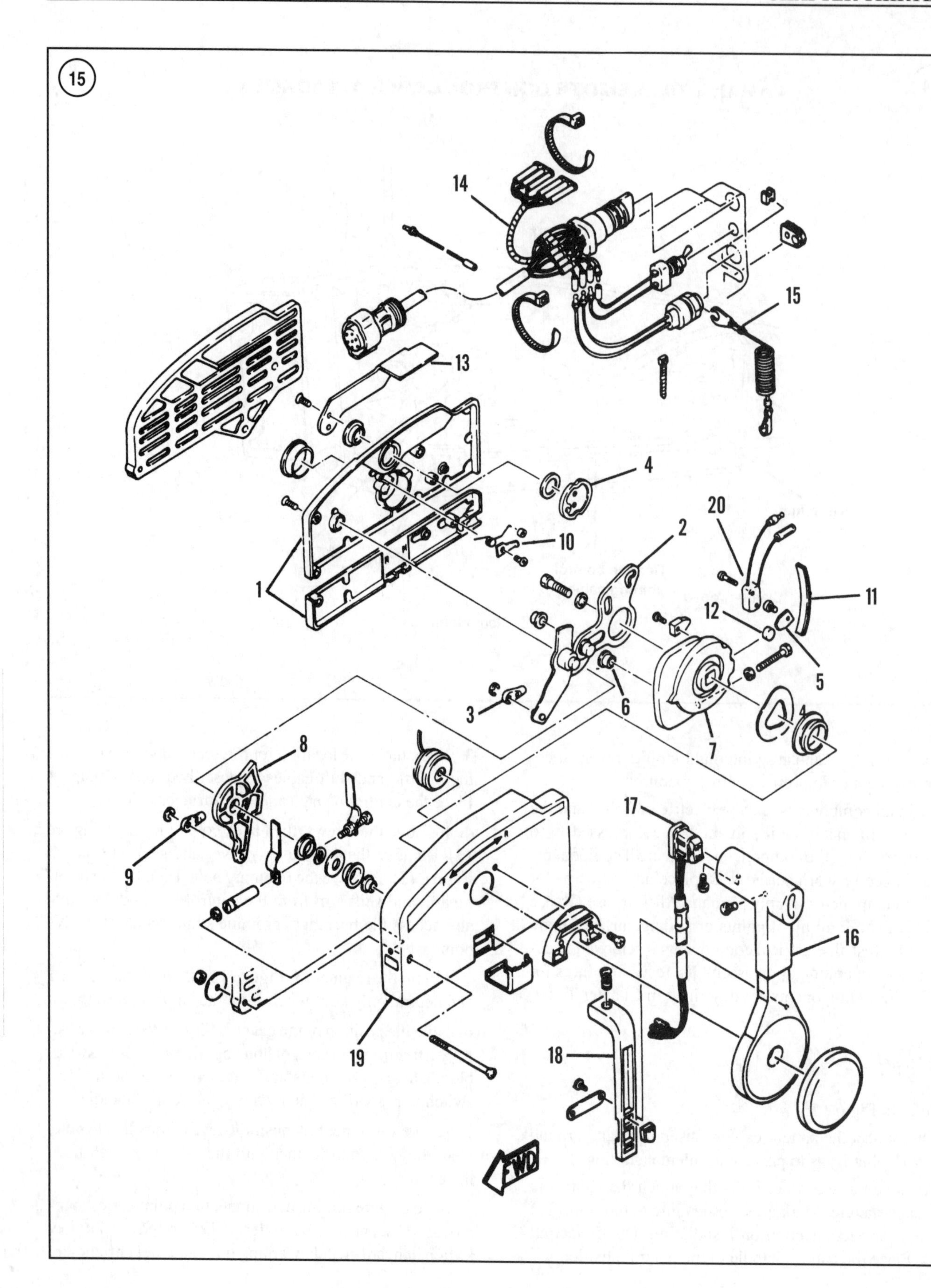
15
14
15
13
4
20
2
10
1
11
12
5
3
6
4
7
8
17
9
16
19
18
FWD

YAMAHA 703 REMOTE CONTROL

1. Back covers
2. Throttle arm/lever
3. Cable connector
4. Throttle only shaft/cam
5. Retainer
6. Bushing
7. Gear
8. Shift arm
9. Cable connector
10. Detent roller
11. Leaf spring
12. Throttle only roller
13. Throttle only lever
14. Start switch leads/connectors
15. lanyard
16. Control handle
17. Tilt/trim switch *
18. Neutral position lever
19. Control housing
20. Neutral only start switch

*Used only on tilt/trim models

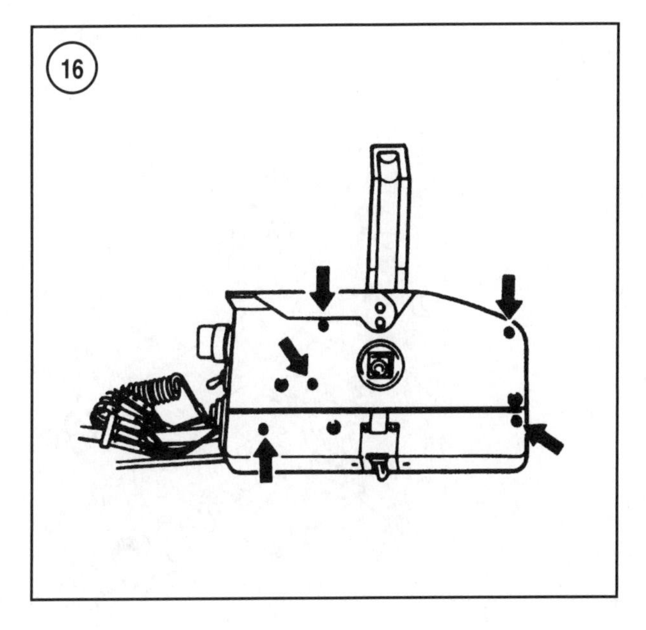

switch. Note the location and orientation of any remaining components. Then remove them from the control housing.
9. Assembly is the reverse of disassembly. Apply Yamaha all purpose grease to all bushings, pivot points and sliding surfaces. Apply Loctite 271 to the threads of all bolts and screws on assembly. Tighten all fasteners to the specifications in **Table 1** or Chapter One.
10. Install the throttle and shift cables as described in this chapter. Install both back covers (**Figure 16**). Reattach all disconnected leads to the control. Connect the cables to the battery. Adjust the cables at the engine as described in Chapter Four. Check for proper remote control operation before putting the engine into service.

13

Disassembly and assembly (705 model)

To assist with component identification and orientation, refer to **Figure 17**. Note the location and orientation of *ALL* components prior to removal.
1. Disconnect the battery cables. Disconnect and ground the spark plug leads to prevent accidental starting.
2. Remove the screws or bolts that attach the control to the boat structure. Place the control handle in the neutral position. Mark the cable location for reference, and then remove the throttle and shift cables as described in this chapter. Place the control on a clean work surface.
3. Mark all wires leading to the control to ensure proper connection during installation. Disconnect all wires.
4. Mark, make notes or drawings indicating the location and orientation of each component prior to removal.
5. Place the control handle in the neutral gear position. Remove the circlip (1, **Figure 18**) from the throttle lever

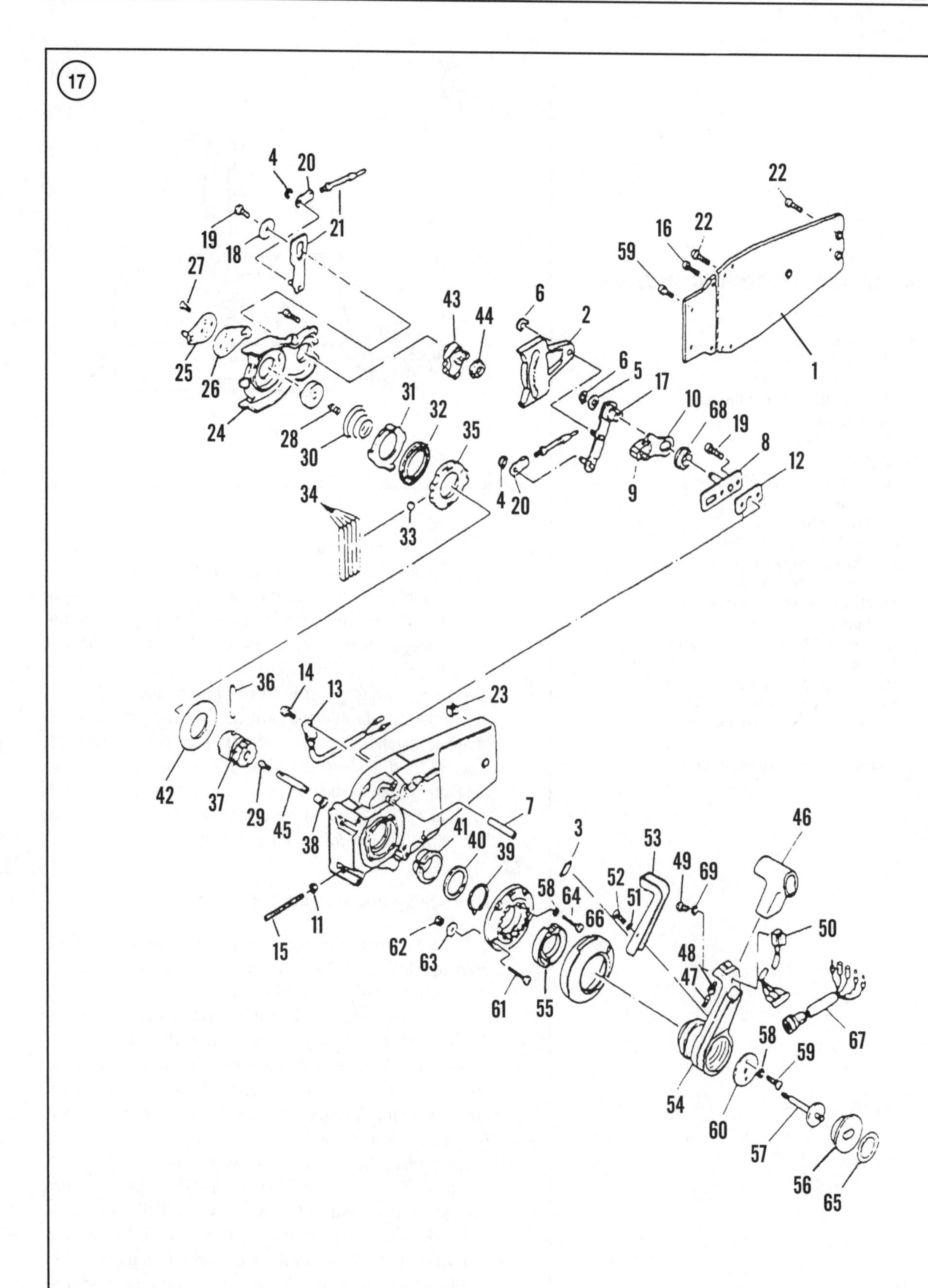
17
1
2
3
4
5
6
7
8
9
10
11
12
13
14
15
16
17
18
19
20
21
22
23
24
25
26
27
28
29
30
31
32
33
34
35
36
37
38
39
40
41
42
43
44
45
46
47
48
49
50
51
52
53
54
55
56
57
58
59
60
61
62
63
64
65
66
67
68
69

YAMAHA 705 REMOTE CONTROL

1. Back cover
2. Cam plate
3. Cap
4. Circlip
5. Washer
6. Circlip
7. Friction shaft
8. Throttle shaft
9. Throttle friction lever
10. Friction adjusting plate
11. Nut
12. Spacer
13. Neutral only start switch
14. Screw
15. Screw
16. Screw
17. Throttle lever
18. Washer
19. Bolt
20. Cable connector
21. Shift cable
22. Screw
23. Cable grommet
24. Gear cover
25. Drive arm
26. Drive arm limiter
27. Screw
28. Spring
29. Screw
30. Spring
31. Locking plate
32. Shifting plate
33. Throttle only roller
34. Spring
35. Gear
36. Pin
37. Shaft
38. Bushing
39. Circlip
40. Washer
41. Bushing
42. Washer
43. Gear
44. Bushing
45. Shaft
46. Handle grip
47. Spring
48. Screw
49. Screw
50. Tilt/trim switch*
51. Washer
52. Screw
53. Neutral lock lever
54. Control handle
55. Neutral lock plate
56. Cover
57. Throttle only shaft
58. Washer
59. Screw
60. Washer
61. Screw
62. Nut
63. Washer
64. Screw
65. Decal cover
66. Decal
67. Wire *
68. Bushing
69. Washer

*Used only on tilt/trim models.

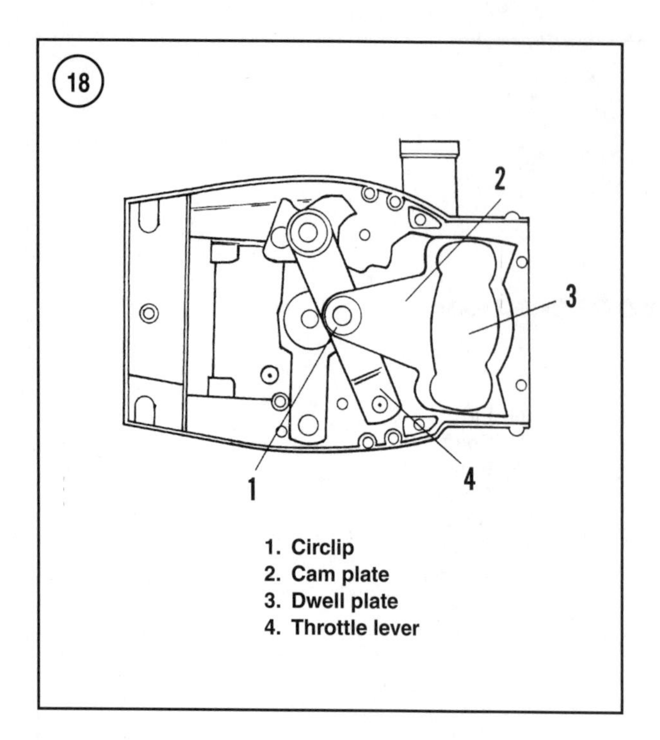

1. Circlip
2. Cam plate
3. Dwell plate
4. Throttle lever

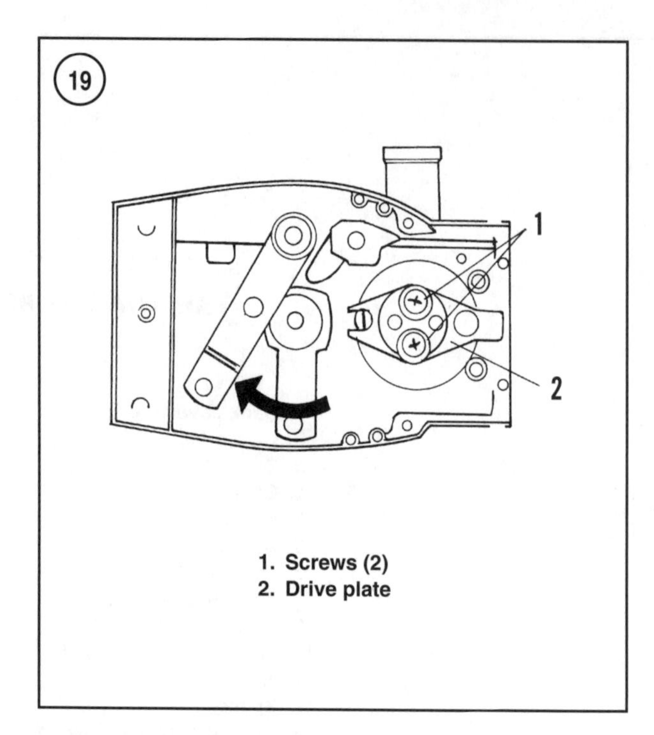

1. Screws (2)
2. Drive plate

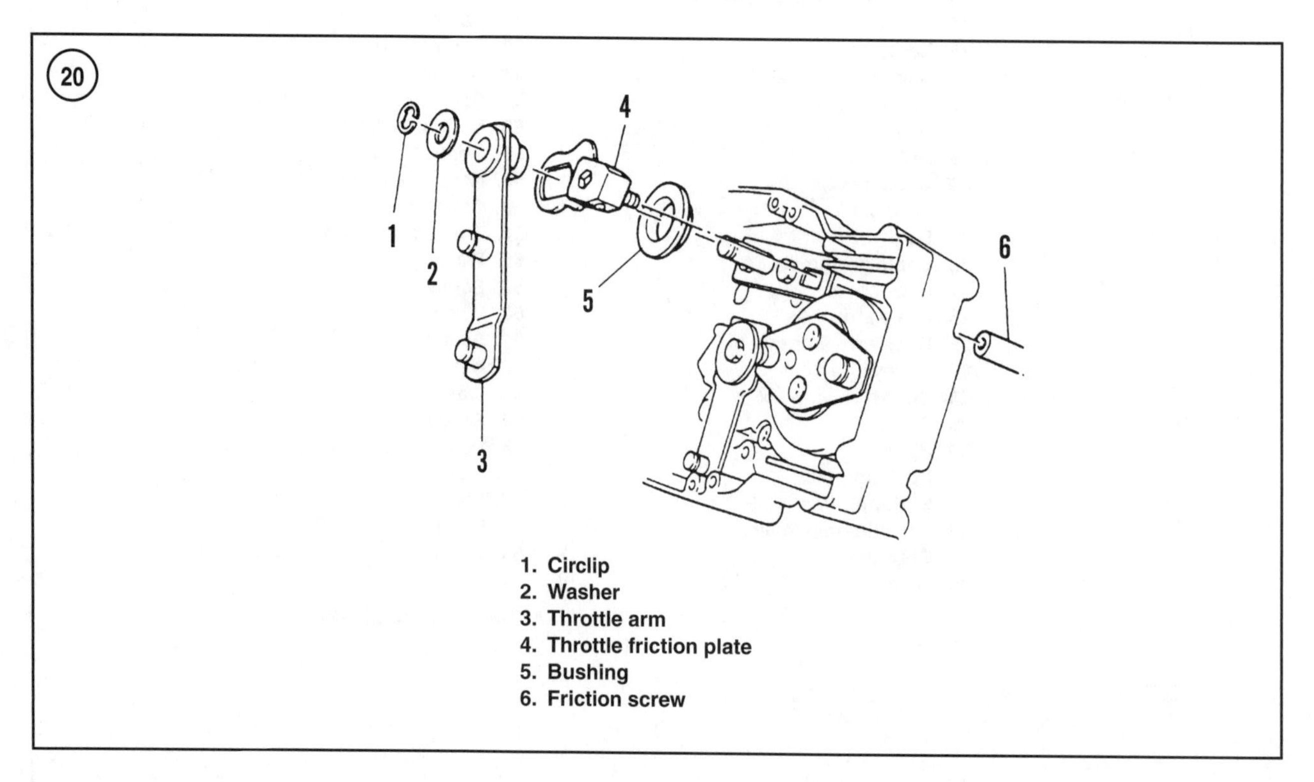

1. Circlip
2. Washer
3. Throttle arm
4. Throttle friction plate
5. Bushing
6. Friction screw

pivot. Lift the cam plate (2, **Figure 18**) and dwell plate (3) from the control.

6. Slide the throttle lever toward the back of the control to access the two screws (1, **Figure 19**) that retain the drive plate (2) to the drive arm limiter (26, **Figure 17**), then lift them from the control housing.

7. Count the number of turns while tightening the throttle friction plate to the fully seated position. Record the num-

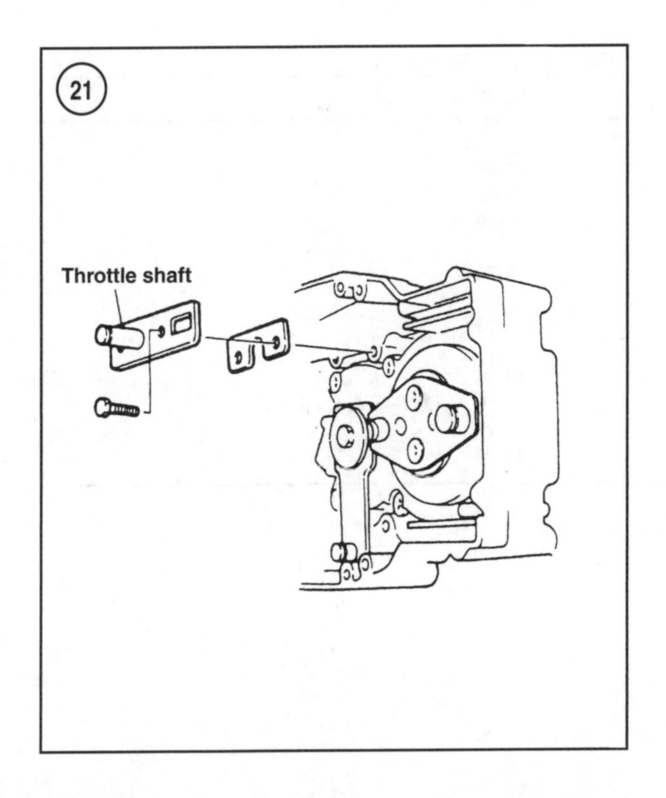

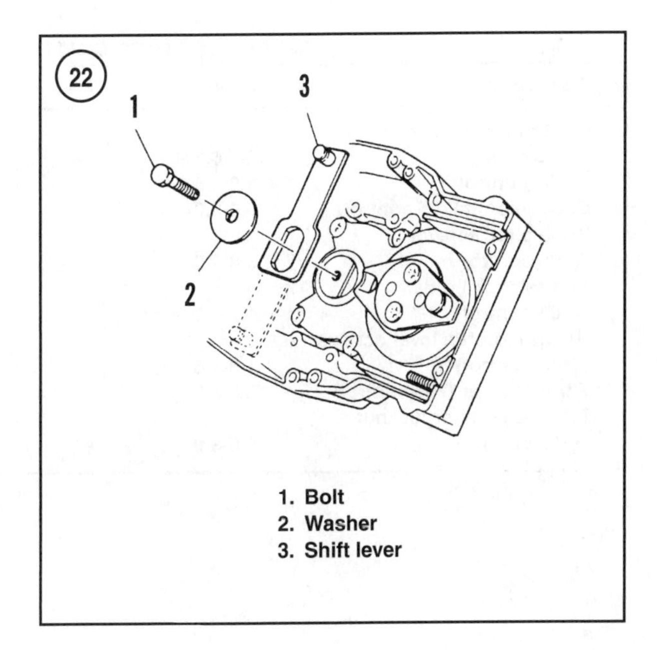

1. Bolt
2. Washer
3. Shift lever

ber of turns. Loosen the circlip (1, **Figure 20**). Note the orientation of the components, then lift the washer, throttle arm, throttle friction plate and bushing (2-5, **Figure 20**) from the control housing. Remove the friction screw (6, **Figure 20**) from the control housing.

8. Remove the two hex bolts, then lift the throttle shaft (**Figure 21**) from the control housing. Note the position of the shift lever (3, **Figure 22**). It must be installed in the same position as removed. Remove the bolt (1, **Figure 22**) and washer (2). Then lift the throttle lever from the control housing.

9. Remove the screws (22, **Figure 17**). Then carefully lift the gear cover (24) from the control housing. Loosen the locking nut (11, **Figure 17**) without turning the stopper screw (15). Count the turns while turning the stopper screw clockwise to the fully seated position. Record the number of turns. Then remove the screw and locking nut.

10. Carefully pry the throttle-only cover (56, **Figure 17**) from the throttle only shaft (57). Use pliers to pull the throttle-only shaft from the control housing. The shaft may break during the removal procedure. Remove the two screws (59, **Figure 17**) from the washer (60). Note the routing of the tilt/trim switch wires, then pull the control handle and trim switch wires from the control. Disconnect the tilt/trim switch wires from the instrument harness. Remove the decal cover (66, **Figure 17**) and neutral lock plate (55).

11. Mark all components prior to removal. Remove the screws (61, **Figure 17**) along with the cover/mount. Remove the circlip (39, **Figure 17**) from the drive shaft (37). Note the orientation of the lock plate, shift plate and gear (31, 32 and 35, **Figure 17**). Then slide them from the drive shaft. Note the location and orientation of all remaining control components. Then remove them.

12. Remove the neutral only key switch (13, **Figure 17**), trim switch and wires (50 and 67) along with the key switch, fuel enrichment switch and warning horn. Refer to Chapter Two and test these components prior to assembly.

13. Assembly is the reverse of disassembly. Note the following: Apply Yamaha All-Purpose Grease to all pivot points and sliding surfaces. Install the throttle friction screw and stopper screw until seated, then back out the recorded number of turns. Tighten the jam nut securely. Apply Loctite 271 to the threads of all fasteners (excluding the throttle friction and stopper screw). Tighten all fasteners to the specification in **Table 1** or Chapter One.

   a. Install the throttle and shift cables as instructed in this chapter. Install the control to the boat and securely tighten the mounting screws.
   b. Adjust the cables at the engine as described in Chapter Four. Check for proper remote control operation and correct as required.

Table 1 is on the following page.

Table 1 REMOTE CONTROL TORQUE SPECIFICATIONS

Fastener	N·m	in.-lb.
Control handle screw		
703 control	6.0-6.5	52-56
705 control	3.0-4.5	26-39
Cover/mount (705 control)	5.0-8.0	44-70
Drive plate screws		
(705 control)	5.0-8.0	44-70
Neutral lock holder		
(703 control)	1.2-1.5	11-13
Neutral throttle lever screws		
(703 control)	1.5-1.8	13-16
Throttle lever (705 control)	5.0-8.0	44-70
Throttle stopper locknut		
(705 control)	5.0-8.0	44-70

Index

V

W

WIRING DIAGRAMS

80 JET, 115 HP AND 130 HP (PREMIX)

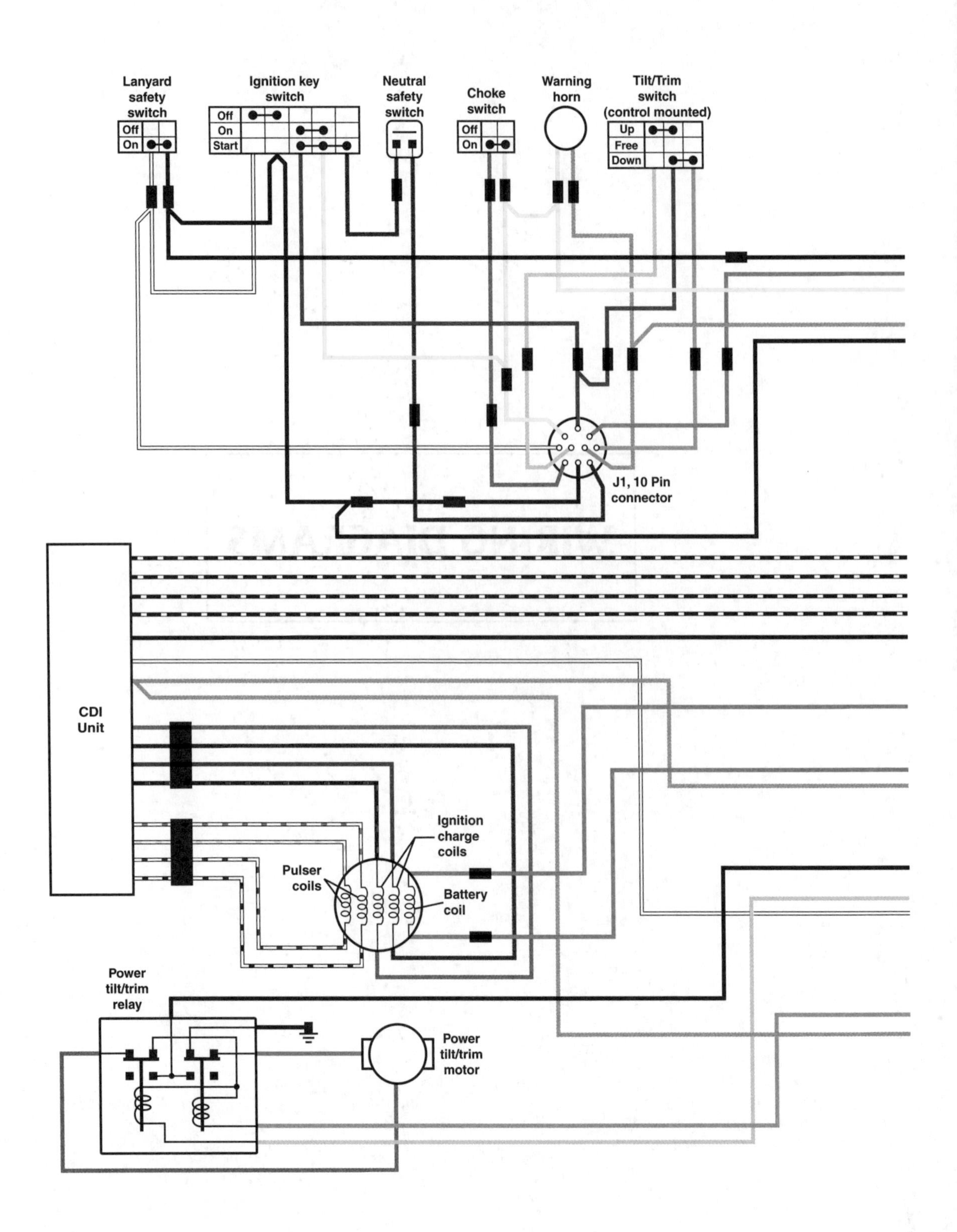

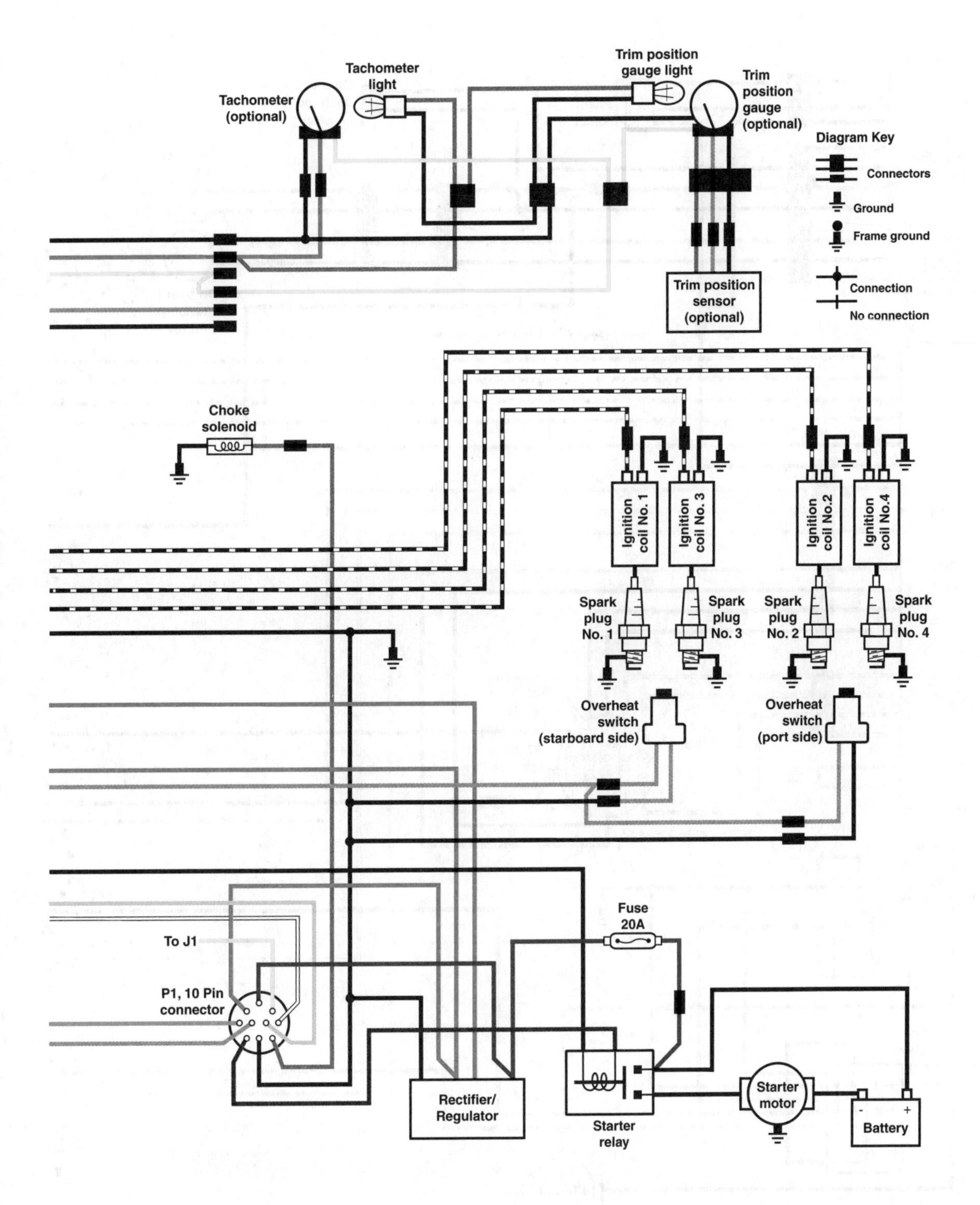
Tachometer (optional)
Tachometer light
Trim position gauge light
Trim position gauge (optional)
Diagram Key
Connectors
Ground
Frame ground
Connection
No connection
Trim position sensor (optional)
Choke solenoid
Ignition coil No. 1
Ignition coil No. 3
Ignition coil No.2
Ignition coil No.4
Spark plug No. 1
Spark plug No. 3
Spark plug No. 2
Spark plug No. 4
Overheat switch (starboard side)
Overheat switch (port side)
Fuse 20A
To J1
P1, 10 Pin connector
Rectifier/ Regulator
Starter relay
Starter motor
Battery

80 JET AND 115-130 HP (OIL INJECTION)

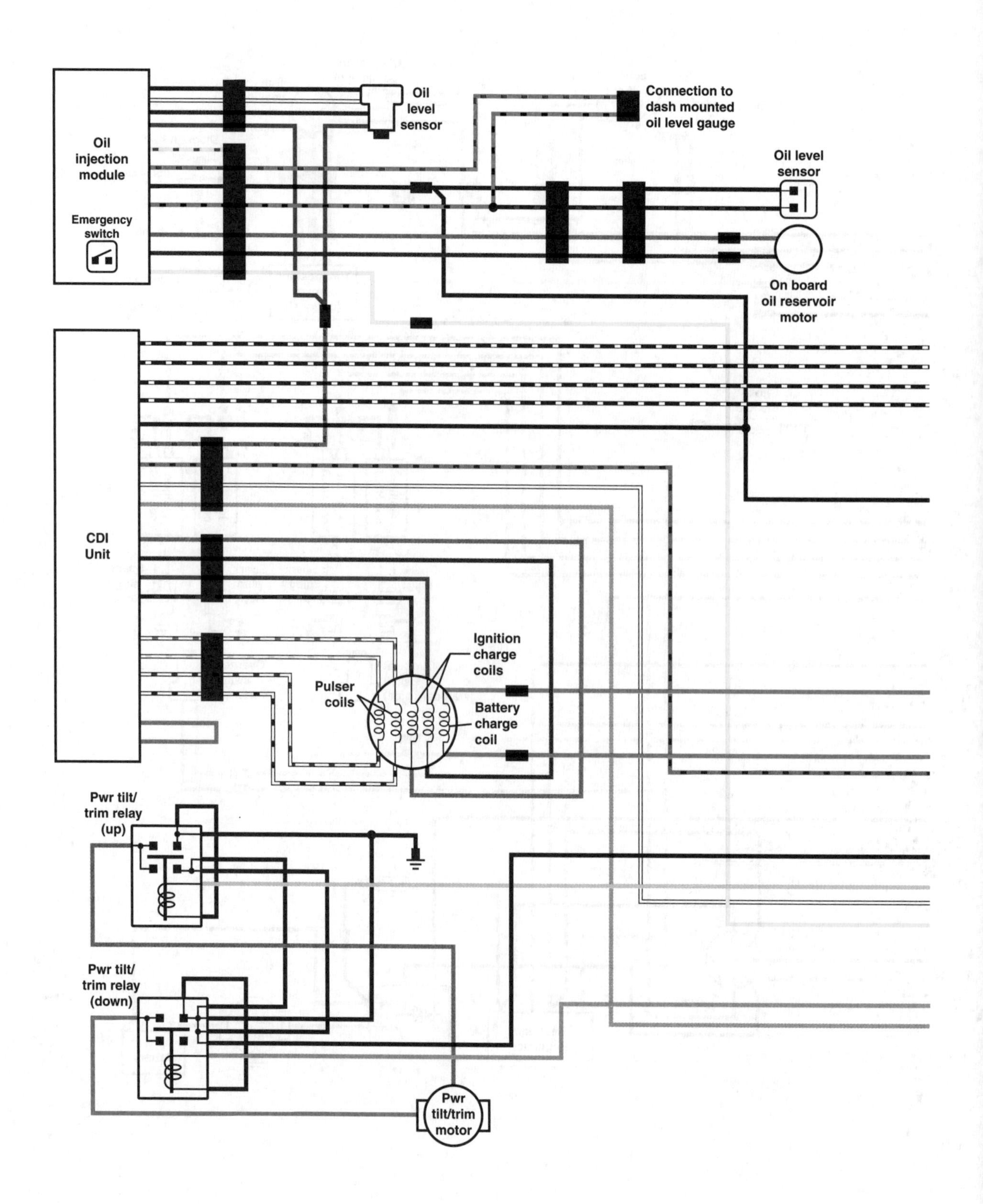

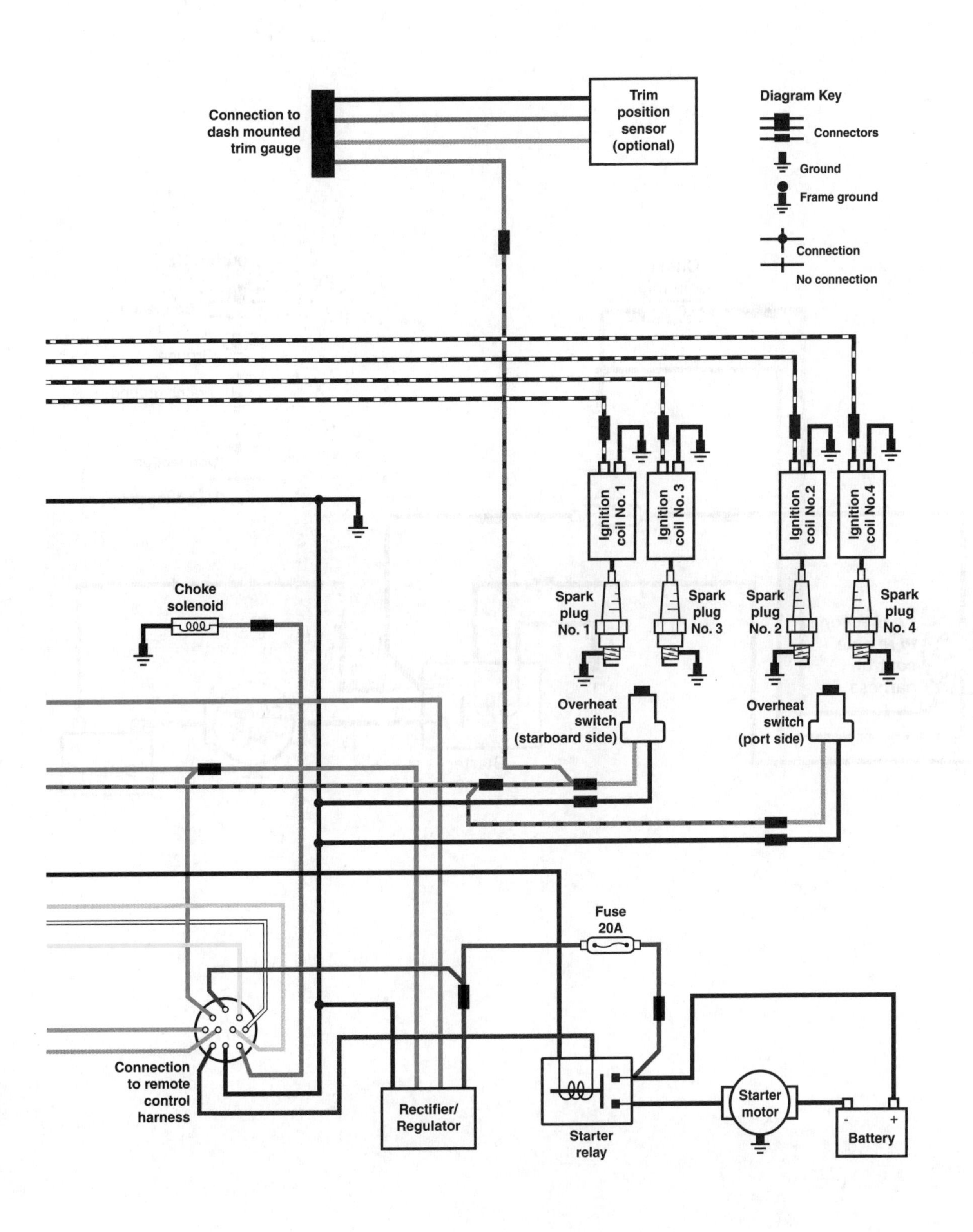
Connection to
dash mounted
trim gauge
Trim
position
sensor
(optional)
Diagram Key
Connectors
Ground
Frame ground
Connection
No connection
Ignition
coil No. 1
Ignition
coil No. 3
Ignition
coil No.2
Ignition
coil No.4
Spark
plug
No. 1
Spark
plug
No. 3
Spark
plug
No. 2
Spark
plug
No. 4
Choke
solenoid
Overheat
switch
(starboard side)
Overheat
switch
(port side)
Fuse
20A
Connection
to remote
control
harness
Rectifier/
Regulator
Starter
relay
Starter
motor
Battery
15

STARTING SYSTEM (105 JET AND 150-250 HP MODELS)

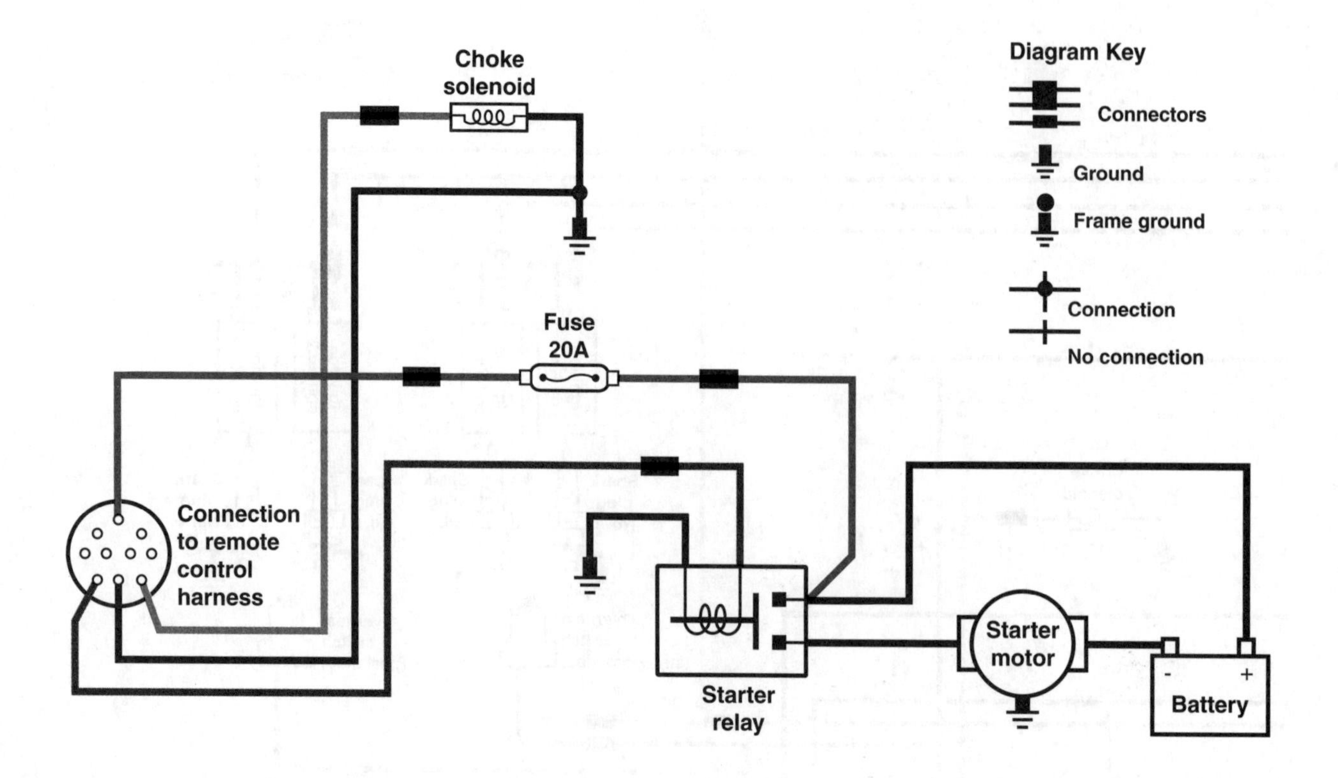

CHARGING SYSTEM (105 JET AND 150-250 HP)

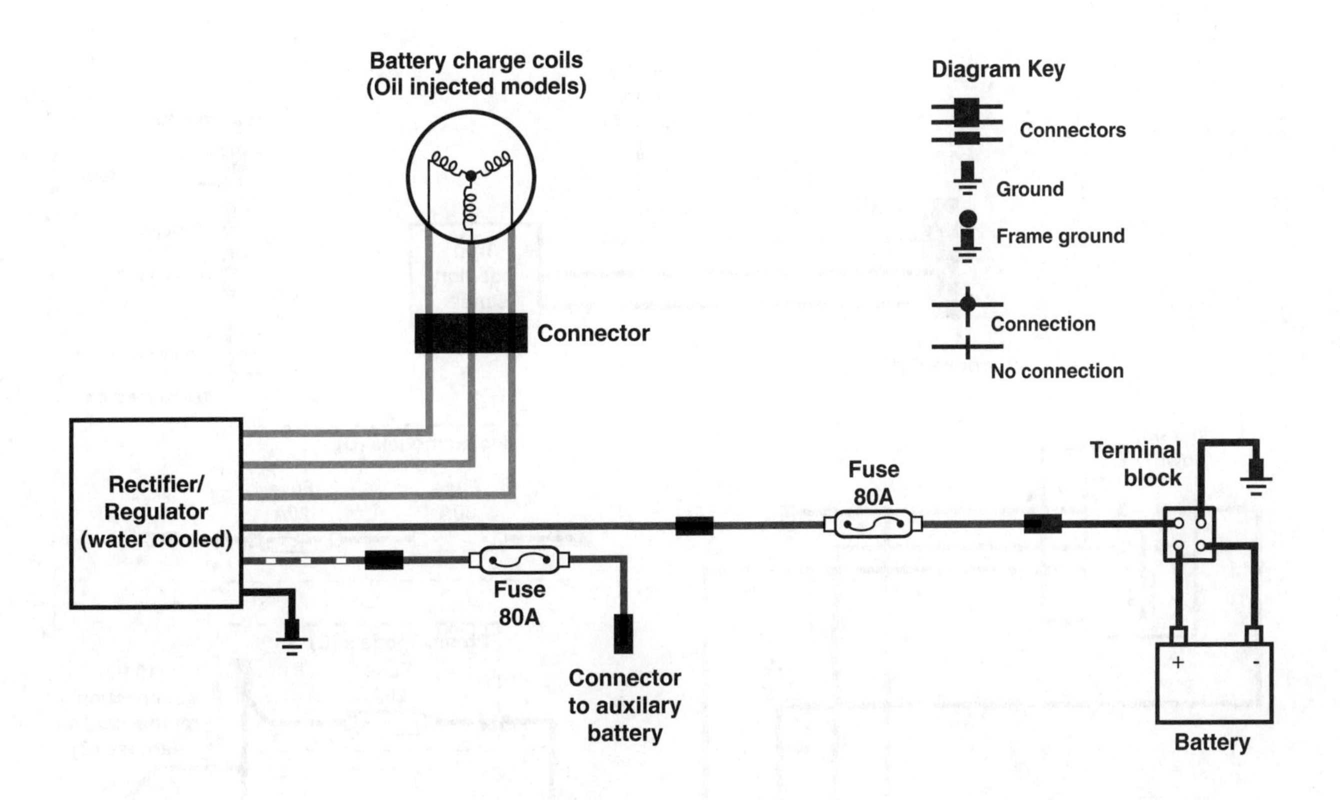

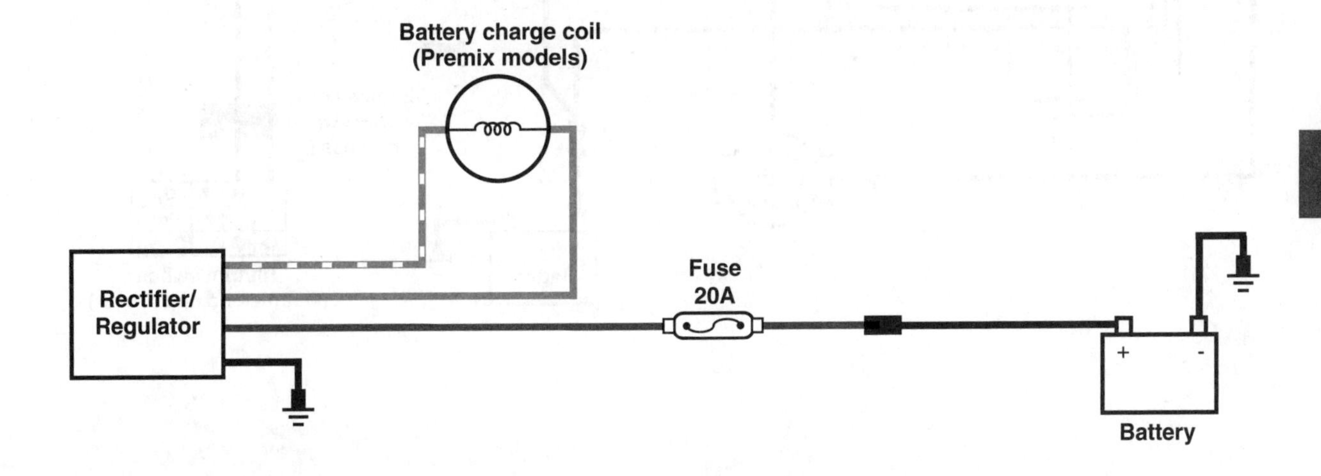

TILT/TRIM SYSTEM (105 JET AND C150 [TWO RELAYS])

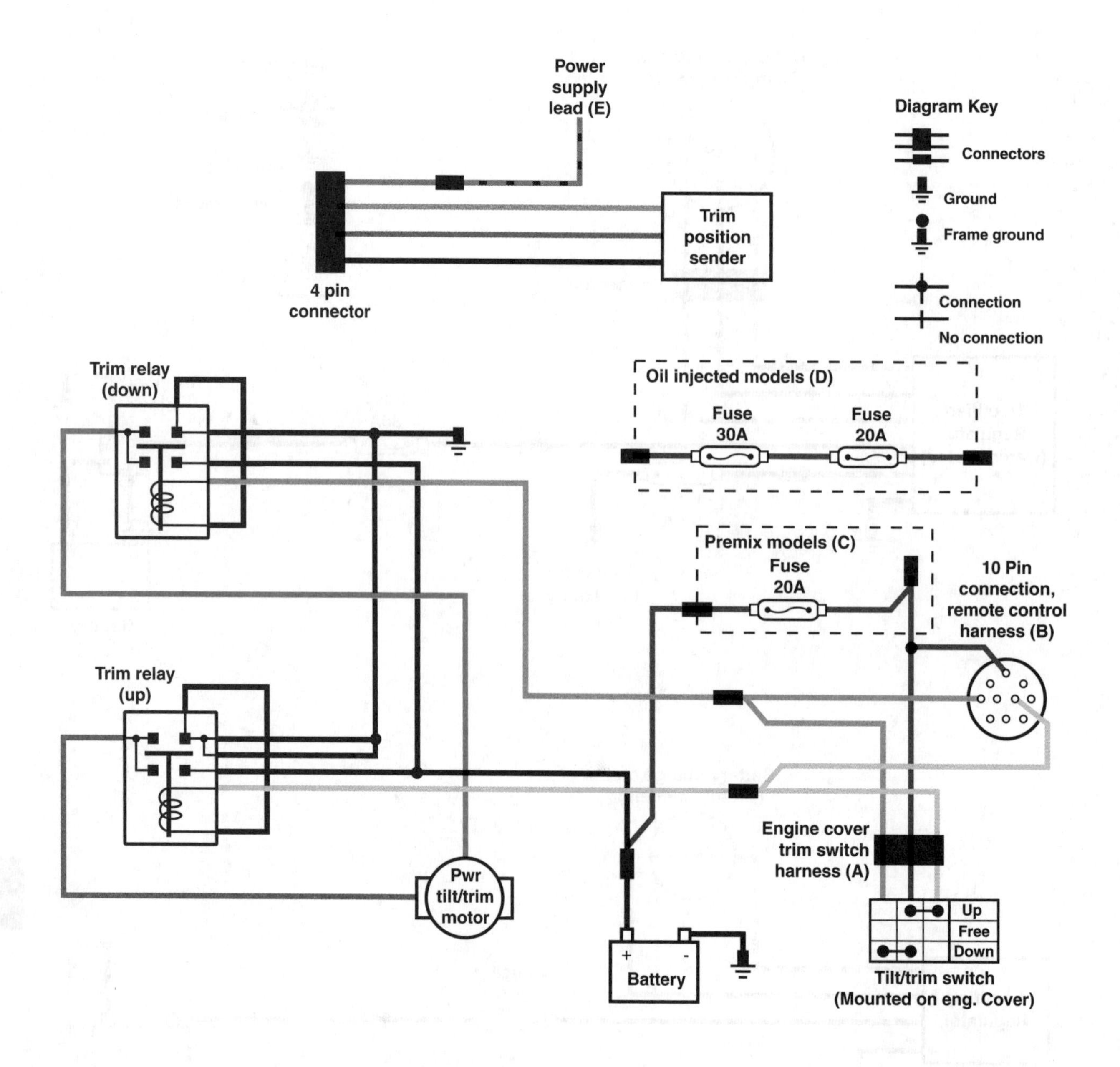

TILT/TRIM SYSTEM (150-250 HP [SINGLE RELAY UNIT])

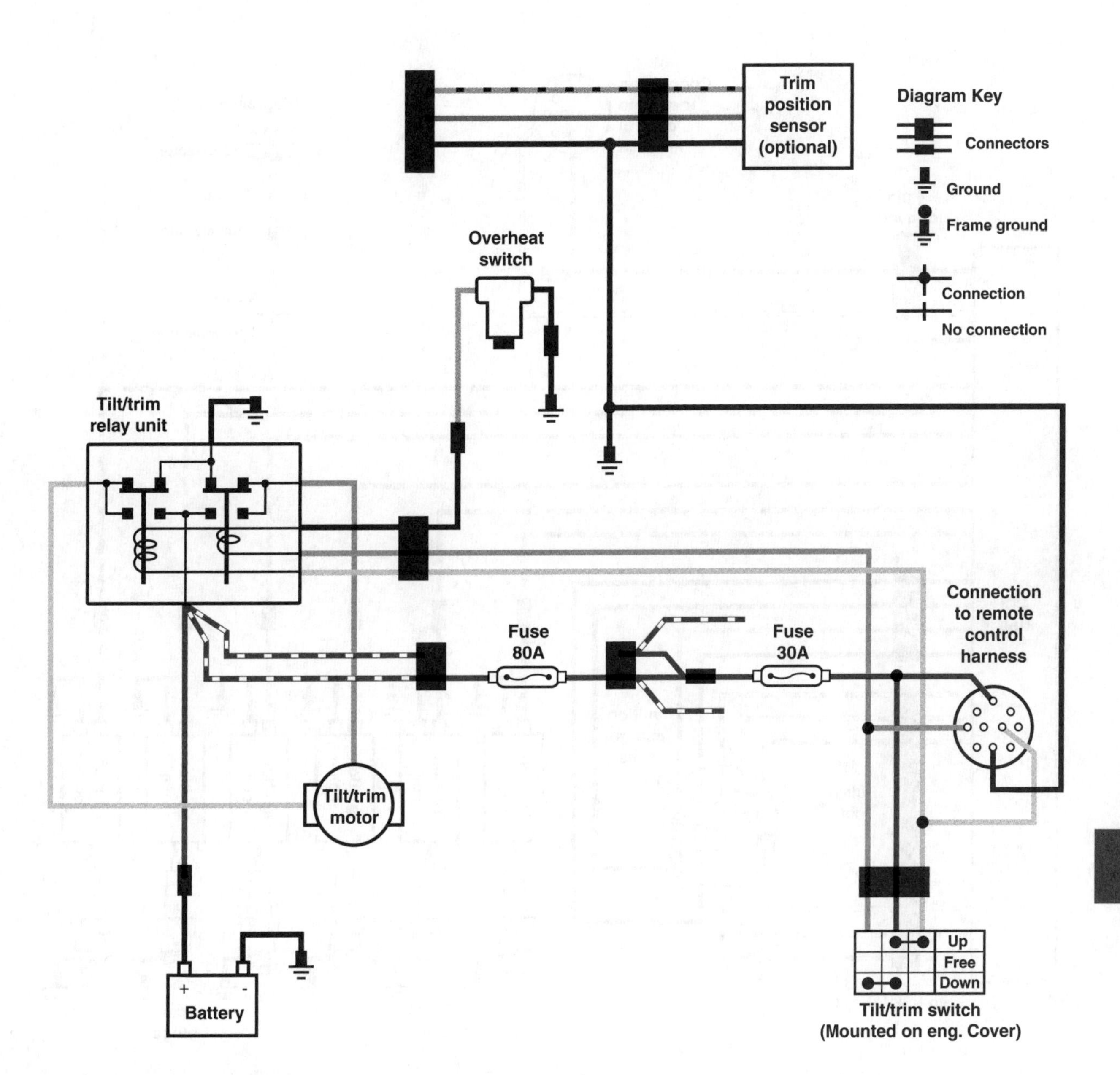

IGNITION SYSTEM (MODEL C150 [PREMIX])

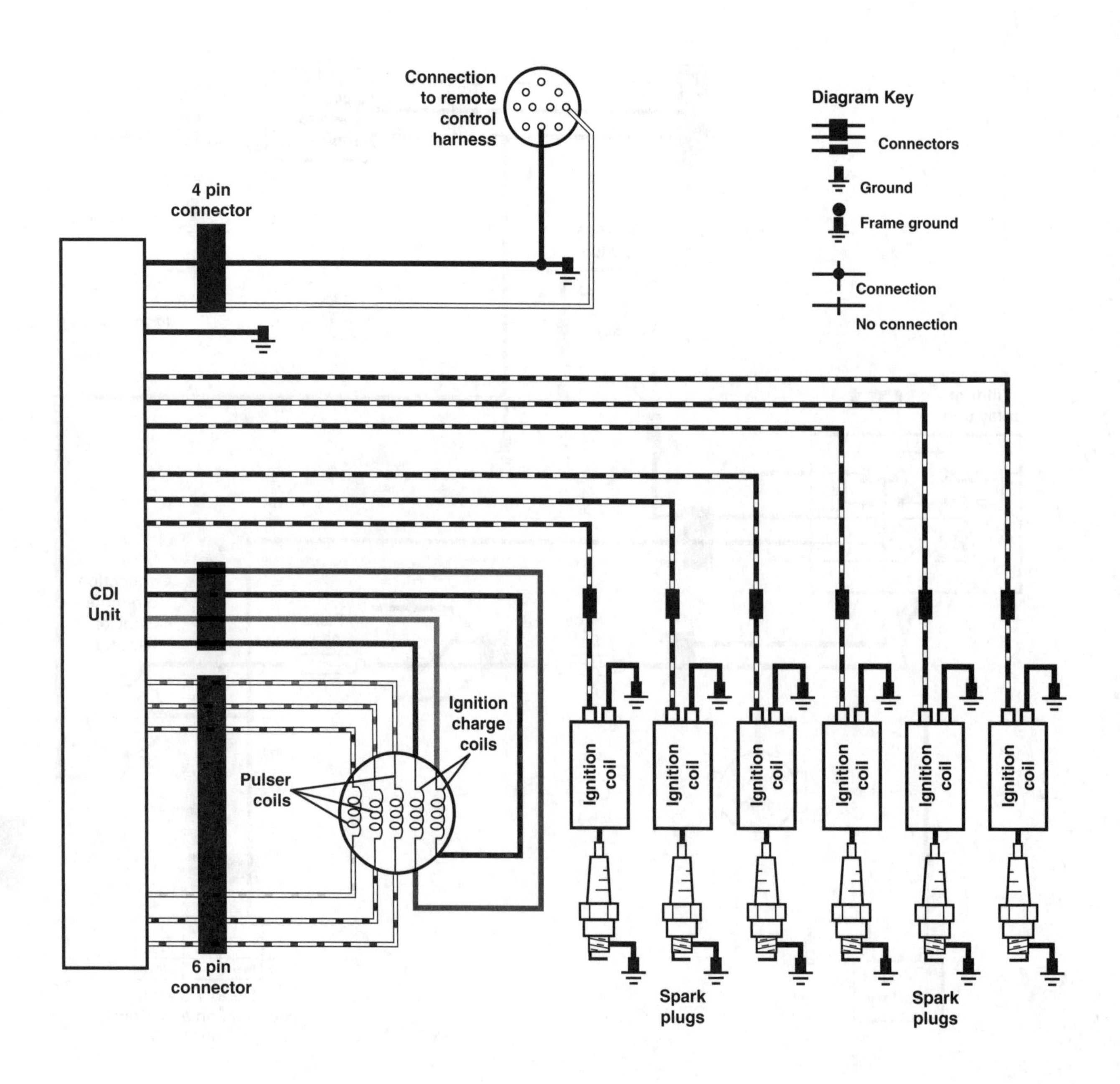

IGNITION SYSTEM (105 JET AND 150-200 HP [CARBURETOR EQUIPPED WITH OIL INJECTION])

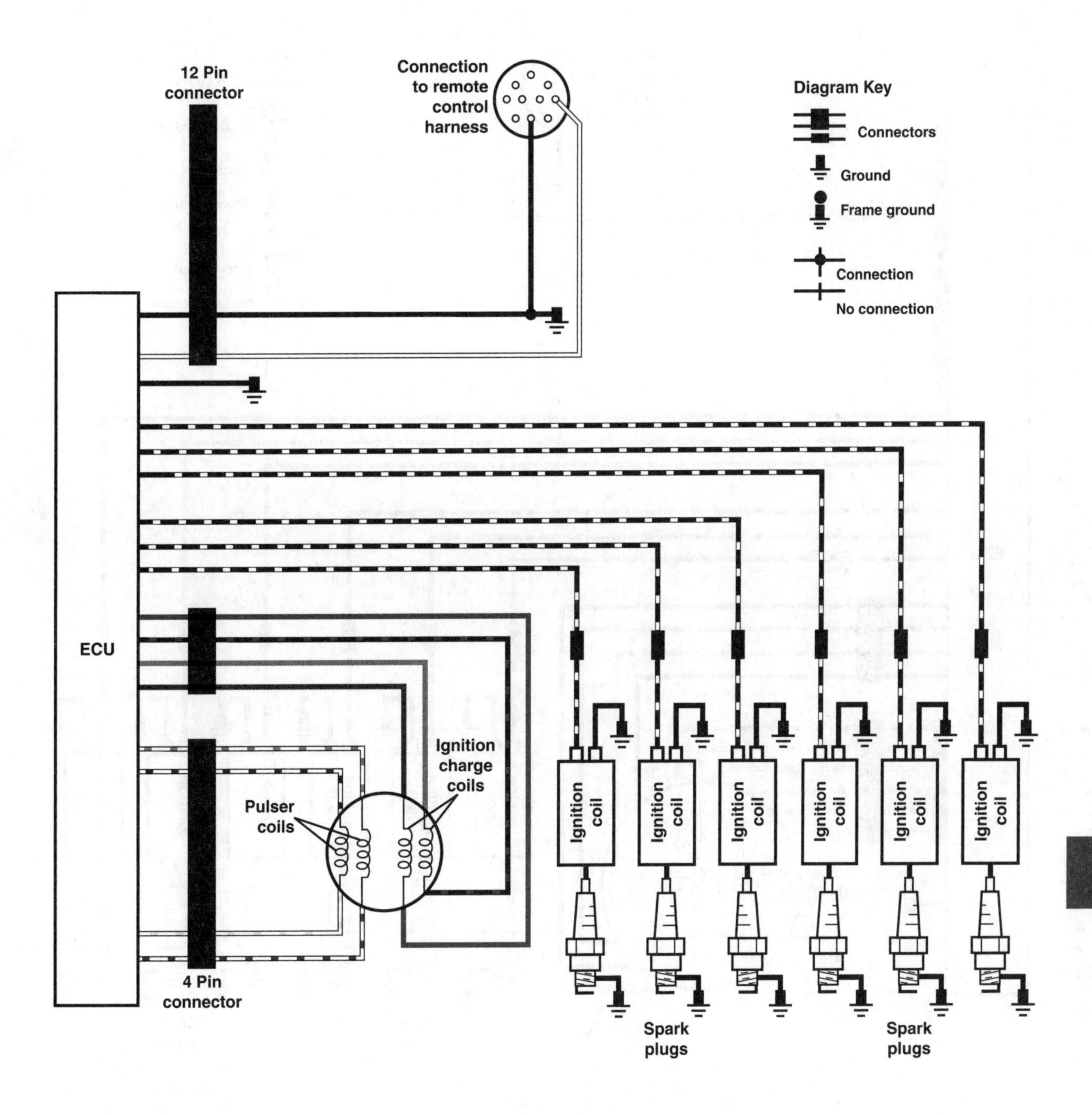

IGNITION SYSTEM (150-250 HP [EFI MODELS])

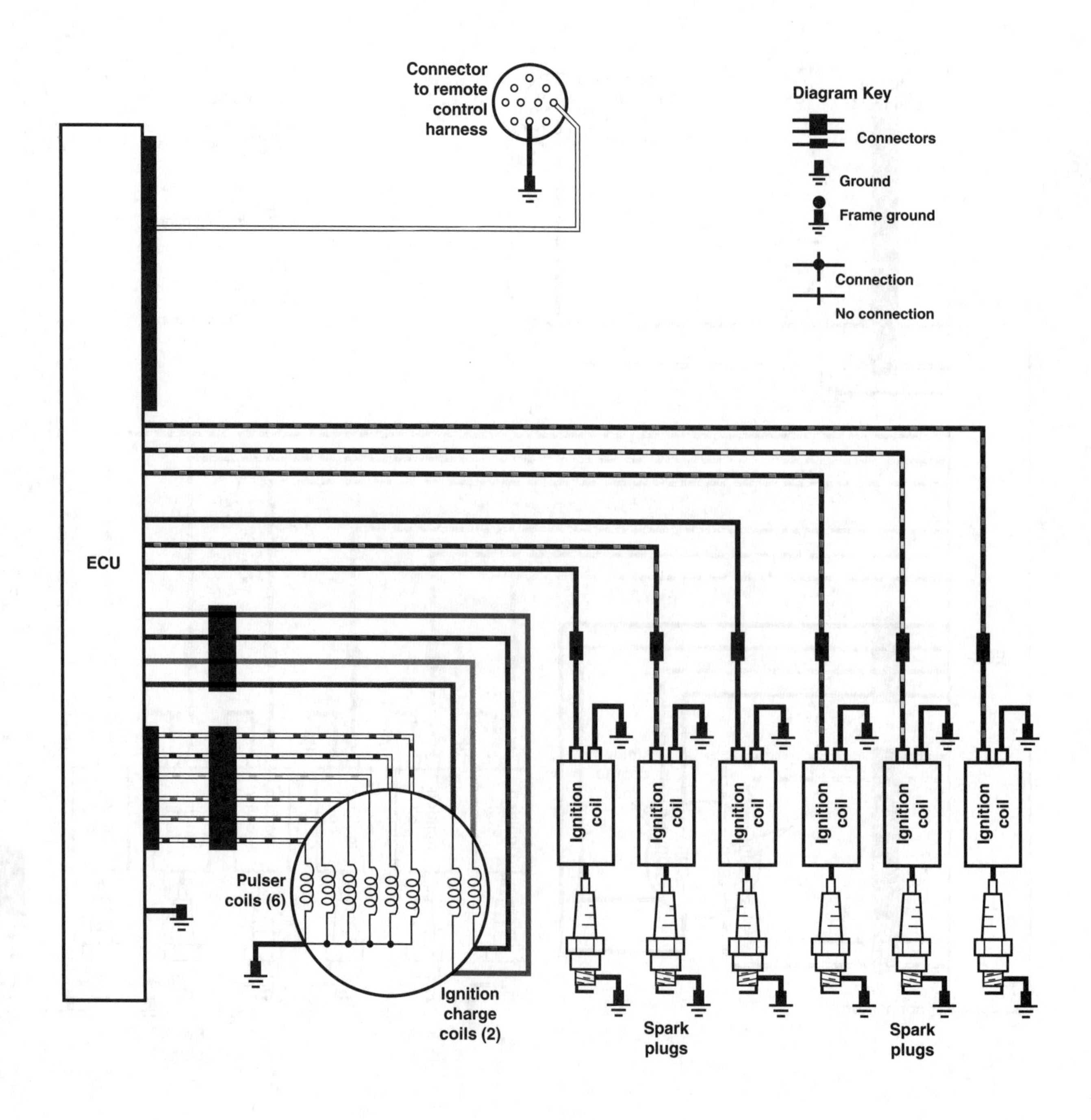

IGNITION SYSTEM (150-250 HP [HPDI MODELS])

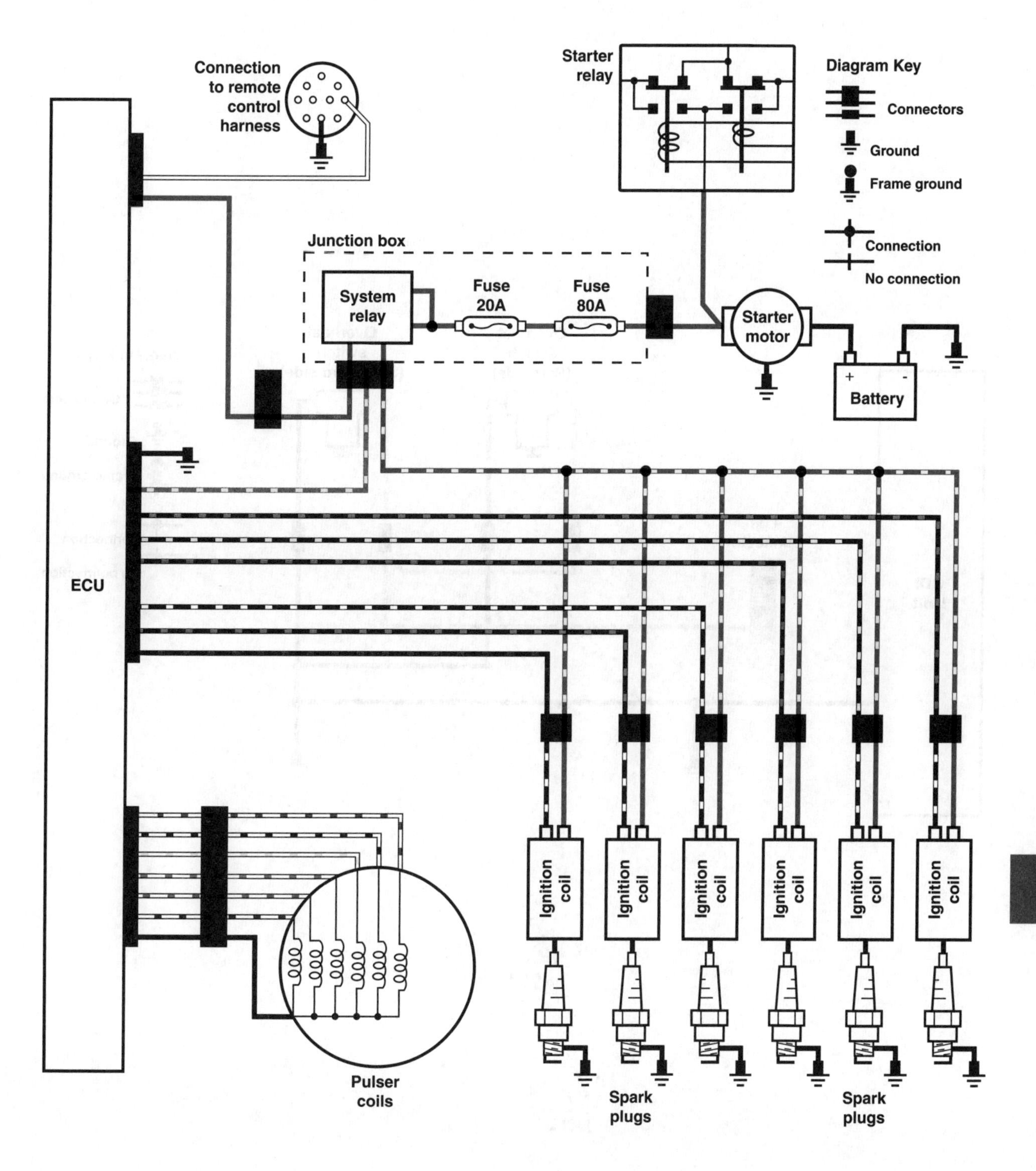

IGNITION CONTROL/WARNING SYSTEM (MODEL C150 [PREMIX])

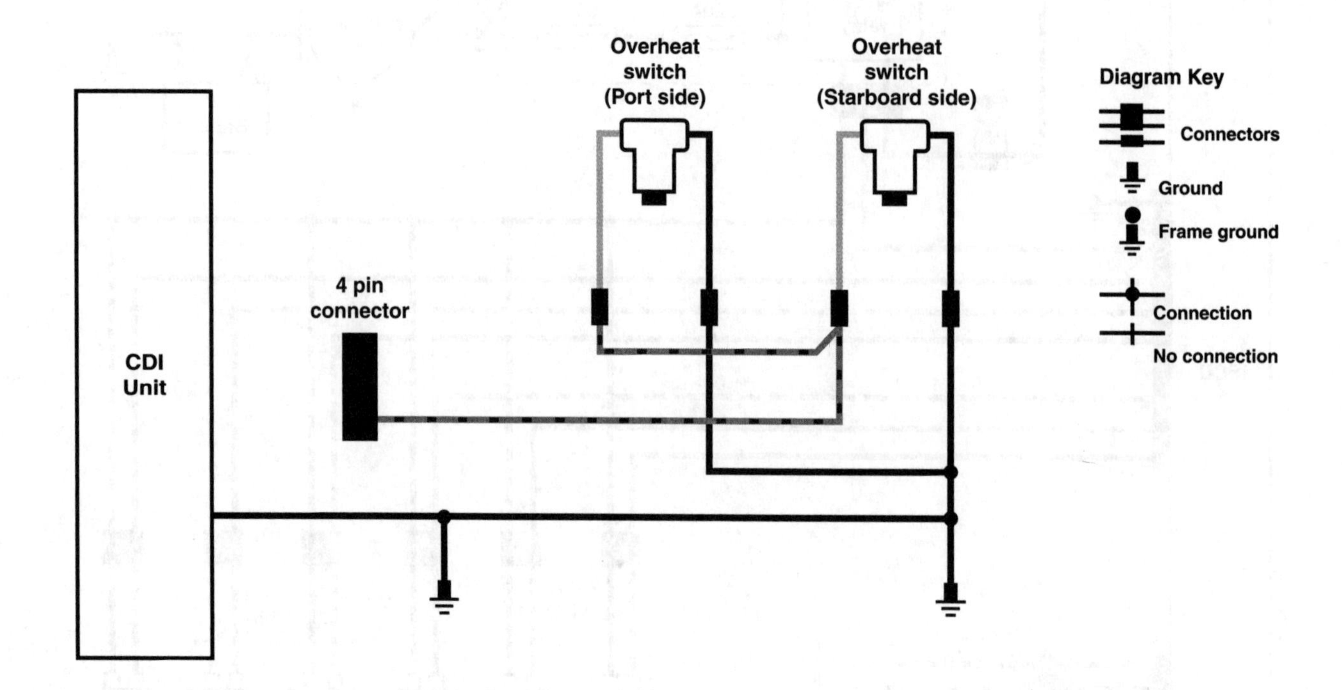

IGNITION CONTROL/WARNING SYSTEM (105 JET AND 150-200 HP PREMIX MODELS [CARBURETOR EQUIPPED])

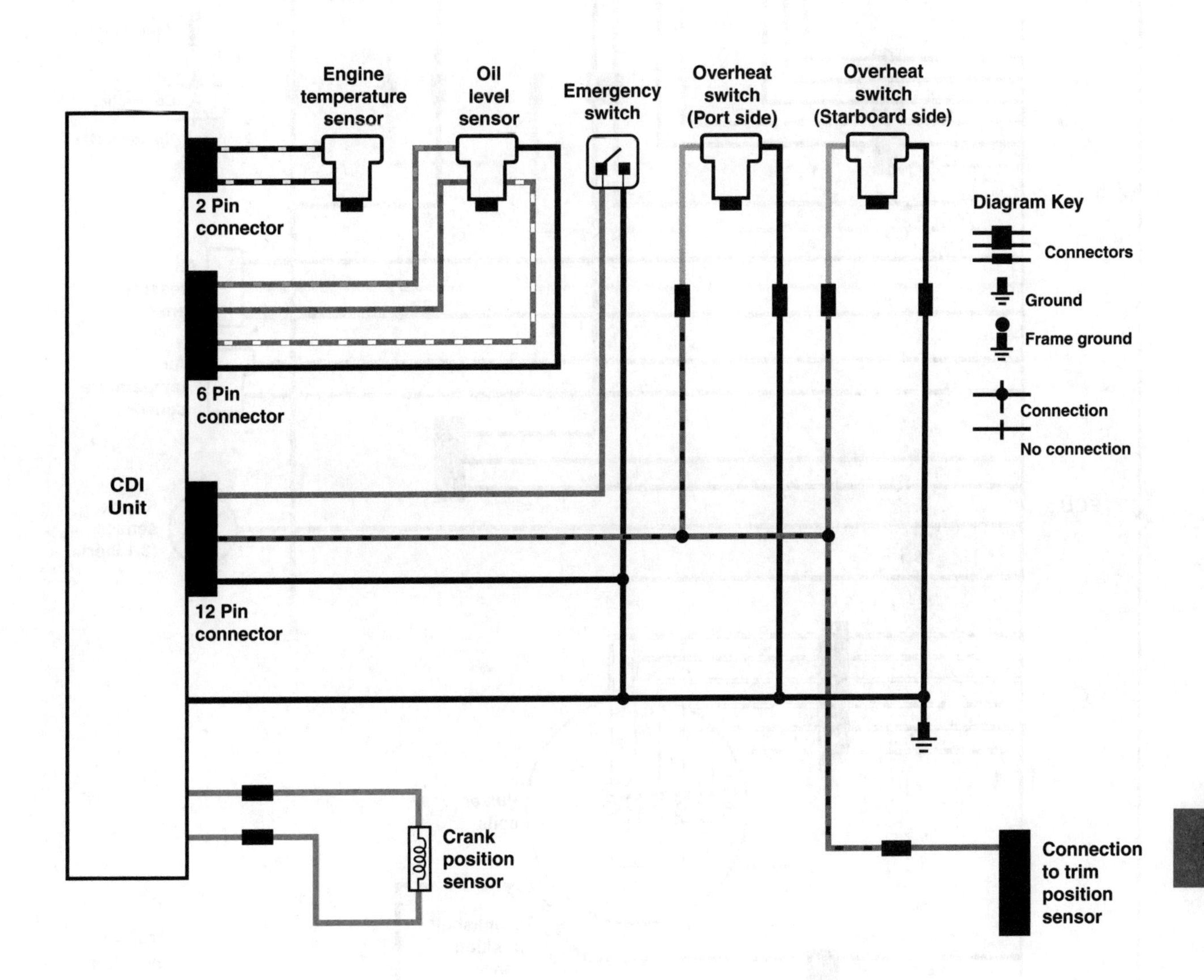

IGNITION CONTROL/WARNING SYSTEM (150-250 HP [EFI MODELS])

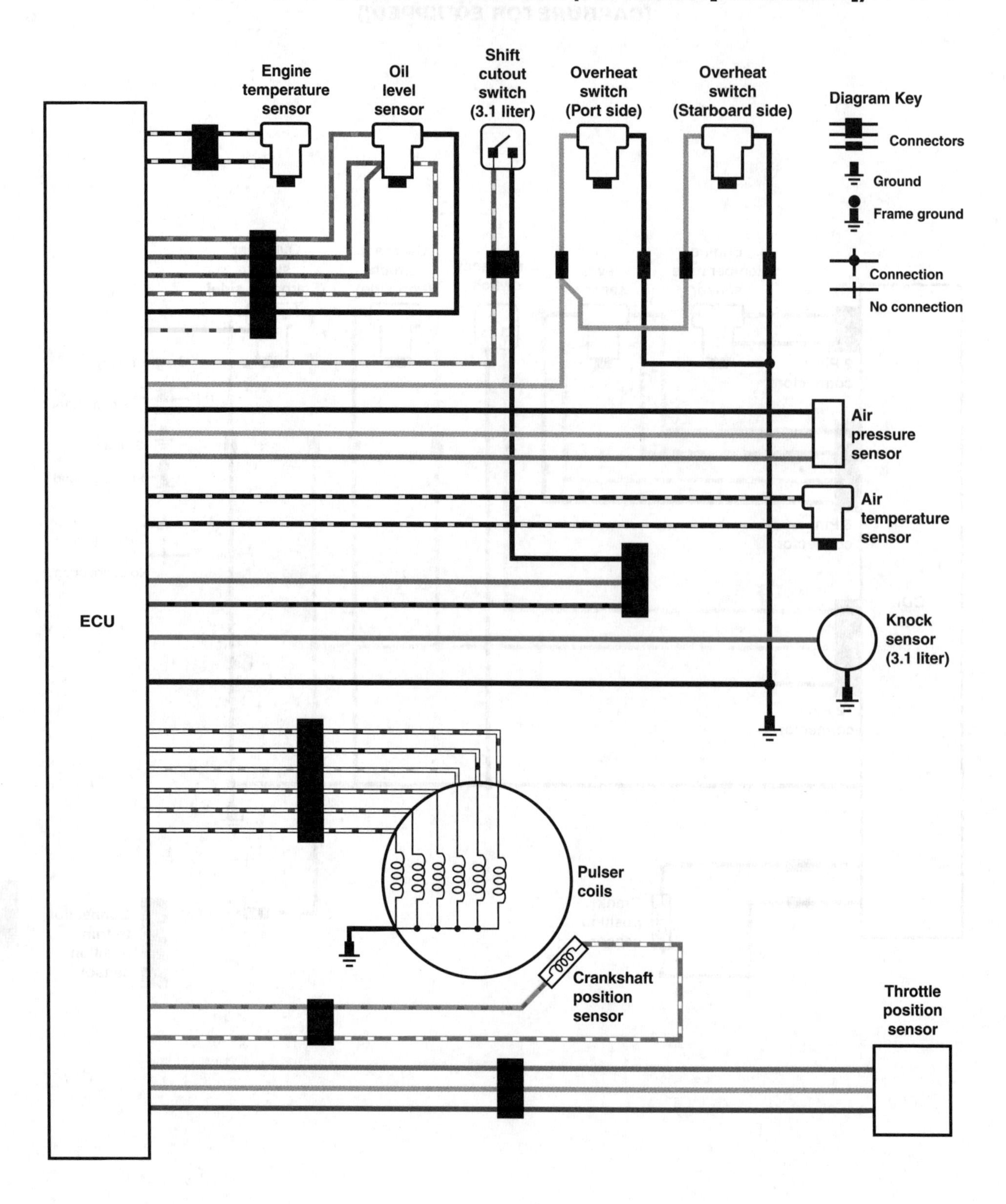

IGNITION CONTROL/WARNING SYSTEM (150-250 HP [HPDI MODELS])

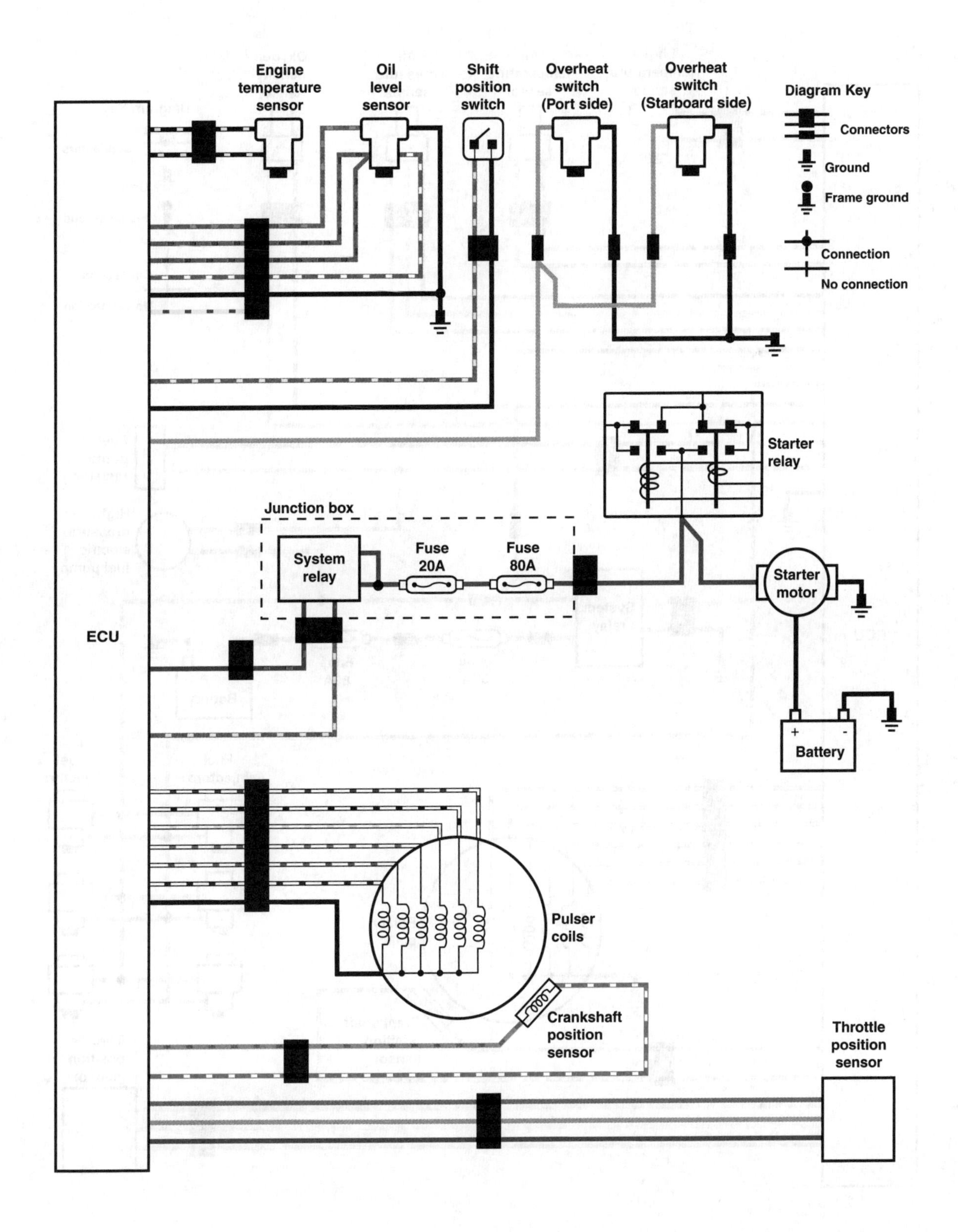

FUEL INJECTION SYSTEM (150-250 HP [EFI MODELS])

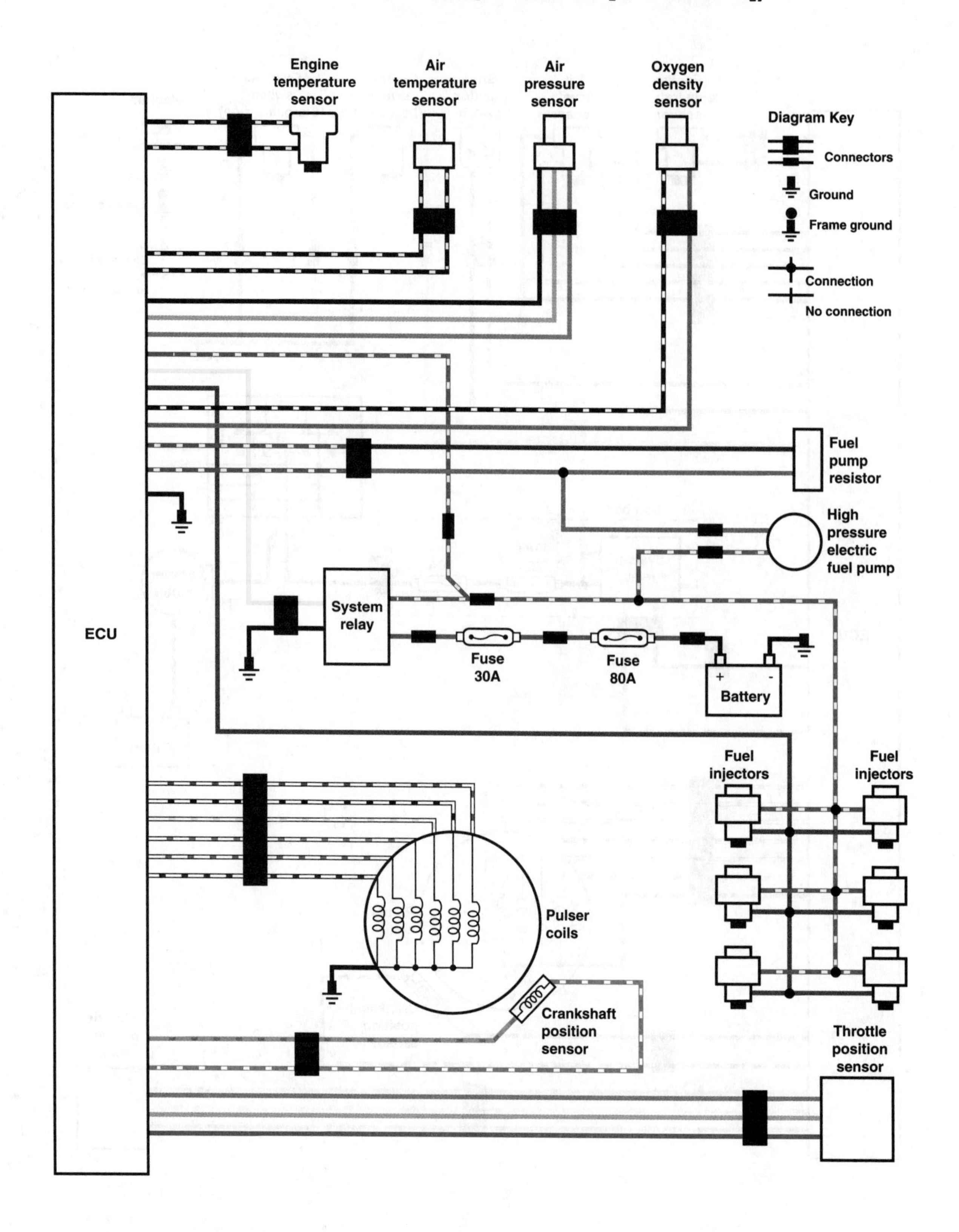

FUEL INJECTION SYSTEM (150-250 HP [HPDI MODELS])

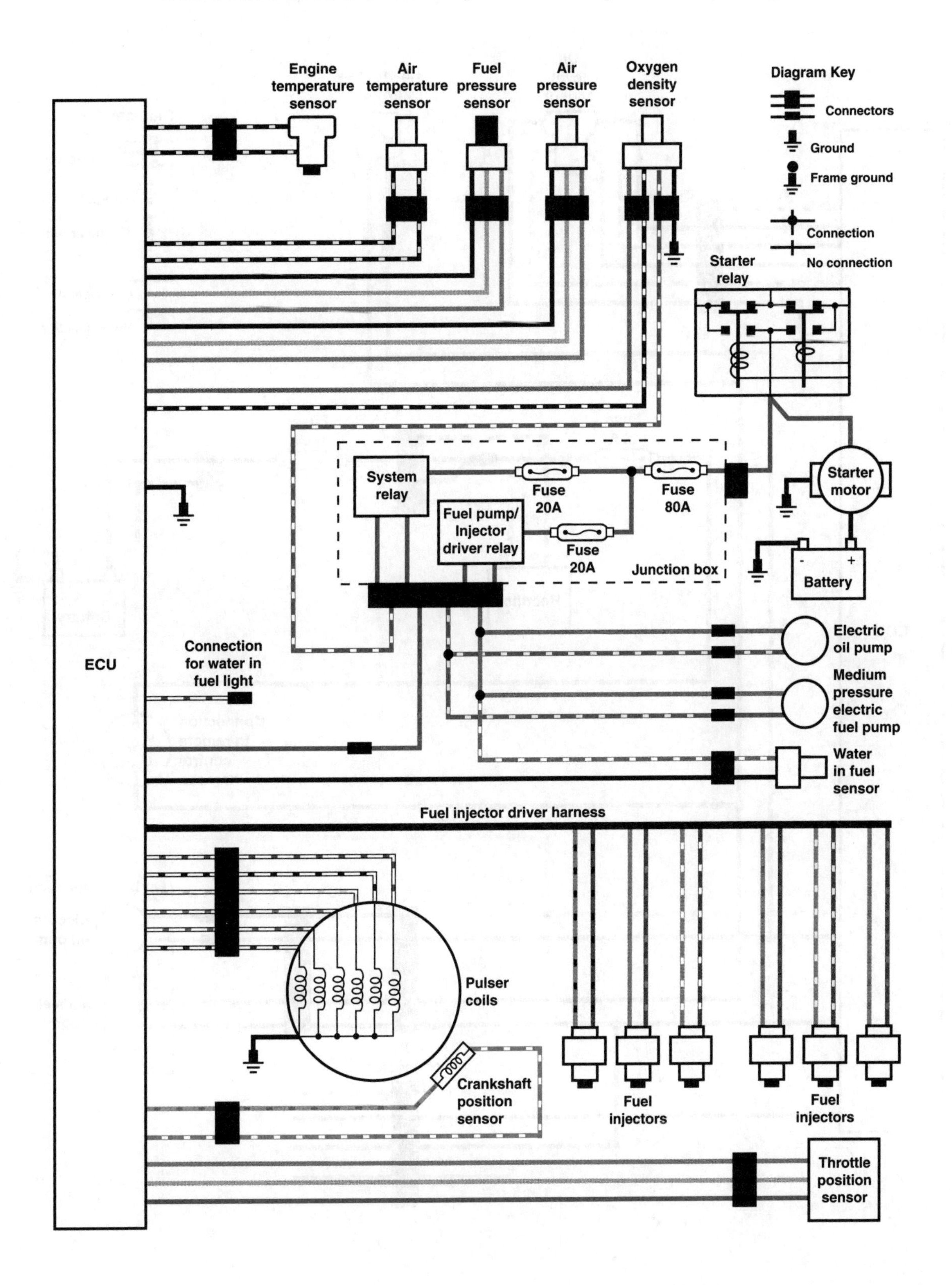

OIL FEED CONTROL SYSTEM (105 JET AND 150-250 HP [OIL INJECTED])

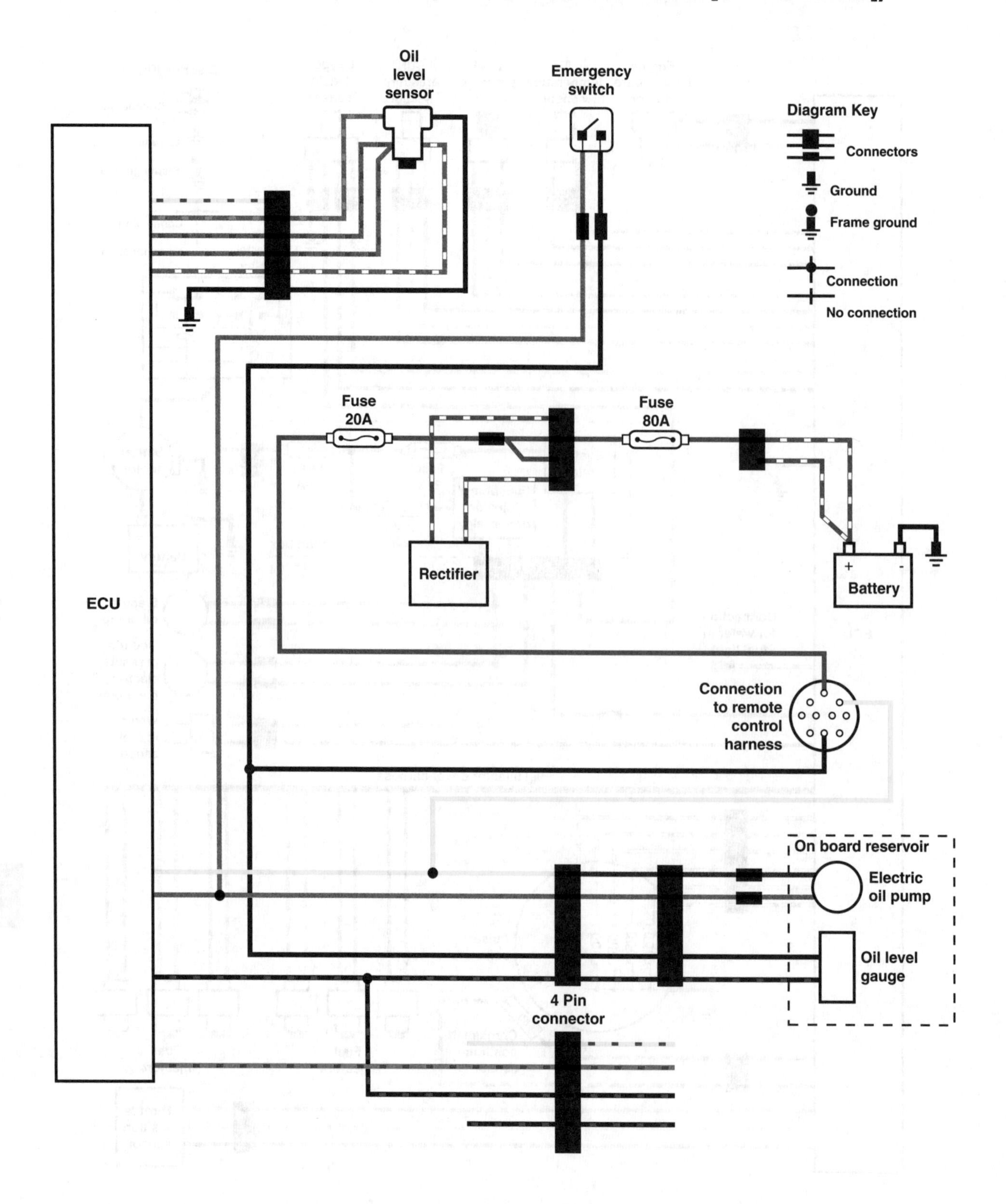

REMOTE CONTROL HARNESS

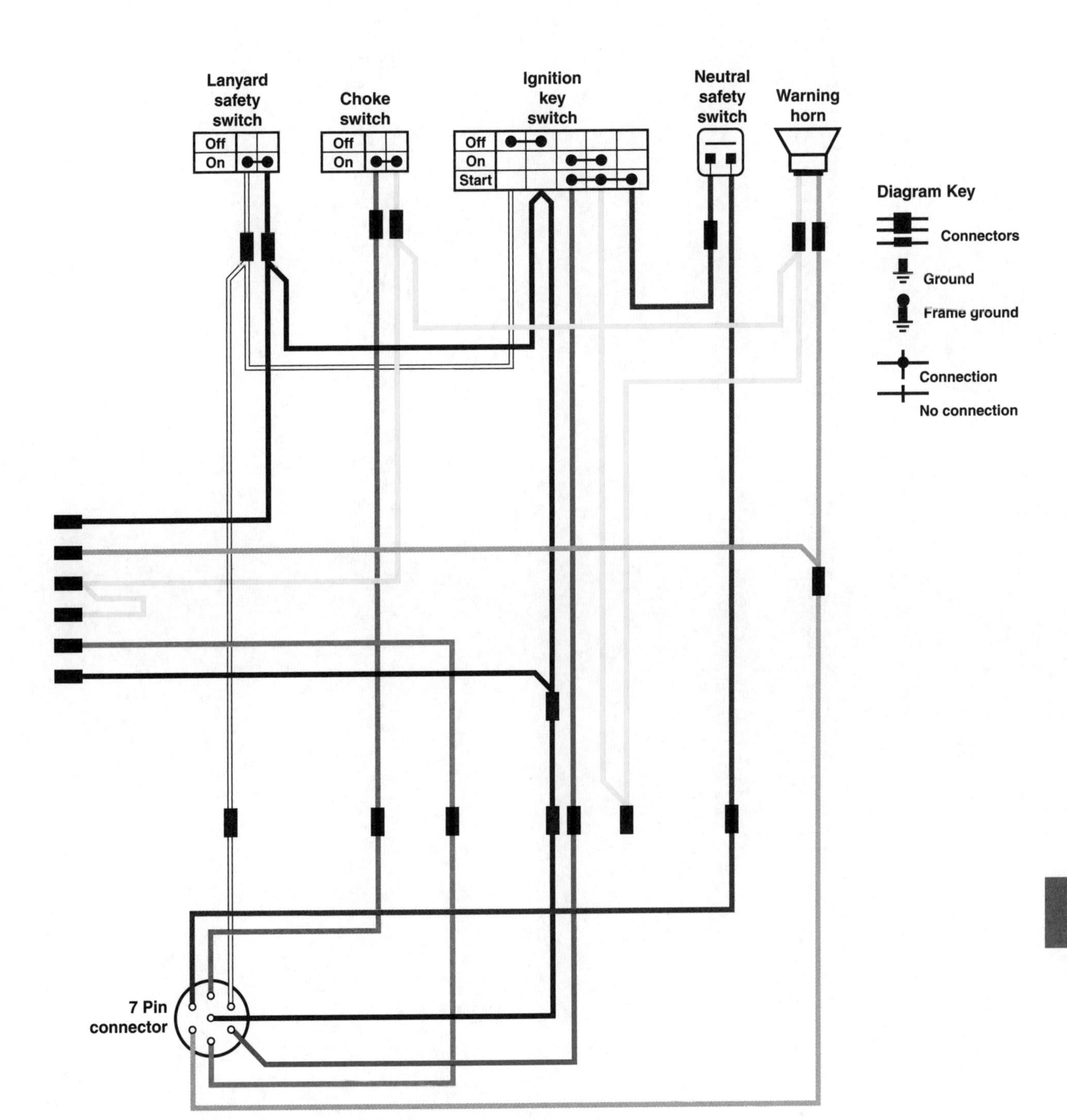

NOTES

NOTES

NOTES

NOTES

NOTES